Calculus

Calculus

Saturnino L. Salas • **Einar Hille**

second edition

one and several variables

with analytic geometry

JOHN WILEY & SONS *New York* • *London* • *Sydney* • *Toronto*

Program Director, Arthur B. Evans
Editor, Marret McCorkle
Design, Mark T. Fowler
Cover Design, Mark T. Fowler

ISBN 0 471 00956 3

Library of Congress Catalog Number: 73-79299

Printed in the United States of America.

10 9 8 7 6 5 4 3 2

Preface

Calculus: One and Several Variables is intended for a three-semester (or five-quarter) course in calculus and analytic geometry:

Chapters 1–9 for the first two semesters (three quarters);
Chapters 12–15 for a third semester (two additional quarters);
Chapters 10 and 11 for either sequence as time permits.

The text is available both as a complete volume and in two parts. In the two-part version, Part 1 deals with functions of one variable, analytic geometry, and infinite series (Chapters 1–11 of the complete volume). Part 2 deals with infinite series, vectors, and functions of several variables (Chapters 10–15 of the complete volume).

The wide acceptance of the first edition has been gratifying and has given us a dividend: a generous flow of reaction from the classroom, a flood of constructive suggestions. As a result, we have added some new material and reworked much of the old. As in the first edition, we have continued to avoid the exotic and tried to stay within the mainstream of calculus—fundamental ideas, basic techniques, standard applications. We have tried to keep the language informal and the presentation uncluttered.

For those of you who are familiar with the first edition, here are some of the changes:

Chapter 1   A more leisurely discussion of composite functions and one-to-one functions.

Chapter 2   More explanation on how to find a δ that "works." More illustrations.

Chapter 3   The differentiability of inverse functions is discussed with more precision; a proof is given in the Appendix. Applications to economics have been added: elasticity, supply and demand curves, isocost lines. More applied max-min problems. More emphasis on differentials and on the *df* notation. A little more on implicit differentiation. The rudiments of partial differentiation.

Chapter 4 Adoption of the increment notation $\Delta x_i = x_i - x_{i-1}$. Emphasis on the *u*-substitution.

Chapter 5 The center stage is now occupied by the notion of exponential growth and decline. The logarithm is motivated by the functional equation $f(xy) = f(x) + f(y)$.

Chapter 6 A discussion of simple harmonic motion has been added. The hyperbolic functions remain, but are now optional.

Chapters 7, 8 A reordering of topics. First, a chapter on the computation of integrals (with less reliance on formulas and more emphasis on general techniques such as the *u*-substitution); then the chapter on analytic geometry.

Chapter 9 Addition of a variant of Bliss's theorem to help with applications of the integral. New sections on fluid pressure and revenue streams.

Chapter 10 Expansion of the discussion of power series, giving emphasis to finding the interval of convergence. More attention to numerical estimation.

Chapter 11 Exploration of some of the common misapplications of L'Hospital's rule.

Chapter 12 Expansion of the discussion of lines and planes in space. More on vector geometry.

Chapter 13 Curvature is now optional.

Chapter 14 Addition of a brief catalog of the quadric surfaces—standard equations, pictures, basic properties.

Chapter 15 More illustrations; more detailed explanations on how to evaluate multiple integrals. More exercises.

Appendix Addition of trig tables, log tables, and exponential tables.

Solutions Manual. Expanded to include detailed solutions to all nonroutine exercises.

Acknowledgments. We are particularly grateful to Victor A. Belfi (Texas Christian University), Fred Gass (Miami University, Ohio), Adam J. Hulin (Louisiana State University, New Orleans), Robert J. Mergener (Moraine Valley Community College), Carl David Minda (University of Cincinnati), C. Stanley Ogilvy (Hamilton College), and B. L. Sanders (Texas Christian University), and to the calculus teaching staffs at the University of Alabama (courtesy of Robert L. Plunkett and Joseph H. Hornback), Clemson University (courtesy of John Kenelly and Stanley M. Lukawecki), and Georgia State University (courtesy of Jan C. Boal).

We owe much to Arthur B. Evans, who directed the project, to Marret McCorkle, who edited the manuscript from beginning to end, and to Dagmar Noll, who assisted throughout.

Special thanks also to Xerox College Publishing for permission to use Tables 1, 4, and 5 in the Appendix and to include exercises from Granville, Smith, and Longley's *Elements of Calculus.*

S.L.S.
E.H.

Contents

Differentiation   75

Integration   174

The Logarithm and Exponential Functions 5   213

The Trigonometric and Hyperbolic Functions 6   268

The Technique of Integration 7   314

L'Hospital's Rule; Improper Integrals 11   537

Vectors 12   559

Vector Calculus 13   609

Functions of Several Variables **14**   651

Multiple Integrals **15**   765

Appendix A   A1

Appendix B. Tables   A15

Answers to Starred Exercises

Index

The Greek Alphabet

A	α	alpha
B	β	beta
Γ	γ	gamma
Δ	δ	delta
E	ϵ	epsilon
Z	ζ	zeta
H	η	eta
Θ	θ	theta
I	ι	iota
K	κ	kappa
Λ	λ	lambda
M	μ	mu
N	ν	nu
Ξ	ξ	xi
O	o	omicron
Π	π	pi
P	ρ	rho
Σ	σ	sigma
T	τ	tau
Υ	υ	upsilon
Φ	ϕ	phi
X	χ	chi
Ψ	ψ	psi
Ω	ω	omega

Introduction

1

1.1 What is Calculus?

To a Roman in the days of the empire a "calculus" was a little pebble that he used in counting and in gambling. Centuries later the verb "calculare" came to mean "to compute," "to reckon," "to figure out." To the engineer and mathematician of today calculus is the branch of mathematics that takes in elementary algebra and geometry and adds one more ingredient: *the limit process*.

Calculus begins where elementary mathematics leaves off. It takes ideas from elementary mathematics and extends them to a much more general situation. Here are some examples. On the left-hand side you will find an idea from elementary mathematics; on the right, this same idea as enriched by calculus.

Elementary Mathematics	*Calculus*

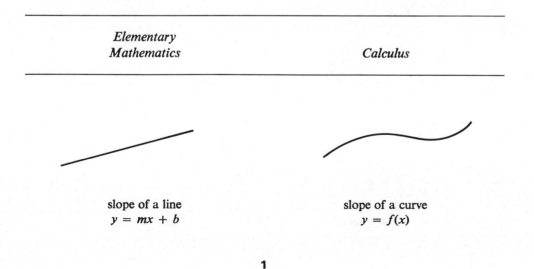

<table>
<tr><td align="center">slope of a line
$y = mx + b$</td><td align="center">slope of a curve
$y = f(x)$</td></tr>
</table>

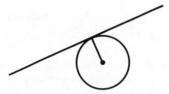

tangent line to
a circle

tangent line to a more
general curve

average velocity,
average acceleration

instantaneous velocity,
instantaneous acceleration

distance moved under
a constant velocity

distance moved under
varying velocity

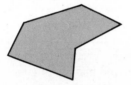

area of a region bounded
by line segments

area of a region bounded
by curves

sum of a finite collection
of numbers
$a_1 + a_2 + \cdots + a_n$

sum of an infinite series
$a_1 + a_2 + \cdots + a_n + \cdots$

average of a finite
collection of numbers

average value of a function
on an interval

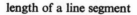

length of a line segment

length of a curve

center of a circle

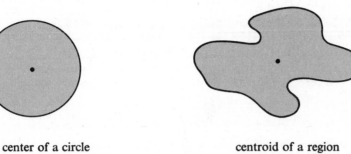

centroid of a region

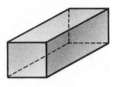

volume of
a rectangular solid

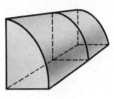

volume of a solid
with a curved boundary

surface area of
a cylinder

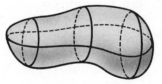

surface area of
a more general solid

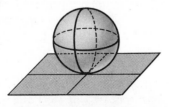

tangent plane to
a sphere

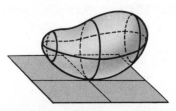

tangent plane to
a more general surface

work done by a constant force	work done by a varying force

mass of an object of constant density	mass of an object of varying density

center of a sphere	center of gravity of a more general solid

It is fitting to say something about the discovery of calculus and about the men who made it. A certain amount of preliminary work was done during the first half of the seventeenth century, but the actual discovery can be credited to Sir Isaac Newton (1642–1727) and to Gottfried Wilhelm Leibniz (1646–1716), an Englishman and a German. Newton's discovery is one of the few good turns that the Great Plague did mankind. The plague forced the closing of Cambridge University in 1665 and young Isaac Newton of Trinity College returned to his home in Lincolnshire for eighteen months of meditation, out of which grew his *method of fluxions*, his *theory of gravitation* (*Philosophiae Naturalis Principia Mathematica*, 1687), and his *theory of light* (*Opticks*, 1704). The method of fluxions is what concerns us here. A treatise with this title was written by Newton in 1672, but it remained unpublished until 1736, nine years after his death. The new method (calculus to us) was first announced in 1687 but in vague general terms without symbolism, formulas, or applications. Newton himself seemed very reluctant to publish anything tangible about his discovery, and it is not surprising that the development on the Continent, in spite of a late start, soon overtook Newton and went beyond him.

Leibniz started his work in 1673, eight years after Newton. In 1675 he initiated the basic notation: dx and $\int$. His first publications appeared in 1684 and 1686. These made little stir in Germany but the two brothers Bernoulli of Basel (Switzerland) took up the ideas and added profusely to them. From 1690 onward calculus grew rapidly. It reached approximately its present state in about one hundred years.

1.2   Notions and Formulas from Elementary Mathematics

The following outline is presented for review and easy reference.

I. Sets

the object x is in the set A (*x* is an element of *A*)	$x \in A$
the object x is not in the set A	$x \notin A$
containment	$A \subseteq B$
union	$A \cup B$
intersection	$A \cap B$
empty set	a set with no elements

(These are the only notions from set theory that you will need for this book. If you are not familiar with them, see Section 1 of the Appendix.)

II. Real Numbers

CLASSIFICATION

integers	$0, 1, -1, 2, -2, 3, -3$, etc.
rational numbers	$\dfrac{p}{q}$ where p and q are integers, $q \neq 0$.
irrational numbers	real numbers which are not rational; for example, $\sqrt{2}$, π.

ORDER PROPERTIES

(i) Either $a < b$, $b < a$, or $a = b$.

(ii) If $a < b$ and $b < c$, then $a < c$.

(iii) If $a < b$, then $a + c < b + c$ for all real numbers c.

(iv) If $a < b$ and $c > 0$, then $ac < bc$.
     If $a < b$ and $c < 0$, then $ac > bc$.

DENSITY

Between any two numbers there is a rational number and an irrational number.

ABSOLUTE VALUE

$$|a| = \sqrt{a^2}; \quad |a| = \max\{a, -a\}; \quad |a| = \begin{cases} a, & \text{if } a \geq 0 \\ -a, & \text{if } a \leq 0 \end{cases}.$$

geometric interpretation $|a|$ = distance between a and 0.

$|a - c|$ = distance between a and c.

properties   (i) $|a| = 0$ iff $a = 0$.†

(ii) $|-a| = |a|$.

(iii) $|ab| = |a|\,|b|$.

(iv) $|a + b| \le |a| + |b|$.   (the triangle inequality)

(v) $\big||a| - |b|\big| \le |a - b|$.   (another form of the triangle inequality)

INTERVALS

The notation

$$\{x: (\ \ )\}$$

will be used to denote the set of all x such that ( ). Suppose now that $a < b$. The *open interval* (a, b) is the set of all numbers between a and b:

$$(a, b) = \{x: a < x < b\}.$$

The *closed interval* $[a, b]$ is the open interval (a, b) together with the endpoints:

$$[a, b] = \{x: a \le x \le b\}.$$

There are seven other types of intervals:

$$(a, b] = \{x: a < x \le b\} \qquad [a, b) = \{x: a \le x < b\}$$
$$(a, \infty) = \{x: a < x\} \qquad [a, \infty) = \{x: a \le x\}$$
$$(-\infty, b) = \{x: x < b\} \qquad (-\infty, b] = \{x: x \le b\}$$
$$(-\infty, \infty) = \mathbf{R} = \text{set of real numbers.}$$

BOUNDEDNESS

A set S of real numbers is said to be

(i) *bounded above* iff there exists a real number M such that

$$x \le M \qquad \text{for all} \quad x \in S.$$

(M is called an *upper bound* for S.)

(ii) *bounded below* iff there exists a real number m such that

$$m \le x \qquad \text{for all} \quad x \in S.$$

(m is called a *lower bound* for S.)

(iii) *bounded* iff it is bounded above and below.

Example.   The intervals $(-\infty, 2]$ and $(-\infty, 2)$ are bounded above but not below.

---

† By "iff" we mean "if and only if." This expression is used so often in mathematics that it is convenient to have an abbreviation for it.

Example.   The set of positive integers is bounded below but not above.

Example.   The intervals $[0, 1]$, $(0, 1)$, and $(0, 1]$ are bounded (both above and below).

III. Algebra and Geometry

GENERAL QUADRATIC FORMULA

The quadratic equation

$$ax^2 + bx + c = 0, \qquad a \neq 0$$

has solutions

$$x = \frac{-b \pm \sqrt{b^2 - 4ac}}{2a}.$$

FACTORIALS

$$0! = 1$$
$$1! = 1$$
$$2! = 2 \cdot 1$$
$$3! = 3 \cdot 2 \cdot 1$$
$$\vdots$$
$$n! = n(n - 1) \cdots (2)(1).$$

ELEMENTARY FIGURES

circle   area $= \pi r^2$; circumference $= 2\pi r$; area of sector $= \frac{1}{2} r^2 \theta$, where θ is the central angle measured in radians; length of circular arc $= r\theta$, where θ is the central angle measured in radians.

right circular cylinder   volume $= \pi r^2 h$; lateral area $= 2\pi rh$; total surface area $= 2\pi r(r + h)$.

cone   volume $= \frac{1}{3}\pi r^2 h$; lateral area $= \pi r\sqrt{r^2 + h^2}$; total surface area $= \pi r(r + \sqrt{r^2 + h^2})$.

sphere   volume $= \frac{4}{3}\pi r^3$; surface area $= 4\pi r^2$.

IV. Analytic Geometry

ABSCISSA AND ORDINATE

The x-coordinate of a point is called the *abscissa*; the y-coordinate is called the *ordinate*.

DISTANCE BETWEEN TWO POINTS $P(x_0, y_0)$ AND $P(x_1, y_1)$:

$$d = \sqrt{(x_1 - x_0)^2 + (y_1 - y_0)^2}.$$

LINES

(i) Slope of a line $= \dfrac{\text{rise}}{\text{run}} = \tan \alpha$, where α is the angle of inclination.

(ii) Equations for lines:

general form	$Ax + By + C = 0$, A and B not both 0.
vertical line	$x = a$.
slope-intercept form	$y = mx + b$.
point-slope form	$y - y_0 = m(x - x_0)$.
two-point form	$y - y_0 = \dfrac{y_1 - y_0}{x_1 - x_0}(x - x_0)$.
two-intercept form	$\dfrac{x}{a} + \dfrac{y}{b} = 1$.

(iii) Angle between two nonvertical lines:

$$\tan \theta = \frac{m_1 - m_0}{1 + m_1 m_0}.$$

(iv) Perpendicular lines:

$$m_1 = -\frac{1}{m_0}.$$

EQUATIONS OF CONIC SECTIONS

circle of radius r centered at the origin	$x^2 + y^2 = r^2$.
circle of radius r centered at $P(x_0, y_0)$	$(x - x_0)^2 + (y - y_0)^2 = r^2$.
parabolas	$x^2 = 4cy, \quad y^2 = 4cx$.
ellipse	$\dfrac{x^2}{a^2} + \dfrac{y^2}{b^2} = 1$.
hyperbolas	$\dfrac{x^2}{a^2} - \dfrac{y^2}{b^2} = 1, \quad xy = c$.

(The conic sections are discussed in detail in Chapter 8.)

V. Trigonometry

ANGLE MEASUREMENT

$$\pi \cong 3.14159.$$

2π radians $= 360$ degrees $=$ one complete revolution.

One radian $= \dfrac{180}{\pi}$ degrees $\cong 57.296$ degrees.

One degree $= \dfrac{\pi}{180}$ radians $\cong 0.0175$ radians.

SECTORS

Length of subtended arc $= r\theta$, where θ is the central angle measured in radians. (Figure 1.2.1)

Area of sector $= \frac{1}{2}r^2\theta$, where θ is the central angle measured in radians
$\qquad\qquad = \frac{1}{2}r \cdot$ length of subtended arc.

SINE AND COSINE

Take θ between 0 and 2π. In the unit circle (Figure 1.2.2) mark off the sector OAP that has central angle θ radians:

$$\cos\theta = x\text{-coordinate of } P, \qquad \sin\theta = y\text{-coordinate of } P.$$

These definitions are then extended periodically:

$$\cos\theta = \cos(\theta + 2n\pi), \qquad \sin\theta = \sin(\theta + 2n\pi).$$

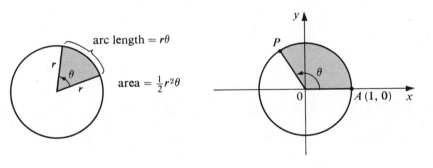

FIGURE 1.2.1   FIGURE 1.2.2

OTHER TRIGONOMETRIC FUNCTIONS

$$\tan\theta = \frac{\sin\theta}{\cos\theta}, \qquad \cot\theta = \frac{\cos\theta}{\sin\theta},$$

$$\sec\theta = \frac{1}{\cos\theta}, \qquad \operatorname{cosec}\theta = \frac{1}{\sin\theta}.$$

IN TERMS OF A RIGHT TRIANGLE

For θ between 0 and $\frac{1}{2}\pi$,

$$\sin \theta = \frac{\text{opposite side}}{\text{hypotenuse}} \qquad \cos \theta = \frac{\text{adjacent side}}{\text{hypotenuse}}$$

$$\tan \theta = \frac{\text{opposite side}}{\text{adjacent side}} \qquad \cot \theta = \frac{\text{adjacent side}}{\text{opposite side}}$$

$$\sec \theta = \frac{\text{hypotenuse}}{\text{adjacent side}} \qquad \operatorname{cosec} \theta = \frac{\text{hypotenuse}}{\text{opposite side}}.$$

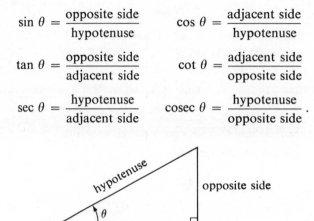

FIGURE 1.2.3

ODDNESS AND EVENNESS

cosine function is even $\cos(-\theta) = \cos \theta$. (The graph is symmetric about the y-axis.)

sine and tangent functions are odd $\sin(-\theta) = -\sin \theta$, $\tan(-\theta) = -\tan \theta$. (The graph is symmetric about the origin.)

IDENTITIES

(i) $\cos^2 \theta + \sin^2 \theta = 1$, $\tan^2 \theta + 1 = \sec^2 \theta$, $\cot^2 \theta + 1 = \operatorname{cosec}^2 \theta$.

(ii) $\sin(\theta_1 + \theta_2) = \sin \theta_1 \cos \theta_2 + \cos \theta_1 \sin \theta_2$.
$\sin(\theta_1 - \theta_2) = \sin \theta_1 \cos \theta_2 - \cos \theta_1 \sin \theta_2$.
$\cos(\theta_1 + \theta_2) = \cos \theta_1 \cos \theta_2 - \sin \theta_1 \sin \theta_2$.
$\cos(\theta_1 - \theta_2) = \cos \theta_1 \cos \theta_2 + \sin \theta_1 \sin \theta_2$.

(iii) $\sin 2\theta = 2 \sin \theta \cos \theta$, $\cos 2\theta = \cos^2 \theta - \sin^2 \theta$.

(iv) $\sin^2 \theta = \frac{1}{2}(1 - \cos 2\theta)$, $\cos^2 \theta = \frac{1}{2}(1 + \cos 2\theta)$.

IN AN ARBITRARY TRIANGLE

law of sines $\dfrac{\sin A}{a} = \dfrac{\sin B}{b} = \dfrac{\sin C}{c}$.

law of cosines $a^2 = b^2 + c^2 - 2bc \cos A$.

(The calculus of the trigonometric functions is discussed in Chapter 6.)

VI. Induction

Axiom   Let S be a set of integers. If
        (A) 1 is in S and   (B) k is in S implies $k + 1$ is in S,
        then all the positive integers are in S.

    In general, we use induction when we want to prove that a certain proposition is true for all positive integers n. To do this, first we prove that the proposition is true for $n = 1$. Then we prove that, if it is true for $n = k$, then it is true for $n = k + 1$. It follows then from the axiom that the proposition is true for all positive integers n. (We will use induction only sparingly. For a discussion of it we refer you to Section 2 of the Appendix.)

1.3 Inequalities; Absolute Value

To solve an inequality is to find the set of numbers that satisfy it. Inequalities play such an important role in calculus, that it is probably a good idea to refresh your memory on the subject. Here are some sample problems.

Problem.   Solve the inequality

$$3(1 - x) \le 6.$$

SOLUTION

$$
\begin{aligned}
3(1 - x) &\le 6, & \\
1 - x &\le 2, & \text{(we divided by 3)} \\
-x &\le 1, & \text{(subtracted 1)} \\
x &\ge -1. & \text{(multiplied by } -1\text{)}
\end{aligned}
$$

As the solution we have the interval $[-1, \infty)$. ☐

Problem.   Solve the inequality

$$\tfrac{1}{5}(x^2 - 4x + 3) > 0.$$

SOLUTION

$$
\begin{aligned}
\tfrac{1}{5}(x^2 - 4x + 3) &> 0, & \\
x^2 - 4x + 3 &> 0, & \text{(multiplied by 5)} \\
(x - 1)(x - 3) &> 0. & \text{(factored)}
\end{aligned}
$$

The product $(x - 1)(x - 3)$ is positive iff both factors are negative:

$$x - 1 < 0, \qquad x - 3 < 0,$$

or both factors are positive:

$$x - 1 > 0, \qquad x - 3 > 0.$$

The first situation arises when

$$x < 1$$

and the second situation arises when

$$x > 3.$$

As the solution we have

$$\{x: x < 1 \text{ or } x > 3\} = (-\infty, 1) \cup (3, \infty). \quad \square$$

Problem.   Solve the inequality

$$\frac{x + 2}{1 - x} > 1.$$

SOLUTION.   Since division by 0 is not defined, $1 - x$ cannot be 0 and therefore x cannot be 1. To clear fractions we will have to multiply by $1 - x$. It is therefore important to keep track of the sign of $1 - x$.

Let's assume first that

$$1 - x > 0, \quad \text{or equivalently, that} \quad x < 1.$$

In this situation

$$\frac{x + 2}{1 - x} > 1$$

gives

$$x + 2 > 1 - x,$$
$$x > -1 - x,$$
$$2x > -1,$$
$$x > -\tfrac{1}{2}.$$

This means that, of all the numbers x that are less than 1, the only ones that satisfy the initial inequality are those that are greater than $-\tfrac{1}{2}$. The set $(-\tfrac{1}{2}, 1)$ is therefore at least a partial solution.

Let's assume now that

$$1 - x < 0, \quad \text{or equivalently, that} \quad x > 1.$$

In this situation,

$$\frac{x + 2}{1 - x} > 1$$

gives

$$x + 2 < 1 - x,$$
$$x < -1 - x,$$
$$2x < -1,$$
$$x < -\tfrac{1}{2}.$$

This means that, of all the numbers that are greater than 1, the only ones that satisfy the initial inequality are those that are less than $-\tfrac{1}{2}$. There are no such numbers.

The set $(-\tfrac{1}{2}, 1)$, which we obtained in the first case, is therefore the total solution. $\square$

We come now to some inequalities that involve absolute values. We begin with the inequality

$$|x| < d.$$

The easiest way to handle this inequality is to think of $|x|$ as the distance between 0 and x. It then becomes immediately obvious that

$$|x| < d \quad \text{iff} \quad -d < x < d.$$

The inequality

$$|x - c| < d$$

can be handled with similar ease. The number $|x - c|$ represents the distance between c and x. It is therefore obvious that

$$|x - c| < d \quad \text{iff} \quad c - d < x < c + d.$$

We illustrate the boxed-in assertions in Figure 1.3.1.

FIGURE 1.3.1

As another example of this type we take the inequality

$$0 < |x - c| < d.$$

Here we have the inequality

$$|x - c| < d$$

with the additional requirement that x cannot be c. Consequently we have

$$0 < |x - c| < d \quad \text{iff} \quad c - d < x < c \quad \text{or} \quad c < x < c + d.$$

One of the fundamental inequalities of calculus is the triangle inequality

$$|a + b| \le |a| + |b|.$$

Problem. Prove the validity of the triangle inequality

$$|a + b| \le |a| + |b|.$$

PROOF. The proof is easy if you think of $|x|$ as $\sqrt{x^2}$. We begin by noting that

$$(a + b)^2 = a^2 + 2ab + b^2 \le |a|^2 + 2|a||b| + |b|^2 = (|a| + |b|)^2.$$

Comparing the extremes of the inequality and taking square roots, we get

$$\sqrt{(a+b)^2} \le |a| + |b|.$$

The result follows from observing that

$$\sqrt{(a+b)^2} = |a+b|. \quad \square$$

Here is another inequality that comes up in calculus:

$$\boxed{\big||x| - |c|\big| \le |x - c|.}$$

Problem. Prove that

$$\big||x| - |c|\big| \le |x - c|.$$

PROOF. Once again we begin by squaring.

$$\begin{aligned}
\big||x| - |c|\big|^2 = (|x| - |c|)^2 &= |x|^2 - 2|x||c| + |c|^2 \\
&= x^2 - 2|x||c| + c^2 \\
&\le x^2 - 2xc + c^2 \qquad \text{(why?)} \\
&= (x - c)^2.
\end{aligned}$$

Comparing the extremes we get

$$\big||x| - |c|\big|^2 \le (x - c)^2.$$

Now we take square roots:

$$\big||x| - |c|\big| \le \sqrt{(x - c)^2} = |x - c|. \quad \square$$

Exercises†

Solve the following equations and inequalities.

*1. $|x - 2| = 0$.

2. $|3x - 5| = 0$.

*3. $|(x - a)(x - b)| = 0$.

*4. $|x - 2| < 1$.

5. $|x - 2| \le 1$.

*6. $|x - 2| > 1$.

*7. $4x - 1 < x$.

8. $4(1 - 2x) < 5$.

*9. $(x - 1)(x + 1) < 0$.

*10. $x \le |x|$.

11. $x - 2 = |2 - x|$.

*12. $|x| \le x$.

*13. $x^3 + x < 0$.

14. $x^4 - 3x^3 + 2x^2 > 0$.

*15. $|x - 2| < x$.

*16. $(x - 1)(x + 1) \le 3$.

17. $\dfrac{1}{x} < x$.

*18. $x + \dfrac{1}{x} \ge 1$.

*19. $\dfrac{x}{x - 5} \ge 0$.

20. $|2x - 3| < \frac{1}{2}$.

---

† The starred exercises have answers at the back of the book.

1.4 Some Comments on Functions

As the last step of our review of elementary mathematics we single out for special attention some properties of functions. In particular we go over some terminology.

Domain and Range

First we need a working definition for the word *function*. Let D be a set of real numbers. By a *function* on D we mean a rule which assigns a unique number to each number in D. The set D is called the *domain* of the function. The set of assignments that the function makes (the set of values that the function takes on) is called the *range* of the function.

Here are some examples. We begin with the squaring function

$$f(x) = x^2, \qquad x \text{ real.}$$

The domain of f is explicitly given as the set of all real numbers. As x runs through all the real numbers, x^2 runs through all the nonnegative numbers. The range is therefore $[0, \infty)$. In abbreviated form we can write

$$\operatorname{dom}(f) = (-\infty, \infty) \quad \text{and} \quad \operatorname{ran}(f) = [0, \infty).$$

Now take the function

$$g(x) = \sqrt{x + 4}, \qquad x \in [0, 5].$$

The domain of g is given as the closed interval $[0, 5]$. At 0, g takes on the value 2:

$$g(0) = \sqrt{0 + 4} = \sqrt{4} = 2.$$

At 5, g takes on the value 3:

$$g(5) = \sqrt{5 + 4} = \sqrt{9} = 3.$$

As x runs through all the numbers from 0 to 5, $g(x)$ runs through all the numbers from 2 to 3. The range of g is therefore the closed interval $[2, 3]$. In abbreviated form

$$\operatorname{dom}(g) = [0, 5], \qquad \operatorname{ran}(g) = [2, 3].$$

The graph of f is the parabola depicted in Figure 1.4.1. The graph of g is the arc of Figure 1.4.2.

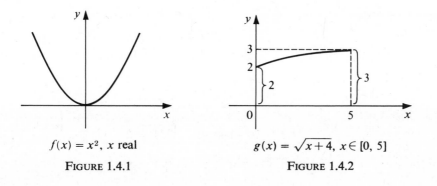

$f(x) = x^2$, x real

FIGURE 1.4.1

$g(x) = \sqrt{x + 4}$, $x \in [0, 5]$

FIGURE 1.4.2

Sometimes the domain is not explicitly given. For example, we might write

$$F(x) = x^3 \quad \text{and} \quad G(x) = \sqrt{x}. \quad \text{(Figures 1.4.3 and 1.4.4)}$$

In such cases take as the domain the maximal set for which the definition makes sense. For F that would be the set of real numbers and for G, the set of nonnegative numbers.

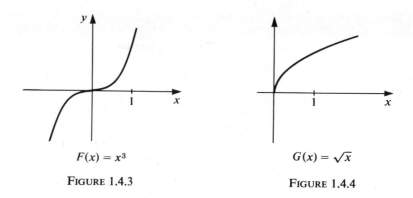

$$F(x) = x^3$$

FIGURE 1.4.3

$$G(x) = \sqrt{x}$$

FIGURE 1.4.4

Problem. Find the domain and range of

$$f(x) = \frac{1}{\sqrt{2 - x}} + 5.$$

SOLUTION. First we look for the domain. To be able to form $\sqrt{2 - x}$ we need $2 - x \geq 0$ and therefore $x \leq 2$. But at $x = 2$, $\sqrt{2 - x} = 0$ and its reciprocal makes no sense. We must therefore restrict x to $x < 2$. The domain is thus $(-\infty, 2)$.

Now we look for the range. As x runs through $(-\infty, 2)$, $\sqrt{2 - x}$ runs through the positive numbers and so does its reciprocal. The range of f consists therefore of the positive numbers plus 5. In short, ran $(f) = (5, \infty)$. $\square$

One-to-Oneness

A function can take on the same value at different points of its domain. Thus, for example, the squaring function takes on the same value at $-c$ as it does at c. In the case of

$$f(x) = (x - 3)(x - 5)$$

we have not only

$$f(3) = 0 \quad \text{but also} \quad f(5) = 0.$$

Functions for which this kind of repetition does not occur are called *one-to-one*.

Definition of One-to-Oneness

A function is said to be *one-to-one* iff there are no two points at which it takes on the same value.

As examples we can take
$$f(x) = x^3 \quad \text{and} \quad g(x) = \sqrt{x}.$$

The cubing function is one-to-one because no two numbers have the same cube. The square root function is one-to-one because no two numbers have the same square root.

In the case of a one-to-one function no horizontal line intersects the graph more than once. (Figure 1.4.5) If a horizontal line intersects the graph more than once, then the function is not one-to-one. (Figure 1.4.6)

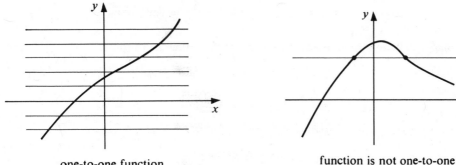

one-to-one function

FIGURE 1.4.5

function is not one-to-one

FIGURE 1.4.6

Combining Functions

Functions with a common domain can be added and subtracted:
$$(f \pm g)(x) = f(x) \pm g(x).$$

With
$$f(x) = x^2 \quad \text{and} \quad g(x) = x + 1,$$

we have
$$(f + g)(x) = f(x) + g(x) = x^2 + x + 1$$

and
$$(f - g)(x) = f(x) - g(x) = x^2 - x - 1.$$

Functions with a common domain can be multiplied:
$$(fg)(x) = f(x)g(x).$$

With
$$f(x) = x^2 \quad \text{and} \quad g(x) = x + 1,$$

we have
$$(fg)(x) = f(x)g(x) = x^2(x + 1) = x^3 + x^2.$$

Functions with a common domain can be divided:

$$\frac{f}{g}(x) = \frac{f(x)}{g(x)} \quad \text{provided that} \quad g(x) \neq 0.$$

With

$$f(x) = x^2 \quad \text{and} \quad g(x) = x^4 + 1,$$

we have

$$\frac{f}{g}(x) = \frac{f(x)}{g(x)} = \frac{x^2}{x^4 + 1}.$$

A function can also be multiplied by a scalar (a real number) α:

$$(\alpha f)(x) = \alpha f(x).$$

With

$$\alpha = 4 \quad \text{and} \quad f(x) = 2x^2 - 1,$$

we have

$$(4f)(x) = 4f(x) = 4(2x^2 - 1) = 8x^2 - 4.$$

Example. There is no reason to always use the letter x. With

$$f(t) = |2 - t| \quad \text{and} \quad g(t) = t^2 + 1,$$

we have

$$(f + g)(t) = f(t) + g(t) = |2 - t| + t^2 + 1,$$
$$(f - g)(t) = f(t) - g(t) = |2 - t| - (t^2 + 1),$$
$$(fg)(t) = f(t)g(t) = |2 - t|(t^2 + 1),$$
$$\frac{f}{g}(t) = \frac{f(t)}{g(t)} = \frac{|2 - t|}{t^2 + 1},$$
$$(2f)(t) = 2f(t) = 2|2 - t|. \quad \square$$

Exercises

Find the domain and the range.

*1. $f(x) = |x|$.

*3. $F(t) = 2t - 1$.

*5. $f(x) = \dfrac{1}{x^2}$.

*7. $f(t) = \sqrt{1 - t}$.

*9. $f(x) = \sqrt{1 - x} - 1$.

*11. $h(x) = \dfrac{1}{\sqrt{1 - x}}$.

2. $g(x) = x^2 - 1$.

4. $G(z) = \sqrt{z} - 1$.

6. $g(x) = \dfrac{1}{x}$.

8. $g(t) = \sqrt{t - 1}$.

10. $g(x) = \sqrt{x - 1} - 1$.

12. $F(x) = \sqrt{1 - x^2}$.

Determine whether the given function is one-to-one.

*13. $f(x) = x + 3$.   14. $g(x) = x^2 + 1$.

*15. $f(x) = x^2 - 3x + 2$.   16. $g(x) = x^3 + 1$.

*17. $f(x) = (x - 1)(x - 2)$.   18. $g(x) = x^{1/3}$.

*19. $f(x) = x^4 + 5$.   20. $g(t) = \sqrt{t - 1}$.

*21. $f(t) = \sqrt{1 - t}$.   22. $g(z) = \sqrt{1 - z^2}$.

*23. $f(t) = \sqrt{1 - t^3}$.   24. $g(x) = x|x|$.

*25. $f(x) = x^2|x|$.   26. $g(x) = x + \dfrac{1}{x}$.

*27. $f(x) = (2 - 3x)^3$.   28. $g(x) = (2 - 3x^2)^3$.

A function f is called *even* iff

$$f(-x) = f(x) \quad \text{for all } x \in \text{dom }(f).$$

It is called *odd* iff

$$f(-x) = -f(x) \quad \text{for all } x \in \text{dom }(f).$$

Test each of the functions below for evenness and oddness.

*29. $f(x) = x^3$.   30. $f(x) = x^2$.

*31. $g(x) = x(x - 1)$.   32. $g(x) = x(x^2 + 1)$.

*33. $f(x) = |x|$.   34. $f(x) = x + \dfrac{1}{x}$.

*35. What can you conclude about the product of two odd functions?

*36. What can you conclude about the product of two even functions?

*37. What can you conclude about the product of an odd function and an even function?

*38. What is the characteristic feature of the graph of an even function?

*39. What is the characteristic feature of the graph of an odd function?

1.5   Composition of Functions

The arithmetical ways of combining functions that we described in the last section should present no difficulties. There is another way of combining functions on which we want to refresh your memory. To describe it, we begin with two functions f and g and a number x in the domain of g. By applying g to x, we get the number $g(x)$. If $g(x)$ is in the domain of f, then we can apply f to $g(x)$ and thereby obtain the number $f(g(x))$.

What is $f(g(x))$? It is the result of first applying g to x and then applying f to $g(x)$. The idea is illustrated in Figure 1.5.1.

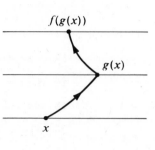

FIGURE 1.5.1

If the range of g is completely contained in the domain of f (namely, if each value that g takes on is in the domain of f), then we can form $f(g(x))$ for *each* x in the domain of g and in this manner create a new function. This new function—it takes each x in the domain of g and assigns to it the value $f(g(x))$—is called the *composition* of f and g and is denoted by the symbol $f \circ g$. What is this function $f \circ g$? It is the function which results from first applying g and then applying f.

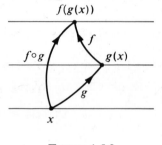

FIGURE 1.5.2

Definition of Composition

If the range of the function g is contained in the domain of the function f, then the composition $f \circ g$ (read "f circle g") is the function defined on the domain of g by setting

$$(f \circ g)(x) = f(g(x)).$$

Examples

(1) Suppose that

$$g(x) = x^2 \qquad \text{(the squaring function)}$$

and

$$f(x) = x + 1. \qquad \text{(the function which adds 1)}$$

Then

$$(f \circ g)(x) = f(g(x)) = g(x) + 1 = x^2 + 1.$$

In other words, $f \circ g$ is the function which *first* squares and *then* adds 1. □

(2) If on the other hand

$$g(x) = x + 1 \qquad\qquad\qquad (g \text{ adds } 1)$$

and

$$f(x) = x^2, \qquad\qquad\qquad (f \text{ squares})$$

then

$$(f \circ g)(x) = f(g(x)) = [g(x)]^2 = (x + 1)^2.$$

Here $f \circ g$ is the function which *first* adds 1 and *then* squares.   □

(3) If

$$g(x) = \frac{1}{x} + 1 \qquad\qquad (g \text{ adds } 1 \text{ to the reciprocal})$$

and

$$f(x) = x^2 - 5, \qquad (f \text{ subtracts } 5 \text{ from the square})$$

then

$$(f \circ g)(x) = f(g(x)) = [g(x)]^2 - 5 = \left(\frac{1}{x} + 1\right)^2 - 5.$$

Here $f \circ g$ *first* adds 1 to the reciprocal and *then* subtracts 5 from the square.   □

(4) If

$$f(x) = \frac{1}{x} + x^2 \quad \text{and} \quad g(x) = \frac{x^2 + 1}{x^4 + 1},$$

then

$$(f \circ g)(x) = f(g(x)) = \frac{1}{g(x)} + [g(x)]^2 = \frac{x^4 + 1}{x^2 + 1} + \left(\frac{x^2 + 1}{x^4 + 1}\right)^2. \quad □$$

It is possible to form the composition of more than two functions. For example, the triple composition $f \circ g \circ h$ consists of first h, then g, and then f:

$$(f \circ g \circ h)(x) = f(g(h(x))).$$

Obviously we can go on in this manner with as many functions as we like.

Example. If

$$f(x) = \frac{1}{x}, \qquad g(x) = x + 2, \qquad h(x) = (x^2 + 1)^2,$$

then

$$(f \circ g \circ h)(x) = f(g(h(x)))$$

$$= \frac{1}{g(h(x))}$$

$$= \frac{1}{h(x) + 2}$$

$$= \frac{1}{(x^2 + 1)^2 + 2}. \quad □$$

To apply some of the techniques of calculus you will need to be able to recognize composites. You will need to be able to start with a function, say $F(x) = (x + 1)^5$, and recognize how it is a composition.

Problem. Find functions f and g such that $f \circ g = F$ if

$$F(x) = (x + 1)^5.$$

A SOLUTION. The function consists of first adding 1 and then taking the fifth power. We can therefore set

$$g(x) = x + 1 \qquad \text{(adding 1)}$$

and

$$f(x) = x^5. \qquad \text{(taking the fifth power)}$$

As you can see,

$$(f \circ g)(x) = f(g(x)) = [g(x)]^5 = (x + 1)^5. \quad \square$$

Problem. Find functions f and g such that $f \circ g = F$ if

$$F(x) = \frac{1}{x} - 6.$$

A SOLUTION. F takes the reciprocal and then subtracts 6. We can therefore set

$$g(x) = \frac{1}{x} \qquad \text{(taking the reciprocal)}$$

and

$$f(x) = x - 6. \qquad \text{(subtracting 6)}$$

As you can check,

$$(f \circ g)(x) = f(g(x)) = g(x) - 6 = \frac{1}{x} - 6. \quad \square$$

Problem. Find three functions f, g, h such that $f \circ g \circ h = F$ if

$$F(x) = \frac{1}{|x| + 3}.$$

A SOLUTION. F takes the absolute value, adds 3, and then inverts. Let h take the absolute value:

$$\text{set} \quad h(x) = |x|.$$

Let g add 3:

$$\text{set} \quad g(x) = x + 3.$$

Let f do the inverting:

$$\text{set} \quad f(x) = \frac{1}{x}.$$

With this choice of f, g, h we have

$$(f \circ g \circ h)(x) = f(g(h(x)))$$

$$= \frac{1}{g(h(x))}$$

$$= \frac{1}{h(x) + 3}$$

$$= \frac{1}{|x| + 3}. \quad \square$$

In the next problem we start with the same function

$$F(x) = \frac{1}{|x| + 3},$$

but this time we are asked to break it up as the composition of only two functions.

Problem. Find two functions f and g such that $f \circ g = F$ if

$$F(x) = \frac{1}{|x| + 3}.$$

SOME SOLUTIONS. F takes the absolute value, adds 3, and then inverts. We can let g do the first two things by setting

$$g(x) = |x| + 3$$

and then let f do the inverting by setting

$$f(x) = \frac{1}{x}.$$

Or, we could let g just take the absolute value,

$$g(x) = |x|,$$

and then have f add 3 and invert:

$$f(x) = \frac{1}{x + 3}.$$

There are still other possible choices for f and g. For example, we could set

$$g(x) = x \quad \text{and} \quad f(x) = \frac{1}{|x| + 3}$$

or

$$g(x) = \frac{1}{|x| + 3} \quad \text{and} \quad f(x) = x,$$

but these last choices of f and g don't advance us at all. They are much like factoring 12 into $12 \cdot 1$ or $1 \cdot 12$. $\quad \square$

Exercises

Form the composition $f \circ g$.

*1. $f(x) = 2x + 5, \quad g(x) = x^2.$ 2. $f(x) = x^2, \quad g(x) = 2x + 5.$

*3. $f(x) = \sqrt{x}, \quad g(x) = x^2 + 5.$ 4. $f(x) = x^2 + x, \quad g(x) = x^{1/3}.$

*5. $f(x) = \dfrac{1}{x}, \quad g(x) = \dfrac{1}{x}.$ 6. $f(x) = x^2, \quad g(x) = \dfrac{1}{x - 1}.$

*7. $f(x) = x - 1, \quad g(x) = \dfrac{1}{x}.$ 8. $f(x) = x^2 - 1, \quad g(x) = x(x - 1).$

*9. $f(x) = \dfrac{1}{\sqrt{x} - 1}, \quad g(x) = (x^2 + 2)^2.$

10. $f(x) = \dfrac{1}{x} - \dfrac{1}{x + 1}, \quad g(x) = \dfrac{1}{x^2}.$

Form the composition $f \circ g \circ h$.

*11. $f(x) = 4x, \quad g(x) = x - 1, \quad h(x) = x^2.$

12. $f(x) = x - 1, \quad g(x) = 4x, \quad h(x) = x^2.$

*13. $f(x) = x^2, \quad g(x) = x - 1, \quad h(x) = x^4.$

14. $f(x) = x - 1, \quad g(x) = x^2, \quad h(x) = 4x.$

*15. $f(x) = \dfrac{1}{x}, \quad g(x) = \dfrac{1}{2x + 1}, \quad h(x) = x^2.$

16. $f(x) = \dfrac{x + 1}{x}, \quad g(x) = \dfrac{1}{2x + 1}, \quad h(x) = x^2.$

17. Find f such that $f \circ g = F$ given that

*(a) $g(x) = \dfrac{1 - x}{1 + x}$ and $F(x) = \dfrac{1 + x}{1 - x}.$

(b) $g(x) = x^2$ and $F(x) = ax^2 + b.$

*(c) $g(x) = -x^2$ and $F(x) = \sqrt{a^2 - x^2}.$

18. Find g such that $f \circ g = F$ given that

(a) $f(x) = x^2$ and $F(x) = \left(1 - \dfrac{1}{x^4}\right)^2.$

*(b) $f(x) = x + \dfrac{1}{x}$ and $F(x) = a^2 x^2 + \dfrac{1}{a^2 x^2}.$

(c) $f(x) = x^2 + 1$ and $F(x) = (2x^3 - 1)^2 + 1.$

1.6   Inverse Functions

I am thinking of a real number and its cube is 8. Can you figure out the number? Obviously yes. It must be 2.

Now I am thinking of another number and its square is 9. Can you figure out the number? No, not unless you are clairvoyant. It could be 3 but it could also be -3.

The difference between the two situations can be phrased this way: In the first case we are dealing with a one-to-one function—no two numbers have the same cube. So, knowing the cube, we can work back to the original number. In the second case we are dealing with a function which is not one-to-one—different numbers can have the same square. Knowing the square we still can't figure out what the original number was.

Throughout the rest of this section we shall be looking at functional values and tracing them back to where they came from. So as to be able to trace these values back, we shall restrict ourselves entirely to functions which are one-to-one.

We begin with a theorem about one-to-one functions.

Theorem

If f is a one-to-one function, then there is one and only one function g which is defined on the range of f and satisfies the equation

$$f(g(x)) = x \quad \text{for all } x \text{ in the range of } f.$$

PROOF. The proof is easy. If x is in the range of f, then f must take on the value x at some number. Since f is one-to-one, there can be only one such number. We've called it $g(x)$. $\square$

The function that we've named g in the theorem is called the inverse of f and is usually denoted by f^{-1}.

Definition of Inverse Function

Let f be a one-to-one function. The *inverse* of f, denoted by f^{-1}, is the unique function which is defined on the range of f and satisfies the equation

$$f(f^{-1}(x)) = x \quad \text{for all } x \text{ in the range of } f.$$

Problem.   Find the inverse of f for

$$f(x) = x^3.$$

SOLUTION.   We begin by setting

$$f(f^{-1}(x)) = x.$$

Since f is the cubing function, we must have

$$f(f^{-1}(x)) = [f^{-1}(x)]^3$$

so that

$$[f^{-1}(x)]^3 = x.$$

Taking cube roots we have

$$f^{-1}(x) = x^{1/3}.$$

What we've found is that the cube root function is the inverse of the cubing function. As you can check,

$$f(f^{-1}(x)) = [f^{-1}(x)]^3 = [x^{1/3}]^3 = x. \quad \square$$

Problem.   Find the inverse of f for

$$f(x) = 3x - 5.$$

Solution.   Set

$$f(f^{-1}(x)) = x.$$

With our choice of f this gives

$$3f^{-1}(x) - 5 = x,$$

which we can now solve for $f^{-1}(x)$:

$$3f^{-1}(x) = x + 5,$$
$$f^{-1}(x) = \tfrac{1}{3}x + \tfrac{5}{3}.$$

As you can check,

$$f(f^{-1}(x)) = 3f^{-1}(x) - 5$$
$$= 3(\tfrac{1}{3}x + \tfrac{5}{3}) - 5$$
$$= (x + 5) - 5$$
$$= x. \quad \square$$

Problem.   Find the inverse of f for

$$f(x) = (1 - x^3)^{1/5} + 2.$$

Solution.   Once again we begin by setting

$$f(f^{-1}(x)) = x.$$

Since

$$f(f^{-1}(x)) = \{1 - [f^{-1}(x)]^3\}^{1/5} + 2,$$

we must have

$$\{1 - [f^{-1}(x)]^3\}^{1/5} + 2 = x.$$

We now solve for $f^{-1}(x)$:

$$\{1 - [f^{-1}(x)]^3\}^{1/5} = x - 2$$
$$1 - [f^{-1}(x)]^3 = (x - 2)^5$$
$$[f^{-1}(x)]^3 = 1 - (x - 2)^5$$
$$f^{-1}(x) = [1 - (x - 2)^5]^{1/3}. \quad \square$$

By definition f^{-1} satisfies the equation

(1.6.1) $$\boxed{f(f^{-1}(x)) = x \quad \text{for all } x \text{ in the range of } f.}$$

Replacing the letter x by the letter t, we have

(1.6.2) $$f(f^{-1}(t)) = t \quad \text{for all } t \text{ in the range of } f.$$

It is also true that

(1.6.3) $$\boxed{f^{-1}(f(x)) = x \quad \text{for all } x \text{ in the domain of } f.}$$

PROOF.   If x is in the domain of f then $f(x)$ is in the range of f and therefore by (1.6.2)

$$f(f^{-1}(f(x))) = f(x). \qquad\qquad (\text{let } t = f(x))$$

This tells us that f takes on the same value at $f^{-1}(f(x))$ as it does at x. With f one-to-one, this can only happen if

$$f^{-1}(f(x)) = x. \quad \square$$

That

$$f^{-1}(f(x)) = x \quad \text{for all } x \text{ in the domain of } f$$

tells us that f^{-1} "undoes" the work of f: f takes x to $f(x)$ and f^{-1} takes $f(x)$ back to x. See Figure 1.6.1.

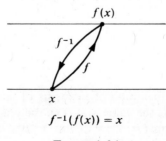

$$f^{-1}(f(x)) = x$$

FIGURE 1.6.1

That

$$f(f^{-1}(x)) = x \quad \text{for all } x \text{ in the range of } f$$

tells us that f "undoes" the work of f^{-1}: f^{-1} takes x to $f^{-1}(x)$ and f takes $f^{-1}(x)$ back to x. See Figure 1.6.2.

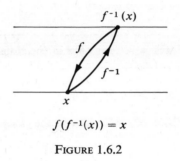

$$f(f^{-1}(x)) = x$$

FIGURE 1.6.2

There is an interesting relation between the graph of a one-to-one function and the graph of its inverse. See Figure 1.6.3. Each graph is the mirror image of the other, the mirror being the line $y = x$. Instead of giving a formal proof we refer you to

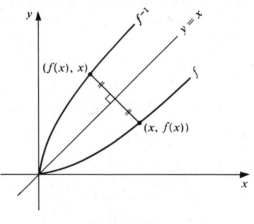

FIGURE 1.6.3

Figure 1.6.4. As you know the graph of f consists of points with coordinates of the form $(x, f(x))$. Since f^{-1} has the value x at $f(x)$, the graph of f^{-1} consists of points with coordinates of the form $(f(x), x)$. These points, $(x, f(x))$ and $(f(x), x)$, are opposite vertices of the shaded square. $\square$

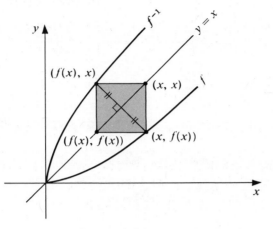

$(f(x), x)$
(x, x)
$(f(x), f(x))$
$(x, f(x))$

FIGURE 1.6.4

Problem. Given that the graph of f is as in Figure 1.6.5, sketch the graph of f^{-1}.

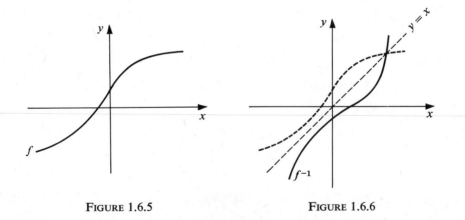

FIGURE 1.6.5   FIGURE 1.6.6

SOLUTION. First we draw in the line $y = x$ and then reproduce the mirror image of the graph of f. See Figure 1.6.6. ☐

Exercises

Find the inverse.

*1. $f(x) = 5x + 3$.   2. $f(x) = 4x - 7$.

*3. $f(x) = mx + b$.   4. $f(x) = x^5$.

*5. $f(x) = x^5 + 1$.   6. $f(x) = (1 - x)^3$.

*7. $f(x) = 1 + 3x^3$.   8. $f(x) = x^3 - 1$.

*9. $f(x) = (x + 1)^3 + 2.$ 10. $f(x) = (4x - 1)^3.$

*11. $f(x) = x^{3/5}.$ 12. $f(x) = 1 + x^5.$

*13. $f(x) = (1 - x^3)^5.$ 14. $f(x) = 1 - (x - 2)^{1/3}.$

*15. $f(x) = \dfrac{1}{x}.$ 16. $f(x) = \dfrac{1}{1 - x}.$

*17. $f(x) = \dfrac{1}{x^3 + 1}.$ 18. $f(x) = \dfrac{1}{1 - x} - 1.$

19. Sketch the graph of f^{-1} given that the graph of f is as in
   (a) Figure 1.6.7.   (b) Figure 1.6.8.   (c) Figure 1.6.9.

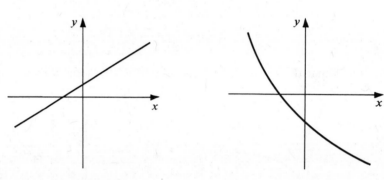

FIGURE 1.6.7   FIGURE 1.6.8

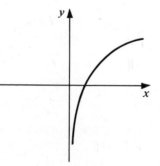

FIGURE 1.6.9

20. (a) (*Optional*) Show that the composition of two one-to-one functions is one-to-one.
   (b) (*Optional*) Express $(f \circ g)^{-1}$ in terms of f^{-1} and g^{-1}.

Limits and Continuity

2.1 The Idea of Limit

Chapter 1 was not calculus. It was just preliminary, just warm-up. Calculus begins with the idea of limit.

To introduce this idea we begin with a number l and a function f defined *near* the number c but not necessarily defined at c itself. A rough translation of

$$\lim_{x \to c} f(x) = l$$

might read

$$\text{as } x \text{ approaches } c, \quad f(x) \text{ approaches } l$$

or

$$\text{for } x \text{ close to } c, \quad f(x) \text{ is close to } l$$

or still

$$\text{for } x \text{ approximately equal to } c, \quad f(x) \text{ is approximately equal to } l.$$

We illustrate the idea in Figure 2.1.1. The curve represents the graph of a function f. The number c appears on the x-axis, the limit l on the y-axis. As x approaches c (along the x-axis), $f(x)$ approaches l (along the y-axis).

In taking the limit as x approaches c, it does not matter whether or not f is defined at c and, if so, how it is defined there. The only thing that matters is how f is defined *near* c. For example, in Figure 2.1.2 the graph of f is a broken curve defined peculiarly at c and yet

$$\lim_{x \to c} f(x) = l$$

because, as suggested in Figure 2.1.3,

$$\text{as } x \text{ approaches } c, \quad f(x) \text{ approaches } l.$$

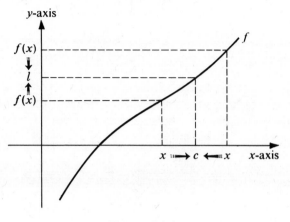

FIGURE 2.1.1

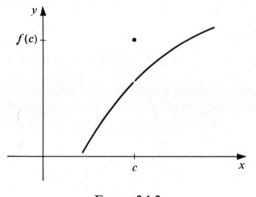

FIGURE 2.1.2

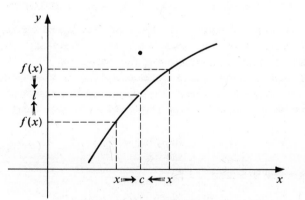

FIGURE 2.1.3

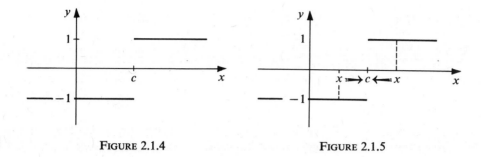

FIGURE 2.1.4   FIGURE 2.1.5

An interesting situation arises in the case of the function

$$f(x) = \begin{cases} -1, & x < c \\ 1, & x > c \end{cases}.$$ (Figure 2.1.4)

As x approaches c from the left (Figure 2.1.5), $f(x)$ is constantly -1. As c approaches c from the right, $f(x)$ is constantly 1. There is *no one* number which $f(x)$ approaches as x approaches c from both sides; namely,

$$\lim_{x \to c} f(x) \quad \text{does not exist.}$$

Other instances in which

$$\lim_{x \to c} f(x) \quad \text{does not exist}$$

are given by

$$f(x) = \frac{1}{x - c}, \quad f(x) = \frac{1}{|x - c|}, \quad f(x) = \begin{cases} 1, & x \text{ rational} \\ 0, & x \text{ irrational} \end{cases}.$$

We begin with

$$f(x) = \frac{1}{x - c}.$$

The graph is displayed in Figure 2.1.6. As x approaches c from the right (Figure 2.1.7), $f(x)$ becomes arbitrarily large—larger than any preassigned positive number.

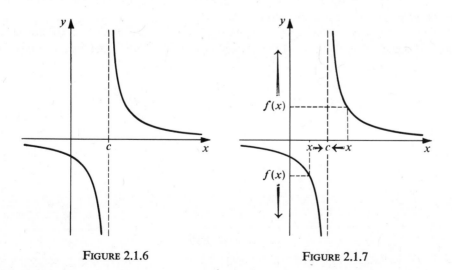

FIGURE 2.1.6   FIGURE 2.1.7

As x approaches c from the left, $f(x)$ becomes arbitrarily large negative—less than any preassigned negative number. Under these circumstances $f(x)$ can't possibly approach a fixed number.

A slightly different situation occurs in the case of

$$f(x) = \frac{1}{|x - c|}. \qquad \text{(Figure 2.1.8)}$$

Here, as x approaches c from either side, $f(x)$ becomes arbitrarily large. (Figure 2.1.9) Becoming arbitrarily large, $f(x)$ cannot stay close to any fixed number l.

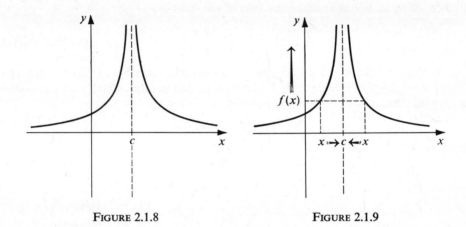

FIGURE 2.1.8   FIGURE 2.1.9

The final example

$$f(x) = \begin{cases} 1, & x \text{ rational} \\ 0, & x \text{ irrational} \end{cases}, \qquad \text{(see Figure 2.1.10)}$$

is somewhat exotic. As x approaches c, x passes through both rational and irrational numbers. As this happens, $f(x)$ jumps wildly back and forth between 0 and 1, and cannot stay close to any fixed number l.

If you've taken the attitude that all this is very imprecise, then you are absolutely right. It is imprecise but it need not remain imprecise. One of the great triumphs of calculus has been its capacity to formulate limit statements with precision. But for

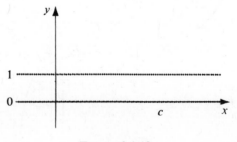

FIGURE 2.1.10

this precision you'll have to wait until Section 2.2. Throughout the rest of the present section we'll go on in an informal, intuitive manner.

Numerical Examples

(1) $$\lim_{x \to 3} (2x + 5) = 11.$$

As x approaches 3, $2x$ approaches 6, and $2x + 5$ approaches 11.   □

(2) $$\lim_{x \to 2} (x^2 - 1) = 3.$$

As x approaches 2, x^2 approaches 4 and $x^2 - 1$ approaches $4 - 1 = 3$.   □

(3) $$\lim_{x \to 3} \frac{1}{x - 1} = \frac{1}{2}.$$

As x approaches 3, $x - 1$ approaches 2 and the reciprocal of $x - 1$ approaches $\frac{1}{2}$.   □

(4) $$\lim_{x \to 2} \frac{1}{x - 2} \quad \text{does not exist.}$$

It was argued before more generally that

$$\lim_{x \to c} \frac{1}{x - c} \quad \text{does not exist}$$

no matter what c is. (Figures 2.1.6 and 2.1.7)   □

Before going to the next examples, remember that, in taking the limit as x approaches a given number c, it does not matter whether or not the function is defined at the number c and, if so, how it is defined there. The only thing that matters is how the function is defined *near* the number c.

(5) $$\lim_{x \to 2} \frac{x^2 - 4x + 4}{x - 2} = 0.$$

At $x = 2$, the function is undefined: both numerator and denominator are 0. But that doesn't matter. For all $x \neq 2$, and therefore for all x near 2,

$$\frac{x^2 - 4x + 4}{x - 2} = x - 2.$$

It follows that

$$\lim_{x \to 2} \frac{x^2 - 4x + 4}{x - 2} = \lim_{x \to 2} (x - 2) = 0. \quad □$$

(6) $$\lim_{x \to 2} \frac{x - 2}{x - 2} = 1.$$

At $x = 2$, the function is undefined: both numerator and denominator are 0. But, as we said before, that doesn't matter. For all $x \neq 2$, and therefore for all x near 2,

$$\frac{x - 2}{x - 2} = 1$$

so that

$$\lim_{x \to 2} \frac{x - 2}{x - 2} = \lim_{x \to 2} 1 = 1. \quad \square$$

(7) $$\lim_{x \to 2} \frac{x - 2}{x^2 - 4x + 4} \quad \text{does not exist.}$$

This result does not follow from the fact the function is not defined at 2 but rather from the fact that for all $x \neq 2$ and therefore for all x near 2,

$$\frac{x - 2}{x^2 - 4x + 4} = \frac{1}{x - 2}$$

and, as you saw before,

$$\lim_{x \to 2} \frac{1}{x - 2} \quad \text{does not exist.} \quad \square$$

(8) If

$$f(x) = \begin{cases} 3x - 1, & x \neq 2 \\ 45, & x = 2 \end{cases},$$

then

$$\lim_{x \to 2} f(x) = 5.$$

It doesn't matter to us that $f(2) = 45$. For $x \neq 2$,

$$f(x) = 3x - 1$$

so that

$$\lim_{x \to 2} f(x) = \lim_{x \to 2} (3x - 1) = 5. \quad \square$$

(9) If

$$f(x) = \begin{cases} -2x, & x < 1 \\ 2x, & x > 1 \end{cases},$$

then

$$\lim_{x \to 1} f(x) \quad \text{does not exist.}$$

As x approaches 1 from the left, $f(x)$ (being $-2x$) approaches -2. As x approaches 1 from the right, $f(x)$ (being $2x$) approaches 2. There is no one number that $f(x)$ approaches as x approaches 2 from both sides. $\square$

(10) If

$$f(x) = \begin{cases} -2x, & x < 0 \\ 2x, & x > 0 \end{cases},$$

then

$$\lim_{x \to 0} f(x) = 0.$$

As x approaches 0 from the left, $f(x)$ (being $-2x$) approaches 0. As x approaches 0 from the right, $f(x)$ (being $2x$) also approaches 0. As x approaches 0 from either side, $f(x)$ approaches 0. □

Exercises

Decide on intuitive grounds whether or not the indicated limit exists. Evaluate the limits which do exist.

*1. $\lim_{x \to 1} 3x$.

2. $\lim_{x \to 0} (2x - 1)$.

*3. $\lim_{x \to 1} (2 - 5x)$.

4. $\lim_{x \to -2} x^2$.

*5. $\lim_{x \to 0} \dfrac{1}{|x|}$.

6. $\lim_{x \to 1} \dfrac{3}{x + 1}$.

*7. $\lim_{x \to 0} \dfrac{4}{x + 1}$.

8. $\lim_{x \to 2} \dfrac{1}{3x - 6}$.

*9. $\lim_{x \to 3} \dfrac{2x - 6}{x - 3}$.

10. $\lim_{x \to 3} \dfrac{x^2 - 6x + 9}{x - 3}$.

*11. $\lim_{x \to 0} \left(x + \dfrac{1}{x} \right)$.

12. $\lim_{x \to 1} \left(x + \dfrac{1}{x} \right)$.

*13. $\lim_{x \to 0} \dfrac{2x - 5x^2}{x}$.

14. $\lim_{x \to 1} \dfrac{x^2 - 1}{x - 1}$.

*15. $\lim_{x \to 1} \dfrac{x^3 - 1}{x - 1}$.

16. $\lim_{x \to 1} \dfrac{x^2 + 1}{x^2 - 1}$.

*17. $\lim_{x \to 0} f(x)$ if $f(x) = \left\{ \begin{array}{ll} -x^2, & x < 0 \\ x^2, & x > 0 \end{array} \right]$.

*18. $\lim_{x \to 0} f(x)$ if $f(x) = \left\{ \begin{array}{ll} x^2, & x < 0 \\ 1 + x, & x > 0 \end{array} \right]$.

19. $\lim_{x \to 4} f(x)$ if $f(x) = \left\{ \begin{array}{ll} x^2, & x \neq 4 \\ 0, & x = 4 \end{array} \right]$.

20. $\lim_{x \to 0} f(x)$ if $f(x) = \left\{ \begin{array}{ll} 2, & x \text{ rational} \\ -2, & x \text{ irrational} \end{array} \right]$.

2.2   Definition of Limit

In Section 2.1 we tried to convey the idea of limit in an intuitive manner. Our remarks, however, were too vague to be called mathematics. In this section we shall be more precise. To say that

$$\lim_{x \to c} f(x) = l$$

is to say that the *difference between $f(x)$ and l can be made arbitrarily small (smaller than any preassigned positive number) simply by requiring that x be sufficiently close to c.*

Think of a positive number and call it ϵ (epsilon). If it is true that

$$\lim_{x \to c} f(x) = l,$$

then you can be sure that

$$|f(x) - l| < \epsilon \quad \text{for all } x \text{ sufficiently close to } c;$$

namely, you can be sure that there exists a positive number δ (delta) such that

$$\text{if} \quad 0 < |x - c| < \delta, \quad \text{then} \quad |f(x) - l| < \epsilon.$$

Definition of Limit

$$\lim_{x \to c} f(x) = l \quad \text{iff} \quad \begin{cases} \text{for each } \epsilon > 0 \text{ there exists } \delta > 0 \text{ such that} \\ \text{if} \quad 0 < |x - c| < \delta, \quad \text{then} \quad |f(x) - l| < \epsilon. \end{cases}$$

Figures 2.2.1 and 2.2.2 illustrate this definition.

REMARK 1.   Note that for f to have a limit at c it must be defined at all numbers sufficiently close to c, but it need not be defined at c. Even if it is defined at c, its value there is irrelevant.

REMARK 2.   In general, the choice of δ depends upon the previous choice of ϵ. We do not require that there exist a number δ which "works" for *all* ϵ but, rather, that for *each* ϵ there exist a δ which "works" for it.

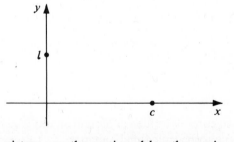

picture c on the x-axis and l on the y-axis

FIGURE 2.2.1

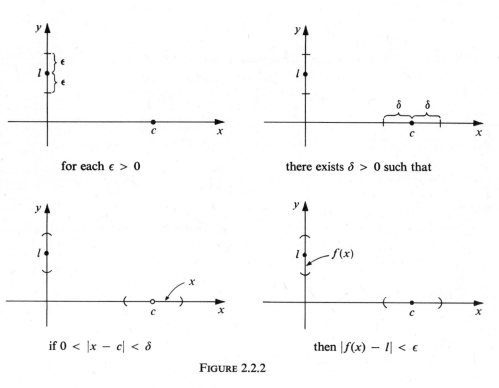

for each $\epsilon > 0$ there exists $\delta > 0$ such that

if $0 < |x - c| < \delta$ then $|f(x) - l| < \epsilon$

FIGURE 2.2.2

In Figure 2.2.3, we give two choices of ϵ and for each we display a suitable δ. For a δ to be suitable, all the points within δ of c (with the possible exception of c itself) must be taken by the function to within ϵ of l. In part (b) of Figure 2.2.3 we began with a smaller ϵ and had to use a smaller δ.

The δ of Figure 2.2.4 is too large for the given ϵ. In particular the points marked x_1 and x_2 in the figure are not taken by the function to within ϵ of l.

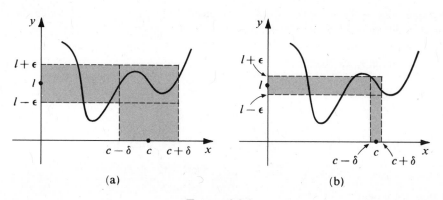

(a)   (b)

FIGURE 2.2.3

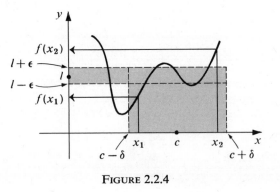

FIGURE 2.2.4

Below we apply the ϵ, δ definition of limit to a wide variety of functions. If you have never run across ϵ, δ arguments before, you may find them confusing at first. It usually takes a little while for the ϵ, δ idea to take hold.

Example. Show that

$$\lim_{x \to 2} (2x - 1) = 3. \hspace{2cm} \text{(Figure 2.2.5)}$$

Finding a δ. Let $\epsilon > 0$. We seek a number $\delta > 0$ such that,

$$\text{if} \quad 0 < |x - 2| < \delta, \quad \text{then} \quad |(2x - 1) - 3| < \epsilon.$$

What we have to do first is establish a connection between

$$|(2x - 1) - 3| \quad \text{and} \quad |x - 2|.$$

The connection is simple:

$$|(2x - 1) - 3| = |2x - 4|$$

so that

$$(1) \hspace{3cm} |(2x - 1) - 3| = 2|x - 2|.$$

To make $|(2x - 1) - 3|$ less than ϵ, we need only make $|x - 2|$ twice as small. This suggests that we choose $\delta = \frac{1}{2}\epsilon$.

   Showing that the δ "works." If $0 < |x - 2| < \frac{1}{2}\epsilon$, then $2|x - 2| < \epsilon$ and, by (1), $|(2x - 1) - 3| < \epsilon$. □

Example. Show that

$$\lim_{x \to -1} (2 - 3x) = 5. \hspace{2cm} \text{(Figure 2.2.6)}$$

Finding a δ. Let $\epsilon > 0$. We seek a number $\delta > 0$ such that,

$$\text{if} \quad 0 < |x - (-1)| < \delta, \quad \text{then} \quad |(2 - 3x) - 5| < \epsilon.$$

To find a connection between

$$|x - (-1)| \quad \text{and} \quad |(2 - 3x) - 5|,$$

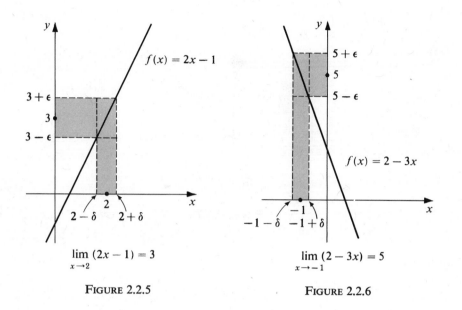

$$\lim_{x \to 2} (2x - 1) = 3$$

FIGURE 2.2.5

$$\lim_{x \to -1} (2 - 3x) = 5$$

FIGURE 2.2.6

we simplify both expressions:

$$|x - (-1)| = |x + 1|$$

and

$$|(2 - 3x) - 5| = |-3x - 3| = 3|-x - 1| = 3|x + 1|.$$

We can conclude that

(2) $$|(2 - 3x) - 5| = 3|x - (-1)|.$$

We can make the expression on the left less than ϵ by making $|x - (-1)|$ three times as small. This suggests that we set $\delta = \frac{1}{3}\epsilon$.

Showing that the δ "works." If $0 < |x - (-1)| < \frac{1}{3}\epsilon$, then $3|x - (-1)| < \epsilon$ and, by (2), $|(2 - 3x) - 5| < \epsilon$. □

Example

(2.2.1) $$\lim_{x \to c} |x| = |c|.$$ (Figure 2.2.7)

PROOF.   Let $\epsilon > 0$. We seek $\delta > 0$ such that

$$\text{if} \quad 0 < |x - c| < \delta, \quad \text{then} \quad \big||x| - |c|\big| < \epsilon.$$

Since

$$\big||x| - |c|\big| \le |x - c|$$

we can choose $\delta = \epsilon$; for

$$\text{if} \quad 0 < |x - c| < \epsilon, \quad \text{then} \quad \big||x| - |c|\big| < \epsilon. □$$

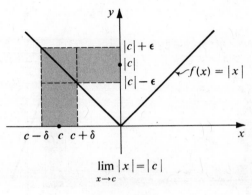

$$\lim_{x \to c} |x| = |c|$$

FIGURE 2.2.7

Example

(2.2.2) $$\lim_{x \to c} x = c.$$ (Figure 2.2.8)

PROOF.   Let $\epsilon > 0$. Here we must find $\delta > 0$ such that

$$\text{if} \quad 0 < |x - c| < \delta, \quad \text{then} \quad |x - c| < \epsilon.$$

Obviously we can choose $\delta = \epsilon$. $\square$

Example

(2.2.3) $$\lim_{x \to c} a = a.$$ (Figure 2.2.9)

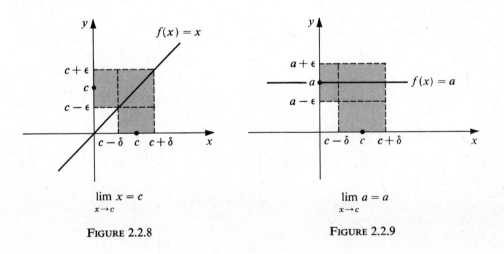

$$\lim_{x \to c} x = c$$

FIGURE 2.2.8

$$\lim_{x \to c} a = a$$

FIGURE 2.2.9

PROOF. Here we are dealing with the constant function

$$f(x) = a.$$

Let $\epsilon > 0$. We must find $\delta > 0$ such that

$$\text{if} \quad 0 < |x - c| < \delta, \quad \text{then} \quad |a - a| < \epsilon.$$

Since

$$|a - a| = 0,$$

we always have

$$|a - a| < \epsilon$$

no matter how δ is chosen; in short any positive number will do for δ.   □

The remaining examples are more complicated. The ϵ, δ proofs that accompany the examples can certainly be skipped at this stage.

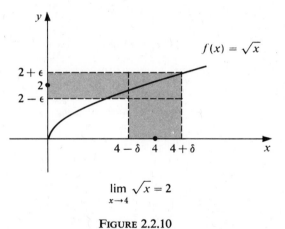

$$\lim_{x \to 4} \sqrt{x} = 2$$

FIGURE 2.2.10

Example

$$\lim_{x \to 4} \sqrt{x} = 2. \qquad\qquad \text{(Figure 2.2.10)}$$

PROOF. *(Optional)* Let $\epsilon > 0$. We seek $\delta > 0$ such that

$$\text{if} \quad 0 < |x - 4| < \delta, \quad \text{then} \quad |\sqrt{x} - 2| < \epsilon.$$

First we want a relation between

$$|x - 4| \quad \text{and} \quad |\sqrt{x} - 2|.$$

To be able to form $\sqrt{x}$ at all we need $x \geq 0$. To insure this we must have $\delta \leq 4$. (Why?)

Remembering that we must have $\delta \leq 4$, let's go on. If $x \geq 0$, then we can form $\sqrt{x}$ and write

$$x - 4 = (\sqrt{x})^2 - 2^2 = (\sqrt{x} + 2)(\sqrt{x} - 2).$$

Taking absolute values, we have

$$|x - 4| = |\sqrt{x} + 2| \, |\sqrt{x} - 2|.$$

Since

$$|\sqrt{x} + 2| > 1,$$

we have

$$|\sqrt{x} - 2| < |x - 4|.$$

This last inequality suggests that we simply set $\delta = \epsilon$. But remember now the previous requirement $\delta \le 4$. We can meet all requirements by setting $\delta =$ minimum of 4 and ϵ.

Let's make sure that this choice of δ "works." We begin by assuming that

$$0 < |x - 4| < \delta.$$

Since $\delta \le 4$, we have $x \ge 0$ and can write

$$|x - 4| = |\sqrt{x} + 2| \, |\sqrt{x} - 2|.$$

Since

$$|\sqrt{x} + 2| > 1,$$

we can conclude that

$$|\sqrt{x} - 2| < |x - 4|.$$

Since

$$|x - 4| < \delta \quad \text{and} \quad \delta \le \epsilon,$$

it does follow that

$$|\sqrt{x} - 2| < \epsilon. \quad \square$$

Example 2.2.4. If

$$h(x) = \begin{cases} -1, & x < 0 \\ 1, & x > 0 \end{cases}, \qquad \text{(Figure 2.2.11)}$$

then

$$\lim_{x \to 0} h(x) \quad \text{does not exist.}$$

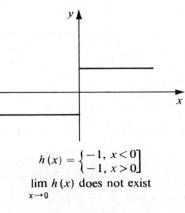

$$h(x) = \begin{cases} -1, & x < 0 \\ -1, & x > 0 \end{cases}$$
$$\lim_{x \to 0} h(x) \text{ does not exist}$$

FIGURE 2.2.11

PROOF. (*Optional*) Suppose on the contrary that

$$\lim_{x \to 0} h(x) = l.$$

Then there must exist $\delta > 0$ such that

if $\quad 0 < |x - 0| < \delta, \qquad$ then $\quad |h(x) - l| < 1. \qquad$ (take $\epsilon = 1$)

Since

$$0 < |\tfrac{1}{2}\delta - 0| < \delta \quad \text{and} \quad h(\tfrac{1}{2}\delta) = 1,$$

we must have

$$|1 - l| < 1.$$

Since

$$0 < |-\tfrac{1}{2}\delta - 0| < \delta \quad \text{and} \quad h(-\tfrac{1}{2}\delta) = -1,$$

we must have

$$|-1 - l| < 1.$$

From $|1 - l| < 1$ we conclude that $l > 0$. From $|-1 - l| < 1$ we conclude that $l < 0$. Clearly no such number l exists. $\square$

Example 2.2.5. If

$$g(x) = \begin{cases} x, & x \text{ rational} \\ 0, & x \text{ irrational} \end{cases},$$

then

$$\lim_{x \to 0} g(x) = 0.$$

PROOF. (*Optional*) Let $\epsilon > 0$. We shall show that we can use ϵ itself as δ; namely, we shall show that

if $\quad 0 < |x - 0| < \epsilon, \qquad$ then $\quad |g(x) - 0| < \epsilon.$

For suppose that

$$0 < |x - 0| < \epsilon.$$

If x is rational, then $g(x) = x$. If x is irrational, then $g(x) = 0$. In either case,

$$|g(x) - 0| < \epsilon. \quad \square$$

Example

If $\quad c \neq 0, \qquad$ then $\quad \lim_{x \to c} \dfrac{1}{x} = \dfrac{1}{c}. \qquad$ (Figure 2.2.12)

PROOF. (*Optional*) Let $\epsilon > 0$. Our task is to show that there exists $\delta > 0$ such that

if $\quad 0 < |x - c| < \delta, \qquad$ then $\quad \left| \dfrac{1}{x} - \dfrac{1}{c} \right| < \epsilon.$

Note that, for $x \neq 0$,

$$\left| \frac{1}{x} - \frac{1}{c} \right| = \left| \frac{x - c}{xc} \right| = \frac{|x - c|}{|x| \, |c|}.$$

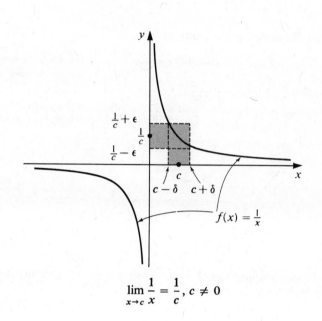

$$\lim_{x \to c} \frac{1}{x} = \frac{1}{c}, \, c \neq 0$$

FIGURE 2.2.12

Let δ be any positive number satisfying

$$\delta < \tfrac{1}{2}|c| \quad \text{and} \quad \delta < \tfrac{1}{2}|c|^2\,\epsilon.$$

(Why this choice of δ will be clear in a moment.) Now if

$$0 < |x - c| < \delta,$$

then

$$|x| > \tfrac{1}{2}|c| \quad \text{and} \quad |x - c| < \tfrac{1}{2}|c|^2\epsilon. \qquad \text{(explain)}$$

It follows then that

$$\frac{1}{|x|} < \frac{1}{\tfrac{1}{2}|c|} \quad \text{and} \quad \frac{|x - c|}{\tfrac{1}{2}|c|\,|c|} < \epsilon$$

and therefore that

$$\frac{|x - c|}{|x|\,|c|} < \epsilon.$$

Since, for such x,

$$\left| \frac{1}{x} - \frac{1}{c} \right| = \frac{|x - c|}{|x|\,|c|},$$

we have

$$\left| \frac{1}{x} - \frac{1}{c} \right| < \epsilon. \quad \square$$

Example 2.2.6. If

$$f(x) = \begin{cases} 1, & x \text{ rational} \\ 0, & x \text{ irrational} \end{cases}, \qquad \text{(the Dirichlet function)}$$

then at no number c does

$$\lim_{x \to c} f(x)$$

exist.

PROOF.   (*Optional*) Suppose, on the contrary, that

$$\lim_{x \to c} f(x) = l \quad \text{for some particular } c.$$

Taking $\epsilon = \frac{1}{2}$, there must exist $\delta > 0$ such that,

$$\text{if}\quad 0 < |x - c| < \delta, \qquad \text{then}\quad |f(x) - l| < \tfrac{1}{2}.$$

Let x_1 be a rational number satisfying

$$0 < |x_1 - c| < \delta$$

and x_2 an irrational number satisfying

$$0 < |x_2 - c| < \delta.$$

(That such numbers exist follows from the fact that every interval contains both rational and irrational numbers.) Now

$$f(x_1) = 1 \quad \text{and} \quad f(x_2) = 0.$$

Thus we must have both

$$|1 - l| < \tfrac{1}{2} \quad \text{and} \quad |0 - l| < \tfrac{1}{2}.$$

From the first inequality we conclude that $l > \frac{1}{2}$. From the second, we conclude that $l < \frac{1}{2}$. Clearly no such number l exists. $\square$

Exercises

Decide in the manner of Section 2.1 whether or not the indicated limit exists. Evaluate the limits which do exist.

*1. $\displaystyle \lim_{x \to 1} \frac{x}{x + 1}$.

2. $\displaystyle \lim_{x \to 16} \sqrt{x}$.

*3. $\displaystyle \lim_{x \to 0} \frac{x^2(1 + x)}{2x}$.

4. $\displaystyle \lim_{x \to 0} \frac{x(1 + x)}{2x^2}$.

*5. $\displaystyle \lim_{x \to 4} \frac{x}{\sqrt{x} + 1}$.

6. $\displaystyle \lim_{x \to 4} \frac{x}{\sqrt{x} + 1}$.

*7. $\displaystyle \lim_{x \to -1} \frac{1 - x}{x + 1}$.

8. $\displaystyle \lim_{x \to 1} \frac{x^4 - 1}{x - 1}$.

*9. $\displaystyle \lim_{x \to 0} \frac{x}{|x|}$.

10. $\displaystyle \lim_{x \to 2} \frac{x}{|x|}$.

*11. Referring to Figure 2.2.13, which of the δ's displayed "works" for the given ϵ?

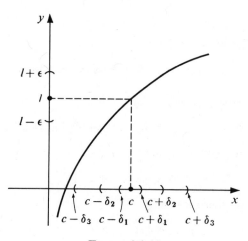

FIGURE 2.2.13

For the following limits, find the largest δ which "works" for a given arbitrary ϵ.

12. $\lim_{x \to 1} 2x = 2$.

*13. $\lim_{x \to 4} 5x = 20$.

14. $\lim_{x \to 2} \frac{1}{2}x = 1$.

*15. $\lim_{x \to 2} \frac{1}{5}x = \frac{2}{5}$.

Give an ϵ, δ proof of the following limits.

16. $\lim_{x \to 4} (2x - 5) = 3$.

*17. $\lim_{x \to 2} (3x - 1) = 5$.

18. $\lim_{x \to 3} (6x - 7) = 11$.

*19. $\lim_{x \to 0} (2 - 5x) = 2$.

20. $\lim_{x \to 2} |1 - 3x| = 5$.

*21. $\lim_{x \to 2} |x - 2| = 0$.

(*Optional*) Give an ϵ, δ proof of the following limits.

22. $\lim_{x \to 1} \sqrt{x} = 1$.

*23. $\lim_{x \to 9} \sqrt{x} = 3$.

24. $\lim_{x \to 2} x^2 = 4$.

*25. $\lim_{x \to 5} \sqrt{x - 1} = 2$.

2.3   Some Limit Theorems

As you saw in the last section it can be rather tedious to apply the ϵ, δ limit test to individual functions. By proving some general theorems about limits we can avoid some of this repetitive work. The theorems themselves of course (at least the first ones) will have to be proved by ϵ, δ methods.

We begin by showing that if a limit exists, it is unique.

Uniqueness Theorem

If
$$\lim_{x \to c} f(x) = l \quad \text{and} \quad \lim_{x \to c} f(x) = m,$$
then
$$l = m.$$

PROOF.　We shall show that $l = m$ by proving that the assumption $l \neq m$ leads to the absurd result $|l - m| < |l - m|$.

Let's assume that $l \neq m$. It follows that $\frac{1}{2}|l - m| > 0$. Since

$$\lim_{x \to c} f(x) = l,$$

we know that there exists $\delta_1 > 0$ such that,

$$\text{if} \quad 0 < |x - c| < \delta_1, \quad \text{then} \quad |f(x) - l| < \tfrac{1}{2}|l - m|.$$

Since

$$\lim_{x \to c} f(x) = m,$$

we know that there exists $\delta_2 > 0$ such that

$$\text{if} \quad 0 < |x - c| < \delta_2, \quad \text{then} \quad |f(x) - m| < \tfrac{1}{2}|l - m|.$$

For x_1 satisfying

$$0 < |x_1 - c| < \text{minimum of } \delta_1 \text{ and } \delta_2,$$

we find that

$$|f(x_1) - l| < \tfrac{1}{2}|l - m| \quad \text{and} \quad |f(x_1) - m| < \tfrac{1}{2}|l - m|,$$

which for convenience we rewrite as

$$|l - f(x_1)| < \tfrac{1}{2}|l - m| \quad \text{and} \quad |f(x_1) - m| < \tfrac{1}{2}|l - m|.$$

It follows now that

$$|l - m| = |[l - f(x_1)] + [f(x_1) - m]| \leq |l - f(x_1)| + |f(x_1) - m|$$
$$< \tfrac{1}{2}|l - m| + \tfrac{1}{2}|l - m| = |l - m|.$$

In short, we have arrived at the absurdity

$$|l - m| < |l - m|. \quad \square$$

Theorem 2.3.1

If

$$\lim_{x \to c} f(x) = l \quad \text{and} \quad \lim_{x \to c} g(x) = m,$$

then

(i) $$\lim_{x \to c} [f(x) + g(x)] = l + m,$$

(ii) $$\lim_{x \to c} [\alpha f(x)] = \alpha l \quad \text{for each real } \alpha,$$

(iii) $$\lim_{x \to c} [f(x)g(x)] = lm.$$

PROOF.   Let $\epsilon > 0$. To prove (i) we must show that there exists $\delta > 0$ such that

$$\text{if} \quad 0 < |x - c| < \delta, \quad \text{then} \quad |[f(x) - g(x)] - [l + m]| < \epsilon.$$

Note that

$$(2.3.2) \qquad |[f(x) + g(x)] - [l + m]| = |[f(x) - l] + [g(x) - m]|$$
$$\leq |f(x) - l| + |g(x) - m|.$$

We will make

$$|[f(x) + g(x)] - [l + m]|$$

less than ϵ by making

$$|f(x) - l| \quad \text{and} \quad |g(x) - m|$$

each less than $\frac{1}{2}\epsilon$. Since $\epsilon > 0$, we know that $\frac{1}{2}\epsilon > 0$. Since

$$\lim_{x \to c} f(x) = l \quad \text{and} \quad \lim_{x \to c} g(x) = m,$$

we know that there exist positive numbers δ_1 and δ_2 such that

$$\text{if} \quad 0 < |x - c| < \delta_1, \quad \text{then} \quad |f(x) - l| < \tfrac{1}{2}\epsilon$$

and

$$\text{if} \quad 0 < |x - c| < \delta_2, \quad \text{then} \quad |g(x) - m| < \tfrac{1}{2}\epsilon.$$

Now we set

$$\delta = \text{minimum of } \delta_1 \text{ and } \delta_2$$

and note that,

$$\text{if} \quad 0 < |x - c| < \delta, \quad \text{then} \quad |f(x) - l| < \tfrac{1}{2}\epsilon \quad \text{and} \quad |g(x) - m| < \tfrac{1}{2}\epsilon,$$

and thus by (2.3.2)

$$|[f(x) + g(x)] - [l + m]| < \epsilon.$$

In summary, by setting $\delta = \min \{\delta_1, \delta_2\}$, we found that,

$$\text{if} \quad 0 < |x - c| < \delta, \quad \text{then} \quad |[f(x) + g(x)] - [l + m]| < \epsilon.$$

Thus (i) is proved.

To prove (ii) we consider two cases: $\alpha \neq 0$ and $\alpha = 0$. If $\alpha \neq 0$, then $\epsilon/|\alpha| > 0$ and since

$$\lim_{x \to c} f(x) = l,$$

we know that there exists $\delta > 0$ such that,

$$\text{if} \quad 0 < |x - c| < \delta, \quad \text{then} \quad |f(x) - l| < \frac{\epsilon}{|\alpha|}.$$

From the last inequality we obtain

$$|\alpha| \, |f(x) - l| < \epsilon,$$

and thus

$$|\alpha f(x) - \alpha l| < \epsilon.$$

The case $\alpha = 0$ was treated before in (2.2.3). For a proof of (iii), see the supplement to this section. $\square$

The results of Theorem 2.3.1 are easily extended to apply to any finite number of functions; namely, if

$$\lim_{x \to c} f_1(x) = l_1, \quad \lim_{x \to c} f_2(x) = l_2, \ldots, \quad \lim_{x \to c} f_n(x) = l_n$$

then

(2.3.3) $$\lim_{x \to c} [\alpha_1 f_1(x) + \alpha_2 f_2(x) + \cdots + \alpha_n f_n(x)] = \alpha_1 l_1 + \alpha_2 l_2 + \cdots + \alpha_n l_n$$

and

(2.3.4) $$\lim_{x \to c} [f_1(x) f_2(x) \cdots f_n(x)] = l_1 l_2 \cdots l_n.$$

From this it is easy to see that every polynomial

$$\boxed{P(x) = \alpha_n x^n + \cdots + \alpha_1 x + \alpha_0}$$

satisfies

$$\boxed{\lim_{x \to c} P(x) = P(c).}$$

PROOF. We already know that

$$\lim_{x \to c} x = c.$$

Applying (2.3.4) we have

$$\lim_{x \to c} x^k = c^k \quad \text{for each positive integer } k.$$

We also know that

$$\lim_{x \to c} \alpha_0 = \alpha_0.$$

It follows now from (2.3.3) that

$$\lim_{x \to c} (\alpha_n x^n + \cdots + \alpha_1 x + \alpha_0) = \alpha_n c^n + \cdots + \alpha_1 c + \alpha_0;$$

that is,

$$\lim_{x \to c} P(x) = P(c). \quad \square$$

Examples

$$\lim_{x \to 1} (5x^2 - 12x + 2) = 5(1)^2 - 12(1) + 2 = -5,$$

$$\lim_{x \to 0} (14x^5 - 7x^2 + 2x + 8) = 8,$$

$$\lim_{x \to -1} (2x^3 + x^2 - 2x - 3) = 2(-1)^3 + (-1)^2 - 2(-1) - 3 = -2. \quad \square$$

We come now to reciprocals and quotients.

Theorem 2.3.5

If

$$\lim_{x \to c} g(x) = m \quad \text{and} \quad m \neq 0, \qquad \text{then} \qquad \lim_{x \to c} \frac{1}{g(x)} = \frac{1}{m}.$$

PROOF.   See the supplement to this section. $\square$

Once you know that reciprocals present no trouble, quotients become easy to handle.

Theorem

If

$$\lim_{x \to c} f(x) = l \quad \text{and} \quad \lim_{x \to c} g(x) = m \quad \text{with} \quad m \neq 0,$$

then

$$\lim_{x \to c} \frac{f(x)}{g(x)} = \frac{l}{m}.$$

PROOF.   The key here is to observe that the quotient can be written as a product

$$\frac{f(x)}{g(x)} = f(x) \, \frac{1}{g(x)}.$$

With

$$\lim_{x \to c} f(x) = l \quad \text{and} \quad \lim_{x \to c} \frac{1}{g(x)} = \frac{1}{m},$$

the product rule [part (iii) of Theorem 2.3.1] gives

$$\lim_{x \to c} \frac{f(x)}{g(x)} = l\,\frac{1}{m} = \frac{l}{m}. \quad \Box$$

As an immediate consequence of this theorem on quotients you can see that, if P and Q are polynomials and $Q(c) \neq 0$, then

$$\boxed{\lim_{x \to c} \frac{P(x)}{Q(x)} = \frac{P(c)}{Q(c)}.}$$

Examples

$$\lim_{x \to 2} \frac{3x - 5}{x^2 + 1} = \frac{6 - 5}{4 + 1} = \frac{1}{5}, \qquad \lim_{x \to -3} \left(x + \frac{1}{x} \right) = -3 + \frac{1}{-3} = -3\frac{1}{3},$$

$$\lim_{x \to 3} \frac{x^3 - 3x^2}{1 - x^2} = \frac{27 - 27}{1 - 9} = 0.$$

There is no point looking for a limit that does not exist. The last theorem of this section gives a condition under which a quotient does not have a limit.

Theorem 2.3.6

If

$$\lim_{x \to c} f(x) = l \quad \text{with} \quad l \neq 0, \quad \text{and} \quad \lim_{x \to c} g(x) = 0,$$

then

$$\lim_{x \to c} \frac{f(x)}{g(x)} \quad \text{does not exist.}$$

PROOF. Suppose on the contrary that there exists a real number L such that

$$\lim_{x \to c} \frac{f(x)}{g(x)} = L.$$

Then

$$l = \lim_{x \to c} f(x) = \lim_{x \to c} \left[g(x) \cdot \frac{f(x)}{g(x)} \right] = \lim_{x \to c} g(x) \cdot \lim_{x \to c} \frac{f(x)}{g(x)} = 0 \cdot L = 0,$$

which contradicts our assumption that $l \neq 0$. $\quad \Box$

Examples. Applying Theorem 2.3.6, you can see that

$$\lim_{x \to 1} \frac{x^2}{x - 1}, \qquad \lim_{x \to 2} \frac{3x - 7}{x^2 - 4}, \qquad \text{and} \qquad \lim_{x \to 0} \frac{5}{x}$$

all fail to exist. $\quad \Box$

Exercises

1. Evaluate the limits that exist.

   (a) $\lim\limits_{x \to 2} \dfrac{x^2 + x + 1}{x^2 + 2x}$.

   *(b) $\lim\limits_{x \to 2} \dfrac{x - 2}{x^2 - 4}$.

   *(c) $\lim\limits_{x \to 2} \dfrac{x^2 - 4}{x - 2}$.

   (d) $\lim\limits_{x \to -1} (\sqrt{2x - 6x + 1})$.

   *(e) $\lim\limits_{x \to 0} \left(x - \dfrac{4}{x} \right)$.

   (f) $\lim\limits_{x \to -1} \dfrac{|x| - x^2}{(3x - 1)(x^4 - 2)}$.

   (g) $\lim\limits_{x \to 5} \dfrac{2 - x^2}{4x}$.

   *(h) $\lim\limits_{x \to 1} \dfrac{x - x^2}{1 - x}$.

2. Evaluate the limits that exist.

   (a) $\lim\limits_{x \to -3} \dfrac{5x + 15}{x^3 + 3x^2 - x - 3}$.

   *(b) $\lim\limits_{x \to -4} \left(\dfrac{2x}{x + 4} + \dfrac{8}{x + 4} \right)$.

   *(c) $\lim\limits_{t \to 1} \dfrac{t}{t^2 - 1}$.

   (d) $\lim\limits_{h \to 0} |h - |h||$.

3. Given that

   $$\lim_{x \to c} f(x) = 2, \quad \lim_{x \to c} g(x) = -1, \quad \text{and} \quad \lim_{x \to c} h(x) = 0,$$

   evaluate the limits that exist.

   *(a) $\lim\limits_{x \to c} [f(x) - g(x)]$.

   (b) $\lim\limits_{x \to c} [f(x)]^2$.

   (c) $\lim\limits_{x \to c} \dfrac{f(x)}{g(x)}$.

   *(d) $\lim\limits_{x \to c} \dfrac{h(x)}{f(x)}$.

   *(e) $\lim\limits_{x \to c} \dfrac{f(x)}{h(x)}$.

   (f) $\lim\limits_{x \to c} \dfrac{1}{f(x) - g(x)}$.

4. Given that $f(x) = x^3$, evaluate the limits that exist.

   *(a) $\lim\limits_{x \to 3} \dfrac{f(x) - f(3)}{x - 3}$.

   (b) $\lim\limits_{x \to 3} \dfrac{f(x) - f(2)}{x - 3}$.

   *(c) $\lim\limits_{x \to 3} \dfrac{f(x) - f(3)}{x - 2}$.

   (d) $\lim\limits_{x \to 1} \dfrac{f(x) - f(1)}{x - 1}$.

5. Given that

   $$\lim_{x \to c} g(x) = 0 \quad \text{and} \quad f(x)g(x) = 1 \quad \text{for all real } x,$$

   prove that $\lim\limits_{x \to c} f(x)$ does not exist.

Supplement to Section 2.3

PROOF OF THEOREM 2.3.1 (iii)

$$|f(x)g(x) - lm| \le |f(x)g(x) - f(x)m| + |f(x)m - lm|$$
$$= |f(x)| \, |g(x) - m| + |m| \, |f(x) - l|$$
$$\le |f(x)| \, |g(x) - m| + (1 + |m|)|f(x) - l|.$$

Let $\epsilon > 0$. Since

$$\lim_{x \to c} f(x) = l \quad \text{and} \quad \lim_{x \to c} g(x) = m,$$

we know

(1) that there exists $\delta_1 > 0$ such that, if $0 < |x - c| < \delta_1$, then

$$|f(x) - l| < 1 \quad \text{and thus} \quad |f(x)| < 1 + |l|;$$

(2) that there exists $\delta_2 > 0$ such that

$$\text{if} \quad 0 < |x - c| < \delta_2, \quad \text{then} \quad |g(x) - m| < \frac{\epsilon}{2}\left(\frac{1}{1 + |l|}\right);$$

(3) that there exists $\delta_3 > 0$ such that

$$\text{if} \quad 0 < |x - c| < \delta_3, \quad \text{then} \quad |f(x) - l| < \frac{\epsilon}{2}\left(\frac{1}{1 + |m|}\right).$$

We can now set $\delta = \min\{\delta_1, \delta_2, \delta_3\}$ and observe that if $0 < |x - c| < \delta$, then

$$|f(x)g(x) - lm| < (1 + |l|)\frac{\epsilon}{2}\left(\frac{1}{1 + |l|}\right) + (1 + |m|)\frac{\epsilon}{2}\left(\frac{1}{1 + |m|}\right) = \epsilon. \quad \square$$

PROOF OF THEOREM 2.3.5.   For $g(x) \ne 0$,

$$\left|\frac{1}{g(x)} - \frac{1}{m}\right| = \frac{|g(x) - m|}{|m| \, |g(x)|}.$$

Choose $\delta_1 > 0$ such that

$$\text{if} \quad 0 < |x - c| < \delta_1, \quad \text{then} \quad |g(x) - m| < \frac{|m|}{2}.$$

For such x,

$$|g(x)| > \frac{|m|}{2}, \quad \frac{1}{|g(x)|} < \frac{2}{|m|}$$

and thus

$$\left|\frac{1}{g(x)} - \frac{1}{m}\right| \le \frac{2}{|m|^2} |g(x) - m|.$$

Now let $\epsilon > 0$ and choose $\delta_2 > 0$ such that

$$\text{if} \quad 0 < |x - c| < \delta_2, \quad \text{then} \quad |g(x) - m| < \frac{|m|^2}{2}\epsilon.$$

Setting $\delta = \min\{\delta_1, \delta_2\}$, we find that

$$\text{if} \quad 0 < |x - c| < \delta, \quad \text{then} \quad \left|\frac{1}{g(x)} - \frac{1}{m}\right| < \epsilon. \quad \square$$

2.4   More on Limits

As we have indicated before, whether or not

$$\lim_{x \to c} f(x)$$

exists and, if so, what it is, depends only on the behavior of f for x close to c. If, for x close to c, f is trapped between two functions both of which tend to the same limit l (see Figure 2.4.1), then $f(x)$ must also tend to l. This idea is the substance of what we call the "pinching theorem."

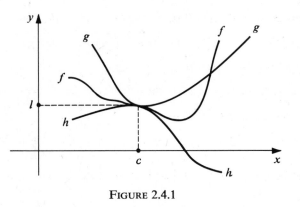

FIGURE 2.4.1

The Pinching Theorem 2.4.1

Suppose that there exists a number $p > 0$ such that

$$h(x) \leq f(x) \leq g(x) \quad \text{for all } x \text{ satisfying } 0 < |x - c| < p.$$

If

$$\lim_{x \to c} h(x) = l \quad \text{and} \quad \lim_{x \to c} g(x) = l,$$

then

$$\lim_{x \to c} f(x) = l.$$

PROOF.   Let $\epsilon > 0$. Let $p > 0$ be such that,

$$\text{if} \quad 0 < |x - c| < p, \quad \text{then} \quad h(x) \leq f(x) \leq g(x).$$

Choose $\delta_1 > 0$ such that

$$\text{if} \quad 0 < |x - c| < \delta_1, \quad \text{then} \quad l - \epsilon < h(x) < l + \epsilon.$$

Choose $\delta_2 > 0$ such that

$$\text{if} \quad 0 < |x - c| < \delta_2, \quad \text{then} \quad l - \epsilon < g(x) < l + \epsilon.$$

Let $\delta = \min(p, \delta_1, \delta_2)$. For x satisfying $0 < |x - c| < \delta$, we have
$$l - \epsilon < h(x) \leq f(x) \leq g(x) < l + \epsilon$$
and thus
$$|f(x) - l| < \epsilon. \quad \square$$

Example. Suppose that
$$-x^2 \leq f(x) \leq x^2 \qquad \text{for all } x \neq 0.$$
Since
$$\lim_{x \to 0} -x^2 = 0 \quad \text{and} \quad \lim_{x \to 0} x^2 = 0,$$
we know that
$$\lim_{x \to 0} f(x) = 0. \quad \square$$

Example. Suppose that
$$x^2 + 12x - 9 \leq f(x) \leq 5x^2 \qquad \text{for all } x \neq \tfrac{3}{2}.$$
Since
$$\lim_{x \to 3/2} (x^2 + 12x - 9) = (\tfrac{3}{2})^2 + 12(\tfrac{3}{2}) - 9 = \tfrac{45}{4}$$
and
$$\lim_{x \to 3/2} 5x^2 = 5(\tfrac{3}{2})^2 = \tfrac{45}{4},$$
we know that
$$\lim_{x \to 3/2} f(x) = \tfrac{45}{4}. \quad \square$$

Example. If
$$f(x) = \begin{cases} x^2, & x \text{ rational} \\ x^4, & x \text{ irrational} \end{cases},$$
then
$$\lim_{x \to 0} f(x) = 0.$$
To verify this, we note that for each x satisfying $|x| < 1$, we have
$$0 \leq f(x) \leq x^2. \qquad \text{(explain)}$$
Since
$$\lim_{x \to 0} x^2 = 0 \qquad (\text{and, of course, } \lim_{x \to 0} 0 = 0),$$
the pinching theorem gives us
$$\lim_{x \to 0} f(x) = 0. \quad \square$$

There are many different ways of formulating the same limit statement. Sometimes one formulation is more convenient, sometimes another. In any case, it is useful to recognize that the following are equivalent:

(2.4.2)

(i) $\lim_{x \to c} f(x) = l.$ (ii) $\lim_{x \to c} (f(x) - l) = 0.$

(iii) $\lim_{x \to c} |f(x) - l| = 0.$ (iv) $\lim_{h \to 0} f(c + h) = l.$

The equivalence of these four statements is obvious. It is, however, a good exercise in ϵ, δ technique to show that (i) is equivalent to (iv).

Example.   For $f(x) = x^3$, we have

$$\lim_{x \to 2} x^3 = 8, \qquad\qquad \lim_{x \to 2} (x^3 - 8) = 0,$$

$$\lim_{x \to 2} |x^3 - 8| = 0, \qquad \lim_{h \to 0} (2 + h)^3 = 8. \quad \square$$

While we are on the subject of equivalent statements, note that

$$\lim_{x \to c} f(x) = l \quad \text{and} \quad \lim_{x \to c} |f(x)| = |l|$$

are *not* equivalent. While

$$\lim_{x \to c} f(x) = l \quad \text{implies} \quad \lim_{x \to c} f(x) = |l|, \qquad\qquad \text{(Exercise 2)}$$

the converse is false. There are instances in which

$$\lim_{x \to c} |f(x)| \text{ exists,} \quad \text{but} \quad \lim_{x \to c} f(x) \text{ does not.} \qquad \text{(Exercise 3)}$$

Exercises

1. Find the limits that exist.

   *(a) $\lim\limits_{x \to 0} f(x)$ if $-|x| \le f(x) \le x$ for $x \ne 0$.

   (b) $\lim\limits_{x \to 3} f(x)$ if $1 \le f(x) \le (x - 3)^2 + 1$ for $x \ne 3$.

   *(c) $\lim\limits_{x \to 2} f(x)$ if $-(x - 2)^2 \le f(x) \le 0$ for $x \ne 2$.

   (d) $\lim\limits_{x \to 5} f(x)$ if $5 + x \le f(x) \le x^2 - 9x + 30$ for $x \ne 5$.

   *(e) $\lim\limits_{x \to 0} f(x)$ if $\left| \dfrac{f(x)}{x} \right| \le 1$ for $x \ne 0$.

   (f) $\lim\limits_{h \to 0} (3 + h)^2$.
   $\qquad\qquad\qquad$ (g) $\lim\limits_{h \to 0} h \left(1 + \dfrac{1}{h} \right)$.

   *(h) $\lim\limits_{h \to 0} h \left(1 + \dfrac{1}{h^2} \right)$.
   $\qquad\qquad$ (i) $\lim\limits_{h \to 1} \dfrac{h}{|h - 1|}$.

   *(j) $\lim\limits_{h \to 0} \dfrac{1 - 1/h}{1 + 1/h}$.
   $\qquad\qquad$ (k) $\lim\limits_{h \to 0} \dfrac{1 - 1/h^2}{1 - 1/h}$.

   *(l) $\lim\limits_{x \to 1} \left(1 - \dfrac{1}{x} \right)\left(\dfrac{1}{x - 1} \right)$.
   $\qquad$ (m) $\lim\limits_{x \to c} \dfrac{x - c}{|x - c|}$.

2. Show that

   $$\text{if} \quad \lim_{x \to c} f(x) = l \quad \text{then} \quad \lim_{x \to c} |f(x)| = |l|.$$

*3. Given that

$$f(x) = \begin{cases} -1, & x < 0 \\ 1, & x > 0 \end{cases},$$

find

(a) $\lim_{x \to 0} f(x)$ and   (b) $\lim_{x \to 0} |f(x)|$.

4. Show that, for each $c > 0$

$$\lim_{x \to c} \sqrt{x} = \sqrt{c}.$$

HINT: If x and c are positive, then

$$0 \le |\sqrt{x} - \sqrt{c}| = \frac{|x - c|}{\sqrt{x} + \sqrt{c}} \le \frac{1}{\sqrt{c}} |x - c|.$$

5.   (a) Verify that

$$\max \{f(x), g(x)\} = \tfrac{1}{2}\{[f(x) + g(x)] + |f(x) - g(x)|\}.$$

*(b) Find a similar expression for

$$\min \{f(x), g(x)\}.$$

6. Let

$$h(x) = \min \{f(x), g(x)\} \quad \text{and} \quad H(x) = \max \{f(x), g(x)\}.$$

Show that if

$$\lim_{x \to c} f(x) = l \quad \text{and} \quad \lim_{x \to c} g(x) = l$$

then

$$\lim_{x \to c} h(x) = l \quad \text{and} \quad \lim_{x \to c} H(x) = l.$$

HINT: Use Exercise 5.

7. (*Optional*)   Give an ϵ, δ proof that statement (i) of 2.4.2 is equivalent to statement (iv).

2.5   One-Sided Limits

Whether or not $\lim_{x \to c} f(x)$ exists, and, if so, what it is, depends on the behavior of f on both sides of c. The situation is simpler in the case of one-sided limits: the *left-hand* limit, $\lim_{x \uparrow c} f(x)$, and the *right-hand* limit, $\lim_{x \downarrow c} f(x)$.

You can think of the left-hand limit

$$\lim_{x \uparrow c} f(x)$$

as the number that $f(x)$ approaches as x comes into c from the left, and the right-hand limit

$$\lim_{x \downarrow c} f(x)$$

as the number that $f(x)$ approaches as x comes into c from the right. See Figure 2.5.1.

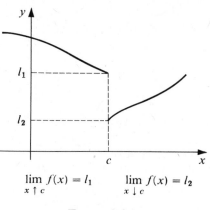

$$\lim_{x \uparrow c} f(x) = l_1 \qquad \lim_{x \downarrow c} f(x) = l_2$$

FIGURE 2.5.1

Here are the appropriate ϵ, δ definitions:

Definition of Left-Hand Limit

Let f be a function defined at least on an interval of the form (a, c).

$$\lim_{x \uparrow c} f(x) = l \quad \text{iff} \quad \begin{cases} \text{for each } \epsilon > 0 \text{ there exists } \delta > 0 \text{ such that} \\ \text{if} \quad c - \delta < x < c, \qquad \text{then} \quad |f(x) - l| < \epsilon. \end{cases}$$

Definition of Right-Hand Limit

Let f be a function defined at least on an interval of the form (c, d).

$$\lim_{x \downarrow c} f(x) = l \quad \text{iff} \quad \begin{cases} \text{for each } \epsilon > 0 \text{ there exists } \delta > 0 \text{ such that} \\ \text{if} \quad c < x < c + \delta, \qquad \text{then} \quad |f(x) - l| < \epsilon. \end{cases}$$

Example.   If

$$h(x) = \begin{cases} -1, & x < 0 \\ 1, & x > 0 \end{cases},$$

then

$$\lim_{x \uparrow 0} h(x) = -1 \quad \text{and} \quad \lim_{x \downarrow 0} h(x) = 1$$

although, as you saw before,

$$\lim_{x \to 0} h(x)$$

does not exist.   □

Example.   Since the square-root function is not defined on both sides of 0, we cannot consider

$$\lim_{x \to 0} \sqrt{x}.$$

However, it is true that

$$\lim_{x \downarrow 0} \sqrt{x} = 0.$$ (Figure 2.5.2)

For a formal proof, begin with $\epsilon > 0$ and take ϵ^2 as δ:

if $0 < x < \epsilon^2$ then $|\sqrt{x} - 0| = \sqrt{x} < \epsilon.$ $\square$

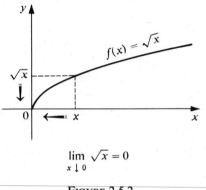

$$\lim_{x \downarrow 0} \sqrt{x} = 0$$

FIGURE 2.5.2

Example.   An interesting function to look at when dealing with one-sided limits is the *greatest-integer function*:

(2.5.1) $[x] = \text{greatest integer} \leq x.$

For the graph we refer to Figure 2.5.3. At 0 the function is 0 and it remains 0 throughout the interval $[0, 1)$. At 1 the function jumps up to 1 and it remains 1 throughout

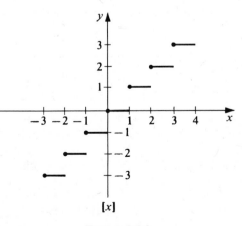

FIGURE 2.5.3

the interval $[1, 2)$. At 2 the function jumps up to 2 and so on. It is easy to see that for each integer n,

$$\lim_{x \uparrow n} [x] = n - 1 \quad \text{and} \quad \lim_{x \downarrow n} [x] = n.$$

For example,

$$\lim_{x \uparrow 2} [x] = 1 \quad \text{and} \quad \lim_{x \downarrow 2} [x] = 2.$$

If c is a number which is not an integer, then c lies between consecutive integers n and $n + 1$. For such c, we have both

$$\lim_{x \uparrow c} [x] = n \quad \text{and} \quad \lim_{x \downarrow c} [x] = n.$$

For example,

$$\lim_{x \uparrow 3/2} [x] = 1 \quad \text{and} \quad \lim_{x \downarrow 3/2} [x] = 1. \quad \square$$

Results like those proved in Sections 2.3 and 2.4 for two-sided limits also hold for one-sided limits. Thus, for instance, if

$$\lim_{x \uparrow c} f(x) = l \quad \text{and} \quad \lim_{x \uparrow c} g(x) = m,$$

then

$$\lim_{x \uparrow c} [f(x) + g(x)] = l + m,$$

$$\lim_{x \uparrow c} [\alpha f(x)] = \alpha l,$$

$$\lim_{x \uparrow c} [f(x) g(x)] = lm.$$

Moreover, if $m \neq 0$, then

$$\lim_{x \uparrow c} \frac{1}{g(x)} = \frac{1}{m} \quad \text{and} \quad \lim_{x \uparrow c} \frac{f(x)}{g(x)} = \frac{l}{m}.$$

Similar results hold for right-hand limits. Proofs can be obtained by mimicking the proofs given before.

One-sided limits give us a simple way of determining whether or not a (two-sided) limit exists:

(2.5.2)

$$\boxed{\lim_{x \to c} f(x) = l \quad \text{iff} \quad \lim_{x \uparrow c} f(x) = l \quad \text{and} \quad \lim_{x \downarrow c} f(x) = l.}$$

The proof is left as an exercise. It is easy because any δ that "works" for the limit will work for the two one-sided limits and any δ that "works" for the two one-sided limits will work for the limit.

We illustrate the effectiveness of (2.5.2) by some examples.

Example. For

$$h(x) = \begin{cases} -1, & x < 0 \\ 1, & x > 0 \end{cases},$$

we have

$$\lim_{x \uparrow 0} h(x) = -1 \quad \text{and} \quad \lim_{x \downarrow 0} h(x) = 1.$$

Since the two one-sided limits are not equal,

$$\lim_{x \to 0} h(x) \quad \text{does not exist.} \quad \square$$

Example.   With

$$g(x) = \begin{cases} x^2, & x < 0 \\ \sqrt{x}, & x > 0 \end{cases},$$

we have both

$$\lim_{x \uparrow 0} g(x) = 0 \quad \text{and} \quad \lim_{x \downarrow 0} g(x) = 0.$$

Consequently,

$$\lim_{x \to 0} g(x) = 0. \quad \square$$

Example.   Set

$$k(x) = \begin{cases} 1, & \text{negative rational } x \\ 0, & \text{negative irrational } x \\ x, & \text{positive } x \end{cases}.$$

Here

$$\lim_{x \uparrow 0} k(x) \quad \text{does not exist}$$

and therefore

$$\lim_{x \to 0} k(x) \quad \text{does not exist either.} \quad \square$$

Exercises

1.   (a) Sketch the graph of

$$f(x) = \begin{cases} x - 1, & x < 0 \\ 1 + x, & x > 0 \end{cases}.$$

(b) Evaluate the limits that exist.

     *(i) $\lim_{x \uparrow 0} f(x)$.   (ii) $\lim_{x \downarrow 0} f(x)$.   *(iii) $\lim_{x \to 0} f(x)$.

2.   (a) Sketch the graph of

$$g(x) = \begin{cases} x^2, & x < 1 \\ x, & 1 < x < 4 \\ 4 - x, & 4 < x \end{cases}.$$

(b) Evaluate the limits that exist.

     *(i) $\lim_{x \uparrow 1} g(x)$.   (ii) $\lim_{x \downarrow 1} g(x)$.   *(iii) $\lim_{x \to 1} g(x)$.

     (iv) $\lim_{x \uparrow 2} g(x)$.   *(v) $\lim_{x \downarrow 2} g(x)$.   (vi) $\lim_{x \to 2} g(x)$.

     *(vii) $\lim_{x \uparrow 4} g(x)$.   (viii) $\lim_{x \downarrow 4} g(x)$.   *(ix) $\lim_{x \to 4} g(x)$.

3. Evaluate the limits that exist.

* (a) $\lim\limits_{x \uparrow 2} \dfrac{x^2}{x + 2}$.

(b) $\lim\limits_{x \downarrow 2} \dfrac{x^2}{x + 2}$.

* (c) $\lim\limits_{x \uparrow 2} \dfrac{x - 2}{|x - 2|}$.

(d) $\lim\limits_{x \downarrow 2} \dfrac{x - 2}{|x - 2|}$.

* (e) $\lim\limits_{x \to 1} \dfrac{x^3 - 1}{|x^3 - 1|}$.

(f) $\lim\limits_{x \downarrow 0} \dfrac{1 + \sqrt{x}}{1 - \sqrt{x}}$.

* (g) $\lim\limits_{x \uparrow 1} \sqrt{|x|} - x$.

(h) $\lim\limits_{x \downarrow 1} \sqrt{|x|} - x$.

* (i) $\lim\limits_{x \uparrow 1} \sqrt{x - [x]}$.

(j) $\lim\limits_{x \downarrow 1} \sqrt{x - [x]}$.

* (k) $\lim\limits_{x \downarrow 0} \dfrac{[x] - x}{x}$.

(l) $\lim\limits_{x \uparrow 1} (x^2 + 2)^{[x]}$.

* (m) $\lim\limits_{x \uparrow 2} \left(\dfrac{1}{x - 2} - \dfrac{1}{|x - 2|} \right)$.

(n) $\lim\limits_{x \downarrow 2} \left(\dfrac{1}{x - 2} - \dfrac{1}{|x - 2|} \right)$.

4. (a) Sketch the graph of

$$f(x) = [2x].$$

(b) Evaluate the limits that exist.

* (i) $\lim\limits_{x \uparrow 1/2} [2x]$.   * (ii) $\lim\limits_{x \downarrow 1/2} [2x]$.   * (iii) $\lim\limits_{x \uparrow 3/4} [2x]$.

* (iv) $\lim\limits_{x \downarrow 3/4} [2x]$.   * (v) $\lim\limits_{x \to 1/2} [2x]$.   * (vi) $\lim\limits_{x \to 3/4} [2x]$.

5. Give an ϵ, δ proof of (2.5.2).

6. (*Optional*) Give an ϵ, δ proof that

(a) $\lim\limits_{x \uparrow c} f(x) = l$ iff $\lim\limits_{h \to 0} f(c - |h|) = l$.

(b) $\lim\limits_{x \downarrow c} f(x) = l$ iff $\lim\limits_{h \to 0} f(c + |h|) = l$.

7. (*Optional*) State the pinching theorem (2.4.1)

(a) for left-hand limits.   (b) for right-hand limits.

8. (*Optional*) Evaluate the limits that exist.

* (a) $\lim\limits_{x \to 1} [2x](x - 1)$.

(b) $\lim\limits_{x \to 0} [x][x + 1]$.

* (c) $\lim\limits_{x \to 0} x [1/x]$.

(d) $\lim\limits_{x \to 0} x^3 [1/x]$.

HINT: Use one-sided limits. For (c) and (d) note that $t - 1 \leq [t] \leq t$ and use the appropriate pinching theorem.

2.6 Continuity

In ordinary language, to say that a certain process is "continuous" is to say that it goes on without interruption and without abrupt changes. In mathematics the word "continuous" has much the same meaning.

The idea of continuity is so important to calculus that we will have to discuss it with some care. First we introduce the idea of continuity at a point (or number) c and then we talk about continuity on an interval.

Continuity at a Point

Definition

A function f is said to be *continuous* at c iff $\lim\limits_{x \to c} f(x) = f(c)$.

If c is in the domain of f, then f can fail to be continuous at c only for one of two reasons: either $f(x)$ does not have a limit as x tends to c, or it does have a limit, but the limit is not $f(c)$. These possibilities are illustrated in Figures 2.6.1, 2.6.2, and 2.6.3.

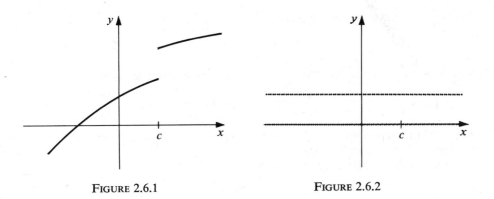

FIGURE 2.6.1   FIGURE 2.6.2

The function depicted in Figure 2.6.1 does not have a limit at c, and hence is not continuous at c. In Figure 2.6.2 we have tried to suggest the Dirichlet function

$$f(x) = \begin{cases} 1, & x \text{ rational} \\ 0, & x \text{ irrational} \end{cases}.$$

At no point c does f have a limit. It is therefore everywhere discontinuous. The function depicted in Figure 2.6.3 does have a limit at c. It is discontinuous at c because the limit at c is not the value at c.

The discontinuity displayed in Figure 2.6.3 is not serious in the sense that it can be removed simply by redefining the function at c (by lowering the dot into place). See Figure 2.6.4.

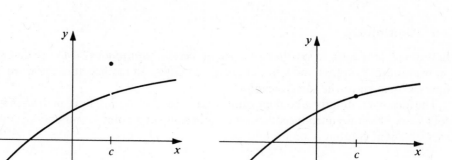

FIGURE 2.6.3   FIGURE 2.6.4

Most of the functions that you have encountered so far are continuous at each point of their domains. In particular, this is true for polynomials,

$$\lim_{x \to c} P(x) = P(c),$$

for rational functions, by which we mean quotients of polynomials,

$$\lim_{x \to c} \frac{P(x)}{Q(x)} = \frac{P(c)}{Q(c)} \quad \text{if} \quad Q(c) \neq 0,$$

and for the absolute-value function,

$$\lim_{x \to c} |x| = |c|.$$

In one of the exercises you showed (or at least were asked to show) that

$$\lim_{x \to c} \sqrt{x} = \sqrt{c} \quad \text{for each } c > 0.$$

This makes the square-root function continuous at each positive number. What happens at 0, we discuss later.

With f and g continuous at c, we have

$$\lim_{x \to c} f(x) = f(c), \qquad \lim_{x \to c} g(x) = g(c)$$

and thus, by the limit theorems,

$$\lim_{x \to c} [f(x) + g(x)] = f(c) + g(c)$$

$$\lim_{x \to c} [\alpha f(x)] = \alpha f(c)$$

$$\lim_{x \to c} [f(x)g(x)] = f(c)g(c)$$

and, if $g(c) \neq 0$,

$$\lim_{x \to c} \frac{f(x)}{g(x)} = \frac{f(c)}{g(c)}.$$

This tells us that

> if f and g are continuous at c, then $f + g$, αf, and fg are continuous at c; moreover, if $g(c) \neq 0$, then the quotient f/g is also continuous at c.

These results can obviously be extended to any finite number of functions.

Our next topic is the continuity of composite functions. Before getting into this, however, let's take a look at continuity in terms of ϵ, δ. A direct translation of

$$\lim_{x \to c} f(x) = f(c)$$

into ϵ, δ terms reads like this:

> for each $\epsilon > 0$ there exists $\delta > 0$ such that
> if $\ 0 < |x - c| < \delta, \quad$ then $\ |f(x) - f(c)| < \epsilon.$

Here, however, the restriction

$$0 < |x - c|$$

is unnecessary; namely, we can allow $|x - c| = 0$. We can allow it because, if $|x - c| = 0$, then $x = c$ and $f(x) = f(c)$ and thus $|f(x) - f(c)| = 0$. Being 0, $|f(x) - f(c)|$ is certainly less than ϵ.

In summary, an ϵ, δ characterization of continuity at c reads as follows:

> f is continuous at c $\quad$ iff $\quad \begin{cases} \text{for each } \epsilon > 0 \text{ there exists } \delta > 0 \text{ such that} \\ \text{if } \ |x - c| < \delta \quad \text{ then } \ |f(x) - f(c)| < \epsilon. \end{cases}$

In simple intuitive language

> f is continuous at c $\quad$ iff $\quad$ for x close to c, $f(x)$ is close to $f(c)$.

We are now ready to take up the continuity of composites.

Theorem

If g is continuous at c and f is continuous at $g(c)$ then the composition $f \circ g$ is continuous at c.

The idea here is simple: with g continuous at c, we know that

$$\text{for } x \text{ close to } c, \quad g(x) \text{ is close to } g(c);$$

from the continuity of f at $g(c)$, we know that

with $g(x)$ close to $g(c)$, $f(g(x))$ is close to $f(g(c))$.

In summary,

with x close to c, $f(g(x))$ is close to $f(g(c))$.

The argument we just gave is too vague to be a proof. Here in contrast is a proof:

We begin with $\epsilon > 0$ and must show that there exists some number $\delta > 0$ such that

if $|x - c| < \delta$,   then $|f(g(x)) - f(g(c))| < \epsilon$.

In the first place, we observe that since f is continuous at $g(c)$ there does exist a number $\delta_1 > 0$ such that

if $|t - g(c)| < \delta_1$,   then $|f(t) - f(g(c))| < \epsilon$.

With $\delta_1 > 0$ we know from the continuity of g at c that there exists a number $\delta > 0$ such that

if $|x - c| < \delta$,   then $|g(x) - g(c)| < \delta_1$.

Putting these two statements together, we see that

if $|x - c| < \delta$,   then $|g(x) - g(c)| < \delta_1$

and thus

$$|f(g(x)) - f(g(c))| < \epsilon. \quad \square$$

This proof is illustrated in Figure 2.6.5. The numbers within δ of c are taken by g to within δ_1 of $g(c)$ and then by f to within ϵ of $f(g(c))$.

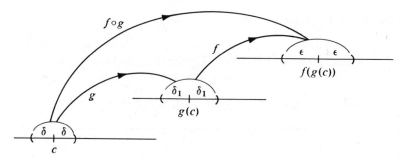

FIGURE 2.6.5

It is time to look at some examples.

Example.   The function

$$F(x) = \sqrt{x^2 + 1}$$

is everywhere continuous. To see this, note that

$$F = f \circ g,$$

where

$$g(x) = x^2 + 1 \quad \text{and} \quad f(x) = \sqrt{x}. \quad \square$$

Example. The function

$$F(x) = \left| \frac{x^2 + 1}{(x - 8)^4} \right|$$

is continuous on its domain. This of course excludes $x = 8$. To see this, note that

$$F = f \circ g,$$

where

$$g(x) = \frac{x^2 + 1}{(x - 8)^4} \quad \text{and} \quad f(x) = |x|. \quad \square$$

The continuity of composites holds for any finite number of functions. The only requirement is that each function be continuous *where it is applied.*

Example. The function

$$F(x) = \frac{1}{5 - \sqrt{x^2 + 16}}$$

is continuous everywhere except at $x = \pm 3$, where it is not defined. To see this, note that

$$F = f \circ g \circ k \circ h,$$

where

$$f(x) = \frac{1}{x}, \qquad g(x) = 5 - x, \qquad k(x) = \sqrt{x}, \qquad h(x) = x^2 + 16$$

and observe that each of these functions is being applied only where it is continuous. In particular f is being applied only to nonzero numbers and k is being applied only to positive numbers. $\square$

Just as we considered one-sided limits, we can consider one-sided continuity.

Definition of One-Sided Continuity

A function f is called

         continuous from the left at c iff $\lim\limits_{x \uparrow c} f(x) = f(c).$

It is called

         continuous from the right at c iff $\lim\limits_{x \downarrow c} f(x) = f(c).$

It is clear from (2.5.2) that a function f is continuous at c iff it is continuous there from both sides.

At each integer the greatest-integer function

$$f(x) = [x] \qquad\qquad \text{(Figure 2.6.6)}$$

is continuous from the right but discontinuous from the left. This situation is reversed in the case of

$$g(x) = [x) = \text{greatest integer less than } x.$$

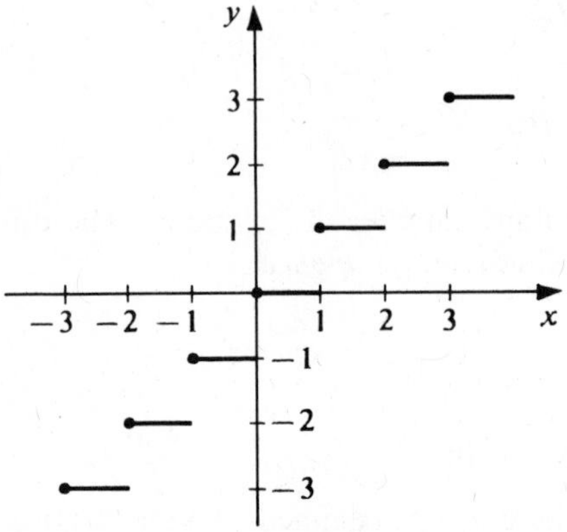

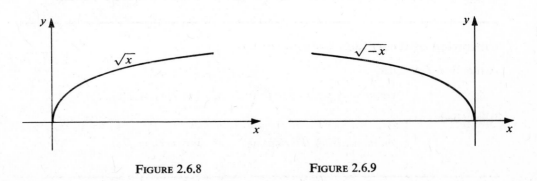

$$f(x) = [x]$$

FIGURE 2.6.6

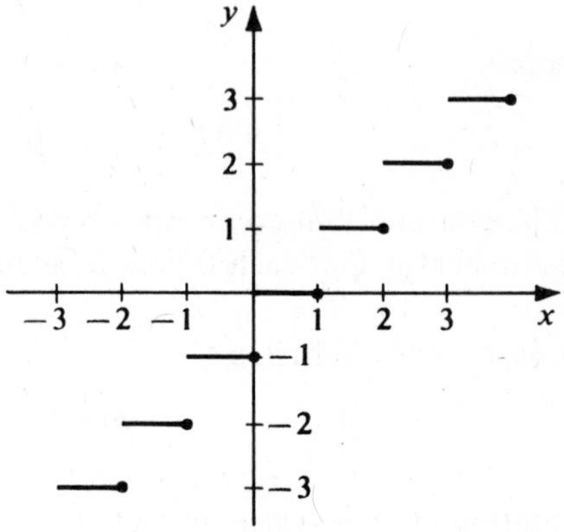

$$g(x) = [x)$$

FIGURE 2.6.7

For the graph, see Figure 2.6.7. At each integer, g is continuous from the left but discontinuous from the right.

For more examples of one-sided continuity we refer to Figures 2.6.8 and 2.6.9. In Figure 2.6.8 we have continuity from the right at 0. In Figure 2.6.9 we have continuity from the left at 0.

FIGURE 2.6.8   FIGURE 2.6.9

Continuity on $[a, b]$

From a function defined on a closed interval $[a, b)$ the most continuity that we can possibly expect is

(1) continuity at each point c of the open interval (a, b),
(2) continuity from the right at a, and
(3) continuity from the left at b.

Any function f that fulfills these requirements is called *continuous* on $[a, b]$.
As an example we take the function

$$f(x) = \sqrt{1 - x^2}.$$

Its graph is the semicircle displayed in Figure 2.6.10. The function is continuous on $[-1, 1]$ because it is continuous at each number c in $(-1, 1)$, continuous from the right at -1 and continuous from the left at 1.

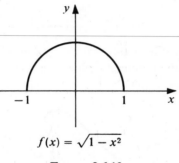

$$f(x) = \sqrt{1 - x^2}$$

FIGURE 2.6.10

Functions which are continuous on a closed interval $[a, b]$ have special properties that, in general, discontinuous functions do not have. We close this section by emphasizing two of these properties.

In the first place, a function which is continuous on an interval does not "skip" any values and thus its graph is an "unbroken curve." There are no "holes" in it and no "gaps." This is the idea behind the *intermediate-value theorem*.

The Intermediate-Value Theorem 2.6.1

If f is continuous on $[a, b]$ and C is a number between $f(a)$ and $f(b)$, then there is at least one number c between a and b such that $f(c) = C$.

We illustrate the theorem in Figure 2.6.11. What can happen in the discontinuous case is illustrated in Figure 2.6.12. There the number C has been "skipped."

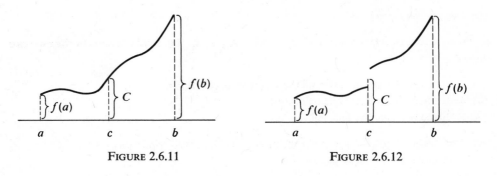

FIGURE 2.6.11   FIGURE 2.6.12

We come now to the maximum-minimum property of continuous functions.

Maximum-Minimum Theorem 2.6.2

If f is continuous on $[a, b]$ then, somewhere on $[a, b]$, f takes on a maximum value M and a minimum value m.

The situation is illustrated in Figure 2.6.13. The maximum value M is taken on at the point marked x_2 and the minimum value m is taken on at the point marked x_1.

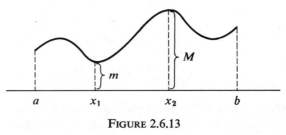

FIGURE 2.6.13

It is interesting to note that in the maximum-minimum theorem the full hypothesis is needed. If we drop the continuity requirement, the conclusion does not follow. As an example, we can take the function

$$f(x) = \left\{ \begin{array}{ll} \frac{1}{2}(a + b), & x = a \\ x, & a < x < b \\ \frac{1}{2}(a + b), & x = b \end{array} \right].$$

The graph of this function is pictured in Figure 2.6.14. The function is defined on $[a, b]$ but it takes on neither a maximum nor a minimum value. If, instead of dropping the continuity requirement, we drop the requirement that the interval be of the form $[a, b]$, then again the result fails. For example, in the case of

$$g(x) = x \qquad \text{for } x \in (a, b),$$

g is continuous at each point of (a, b) but again there is no maximum value and no minimum value. See Figure 2.6.15. (For proofs of the intermediate-value theorem and the maximum-minimum theorem see the Appendix at the end of the book.)

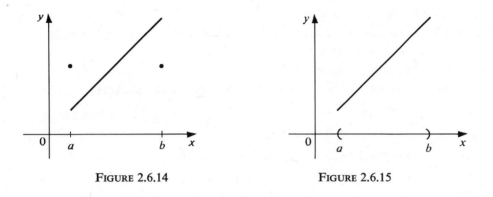

FIGURE 2.6.14   FIGURE 2.6.15

Exercises

*1. In each case, state whether or not the function is continuous at 2.

(a) $f(x) = 6x^7 - 100x + 1$.

(b) $g(x) = \sqrt{(x - 2)^3} + 5$.

(c) $h(x) = \begin{cases} x^2 + 4, & x < 2 \\ x^3, & x \geq 2 \end{cases}$.

(d) $l(x) = \sqrt{1 - x}$.

(e) $m(x) = (x - 2)[x]$.

(f) $n(x) = \begin{cases} \dfrac{1}{x - 2}, & x \neq 2 \\ 0, & x = 2 \end{cases}$.

(g) $v(x) = \dfrac{\sqrt{x^3 - 1} - x^4}{x}$.

(h) $w(x) = \begin{cases} 0, & 1.99 < x < 2.001 \\ -1, & x \notin (1.99, 2.001) \end{cases}$.

(i) $z(x) = \begin{cases} 0, & x < 2 \\ 0.01, & x \geq 2 \end{cases}$.

(j) $q(x) = [x]$.

*2. For each of the following functions sketch the graph and find the discontinuities (if any).

(a) $f(x) = |x - 1|$.

(b) $f(x) = |x^2 - 1|$.

(c) $f(x) = \begin{cases} 2x - 1, & x < 1 \\ x^2, & x \geq 1 \end{cases}$.

(d) $f(x) = \max \{x, x^2\}$.

(e) $f(x) = \dfrac{1}{x - 3}$.

(f) $f(x) = \begin{cases} -1, & x < 2 \\ \frac{1}{2}x, & 2 \leq x < 3 \\ x^{1/2}, & x > 3 \end{cases}$.

3. Find a function which is defined everywhere and discontinuous
   (a) only at 0.
   (b) only at 0 and 1.
   (c) from the left only at 0.
   (d) from the right only at 0.
4. Find a function which is continuous on the open interval $(0, 1)$ and has
   (a) no maximum value there.
   (b) a maximum value there.
5. At what points (if any) are the following functions continuous?

   *(a) $f(x) = \begin{cases} 1, & x \text{ rational} \\ 0, & x \text{ irrational} \end{cases}$.

   *(b) $g(x) = \begin{cases} x, & x \text{ rational} \\ 0, & x \text{ irrational} \end{cases}$.

6. Prove that there exists a real number x_0 such that

$$x_0^5 - 4x_0 + 1 = 7.21.$$

   HINT: Apply the intermediate-value theorem to the polynomial

$$P(x) = x^5 - 4x + 1.$$

7. Prove that f is continuous at c iff $\lim_{h \to 0} f(c + h) = f(c)$.

8. Let f and g be continuous at c. Prove that if
   (a) $f(c) > 0$, then there exists $\delta > 0$ such that

$$f(x) > 0 \quad \text{for all } x \in (c - \delta, c + \delta).$$

   (b) $f(c) < 0$, then there exists $\delta > 0$ such that

$$f(x) < 0 \quad \text{for all } x \in (c - \delta, c + \delta).$$

   (c) $f(c) < g(c)$, then there exists $\delta > 0$ such that

$$f(x) < g(x) \quad \text{for all } x \in (c - \delta, c + \delta).$$

9. (Optional) Let n be a positive integer.

   (a) Prove that, if $0 \le a < b$, then $a^n < b^n$.

   (b) Prove that every nonnegative real number x has a unique nonnegative nth root $x^{1/n}$. HINT: The existence of $x^{1/n}$ can be seen by applying the intermediate-value theorem to the function $f(t) = t^n$ for $t \ge 0$. The uniqueness follows from part (a).

Differentiation

3

3.1 The Idea of Derivative

Informal Beginnings

We begin with a function f and on its graph choose a point $(x, f(x))$. See Figure 3.1.1. Through the point $(x, f(x))$ we mark the line which to the naked eye seems to best approximate the graph of f near the chosen point. We call this line the *tangent line* at $(x, f(x))$. Our question is this: How can we determine the tangent line analytically?

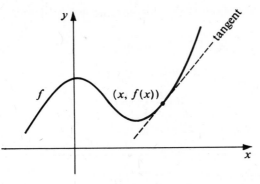

FIGURE 3.1.1

We already know that the tangent line passes through $(x, f(x))$ and hence all we need to determine is the slope. To do this, we choose a small number $h \neq 0$ and on the graph of f mark the point $(x + h, f(x + h))$. See Figure 3.1.2. The secant line that passes through $(x, f(x))$ and $(x + h, f(x + h))$ has slope

$$\frac{f(x + h) - f(x)}{h}.$$

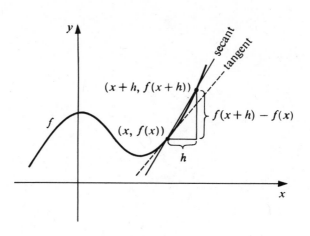

FIGURE 3.1.2

This is certainly not the slope of the tangent line but it is an approximation to it. To get the slope of the tangent line we will let h tend to zero. If all goes well (this is all very informal), as h tends to 0, the point $(x + h, f(x + h))$ will slide along the curve toward the point $(x, f(x))$ and, as this happens, the secant line will have the tangent as a limiting position. With

$$\frac{f(x + h) - f(x)}{h} \quad \text{as the slope of the secant line,}$$

we can view

$$\lim_{h \to 0} \frac{f(x + h) - f(x)}{h} \quad \text{as the slope of the tangent line.}$$

This in fact is how the slope of the tangent line will be defined.

Definitions and Examples

Not all functions possess limits of the form

$$\lim_{h \to 0} \frac{f(x + h) - f(x)}{h}.$$

Those which do are called *differentiable*.

Definition of Derivative

A function f is said to be *differentiable* at x iff

$$\lim_{h \to 0} \frac{f(x + h) - f(x)}{h} \quad \text{exists.}$$

If this limit exists, it is called the *derivative* of f at x and is denoted by $f'(x)$.†

---

† This prime notation goes back to the French mathematician Joseph Louis Lagrange (1736–1813). Another will be introduced later.

The number

$$f'(x) = \lim_{h \to 0} \frac{f(x + h) - f(x)}{h}$$

can be interpreted as the *slope* of the graph of f at the point $(x, f(x))$. The line through $(x, f(x))$ which has this slope is called the *tangent line* at $(x, f(x))$.

It is time we looked at some examples.

Example.   We begin with the squaring function

$$f(x) = x^2. \qquad\qquad \text{(Figure 3.1.3)}$$

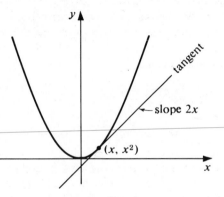

squaring function

FIGURE 3.1.3

To find $f'(x)$ we form the *difference quotient*

$$\frac{f(x + h) - f(x)}{h} = \frac{(x + h)^2 - x^2}{h}.$$

Since

$$\frac{(x + h)^2 - x^2}{h} = \frac{x^2 + 2xh + h^2 - x^2}{h} = \frac{2xh + h^2}{h} = 2x + h,$$

we have

$$\frac{f(x + h) - f(x)}{h} = 2x + h.$$

Obviously then

$$f'(x) = \lim_{h \to 0} \frac{f(x + h) - f(x)}{h} = 2x. \quad \square$$

Example.   In the case of a linear function

$$f(x) = mx + b$$

we have

$$f'(x) = m.$$

In other words, the slope is constantly m. To verify this, note that

$$\frac{f(x + h) - f(x)}{h} = \frac{[m(x + h) + b] - [mx + b]}{h} = \frac{mh}{h} = m.$$

It follows that

$$f'(x) = \lim_{h \to 0} \frac{f(x + h) - f(x)}{h} = m. \quad \square$$

The derivative

$$f'(x) = \lim_{h \to 0} \frac{f(x + h) - f(x)}{h}$$

is a two-sided limit. It can therefore not be taken at an endpoint of the domain. In our next example we will be dealing with the square-root function. Although this is defined for all $x \geq 0$, you can expect a derivative only for $x > 0$.

Example.   The square-root function

$$f(x) = \sqrt{x}, \qquad x \geq 0 \qquad\qquad \text{(Figure 3.1.4)}$$

has derivative

$$f'(x) = \frac{1}{2\sqrt{x}}, \qquad \text{for } x > 0.$$

To verify this, we begin with $x > 0$ and form the difference quotient

$$\frac{f(x + h) - f(x)}{h} = \frac{\sqrt{x + h} - \sqrt{x}}{h}.$$

To remove the radicals from the numerator we multiply both the numerator and the denominator by $\sqrt{x + h} + \sqrt{x}$. This gives

$$\frac{f(x + h) - f(x)}{h} = \left(\frac{\sqrt{x + h} - \sqrt{x}}{h}\right)\left(\frac{\sqrt{x + h} + \sqrt{x}}{\sqrt{x + h} + \sqrt{x}}\right) = \frac{1}{\sqrt{x + h} + \sqrt{x}}$$

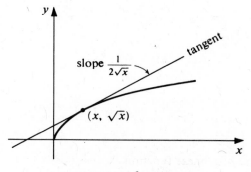

square root function

FIGURE 3.1.4

and thus

$$f'(x) = \lim_{h \to 0} \frac{f(x + h) - f(x)}{h} = \frac{1}{2\sqrt{x}}. \quad \square$$

To *differentiate* a function is to find its derivative.

Example.   Here we differentiate the function

$$f(x) = \sqrt{1 - x^2}.$$

Its graph is the upper half of the unit circle. (Figure 3.1.5)

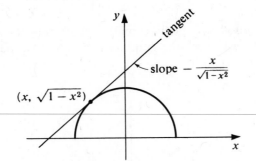

upper half of unit circle

FIGURE 3.1.5

The domain of f is the closed interval $[-1, 1]$ but we can expect differentiability at most on the open interval $(-1, 1)$. For x in $(-1, 1)$ we form the difference quotient

$$\frac{f(x - h) - f(x)}{h} = \frac{\sqrt{1 - (x + h)^2} - \sqrt{1 - x^2}}{h}.$$

Once again we remove the radicals from the numerator, this time by multiplying both numerator and denominator by

$$\sqrt{1 - (x + h)^2} + \sqrt{1 - x^2}.$$

After simplification these calculations yield

$$\frac{f(x + h) - f(x)}{h} = \frac{-2x - h}{\sqrt{1 - (x + h)^2} + \sqrt{1 - x^2}}$$

and thus

$$f'(x) = \lim_{h \to 0} \frac{f(x + h) - f(x)}{h} = \frac{-x}{\sqrt{1 - x^2}}. \quad \square$$

Differentiability versus Continuity

A function can be continuous at some number x without being differentiable there. For example, the absolute-value function

$$f(x) = |x|$$

is continuous at 0 (it is everywhere continuous) but it is not differentiable at 0:

$$\frac{f(0 + h) - f(0)}{h} = \frac{|0 + h| - |0|}{h} = \frac{|h|}{h} = \begin{cases} -1, & h < 0 \\ 1, & h > 0 \end{cases},$$

so that

$$\lim_{h \to 0} \frac{f(0 + h) - f(0)}{h} \quad \text{does not exist.}$$

The failure of the absolute-value function to be differentiable at 0 is reflected by its graph. (Figure 3.1.6) At $(0, 0)$, the graph comes to a sharp point and there is no tangent line.

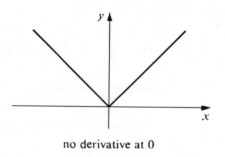

no derivative at 0

FIGURE 3.1.6

A similar situation occurs with

$$f(x) = |x^2 - 1|.$$

The graph of f (pictured in Figure 3.1.7) can be obtained by starting with the graph of

$$g(x) = x^2 - 1 \qquad \text{(Figure 3.1.8)}$$

and replacing the portion which lies below the x-axis by its mirror image above the x-axis. Both functions f and g are everywhere continuous, but while g is everywhere differentiable, f is not. The graph of f comes to a sharp point at $(-1, 0)$ and $(1, 0)$. The function is not differentiable at -1 and it is not differentiable at 1.

Although not every continuous function is differentiable, every differentiable function is continuous. In particular,

(3.1.1) | if f is differentiable at x,   then f is continuous at x.

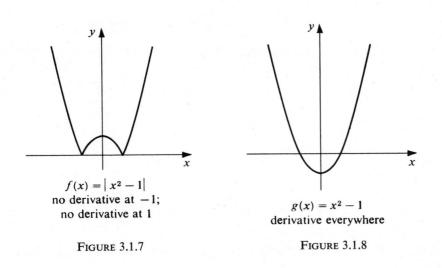

$f(x) = |x^2 - 1|$
no derivative at -1;
no derivative at 1

FIGURE 3.1.7

$g(x) = x^2 - 1$
derivative everywhere

FIGURE 3.1.8

To prove this, we need only observe that for $h \neq 0$ and $x + h$ in the domain of f,

$$f(x + h) - f(x) = \frac{f(x + h) - f(x)}{h} \cdot h.$$

With f differentiable at x,

$$\lim_{h \to 0} \frac{f(x + h) - f(x)}{h} = f'(x).$$

Since

$$\lim_{h \to 0} h = 0,$$

we have

$$\lim_{h \to 0} [f(x + h) - f(x)] = \left[\lim_{h \to 0} \frac{f(x + h) - f(x)}{h} \right] \cdot \left[\lim_{h \to 0} h \right] = f'(x) \cdot 0 = 0.$$

This tells us that

$$\lim_{h \to 0} f(x + h) = f(x)$$

and thus (by Exercise 7, Section 2.6) that f is continuous at x.   □

Exercises

Differentiate each of the following functions by forming a suitable difference quotient and taking the limit as h tends to 0.

* 1. $f(x) = \alpha.$ 2. $f(x) = 3x.$ * 3. $f(x) = 3x + 5.$
  4. $f(x) = mx + b.$ * 5. $f(x) = 1/x.$ 6. $f(x) = 1/x^2.$
* 7. $f(x) = \frac{1}{2}x^2.$ 8. $f(x) = -x^2.$ * 9. $f(x) = 2x^3.$
  10. $f(x) = x^4.$ * 11. $f(x) = (x - 1)^2.$ 12. $f(x) = (2x - 3)^2.$
* 13. $f(x) = 1/\sqrt{x}.$ 14. $f(x) = \sqrt{x - 1}.$ * 15. $f(x) = 1/\sqrt{x - 1}.$

For each of the functions below find $f'(2)$ by forming the difference quotient

$$\frac{f(2 + h) - f(2)}{h}$$

and taking the limit as $h \to 0$.

16. $f(x) = x^2$. $\left[\dfrac{f(2 + h) - f(2)}{h} = \dfrac{(2 + h)^2 - 2^2}{h} = 4 + h, \text{ etc.} \right]$

*17. $f(x) = \dfrac{1}{x + 1}$.

18. $f(x) = \dfrac{1}{2x}$.

*19. $f(x) = \sqrt{x + 2}$.

20. $f(x) = \dfrac{1}{\sqrt{x + 2}}$.

Draw the graph of each of the following functions and indicate where it is not differentiable.

21. $f(x) = |2x - 5|$.

22. $f(x) = |x^2 - 4|$.

23. $f(x) = \sqrt{|x|}$.

24. $f(x) = |x^3 - 1|$.

3.2   Some Differentiation Formulas

Calculating the derivative of

$$f(x) = (x^3 - 2)(4x + 1) \quad \text{or} \quad f(x) = \frac{6x^2 - 1}{x^4 + 5x + 1}$$

by forming the proper difference quotient

$$\frac{f(x + h) - f(x)}{h}$$

and then taking the limit as h tends to 0 is rather time consuming. Here we derive some general formulas which make such calculations quite simple.

For future reference we first point out that the constant functions have derivative 0:

$$\text{if} \quad f(x) = \alpha, \qquad \text{then} \quad f'(x) = 0$$

and the identity function has derivative 1:

$$\text{if} \quad f(x) = x, \qquad \text{then} \quad f'(x) = 1.$$

PROOF.   For $f(x) = \alpha$,

$$f'(x) = \lim_{h \to 0} \frac{f(x + h) - f(x)}{h} = \lim_{h \to 0} \frac{\alpha - \alpha}{h} = \lim_{h \to 0} 0 = 0.$$

For $f(x) = x$,

$$f'(x) = \lim_{h \to 0} \frac{f(x + h) - f(x)}{h} = \lim_{h \to 0} \frac{(x + h) - x}{h} = \lim_{h \to 0} \frac{h}{h} = \lim_{h \to 0} 1 = 1. \quad \square$$

Derivatives of Sums and Scalar Multiples

Let α be a real number. If f and g are differentiable at x, then $f + g$ and αf are differentiable at x. Moreover,

(i) $(f + g)'(x) = f'(x) + g'(x)$,

(ii) $(\alpha f)'(x) = \alpha f'(x)$.

PROOF.   To prove (i) we note that

$$\frac{(f + g)(x + h) - (f + g)(x)}{h} = \frac{[f(x + h) + g(x + h)] - [f(x) + g(x)]}{h}$$

$$= \frac{f(x + h) - f(x)}{h} + \frac{g(x + h) - g(x)}{h}.$$

By definition,

$$\lim_{h \to 0} \frac{f(x + h) - f(x)}{h} = f'(x), \qquad \lim_{h \to 0} \frac{g(x + h) - g(x)}{h} = g'(x).$$

Thus

$$\lim_{h \to 0} \frac{(f + g)(x + h) - (f + g)(x)}{h} = f'(x) + g'(x),$$

which means that

$$(f + g)'(x) = f'(x) + g'(x).$$

To show that (ii) holds we must show that

$$\lim_{h \to 0} \frac{(\alpha f)(x + h) - (\alpha f)(x)}{h} = \alpha f'(x).$$

This follows directly from the fact that

$$\frac{(\alpha f)(x + h) - (\alpha f)(x)}{h} = \alpha \left[\frac{f(x + h) - f(x)}{h} \right]. \quad \square$$

Product Rule

If f and g are differentiable at x, then so is their product, and

$$(fg)'(x) = f(x)g'(x) + g(x)f'(x).$$

PROOF.   In the first place

$$\frac{(fg)(x + h) - (fg)(x)}{h} = \frac{f(x + h)g(x + h) - f(x)g(x)}{h}.$$

Subtracting and adding $f(x + h)g(x)$ to the numerator, we can rewrite the above difference quotient as

$$f(x + h) \left[\frac{g(x + h) - g(x)}{h} \right] + g(x) \left[\frac{f(x + h) - f(x)}{h} \right].$$

Since f is differentiable at x, we know that f is continuous at x (3.1.1) and thus that

$$\lim_{h \to 0} f(x + h) = f(x).$$

Since

$$\lim_{h \to 0} \frac{g(x + h) - g(x)}{h} = g'(x) \quad \text{and} \quad \lim_{h \to 0} \frac{f(x + h) - f(x)}{h} = f'(x),$$

we obtain

$$\lim_{h \to 0} \frac{(fg)(x + h) - (fg)(x)}{h} = f(x)g'(x) + g(x)f'(x). \quad \square$$

Using the product rule it is not hard to prove that

(3.2.1)

$$\boxed{\begin{array}{l} \text{for each positive integer } n \\ p(x) = x^n \text{ has derivative } p'(x) = nx^{n-1}. \end{array}}$$

In particular

$$p(x) = x \quad \text{has derivative} \quad p'(x) = 1 \cdot x^0 = 1,$$
$$p(x) = x^2 \quad \text{has derivative} \quad p'(x) = 2x,$$
$$p(x) = x^3 \quad \text{has derivative} \quad p'(x) = 3x^2,$$
$$p(x) = x^4 \quad \text{has derivative} \quad p'(x) = 4x^3,$$

and so on.

PROOF OF (3.2.1). We proceed by induction on n. If $n = 1$, then we have the identity function

$$p(x) = x,$$

which we know satisfies

$$p'(x) = 1 = 1 \cdot x^0.$$

This means that the formula holds for $n = 1$.

We suppose now that the result holds for $n = k$, and show that it holds for $n = k + 1$. We let

$$p(x) = x^{k+1}$$

and note that

$$p(x) = x \cdot x^k.$$

Applying the product rule and our inductive hypothesis, we obtain

$$p'(x) = x \cdot kx^{k-1} + 1 \cdot x^k = (k + 1)x^k,$$

which shows that the formula holds for $k + 1$. $\quad \square$

The formula for differentiating polynomials should now be obvious:

$$\text{if} \quad P(x) = a_n x^n + a_{n-1} x^{n-1} + \cdots + a_1 x + a_0,$$
$$\text{then} \quad P'(x) = n a_n x^{n-1} + (n-1) a_{n-1} x^{n-2} + \cdots + a_1.$$

For example,

$$P(x) = 12x^3 - 6x - 2 \quad \text{has derivative} \quad P'(x) = 36x^2 - 6$$

and

$$P(x) = \tfrac{1}{4}x^4 - 2x^2 + x + 5 \quad \text{has derivative} \quad P'(x) = x^3 - 4x + 1.$$

Problem. Differentiate

$$F(x) = (x^3 - 2)(4x + 1).$$

SOLUTION. We have a product

$$F(x) = f(x)g(x)$$

with

$$f(x) = x^3 - 2 \quad \text{and} \quad g(x) = 4x + 1.$$

The product rule gives

$$
\begin{aligned}
F'(x) &= f(x)g'(x) + g(x)f'(x) \\
&= (x^3 - 2) \cdot 4 + (4x + 1)(3x^2) \\
&= 4x^3 - 8 + 12x^3 + 3x^2 \\
&= 16x^3 + 3x^2 - 8. \quad \square
\end{aligned}
$$

Problem. Differentiate

$$F(x) = (ax + b)(cx + d).$$

SOLUTION. We have a product

$$F(x) = f(x)g(x)$$

with

$$f(x) = ax + b \quad \text{and} \quad g(x) = cx + d.$$

Again we use the product rule

$$F'(x) = f(x)g'(x) + g(x)f'(x).$$

In this case

$$
\begin{aligned}
F'(x) &= (ax + b)c + (cx + d)a \\
&= 2acx + bc + ad. \quad \square
\end{aligned}
$$

We can also do this problem without using the product rule by first carrying out the multiplication:

$$F(x) = acx^2 + bcx + adx + bd$$

and then differentiating:

$$F'(x) = 2acx + bc + ad. \quad \square$$

We come now to reciprocals.

Reciprocal Rule

If g is differentiable at x and $g(x) \neq 0$, then $1/g$ is differentiable at x and

$$\left(\frac{1}{g}\right)'(x) = -\frac{g'(x)}{[g(x)]^2}.$$

PROOF.   Since g is differentiable at x, g is continuous at x. (3.1.1) Since $g(x) \neq 0$, we know that $1/g$ is continuous at x, and thus that

$$\lim_{h \to 0} \frac{1}{g(x+h)} = \frac{1}{g(x)}.$$

Let $h \neq 0$. If $g(x + h) \neq 0$, then

$$\frac{1}{h}\left[\frac{1}{g(x+h)} - \frac{1}{g(x)}\right] = -\left[\frac{g(x+h) - g(x)}{h}\right]\frac{1}{g(x+h)}\frac{1}{g(x)}.$$

Taking the limit as h tends to zero, we see that the right-hand side (and thus the left) tends to

$$-\frac{g'(x)}{[g(x)]^2}. \quad \square$$

From this last result it is easy to see that the formula for the derivative of a power, x^n, also applies to negative powers; namely,

(3.2.2)

> for each negative integer n
> $p(x) = x^n$ has derivative $p'(x) = nx^{n-1}$

except of course at $x = 0$ where no negative power is even defined. In particular

$$p(x) = x^{-1} \quad \text{has derivative} \quad p'(x) = (-1)x^{-2} = -x^{-2},$$
$$p(x) = x^{-2} \quad \text{has derivative} \quad p'(x) = -2x^{-3},$$
$$p(x) = x^{-3} \quad \text{has derivative} \quad p'(x) = -3x^{-4},$$

and so on.

PROOF OF (3.2.2)

$$p(x) = \frac{1}{g(x)} \quad \text{with} \quad g(x) = x^{-n}$$

and $-n$ is a positive integer. The rule for reciprocals gives

$$p'(x) = -\frac{g'(x)}{[g(x)]^2} = -\frac{(-nx^{-n-1})}{x^{-2n}} = nx^{n-1}. \quad \square$$

Problem. Differentiate

$$f(x) = 5x^2 - \frac{6}{x}.$$

SOLUTION

$$f(x) = 5x^2 - 6x^{-1},$$

so that

$$f'(x) = 10x + 6x^{-2},$$

which, if you don't like negative exponents, you can rewrite as

$$f'(x) = 10x + \frac{6}{x^2}. \quad \square$$

Problem. Differentiate

$$f(x) = \frac{1}{ax^2 + bx + c}.$$

SOLUTION. Here we have a reciprocal

$$f(x) = \frac{1}{g(x)}$$

with

$$g(x) = ax^2 + bx + c.$$

The reciprocal rule gives

$$f'(x) = -\frac{g'(x)}{[g(x)]^2}$$

$$= -\frac{2ax + b}{[ax^2 + bx + c]^2}. \quad \square$$

Finally we come to quotients in general.

Quotient Rule

If f and g are differentiable at x and $g(x) \neq 0$, then the quotient f/g is differentiable at x and

$$\left(\frac{f}{g}\right)'(x) = \frac{g(x)f'(x) - f(x)g'(x)}{[g(x)]^2}.$$

The proof is left to you. HINT: $f/g = f \cdot 1/g$.

From the quotient rule you can see that all rational functions (quotients of polynomials) are differentiable wherever they are defined.

Problem. Differentiate

$$F(x) = \frac{ax + b}{cx + d}.$$

SOLUTION.   We are dealing with a quotient

$$F(x) = \frac{f(x)}{g(x)}.$$

The quotient rule,

$$F'(x) = \frac{g(x)f'(x) - f(x)g'(x)}{[g(x)]^2},$$

gives

$$F'(x) = \frac{(cx + d) \cdot a - (ax + b) \cdot c}{(cx + d)^2}$$

$$= \frac{ad - bc}{(cx + d)^2}. \quad \square$$

Problem. Differentiate

$$F(x) = \frac{6x^2 - 1}{x^4 + 5x + 1}.$$

SOLUTION.   We are dealing with a quotient

$$F(x) = \frac{f(x)}{g(x)}.$$

The quotient rule,

$$F'(x) = \frac{g(x)f'(x) - f(x)g'(x)}{[g(x)]^2},$$

gives

$$F'(x) = \frac{(x^4 + 5x + 1)(12x) - (6x^2 - 1)(4x^3 + 5)}{(x^4 + 5x + 1)^2}. \quad \square$$

Problem. Find $f'(0)$, $f'(1)$, and $f'(2)$ for the function

$$f(x) = \frac{5x}{1 + x}.$$

SOLUTION.   First we find a general expression for $f'(x)$ and then evaluate it at 0, 1, and 2. Using the quotient rule, we get

$$f'(x) = \frac{(1 + x)5 - 5x(1)}{(1 + x)^2} = \frac{5}{(1 + x)^2}.$$

This gives

$$f'(0) = \frac{5}{(1 + 0)^2} = 5, \quad f'(1) = \frac{5}{(1 + 1)^2} = \frac{5}{4},$$

$$f'(2) = \frac{5}{(1 + 2)^2} = \frac{5}{9}. \quad \square$$

Problem.   Find $f'(-1)$ for

$$f(x) = \frac{x^2}{ax^2 + b}.$$

SOLUTION.   We first find $f'(x)$ in general by the quotient rule:

$$f'(x) = \frac{(ax^2 + b)2x - x^2(2ax)}{(ax^2 + b)^2}.$$

Now we evaluate f' at -1:

$$f'(-1) = \frac{(a + b)(-2) - (-2a)}{(a + b)^2}$$

$$= \frac{-2a - 2b + 2a}{(a + b)^2}$$

$$= -\frac{2b}{(a + b)^2}. \quad \square$$

Exercises

Differentiate the following functions.

*1. $F(x) = 1 - x.$

2. $F(x) = 2(1 + x).$

*3. $F(x) = 11x^5 - 6x^3 + 8.$

4. $F(x) = \frac{x^4}{4} - \frac{x^3}{3} + \frac{x^2}{2} - \frac{x}{1}.$

*5. $F(x) = ax^2 + bx + c.$

6. $F(x) = \frac{3}{x^2}.$

*7. $F(x) = -\frac{1}{x^2}.$

8. $F(x) = \frac{(x^2 + 2)}{x^3}.$

*9. $F(x) = \frac{ax - b}{cx - d}.$

10. $F(x) = x - \frac{1}{x}.$

*11. $G(x) = \frac{x^3}{1 - x}.$

12. $G(x) = (x^2 + 1)^2.$

*13. $G(x) = (x^2 - 1)(x - 3)$.

14. $G(x) = \dfrac{7x^4 + 11}{x + 1}$.

*15. $G(x) = (x - 1)(x - 2)$.

16. $G(x) = \dfrac{2x^2 + 1}{x + 2}$.

*17. $G(x) = \dfrac{6 - 1/x}{x - 2}$.

18. $G(x) = \dfrac{1 + x^4}{x^2}$.

*19. $G(x) = (9x^8 - 8x^9)\left(x + \dfrac{1}{x}\right)$.

20. $G(x) = \left(1 - \dfrac{1}{x}\right)^2$.

Find $f'(0)$ and $f'(1)$.

*21. $f(x) = \dfrac{1}{x - 2}$.

22. $f(x) = x^2(x + 1)$.

*23. $f(x) = \dfrac{1 - x^2}{1 + x^2}$.

24. $f(x) = \dfrac{2x^2 + x + 1}{x^2 + 2x + 1}$.

*25. $f(x) = \dfrac{ax + b}{cx + d}$.

26. $f(x) = \dfrac{ax^2 + bx + c}{cx^2 + bx + a}$.

Given that $h(0) = 1$ and $h'(0) = 2$, find $f'(0)$.

*27. $f(x) = xh(x)$.

28. $f(x) = 3[h(x)]^2 - 5x$.

*29. $f(x) = \dfrac{1}{[h(x)]^2}$.

30. $f(x) = h(x) - \dfrac{1}{h(x)}$.

31. Prove the validity of quotient rule.

3.3   The *d/dx* Notation

So far we have indicated the derivative by a prime but there are other notations which are widely used, particularly in science and engineering. The most popular of these is the double-*d* notation of Leibniz. In the Leibniz notation the derivative of a function *y* is indicated by writing

$$\frac{dy}{dx}, \frac{dy}{dt}, \quad \text{or} \quad \frac{dy}{dz}, \text{ etc.,}$$

depending upon whether the letter x, t, or z, etc., is being used for the elements of the domain of y. For instance, if y is initially defined by

$$y(x) = x^3,$$

then the Leibniz notation gives

$$\frac{dy}{dx}(x) = 3x^2.$$

Usually writers drop the x on the left and simply write

$$y = x^3$$

and

$$\frac{dy}{dx} = 3x^2.$$

The symbols

$$\frac{d}{dx}, \frac{d}{dt}, \frac{d}{dz}, \text{ etc.,}$$

are also used as prefixes before expressions to be differentiated. For example,

$$\frac{d}{dx}(x^3 - 4x) = 3x^2 - 4, \qquad \frac{d}{dt}(t^2 + 3t + 1) = 2t + 3, \qquad \frac{d}{dz}(z^5 - 1) = 5z^4.$$

In the Leibniz notation the differentiation formulas look like this:

I.
$$\frac{d}{dx}[f(x) + g(x)] = \frac{d}{dx}[f(x)] + \frac{d}{dx}[g(x)].$$

II.
$$\frac{d}{dx}[\alpha f(x)] = \alpha \frac{d}{dx}[f(x)].$$

III.
$$\frac{d}{dx}[f(x)g(x)] = f(x)\frac{d}{dx}[g(x)] + g(x)\frac{d}{dx}[f(x)].$$

IV.
$$\frac{d}{dx}\left[\frac{1}{g(x)}\right] = -\frac{1}{[g(x)]^2}\frac{d}{dx}[g(x)].$$

V.
$$\frac{d}{dx}\left[\frac{f(x)}{g(x)}\right] = \frac{g(x)\dfrac{d}{dx}[f(x)] - f(x)\dfrac{d}{dx}[g(x)]}{[g(x)]^2}.$$

Often the functions f and g are replaced by u and v and the x is left out altogether. Formulas I–V then look like this:

I.
$$\frac{d}{dx}(u + v) = \frac{du}{dx} + \frac{dv}{dx}.$$

II.
$$\frac{d}{dx}(\alpha u) = \alpha \frac{du}{dx}.$$

III.
$$\frac{d}{dx}(uv) = u\frac{dv}{dx} + v\frac{du}{dx}.$$

IV.
$$\frac{d}{dx}\left(\frac{1}{v}\right) = -\frac{1}{v^2}\frac{dv}{dx}.$$

V.
$$\frac{d}{dx}\left(\frac{u}{v}\right) = \frac{v\dfrac{du}{dx} - u\dfrac{dv}{dx}}{v^2}.$$

Probably the only way to develop a feeling for this new notation is to use it. Below we work out some practice problems.

Problem. Find

$$\frac{dy}{dx} \quad \text{if} \quad y = \frac{x - 1}{x + 2}.$$

SOLUTION. Here we use the quotient rule (V):

$$\frac{dy}{dx} = \frac{(x + 2)\dfrac{d}{dx}(x - 1) - (x - 1)\dfrac{d}{dx}(x + 2)}{(x + 2)^2}$$

$$= \frac{(x + 2) - (x - 1)}{(x + 2)^2}$$

$$= \frac{3}{(x + 2)^2}. \quad \square$$

Problem. Find

$$\frac{dy}{dx} \quad \text{if} \quad y = (x^3 + 1)(3x^5 + 2x - 1).$$

SOLUTION. Here we can use the product rule (III):

$$\frac{dy}{dx} = (x^3 + 1)\frac{d}{dx}(3x^5 + 2x - 1) + (3x^5 + 2x - 1)\frac{d}{dx}(x^3 + 1)$$

$$= (x^3 + 1)(15x^4 + 2) + (3x^5 + 2x - 1)(3x^2)$$

$$= (15x^7 + 15x^4 + 2x^3 + 2) + (9x^7 + 6x^3 - 3x^2)$$

$$= 24x^7 + 15x^4 + 8x^3 - 3x^2 + 2.$$

We could have achieved the same result by carrying out the initial multiplication:

$$y = (x^3 + 1)(3x^5 + 2x - 1)$$
$$= (3x^8 + 2x^4 - x^3) + (3x^5 + 2x - 1)$$
$$= 3x^8 + 3x^5 + 2x^4 - x^3 + 2x - 1$$

and then differentiating:

$$\frac{dy}{dx} = 24x^7 + 15x^4 + 8x^3 - 3x^2 + 2. \quad \square$$

Problem. Find

$$\frac{d}{dt}\left(t^3 - \frac{t}{t^2 - 1}\right).$$

SOLUTION

$$\frac{d}{dt}\left(t^3 - \frac{t}{t^2 - 1}\right) = \frac{d}{dt}(t^3) - \frac{d}{dt}\left(\frac{t}{t^2 - 1}\right)$$

$$= 3t^2 - \left[\frac{(t^2 - 1)\cdot 1 - t(2t)}{(t^2 - 1)^2}\right]$$

$$= 3t^2 + \frac{t^2 + 1}{(t^2 - 1)^2}. \quad \Box$$

Problem.   Find

$$\frac{du}{dx} \quad \text{if} \quad u = \frac{ax + b}{cx + d} - \frac{cx + d}{ax + b}.$$

SOLUTION

$$\frac{du}{dx} = \frac{d}{dx}\left(\frac{ax + b}{cx + d}\right) - \frac{d}{dx}\left(\frac{cx + d}{ax + b}\right)$$

$$= \frac{(cx + d)a - (ax + b)c}{(cx + d)^2} - \frac{(ax + b)c - (cx + d)a}{(ax + b)^2}$$

$$= \frac{ad - bc}{(cx + d)^2} - \frac{bc - ad}{(ax + b)^2}$$

$$= (ad - bc)\left[\frac{1}{(cx + d)^2} + \frac{1}{(ax + b)^2}\right]. \quad \Box$$

Problem.   Evaluate dy/dx at $x = 0$ and $x = 1$ if

$$y = \frac{x^2}{x^2 - 4}.$$

SOLUTION

$$\frac{dy}{dx} = \frac{(x^2 - 4)2x - x^2(2x)}{(x^2 - 4)^2} = -\frac{8x}{(x^2 - 4)^2}.$$

At $x = 0$, $dy/dx = 0$; at $x = 1$, $dy/dx = -\frac{8}{9}$. $\Box$

Exercises

Find dy/dx.

*1. $y = 3x^4 - x^2 + 1$.

2. $y = x^2 + 2x^{-4}$.

*3. $y = x - \dfrac{1}{x}$.

4. $y = \dfrac{2x}{1 - x}$.

*5. $y = \dfrac{x}{1 + x^2}$.

6. $y = x(x - 2)(x + 1)$.

*7. $y = \dfrac{x^2}{1 - x}$.

8. $y = \left(\dfrac{x}{1 + x}\right)\left(\dfrac{2 - x}{3}\right)$.

*9. $y = \dfrac{x^3 + 1}{x^3 - 1}$.

10. $y = \dfrac{x^2}{(1 + x)^2}$.

Find the indicated derivative.

*11. $\dfrac{d}{dx}(2x - 5)$.

12. $\dfrac{d}{dx}(5x + 2)$.

*13. $\dfrac{d}{dx}(7x^3 - 1)$.

14. $\dfrac{d}{dx}[(1 - x)^2(1 + x)]$.

*15. $\dfrac{d}{dt}\left(\dfrac{t^2 + 1}{t^2 - 1}\right)$.

16. $\dfrac{d}{dt}\left(\dfrac{2t^3 + 1}{t^4}\right)$.

*17. $\dfrac{d}{dt}\left(\dfrac{t^4}{2t^3 - 1}\right)$.

18. $\dfrac{d}{dt}\left[\dfrac{t}{(1 + t)^2}\right]$.

*19. $\dfrac{d}{du}\left(\dfrac{2u}{1 - 2u}\right)$.

20. $\dfrac{d}{du}\left(\dfrac{u^2}{u^3 + 1}\right)$.

*21. $\dfrac{d}{du}\left(\dfrac{u}{u - 1} - \dfrac{u}{u + 1}\right)$.

22. $\dfrac{d}{du}[u^2(1 - u^2)(1 - u^3)]$.

*23. $\dfrac{d}{dx}\left(\dfrac{x^2}{1 - x^2} - \dfrac{1 - x^2}{x^2}\right)$.

24. $\dfrac{d}{dx}\left(\dfrac{3x^4 + 2x + 1}{x^4 + x - 1}\right)$.

*25. $\dfrac{d}{dx}\left(\dfrac{x^3 + x^2 + x + 1}{x^3 - x^2 + x - 1}\right)$.

26. $\dfrac{d}{dx}\left(\dfrac{x^3 + x^2 + x - 1}{x^3 - x^2 + x + 1}\right)$.

Evaluate dy/dx at $x = 2$.

*27. $y = (x + 1)(x + 2)(x + 3)$.

28. $y = (x + 1)(x^2 + 2)(x^3 + 3)$.

*29. $y = \dfrac{(x - 1)(x - 2)}{(x + 2)}$.

30. $y = \dfrac{(x^2 + 1)(x^2 - 2)}{x^2 + 2}$.

3.4 The Derivative as a Rate of Change

In the case of a linear function

$$y = mx + b$$

the graph is a straight line and the slope m gives *the rate of change of y with respect to x*:

$$m = \frac{\text{change in } y}{\text{change in } x}.$$ (Figure 3.4.1)

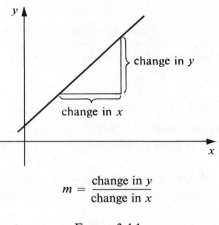

$$m = \frac{\text{change in } y}{\text{change in } x}$$

FIGURE 3.4.1

In other words, on a line of slope m, y changes m times as fast as x.

In the more general case of a differentiable function

$$y = f(x)$$

the graph is a curve. The slope

$$\frac{dy}{dx} = f'(x)$$

still gives *the rate of change of y with respect to x* but this rate can vary from point to point. At x_1 (see Figure 3.4.2) the rate of change is $f'(x_1)$; at x_2, the rate of change is $f'(x_2)$; at x_3, the rate of change is $f'(x_3)$.

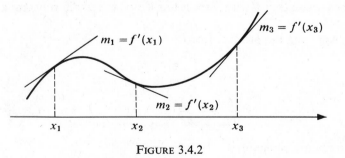

FIGURE 3.4.2

This rate of change idea is one of the fundamental ideas of the calculus. Keep it in mind whenever you see a derivative.

Problem. Find the rate of change of the area of a circle with respect to its radius r. Evaluate this rate of change when $r = 2$.

SOLUTION.   Since

$$A = \pi r^2$$

the rate of change of A with respect to r is given by

$$\frac{dA}{dr} = 2\pi r,$$

which, by the way, is the circumference. When $r = 2$, the rate of change is 4π. This means that when the radius is 2, the area is changing 4π times as fast as the radius.   □

Problem.   Find the rate of change of the area of an equilateral triangle with respect to the length of a side. Evaluate this rate of change when the side has length $\sqrt{3}$.

SOLUTION.   If the side has length s, the area is given by the formula

$$A = \frac{\sqrt{3}}{4} s^2. \qquad\qquad \text{(check this out)}$$

The rate of change is therefore

$$\frac{dA}{ds} = \frac{\sqrt{3}}{2} s.$$

When $s = \sqrt{3}$, the rate of change is $\frac{3}{2}$. In other words, when the side has length $\sqrt{3}$, the area is changing $\frac{3}{2}$ times as fast as the length of the side.   □

Problem.   The volume of a circular cylinder of height h and base radius r is given by the formula

$$V = \pi r^2 h.$$

Find the rate of change of h with respect to r if V is to remain constant as r increases.

SOLUTION.   Applying the product rule to

$$V = \pi r^2 h$$

we get

$$\frac{dV}{dr} = \pi r^2 \frac{dh}{dr} + h \frac{d}{dr}(\pi r^2) = \pi r^2 \frac{dh}{dr} + 2\pi rh.$$

Since V is to remain constant,

$$\frac{dV}{dr} = 0$$

and so

$$\pi r^2 \frac{dh}{dr} + 2\pi rh = 0.$$

Dividing through by πr, we get

$$r \frac{dh}{dr} + 2h = 0$$

and thus

$$\frac{dh}{dr} = -\frac{2h}{r}.$$

For the volume to remain constant, h must decrease $2h/r$ times as fast as r increases. $\square$

REMARK. Rates of change per unit time are discussed in Section 3.17.

Exercises

*1. Find the rate of change of the area of a square with respect to the length s of one of its sides. Evaluate this rate of change when $s = 4$.

2. Find the rate of change of the volume of a cube with respect to the length s of one of its sides. Evaluate this rate of change when $s = 4$.

*3. Find the rate of change of the area of a square with respect to the length z of one of its diagonals. Evaluate this rate of change when $z = 4$.

4. Find the rate of change of

$$y = \frac{1}{x}$$

with respect to x when $x = -1$.

*5. Find the rate of change of

$$y = \frac{1}{x(x + 1)}$$

with respect to x when $x = 2$.

6. Find the values of x for which the rate of change of

$$y = x^3 - 12x^2 + 45x - 13$$

is zero.

7. Find the rate of change of the volume of a ball with respect to its radius, given that $V = \frac{4}{3}\pi r^3$.

*8. Find the rate of change of the surface area of a ball with respect to its radius given that $A = 4\pi r^2$. What is this rate of change when $r = r_0$? How must r_0 be chosen so that the rate of change is 1?

9. Find x_0 given that the rate of change of

$$y = 2x^2 + x - 1$$

with respect to x is 4 when $x = x_0$.

*10. The dimensions of a rectangle are changing in such a way that the area of the rectangle remains constant. Find the rate of change of the height h with respect to the base b.

11. For what value of x is the rate of change of

$$y = ax^2 + bx + c \quad \text{with respect to } x$$

the same as the rate of change of

$$z = bx^2 + ax + c \quad \text{with respect to } x?$$

Assume $a \neq b$.

*12. Find the rate of change of the product $f(x)g(x)h(x)$ with respect to x when $x = 1$ given that

$$f(1) = 0, \qquad g(1) = 2, \qquad h(1) = -2,$$

$$f'(1) = 1, \qquad g'(1) = -1, \qquad h'(1) = 0.$$

3.5   The Chain Rule

Here we take up the differentiation of composite functions. Until we get to Theorem 3.5.5 our approach is completely intuitive—no real definitions, no proofs, just informal discussion ("arm waving," if you like). Our purpose is to give you some experience with the standard computational procedures and to give you some insight into why these procedures work. Theorem 3.5.5 puts it all on a sound footing.

Assume for the moment that the composition of differentiable functions is differentiable. Knowing the derivative of f and the derivative of g, how can we find the derivative of $f \circ g$? That is, how can we find

$$\frac{d}{dx}[f(g(x))]?$$

The key to answering this question lies in the following two ideas:

I. For any differentiable function h, $h(t)$ changes $h'(t)$ times as fast as t.
   (This comes from $h'(t)$ being the rate of change of $h(t)$ with respect to t.)

II. If A changes m times as fast as B and B changes n times as fast as C, then A changes mn times as fast as C.

Once you accept I and II, the computation of $\dfrac{d}{dx}[f(g(x))]$ is easy. By I,

$$f(g(x)) \text{ changes } f'(g(x)) \text{ times as fast as } g(x)$$

and

$$g(x) \text{ changes } g'(x) \text{ times as fast as } x.$$

It follows by II that

$$f(g(x)) \text{ changes } f'(g(x))g'(x) \text{ times as fast as } x.$$

This makes

$$f'(g(x))g'(x) \text{ the rate of change of } f(g(x)) \text{ with respect to } x,$$

which means that

(3.5.1)
$$\frac{d}{dx}[f(g(x))] = f'(g(x))g'(x).†$$

Formula (3.5.1) is called the *chain rule*. The name is particularly appropriate when three or more functions are involved. For the composition of three differentiable functions we have

$$\frac{d}{dx}[f(g(h(x)))] = f'(g(h(x)))g'(h(x))h'(x.)$$

If four functions are involved then a new link is added to the chain. The pattern must be obvious.

As an application of the chain rule we show how to differentiate integral powers of a differentiable function. The formula is this:

(3.5.2)
$$\frac{d}{dx}[g(x)]^n = n[g(x)]^{n-1}g'(x).$$

To derive this formula we set

$$f(x) = x^n$$

so that

$$\frac{d}{dx}[g(x)]^n \quad \text{becomes} \quad \frac{d}{dx}[f(g(x))].$$

The chain rule now gives

(1)
$$\frac{d}{dx}[g(x)]^n = f'(g(x))g'(x).$$

Since

$$f(x) = x^n \quad \text{has derivative} \quad f'(x) = nx^{n-1}$$

we have

$$f'(g(x)) = n[g(x)]^{(n-1)}.$$

Substituting this into (1), we get the desired formula. □

Formula (3.5.2) can also be written

(3.5.3)
$$\frac{d}{dx}[g(x)]^n = n[g(x)]^{(n-1)}\frac{d}{dx}[g(x)].$$

---

† If you wish to use the prime notation throughout, then the left-hand side becomes $(f \circ g)'(x)$ and the formula reads

$$(f \circ g)'(x) = f'(g(x))g'(x).$$

Example

$$\frac{d}{dx}(x^2 + 1)^3 = 3(x^2 + 1)^2 \frac{d}{dx}(x^2 + 1)$$

$$= 3(x^2 + 1)^2 2x$$

$$= 6x(x^2 + 1)^2. \quad \square$$

Example

$$\frac{d}{dx}\left(x + \frac{1}{x}\right)^{-3} = -3\left(x + \frac{1}{x}\right)^{-4} \frac{d}{dx}\left(x + \frac{1}{x}\right)$$

$$= -3\left(x + \frac{1}{x}\right)^{-4}\left(1 - \frac{1}{x^2}\right). \quad \square$$

Example

$$\frac{d}{dx}[1 - (2 + 3x)^2]^3 = 3[1 - (2 + 3x)^2]^2 \frac{d}{dx}[1 - (2 + 3x)^2].$$

Since

$$\frac{d}{dx}[1 - (2 + 3x)^2] = -2(2 + 3x)\frac{d}{dx}(3x) = -6(2 + 3x),$$

we have

$$\frac{d}{dx}[1 - (2 + 3x)^2]^3 = -18(2 + 3x)[1 - (2 + 3x)^2]^2. \quad \square$$

Another Formulation of the Chain Rule

If y is expressed in terms of u:

$$y = f(u),$$

and u is expressed in terms of x:

$$u = g(x),$$

then y can be expressed in terms of x:

$$y = f(u) = f(g(x)).$$

In this situation we have

$$\frac{d}{dx}[f(g(x))] = \frac{dy}{dx}$$

$$f'(g(x)) = f'(u) = \frac{dy}{du}$$

$$g'(x) = \frac{du}{dx}$$

so that Formula 3.5.1 becomes

(3.5.4)
$$\boxed{\frac{dy}{dx} = \frac{dy}{du}\frac{du}{dx}.}$$

This formulation of the chain rule is particularly easy to understand. It simply points out that

with y changing $\dfrac{dy}{du}$ times as fast as u,

and u changing $\dfrac{du}{dx}$ times as fast as x,

y must change $\dfrac{dy}{du}\dfrac{du}{dx}$ times as fast as x.

Problem. Find dy/dx given that

$$y = 3u^2 + 2 \quad \text{and} \quad u = \frac{1}{x-1}.$$

SOLUTION

$$\frac{dy}{du} = 6u = \frac{6}{x-1}, \qquad \frac{du}{dx} = -\frac{1}{(x-1)^2}$$

so that

$$\frac{dy}{dx} = \frac{dy}{du}\frac{du}{dx} = \frac{6}{x-1}\left[-\frac{1}{(x-1)^2}\right] = -\frac{6}{(x-1)^3}. \quad \square$$

ALTERNATIVE SOLUTION. We can achieve the same result by first expressing y in terms of x and then differentiating. With

$$y = 3u^2 + 2 \quad \text{and} \quad u = \frac{1}{x-1}$$

we have

$$y = \frac{3}{(x-1)^2} + 2$$

so that

$$\frac{dy}{dx} = -\frac{6}{(x-1)^3}. \quad \square$$

Problem. Find dy/dx for $x = 2$ given that

$$y = \frac{u-1}{u+1} \quad \text{and} \quad u = x^2.$$

SOLUTION

$$\frac{dy}{du} = \frac{(u+1)1 - (u-1)1}{(u+1)^2} = \frac{2}{(u+1)^2} = \frac{2}{(x^2+1)^2} \quad \text{and} \quad \frac{du}{dx} = 2x$$

so that

$$\frac{dy}{dx} = \frac{dy}{du}\frac{du}{dx} = \frac{2}{(x^2+1)^2}(2x) = \frac{4x}{(x^2+1)^2}.$$

At $x = 2$,

$$\frac{dy}{dx} = \frac{4(2)}{(2^2 + 1)^2} = \frac{8}{25} \, . \quad \square$$

Alternative Solution.   With

$$y = \frac{u - 1}{u + 1} \quad \text{and} \quad u = x^2$$

we have

$$y = \frac{x^2 - 1}{x^2 + 1}$$

so that

$$\frac{dy}{dx} = \frac{(x^2 + 1)2x - (x^2 - 1)2x}{(x^2 + 1)^2} = \frac{4x}{(x^2 + 1)^2} \, .$$

At $x = 2$,

$$\frac{dy}{dx} = \frac{4(2)}{(2^2 + 1)^2} = \frac{8}{25} \, . \quad \square$$

The formula

$$\frac{dy}{dx} = \frac{dy}{du}\frac{du}{dx}$$

can easily be extended to more variables. For example, if x itself depends on s, then we have

$$\boxed{\frac{dy}{ds} = \frac{dy}{du}\frac{du}{dx}\frac{dx}{ds}} \, .$$

If, in addition, s depends on t, then

$$\boxed{\frac{dy}{dt} = \frac{dy}{du}\frac{du}{dx}\frac{dx}{ds}\frac{ds}{dt}}$$

and so on. Each new dependence adds a new link to the chain.

A More Rigorous Approach

So far our approach to the differentiation of composite functions has been entirely intuitive. As we said at the beginning, our purpose was to give you some familiarity with the standard formulas and some insight into why these formulas "work."

It is time to be more careful. How do we know that the composition of differentiable functions is differentiable? What assumptions do we need? Precisely under what circumstances does

$$(f \circ g)'(x) = f'(g(x))g'(x)?$$

The following theorem takes care of these loose ends.

Theorem 3.5.5   Chain Rule Theorem

If g is differentiable at x and f is differentiable at $g(x)$, then the composition $f \circ g$ is differentiable at x and

$$(f \circ g)'(x) = f'(g(x))g'(x).$$

For a proof see the supplement to this section. We have made the proof optional because it is somewhat tricky. □

Exercises

Differentiate.

*1. $f(x) = (1 - 2x)^{-1}$.

2. $f(x) = (1 + 2x)^5$.

*3. $f(x) = (x^5 - x^{10})^{20}$.

4. $f(x) = \left(x^2 + \dfrac{1}{x^2}\right)^3$.

*5. $f(x) = \left(x - \dfrac{1}{x}\right)^4$.

6. $f(x) = \left(x + \dfrac{1}{x}\right)^3$.

*7. $f(x) = (x - x^3 - x^5)^4$.

8. $f(t) = \left(\dfrac{1}{1 + t}\right)^4$.

*9. $f(t) = (t^2 - 1)^{100}$.

10. $f(t) = (c^2 + t^2)^3$.

*11. $f(t) = (t^{-1} + t^{-2})^4$.

12. $f(x) = \left(\dfrac{ax + b}{cx + d}\right)^3$.

*13. $f(x) = \left(\dfrac{3x}{x^2 + 1}\right)^4$.

14. $f(x) = [(2x + 1)^2 + (x + 1)^2]^3$.

*15. $f(x) = (x^4 + x^2 + x)^2$.

16. $f(x) = (x^2 + 2x + 1)^3$.

*17. $f(x) = \left(\dfrac{x^3}{3} + \dfrac{x^2}{2} + \dfrac{x}{1}\right)^{-1}$.

18. $f(x) = \left(\dfrac{x^2 + 2}{x^2 + 1}\right)^5$.

*19. $f(x) = \left(\dfrac{1}{x + 2} - \dfrac{1}{x - 2}\right)^3$.

20. $f(x) = [(6x + x^5)^{-1} + x]^2$.

Find dy/dx.

*21. $y = \dfrac{1}{1 + u^2}$, $u = 2x + 1$.

22. $y = u + \dfrac{1}{u}$, $u = (3x + 1)^4$.

*23. $y = \dfrac{2u}{1 - 4u}$, $u = (5x^2 + 1)^4$.

24. $y = u^3 - u + 1$, $u = \dfrac{1 - x}{1 + x}$.

Find dy/dt.

25. $y = \dfrac{1 - 7u}{1 + u^2}$, $\quad u = 1 + x^2$, $\quad x = 2t - 5$.

26. $y = 1 + u^2$, $\quad u = \dfrac{1 - 7x}{1 + x^2}$, $\quad x = 5t + 2$.

Given that

$$f(0) = 1, \quad f'(0) = 2, \quad f(1) = 0, \quad f'(1) = 1, \quad f(2) = 1, \quad f'(2) = 1,$$
$$g(0) = 2, \quad g'(0) = 1, \quad g(1) = 1, \quad g'(1) = 0, \quad g(2) = 2, \quad g'(2) = 1,$$
$$h(0) = 1, \quad h'(0) = 2, \quad h(1) = 2, \quad h'(1) = 1, \quad h(2) = 0, \quad h'(2) = 2,$$

evaluate the following.

*27. $(f \circ g)'(0)$.	28. $(f \circ g)'(1)$.	*29. $(f \circ g)'(2)$.
30. $(g \circ f)'(0)$.	*31. $(g \circ f)'(1)$.	32. $(g \circ f)'(2)$.
*33. $(f \circ h)'(0)$.	34. $(h \circ f)'(1)$.	*35. $(h \circ f)'(0)$.
36. $(f \circ h \circ g)'(1)$.	*37. $(g \circ f \circ h)'(2)$.	38. $(g \circ h \circ f)'(0)$.

Find the following derivatives.

*39. $\dfrac{d}{dx} [f(x^2 + 1)]$.

40. $\dfrac{d}{dx} \left[f\left(\dfrac{x - 1}{x + 1}\right) \right]$.

*41. $\dfrac{d}{dx} \left[[f(x)]^2 + 1 \right]$.

42. $\dfrac{d}{dx} \left[\dfrac{f(x) - 1}{f(x) + 1} \right]$.

Supplement to Section 3.5

To prove Theorem 3.5.5 it is convenient to use a slightly different formulation of derivative.

Theorem 3.5.6

The function f is differentiable at x iff

$$\lim_{t \to x} \frac{f(t) - f(x)}{t - x} \quad \text{exists.}$$

If this limit exists, it is $f'(x)$.

PROOF. For each t in the domain of f, $t \neq x$, define

$$G(t) = \frac{f(t) - f(x)}{t - x}.$$

The theorem follows from noting that f is differentiable at x iff

$$\lim_{h\to 0} G(x + h) \quad \text{exists,}$$

and recalling that

$$\lim_{h\to 0} G(x + h) = l \quad \text{iff} \quad \lim_{t\to x} G(t) = l. \qquad (2.4.2) \quad \square$$

PROOF OF THEOREM 3.5.5.   By Theorem 3.5.6 it is enough to show that

$$\lim_{t\to x} \frac{f(g(t)) - f(g(x))}{t - x} = f'(g(x))g'(x).$$

We begin by defining an auxiliary function F on the domain of f by setting

$$F(y) = \left\{ \begin{array}{ll} \dfrac{f(y) - f(g(x))}{y - g(x)}, & y \neq g(x) \\ f'(g(x)), & y = g(x) \end{array} \right].$$

F is continuous at $g(x)$ since

$$\lim_{y\to g(x)} F(y) = \lim_{y\to g(x)} \frac{f(y) - f(g(x))}{y - g(x)},$$

and the right-hand side is, by Theorem 3.5.6, $f'(g(x))$, which is the value of F at $g(x)$. For $t \neq x$,

(1) $$\frac{f(g(t)) - f(g(x))}{t - x} = F(g(t)) \left[\frac{g(t) - g(x)}{t - x} \right].$$

To see this we note that if $g(t) = g(x)$, then both sides are 0. If $g(t) \neq g(x)$, then

$$F(g(t)) = \frac{f(g(t)) - f(g(x))}{g(t) - g(x)},$$

so that again we have equality.

Since g, being differentiable at x, is continuous at x and since F is continuous at $g(x)$, we know that the composition $F \circ g$ is continuous at x. Thus

$$\lim_{t\to x} F(g(t)) = F(g(x)) = f'(g(x)).$$

This, together with Equation (1), gives

$$\lim_{t\to x} \frac{f(g(t)) - f(g(x))}{t - x} = f'(g(x))g'(x). \quad \square$$

3.6 Derivatives of Higher Order

If a function f is differentiable, then we can form a new function f'. If f' is itself differentiable, then we can form its derivative, called the *second derivative of f*, and denoted by f''. So long as we have differentiability, we can continue in this manner, forming f''', etc. The prime notation is not used beyond the third order. For the fourth derivative we write

$$f^{(4)}$$

and, more generally, for the nth derivative

$$f^{(n)}.$$

As an example, we can take

$$f(x) = x^3 - 5x^2 + 3x - 7$$

for which

$$f'(x) = 3x^2 - 10x + 3 \quad \text{and} \quad f''(x) = 6x - 10.$$

There is a third derivative

$$f'''(x) = 6.$$

All higher derivatives are identically zero.

Since the derivative of a polynomial is again a polynomial and since the derivative of a rational function is again a rational function, you can see that polynomials and rational functions have derivatives of all orders. In the case of a polynomial, the derivatives of order greater than the degree of the polynomial are all identically zero. (Explain.)

In the Leibniz notation the derivatives of higher order are written

$$\frac{d^2y}{dx^2} = \frac{d}{dx}\left(\frac{dy}{dx}\right), \frac{d^3y}{dx^3} = \frac{d}{dx}\left(\frac{d^2y}{dx^2}\right), \ldots, \frac{d^ny}{dx^n} = \frac{d}{dx}\left(\frac{d^{n-1}y}{dx^{n-1}}\right), \ldots$$

or

$$\frac{d^2}{dx^2}[f(x)] = \frac{d}{dx}\left[\frac{d}{dx}[f(x)]\right], \frac{d^3}{dx^3}[f(x)] = \frac{d}{dx}\left[\frac{d^2}{dx^2}[f(x)]\right], \ldots,$$

$$\frac{d^n}{dx^n}[f(x)] = \frac{d}{dx}\left[\frac{d^{n-1}}{dx^{n-1}}[f(x)]\right], \ldots$$

Below we work out some examples.

Example. If

$$y = x^{-1}$$

then

$$\frac{dy}{dx} = -x^{-2}, \quad \frac{d^2y}{dx^2} = 2x^{-3}, \quad \frac{d^3y}{dx^3} = -6x^{-4}, \quad \frac{d^4y}{dx^4} = 24x^{-5}. \quad \square$$

Example

$$\frac{d}{dx}\left(\frac{x}{1+x}\right) = \frac{(1+x)(1) - x(1)}{(1+x)^2} = \frac{1}{(1+x)^2}$$

so that

$$\frac{d^2}{dx^2}\left(\frac{x}{1+x}\right) = \frac{d}{dx}\left[\frac{1}{(1+x)^2}\right] = -\frac{2}{(1+x)^3}$$

and

$$\frac{d^3}{dx^3}\left(\frac{x}{1+x}\right) = \frac{d}{dx}\left[-\frac{2}{(1+x)^3}\right] = \frac{6}{(1+x)^4}. \quad \square$$

Example. If

$$f(x) = \frac{3x^2}{1 + x^2}$$

then

$$f'(x) = \frac{(1 + x^2)6x - 3x^2(2x)}{(1 + x^2)^2} = \frac{6x}{(1 + x^2)^2}$$

and

$$f''(x) = \frac{(1 + x^2)^2 6 - 6x(2)(1 + x^2)2x}{(1 + x^2)^4} = \frac{(1 + x^2)(6 - 24x^2)}{(1 + x^2)^4} = \frac{6(1 - 4x^2)}{(1 + x^2)^3}. \quad \square$$

Exercises

Find the following derivatives.

*1. $f''(x)$ for $f(x) = 7x^3 - 6x^5$.

2. $f''(x)$ for $f(x) = \dfrac{x}{1 + x}$.

*3. $\dfrac{d^2y}{dx^2}$ for $y = x^2 - \dfrac{1}{x^2}$.

4. $\dfrac{d^4y}{dx^4}$ for $y = ax^4$.

*5. $\dfrac{d^3y}{dx^3}$ for $y = (1 + 2x)^3$.

6. $\dfrac{d^3y}{dx^3}$ for $y = (1 + 5x)^2$.

*7. $\dfrac{d^2}{dx^2}\left(\dfrac{1 - x}{1 + x}\right)$.

8. $\dfrac{d^2}{dx^2}\left(x^3 + \dfrac{1}{x^3}\right)$.

*9. $\dfrac{d^2}{dx^2}\left(\dfrac{ax + b}{cx + d}\right)$.

10. $\dfrac{d}{dx}\left[x\,\dfrac{d}{dx}(1 + x^2)\right]$.

*11. $\dfrac{d}{dx}\left[x\,\dfrac{d^2}{dx^2}\left(\dfrac{1}{1 + x}\right)\right]$.

12. $\dfrac{d^2}{dx^2}\left(\dfrac{ax^2 - b}{cx^2 - d}\right)$.

*13. $\dfrac{d^{100}}{dx^{100}}(x^9 - 20x^7 + x^5 + 1)$.

14. $\dfrac{d^5}{dx^5}(x^5 + c^5)$.

*15. $f'''(x)$ for $f(x) = \dfrac{x^3}{3} + \dfrac{x^2}{2} + \dfrac{x}{1} + 1$.

16. $f'''(x)$ for $f(x) = (7x^2 - x)^{-1}$.

*17. $\dfrac{d^2}{dx^2}\left[\left(\dfrac{6 - x}{x^2 - 1}\right)^2\right]$.

18. $\dfrac{d^4}{dx^4}[(1 - x)^4]$.

*19. $\dfrac{d^ny}{dx^n}$ for $y = (1 + x)^n$.

20. $\dfrac{d^ny}{dx^n}$ for $y = \dfrac{1}{x + 1}$.

3.7 Differentiating Inverses; Fractional Exponents

Differentiating Inverses

If f is a one-to-one function, then f has an inverse f^{-1}. This you already know. Suppose now that f is differentiable. Is f^{-1} necessarily differentiable? Yes, if f' does not take on the value 0.

You'll find a proof of this in the Appendix at the end of the book. Right now we assume the result and go on to describe how to calculate the derivative of f^{-1}.

We begin with the equation

$$f(f^{-1}(x)) = x$$

which, as you know, holds for all x in the domain of f^{-1}. Differentiating both sides of the equation, we get

$$\frac{d}{dx}[f(f^{-1}(x))] = 1.$$

With f and f^{-1} both differentiable, we can apply the chain rule to the left side and thereby get

$$f'(f^{-1}(x))(f^{-1})'(x) = 1.$$

Since we are assuming that $f'(f^{-1}(x))$ is not 0, we can divide by it and get

(3.7.1)
$$\boxed{(f^{-1})'(x) = \frac{1}{f'(f^{-1}(x))}.}$$

This is the standard formula for the derivative of an inverse.

nth Roots

For $x > 0$, x^n is one-to-one and its derivative, nx^{n-1}, does not take on the value 0. The inverse $x^{1/n}$ is therefore differentiable and we can use the method just described to find its derivative.

$$(x^{1/n})^n = x,$$

$$\frac{d}{dx}[(x^{1/n})]^n = 1,$$

$$n(x^{1/n})^{n-1}\frac{d}{dx}[x^{1/n}] = 1,$$

$$\frac{d}{dx}[x^{1/n}] = \frac{1}{n}(x^{1/n})^{1-n}.$$

Since

$$(x^{1/n})^{1-n} = x^{(1-n)/n} = x^{(1/n)-1},$$

we can write the differentiation formula as

(3.7.2)
$$\boxed{\frac{d}{dx}[x^{1/n}] = \frac{1}{n}x^{(1/n)-1}.}$$

We derived (3.7.2) for $x > 0$. If n is odd, the formula also holds for $x < 0$. Same argument.　□

Here are some simple examples:

$$\frac{d}{dx}(x^{1/2}) = \frac{1}{2}x^{-1/2}, \qquad \frac{d}{dx}(x^{1/3}) = \frac{1}{3}x^{-2/3}, \qquad \frac{d}{dx}(x^{1/4}) = \frac{1}{4}x^{-3/4}.$$

Rational Powers

We come now to rational exponents p/q. No matter what sign p/q has, we can take q as positive. (Explain.) By definition

$$x^{p/q} = (x^{1/q})^p,$$

so that

$$\frac{d}{dx}[x^{p/q}] = \frac{d}{dx}[(x^{1/q})^p].$$

From the power formula

$$\frac{d}{dx}[(g(x))^p] = p[g(x)]^{p-1}\frac{d}{dx}[g(x)],$$

we see that

$$\frac{d}{dx}[(x^{1/q})^p] = p(x^{1/q})^{p-1}\frac{d}{dx}[x^{1/q}].$$

Since

$$\frac{d}{dx}[x^{1/q}] = \frac{1}{q}x^{(1/q)-1}, \qquad\qquad \text{(Formula 3.7.2)}$$

we have

$$\frac{d}{dx}[x^{p/q}] = \frac{d}{dx}[(x^{1/q})^p] = p(x^{1/q})^{p-1}\frac{1}{q}x^{(1/q)-1} = \frac{p}{q}x^{(p/q)-1}.$$

In short

(3.7.3)
$$\boxed{\frac{d}{dx}[x^{p/q}] = \frac{p}{q}x^{(p/q)-1}.}$$

Here are some simple examples:

$$\frac{d}{dx}(x^{2/3}) = \frac{2}{3}x^{-1/3}, \qquad \frac{d}{dx}(x^{5/2}) = \frac{5}{2}x^{3/2}, \qquad \frac{d}{dx}(x^{-7/9}) = -\frac{7}{9}x^{-16/9}. \quad \square$$

It is easy to go from (3.7.3) to

(3.7.4)
$$\boxed{\frac{d}{dx}[g(x)]^{p/q} = \frac{p}{q}[g(x)]^{(p/q)-1}\frac{d}{dx}[g(x)].}$$

To do this, set

$$f(x) = x^{p/q}$$

and apply the chain rule to

$$f(g(x)) = [g(x)]^{p/q}.$$

The details are left as an exercise. $\quad \square$

The question of the applicability of (3.7.4) remains. If q is odd, then (3.7.4) is valid for all $g(x) \neq 0$. If q is even, then (3.7.4) holds only for $g(x) > 0$. In both instances, of course, g must be differentiable. It is time for some examples.

Example

$$\frac{d}{dx}\left[(x^2 + 1)^{1/5}\right] = \frac{1}{5}(x^2 + 1)^{-4/5} \cdot 2x = \frac{2}{5}x(x^2 + 1)^{-4/5}. \quad \square$$

Example

$$\frac{d}{dx}\left[\sqrt[3]{3x^2 + 7}\right] = \frac{d}{dx}\left[(3x^2 + 7)^{1/3}\right]$$

$$= \frac{1}{3}(3x^2 + 7)^{-2/3} \cdot 6x$$

$$= \frac{2x}{\sqrt[3]{(3x^2 + 7)^2}} \, . \quad \square$$

Example

$$\frac{d}{dx}\left[\sqrt{\frac{x^2}{1 + x^2}}\right] = \frac{d}{dx}\left[\left(\frac{x^2}{1 + x^2}\right)^{1/2}\right]$$

$$= \frac{1}{2}\left(\frac{x^2}{1 + x^2}\right)^{-1/2}\frac{d}{dx}\left(\frac{x^2}{1 + x^2}\right)$$

$$= \frac{1}{2}\left(\frac{x^2}{1 + x^2}\right)^{-1/2}\frac{(1 + x^2)2x - x^2(2x)}{(1 + x^2)^2}$$

$$= \frac{1}{2}\left(\frac{1 + x^2}{x^2}\right)^{1/2}\frac{2x}{(1 + x^2)^2}$$

$$= \frac{x}{(1 + x^2)^2}\sqrt{\frac{1 + x^2}{x^2}} \, . \quad \square$$

Some Further Comments

Let's take a look at formula 3.7.1 from a geometric point of view. The graphs of f and f^{-1} are reflections of one another in the line $y = x$. The tangent lines l_1 and l_2 are also reflections of one another. Figure 3.7.1 illustrates that

$$(f^{-1})'(x) = \text{slope of } l_1 = \frac{f^{-1}(x) - b}{x - b}$$

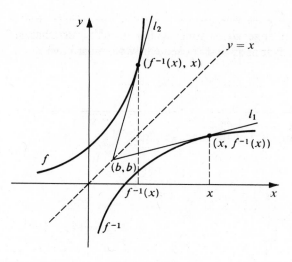

FIGURE 3.7.1

and

$$f'(f^{-1}(x)) = \text{slope of } l_2 = \frac{x - b}{f^{-1}(x) - b}$$

are reciprocals of one another.

In Leibniz's notation formula (3.7.1) reads simply

(3.7.5)
$$\boxed{\frac{dx}{dy} = \frac{1}{dy/dx}.}$$

You can see how this formula evolves by beginning with

$$y = f(x), \qquad x = f^{-1}(y).$$

The derivatives are then

$$\frac{dy}{dx} = f'(x), \qquad \frac{dx}{dy} = (f^{-1})'(y).$$

Differentiation of the identity

$$y = f(f^{-1}(y))$$

yields

$$1 = f'(f^{-1}(y))(f^{-1})'(y) = f'(x)(f^{-1})'(y) = \frac{dy}{dx}\frac{dx}{dy}$$

and finally

$$\frac{dx}{dy} = \frac{1}{dy/dx}. \quad \square$$

One last remark. You can get an intuitive grasp of (3.7.5) by looking at the matter this way:

Since y changes dy/dx times as fast as x, x must change $1/(dy/dx)$ times as fast as y, and consequently we must have

$$\frac{dx}{dy} = \frac{1}{dy/dx}.$$

This is hardly a proof, but the idea is right.

Exercises

Find dy/dx.

*1. $y = \sqrt{x^3 + 1}$.

2. $y = (x + 1)^{1/3}$.

*3. $y = x\sqrt{x^2 + 1}$.

4. $y = x^2\sqrt{x^2 + 1}$.

*5. $y = \sqrt[4]{2x^2 + 1}$.

6. $y = (x + 1)^{1/3}(x + 2)^{2/3}$.

*7. $y = \sqrt{2 - x^2}\sqrt{3 - x^2}$.

8. $y = \sqrt{(x^4 - x + 1)^3}$.

Compute.

*9. $\dfrac{d}{dx}\left(\sqrt{x} + \dfrac{1}{\sqrt{x}}\right)$.

10. $\dfrac{d}{dx}\left(\sqrt{\dfrac{3x + 1}{2x + 5}}\right)$.

*11. $\dfrac{d}{dx}\left(\dfrac{x}{\sqrt{x^2 + 1}}\right)$.

12. $\dfrac{d}{dx}\left(\dfrac{\sqrt{x^2 + 1}}{x}\right)$.

*13. $\dfrac{d}{dx}\left(\sqrt[3]{x} + \dfrac{1}{\sqrt[3]{x}}\right)$.

14. $\dfrac{d}{dx}\left(\sqrt{\dfrac{ax + b}{cx + d}}\right)$.

*15. $\dfrac{d}{dx}\left(\sqrt{\dfrac{ax^2 + b}{cx^2 + d}}\right)$.

16. $\dfrac{d}{dx}\left(\sqrt{\dfrac{1}{x} + \sqrt{x^4 + 1}}\right)$.

Compute.

*17. $\dfrac{d}{dx}[f(\sqrt{x} + 1)]$.

18. $\dfrac{d}{dx}[\sqrt{[f(x)]^2 + 1}]$.

*19. $\dfrac{d}{dx}[\sqrt{f(x^2 + 1)}]$.

20. $\dfrac{d}{dx}[f(\sqrt{x^2 + 1})]$.

Find a formula for $(f^{-1})'(x)$ given that f is one-to-one and its derivative satisfies the indicated equation.

*21. $f'(x) = f(x)$.

22. $f'(x) = 1 + [f(x)]^2$.

*23. $f'(x) = \sqrt{1 - [f(x)]^2}$.

24. In economics the *elasticity* of demand is given by the formula

$$\epsilon = \frac{P}{Q}\left|\frac{dQ}{dP}\right|,$$

where P is price and Q quantity. The demand is said to be

$$\begin{cases} \text{inelastic} & \text{where } \epsilon < 1 \\ \text{unitary} & \text{where } \epsilon = 1 \\ \text{elastic} & \text{where } \epsilon > 1 \end{cases}.$$

Describe the elasticity of each of the following demand curves. Keep in mind that Q and P must remain positive.

*(a) $Q = \dfrac{1}{P^2}$.   (b) $Q = \dfrac{1}{\sqrt{P}}$.   *(c) $Q = \dfrac{1}{P}$.

(d) $Q = 200 - \frac{1}{5}P$.   *(e) $Q = 200 - 5P$.   (f) $Q = (300 - P)^2$.

3.8   Some Tangent-Line Problems

In the case of a differentiable function f, the tangent to the graph of f at the point $(x_0, f(x_0))$ has slope

$$m = f'(x_0).$$

To get an equation for this tangent line we use the point slope formula

$$y - y_0 = m(x - x_0).$$

Using

$$(x_0, f(x_0)) \quad \text{for} \quad (x_0, y_0)$$

and

$$f'(x_0) \quad \text{for} \quad m$$

we get the equation

$$\boxed{y - f(x_0) = f'(x_0)(x - x_0).}$$

Problem.   Find an equation for the line tangent to the graph of

$$f(x) = \frac{2x + 1}{3 - x}$$

at the point $(2, 5)$.

SOLUTION

$$f'(x) = \frac{(3 - x)2 - (2x + 1)(-1)}{(3 - x)^2} = \frac{7}{(3 - x)^2}.$$

At $(2, 5)$ the slope is $f'(2) = 7$. As an equation for tangent we have

$$y - 5 = 7(x - 2).$$

This simplifies to

$$y = 7x - 9. \quad \square$$

Problem. Find an equation for the line tangent to the circle

$$x^2 + y^2 = 25$$

at the point (3, 4).

SOLUTION. The point (3, 4) lies on the upper semicircle

$$y = \sqrt{25 - x^2}.$$

In general

$$\frac{dy}{dx} = \frac{d}{dx}[\sqrt{25 - x^2}]$$

$$= \frac{d}{dx}[(25 - x^2)^{1/2}]$$

$$= \frac{1}{2}(25 - x^2)^{-1/2}(-2x)$$

$$= -\frac{x}{\sqrt{25 - x^2}}.$$

The slope m at (3, 4) is dy/dx evaluated at $x = 3$. With

$$m = -\frac{3}{\sqrt{25 - 9}} = -\frac{3}{4},$$

the tangent line has equation

$$y - 4 = -\frac{3}{4}(x - 3).$$

Multiplying both sides by 4 and collecting terms, we get

$$3x + 4y - 25 = 0. \quad \square$$

If $f'(x_0) = 0$, then the tangent at $(x_0, f(x_0))$ has slope 0. This means that it is horizontal. In Figure 3.8.1 we present several instances of a horizontal tangent. In each case the tangent line is the x-axis, and the point of tangency is the origin.

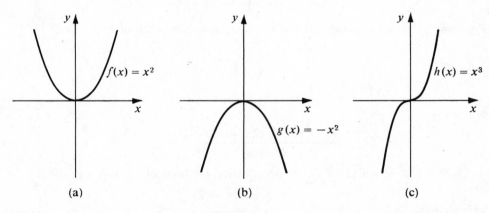

(a)   (b)   (c)

FIGURE 3.8.1

Exercises

1. Find an equation for the tangent at the point $(x_0, f(x_0))$.

   *(a) $f(x) = x^3 - x$, $x_0 = -2$.   (b) $f(x) = x + \dfrac{1}{x}$, $x_0 = 5$.

   *(c) $f(x) = \dfrac{1}{x - 1}$, $x_0 = \tfrac{1}{2}$.   (d) $f(x) = \dfrac{x}{1 + x^2}$, $x_0 = 1$.

2. Find the points (if any) at which the tangent is parallel to the line $y = x$.
   *(a) $y = \sqrt{x}$.   (b) $y = (x - 1)^3$.   *(c) $y = x(5 - x)$.

3. Find the points (if any) at which the tangent to
   *(a) $y = -x^2 - 6$ is parallel to the line $y = 4x - 1$.
   (b) $xy = 1$ is perpendicular to the line $y = x$.
   *(c) $x + y^2 = 1$ is parallel to the line $x + 2y = 0$.
   (d) $x + y^2 = 1$ is perpendicular to the line $x + 2y = 0$.

4. Find the points (if any) on the curve

$$y = \tfrac{2}{3}x^{3/2}$$

   where the inclination of the tangent line is
   *(a) 45°.   (b) 60°.   *(c) 30°.

5. Find an equation for the line which is tangent to the curve $y = x^3$ and passes through the point $(0, 2)$.

6. Find an equation for the line tangent to the graph of f^{-1} at the point (b, a) given that the slope of the graph of f at (a, b) is m.

*7. Isocost lines and indifference curves are studied in economics. For what value of C is the isocost line

$$Ax + By = 1 \qquad (A > 0, B > 0)$$

   tangent to the indifference curve

$$y = \frac{1}{x} + C?$$

   Find the point of tangency (called the equilibrium point).

*8. Determine the coefficients A, B, C so that the curve $y = Ax^2 + Bx + C$ will pass through the point $(1, 3)$ and be tangent to the line $4x + y = 8$ at the point $(2, 0)$.

9. Determine the coefficients A, B, C, D so that the curve $y = Ax^3 + Bx^2 + Cx + D$ will be tangent to the line $y = 3x - 3$ at the point $(1, 0)$ and tangent to the line $y = 18x - 27$ at the point $(2, 9)$.

10. Let

$$P = f(Q) \qquad (P = \text{price}, \quad Q = \text{output})$$

be the supply function of Figure 3.8.2. Compare angles θ and ϕ at points where the elasticity

$$\epsilon = \frac{f(Q)}{Q|f'(Q)|}$$

(a) is 1.   (b) is less than 1.   (c) is greater than 1.

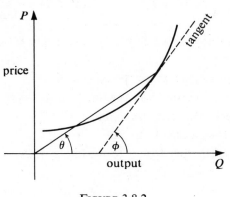

FIGURE 3.8.2

3.9 Additional Practice in Differentiation

Differentiate.

1. $y = x^{2/3} - a^{2/3}$.

* 2. $y = 2x^{3/4} + 4x^{-1/4}$.

3. $y = \dfrac{a + bx + cx^2}{x}$.

* 4. $y = \dfrac{\sqrt{x}}{2} - \dfrac{2}{\sqrt{x}}$.

5. $y = \sqrt{ax} + \dfrac{a}{\sqrt{ax}}$.

* 6. $s = \dfrac{a + bt + ct^2}{\sqrt{t}}$.

7. $r = \sqrt{1 - 2\theta}$.

* 8. $f(t) = (2 - 3t^2)^3$.

9. $f(x) = \dfrac{1}{\sqrt{a^2 - x^2}}$.

* 10. $y = \left(a - \dfrac{b}{x}\right)^2$.

11. $y = \left(a + \dfrac{b}{x^2}\right)^3$.

* 12. $y = x\sqrt{a + bx}$.

13. $s = t\sqrt{a^2 + t^2}$.

* 14. $y = \dfrac{a - x}{a + x}$.

15. $y = \dfrac{a^2 + x^2}{a^2 - x^2}$.

* 16. $y = \dfrac{\sqrt{a^2 + x^2}}{x}$.

17. $y = \dfrac{2 - x}{1 + 2x^2}$.

* 18. $y = \dfrac{x}{\sqrt{a^2 - x^2}}$.

19. $y = \dfrac{x}{\sqrt{a - bx}}$.

*20. $r = \theta^2\sqrt{3 - 4\theta}$.

21. $y = \sqrt{\dfrac{1 - cx}{1 + cx}}$.

*22. $f(x) = x\sqrt[3]{2 + 3x}$.

23. $y = \sqrt{\dfrac{a^2 + x^2}{a^2 - x^2}}$.

*24. $s = \sqrt[3]{\dfrac{2 + 3t}{2 - 3t}}$.

25. $r = \dfrac{\sqrt[3]{a + b\theta}}{\theta}$.

*26. $s = \sqrt{2t - \dfrac{1}{t^2}}$.

27. $y = \dfrac{b}{a}\sqrt{a^2 - x^2}$.

*28. $y = (x + 2)^2\sqrt{x^2 + 2}$.

29. $y = (a^{2/3} - x^{2/3})^{3/2}$.

*30. $y = (a^{3/5} - x^{3/5})^{5/3}$.

Evaluate dy/dx at the given x.

*31. $y = (x^2 - x)^3$, $x = 3$.

32. $y = (4 - x^2)^3$, $x = 3$.

*33. $y = \sqrt[3]{x} + \sqrt{x}$, $x = 64$.

34. $y = x\sqrt{3 + 2x}$, $x = 3$.

*35. $y = (2x)^{1/3} + (2x)^{2/3}$, $x = 4$.

36. $y = \sqrt{9 + 4x^2}$, $x = 2$.

*37. $y = \dfrac{1}{\sqrt{25 - x^2}}$, $x = 3$.

38. $y = \dfrac{x^2 + 2}{2 - x^2}$, $x = 2$.

*39. $y = \dfrac{\sqrt{16 + 3x}}{x}$, $x = 3$.

40. $y = \dfrac{\sqrt{5 - 2x}}{2x + 1}$, $x = \frac{1}{2}$.

*41. $y = x\sqrt{8 - x^2}$, $x = 2$.

42. $y = \sqrt{\dfrac{4x + 1}{5x - 1}}$, $x = 2$.

*43. $y = x^2\sqrt{1 + x^3}$, $x = 2$.

44. $y = \sqrt{\dfrac{x^2 - 5}{10 - x^2}}$, $x = 3$.

3.10   The Mean-Value Theorem

The main object of this section is to state and prove a result which is known as *the mean-value theorem*. At first glance the theorem may appear innocuous. It is certainly easy to understand. Its remarkable feature is its immense usefulness.

The Mean-Value Theorem

If f is differentiable on (a, b) and continuous on $[a, b]$, then there is at least one number c in (a, b) at which

$$f'(c) = \frac{f(b) - f(a)}{b - a}.$$

To explain what this means geometrically we refer to Figure 3.10.1. The number

$$\frac{f(b) - f(a)}{b - a}$$

is the slope of the line l which passes through the points $(a, f(a))$ and $(b, f(b))$. To say that there is at least one number c at which

$$f'(c) = \frac{f(b) - f(a)}{b - a}$$

is to say that the graph of f has at least one point $(c, f(c))$ at which the tangent is parallel to l.

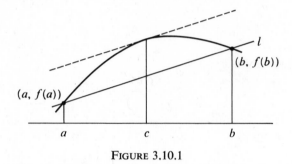

FIGURE 3.10.1

To prove the mean-value theorem we need first some other results.

Theorem 3.10.1

Let f be differentiable at x_0. If $f'(x_0) > 0$, then

$$f(x_0 - h) < f(x_0) < f(x_0 + h)$$

for all positive h sufficiently small. If $f'(x_0) < 0$, then

$$f(x_0 + h) < f(x_0) < f(x_0 - h)$$

for all positive h sufficiently small.

PROOF.   We take the case $f'(x_0) > 0$ and leave the other case to you. By definition of derivative,

$$\lim_{k \to 0} \frac{f(x_0 + k) - f(x_0)}{k} = f'(x_0).$$

With $f'(x_0) > 0$ we can use $f'(x_0)$ itself as ϵ and conclude that there exists $\delta > 0$ such that

$$\text{if}\quad 0 < |k| < \delta, \quad\text{then}\quad \left| \frac{f(x_0 + k) - f(x_0)}{k} - f'(x_0) \right| < f'(x_0).$$

For such k we have

$$\frac{f(x_0 + k) - f(x_0)}{k} > 0. \qquad \text{(why?)}$$

If now $0 < h < \delta$, then

$$\frac{f(x_0 + h) - f(x_0)}{h} > 0 \quad \text{and} \quad \frac{f(x_0 - h) - f(x_0)}{-h} > 0.$$

The first inequality gives us

$$f(x_0) < f(x_0 + h)$$

and the second gives us

$$f(x_0 - h) < f(x_0). \quad \Box$$

Next we prove a special case of the mean-value theorem, known as Rolle's theorem. In Rolle's theorem we make the additional assumption that $f(a)$ and $f(b)$ are both 0. (See Figure 3.10.2.) In this case the line joining $(a, f(a))$ to $(b, f(b))$ is horizontal. (It is the x-axis.) The conclusion is that there is a point $(c, f(c))$ at which the tangent is horizontal.

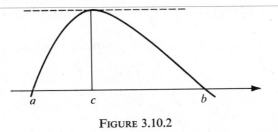

FIGURE 3.10.2

Rolle's Theorem

Let f be differentiable on (a, b) and continuous on $[a, b]$. If $f(a)$ and $f(b)$ are both 0, then there is at least one number c in (a, b) at which

$$f'(c) = 0.$$

PROOF.   If f is constantly 0 on $[a, b]$, the result is obvious. If f is not constantly 0 on $[a, b]$, then either f takes on some positive values or some negative ones. We assume the former and leave the other case to you.

Since f is continuous on $[a, b]$, at some point c of $[a, b]$, f must take on a maximum value. (Theorem 2.6.2.) This maximum value, $f(c)$, must be positive. Since $f(a)$ and $f(b)$ are both 0, c cannot be a and it cannot be b. This means that c must be in the open interval (a, b) and therefore $f'(c)$ exists. Now $f'(c)$ cannot be greater than 0 and it cannot be less than 0 because each of these conditions would imply that f takes on values greater than $f(c)$. (This follows from Theorem 3.10.1.) We conclude therefore that $f'(c) = 0$. $\Box$

We are now ready to give a proof of the mean-value theorem. The idea of the proof is to create a function g which satisfies the conditions of Rolle's theorem and is so related to f that the conclusion $g'(c) = 0$ leads to the conclusion

$$f'(c) = \frac{f(b) - f(a)}{b - a}.$$

It is not hard to see that

$$g(x) = f(x) - \left[\frac{f(b) - f(a)}{b - a}(x - a) + f(a)\right]$$

is exactly such a function. Geometrically $g(x)$ is represented in Figure 3.10.3. The line that passes through $(a, f(a))$ and $(b, f(b))$ has equation

$$y = \frac{f(b) - f(a)}{b - a}(x - a) + f(a).$$

(This is not hard to verify. The slope is right and when $x = a$, $y = f(a)$.) The difference

$$g(x) = f(x) - \left[\frac{f(b) - f(a)}{b - a}(x - a) + f(a)\right]$$

is simply the vertical separation between the graph of f and the line in question.

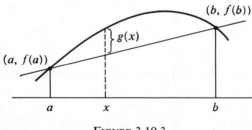

FIGURE 3.10.3

If f is differentiable on (a, b) and continuous on $[a, b]$, then so is g. As you can check, $g(a)$ and $g(b)$ are both 0. Therefore by Rolle's theorem there is at least one point c in (a, b) at which $g'(c) = 0$. Since in general

$$g'(x) = f'(x) - \frac{f(b) - f(a)}{b - a},$$

in particular

$$g'(c) = f'(c) - \frac{f(b) - f(a)}{b - a}.$$

With $g'(c) = 0$, we have

$$f'(c) = \frac{f(b) - f(a)}{b - a}. \quad \square$$

Exercises

With f, a, and b as below find those numbers c which satisfy the condition of the mean-value theorem.

*1. $f(x) = x^2$, $a = 1$, $b = 2$.
2. $f(x) = 3\sqrt{x} - 4x$, $a = 1$, $b = 4$.

*3. $f(x) = x^3$, $a = 1$, $b = 3$.
4. $f(x) = x^{2/3}$, $a = 1$, $b = 8$.

5. Given that

$$f(x) = \frac{1}{x}, \qquad a = -1, \quad b = 1,$$

verify that there is no number c such that

$$f'(c) = \frac{f(b) - f(a)}{b - a}.$$

Explain how this does not violate the mean-value theorem.

6. Graph the function

$$f(x) = |2x - 1| - 3.$$

Verify that $f(-1)$ and $f(2)$ are both 0 and yet f' does not take on the value 0. Explain how this does not contradict Rolle's theorem.

7. Given that $|f'(x)| \le 1$ for all numbers x, show that

$$|f(x_1) - f(x_2)| \le |x_1 - x_2| \quad \text{for all numbers } x_1, x_2.$$

8. Let P be a nonconstant polynomial

$$P(x) = a_n x^n + \cdots + a_1 x + a_0.$$

Show that between any two consecutive roots of the equation $P'(x) = 0$ there is at most one root of the equation $P(x) = 0$.

3.11   Increasing and Decreasing Functions

It is often useful to know on what intervals a given function is "increasing" and on what intervals it is "decreasing." If the function in question is differentiable, we can answer such questions by looking at the sign of the derivative. As we shall see (and as one would expect from the interpretation of the derivative as the slope of the graph), a function is

(a) "increasing" on any interval in which its derivative is positive,
(b) "decreasing" on any interval in which its derivative is negative, and
(c) constant on any interval in which its derivative is identically 0.

Before attempting to prove these results, we will explain exactly what we mean by a function "increasing" or "decreasing" on an interval.

Definition 3.11.1

A function f is said to be

(i) *increasing* on the interval I iff for every two numbers x_1, x_2 in I

$$x_1 < x_2 \quad \text{implies} \quad f(x_1) < f(x_2).$$

(ii) *decreasing* on the interval I iff for every two numbers x_1, x_2 in I

$$x_1 < x_2 \quad \text{implies} \quad f(x_1) > f(x_2).$$

Preliminary Examples

1. The squaring function

$$f(x) = x^2 \qquad\qquad \text{(Figure 3.11.1)}$$

   is decreasing on $(-\infty, 0]$ and increasing on $[0, \infty)$.
2. The function

$$f(x) = \begin{cases} 1, & x < 0 \\ x, & x \geq 0 \end{cases}, \qquad \text{(Figure 3.11.2)}$$

   is constant on $(-\infty, 0)$ and increasing on $[0, \infty)$.

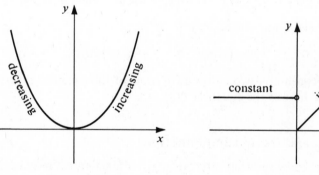

FIGURE 3.11.1   FIGURE 3.11.2

3. The cubing function

$$f(x) = x^3 \qquad\qquad \text{(Figure 3.11.3)}$$

   is everywhere increasing.
4. In the case of the Dirichlet function

$$g(x) = \begin{cases} 1, & x \text{ rational} \\ 0, & x \text{ irrational} \end{cases}, \qquad \text{(Figure 3.11.4)}$$

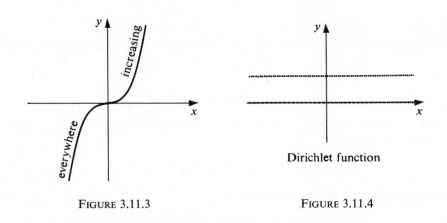

FIGURE 3.11.3   FIGURE 3.11.4

there is no interval on which the function increases and no interval on which the function decreases. On every interval it jumps back and forth between 0 and 1 an infinite number of times. □

Now we start proving some theorems. As you will see, all of them are consequences of the mean-value theorem.

Theorem 3.11.2

Let I be an open interval.

(i) If $f'(x) > 0$ for all x in I, then f is increasing on I.
(ii) If $f'(x) < 0$ for all x in I, then f is decreasing on I.
(iii) If $f'(x) = 0$ for all x in I, then f is constant on I.

PROOF.   We take x_1 and x_2 in I with $x_1 < x_2$. Since f is differentiable on (x_1, x_2) and continuous on $[x_1, x_2]$, we see from the mean-value theorem that there is a number c in (x_1, x_2) such that

$$f'(c) = \frac{f(x_2) - f(x_1)}{x_2 - x_1}.$$

In (i), we have

$$\frac{f(x_2) - f(x_1)}{x_2 - x_1} > 0 \quad \text{and thus} \quad f(x_1) < f(x_2).$$

In (ii), we have

$$\frac{f(x_2) - f(x_1)}{x_2 - x_1} < 0 \quad \text{and thus} \quad f(x_1) > f(x_2).$$

In (iii), we have

$$\frac{f(x_2) - f(x_1)}{x_2 - x_1} = 0 \quad \text{and thus} \quad f(x_1) = f(x_2). \quad \square$$

The theorem we just proved is useful but it has some deficiencies. For example, in the case of the squaring function

$$f(x) = x^2$$

the derivative

$$f'(x) = 2x$$

is negative for x in $(-\infty, 0)$, 0 for $x = 0$, and positive for x in $(0, \infty)$. Theorem 3.11.2 assures us that

> f decreases on $(-\infty, 0)$ and increases on $(0, \infty)$,

but actually stronger results are true:

> f decreases on $(-\infty, 0]$ and increases on $[0, \infty)$.

To get these stronger results we need a theorem that works for closed intervals.

To extend Theorem 3.11.2 so that it works for a closed interval I, the only thing we need is continuity at the endpoint(s).

Theorem 3.11.3

Let f be continuous on a closed interval I.

(i) If $f'(x) > 0$ for all x in the interior of I, then f is increasing on all of I.
(ii) If $f'(x) < 0$ for all x in the interior of I, then f is decreasing on all of I.
(iii) If $f'(x) = 0$ for all x in the interior of I, then f is constant on all of I.

A proof of this result is not hard to construct. As you can verify, a word-for-word copy of our proof of Theorem 3.11.2 also works here. □

It is time to give more examples.

Example. The function

$$f(x) = \sqrt{1 - x^2}$$

has as a graph the upper semicircle of Figure 3.11.5. It is therefore obvious that f increases on $[-1, 0]$ and decreases on $[0, 1]$. We can get this information without reference to Figure 3.11.5.

That f increases on $[-1, 0]$ can be seen from the fact that the derivative

$$f'(x) = -\frac{x}{\sqrt{1 - x^2}}$$

is positive for x in $(-1, 0)$ and f itself is continuous on $[-1, 0]$. That f decreases on $[0, 1]$ can be seen from the fact that $f'(x)$ is negative for x in $(0, 1)$ and f is continuous on $[0, 1]$. □

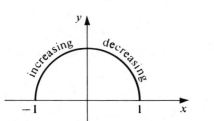

FIGURE 3.11.5

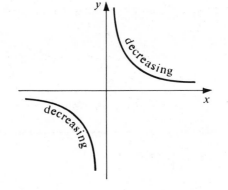

FIGURE 3.11.6

Example. The function

$$f(x) = \frac{1}{x}$$

has derivative

$$f'(x) = -\frac{1}{x^2}.$$

Since $f'(x) < 0$ for all $x \neq 0$ and f is not defined at 0, f decreases both on $(-\infty, 0)$ and on $(0, \infty)$. (Figure 3.11.6) As x approaches 0, the graph becomes steeper; it flattens out as x tends away from 0. ☐

Example. If

$$g(x) = 4x^5 - 15x^4 - 20x^3 + 110x^2 - 120x + 40,$$

then g is everywhere differentiable and

$$g'(x) = 20x^4 - 60x^3 - 60x^2 + 220x - 120$$
$$= 20(x^4 - 3x^3 - 3x^2 + 11x - 6)$$
$$= 20(x + 2)(x - 1)^2(x - 3).$$

The derivative g' takes on the value 0 at -2, at 1, and at 3. Moreover it is not hard to see that

$$g'(x) \text{ is } \begin{cases} \text{positive} & \text{for } x \in (-\infty, -2) \\ \text{negative} & \text{for } x \in (-2, 1) \\ \text{negative} & \text{for } x \in (1, 3) \\ \text{positive} & \text{for } x \in (3, \infty) \end{cases}.$$

Since g' is positive on $(-\infty, -2)$ we can conclude that g increases on $(-\infty, -2]$. Since g' is negative on $(-2, 1)$ and on $(1, 3)$, we can conclude that g decreases both on $[-2, 1]$ and on $[1, 3]$. This tells us that g actually decreases on $[-2, 3]$. (Explain.) Finally, since g' is positive on $(3, \infty)$, g must increase on $[3, \infty)$. ☐

If two differentiable functions differ by a constant,

$$f(x) = g(x) + C,$$

then their derivatives are obviously equal:

$$f'(x) = g'(x).$$

The converse is also true. In fact, we have the following theorem.

Theorem 3.11.4

I. If

$$f'(x) = g'(x) \quad \text{for all } x \text{ in an open interval } I,$$

then

$$f \text{ and } g \text{ differ by a constant on } I.$$

II. If

$$f'(x) = g'(x) \quad \text{for all } x \text{ in the interior of a closed interval } I$$

and

$$f \text{ and } g \text{ are continuous at the endpoints of } I,$$

then

$$f \text{ and } g \text{ differ by a constant on } I.$$

PROOF. Set $h = f - g$. For the first assertion apply (iii) of Theorem 3.11.2 to h. For the second assertion apply (iii) of Theorem 3.11.3 to h. We leave the details as an exercise. □

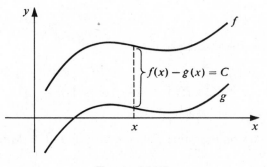

FIGURE 3.11.7

For an illustration of this theorem see Figure 3.11.7. At points with the same first coordinate the slopes are equal and therefore the curves have the same steepness. The separation between the curves remains constant.

Problem. Find a general expression for $f(x)$ given that

$$f'(x) = x^2 - x \quad \text{for all real } x \quad \text{and} \quad f(0) = 1.$$

SOLUTION.   Since

$$\frac{d}{dx}\left(\frac{x^3}{3} - \frac{x^2}{2}\right) = x^2 - x,$$

we know that $f(x)$ has the form

$$f(x) = \frac{x^3}{3} - \frac{x^2}{2} + C.$$

On the one hand

$$f(0) = C, \qquad\qquad\qquad \text{(set } x = 0\text{)}$$

and on the other

$$f(0) = 1. \qquad\qquad\qquad \text{(given)}$$

This forces $C = 1$ and

$$f(x) = \frac{x^3}{3} - \frac{x^2}{2} + 1. \quad \square$$

Problem.   Find a general expression for $g(x)$ given that

$$g'(x) = \frac{3}{(x + 1)^2} \qquad \text{for } x > -1 \quad \text{and} \quad g(1) = 2.$$

SOLUTION.   It is not hard to see that

$$\frac{d}{dx}\left(\frac{1}{x + 1}\right) = \frac{d}{dx}[(x + 1)^{-1}] = -\frac{1}{(x + 1)^2}$$

and therefore

$$\frac{d}{dx}\left(-\frac{3}{x + 1}\right) = \frac{3}{(x + 1)^2}.$$

This means that

$$g(x) = -\frac{3}{x + 1} + C.$$

To require $g(1) = 2$ is to require

$$-\tfrac{3}{2} + C = 2.$$

This leads to

$$C = \tfrac{7}{2}.$$

The solution is therefore

$$g(x) = -\frac{3}{x + 1} + \frac{7}{2}. \quad \square$$

Exercises

Find the intervals on which f is increasing and those on which it is decreasing, if $f(x)$ is as follows.

*1. $x^3 - 3x + 2$.   2. $x^3 - 3x^2 + 6$.   *3. $x + \dfrac{1}{x}$.

4. $x^3(1 + x)$. * 5. $x(x + 1)(x + 2)$. 6. $(x + 1)^4$.

*7. $\dfrac{1}{|x - 2|}$. 8. $\dfrac{x}{1 + x^2}$. *9. $\dfrac{x^2 + 1}{x^2 - 1}$.

10. $\dfrac{x^2}{x^2 + 1}$. *11. $|x^2 - 5|$. 12. $x^2(1 + x)^2$.

*13. $\dfrac{x - 1}{x + 1}$. 14. $x^2 + \dfrac{16}{x^2}$. *15. $\left(\dfrac{1 - \sqrt{x}}{1 + \sqrt{x}}\right)^7$.

16. $\sqrt{\dfrac{2 + x}{1 + x}}$. *17. $\sqrt{\dfrac{1 + x^2}{2 + x^2}}$. 18. $|x + 1||x - 2|$.

Find a general expression for $f(x)$ given the following.

*19. $f'(x) = x^2 - 1$ for all real x and $f(0) = 1$.
20. $f'(x) = x^2 - 1$ for all real x and $f(1) = 2$.
*21. $f'(x) = 2x - 5$ for all real x and $f(2) = 4$.
22. $f'(x) = 5x^4 + 4x^3 + 3x^2 + 2x + 1$ for all real x and $f(0) = 5$.
*23. $f'(x) = 4x^{-3}$ for $x \neq 0$ and $f(1) = 0$.
24. $f'(x) = x(x^2 + 1)^4$ for all real x and $f(0) = 1$.
*25. $f'(x) = 3x^2(x^3 - 1)^4$ for all real x and $f(1) = 0$.

26. Given that

$$f'(x) > g'(x) \quad \text{for all real } x \quad \text{and} \quad f(0) = g(0)$$

compare $f(x)$ with $g(x)$

(a) on $(-\infty, 0)$. (b) on $(0, \infty)$.

Justify your answers.
27. Prove Theorem 3.11.4.

3.12 Maxima and Minima

Closely related to the material just discussed is the problem of determining the maximum and minimum values of functions. There is a practical aspect to this matter. In many problems of economics, engineering, and physics it is important to find out how large or how small a certain quantity may become. If the problem admits a mathematical formulation, it is often reducible to the problem of finding the extreme values of some function.

In this section we consider functions which are defined on an open interval or on the union of open intervals. We begin with a definition.

A function f is said to have a *local* (or *relative*) *maximum* at c iff

$$f(c) \geq f(x) \quad \text{for all } x \text{ sufficiently close to } c.$$

It is said to have a *local* (or *relative*) *minimum* at c iff

$$f(c) \leq f(x) \quad \text{for all } x \text{ sufficiently close to } c.$$

We illustrate these notions in Figure 3.12.1 and Figure 3.12.2. A careful look at the figures suggests that local maxima and minima occur only at points where the tangent is horizontal $[f'(c) = 0]$ or where there is no tangent line $[f'(c)$ does not exist]. This is indeed the case.

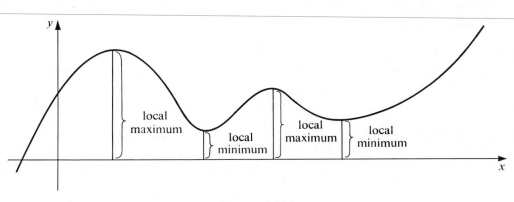

FIGURE 3.12.1

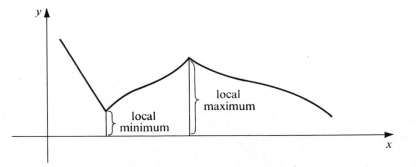

Theorem 3.12.1

If f has a local maximum or minimum at c then

$$\text{either } f'(c) = 0 \text{ or } f'(c) \text{ does not exist.}$$

PROOF. If $f'(c) > 0$ or $f'(c) < 0$, then, by Theorem 3.10.1, there must be numbers x_1 and x_2 which are arbitrarily close to c and yet satisfy

$$f(x_1) < f(c) < f(x_2).$$

This makes it impossible for a local maximum or minimum to occur at c. □

In view of this result, in searching for local maxima and local minima of a function f, the only points we need to test are those points c at which $f'(c) = 0$ or $f'(c)$ does not exist. Such points are called *critical points*.

We illustrate the technique for finding local maxima and local minima by some examples. In each example the first step will be to find the critical points. We begin with very simple cases.

Example. In the case of

$$f(x) = 3 - x^2 \qquad \text{(Figure 3.12.3)}$$

the derivative

$$f'(x) = -2x$$

exists everywhere. Since $f'(x) = 0$ only at $x = 0$, 0 is the only critical point. The number $f(0) = 3$ is obviously a local maximum. □

Example. In the case of

$$f(x) = |x + 1| + 2 \qquad \text{(Figure 3.12.4)}$$

the derivative is

$$f'(x) = \begin{cases} 1, & x > -1 \\ -1, & x < -1 \end{cases}.$$

This derivative is never 0. It fails to exist only at -1. The number -1 is the only critical point. The value $f(-1) = 2$ is a local minimum. □

Example. In the case of

$$f(x) = \frac{1}{x - 1} \qquad \text{(Figure 3.12.5)}$$

the derivative

$$f'(x) = -\frac{1}{(x - 1)^2}$$

exists throughout the domain of f and is never 0. There are therefore no critical points and no local extreme values. □

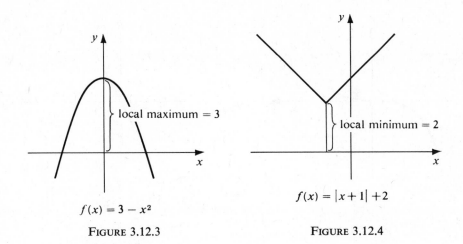

local maximum = 3

$f(x) = 3 - x^2$

FIGURE 3.12.3

local minimum = 2

$f(x) = |x + 1| + 2$

FIGURE 3.12.4

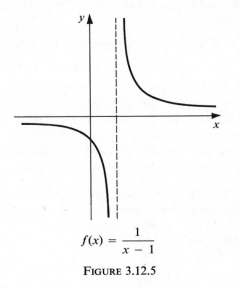

$$f(x) = \frac{1}{x - 1}$$

FIGURE 3.12.5

CAUTION.   The fact that c is a critical point of f does not guarantee that $f(c)$ is a local extreme value. This is illustrated in the next two examples.

Example.   In the case of the cubing function

$$f(x) = x^3 \qquad \text{(Figure 3.12.6)}$$

the derivative $f'(x) = 3x^2$ is 0 at 0 but $f(0) = 0$ is not a local extreme value. The cubing function is everywhere increasing.   □

Example.   The function $f(x) = \begin{cases} 2x, & x < 1 \\ \frac{1}{2}x + \frac{3}{2}, & x \geq 1 \end{cases}$,   (Figure 3.12.7)

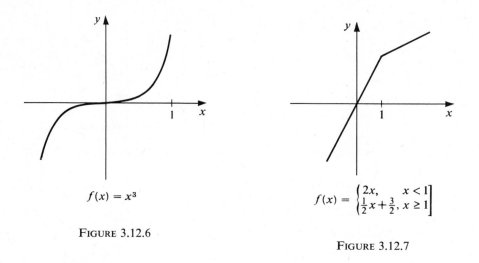

$$f(x) = x^3$$

FIGURE 3.12.6

$$f(x) = \begin{cases} 2x, & x < 1 \\ \frac{1}{2}x + \frac{3}{2}, & x \geq 1 \end{cases}$$

FIGURE 3.12.7

is everywhere increasing. Although 1 is a critical point $[f'(1)$ does not exist], $f(1) = 2$ is not a local extreme value. □

The following test is often useful in determining the behavior of a function at a critical point.

Test 1(*The first-derivative test*).   Suppose that c is a critical point for f and that f is continuous at c. If there exists an interval $(c - \delta, c + \delta)$ such that $f'(x) > 0$ for all x in $(c - \delta, c)$ and $f'(x) < 0$ for all x in $(c, c + \delta)$, then $f(c)$ is a local maximum. If both inequalities are reversed, then $f(c)$ is a local minimum.

PROOF.   The proof is easy. In the first case f increases on $(c - \delta, c]$ and decreases on $[c, c + \delta)$. This makes $f(c)$ a local maximum. (Figures 3.12.8 and 3.12.9) In the second case f decreases on $(c - \delta, c]$ and increases on $[c, c + \delta)$. This makes $f(c)$ a local minimum. (Figures 3.12.10 and 3.12.11) □

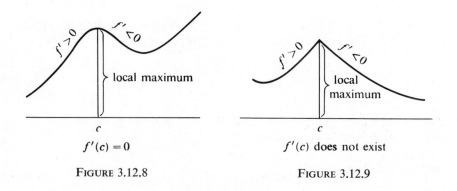

local maximum

$f'(c) = 0$

FIGURE 3.12.8

local maximum

$f'(c)$ does not exist

FIGURE 3.12.9

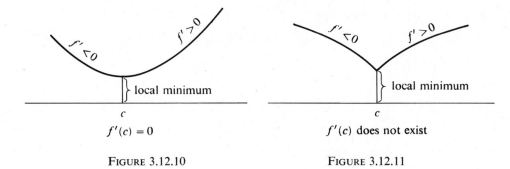

FIGURE 3.12.10   FIGURE 3.12.11

Example.   In the case of

$$f(x) = |x^2 - 1|$$ (Figure 3.12.12)

the derivative is given by

$$f'(x) = \begin{cases} 2x, & x < -1 \\ -2x, & -1 < x < 1 \\ 2x, & 1 < x \end{cases}.$$ (check this out)

The critical points are 0, -1, and 1. At 0, $f'(x) = 0$. At -1 and 1, $f'(x)$ does not exist. To analyze the behavior of f at the critical points we use Test 1. Since

$$f'(x) \text{ is } \begin{cases} \text{negative,} & \text{for} & x < -1 \\ \text{positive,} & \text{for} & -1 < x < 0 \\ \text{negative,} & \text{for} & 0 < x < 1 \\ \text{positive,} & \text{for} & 1 < x \end{cases},$$

we see that $f(-1) = 0$ is a local minimum, $f(0) = 1$ is a local maximum, and $f(1) = 0$ is a local minimum.   □

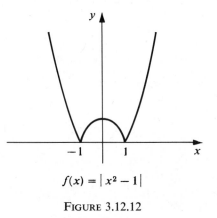

$$f(x) = |x^2 - 1|$$

FIGURE 3.12.12

Example.   Here we begin with

$$f(x) = (x - 2)(x - 1)^4.$$

Differentiating we obtain

$$\begin{aligned} f'(x) &= (x - 2)4(x - 1)^3 + (x - 1)^4 \\ &= (x - 1)^3(4x - 8 + x - 1) \\ &= (x - 1)^3(5x - 9). \end{aligned}$$

The critical points are 1 and $\frac{9}{5}$. Since

$$f'(x) \text{ is } \begin{cases} \text{positive,} & \text{for} \quad x < 1 \\ \text{negative,} & \text{for} \quad 1 < x < \frac{9}{5} \\ \text{positive,} & \text{for} \quad \frac{9}{5} < x \end{cases},$$

we see that $f(1) = 0$ is a local maximum and $f(\frac{9}{5}) = -4^4/5^5$ is a local minimum.   □

In certain cases it may be difficult to determine the sign of the first derivative both to the left of c and to the right of c. If f is twice differentiable at c, the following test may be easier to apply.

Test 2 (*The second-derivative test*).   Suppose that $f'(c) = 0$.
If $f''(c) > 0$,   then $f(c)$ is a local minimum value.
If $f''(c) < 0$,   then $f(c)$ is a local maximum value.

PROOF.   We handle the case $f''(c) > 0$. The other is left as an exercise. Since f'' is the derivative of f', we see from Theorem 3.10.1 that there exists $\delta > 0$ such that, if

$$c - \delta < x_1 < c < x_2 < c + \delta,$$

then

$$f'(x_1) < f'(c) < f'(x_2).$$

Since $f'(c) = 0$, we have

$$f'(x) < 0 \quad \text{for } x \text{ in } (c - \delta, c)$$

and

$$f'(x) > 0 \quad \text{for } x \text{ in } (c, c + \delta).$$

By Test 1 this shows that $f(c)$ is a local minimum.   □

Now we illustrate the second-derivative test:

Example.   For

$$f(x) = x^3 - x$$

we have

$$f'(x) = 3x^2 - 1.$$

The critical points are $-1/\sqrt{3}$ and $1/\sqrt{3}$. At both of these points the derivative is 0. As a second derivative we have

$$f''(x) = 6x.$$

Since $f''(-1/\sqrt{3}) < 0$ and $f''(1/\sqrt{3}) > 0$ we can conclude from the second-derivative test that $f(-1/\sqrt{3}) = 2/3\sqrt{3}$ is a local maximum and $f(1/\sqrt{3}) = -2/3\sqrt{3}$ is a local minimum. □

Exercises

Find the critical points and the local extreme values.

*1. $x^3 + 3x - 2$.

2. $2x^4 - 4x^2 + 6$.

*3. $x + \dfrac{1}{x}$.

4. $x^2(1 - x)$.

*5. $x(x + 1)(x + 2)$.

6. $(1 - x)^2(1 + x)$.

*7. $\dfrac{1}{|x - 2|}$.

8. $\dfrac{1 + x}{1 - x}$.

*9. $\dfrac{2 - 3x}{2 + x}$.

10. $\dfrac{2}{x(x + 1)}$.

*11. $|x^2 - 16|$.

12. $x^3(1 - x)^2$.

*13. $\left(\dfrac{x - 2}{x + 2}\right)^3$.

14. $(1 - 2x)(x - 1)^3$.

*15. $(1 - x)(1 + x)^3$.

16. $\dfrac{x^2}{1 + x}$.

*17. $\dfrac{|x|}{1 + |x|}$.

18. $(3x - 5)^3$.

*19. $|x - 1||x + 2|$.

20. $x\sqrt[3]{1 - x}$.

*21. $-\dfrac{x^3}{x + 1}$.

22. $\dfrac{1}{x + 1} - \dfrac{1}{x - 2}$.

*23. $\dfrac{1}{x + 1} - \dfrac{1}{x + 2}$.

24. $|x - 3| + |2x + 1|$.

25. Prove the validity of the second-derivative test in the case that $f''(c) < 0$.

3.13 More on Maxima and Minima

Endpoint Maxima and Minima

For functions defined on an open interval or on the union of open intervals the critical points are those at which the derivative is 0 or the derivative does not exist. For functions defined on a closed interval

$$[a, b], \quad [a, \infty), \quad \text{or} \quad (-\infty, b]$$

the *endpoints* of the domain (a and b in the case of $[a, b]$, a in the case of $[a, \infty)$, and b in the case of $(-\infty, b]$) are also called *critical points*.

Endpoints can give rise to what are called *endpoint maxima* and *endpoint minima*. See Figures 3.13.1, 3.13.2, and 3.13.3.

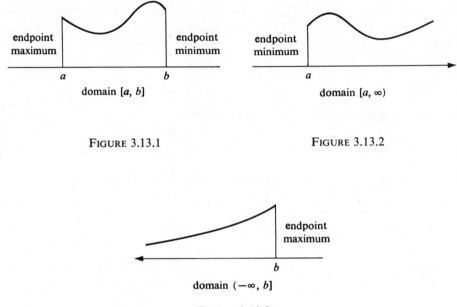

FIGURE 3.13.1   FIGURE 3.13.2

FIGURE 3.13.3

The figures render the general idea so obvious that formal definitions seem unnecessary. Nevertheless, here they are:

If c is an endpoint of the domain of f, then f is said to have an *endpoint maximum* at c iff

$\quad f(c) \geq f(x)$ for all x in the domain of f which are sufficiently close to c.

It is said to have an *endpoint minimum* at c iff

$\quad f(c) \leq f(x)$ for all x in the domain of f which are sufficiently close to c.

Absolute Maxima and Minima

Whether or not a function f has a local or endpoint extreme at some point depends entirely on the behavior of f for x close to that point. Absolute extreme values, which we define below, depend on the behavior of the function on its entire domain.

We begin with a number d in the domain of f. Here d can be an interior point or an endpoint.

$f(d)$ is called *the absolute maximum value* of f (or simply *the maximum value* of f) iff

$$f(d) \geq f(x) \quad \text{for all } x \text{ in the domain of } f.$$

$f(d)$ is called *the absolute minimum value* of f (or simply *the minimum value* of f) iff

$$f(d) \leq f(x) \quad \text{for all } x \text{ in the domain of } f.$$

Usually the most practical way of determining the absolute extremes of a function is to gather together the local extremes and the endpoint extremes. The largest of these numbers (if there is a largest) is obviously the absolute maximum, and the smallest (if there is a smallest) is the absolute minimum.

It is worth repeating here that in the case of a function continuous on a bounded closed interval $[a, b]$ there is both an absolute maximum and an absolute minimum. (Theorem 2.6.2)

Summary

We summarize the central ideas of this section and of the last one by outlining a step-by-step procedure for finding the extreme values (local, endpoint, and absolute) of a function f.

1. Find the critical points. These are the points at which $f'(x) = 0$, the points at which $f'(x)$ does not exist, and the endpoints of the domain.
2. Test each interior critical point for local extremes
   (a) by inspection (some graphing may help),
   (b) by applying Test 1 (the first-derivative test),
   (c) or, if $f'(c) = 0$, by applying Test 2 (the second-derivative test).
3. Test each endpoint for endpoint extremes.
4. Test for absolute extremes by examining the local extremes and the endpoint extremes.

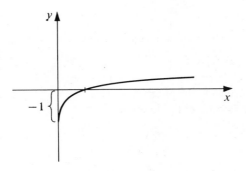

FIGURE 3.13.4

Problem. Find the critical points and classify the extreme values of

$$f(x) = \frac{\sqrt{x} - 1}{\sqrt{x} + 1}.$$

SOLUTION. The domain here is $[0, \infty)$. First we check the interior points $x \in (0, \infty)$. On $(0, \infty)$ f is differentiable and

$$f'(x) = \frac{(\sqrt{x} + 1)\dfrac{1}{2\sqrt{x}} - (\sqrt{x} - 1)\dfrac{1}{2\sqrt{x}}}{(\sqrt{x} + 1)^2}$$

$$= \frac{1}{\sqrt{x}(\sqrt{x} + 1)^2}.$$

Since the derivative $f'(x)$ is never 0, there are no interior critical points.

The number 0 is a critical point because it is an endpoint. Since $f'(x) > 0$ for all x in $(0, \infty)$ and f is continuous on $[0, \infty)$, f must be increasing on $[0, \infty)$. This makes $f(0) = -1$ an endpoint minimum. It is also the absolute minimum. For the graph see Figure 3.13.4. □

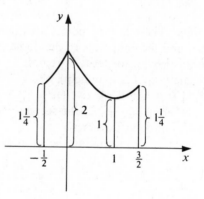

FIGURE 3.13.5

Problem.   Find the critical points and classify the extreme values of

$$f(x) = x^2 - 2|x| + 2, \qquad x \in [-\tfrac{1}{2}, \tfrac{3}{2}].$$

SOLUTION.   First we check the interior points $x \in (-\tfrac{1}{2}, \tfrac{3}{2})$. On the open interval $(-\tfrac{1}{2}, \tfrac{3}{2})$ the function is differentiable except at 0:

$$f'(x) = \begin{cases} 2x + 2, & -\tfrac{1}{2} < x < 0 \\ 2x - 2, & 0 < x < \tfrac{3}{2} \end{cases}. \qquad \text{(check this out)}$$

This makes 0 a critical point. Since $f'(x) = 0$ at $x = 1$, 1 is also a critical point. The endpoints $-\tfrac{1}{2}$ and $\tfrac{3}{2}$ are also critical points.

A little to the left of 0, $f' > 0$ and a little to the right of 0, $f' < 0$. By Test 1, this means that $f(0) = 2$ is a local maximum.

A little to the left of 1, $f' < 0$ and a little to the right of 1, $f' > 0$. This makes $f(1) = 1$ a local minimum. (Again by Test 1.)

Now we check the endpoints. Since $f' > 0$ a little to the right of $-\tfrac{1}{2}$,

$$f(-\tfrac{1}{2}) = (-\tfrac{1}{2})^2 - 2|-\tfrac{1}{2}| + 2 = \tfrac{1}{4} - 1 + 2 = 1\tfrac{1}{4}$$

is an endpoint minimum. (Explain.) Since $f' > 0$ a little to the left of $\tfrac{3}{2}$,

$$f(\tfrac{3}{2}) = (\tfrac{3}{2})^2 - 2|\tfrac{3}{2}| + 2 = \tfrac{9}{4} - 3 + 2 = 1\tfrac{1}{4}$$

is an endpoint maximum. (Explain.)

In short, the critical points are $-\frac{1}{2}$, 0, 1, and $\frac{3}{2}$.

$f(-\frac{1}{2}) = 1\frac{1}{4}$ is an endpoint minimum.

$f(0) = 2$ is a local maximum; it is also the absolute maximum.

$f(1) = 1$ is a local minimum; it is also the absolute minimum.

$f(\frac{3}{2}) = 1\frac{1}{4}$ is an endpoint maximum.

For the graph see Figure 3.13.5. □

Problem. Find the critical points and classify the extreme values of

$$f(x) = \sqrt{1 - x^2} + \frac{x}{2}.$$

SOLUTION. The domain is the closed interval $[-1, 1]$. The endpoints -1 and 1 are critical points. To check the interior points $x \in (-1, 1)$, we differentiate. This gives

$$f'(x) = -\frac{x}{\sqrt{1 - x^2}} + \frac{1}{2}.$$

Setting $f'(x) = 0$ gives

$$\frac{x}{\sqrt{1 - x^2}} = \frac{1}{2}.$$

Squaring both sides, we get

$$\frac{x^2}{1 - x^2} = \frac{1}{4},$$

$$4x^2 = 1 - x^2,$$

$$5x^2 = 1,$$

and thus

$$x = \pm \frac{1}{\sqrt{5}}.$$

Of these two numbers only $1/\sqrt{5}$ is a root of the equation $f'(x) = 0$. The other, $-1/\sqrt{5}$, is an extraneous root that we introduced when we did some squaring. In short there are three critical points to examine: -1, $1/\sqrt{5}$, and 1. Since $f' > 0$ a little to the right of -1, $f(-1) = -\frac{1}{2}$ is an endpoint minimum. To test $f(1/\sqrt{5})$ we apply the second-derivative test:

$$f''(x) = -\left[\frac{\sqrt{1 - x^2} \cdot 1 - x\left(-\dfrac{x}{\sqrt{1 - x^2}}\right)}{1 - x^2}\right]$$

$$= -(1 - x^2)^{-3/2}.$$

It is not hard to see that $f''(1/\sqrt{5})$ is negative and therefore

$$f\left(\frac{1}{\sqrt{5}}\right) = \sqrt{1 - \frac{1}{5}} + \frac{1}{2\sqrt{5}} = 2\frac{1}{\sqrt{5}} + \frac{1}{2\sqrt{5}} = \frac{5}{2\sqrt{5}} = \frac{\sqrt{5}}{2}$$

is a local maximum value. Since $f' < 0$ a little to the left of $1, f(1) = \frac{1}{2}$ is an endpoint minimum.

To summarize, the critical points are $-1, 1/\sqrt{5}$, and 1.

$$f(-1) = -\frac{1}{2} \quad \text{is an endpoint minimum; it is also the absolute minimum.}$$

$$f\left(\frac{1}{\sqrt{5}}\right) = \frac{\sqrt{5}}{2} \quad \text{is a local maximum; it is also the absolute maximum.}$$

$$f(1) = \frac{1}{2} \quad \text{is an endpoint minimum.} \quad \square$$

Exercises

Find the critical points and classify the extreme values.

*1. $f(x) = \sqrt{x + 2}$.

2. $f(x) = (x - 1)(x - 2)$.

*3. $f(x) = x^2 - 4x + 1, \quad x \in [0, 3]$.

4. $f(x) = 2x^2 + 5x - 1, \quad x \in [-2, 0]$.

*5. $f(x) = x^2 + \frac{1}{x}$.

6. $f(x) = x + \frac{1}{x^2}$.

*7. $f(x) = x^2 + \frac{1}{x}, \quad x \in [\frac{1}{10}, 2]$.

8. $f(x) = x + \frac{1}{x^2}, \quad x \in (-1, 0)$.

*9. $f(x) = (x - 1)(x - 2), \quad x \in [0, 2]$.

10. $f(x) = (x - 1)^2(x - 2)^2, \quad x \in [0, 4]$.

*11. $f(x) = \frac{1 - 3\sqrt{x}}{3 - \sqrt{x}}$.

12. $f(x) = \frac{x^2}{1 + x^2}, \quad x \in [-1, 2]$.

*13. $f(x) = (x - \sqrt{x})^2$.

14. $f(x) = x\sqrt{4 - x^2}$.

*15. $f(x) = x\sqrt{3 - x}$.

16. $f(x) = \sqrt{x} - \frac{1}{\sqrt{x}}$.

*17. $f(x) = 1 - \sqrt[3]{x - 1}$.

18. $f(x) = (4x - 1)^{1/3}(2x - 1)^{2/3}$.

*19. $f(x) = \begin{cases} -2x, & x < 1 \\ x - 3, & 1 \le x \le 4 \\ 5 - x, & 4 < x \end{cases}$.

20. $f(x) = \begin{cases} -x^2, & 0 \le x < 1 \\ -2x, & 1 < x < 2 \\ -\frac{1}{2}x^2, & 2 \le x \le 3 \end{cases}$.

3.14   Additional Maximum-Minimum Problems

The techniques of the last two sections can be applied to a wide variety of maximum-minimum problems. Here are some examples.

Problem.   A window in the shape of a rectangle capped by a semicircle is to be surrounded by p inches of metal border. Find the radius of the semicircular part if the total area of the window is to be a maximum.

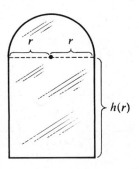

FIGURE 3.14.1

SOLUTION.   As in Figure 3.14.1 we let r be the radius of the semicircular part and let $h(r)$ be the height of the rectangular part. The total area of the window is then given by the function

$$A(r) = 2rh(r) + \frac{\pi r^2}{2}.$$

Since the total perimeter of the window is p, we have

$$p = 2r + 2h(r) + \pi r \quad \text{and thus} \quad h(r) = \frac{p - 2r - \pi r}{2}.$$

Substituting

$$h(r) = \frac{p - 2r - \pi r}{2}$$

into the area formula, we get

$$A(r) = 2r\left(\frac{p - 2r - \pi r}{2}\right) + \frac{\pi r^2}{2} = pr - 2r^2 - \frac{\pi r^2}{2}.$$

Our problem now is simply to find out what value of r maximizes the function A. Differentiating, we get

$$A'(r) = p - 4r - \pi r.$$

Setting $A'(r) = 0$ gives

$$p - 4r - \pi r = 0 \quad \text{and consequently} \quad r = \frac{p}{4 + \pi}.$$

Since the second derivative

$$A''(r) = -4 - \pi$$

is constantly negative, we can conclude that $r = p/(4 + \pi)$ does give the desired maximum. □

Problem.   Find the dimensions of the base of the rectangular box of greatest volume that can be constructed from 200 square inches of cardboard if the base is to be three times as long as it is wide.

SOLUTION.   If the dimensions of the base are x and $3x$ and the height is $h(x)$ then the volume is given by

$$V(x) = 3x^2 h(x). \hspace{2cm} \text{(Figure 3.14.2)}$$

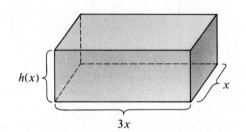

FIGURE 3.14.2

To obtain $h(x)$ in terms of x, we use the fact that the total surface area is 200 square inches. This gives

$$200 = 6xh(x) + 6x^2 + 2xh(x)$$

and thus

$$h(x) = \frac{200 - 6x^2}{8x} = \frac{100 - 3x^2}{4x}.$$

Substitution of this last expression for $h(x)$ into the volume formula gives

$$V(x) = 3x^2 \left(\frac{100 - 3x^2}{4x} \right),$$

which simplifies to

$$V(x) = \tfrac{3}{4}(100x - 3x^3).$$

Differentiation yields

$$V'(x) = \tfrac{3}{4}(100 - 9x^2) \quad \text{and} \quad V''(x) = -\tfrac{27}{2}x.$$

Setting

$$V'(x) = 0,$$

we get

$$100 - 9x^2 = 0 \quad \text{and thus} \quad x = \pm \tfrac{10}{3}.$$

Of these two answers the only one that fits the conditions of the problem is $x = \frac{10}{3}$. Since the second derivative

$$V''(x) = -\frac{27}{2}x$$

is negative at $x = \frac{10}{3}$, we can conclude that $x = \frac{10}{3}$ gives rise to a maximum. The rectangle must be $\frac{10}{3}$ inches wide and 10 inches long. $\square$

Problem. A manufacturing plant has a capacity of 25 articles per week. Experience has shown that n articles per week can be sold at a price of p dollars each, where $p = 110 - 2n$, and the cost of producing n articles is $(600 + 10n + n^2)$ dollars. How many articles should be made each week to give the largest profit?

SOLUTION. The profit (P dollars) on the sale of n articles is

$$P = np - (600 + 10n + n^2),$$

which simplifies to

$$P = 100n - 600 - 3n^2.$$

In this problem n must be an integer and it therefore doesn't make any sense to differentiate P with respect to n. The formula shows that P is negative if n is less than 8. By direct calculation we construct Table 3.14.1. The table shows that the largest profit is obtained when 17 articles per week are manufactured.

TABLE 3.14.1

n	P	n	P	n	P
8	8	14	212	20	200
9	57	15	225	21	117
10	100	16	232	22	148
11	137	17	233	23	113
12	168	18	228	24	72
13	193	19	217	25	25

We can avoid such detailed computation by considering the function

$$f(x) = 100x - 600 - 3x^2, \quad 0 \le x \le 25.$$

This function is differentiable with respect to x and for integral values of x it agrees with P. Differentiation of f gives

$$f'(x) = 100 - 6x.$$

Setting $f'(x) = 0$, we find $x = 16\frac{2}{3}$. It is now apparent that the largest value of f corresponding to an integral value of x will occur for $x = 16$ or $x = 17$. Direct calculation of $f(16)$ and $f(17)$ shows that the choice of $x = 17$ is correct. $\square$

Exercises

* 1. Find the two positive numbers whose sum is 40 and whose product is a maximum.

2. Find the dimensions of the rectangle of perimeter p that has the largest area.

* 3. A rectangular garden 200 square feet in area is to be fenced off against rabbits. Find the dimensions that will require the least amount of fencing if one side of the garden is already protected by a barn.

4. Find the largest area that a rectangle can have if its base lies on the x-axis and its upper vertices lie on the curve

$$y = 4 - x^2.$$

* 5. Find the largest possible area for a rectangle inscribed in a circle of radius r.

6. Let Q denote output, R revenue, C cost, and P profit. Given that

$$P = R - C,$$

what is the relation between marginal revenue

$$MR = \frac{dR}{dQ}$$

and marginal cost

$$MC = \frac{dC}{dQ}$$

when profits are at a maximum?

* 7. A manufacturer finds that the total cost of producing Q tons is $aQ^2 + bQ + c$ dollars and the price at which each ton can be sold is $\beta - \alpha Q$. Assuming that a, b, c, α, β are all positive, what is the output for maximum profit?

8. Davis Rent-A-TV derives an average net profit of $15 per customer if it services 1000 customers or less. If it services over 1000 customers, then the average profit decreases 1¢ for each customer above that number. How many customers give the maximum net profit?

* 9. The cross section of a beam is in the form of a rectangle of length l and width w. Assuming that the strength of the beam varies directly with $w^2 l$, what are the dimensions of the strongest beam that can be sawed from a round log of diameter d?

10. Let

$$P = f(Q) \qquad (P = \text{price}, \ Q = \text{output})$$

be a demand curve. Show that at an output which maximizes total revenue (price × output) the elasticity

$$\epsilon = \frac{f(Q)}{Q|f'(Q)|}$$

is 1.

* 11. Of all the isosceles triangles with a given perimeter, which has the largest area?

12. What are the dimensions of the base of the rectangular box of greatest volume that can be constructed from 100 square inches of cardboard if the base is to be twice as long as it is wide?
    (a) Assume that the box has a top.
    (b) Assume that the box has no top.

*13. Find the maximum area that a parallelogram can have and be inscribed in a triangle ABC in such a way that one vertex coincides with A while the others fall one on each side of the triangle.

14. Find the dimensions of the triangle of least area that can circumscribe a circle of radius r.

*15. From a rectangular piece of cardboard of dimensions $a \times b$ four congruent squares are to be cut out, one at each corner. The remaining crosslike piece is then to be folded into an open box. What size squares should be cut out if the volume of the resulting box is to be a maximum?

*16. What is the maximum area possible for a triangle inscribed in a circle of radius r?

17. A lighting fixtures manufacturer finds that he can sell x standing lamps per week at p dollars each, where $5x = 375 - 3p$. The cost of production is $(500 + 15x + \frac{1}{5}x^2)$ dollars. Show that maximum profit is obtained when production is about 30 units per week.

18. In Exercise 17 suppose that the relation between x and p is

$$x = 100 - 20\sqrt{\frac{p}{5}}.$$

Show that for maximum profit the manufacturer should produce only about 25 units per week.

19. In Exercise 17 suppose that the relation between x and p is

$$x^2 = 2500 - 20p.$$

What production produces maximum profit in this instance?

20. Show that a conical tent of a given volume will require the least amount of canvas when the height is $\sqrt{2}$ times the radius of the base. Assume no canvas flooring.

*21. A string of length l is to be cut into two pieces, one piece to form an equilateral triangle and the other to form a circle. How should the string be cut so as to (a) maximize the sum of the two areas? (b) minimize the sum of the two areas?

22. Given a right circular cone of height h and base radius r, find the inscribed cylinder:
    *(a) with the greatest volume.   (b) with the greatest curved surface.

23. Cylinders and right circular cones are inscribed in a sphere of radius R. Find:
    *(a) the cylinder with the greatest volume.
    (b) the cylinder with the greatest curved surface.
    *(c) the cone with the greatest volume.

24. What is the maximum volume of a right circular cone of slant height a?

*25. (*Optional*)   The cost of erecting an office building is $500,000 for the first story, $525,000 for the second, $550,000 for the third, and so on. Other expenses (lot, basement, etc.) are $3,500,000. Assume that the net annual income is $50,000 per story. How many stories will provide the greatest rate of return on investment?

3.15   Concavity and Points of Inflection

If you look at the graph of the cubing function (Figure 3.15.1), you will see that for x in $(-\infty, 0]$ the graph is "curving down" and for x in $[0, \infty)$ the graph is "curving up." In the language of calculus we say that the graph is *concave down* on $(-\infty, 0]$ and *concave up* on $[0, \infty)$.

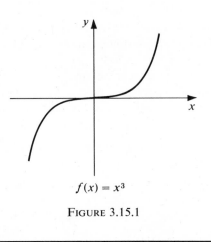

$$f(x) = x^3$$

FIGURE 3.15.1

Definition of Concavity

The graph of a differentiable function f is said to be *concave up* on the interval I iff f' is increasing on I; it is said to be *concave down* on I iff f' is decreasing on I.

If f'' exists, it serves well as a test for concavity. If $f''(x) > 0$ for all x in I, then f' is increasing there, so that the graph of f is concave up on I. If $f''(x) < 0$ for all x in I, then f' is decreasing there, so that the graph of f is concave down on I.

Of special interest in curve sketching are the points which separate arcs of opposite concavity. These are called *points of inflection.*

Definition of Point of Inflection

The point $(c, f(c))$ is called a *point of inflection* iff there exists $\delta > 0$ such that the graph of f is concave in one sense on $(c - \delta, c]$ and concave in the opposite sense on $[c, c + \delta)$.

It is not hard to see that if $(c, f(c))$ is a point of inflection then the derivative f' either has a local maximum at c (Figure 3.15.2) or it has a local minimum at c (Figure 3.15.3). It follows therefore that

if $(c, f(c))$ is a point of inflection, then either $f''(c) = 0$ or $f''(c)$ does not exist.

(To show this, all you have to do is apply Theorem 3.12.1, not to f, but to f'.)

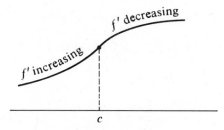

f' has a local maximum at c

FIGURE 3.15.2

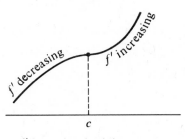

f' has a local minimum at c

FIGURE 3.15.3

Example. For
$$f(x) = x^3 + \tfrac{1}{2}x^2 - 2x + 1$$
we have
$$f'(x) = 3x^2 + x - 2 \quad \text{and} \quad f''(x) = 6x + 1.$$
Since
$$f''(x) \text{ is } \begin{cases} \text{negative}, & \text{for } x < -\tfrac{1}{6} \\ 0, & \text{for } x = -\tfrac{1}{6} \\ \text{positive}, & \text{for } x > -\tfrac{1}{6} \end{cases},$$

the graph of f is concave down on $(-\infty, -\tfrac{1}{6}]$ and concave up on $[-\tfrac{1}{6}, \infty)$. The point $(-\tfrac{1}{6}, f(-\tfrac{1}{6}))$ is a point of inflection. □

Example. For
$$f(x) = 3x^{5/3} - x$$
we have
$$f'(x) = 5x^{2/3} - 1 \quad \text{and} \quad f''(x) = \tfrac{10}{3}x^{-1/3}.$$
The second derivative fails to exist at 0. Since
$$f''(x) \text{ is } \begin{cases} \text{negative,} & \text{for } x < 0 \\ \text{positive,} & \text{for } x > 0 \end{cases},$$
the point $(0, f(0)) = (0, 0)$ is a point of inflection. The graph is concave down on $(-\infty, 0]$ and concave up on $[0, \infty)$. □

Exercises

Describe the concavity of the graph and find the points of inflection (if any).

*1. $\dfrac{1}{x}$.

2. $x + \dfrac{1}{x}$.

*3. $x^3 - 3x + 2$.

4. $2x^2 - 5x + 2$.

*5. $\tfrac{1}{4}x^4 - \tfrac{1}{2}x^2$.

6. $x^3(1 - x)$.

*7. $\dfrac{x}{x^2 - 1}$.

8. $\dfrac{x + 2}{x - 2}$.

*9. $(1 - x)^2(1 + x)^2$.

10. $\dfrac{6x}{x^2 + 1}$.

*11. $\dfrac{1 - \sqrt{x}}{1 + \sqrt{x}}$.

12. $(x - 3)^{1/5}$.

*13. Find d given that $(d, f(d))$ is a point of inflection of
$$f(x) = (x - a)(x - b)(x - c).$$

14. Find c given that the graph of
$$f(x) = cx^2 + \frac{1}{x^2}$$
has an inflection point at $(1, f(1))$.

*15. Find a and b given that the graph of
$$f(x) = ax^3 + bx^2$$
passes through $(-1, 1)$ and has an inflection point when $x = \tfrac{1}{3}$.

16. Determine A and B so that the curve
$$y = Ax^{1/2} + Bx^{-1/2}$$
will have a point of inflection at $(1, 4)$.

3.16   Some Curve Sketching

During the course of the last few sections you have seen how to determine the extreme values of a function and how to find where a function is increasing and where it is decreasing. Now you have seen how to determine the concavity of the graph and how to find the points of inflection. By putting all this information together you can get a good idea of what the graph of a given function looks like.

Example.   In the case of

$$f(x) = \tfrac{1}{4}(x^3 - \tfrac{3}{2}x^2 - 6x + 2),$$

$f(0) = \tfrac{1}{2}$ and therefore the graph intersects the y-axis at the point $(0, \tfrac{1}{2})$.† Differentiation gives

$$f'(x) = \tfrac{1}{4}(3x^2 - 3x - 6) = \tfrac{3}{4}(x + 1)(x - 2)$$

and

$$f''(x) = \tfrac{1}{4}(6x - 3) = \tfrac{3}{4}(2x - 1).$$

The first derivative is 0 at $x = -1$ and $x = 2$. These are the only critical points. Since $f''(-1) < 0$,

$$f(-1) = \tfrac{1}{4}[(-1)^3 - \tfrac{3}{2}(-1)^2 - 6(-1) + 2] = \tfrac{1}{4}[-1 - \tfrac{3}{2} + 6 + 2] = 1\tfrac{3}{8}$$

is a local maximum. Since $f''(2) > 0$,

$$f(2) = \tfrac{1}{4}[(2)^3 - \tfrac{3}{2}(2)^2 - 6(2) + 2] = -2$$

is a local minimum. Since

$$f''(x) \text{ is } \begin{cases} \text{negative,} & \text{for } x < \tfrac{1}{2} \\ \qquad 0, & \text{for } x = \tfrac{1}{2} \\ \text{positive,} & \text{for } x > \tfrac{1}{2} \end{cases},$$

the point $(\tfrac{1}{2}, f(\tfrac{1}{2})) = (\tfrac{1}{2}, -\tfrac{5}{16})$ is the only point of inflection. The graph is concave down on $(-\infty, \tfrac{1}{2}]$ and concave up on $[\tfrac{1}{2}, \infty)$.

If you now put together all this information and plot a few points, you will not find it hard to come up with a graph that looks like the one displayed in Figure 3.16.1.   □

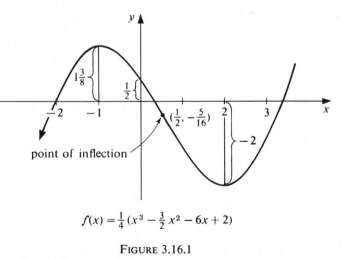

$$f(x) = \tfrac{1}{4}(x^3 - \tfrac{3}{2}x^2 - 6x + 2)$$

FIGURE 3.16.1

---

† It would be nice to find out exactly where the graph intersects the x-axis, but this would require that we solve the equation

$$x^3 - \tfrac{3}{2}x^2 - 6x + 2 = 0.$$

Example.   The function

$$g(x) = \frac{1}{1 + x^2}$$

remains positive throughout. Since $g(-x) = g(x)$, the function is even and its graph is symmetric with respect to the y-axis. When $x = 0$, $g(x) = 1$. This is clearly the maximum value of g. As x tends away from zero

$$1 + x^2 \quad \text{increases without bound}$$

and therefore

$$g(x) = \frac{1}{1 + x^2} \quad \text{decreases toward 0.}$$

So far we have not used calculus in this example but we need it now to determine the concavity of the graph. Differentiation gives

$$g'(x) = -\frac{2x}{(1 + x^2)^2} \quad \text{and} \quad g''(x) = \frac{2(3x^2 - 1)}{(1 + x^2)^3}.$$

Since

$$g''(x) \text{ is} \begin{cases} \text{positive,} & \text{for } x < -1/\sqrt{3} \\ 0, & \text{for } x = -1/\sqrt{3} \\ \text{negative,} & \text{for } -1/\sqrt{3} < x < 1/\sqrt{3} \\ 0, & \text{for } x = 1/\sqrt{3} \\ \text{positive,} & \text{for } x > 1/\sqrt{3} \end{cases},$$

the graph is concave down on $[-1/\sqrt{3}, 1/\sqrt{3}]$ and concave up on $(-\infty, -1/\sqrt{3}]$ and $[1/\sqrt{3}, \infty)$. The points $(-1/\sqrt{3}, 3/4)$ and $(1/\sqrt{3}, 3/4)$ are points of inflection.

It is now easy to see that the graph is as in Figure 3.16.2. $\square$

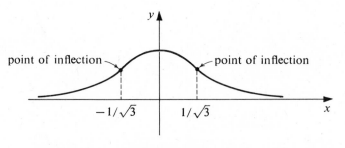

FIGURE 3.16.2

Example.   As our final example we take the function

$$f(x) = 3(x^{5/3} - x^{4/3}).$$

We can rewrite this as

$$f(x) = 3x^{4/3}(x^{1/3} - 1).$$

The function is zero at $x = 0$ and at $x = 1$. The graph crosses the x-axis at the origin and at the point $(1, 0)$.

$$f'(x) = 5x^{2/3} - 4x^{1/3} = x^{1/3}(5x^{1/3} - 4),$$

$$f''(x) = \tfrac{10}{3}x^{-1/3} - \tfrac{4}{3}x^{-2/3} = \tfrac{2}{3}x^{-2/3}(5x^{1/3} - 2).$$

The first derivative is zero at $x = 0$ and at $x = 4^3/5^3 = \tfrac{64}{125}$. These are the only critical points. Note that

$$f'(x) \text{ is} \begin{cases} \text{positive,} & \text{for } x < 0 \\ 0, & \text{for } x = 0 \\ \text{negative,} & \text{for } 0 < x < \tfrac{64}{125} \\ 0, & \text{for } x = \tfrac{64}{125} \\ \text{positive,} & \text{for } x > \tfrac{64}{125} \end{cases}.$$

This shows that

$$f(0) = 0 \quad \text{is a local maximum}$$

and

$$f(\tfrac{64}{125}) = -\tfrac{768}{3125} \overset{\text{approx}}{=} -\tfrac{1}{4} \quad \text{is a local minimum.}$$

The second derivative does not exist at $x = 0$. It is zero at $x = 2^3/5^3 = \tfrac{8}{125}$. Since

$$f''(x) \text{ is} \begin{cases} \text{negative,} & \text{for } x < 0 \\ \text{nonexistent,} & \text{for } x = 0 \\ \text{negative,} & \text{for } 0 < x < \tfrac{8}{125} \\ 0, & \text{for } x = \tfrac{8}{125} \\ \text{positive,} & \text{for } x > \tfrac{8}{125} \end{cases},$$

the graph is concave down on all of $(-\infty, \tfrac{8}{125}]$ (explain) and concave up on $[\tfrac{8}{125}, \infty)$. The point

$$(\tfrac{8}{125}, f(\tfrac{8}{125})) \overset{\text{approx}}{=} (0.06, -0.04)$$

is the only point of inflection. Note that this is close to the origin.

Before sketching the graph we gather a little more information. The easiest numbers to work with are the perfect cubes:

at $x = -\tfrac{64}{125}$, $f(x)$ is about -2.2 and the slope is 6.4;

at $x = -\tfrac{8}{27}$, $f(x)$ is about -0.1 and the slope is about 5;

at $x = 1$, the slope is 1;

at $x = \tfrac{125}{64}$, $f(x)$ is about 1.8 and the slope is about 3.

Use this information and you will come up with a curve that looks like the one displayed in Figure 3.16.3. □

REMARK. So far we have paid little attention to questions of asymptotic behavior. Asymptotes, infinite limits, limits as $x \to \pm\infty$ are discussed in Chapter 11.

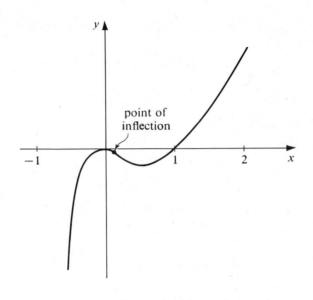

FIGURE 3.16.3

Exercises

For each of the following functions (i) find the critical points, (ii) find and classify the extreme values (local, endpoint, absolute), (iii) indicate where the function is increasing and where it is decreasing, (iv) indicate the concavity of the graph, (v) specify the points of inflection, and then (vi) sketch the graph.

*1. $f(x) = (x - 2)^2$.

2. $f(x) = 1 - (x - 2)^2$.

*3. $f(x) = x^3 - 2x^2 + x + 1$.

4. $f(x) = \frac{2}{3}x^3 - \frac{1}{2}x^2 - 10x - 1$.

*5. $f(x) = \dfrac{x}{3x - 1}$.

6. $f(x) = \dfrac{2x}{4x - 3}$.

*7. $f(x) = \dfrac{x^2}{3x + 1}$.

8. $f(x) = \dfrac{2x^2}{x + 1}$.

*9. $f(x) = \dfrac{x}{(3x + 1)^2}$.

10. $f(x) = \dfrac{2x}{(x + 1)^2}$.

*11. $f(x) = 3x^5 + 5x^3$.

12. $f(x) = 3x^4 + 4x^3$.

*13. $f(x) = 1 + (x - 2)^{4/3}$.

14. $f(x) = 1 + (x - 2)^{5/3}$.

*15. $f(x) = x^2(1 + x)^2$.

16. $f(x) = x^2(1 + x)^3$.

*17. $f(x) = x\sqrt{1 - x}$.

18. $f(x) = \sqrt{x - x^2}$.

*19. $f(x) = \dfrac{2x}{x^2 + 2}$.

20. $f(x) = \dfrac{2x^2}{x^2 + 2}$.

*21. $f(x) = 2 + (x + 1)^{6/5}$.

22. $f(x) = 2 + (x + 1)^{7/5}$.

3.17   Rates of Change per Unit Time

You've already seen that the derivative of a function gives its rate of change. Suppose now that $f(t)$ represents some quantity at time t. The derivative $f'(t)$ then gives the *rate of change* of that quantity *per unit time.*

Example.   A spherical balloon (Figure 3.17.1) is expanding under the influence of solar radiation. If its radius is increasing at the rate of 2 inches per minute, how fast is the volume increasing when the radius is 5 inches?

FIGURE 3.17.1

SOLUTION.   We can express the volume at time t in terms of the radius at time t by the formula

$$V(t) = \tfrac{4}{3}\pi[r(t)]^3.$$

Differentiation gives

$$V'(t) = 4\pi[r(t)]^2 r'(t).$$

If at time t_0 the radius is 5 inches, then

$$V'(t_0) = 4\pi(25)2 = 200\pi.$$

This means that the volume is increasing at the rate of 200π cubic inches per minute. ☐

Example.   A point is moving along the unit circle. (Figure 3.17.2) Each time that it has coordinates $(\tfrac{1}{2}, \tfrac{1}{2}\sqrt{3})$ its ordinate is decreasing at the rate of 3 units per second. At what rate is the abscissa changing?

SOLUTION.   At each time t,

$$[x(t)]^2 + [y(t)]^2 = 1,$$

and thus

$$2x(t)x'(t) + 2y(t)y'(t) = 0.$$

Canceling the 2, we have

$$x(t)x'(t) + y(t)y'(t) = 0.$$

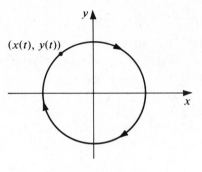

FIGURE 3.17.2

If we let t_0 be a time at which the moving point has coordinates $(\frac{1}{2}, \frac{1}{2}\sqrt{3})$, we then have

$$\tfrac{1}{2}x'(t_0) + \tfrac{1}{2}\sqrt{3}\,(-3) = 0,$$

which implies that

$$x'(t_0) = 3\sqrt{3}.$$

The abscissa is thus increasing at the rate of $3\sqrt{3}$ units per second.   □

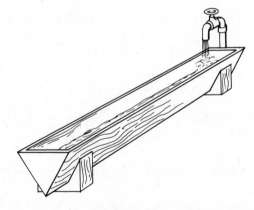

FIGURE 3.17.3

Example.   A water trough (Figure 3.17.3) with vertical cross section in the shape of an equilateral triangle is being filled at the rate of 4 cubic feet per minute. Given that the trough is 12 feet long, how fast is the water level rising when the water reaches a depth of $1\frac{1}{2}$ feet?

SOLUTION.   When the water is x feet deep, a vertical cross section of water has area $x^2/\sqrt{3}$. Since the trough is 12 feet long, the volume of the water is then $12x^2/\sqrt{3}$. If we denote by $x(t)$ the depth of the water at time t, then we obtain

$$V(t) = \frac{12}{\sqrt{3}}\,[x(t)]^2$$

and thus

$$V'(t) = \frac{24}{\sqrt{3}} x(t)x'(t).$$

If it is at time t_0 that the water is $1\frac{1}{2}$ feet deep, then we have

$$4 = \frac{24}{\sqrt{3}} (1\tfrac{1}{2})x'(t_0),$$

so that

$$x'(t_0) = \tfrac{1}{9}\sqrt{3}.$$

This tells us that the water is rising at the rate of $\frac{1}{9}\sqrt{3}$ feet per minute. □

Velocity and Acceleration

Suppose now that an object is moving along a straight line and that for each time t during a certain time interval the object has position (coordinate) $x(t)$. If $x'(t)$ exists, then $x'(t)$ gives the rate of change of position at time t. This rate of change of position is called the *velocity* of the object at time t; in symbols,

$$\boxed{v(t) = x'(t).}$$

If the velocity function is itself differentiable, then its rate of change at time t is called the *acceleration* at time t; in symbols,

$$\boxed{a(t) = v'(t) = x''(t).}$$

In the Leibniz notation,

$$\boxed{v = \frac{dx}{dt} \quad \text{and} \quad a = \frac{dv}{dt} = \frac{d^2x}{dt^2}.}$$

A positive velocity indicates an increase in coordinate and thus a motion in the positive direction. A negative velocity indicates a decrease in coordinate and thus a motion in the negative direction. When the direction of motion is immaterial, we speak of *speed*. This is by definition the absolute value of the velocity:

$$\boxed{\text{speed at time } t = |v(t)|.}$$

A word about units is in order. If distance is measured in feet and time in seconds, then velocity and speed are given in feet per second and acceleration is given in feet per second per second.

We can illustrate these ideas by considering the motion of an object which falls freely under the influence of gravity. If we neglect air resistance, then the height of the object at time t is given by the equation

$$x(t) = \tfrac{1}{2}gt^2 + v_0 t + x_0.$$

Here v_0 represents the velocity at time $t = 0$ (called the *initial velocity*) and x_0 represents the height at time $t = 0$ (called the *initial position*). The constant g, called the gravitational constant, is the acceleration due to gravity. If we measure time in seconds and distances in feet, then g is approximately -32 feet per second per second. (Why is it negative?) In making numerical calculations we shall take g as -32 feet per second per second. Doing so we obtain

$$x(t) = -16t^2 + v_0 t + x_0.$$

Problem. A stone is dropped from a height of 1600 feet. How long does it take to reach the ground and what is its speed at impact?

SOLUTION. Here $x_0 = 1600$ and $v_0 = 0$. Consequently

$$x(t) = -16t^2 + 1600.$$

To find t at the moment of impact we set $x(t) = 0$. This gives

$$-16t^2 + 1600 = 0$$
$$t^2 = 100$$
$$t = \pm 10.$$

We can disregard the negative answer because the stone had not even been dropped before $t = 0$. We conclude that it takes 10 seconds for the stone to reach the ground.

The velocity at impact is the velocity when $t = 10$. Since

$$v(t) = x'(t) = -32t$$

we have

$$v(10) = -320.$$

The speed at impact is therefore 320 feet per second. $\square$

Problem. A stone is projected from ground level vertically upward with an initial speed of 72 feet per second.

(a) How many seconds later does it attain its maximum height?
(b) What is this maximum height?
(c) How fast is it traveling when it reaches a height of 32 feet?
    (i) going up?   (ii) going down?

SOLUTION.   The basic equation is again

$$x(t) = -16t^2 + v_0 t + x_0.$$

Here $x_0 = 0$ (it starts at ground level) and $v_0 = 72$. (The initial velocity is 72 feet per second.) The equation of motion is therefore

$$x(t) = -16t^2 + 72t.$$

Differentiation gives

$$v(t) = x'(t) = -32t + 72.$$

The maximum height is obviously reached when the velocity is 0. This occurs when $t = \frac{72}{32} = \frac{9}{4}$. To get the maximum height attained we need only evaluate $x(t)$ at $t = \frac{9}{4}$. Since

$$x(\tfrac{9}{4}) = 81,$$

we see that the maximum height attained is 81 feet.

To answer part (c) we must find those numbers t for which

$$x(t) = 32.$$

Since

$$x(t) = -16t^2 + 72t,$$

we are led to the quadratic equation

$$16t^2 - 72t + 32 = 0.$$

This quadratic has two solutions $t = \frac{1}{2}$ and $t = 4$. For (i) we have $v(\frac{1}{2}) = 56$; and for (ii), $v(4) = -56$. The speed in each case is 56 feet per second. $\square$

Exercises

*1. A point moves along the straight line $x + 2y = 2$. Find: (a) the rate of change of the y-coordinate, given that the x-coordinate increases 4 units per second; (b) the rate of change of the x-coordinate, given that the y-coordinate decreases 2 units per second.

2. A rubbish heap in the shape of a cube is being compacted. Given that the volume decreases at the rate of 2 cubic inches per minute, find (a) the rate of change of an edge and (b) the rate of change of the total surface area when the volume of the cube is 64 cubic inches.

*3. A point P is moving in the circular orbit $x^2 + y^2 = 25$. As it passes through the point (3, 4), its y-coordinate is decreasing at the rate of 2 units per second. How is the x-coordinate changing?

4. A barge, the deck of which is 12 feet below the level of a dock, is drawn to it by means of a cable attached to a ring in the floor of the dock, the cable being hauled in by a windlass on deck at the rate of 8 feet per minute. How fast is the barge moving toward the dock when 16 feet away?

*5. A boat is fastened to a rope which is wound about a windlass 20 feet above the level at which the rope is attached to the boat. The boat is drifting away at the rate of 8 feet per second. How fast is it unwinding the rope when 30 feet from the point directly under the windlass?

6. At a certain instant the dimensions of a rectangle are a and b. These dimensions are changing at the rates m, n, respectively. Find the rate at which the area is changing.

7. A ladder 13 feet long is leaning against a wall. If the base of the ladder is being pulled away from the wall at the rate of $\frac{1}{2}$ foot per second, how fast is the top of the ladder being lowered when the base is 5 feet from the wall?

*8. Taking

$$x = -t\sqrt{t + 1}, \qquad t \geq 0$$

as the equation of motion, find
(a) the velocity when $t = 1$,
(b) the acceleration when $t = 1$,
(c) the speed when $t = 1$,
(d) the rate of change of the speed when $t = 1$,
(e) the time t, if it exists, when the velocity is a maximum,
(f) the time t, if it exists, when the speed is a maximum.

*9. The height of a cylinder is being increased at the rate of 4 inches per minute. If the volume of the cylinder is to be kept constant, at what rate must the radius be diminished at each instant?

10. In the special theory of relativity the mass of a particle moving at velocity v is

$$m\left(1 - \frac{v^2}{c^2}\right)^{-1/2},$$

where m is the mass at rest and c is the speed of light. At what rate is the mass changing when the particle's velocity is $\frac{1}{2}c$ and the rate of change of the velocity is $0.01c$ per second?

*11. A conical paper cup of radius 2 inches and height 6 inches is leaking water at the rate of 1 cubic inch per minute. At what rate is the level of the water being lowered: (a) when the water is 3 inches deep, (b) when the cup is half full?

12. The shadow cast by a man standing 3 feet from a lamp post is 4 feet long. If the man is 6 feet tall and walks away from the lamp post at a speed of 400 feet per minute, at what rate will his shadow be lengthening: (a) a quarter of a minute later, (b) when he is 20 feet from the lamp post?

*13. A stone is thrown upward from ground level with an initial speed of 32 feet per second. (a) How many seconds later will it hit the ground? (b) What will be the maximum height attained? (c) With what initial speed should it be thrown if it is to reach a maximum height of 36 feet?

14. To estimate the height of a bridge a man drops a stone into the water below. How high is the bridge (a) if the stone hits the water 3 seconds later? (b) if the man hears the splash 3 seconds later? (Use 1080 feet per second as the speed of sound.)

*15. A falling stone is observed to be at a height of 100 feet. Two seconds later it is observed to be at a height of 16 feet. (a) From what height was it dropped? (b) If it was thrown down with an initial speed of 5 feet per second, from what height was it thrown? (c) If it was thrown upward with an initial speed of 10 feet per second, from what height was it thrown?

3.18   The Little-*o*(*h*) Idea; Differentials

For h small, πh^2 is small compared to h, in the sense that the ratio of πh^2 to h tends to 0 as h tends to 0:

$$\lim_{h \to 0} \frac{\pi h^2}{h} = 0.$$

The same holds true for $h^{5/3}\sqrt{1 - h}$:

$$\lim_{h \to 0} \frac{h^{5/3}\sqrt{1 - h}}{h} = 0.$$

Expressions $g(h)$, such as πh^2 and $h^{5/3}\sqrt{1 - h}$ that tend to 0 with h when divided by h, are called *little-o(h)*. More briefly,

$$g(h) \text{ is } little\text{-}o(h) \quad \text{iff} \quad \lim_{h \to 0} \frac{g(h)}{h} = 0.$$

To indicate that $g(h)$ is *little-o(h)* we write

$$g(h) = o(h).$$

As you can check,

$$h^3 = o(h) \quad \text{and} \quad \frac{h^2}{h - 1} = o(h),$$

but

$$h^{1/3} \neq o(h) \quad \text{and} \quad \frac{h}{h - 1} \neq o(h).$$

We can use the *little-o(h)* notation to advantage in connection with differentiation. If f is differentiable at x, then

$$\lim_{h \to 0} \frac{f(x + h) - f(x)}{h} = f'(x).$$

Since obviously

$$f'(x) = \lim_{h \to 0} \frac{f'(x)h}{h},$$

we have

$$\lim_{h \to 0} \frac{[f(x + h) - f(x)] - f'(x)h}{h} = 0.$$

This says that

$$[f(x + h) - f(x)] - f'(x)h = o(h)$$

and therefore

(3.18.1)
$$f(x + h) - f(x) = f'(x)h + o(h).$$

The difference $f(x + h) - f(x)$ is called the *increment* (of f from x to $x + h$) and is usually denoted by Δf:

$$\Delta f = f(x + h) - f(x).$$

The product $f'(x)h$ is called the *differential* (at x with increment h) and is usually denoted by df:

$$df = f'(x)h.$$

Equation (3.18.1) says that for small h

$$\Delta f \quad \text{and} \quad df$$

are approximately equal:

$$\boxed{\Delta f \cong df.}$$

This approximate equality is illustrated in Figure 3.18.1. While Δf gives the precise change in f from x to $x + h$, df gives what the change in f would have been had the graph of f proceeded from x to $x + h$ with constant slope $f'(x)$. The difference between Δf and df is the vertical separation between the graph of f and the tangent line as measured at $x + h$. The less the graph curves, the more accurate the differential estimate becomes.

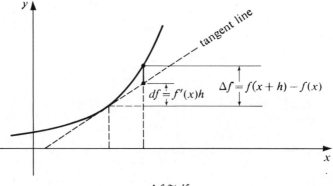

$$\Delta f \cong df$$

FIGURE 3.18.1

To illustrate the use of differentials we begin with a square of side x. The area is

$$f(x) = x^2.$$

An increase h in the length of each side produces a change in area

$$\begin{aligned}
\Delta f &= f(x + h) - f(x) \\
&= (x + h)^2 - x^2 \\
&= (x^2 + 2xh + h^2) - x^2 \\
&= 2xh + h^2.
\end{aligned}$$

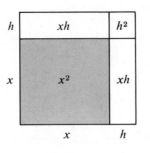

$$\text{FIGURE 3.18.2}$$

As an estimate for this change we can use the differential

$$df = f'(x)h \doteq 2xh. \qquad\qquad \text{(Figure 3.18.2)}$$

The error of our estimate — the difference between the actual change and the estimated change — is the difference

$$\Delta f - df = h^2.$$

Problem. Given that

$$f(x) = x^{2/5}$$

estimate the change in f

(a) if x is increased from 32 to 34.

(b) if x is decreased from 1 to $\frac{9}{10}$.

SOLUTION. Since

$$f'(x) = \tfrac{2}{5}(1/x)^{3/5}$$

we have

$$df = f'(x)h = \tfrac{2}{5}(1/x)^{3/5}h.$$

For part (a) we set $x = 32$ and $h = 2$. The differential then becomes

$$\tfrac{2}{5}(\tfrac{1}{32})^{3/5}2 = \tfrac{1}{10} = 0.10.$$

A change in x from 32 to 34 increases the value of f by approximately 0.10.

For part (b) we set $x = 1$ and $h = -\frac{1}{10}$. The differential then becomes

$$\tfrac{2}{5}(\tfrac{1}{1})^{3/5}(-\tfrac{1}{10}) = -\tfrac{2}{50} = -\tfrac{1}{25} = -0.04.$$

A change in x from 1 to $\frac{9}{10}$ decreases the value of f by approximately 0.04. $\square$

Problem. Use the differential to estimate $\sqrt{104}$.

SOLUTION. We know $\sqrt{100}$. What we need is an estimate for the increase of

$$f(x) = \sqrt{x}$$

from 100 to 104. Differentiation gives

$$f'(x) = \frac{1}{2\sqrt{x}}$$

so that in general

$$df = f'(x)h = \frac{h}{2\sqrt{x}}.$$

With $x = 100$ and $h = 4$, df becomes

$$\frac{4}{2\sqrt{100}} = \frac{1}{5} = 0.2.$$

A change in x from 100 to 104 increases the value of the square root by approximately 0.2. It follows that

$$\sqrt{104} \cong \sqrt{100} + 0.2 = 10 + 0.2 = 10.2.$$

As you can check, $(10.2)^2 = 104.04$, so that we are not far off. $\square$

Let's go back to the little-$o(h)$ idea. We have used

$$df = f'(x)h$$

to approximate

$$\Delta f = f(x + h) - f(x)$$

and in the last problem we used

$$f(x) + f'(x)h \qquad\qquad (\sqrt{100} + 0.2)$$

to approximate

$$f(x + h). \qquad\qquad (\sqrt{104})$$

Now we want to talk about

$$f(x) + f'(x)h.$$

The expression is linear in h:

$$\text{it is of the form } \quad A + Bh$$

and it is an $o(h)$ approximation to $f(x + h)$: you know that

$$f(x + h) - f(x) = f'(x)h + o(h)$$

and therefore

$$f(x + h) = f(x) + f'(x)h + o(h).$$

We now assert that $f(x) + f'(x)h$ is the best linear approximation to $f(x + h)$—the best in that it is the only one which is $o(h)$.

Theorem

If f is differentiable at x, then

$$f(x) + f'(x)h$$

is the only $o(h)$ linear approximation to $f(x + h)$.

PROOF.   All we have to show is that

$$\text{if } \quad f(x + h) = A + Bh + o(h), \qquad \text{then} \quad A = f(x) \quad \text{and} \quad B = f'(x).$$

Let's assume therefore that

(1) $$f(x + h) = A + Bh + o(h).$$

Taking the limit of both sides as h tends to 0, we get on the left, $f(x)$, and on the right, A. (Explain.) This makes

$$A = f(x).$$

Substituting $f(x)$ for A in (1) we have

$$f(x + h) = f(x) + Bh + o(h).$$

This gives

$$f(x + h) - f(x) = Bh + o(h)$$

$$\frac{f(x + h) - f(x)}{h} = B + \frac{o(h)}{h}.$$

Once again we take the limit of both sides as h tends to 0. The left-hand side tends to $f'(x)$ and the right-hand side tends to B. (Explain.) It follows that

$$B = f'(x). \quad \square$$

Exercises

1. Use a differential to estimate the change in the volume of a cube caused by an increase h in the length of each side. Interpret the error of your estimate $\Delta V - dV$ geometrically.
2. Use a differential to estimate the area of a ring of inner radius r and width h. What is the exact area?

Estimate the following by differentials.

*3. $\sqrt[3]{1010}$.
4. $\sqrt{125}$.
*5. $\dfrac{1}{\sqrt{22}}$.
6. $\dfrac{1}{\sqrt{24}}$.

*7. $\sqrt[5]{30}$.
8. $(26)^{2/3}$.
*9. $(33)^{3/5}$.
10. $(33)^{-1/5}$.

11. Show that

$$\text{if} \quad g(h) = o(h) \quad \text{then} \quad \lim_{h \to 0} g(h) = 0.$$

12. Estimate $f(5.4)$ given that

$$f(5) = 1 \quad \text{and} \quad f'(x) = \sqrt[3]{x^2 + 2}.$$

*13. Find the approximate volume of a thin cylindrical sheet with open ends if the inner radius is r, the height is h, and the thickness is t.

*14. A box is to be constructed in the form of a cube to hold 1000 cubic feet. Use a differential to estimate how accurately the inner edge must be made so that the volume will be correct to within 3 cubic feet.

15. Use differentials to estimate for what values of x

*(a) $\sqrt{x + 1} - \sqrt{x} < 0.01$.
(b) $\sqrt[4]{x + 1} - \sqrt[4]{x} < 0.002$.

*16. The time of one vibration of a pendulum is given by the formula

$$t = \pi \sqrt{\frac{l}{g}},$$

where t is measured in seconds, $g = 32.2$, and l is the length of the pendulum measured in feet. Taking $\pi = 3.14$, and given that a pendulum of length 3.26 feet vibrates once a second, find the approximate change in t if the pendulum is lengthened 0.01 feet.

3.19 Implicit Differentiation

Suppose you know that y is a function of x and that y satisfies the equation

$$3x^3y - 4y - 2x + 1 = 0.$$

One way to find dy/dx is to solve first for y:

$$(3x^3 - 4)y - 2x + 1 = 0,$$

$$(3x^3 - 4)y = 2x - 1,$$

$$y = \frac{2x - 1}{3x^3 - 4},$$

and then differentiate:

$$\frac{dy}{dx} = \frac{(3x^3 - 4)2 - (2x - 1)(9x^2)}{(3x^3 - 4)^2}$$

$$= \frac{6x^3 - 8 - 18x^3 + 9x^2}{(3x^3 - 4)^2}$$

(1) $$= -\frac{12x^3 - 9x^2 + 8}{(3x^3 - 4)^2}.$$

It is also possible to find dy/dx without first solving the equation for y. The technique is called *implicit differentiation*.

Let's return to the equation

$$3x^3y - 4y - 2x + 1 = 0.$$

Differentiating both sides of this equation (and remembering that y is a function of x), we have

$$3x^3\frac{dy}{dx} + 9x^2y - 4\frac{dy}{dx} - 2 = 0,$$

$$(3x^3 - 4)\frac{dy}{dx} = 2 - 9x^2y,$$

(2) $$\frac{dy}{dx} = \frac{2 - 9x^2y}{3x^3 - 4}.$$

The answer looks different from what we obtained before because this time y appears on the right-hand side. This is generally no disadvantage. To satisfy yourself that the two answers are really the same all you have to do is substitute

$$y = \frac{2x - 1}{3x^3 - 4}$$

in (2). This gives

$$\frac{dy}{dx} = \frac{2 - 9x^2\left(\dfrac{2x - 1}{3x^3 - 4}\right)}{3x^3 - 4}$$

$$= \frac{6x^3 - 8 - 18x^3 + 9x^2}{(3x^3 - 4)^2}$$

$$= -\frac{12x^3 - 9x^2 + 8}{(3x^3 - 4)^2}.$$

This is the answer we obtained before. □

Implicit differentiation is particularly useful where it is inconvenient (or impossible) to first solve the given equation for y.

Problem. Find dy/dx given that

$$4x^2 + 2xy - xy^3 = 0.$$

SOLUTION. The relation

$$4x^2 + 2xy - xy^3 = 0$$

gives

$$\frac{d}{dx}(4x^2 + 2xy - xy^3) = 0,$$

$$8x + \left(2x\frac{dy}{dx} + 2y\right) - \left(x3y^2\frac{dy}{dx} + y^3\right) = 0,$$

$$8x + 2y - y^3 + x(2 - 3y^2)\frac{dy}{dx} = 0,$$

$$\frac{dy}{dx} = \frac{y^3 - 2y - 8x}{x(2 - 3y^2)}. \quad \square$$

Problem. Find the slope of the curve

$$x^3 - 3xy^2 + y^3 = 0 \quad \text{at } (2, -1).$$

SOLUTION. Differentiation gives

$$3x^2 - 3x\left(2y\frac{dy}{dx}\right) - 3y^2 + 3y^2\frac{dy}{dx} = 0,$$

$$3x^2 - 6xy\frac{dy}{dx} - 3y^2 + 3y^2\frac{dy}{dx} = 0.$$

At $x = 2$ and $y = -1$, the equation becomes

$$12 + 12 \frac{dy}{dx} - 3 + 3 \frac{dy}{dx} = 0$$

$$15 \frac{dy}{dx} = -9$$

$$\frac{dy}{dx} = -\frac{3}{5}.$$

The slope is $-3/5$. ☐

The angle between two curves is the angle between their tangents at the point of intersection. If the slopes are m_1 and m_2, then the angle of intersection can be obtained from the formula

$$\tan \theta = \frac{m_1 - m_2}{1 + m_1 m_2}.$$

Problem.   Find the angle of intersection of the circles

$$C_1: \quad (x - 1)^2 + y^2 = 10,$$

$$C_2: \quad x^2 + (y - 2)^2 = 5.$$

SOLUTION.   If you solve the equations simultaneously, you will find that $(2, 3)$ and $(-2, 1)$ are the points of intersection. Differentiating the first equation implicitly, we get

$$2(x - 1) + 2y \frac{dy}{dx} = 0$$

and therefore

$$m_1 = \frac{dy}{dx} = \frac{1 - x}{y}.$$

Differentiating the second equation, we get

$$2x + 2(y - 2) \frac{dy}{dx} = 0$$

and therefore

$$m_2 = \frac{dy}{dx} = \frac{x}{2 - y}.$$

At $(2, 3)$,

$$m_1 = \frac{1 - 2}{3} = -\frac{1}{3}, \qquad m_2 = \frac{2}{2 - 3} = -2,$$

so that

$$\tan \theta = \frac{m_1 - m_2}{1 + m_1 m_2} = \frac{-\frac{1}{3} + 2}{1 + \frac{2}{3}} = 1.$$

As a value of θ, we can take $\pi/4$ radians (or 45°). At $(-2, 1)$,

$$m_1 = \frac{1 + 2}{1} = 3, \qquad m_2 = \frac{-2}{2 - 1} = -2$$

so that

$$\tan \theta = \frac{m_1 - m_2}{1 + m_1 m_2} = \frac{3 + 2}{1 - 6} = -1.$$

As a value of θ we could take $\theta = -\pi/4$ radians ($-45°$). However, it is customary to take θ positive. So doing, we once again have $\theta = \pi/4$ radians (45°). $\square$

We can also find higher derivatives by implicit differentiation.

Problem. Find d^2y/dx^2 given that

$$b^2x^2 - a^2y^2 = a^2b^2.$$

SOLUTION

$$2b^2x - 2a^2y \frac{dy}{dx} = 0$$

(3) $$\frac{dy}{dx} = \frac{b^2x}{a^2y}.$$

We now differentiate again:

$$\frac{d^2y}{dx^2} = \frac{a^2y(b^2) - (b^2x)a^2 \, dy/dx}{(a^2y)^2}.$$

Substituting for dy/dx its value from (3), we obtain

$$\frac{d^2y}{dx^2} = \frac{a^2b^2y - (b^2x)a^2(b^2x/a^2y)}{(a^2y)^2},$$

which simplifies to

$$\frac{d^2y}{dx^2} = \frac{b^2(a^2y^2 - b^2x^2)}{a^4y^3}.$$

We could leave the answer as it stands, but since we know that

$$b^2x^2 - a^2y^2 = a^2b^2, \qquad \text{(our initial equation)}$$

we can write our answer more neatly as

$$\frac{d^2y}{dx^2} = -\frac{b^4}{a^2y^3}. \quad \square$$

Exercises

Find dy/dx in terms of x and y by implicit differentiation.

*1. $x^2 + y^2 = r^2$.

2. $x^3 + y^3 - 3axy = 0$.

*3. $b^2x^2 + a^2y^2 = a^2b^2$.

4. $\sqrt{x} + \sqrt{y} = \sqrt{a}$.

* 5. $x^{2/3} + y^{2/3} = a^{2/3}$.

6. $y^2 = 4cx$.

* 7. $x^4 + 4x^3y + y^4 = 1$.

8. $(2y)^{1/2} + (3y)^{1/3} = x$.

* 9. $x + 2xy + y = 1$.

10. $x^2 + axy + y^2 = b^2$.

Find the slope at the indicated point.

* 11. $2x + 3y = 5$; $(-2, 3)$.

12. $9x^2 + 4y^2 = 72$; $(2, 3)$.

* 13. $x^2 + xy + 2y^2 = 28$; $(-2, -3)$.

14. $x^3 - axy + 3ay^2 = 3a^3$; (a, a).

* 15. At what angles do the parabolas

$$y^2 = 2px + p^2 \quad \text{and} \quad y^2 = p^2 - 2px$$

intersect?

16. At what angle does the line $y = 2x$ cut the curve

$$x^2 - xy + 2y^2 = 28?$$

17. Show that the hyperbola $x^2 - y^2 = 5$ and the ellipse $4x^2 + 9y^2 = 72$ intersect at right angles.

Find d^2y/dx^2 in terms of x and y.

* 18. $y^2 + 2xy = 16$.

19. $ax^2 + 2xy + by^2 = 1$.

Express d^2y/dx^2 in terms of y alone.

20. $x^2 + y^2 = r^2$.

* 21. $y^2 = 4ax$.

22. $\sqrt{x} + \sqrt{y} = 1$.

* 23. $b^2x^2 + a^2y^2 = a^2b^2$.

Evaluate dy/dx and d^2y/dx^2 at the given point.

* 24. $x^2 - 4y^2 = 9$; $(5, 2)$.

* 25. $x^2 + 4xy + y^2 + 3 = 0$; $(2, -1)$.

26. Find equations for the tangents to the ellipse $4x^2 + y^2 = 72$ that pass through the point $(4, 4)$.

3.20   Additional Exercises; Partial Differentiation

First we give some exercises (40 of them) for extra practice and review. Then we talk briefly about partial differentiation and give you some exercises on that.

* 1. The rate of change of f at x_0 is twice its rate of change at 1. Find x_0 if

(a) $f(x) = x^2$.   (b) $f(x) = 2x^3$.   (c) $f(x) = \sqrt{x}$.

2. An object moves on a line according to the law

$$x = \sqrt{t + 1}.$$

(a) Show that the acceleration is negative and proportional to the cube of the velocity.

(b) Use differentials to find numerical estimates for the position, velocity, and acceleration of the object at time $t = 17$. Base your estimate on $t = 15$.

* 3. A rectangular banner has a red border and a white center. The width of the border at top and bottom is 8 inches and along the sides is 6 inches. The total area is 27 square feet. What should be the dimensions of the banner if the area of the white center is to be a maximum?

* 4. Determine the coefficients A, B, C so that the curve $y = Ax^2 + Bx + C$ will pass through the point $(1, 3)$ and be tangent to the line $x - y + 1 = 0$ at the point $(2, 3)$.

5. Determine the coefficients A, B, C, D so that the curve $y = Ax^3 + Bx^2 + Cx + D$ will be tangent to the line $y = 5x - 4$ at the point $(1, 1)$ and tangent to the line $y = 9x$ at the point $(-1, -9)$.

* 6. Determine the coefficients A, B so that the function $y = Ax^{-1/2} + Bx^{1/2}$ has a minimum value of 6 at $x = 9$.

7. Find a relation between A, B, C given that the function $y = Ax^3 + 3Bx^2 + 3Cx$ has a critical point in common with its derivative.

Test the following functions for extreme values.

* 8. $f(x) = \dfrac{x^2 + x + 4}{x^2 + 2x + 4}$.

9. $f(x) = \dfrac{1}{x} + \dfrac{1}{1 - x}$.

*10. $f(x) = \dfrac{(x - 1)(2 - x)}{x^2}$.

11. $f(x) = \dfrac{x^2}{x^2 + a^2}$.

* 12. If three sides of a trapezoid are each 10 inches long, how long must the fourth side be if the area is a maximum?

13. A rectangular box with a square base and an open top is to be made. Find the volume of the largest box that can be made from 1200 square feet of material.

* 14. A point moves along the parabola $y^2 = 12x$ in such a way that its x-coordinate increases uniformly at the rate of 2 inches per second. At what point do the x-coordinate and the y-coordinate increase at the same rate?

15. A circular plate of metal expands by heat so that its radius increases at the rate of 0.02 inches per second. At what rate is the surface area increasing when the radius is 2 inches?

* 16. A light is hung 12 feet directly above a straight horizontal walk on which a boy 5 feet tall is walking. How fast is the boy's shadow lengthening if he is walking away from the light at the rate of 168 feet per minute?

* 17. One ship was sailing south at 6 miles per hour; another east at 8 miles per hour. At 4 p.m. the second crossed the track of the first, where the first had been 2 hours before.
(a) How was the distance between the ships changing at 3 p.m.?
(b) How at 5 p.m.?

* 18. A railroad track crosses a highway at an angle of 60°. A locomotive is 500 feet from the intersection and moving away from it at the rate of 60 miles per hour. An automobile is 500 feet from the intersection and moving toward it at the rate of 30 miles per hour. What is the rate of change of the distance between them?
HINT: Use the law of cosines.

19. Find the dimensions of the rectangular solid of maximum volume that can be cut from a solid sphere of radius r.

* 20. Find the base and altitude of the isosceles triangle of minimum area that circumscribes the ellipse $b^2x^2 + a^2y^2 = a^2b^2$ and whose base is parallel to the x-axis.

21. At what point of the first quadrant on the ellipse $b^2x^2 + a^2y^2 = a^2b^2$ does the tangent form with the coordinate axes a triangle of minimum area?

* 22. The diameter and altitude of a right circular cylinder are found at a certain instant to be 10 inches and 20 inches, respectively. If the diameter is increasing at the rate of 1 inch per minute, what change in the altitude will keep the volume constant?

* 23. The total cost of producing Q units per week is

$$C(Q) = \tfrac{1}{3}Q^3 - 20Q^2 + 600Q + 1000.$$

Given that the total revenues are

$$R(Q) = 420Q - 2Q^2,$$

find the output which maximizes profit.

24. A steel plant is capable of producing Q_1 tons per day of low-grade steel and Q_2 tons per day of high-grade steel, where

$$Q_2 = \frac{40 - 5Q_1}{10 - Q_1}.$$

If the market price of low-grade steel is half that of high-grade steel, show that about $5\tfrac{1}{2}$ tons of low-grade steel should be produced per day for maximum receipts.

25. The total cost of producing Q articles per week is $aQ^2 + bQ + c$ dollars and the price at which each can be sold is $p = \beta - \alpha Q^2$. Show that the output for maximum profit is

$$Q = \frac{\sqrt{a^2 + 3\alpha(\beta - b)} - a}{3\alpha}.$$

(Take a, b, c, α, β as positive.)

* 26. In a certain industry an output of Q units can be done at total cost of

$$C(Q) = aQ^2 + bQ + c \quad \text{dollars,}$$

bringing in total revenues of

$$R(Q) = \beta Q - \alpha Q^2 \quad \text{dollars.}$$

The government decides to impose an excise tax on the product. What tax rate (dollars per unit) will maximize the government's revenues from this tax? (Take a, b, c, α, β as positive.)

27. A tank contains 1000 cubic feet of natural gas at a pressure of 5 pounds per square inch. If the pressure is decreasing at the rate of 0.05 pounds per square inch per hour, find the rate of increase of the volume. (Assume Boyle's law: $PV = C$.)

* 28. The adiabatic law for the expansion of air is $PV^{1.4} = C$. If at a given time the volume is observed to be 10 cubic feet and the pressure is 50 pounds per square inch, at what rate is the pressure changing if the volume is decreasing 1 cubic foot per second?

* 29. Find the absolute maximum value of

$$y = \frac{x}{(r^2 + x^2)^{3/2}}.$$

30. Given that PQ is the longest or shortest line segment that can be drawn from $P(a, b)$ to the differentiable curve $y = f(x)$, show that PQ is perpendicular to the tangent to the curve at Q.

* 31. What point on the curve $y = x^{3/2}$ is closest to $P(\frac{1}{2}, 0)$?

* 32. The equation of the path of a ball is

$$y = mx - \frac{(m^2 + 1)x^2}{800},$$

where the origin is taken at the point from which the ball is thrown and m is the slope of the curve at the origin. For what value of m will the ball strike (a) at the greatest distance along the same horizontal level? (b) at the greatest height on a vertical wall 300 feet away?

* 33. A horizontal trough 12 feet long has a vertical cross section in the shape of a trapezoid, the bottom being 3 feet wide and the sides inclined to the vertical at an angle whose sine is $\frac{4}{5}$. Water is being poured into it at the rate of 10 cubic feet per minute. How fast is the water level rising when the water is 2 feet deep?

34. In Exercise 33, at what rate is the water being drawn from the trough if the level is falling 0.1 foot per minute when the water is 3 feet deep?

* 35. A point P moves along the parabola $y = x^2$ so that its x-coordinate increases at the constant rate of k units per second. The projection of P on the x-axis is M. At what rate is the area of triangle OMP changing when P is at the point where $x = a$?

* 36. If $y = 4x - x^3$ and x is increasing steadily at the rate of $\frac{1}{3}$ unit per second, find how fast the slope of the graph is changing at the instant when $x = 2$.

37. The sum of the surface areas of a sphere and a cube being given, show that the sum of the volumes will be least when the diameter of the sphere is equal to the edge of the cube. When will the sum of the volumes be greatest?

* 38. A miner wishes to dig a tunnel from a point A to a point B 200 feet below and 600 feet to the east of A. Below the level of A it is bedrock and above A is soft earth. If the cost of tunneling through earth is \$5 per linear foot and through rock \$13, find the minimum cost of a tunnel.

* 39. The distance between two sources of heat A and B, with intensities a and b, respectively, is l. The intensity of heat at a point P between A and B is given by the formula

$$I = \frac{a}{x^2} + \frac{b}{(l - x)^2},$$

where x is the distance between P and A. For what position P will the temperature be lowest?

40. Let $P(x_0, y_0)$ be a point in the first quadrant. Draw a line through P which cuts the positive x-axis at $A(a, 0)$ and the positive y-axis at $B(0, b)$. Find a and b in each of the following cases:
   (a) when the area of $\triangle OAB$ is a minimum.
   (b) when the length of AB is a minimum.
   (c) when $a + b$ is a minimum.
   (d) when the perpendicular distance from O to AB is a maximum.

Partial Differentiation

The idea of partial differentiation is a very simple one. Let's start with a function f of two variables, say

$$f(x, y) = x^2y^3 + 4y + x + 2.$$

The *partial derivative of f with respect to x* is obtained by differentiating f with respect to x, treating y as a constant; in this case

$$\frac{\partial f}{\partial x} = 2xy^3 + 1.$$

The *partial derivative of f with respect to y* is obtained by differentiating f with respect to y, treating x as a constant; in this case

$$\frac{\partial f}{\partial y} = 3x^2y^2 + 4.$$

In the case of a function of three variables x, y, z you can look for three partial derivatives: the partial with respect to x, the partial with respect to y, and also the partial with respect to z. These partials

$$\frac{\partial f}{\partial x}, \quad \frac{\partial f}{\partial y}, \quad \frac{\partial f}{\partial z}$$

are obtained by differentiating with respect to the bottom variable keeping the other two variables constant. Thus for

$$f(x, y, z) = xy^2z^3$$

we have

$$\frac{\partial f}{\partial x} = y^2z^3, \qquad \frac{\partial f}{\partial y} = 2xyz^3, \qquad \frac{\partial f}{\partial z} = 3xy^2z^2.$$

Each partial derivative gives a rate of change:

$\dfrac{\partial f}{\partial x}$ gives the rate of change with respect to x,

$\dfrac{\partial f}{\partial y}$ gives the rate of change with respect to y,

$\dfrac{\partial f}{\partial z}$ gives the rate of change with respect to z.

Find the partial derivatives.

*1. $f(x, y) = 3x^2 - xy + y.$

*2. $g(x, y) = x^2\sqrt{y}.$

*3. $h(x, y) = Ax^2 + Bxy + Cy^2 + Dx + Ey + F.$

*4. $f(x, y) = \sqrt{x^2 - 3y}.$

*5. $g(x, y) = \dfrac{Ax + By}{Cx + Dy}.$

*6. $f(x, y, z) = xy + yz + zx.$

*7. $g(x, y, z) = x^2 + y^2 + z^2.$

*8. $f(x, y, z) = \sqrt{x(1 - y)(1 - z^2)}.$

Integration

4.1 Motivation

We were led to the notion of derivative by considering a tangent problem. By considering two new problems, we will be led to the notion of integral.

An Area Problem

In Figure 4.1.1 we display a region Ω which is bounded above by the graph of a continuous function f, bounded below by the x-axis, bounded to the left by $x = a$, and bounded to the right by $x = b$.

What number, if any, should be called the area of Ω? This is the question we want to resolve.

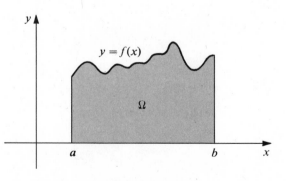

FIGURE 4.1.1

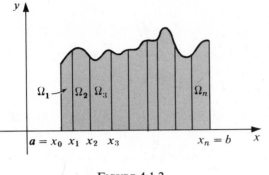

FIGURE 4.1.2

To do this, we begin by splitting up the interval $[a, b]$ into a finite number of non-overlapping intervals.

$$[x_0, x_1], [x_1, x_2], \ldots, [x_{n-1}, x_n] \quad \text{with } a = x_0 < x_1 < \cdots < x_n = b.$$

This breaks up the region Ω into n subregions:

$$\Omega_1, \Omega_2, \ldots, \Omega_n. \qquad \text{(Figure 4.1.2)}$$

We can estimate the total area of Ω by estimating the area of each subregion Ω_j and adding up the results. Let's denote by M_j the maximum value of f on $[x_{j-1}, x_j]$ and by m_j the minimum value. Consider now the rectangles r_j and R_j of Figure 4.1.3.

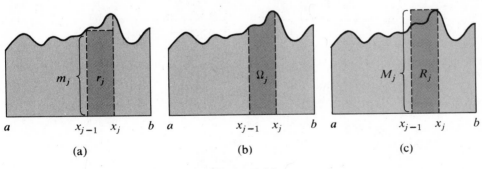

FIGURE 4.1.3

Since

$$r_j \subseteq \Omega_j \subseteq R_j,$$

we must have

$$\text{area of } r_j \leq \text{area of } \Omega_j \leq \text{area of } R_j.$$

Taking the area of rectangles to be given by length times width, we obtain

$$m_j(x_j - x_{j-1}) \leq \text{area of } \Omega_j \leq M_j(x_j - x_{j-1}).$$

Setting
$$x_j - x_{j-1} = \Delta x_j$$
we have
$$m_j \, \Delta x_j \leq \text{area of } \Omega_j \leq M_j \, \Delta x_j.$$
This is presumably true for $j = 1, j = 2, \ldots$ and so on. Adding up these results, we get on the one hand

(4.1.1) $m_1 \, \Delta x_1 + m_2 \, \Delta x_2 + \cdots + m_n \, \Delta x_n \leq \text{area of } \Omega$

and on the other hand

(4.1.2) $\text{area of } \Omega \leq M_1 \, \Delta x_1 + M_2 \, \Delta x_2 + \cdots + M_n \, \Delta x_n.$

A sum of the form
$$m_1 \, \Delta x_1 + m_2 \, \Delta x_2 + \cdots + m_n \, \Delta x_n \qquad \text{(Figure 4.1.4)}$$
is called a *lower sum* for f. A sum of the form
$$M_1 \, \Delta x_1 + M_2 \, \Delta x_2 + \cdots + M_n \, \Delta x_n \qquad \text{(Figure 4.1.5)}$$
is called an *upper sum* for f.

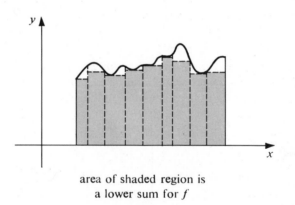

area of shaded region is
a lower sum for f

FIGURE 4.1.4

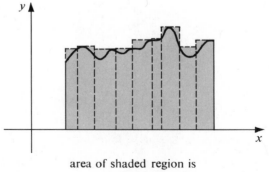

area of shaded region is
an upper sum for f

FIGURE 4.1.5

Inequalities (4.1.1) and (4.1.2) together tell us that for a number to be a candidate for the title of area of Ω it must be greater than or equal to every lower sum for f and less than or equal to every upper sum for f. By an argument that we omit, it can be proven that there is one and only one such number. This number we call the area of Ω.   □

Later we'll return to the subject of area. At this point we turn to a speed-distance problem. As you'll see, this new problem can be solved by the same technique that we just applied to the area problem.

A Speed-Distance Problem

If an object moves at a constant speed for a given amount of time then the total distance traveled is given by the familiar formula

$$\text{distance} = (\text{speed}) \times (\text{time}).$$

Suppose now that during the course of the motion the speed does not remain constant but instead varies continuously. How can the total distance traveled be computed then?

To answer this question, we suppose that the motion begins at time a, ends at time b, and during the time interval $[a, b]$ the speed varies continuously.

As in the case of the area problem we begin by breaking up the interval $[a, b]$ into a finite number of nonoverlapping intervals

$$[t_0, t_1], [t_1, t_2], \ldots, [t_{n-1}, t_n].$$

On each subinterval $[t_{j-1}, t_j]$ the object attains a certain maximum speed M_j and a certain minimum speed m_j. (How do we know this?) If throughout the time interval $[t_{j-1}, t_j]$ the object were to move constantly at its minimum speed, m_j, then it would cover a distance of $m_j \, \Delta t_j$ units. If instead it were to move constantly at its maximum speed, M_j, then it would cover a distance of $M_j \, \Delta t_j$ units. As it is, the actual distance traveled, call it s_j, must lie somewhere in between; namely, we must have

$$m_j \, \Delta t_j \le s_j \le M_j \, \Delta t_j.$$

The total distance traveled during the time interval $[a, b]$, call it s, must be the sum of the distances traveled during the subintervals $[t_{j-1}, t_j]$. In other words we must have

$$s = s_1 + s_2 + \cdots + s_n.$$

Since

$$m_1 \, \Delta t_1 \le s_1 \le M_1 \, \Delta t_1$$
$$m_2 \, \Delta t_2 \le s_2 \le M_2 \, \Delta t_2$$
$$\vdots$$
$$m_n \, \Delta t_n \le s_n \le M_n \, \Delta t_n,$$

it follows by the addition of these inequalities that

$$m_1 \, \Delta t_1 + m_2 \, \Delta t_2 + \cdots + m_n \, \Delta t_n \le s \le M_1 \, \Delta t_1 + M_2 \, \Delta t_2 + \cdots + M_n \, \Delta t_n.$$

A sum of the form

$$m_1 \, \Delta t_1 + m_2 \, \Delta t_2 + \cdots + m_n \, \Delta t_n$$

is called a *lower sum* for the speed function. A sum of the form

$$M_1 \, \Delta t_1 + M_2 \, \Delta t_2 + \cdots + M_n \, \Delta t_n$$

is called an *upper sum* for the speed function. The inequality we just obtained for s tells us that s must be greater than or equal to every lower sum for the speed function and less than or equal to every upper sum for the speed function. As in the case of the area problem, it turns out that there is one and only one such number and it is the total distance traveled. □

4.2   Definition of the Definite Integral

The procedure applied to the two problems of the last section is called integration and the numbers so obtained are called definite integrals. Our purpose here is to establish these notions more precisely. At this stage we restrict ourselves to continuous functions.

> By a *partition* of $[a, b]$ we mean a finite subset of $[a, b]$ which contains the points a and b.

It is convenient to index the elements of a partition according to their natural order. Thus if we write

$$P = \{x_0, x_1, \ldots, x_n\} \text{ is a partition of } [a, b],$$

you can conclude that

$$a = x_0 < x_1 < \cdots < x_n = b.$$

Example.   The sets

$$\{0, 1\}, \{0, \tfrac{1}{2}, 1\}, \{0, \tfrac{1}{4}, \tfrac{1}{2}, 1\}, \quad \text{and} \quad \{0, \tfrac{1}{4}, \tfrac{1}{3}, \tfrac{1}{2}, \tfrac{5}{8}, 1\}$$

are all partitions of the interval $[0, 1]$. □

Suppose now that f is continuous on $[a, b]$. If $P = \{x_0, x_1, \ldots, x_n\}$ is a partition of $[a, b]$, then P breaks up $[a, b]$ into a finite number of nonoverlapping intervals

$$[x_0, x_1], [x_1, x_2], \ldots, [x_{n-1}, x_n] \quad \text{of lengths } \Delta x_1, \Delta x_2, \ldots, \Delta x_n \text{ respectively.}$$

On each such interval $[x_{j-1}, x_j]$ f takes on a maximum value, M_j, and a minimum value, m_j.

> The number
> $$U_f(P) = M_1 \, \Delta x_1 + M_2 \, \Delta x_2 + \cdots + M_n \, \Delta x_n$$
> is called the *P upper sum* for f and the number
> $$L_f(P) = m_1 \, \Delta x_1 + m_2 \, \Delta x_2 + \cdots + m_n \, \Delta x_n$$
> is called the *P lower sum* for f.

Example. If $[a, b] = [0, 1]$, $P = \{0, \frac{1}{4}, \frac{1}{2}, 1\}$, and $f(x) = x^2$, then

$$U_f(P) = \tfrac{1}{16}(\tfrac{1}{4}) + \tfrac{1}{4}(\tfrac{1}{4}) + 1(\tfrac{1}{2}) = \tfrac{37}{64} \quad \text{and} \quad L_f(P) = 0(\tfrac{1}{4}) + \tfrac{1}{16}(\tfrac{1}{4}) + \tfrac{1}{4}(\tfrac{1}{2}) = \tfrac{9}{64}. \quad \square$$

Example. If $[a, b] = [-1, 0]$, $P = \{-1, -\frac{1}{4}, 0\}$ and $f(x) = -(x + 1)$, then

$$U_f(P) = (0)(\tfrac{3}{4}) + (-\tfrac{3}{4})(\tfrac{1}{4}) = -\tfrac{3}{16} \quad \text{and} \quad L_f(P) = (-\tfrac{3}{4})(\tfrac{3}{4}) + (-1)(\tfrac{1}{4}) = -\tfrac{13}{16}. \quad \square$$

By an argument that we omit here (it appears in the Appendix at the end of the book) it can be proved that, with f continuous on $[a, b]$, there is one and only one number I which satisfies the inequality

$$L_f(P) \leq I \leq U_f(P) \quad \text{for } \textit{all} \text{ partitions } P \text{ of } [a, b].$$

This is the number we want.

Definition of the Definite Integral

The unique number I which satisfies the inequality

$$L_f(P) \leq I \leq U_f(P) \quad \text{for all partitions } P \text{ of } [a, b]$$

is called the *definite integral* (or more simply the *integral*) of f on $[a, b]$ and is denoted by

$$\int_a^b f(x)\, dx.$$

The symbol $\int$ dates back to Leibniz and is called an *integral sign*. It is really an elongated S—as in *Sum*. The numbers a and b are called the *limits of integration* and we often speak of *integrating f from a to b.*†

In the expression

$$\int_a^b f(x)\, dx$$

the letter x is a "dummy variable"; in other words, it may be replaced by any other letter not already engaged. Thus, for example, there is no difference between

$$\int_a^b f(x)\, dx, \quad \int_a^b f(t)\, dt, \quad \text{and} \quad \int_a^b f(z)\, dz.$$

All of these mean the definite integral of f on $[a, b]$.

Section 4.1 gives two immediate applications of the definite integral:

   I. If f is nonnegative on $[a, b]$, then

$$A = \int_a^b f(x)\, dx$$

gives the area below the graph of f.

---

† The word "limit" in this setting has no connection with the functional limits of Chapter 2.

II. If $|v(t)|$ is the speed of an object at time t, then

$$s = \int_a^b |v(t)|\, dt$$

gives the distance traveled from time a to time b.

We'll come back to these ideas later. Right now, some simple computations.

Example. If $f(x) = \alpha$ for all x in $[a, b]$, then

$$\int_a^b f(x)\, dx = \alpha(b - a).$$

To see this, we take $P = \{x_0, x_1, \ldots, x_n\}$ as an arbitrary partition of $[a, b]$. Since on each subinterval $[x_{j-1}, x_j]$ f has the constant value α, we see that M_j and m_j are both α. Thus

$$U_f(P) = \alpha\, \Delta x_1 + \alpha\, \Delta x_2 + \cdots + \alpha\, \Delta x_n$$

$$= \alpha(\Delta x_1 + \Delta x_2 + \cdots + \Delta x_n) = \alpha(b - a),$$

and, similarly,

$$L_f(P) = \alpha(b - a).$$

Since

$$L_f(P) \le \alpha(b - a) \le U_f(P)$$

for all partitions P of $[a, b]$, we do find that

$$\int_a^b f(x)\, dx = \alpha(b - a). \quad \square$$

The result of this last example can be written more simply as

$$\int_a^b \alpha\, dx = \alpha(b - a).$$

If $\alpha > 0$ then the region below the graph is simply a rectangle of height α erected on the interval $[a, b]$. (Figure 4.2.1) The integral gives the area of the rectangle.

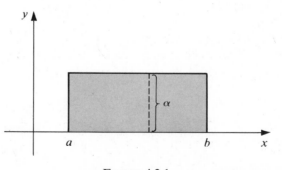

FIGURE 4.2.1

Example

$$\int_a^b x \, dx = \tfrac{1}{2}(b^2 - a^2).$$

To see this, we take $P = \{x_0, x_1, \ldots, x_n\}$ as an arbitrary partition of $[a, b]$. On each subinterval $[x_{j-1}, x_j]$ the function

$$f(x) = x$$

has a maximum

$$M_j = x_j$$

and a minimum

$$m_j = x_{j-1}.$$

It follows that

$$U_f(P) = x_1 \, \Delta x_1 + x_2 \, \Delta x_2 + \cdots + x_n \, \Delta x_n$$
$$= x_1(x_1 - x_0) + x_2(x_2 - x_1) + \cdots + x_n(x_n - x_{n-1})$$

and

$$L_f(P) = x_0 \, \Delta x_1 + x_1 \, \Delta x_2 + \cdots + x_{n-1} \, \Delta x_n$$
$$= x_0(x_1 - x_0) + x_1(x_2 - x_1) + \cdots + x_{n-1}(x_n - x_{n-1}).$$

For each index j,

$$x_{j-1} \leq \tfrac{1}{2}(x_j + x_{j-1}) \leq x_j,$$

and therefore

$$L_f(P) \leq \tfrac{1}{2}(x_1 + x_0)(x_1 - x_0) + \tfrac{1}{2}(x_2 + x_1)(x_2 - x_1) + \cdots$$
$$+ \tfrac{1}{2}(x_n + x_{n-1})(x_n - x_{n-1}) \leq U_f(P).$$

Since the middle expression can be rewritten as

$$\tfrac{1}{2}(x_1^2 - x_0^2 + x_2^2 - x_1^2 + \cdots + x_n^2 - x_{n-1}^2) = \tfrac{1}{2}(x_n^2 - x_0^2) = \tfrac{1}{2}(b^2 - a^2),$$

we do have

$$L_f(P) \leq \tfrac{1}{2}(b^2 - a^2) \leq U_f(P).$$

Since P was chosen arbitrarily, we can conclude that this inequality holds for all partitions P of $[a, b]$. It follows therefore that

$$\int_a^b x \, dx = \tfrac{1}{2}(b^2 - a^2). \quad \square$$

If the interval $[a, b]$ lies to the right of the origin then the region below the graph of

$$f(x) = x, \qquad x \in [a, b]$$

is the trapezoid of Figure 4.2.2. The integral

$$\int_a^b x \, dx$$

gives the area of that trapezoid.

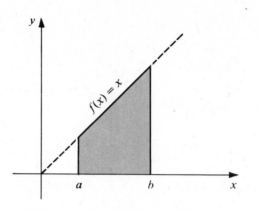

$$f(x) = x$$

FIGURE 4.2.2

Example

$$\int_0^1 x^2 \, dx = \tfrac{1}{3}.$$

To see this, we take $P = \{x_0, x_1, \ldots, x_n\}$ as a partition of $[0, 1]$. On each subinterval $[x_{j-1}, x_j]$ the function

$$f(x) = x^2$$

has a maximum

$$M_j = x_j^2$$

and a minimum

$$m_j = x_{j-1}^2.$$

It follows that

$$U_f(P) = x_1^2 \, \Delta x_1 + \cdots + x_n^2 \, \Delta x_n$$

and

$$L_f(P) = x_0^2 \, \Delta x_1 + \cdots + x_{n-1}^2 \, \Delta x_n.$$

For each index j,

$$x_{j-1}^2 \le \tfrac{1}{3}(x_{j-1}^2 + x_{j-1}x_j + x_j^2) \le x_j^2.$$

If we now multiply this inequality by

$$x_j - x_{j-1},$$

then the middle term reduces to

$$\tfrac{1}{3}(x_j^3 - x_{j-1}^3) \qquad\qquad \text{(check this out)}$$

and consequently we get

$$x_{j-1}^2(x_j - x_{j-1}) \le \tfrac{1}{3}(x_j^3 - x_{j-1}^3) \le x_j^2(x_j - x_{j-1}).$$

Adding up the terms on the left, we get $L_f(P)$. Adding up the middle terms, we get a collapsing sum which reduces to $\tfrac{1}{3}$:

$$\tfrac{1}{3}(x_1^3 - x_0^3 + x_2^3 - x_1^3 + \cdots + x_n^3 - x_{n-1}^3) = \tfrac{1}{3}(x_n^3 - x_0^3) = \tfrac{1}{3}(1^3 - 0^3) = \tfrac{1}{3}.$$

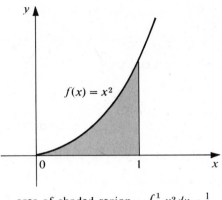

$$f(x) = x^2$$

area of shaded region $= \int_0^1 x^2 dx = \frac{1}{3}$

FIGURE 4.2.3

Adding up the terms on the right, we get $U_f(P)$. It follows therefore that

$$L_f(P) \le \tfrac{1}{3} \le U_f(P).$$

Since P was chosen arbitrarily, we can conclude that this inequality holds for all P. It follows therefore that

$$\int_0^1 x^2 \, dx = \tfrac{1}{3}. \quad \square$$

Exercises

Find $L_f(P)$ and $U_f(P)$ for each of the following:

* 1. $f(x) = 2x$, $x \in [0, 1]$; $P = \{0, \frac{1}{4}, \frac{1}{2}, 1\}$.

2. $f(x) = 1 - x$, $x \in [0, 2]$; $P = \{0, \frac{1}{3}, \frac{3}{4}, 1, 2\}$.

* 3. $f(x) = x^2$, $x \in [-1, 0]$; $P = \{-1, -\frac{1}{2}, -\frac{1}{4}, 0\}$.

4. $f(x) = 1 - x^2$, $x \in [0, 1]$; $P = \{0, \frac{1}{4}, \frac{1}{2}, 1\}$.

* 5. $f(x) = 1 + x^3$, $x \in [0, 1]$; $P = \{0, \frac{1}{2}, 1\}$.

6. Explain why each of the following statements must be false. Take P as a partition of $[-1, 1]$.

(a) $L_f(P) = 3$ and $U_f(P) = 2$.

(b) $L_f(P) = 3$, $U_f(P) = 6$, and $\int_{-1}^1 f(x) \, dx = 2$.

(c) $L_f(P) = 3$, $U_f(P) = 6$, and $\int_{-1}^1 f(x) \, dx = 10$.

7. (a) Given that $P = \{x_0, x_1, \ldots, x_n\}$ is an arbitrary partition of $[a, b]$, find $L_f(P)$ and $U_f(P)$ if

$$f(x) = 1 + 2x.$$

(b) Use your answers to part (a) to evaluate

$$\int_a^b (1 + 2x)\, dx.$$

*8. (a) Given that $P = \{x_0, x_1, \ldots, x_n\}$ is an arbitrary partition of $[a, b]$, find $L_f(P)$ and $U_f(P)$ if

$$f(x) = -3x.$$

(b) Use your answers to part (a) to evaluate

$$\int_a^b (-3x)\, dx.$$

9. (*Optional*) Evaluate

$$\int_0^1 x^3\, dx$$

by the methods of this section.

4.3 The function $F(x) = \int_a^x f(t)\, dt$

The evaluation of

$$\int_a^b f(x)\, dx$$

by its definition as the unique number I satisfying

$$L_f(P) \leq I \leq U_f(P) \quad \text{for all partitions } P \text{ of } [a, b]$$

is at best a laborious process. Try for example to evaluate

$$\int_2^5 \left(x^3 + x^{5/2} - \frac{2x}{1 - x^2} \right) dx \quad \text{or} \quad \int_{-1/2}^{1/4} \frac{x}{1 - x^2}\, dx$$

in this manner. Fortunately there is another way that we can evaluate such integrals. This other way does not work for all functions but it works for many of the elementary functions. This other way of evaluating definite integrals is described in Section 4.4 in what is called the *fundamental theorem of the integral calculus*. Its success depends on a connection between differentiation and integration which is described in Theorem 4.3.2. Before we can prove Theorem 4.3.2 we need some other results.

Theorem

Let P and Q be partitions of the interval $[a, b]$. If $P \subseteq Q$, then

$$L_f(P) \leq L_f(Q) \quad \text{and} \quad U_f(Q) \leq U_f(P).$$

This result is obvious. By adding points to a partition we tend to make the subintervals $[x_{j-1}, x_j]$ smaller. This tends to make the minima, m_j, larger and the maxima, M_j, smaller. Thus the lower sums are made bigger and the upper sums are made smaller. This idea is illustrated in Figures 4.3.1 and 4.3.2.

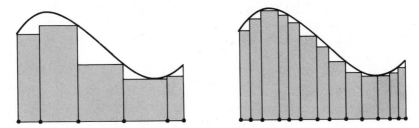

as points are added to a partition, the lower sums tend to get bigger

FIGURE 4.3.1

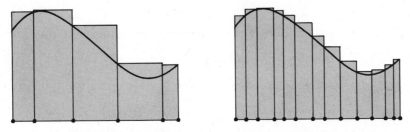

as points are added to a partition, the upper sums tend to get smaller

FIGURE 4.3.2

The next theorem says that the integral is additive on intervals.

Theorem 4.3.1

$$\text{If } a < c < b, \quad \text{then} \quad \int_a^c f(t)\,dt + \int_c^b f(t)\,dt = \int_a^b f(t)\,dt.$$

For nonnegative functions this theorem is easily understood in terms of area. (See Figure 4.3.3.) The area of part I is given by

$$\int_a^c f(t)\,dt.$$

The area of part II is given by

$$\int_c^b f(t)\,dt.$$

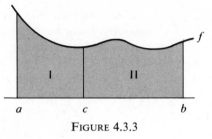

FIGURE 4.3.3

The area of the entire region is given by

$$\int_a^b f(t)\, dt.$$

The theorem says that

> the area of part I + the area of part II = the area of the entire region.

Theorem 4.3.1 can also be viewed in terms of speed and distance. With $f(t) \geq 0$, we can interpret $f(t)$ as the speed of an object at time t. With this interpretation,

$$\int_a^c f(t)\, dt = \text{distance traveled from time } a \text{ to time } c.$$

$$\int_c^b f(t)\, dt = \text{distance traveled from time } c \text{ to time } b.$$

$$\int_a^b f(t)\, dt = \text{distance traveled from time } a \text{ to time } b.$$

It is not surprising that the sum of the first two distances should equal the third.

The fact that the additivity theorem is so easy to understand does not relieve us of the necessity to prove it. Here is a proof.

PROOF OF THEOREM 4.3.1. To prove the theorem we need only show that for each partition P of $[a, b]$

$$L_f(P) \leq \int_a^c f(t)\, dt + \int_c^b f(t)\, dt \leq U_f(P). \qquad \text{(why?)}$$

We begin with

$$P = \{x_0, x_1, \ldots, x_n\}$$

as an arbitrary partition of $[a, b]$. Since the partition $Q = P \cup \{c\}$ contains P, we know from the first theorem that

(1) $$L_f(P) \leq L_f(Q) \quad \text{and} \quad U_f(Q) \leq U_f(P).$$

The sets

$$Q_1 = Q \cap [a, c] \quad \text{and} \quad Q_2 = Q \cap [c, b]$$

are partitions of $[a, c]$ and $[c, b]$ respectively and clearly satisfy

$$L_f(Q_1) + L_f(Q_2) = L_f(Q) \quad \text{and} \quad U_f(Q_1) + U_f(Q_2) = U_f(Q).$$

Since

$$L_f(Q_1) \le \int_a^c f(t)\, dt \le U_f(Q_1) \quad \text{and} \quad L_f(Q_2) \le \int_c^b f(t)\, dt \le U_f(Q_2),$$

we have

$$L_f(Q_1) + L_f(Q_2) \le \int_a^c f(t)\, dt + \int_c^b f(t)\, dt \le U_f(Q_1) + U_f(Q_2).$$

It follows that

$$L_f(Q) \le \int_a^c f(t)\, dt + \int_c^b f(t)\, dt \le U_f(Q)$$

and thus by (1) that

$$L_f(P) \le \int_a^c f(t)\, dt + \int_c^b f(t)\, dt \le U_f(P). \quad \square$$

Until now we have considered the definite integral only over an integral $[a, b]$; namely, in writing

$$\int_a^b f(t)\, dt$$

we have assumed that a was less than b. We can also integrate in the other direction; by definition

$$\int_b^a f(t)\, dt = -\int_a^b f(t)\, dt.$$

Moreover the integral from any point c to itself is defined to be zero:

$$\int_c^c f(t)\, dt = 0.$$

With these two extra conventions, it is not hard to show that, if f is continuous on an interval I, then the additivity condition

$$\int_a^c f(t)\, dt + \int_c^b f(t)\, dt = \int_a^b f(t)\, dt$$

holds for all choices of a, b, c from I, no matter what the order is.

We are now ready to establish the basic link that exists between differentiation and integration. To describe this link, we begin with a function f that is continuous on a closed interval $[a, b]$. For each x in $[a, b]$ the integral

$$\int_a^x f(t)\, dt$$

is a number and consequently we can define a function F on $[a, b]$ by setting

$$F(x) = \int_a^x f(t)\, dt.$$

Theorem 4.3.2

If f is continuous on $[a, b]$, the function F defined on $[a, b]$ by setting

$$F(x) = \int_a^x f(t)\, dt$$

is continuous on $[a, b]$, differentiable on (a, b), and satisfies

$$F'(x) = f(x) \quad \text{for all } x \text{ in } (a, b).$$

PROOF.   We begin with x in the half-open interval $[a, b)$ and show that

$$\lim_{h \downarrow 0} \frac{F(x + h) - F(x)}{h} = f(x).$$

(For a pictorial outline of the proof in the case that f is nonnegative see Figure 4.3.4.)

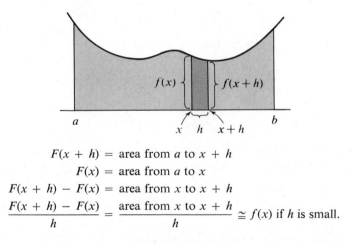

$$F(x + h) = \text{area from } a \text{ to } x + h$$
$$F(x) = \text{area from } a \text{ to } x$$
$$F(x + h) - F(x) = \text{area from } x \text{ to } x + h$$
$$\frac{F(x + h) - F(x)}{h} = \frac{\text{area from } x \text{ to } x + h}{h} \cong f(x) \text{ if } h \text{ is small.}$$

FIGURE 4.3.4

If $x < x + h \leq b$, then

$$F(x + h) - F(x) = \int_a^{x+h} f(t)\, dt - \int_a^x f(t)\, dt.$$

It follows that

$$F(x + h) - F(x) = \int_x^{x+h} f(t)\, dt. \qquad \text{(justify this step)}$$

Now we set

$$M_h = \text{maximum value of } f \text{ on } [x, x + h]$$

and

$$m_h = \text{minimum value of } f \text{ on } [x, x + h].$$

Since

$$M_h[(x + h) - x] = M_h \cdot h$$

is an upper sum for f on $[x, x + h]$ and

$$m_h[(x + h) - x] = m_h \cdot h$$

is a lower sum for f on $[x, x + h]$, we know that

$$m_h \cdot h \leq \int_x^{x+h} f(t)\, dt \leq M_h \cdot h$$

and thus that

$$m_h \leq \frac{F(x + h) - F(x)}{h} \leq M_h.$$

Since f is continuous on $[x, x + h]$, it follows that

(1) $$\lim_{h \downarrow 0} m_h = f(x) = \lim_{h \downarrow 0} M_h$$

and thus that

(2) $$\lim_{h \downarrow 0} \frac{F(x + h) - F(x)}{h} = f(x).$$

In a similar manner, you can verify that if x is in the half-open interval $(a, b]$, then

(3) $$\lim_{h \uparrow 0} \frac{F(x + h) - F(x)}{h} = f(x).$$

If x is in the open interval (a, b), then both (2) and (3) hold, and therefore

$$F'(x) = \lim_{h \to 0} \frac{F(x + h) - F(x)}{h} = f(x).$$

This proves the differentiability condition. Applying (2) to $x = a$, we have

$$\lim_{h \downarrow 0} \frac{F(a + h) - F(a)}{h} = f(a).$$

This implies that

$$\lim_{h \downarrow 0} F(a + h) - F(a) = 0$$

and thus that

$$\lim_{h \downarrow 0} F(a + h) = F(a).$$

This shows that F is continuous from the right at a. That F is continuous from the left at b can be shown by applying (3) to $x = b$. □

Exercises

*1. Given that

$$\int_0^1 f(x)\, dx = 6, \qquad \int_0^2 f(x)\, dx = 4, \qquad \int_2^5 f(x)\, dx = 1,$$

find each of the following:

(a) $\displaystyle\int_0^5 f(x)\, dx.$ (b) $\displaystyle\int_1^2 f(x)\, dx.$ (c) $\displaystyle\int_1^5 f(x)\, dx.$

(d) $\displaystyle\int_0^0 f(x)\, dx.$ (e) $\displaystyle\int_2^0 f(x)\, dx.$ (f) $\displaystyle\int_5^1 f(x)\, dx.$

2. Explain why each of the following statements must be false.
   *(a) $U_f(P_1) = 4$ for the partition $P_1 = \{0, 1, \frac{3}{2}, 2\}$ and
       $U_f(P_2) = 5$ for the partition $P_2 = \{0, \frac{1}{4}, 1, \frac{3}{2}, 2\}$.
   (b) $L_f(P_1) = 5$ for the partition $P_1 = \{0, 1, \frac{3}{2}, 2\}$ and
       $L_f(P_2) = 4$ for the partition $P_2 = \{0, \frac{1}{4}, 1, \frac{3}{2}, 2\}$.

3. For $x > -1$ set $F(x) = \displaystyle\int_0^x t\sqrt{t + 1}\, dt.$

   (a) Find $F(0)$.   *(b) Find $F'(x)$.   (c) Find $F'(2)$.
   *(d) Express $F(2)$ as an integral of $t\sqrt{t + 1}$.
   (e) Express $-F(x)$ as an integral of $t\sqrt{t + 1}$.
4. Use upper and lower sums to show that

$$0.5 < \int_1^2 \frac{dx}{x} < 1.$$

5. (*Optional*) Show that if f is continuous on the interval I then

$$\int_a^c f(t)\, dt + \int_c^b f(t)\, dt = \int_a^b f(t)\, dt$$

   for *every* choice of a, b, c in I.
6. (*Optional*) Show the validity of step (1) in the proof of Theorem 4.3.2.
7. (*Optional*) Complete the proof of Theorem 4.3.2 by showing that

$$\lim_{h \uparrow 0} \frac{F(x + h) - F(x)}{h} = f(x).$$

8. Extend Theorem 4.3.2 by showing that if f is continuous in $[a, b]$ and c is *any* point in $[a, b]$, then

$$F(x) = \int_c^x f(t)\, dt$$

is continuous on $[a, b]$, differentiable on (a, b), and satisfies

$$F'(x) = f(x) \quad \text{for all } x \text{ in } (a, b).$$

HINT: $\int_c^x f(t)\, dt = \int_c^a f(t)\, dt + \int_a^x f(t)\, dt.$

9. Set

$$F(x) = \int_0^x \frac{dt}{t^2 + 9}$$

and find the following:

*(a) $F'(-1)$.   (b) $F'(0)$.   *(c) $F'(\tfrac{1}{2})$.

4.4   The Fundamental Theorem of Integral Calculus

Definition of Antiderivative

A function G is called an *antiderivative* for f on $[a, b]$ iff

(i) G is continuous on $[a, b]$ and
(ii) $G'(x) = f(x)$ for all $x \in (a, b)$.

Theorem 4.3.2 says that, if f is continuous on $[a, b]$, then

$$F(x) = \int_a^x f(t)\, dt$$

is an antiderivative for f on $[a, b]$. This gives us a prescription for constructing anti-derivatives. It tells us that we can construct an antiderivative for f by integrating f.

The so-called "fundamental theorem" goes the other way. It gives us a prescription, not for finding antiderivatives, but for evaluating integrals. It tells us that we can evaluate

$$\int_a^b f(t)\, dt$$

by finding an antiderivative for f.

Theorem 4.4.1　The Fundamental Theorem of Integral Calculus

Let f be continuous on $[a, b]$. If G is an antiderivative of f on $[a, b]$, then

$$\int_a^b f(t)\, dt = G(b) - G(a).$$

PROOF.　From Theorem 4.3.2 we know that the function

$$F(x) = \int_a^x f(t)\, dt$$

is an antiderivative for f on $[a, b]$. If G is also an antiderivative for f on $[a, b]$, then we have both F and G continuous on $[a, b]$ and satisfying $F'(x) = G'(x)$ for all x in (a, b). From Theorem 3.11.4 we can conclude that there exists a constant C such that

$$F(x) = G(x) + C \quad \text{for all } x \text{ in } [a, b].$$

Since $F(a) = 0$, we must have

$$G(a) + C = 0,$$

and thus

$$C = -G(a).$$

This means that

$$F(x) = G(x) - G(a) \quad \text{for all } x \text{ in } [a, b].$$

In particular,

$$F(b) = G(b) - G(a).$$

Since

$$F(b) = \int_a^b f(t)\, dt,$$

the theorem is proved.　□

Applying the Fundamental Theorem

Example.　To evaluate

$$\int_a^b x\, dx,$$

we can use

$$G(x) = \tfrac{1}{2}x^2$$

as an antiderivative. Doing so, we obtain

$$\int_a^b x\, dx = \tfrac{1}{2}(b^2 - a^2). \quad \square$$

Example.   More generally, to evaluate

$$\int_a^b x^n \, dx \quad \text{where } n \text{ is a positive integer}$$

we can use the antiderivative

$$G(x) = \frac{1}{n+1} x^{n+1}.$$

This gives

$$\int_a^b x^n \, dx = \frac{1}{n+1} (b^{n+1} - a^{n+1}). \quad \Box$$

The formula

$$\int_a^b f(x) \, dx = G(b) - G(a),$$

is so useful that we frequently write expressions of the form

$$G(b) - G(a).$$

This difference is often denoted by

$$\left[G(x) \right]_a^b.$$

In this notation

$$\int_a^b x^3 \, dx = \left[\tfrac{1}{4}x^4 \right]_a^b = \tfrac{1}{4}(b^4 - a^4).$$

Example.   For the integral

$$\int_0^1 (2x - 6x^4 + 5) \, dx$$

we choose

$$G(x) = x^2 - \tfrac{6}{5}x^5 + 5x$$

as an antiderivative and obtain

$$\int_0^1 (2x - 6x^4 + 5) \, dx = \left[x^2 - \tfrac{6}{5}x^5 + 5x \right]_0^1 = 4\tfrac{4}{5}. \quad \Box$$

Example.   To evaluate

$$\int_{-1}^1 (x - 1)(x + 2) \, dx$$

we first carry out the indicated multiplication

$$(x - 1)(x + 2) = x^2 + x - 2.$$

It's clear that

$$G(x) = \tfrac{1}{3}x^3 + \tfrac{1}{2}x^2 - 2x$$

serves as an antiderivative. Thus

$$\int_{-1}^{1} (x - 1)(x + 2)\, dx = \left[\tfrac{1}{3}x^3 + \tfrac{1}{2}x^2 - 2x \right]_{-1}^{1} = -\tfrac{10}{3}. \quad \square$$

Example. To evaluate

$$\int_{1}^{2} \frac{dt}{t^2}$$

we need a function with derivative

$$\frac{1}{t^2}.$$

Since

$$\frac{d}{dt}\left(\frac{1}{t}\right) = -\frac{1}{t^2},$$

it's obvious that

$$\frac{d}{dt}\left(-\frac{1}{t}\right) = \frac{1}{t^2}.$$

It follows that

$$\int_{1}^{2} \frac{dt}{t^2} = \left[-\frac{1}{t} \right]_{1}^{2} = \left(-\frac{1}{2}\right) - (-1) = \frac{1}{2}. \quad \square$$

Example. To evaluate

$$\int_{4}^{9} \frac{dt}{\sqrt{t}}$$

we need a function with derivative

$$\frac{1}{\sqrt{t}}.$$

We know that

$$\frac{d}{dt}\left(\sqrt{t}\right) = \frac{1}{2\sqrt{t}}.$$

It follows that

$$\frac{d}{dt}\left(2\sqrt{t}\right) = \frac{1}{\sqrt{t}}$$

and that

$$\int_{4}^{9} \frac{dt}{\sqrt{t}} = \left[2\sqrt{t} \right]_{4}^{9} = 2\sqrt{9} - 2\sqrt{4} = 2. \quad \square$$

Example. To evaluate

$$\int_1^2 \sqrt{x-1}\, dx$$

we need a function with derivative

$$\sqrt{x-1}.$$

It's not hard to see that

$$\frac{d}{dx}[(x-1)^{3/2}] = \tfrac{3}{2}(x-1)^{1/2} = \tfrac{3}{2}\sqrt{x-1}$$

and therefore

$$\frac{d}{dx}[\tfrac{2}{3}(x-1)^{3/2}] = \sqrt{x-1}.$$

This means that

$$\int_1^2 \sqrt{x-1}\, dx = \left[\tfrac{2}{3}(x-1)^{3/2}\right]_1^2 = \tfrac{2}{3}. \quad \square$$

Exercises

Evaluate the following integrals by applying the fundamental theorem.

*1. $\displaystyle\int_0^1 (2x-3)\, dx.$

2. $\displaystyle\int_0^1 (3x+2)\, dx.$

*3. $\displaystyle\int_{-1}^0 5x^4\, dx.$

4. $\displaystyle\int_1^2 (2x+x^2)\, dx.$

*5. $\displaystyle\int_1^4 \sqrt{x}\, dx.$

6. $\displaystyle\int_0^4 \sqrt{x}\, dx.$

*7. $\displaystyle\int_1^5 2\sqrt{x-1}\, dx.$

8. $\displaystyle\int_{-2}^0 (x+1)(x-2)\, dx.$

*9. $\displaystyle\int_1^2 \left(\frac{3}{x^3}+5x\right) dx.$

10. $\displaystyle\int_2^0 \frac{dx}{(x+1)^2}.$

*11. $\displaystyle\int_0^1 (x^{3/2}-x^{1/2})\, dx.$

12. $\displaystyle\int_0^1 (x^{3/4}-2x^{1/2})\, dx.$

*13. $\displaystyle\int_0^1 (x+1)^{17}\, dx.$

14. $\displaystyle\int_0^a (a^2x-x^3)\, dx.$

*15. $\displaystyle\int_0^a (\sqrt{a}-\sqrt{x})^2\, dx.$

16. $\displaystyle\int_{-1}^1 (x-2)^2\, dx.$

*17. $\displaystyle\int_1^2 \frac{6-t}{t^3}\, dt.$

18. $\displaystyle\int_1^2 \frac{2-t}{t^3}\, dt.$

*19. $\displaystyle\int_0^1 x^2(x-1)\, dx.$

20. $\displaystyle\int_1^3 \left(x^2-\frac{1}{x^2}\right) dx.$

4.5   Problems

Problem.   Find the area of the region bounded above by the curve $y = -x^2 + 4$ and below by the x-axis.

SOLUTION.   The curve intersects the x-axis at $x = -2$ and $x = 2$. The region in question is depicted in Figure 4.5.1.

The area is given by the integral

$$\int_{-2}^{2} (-x^2 + 4)\, dx = \left[-\tfrac{1}{3}x^3 + 4x \right]_{-2}^{2} = \tfrac{32}{3}. \quad \square$$

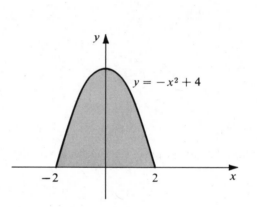

FIGURE 4.5.1

Problem.   Find the area of the crescent depicted in Figure 4.5.2.

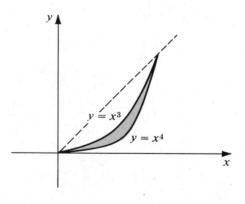

FIGURE 4.5.2

SOLUTION.   The two curves meet at $x = 0$ and at $x = 1$. We can obtain the area of the crescent by finding the area below

$$y = x^3,$$

and subtracting from it the area below

$$y = x^4.$$

This gives

$$A = \int_0^1 x^3 \, dx - \int_0^1 x^4 \, dx = \left[\tfrac{1}{4}x^4\right]_0^1 - \left[\tfrac{1}{5}x^5\right]_0^1 = \tfrac{1}{4} - \tfrac{1}{5} = \tfrac{1}{20}. \quad \square$$

The fundamental theorem gives

$$f(b) - f(a) = \int_a^b f'(x) \, dx$$

and thus

$$f(b) = f(a) + \int_a^b f'(x) \, dx.$$

Problem. Find $f(2)$ given that $f(0) = 1$ and $f'(x) = x^3 + 1$.

SOLUTION. Here

$$f(2) = f(0) + \int_0^2 f'(x) \, dx$$

so that

$$f(2) = 1 + \int_0^2 (x^3 + 1) \, dx$$

$$= 1 + \left[\tfrac{1}{4}x^4 + x\right]_0^2$$

$$= 1 + 6 = 7. \quad \square$$

Problem. An object moves along a coordinate line with velocity

$$v(t) = (1 - t)(2 - t) \text{ units per second.}$$

Its initial position (its position at time $t = 0$) is 2 units to the right of the origin.

(a) Find the position of the object 4 seconds later.
(b) Find the total distance traveled by the object during those first 4 seconds.

SOLUTION. (a) Let $x(t)$ be the position (coordinate) of the object at time t. Then

$$x'(t) = v(t).$$

This means that the position function is an antiderivative for the velocity function. Consequently

$$x(4) - x(0) = \int_0^4 v(t) \, dt.$$

Since the initial position of the object is 2 units to the right of the origin, we have

$$x(0) = 2.$$

From
$$v(t) = (1 - t)(2 - t) = 2 - 3t + t^2,$$
we see that
$$\int_0^4 v(t)\, dt = \int_0^4 (2 - 3t + t^2)\, dt = \left[2t - \tfrac{3}{2}t^2 + \tfrac{1}{3}t^3 \right]_0^4 = 5\tfrac{1}{3}.$$

From
$$x(0) = 2 \quad \text{together with} \quad \int_0^4 v(t)\, dt = 5\tfrac{1}{3},$$
we have
$$x(4) = x(0) + \int_0^4 v(t)\, dt = 2 + 5\tfrac{1}{3} = 7\tfrac{1}{3}.$$

At the end of 4 seconds the object ends up $7\tfrac{1}{3}$ units to the right of the origin.

(b) The total distance traveled by the object during the first 4 seconds is given by the integral
$$s = \int_0^4 |v(t)|\, dt.$$
In this instance
$$s = \int_0^4 |(1 - t)(2 - t)|\, dt.$$

The easiest way to evaluate this integral is to first remove the absolute-value signs. Since
$$|(1 - t)(2 - t)| = \begin{cases} 2 - 3t + t^2, & t \le 1 \\ -2 + 3t - t^2, & 1 \le t \le 2 \\ 2 - 3t + t^2, & 2 \le t \end{cases},$$
we have
$$s = \int_0^4 |(1 - t)(2 - t)|\, dt$$
$$= \int_0^1 (2 - 3t + t^2)\, dt + \int_1^2 (-2 + 3t - t^2)\, dt + \int_2^4 (2 - 3t + t^2)\, dt$$
$$= \left[2t - \tfrac{3}{2}t^2 + \tfrac{1}{3}t^3 \right]_0^1 + \left[-2t + \tfrac{3}{2}t^2 - \tfrac{1}{3}t^3 \right]_1^2 + \left[2t - \tfrac{3}{2}t^2 + \tfrac{1}{3}t^3 \right]_2^4.$$
$$= 5\tfrac{2}{3}.$$

The total distance is $5\tfrac{2}{3}$ units. □

Exercises

*1. In the motion we just discussed an object started out 2 units to the right of the origin. After traveling a distance of $5\tfrac{2}{3}$ units it ended up only $7\tfrac{1}{3}$ units to the right of the origin. Explain how this is possible.

In Exercises 2–8 find the area below the graph.

2. $f(x) = 2 + x^3$, $x \in [0, 1]$.   *3. $f(x) = \dfrac{1}{(x + 1)^2}$, $x \in [0, 2]$.

4. $f(x) = \sqrt{x + 1}$, $x \in [3, 8]$.   *5. $f(x) = \dfrac{1}{2\sqrt{x + 1}}$, $x \in [0, 8]$.

6. $f(x) = (2x^2 + 1)^2$, $x \in [0, 1]$.

*7. Sketch the region which is bounded above by $y = \sqrt{x}$ and below by $y = x^2$. Find the area of this region.

8. Sketch the region which is bounded above by $y = -x^2 + 5$ and below by $y = -x + 3$. Find the area of this region.

*9. Sketch the region which is bounded above by $y = -x^2 + 8$ and below by $y = x^2$. Find the area of this region.

10. Sketch the region bounded above by $y = 6x - x^2$ and below by the x-axis. Find the area of this region.

*11. Find $f(1)$ given that

$$f(0) = 0 \quad \text{and} \quad f'(x) = ax^2 + bx + c.$$

12. Find $f(4)$ given that

$$f(9) = 1 \quad \text{and} \quad f'(x) = (x - 9)^3.$$

*13. An object moves along a coordinate line with velocity

$$v(t) = t(1 - t) \text{ units per second.}$$

Its initial position (its position at time $t = 0$) is 2 units to the left of the origin.
(a) Find the position of the object 10 seconds later.
(b) Find the total distance traveled by the object during the first 10 seconds.

14. An object moves along a coordinate line with acceleration

$$a(t) = \frac{1}{\sqrt{t + 1}} \text{ units per second per second.}$$

(a) Find its velocity at time t_0 given that its initial velocity is 1 unit per second.
(b) Find the position of the object at time t_0 given that its initial velocity is 1 unit per second and its initial position is the origin.

*15. A point moves along the plane in such a manner that its x-coordinate is changing at the rate of $(t^2 + 1)^2$ units per second and its y-coordinate is changing at the rate of $\sqrt{t}$ units per second. Find the position of the point one second after the motion starts if the initial position is the point $P(x_0, y_0)$.

16. An automobile with varying velocity $v(t)$ moves in a fixed direction for 5 minutes and covers a distance of 4 miles. What theorem would you invoke to argue that for at least one instant the speedometer must have read 48 miles per hour?

4.6   The Linearity of the Integral

Here are some simple properties of the integral that are often used in computation.

I. Constants may be factored through the integral sign:

$$\int_a^b \alpha f(x) \, dx = \alpha \int_a^b f(x) \, dx.$$

Example

$$\int_0^{10} \tfrac{3}{7}(x - 5) \, dx = \tfrac{3}{7} \int_0^{10} (x - 5) \, dx = \tfrac{3}{7} \left[\tfrac{1}{2}(x - 5)^2 \right]_0^{10} = 0. \quad \square$$

Example

$$\int_0^1 - (x^2 + 1)^4 x \, dx = -\tfrac{1}{2} \int_0^1 2x(x^2 + 1)^4 \, dx$$

$$= -\tfrac{1}{2} \left[\tfrac{1}{5}(x^2 + 1)^5 \right]_0^1 = -\tfrac{1}{2} \left(\tfrac{32}{5} - \tfrac{1}{5} \right) = -\tfrac{31}{10}. \quad \square$$

II. The integral of a sum is the sum of the integrals:

$$\int_a^b [f(x) + g(x)] \, dx = \int_a^b f(x) \, dx + \int_a^b g(x) \, dx.$$

Example

$$\int_1^2 \left[(x - 1)^2 + \frac{1}{(x + 2)^2} \right] dx = \int_1^2 (x - 1)^2 \, dx + \int_1^2 \frac{dx}{(x + 2)^2}$$

$$= \left[\tfrac{1}{3}(x - 1)^3 \right]_1^2 + \left[-(x + 2)^{-1} \right]_1^2$$

$$= \tfrac{1}{3} - \tfrac{1}{4} + \tfrac{1}{3} = \tfrac{5}{12}. \quad \square$$

Combining I and II, we get the linearity property.

III. The integral of a linear combination is the linear combination of the integrals:

$$\int_a^b [\alpha f(x) + \beta g(x)] \, dx = \alpha \int_a^b f(x) \, dx + \beta \int_a^b g(x) \, dx.$$

Example

$$\int_2^{5/2} \left[\frac{3}{(x-1)^2} - 4x \right] dx = 3 \int_2^{5/2} \frac{dx}{(x-1)^2} - 2 \int_2^{5/2} 2x \, dx$$

$$= 3 \left[-(x-1)^{-1} \right]_2^{5/2} - 2 \left[x^2 \right]_2^{5/2}$$

$$= 3(-\tfrac{2}{3} + 1) - 2(\tfrac{25}{4} - 4) = -\tfrac{7}{2}. \quad \square$$

Properties I and II are just particular instances of property III. To prove III, take *f* and *g* as continuous, and choose for these functions antiderivatives *F* and *G* respectively. The details are left to you.

4.7   The Indefinite Integral Notation

If *F* is an antiderivative for *f* on [a, b], then we have

$$\int_a^b f(x) \, dx = \left[F(x) \right]_a^b.$$

The result is obviously unchanged if we replace $F(x)$ by $F(x) + C$, where *C* is an arbitrary constant:

$$\int_a^b f(x) \, dx = \left[F(x) + C \right]_a^b.$$

If we have no particular interest in the interval [a, b], but wish instead to emphasize that *F* is an antiderivative for *f* on *some* interval, then we can omit the *a* and the *b* and simply write

$$\int f(x) \, dx = F(x) + C.$$

Antiderivatives expressed in this manner are called *indefinite integrals*.

Examples

$$\int x^2 \, dx = \tfrac{1}{3}x^3 + C.$$

$$\int \sqrt{s} \, dx = \tfrac{2}{3}s^{3/2} + C.$$

$$\int \frac{4}{\sqrt{x+2}} \, dx = 8\sqrt{x+2} + C.$$

The following problem will serve to illustrate the notation further.

Problem.   Find the general law of motion of an object which moves in a straight line with constant acceleration *g*.

SOLUTION.   The acceleration is given by the equation

$$a(t) = g.$$

The velocity function is an indefinite integral of the acceleration:

$$v(t) = \int a(t)\, dt = \int g\, dt = gt + C.$$

The constant C is not determined by the acceleration but is arbitrary. Physically C represents the initial velocity v_0. We replace C by v_0 and write

$$v(t) = gt + v_0.$$

The position function is an indefinite integral of the velocity function:

$$x(t) = \int v(t)\, dt = \int (gt + v_0)\, dt = \tfrac{1}{2}gt^2 + v_0 t + K.$$

The constant K (we avoid C because we used it before) is not determined by the velocity function but is arbitrary. Physically it represents the initial position x_0. We therefore write

$$x(t) = \tfrac{1}{2}gt + v_0 t + x_0.$$

This is the general law of motion. It is an equation you saw before in connection with free-falling bodies. □

Exercises

*1. $\int \dfrac{dx}{x^4}.$

2. $\int (x - 1)^2\, dx.$

*3. $\int (ax + b)\, dx.$

4. $\int (ax^2 + b)\, dx.$

*5. $\int \dfrac{dx}{\sqrt{1 + x}}.$

6. $\int \left(\dfrac{x^3 + 1}{x^5}\right) dx.$

*7. $\int \left(\sqrt{x} - \dfrac{1}{\sqrt{x}}\right) dx.$

8. $\int \left(\dfrac{x^3 - 1}{x^2}\right) dx.$

*9. $\int (t - a)(t - b)\, dt.$

10. $\int (t^2 - a)(t^2 - b)\, dt.$

*11. $\int \dfrac{g'(x)}{[g(x)]^2}\, dx.$

12. $\int \dfrac{g(x)f'(x) - g'(x)f(x)}{[g(x)]^2}\, dx.$

*13. Compare

$$\frac{d}{dx}\left[\int f(x)\, dx\right] \quad \text{with} \quad \int \frac{d}{dx}[f(x)]\, dx.$$

14. Find the general law of motion of an object which moves in a straight line with acceleration

*(a) $a(t) = 2A + 6Bt.$

(b) $a(t) = 45t^{1/2} + t^{-1/2}.$

4.8   The *u*-Substitution

Integrals of the form

$$\int f(g(x))g'(x)\ dx$$

can be calculated by computing

$$\int f(u)\ du \qquad \text{and then setting} \quad u = g(x).$$

For suppose that F is an antiderivative for f. Then

$$\int f(g(x))g'(x)\ dx = \int F'(g(x))g'(x)\ dx = F(g(x)) + C$$

by the chain rule ↗

and

$$\int f(u)\ du = F(u) + C = F(g(x)) + C. \quad \square$$

by setting $u = g(x)$ ↗

This relation between the two integrals

$$\int f(g(x))g'(x)\ dx \quad \text{and} \quad \int f(u)\ du$$

is rendered visually obvious by setting

$$u = g(x), \qquad du = g'(x)\ dx.\dagger$$

Problem.   Find

$$\int \frac{dx}{(3 + 2x)^2}.$$

SOLUTION.   Set

$$u = 3 + 2x, \qquad du = 2\ dx.$$

Then

$$\int \frac{dx}{(3 + 2x)^2} = \frac{1}{2} \int \frac{2\ dx}{(3 + 2x)^2}$$

$$= \frac{1}{2} \int \frac{du}{u^2} = -\frac{1}{2u} + C$$

$$= -\frac{1}{2(3 + 2x)} + C. \quad \square$$

Problem.   Find

$$\int x\sqrt{a^2 + b^2 x^2}\ dx.$$

----

† In this setting you can think of $du = g'(x)\ dx$ as a "formal differential." Simply write dx for h.

SOLUTION.   Set

$$u = a^2 + b^2 x^2, \qquad du = 2b^2 x \, dx.$$

$$\int x\sqrt{a^2 + b^2 x^2} \, dx = \frac{1}{2b^2} \int \sqrt{a^2 + b^2 x^2} \, (2b^2 x) \, dx$$

$$= \frac{1}{2b^2} \int \sqrt{u} \, du$$

$$= \frac{1}{2b^2} \left(\frac{2}{3}\right) u^{3/2} + C$$

$$= \frac{1}{3b^2} (a^2 + b^2 x^2)^{3/2} + C. \quad \square$$

For definite integrals we have the following formula:

$$\boxed{\int_a^b f(g(x))g'(x) \, dx = \int_{g(a)}^{g(b)} f(u) \, du.}$$

The formula holds provided that f and g' are both continuous. More precisely, g' must be continuous on an interval that joins a and b, and f must be continuous on the set of values taken on by g.

PROOF.   Let F be an antiderivative for f. Then $F' = f$ and

$$\int_a^b f(g(x))g'(x) \, dx = \int_a^b F'(g(x))g'(x) \, dx$$

$$= \left[F(g(x)) \right]_a^b$$

$$= F(g(b)) - F(g(a))$$

$$= \int_{g(a)}^{g(b)} f(u) \, du. \quad \square$$

Problem.   Evaluate

$$\int_0^1 (x^2 - 1)^4 2x \, dx.$$

SOLUTION.   Set

$$u = x^2 - 1, \qquad du = 2x \, dx.$$

At $x = 0$, $u = -1$. At $x = 1$, $u = 0$.

$$\int_0^1 (x^2 - 1)^4 2x \, dx = \int_{-1}^0 u^4 \, du = \left[\tfrac{1}{5} u^5 \right]_{-1}^0 = \tfrac{1}{5}. \quad \square$$

Problem. Evaluate

$$\int_1^2 \frac{10x^2}{(x^3 + 1)^2} \, dx.$$

SOLUTION. Set

$$u = x^3 + 1, \qquad du = 3x^2 \, dx.$$

At $x = 1$, $u = 2$. At $x = 2$, $u = 9$.

$$\int_1^2 \frac{10x^2}{(x^3 + 1)^2} \, dx = \frac{10}{3} \int_1^2 \frac{3x^2}{(x^3 + 1)^2} \, dx$$

$$= \frac{10}{3} \int_2^9 \frac{du}{u^2}$$

$$= \frac{10}{3} \left[-\frac{1}{u} \right]_2^9$$

$$= \frac{10}{3} \left(\frac{1}{2} - \frac{1}{9} \right)$$

$$= \frac{35}{27}. \quad \square$$

Problem. Evaluate

$$\int_0^2 \frac{x}{\sqrt{4x^2 + 9}} \, dx.$$

SOLUTION. Set

$$u = 4x^2 + 9, \qquad du = 8x \, dx.$$

At $x = 0$, $u = 9$. At $x = 2$, $u = 25$.

$$\int_0^2 \frac{x}{\sqrt{4x^2 + 9}} \, dx = \frac{1}{8} \int_0^2 \frac{8x}{\sqrt{4x^2 + 9}} \, dx$$

$$= \frac{1}{8} \int_9^{25} \frac{du}{\sqrt{u}} = \frac{1}{4} \left[\sqrt{u} \right]_9^{25} = \frac{1}{2}. \quad \square$$

Exercises

Work out the following integrals by a *u*-substitution.

*1. $\displaystyle\int \frac{dx}{(2 - 3x)^2}$.

2. $\displaystyle\int \frac{dx}{\sqrt{2x + 1}}$.

*3. $\displaystyle\int \sqrt{2x + 1} \, dx$.

4. $\displaystyle\int \sqrt{ax + b} \, dx$.

*5. $\displaystyle\int (ax + b)^{3/4} \, dx$.

6. $\displaystyle\int (2ax + b)(ax^2 + bx + c)^4 \, dx$.

*7. $\displaystyle\int \frac{x}{(4x^2 + 9)^2} \, dx.$

8. $\displaystyle\int \frac{3x}{(x^2 + 1)^2} \, dx.$

*9. $\displaystyle\int x^2(5x^3 + 9)^4 \, dx.$

10. $\displaystyle\int t(1 + t^2)^3 \, dt.$

*11. $\displaystyle\int x^2(1 + x^3)^{1/4} \, dx.$

12. $\displaystyle\int x^{n-1}\sqrt{a + bx^n} \, dx.$

*13. $\displaystyle\int \frac{s}{(1 + s^2)^3} \, ds.$

14. $\displaystyle\int \frac{2x}{\sqrt[3]{6 - 5x^2}} \, dx.$

*15. $\displaystyle\int \frac{x}{\sqrt{x^2 + 1}} \, dx.$

16. $\displaystyle\int \frac{3ax^2 - 2bx}{\sqrt{ax^3 - bx^2}} \, dx.$

*17. $\displaystyle\int \frac{b^3x^3}{\sqrt{1 - a^4x^4}} \, dx.$

18. $\displaystyle\int \frac{x^{n-1}}{\sqrt{a + bx^n}} \, dx.$

Evaluate the following integrals by a u-substitution.

*19. $\displaystyle\int_0^1 x(x^2 + 1)^3 \, dx.$

20. $\displaystyle\int_{-1}^0 3x^2(4 + 2x^3)^2 \, dx.$

*21. $\displaystyle\int_0^1 5x(1 + x^2)^4 \, dx.$

22. $\displaystyle\int_1^2 (6 - x)^{-3} \, dx.$

*23. $\displaystyle\int_{-1}^1 \frac{x}{(1 + x^2)^4} \, dx.$

24. $\displaystyle\int_0^3 \frac{r}{\sqrt{r^2 + 16}} \, dr.$

*25. $\displaystyle\int_0^a x\sqrt{a^2 - x^2} \, dx.$

26. $\displaystyle\int_0^a x\sqrt{a^2 + x^2} \, dx.$

4.9   Some Further Properties of the Definite Integral

In this section we feature some important general properties of the integral. The proofs are left mostly to you. You can assume throughout that the functions involved are continuous and that $a < b$.

I. The integral of a nonnegative function is nonnegative:

$$\text{if } f(x) \geq 0 \quad \text{for all } x \in [a, b], \qquad \text{then} \quad \int_a^b f(x) \, dx \geq 0.$$

The integral of a positive function is positive:

$$\text{if } f(x) > 0 \quad \text{for all } x \in [a, b], \qquad \text{then} \quad \int_a^b f(x) \, dx > 0.$$

The next property is an immediate consequence of property I and the linearity property.

II. The integral is order-preserving:

$$\text{if } f(x) \le g(x) \quad \text{for all } x \in [a, b], \qquad \text{then } \int_a^b f(x) \, dx \le \int_a^b g(x) \, dx$$

and

$$\text{if } f(x) < g(x) \quad \text{for all } x \in [a, b], \qquad \text{then } \int_a^b f(x) \, dx < \int_a^b g(x) \, dx.$$

III. Just as the absolute value of a sum is less than or equal to the sum of the absolute values,

$$|x_1 + x_2 + \cdots + x_n| \le |x_1| + |x_2| + \cdots + |x_n|,$$

the absolute value of an integral is less than or equal to the integral of the absolute value:

$$\left| \int_a^b f(x) \, dx \right| \le \int_a^b |f(x)| \, dx.$$

HINT FOR PROOF OF III: Show that

$$\int_a^b f(x) \, dx \quad \text{and} \quad - \int_a^b f(x) \, dx$$

are both less than or equal to

$$\int_a^b |f(x)| \, dx.$$

The next property is one with which you are already familiar.

IV. If m is the minimum value of f on $[a, b]$ and M is the maximum then

$$m(b - a) \le \int_a^b f(x) \, dx \le M(b - a).$$

We come now to a generalization of Theorem 4.3.2. The importance of it will become apparent in Chapter 5.

V. If g is differentiable (and f is continuous), then

$$\frac{d}{dx} \left(\int_a^{g(x)} f(t) \, dt \right) = f(g(x)) g'(x).$$

PROOF.   Set

$$H(x) = \int_a^{g(x)} f(t)\, dt$$

and recognize that H is the composition of differentiable functions:

$$H(x) = F(g(x)) \qquad \text{with} \qquad F(x) = \int_a^x f(t)\, dt.$$

The chain rule gives

$$H'(x) = F'(g(x))g'(x).$$

Theorem 4.3.2 gives

$$F'(x) = f(x).$$

It therefore follows that

$$H'(x) = f(g(x))g'(x). \quad \Box$$

Problem.   Find

$$\frac{d}{dx}\left(\int_0^{x^3} \frac{dt}{1 + t} \right).$$

SOLUTION.   At this stage you'd probably be hard put to carry out the integration. But here that doesn't matter. By Property V you know that

$$\frac{d}{dx}\left(\int_0^{x^3} \frac{dt}{1 + t} \right) = \frac{1}{1 + x^3}\, 3x^2 = \frac{3x^2}{1 + x^3},$$

without carrying out the integration. $\Box$

Exercises

Assume: f and g continuous, $a < b$, and

$$\int_a^b f(x)\, dx > \int_a^b g(x)\, dx.$$

Answer the following questions, giving supporting reasons.

*1. Does it necessarily follow that

$$\int_a^b [f(x) - g(x)]\, dx > 0?$$

*2. Does it necessarily follow that

$$f(x) > g(x) \qquad \text{for all } x \in [a, b]?$$

*3. Does it necessarily follow that

$$f(x) > g(x) \qquad \text{for at least some } x \in [a, b]?$$

*4. Does it necessarily follow that

$$\left| \int_a^b f(x) \, dx \right| > \left| \int_a^b g(x) \, dx \right| ?$$

*5. Does it necessarily follow that

$$\int_a^b |f(x)| \, dx > \int_a^b g(x) \, dx?$$

*6. Does it necessarily follow that

$$\int_a^b |f(x)| \, dx > \int_a^b |g(x)| \, dx?$$

Assume: f continuous, $a < b$, and

$$\int_a^b f(x) \, dx = 0.$$

Answer the following questions, giving supporting reasons.

7. Does it necessarily follow that

$$f(x) = 0 \quad \text{for all} \quad x \in [a, b]?$$

8. Does it necessarily follow that

$$f(x) = 0 \quad \text{for at least some} \quad x \in [a, b]?$$

9. Does it necessarily follow that

$$\left| \int_a^b f(x) \, dx \right| = 0?$$

10. Does it necessarily follow that

$$\int_a^b |f(x)| \, dx = 0?$$

11. Must all upper sums $U_f(P)$ be nonnegative?
12. Must all upper sums $U_f(P)$ be positive?
13. Can a lower sum, $L_f(P)$, be positive?
14. Does it necessarily follow that

$$\int_a^b [f(x)]^2 \, dx = 0?$$

15. Does it necessarily follow that

$$\int_a^b [f(x) + 1] \, dx = b - a?$$

Compute

*16. $\dfrac{d}{dx}\left(\displaystyle\int_1^{x^2} \dfrac{dt}{t}\right)$.

*17. $\dfrac{d}{dx}\left(\displaystyle\int_0^{1+x^2} \dfrac{dt}{\sqrt{2t+5}}\right)$.

18. $\dfrac{d}{dx}\left(\displaystyle\int_0^{x^3} \dfrac{dt}{\sqrt{1+t^2}}\right)$.

19. $\dfrac{d}{dx}\left(\displaystyle\int_x^a f(t)\,dt\right)$.

20. Show that

$$\frac{d}{dx}\left(\int_{g_1(x)}^{g_2(x)} f(t)\,dt\right) = f(g_2(x))g_2'(x) - f(g_1(x))g_1'(x).$$

Here g_1 and g_2 are assumed to be differentiable and f continuous.

HINT: Take a number a from the domain of f. Express the integral as the difference of two integrals each with lower limit a.

Compute using Exercise 20.

*21. $\dfrac{d}{dx}\left(\displaystyle\int_x^{x^2} \dfrac{dt}{t}\right)$.

22. $\dfrac{d}{dx}\left(\displaystyle\int_{1-x}^{1+x} \dfrac{t-1}{t}\,dt\right)$.

23. Prove Property I: (a) by considering lower sums $L_f(P)$; (b) by using an anti-derivative.
24. Prove Property II.
25. Prove Property III.
26. (*Optional*) Prove that if f is continuous on $[a, b]$ and

$$\int_a^b |f(x)|\,dx = 0,$$

then

$$f(x) = 0 \quad \text{for all } x \text{ in } [a, b].$$

4.10 Additional Exercises

Find

*1. $\displaystyle\int (\sqrt{x-a} - \sqrt{x-b})\,dx$.

2. $\displaystyle\int ax\sqrt{1+bx^2}\,dx$.

*3. $\displaystyle\int x^{-1/3}(x^{2/3}-1)^2\,dx$.

4. $\displaystyle\int t^2(1+t^3)^{10}\,dt$.

*5. $\displaystyle\int (1+2\sqrt{x})^2\,dx$.

6. $\displaystyle\int \frac{1}{\sqrt{x}}(1+2\sqrt{x})^5\,dx$.

*7. $\displaystyle\int \frac{(a+b\sqrt{x+1})^2}{\sqrt{x+1}}\,dx$.

8. $\displaystyle\int x\sqrt{x}(1+x^2\sqrt{x})^2\,dx$.

*9. $\displaystyle\int \frac{g'(x)}{[g(x)]^3}\, dx.$

10. $\displaystyle\int \frac{g(x)g'(x)}{\sqrt{1+[g(x)]^2}}\, dx.$

Find the area below the graph.

*11. $y = x\sqrt{2x^2 + 1}, \quad x \in [0, 2].$

12. $y = \dfrac{x}{(2x^2 + 1)^2}, \quad x \in [0, 2].$

*13. Find the area of the portion of the first quadrant that is bounded above by $y = 2x$ and below by $y = x\sqrt{3x^2 + 1}$.

14. At every point of the curve the slope is given by the equation

$$\frac{dy}{dx} = x\sqrt{x^2 + 1}.$$

Find an equation for the curve given that it passes through the point $(0, 1)$.

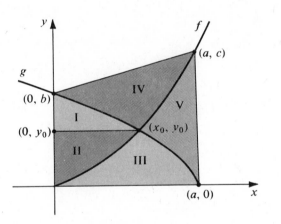

FIGURE 4.10.1

*15. Find a formula for the area of each region depicted in Figure 4.10.1:

(a) I.   (b) II.   (c) III.   (d) IV.   (e) V.

Assume that f and g are continuous, that $a < b$, and that

$$\int_a^b f(x)\, dx > \int_a^b g(x)\, dx.$$

Which of the following statements necessarily hold for all partitions P of $[a, b]$? Support your answers.

*16. $L_g(P) < U_f(P).$

*17. $L_g(P) < L_f(P).$

18. $L_g(P) < \displaystyle\int_a^b f(x)\, dx.$

19. $U_g(P) < U_f(P).$

*20. $U_f(P) > \displaystyle\int_a^b g(x)\, dx.$

21. $U_g(P) < \displaystyle\int_a^b f(x)\, dx.$

A curve which passes through the point $(4, \frac{1}{3})$ has varying slope

$$\frac{dy}{dx} = -\frac{1}{2\sqrt{x}(1 + \sqrt{x})^2} .$$

22. Find an equation for the curve.
23. Sketch the curve.
24. Estimate the area under the curve between $x = 0$ and $x = 1$ by means of the partition $P = \{0, \frac{1}{25}, \frac{4}{25}, \frac{9}{25}, \frac{16}{25}, 1\}$.

The Logarithm and Exponential Functions

5.1 In Search of a Logarithm

The elementary notion of logarithm

$$c = \log_b a \quad \text{iff} \quad b^c = a$$

may be adequate for computational purposes (particularly if you don't ask what b^c means when c is irrational) but it does not lend itself well to the methods of calculus.

From our point of view the advantage of logarithms is that they transform multiplication into addition:

the log of a product = the sum of the logs.

Taking this as the central idea, we make the following definition.

Definition

A *logarithm* function is a nonconstant differentiable function f which has the property that for all positive numbers x and y

$$f(xy) = f(x) + f(y).$$

Let's assume for the moment that such logarithm functions exist and let's see what we can find out about them. In the first place, if f is such a function, then

$$f(1) = f(1 \cdot 1) = f(1) + f(1) = 2f(1)$$

and so

$$f(1) = 0.$$

Taking $y > 0$, we have

$$0 = f(1) = f(y \cdot 1/y) = f(y) + f(1/y)$$

and therefore

$$f(1/y) = -f(y).$$

Taking $x > 0$ and $y > 0$, we have

$$f(x/y) = f(x \cdot 1/y) = f(x) + f(1/y),$$

which, in view of the previous result, means that

$$f(x/y) = f(x) - f(y).$$

We are now ready to look for the derivative. (Remember, we are *assuming* that f is differentiable.) We begin by forming the difference quotient

$$\frac{f(x + h) - f(x)}{h}.$$

From what we have discovered about f,

$$f(x + h) - f(x) = f\left(\frac{x + h}{x}\right) = f(1 + h/x),$$

and therefore

$$\frac{f(x + h) - f(x)}{h} = \frac{f(1 + h/x)}{h}.$$

Remembering that $f(1) = 0$ and multiplying the denominator by x/x, we have

$$\frac{f(x + h) - f(x)}{h} = \frac{1}{x}\left[\frac{f(1 + h/x) - f(1)}{h/x}\right].$$

As h tends to 0, the left side tends to $f'(x)$ and, while $1/x$ remains fixed, the bracketed expression tends to $f'(1)$. In short,

(5.1.1)

$$\boxed{f'(x) = \frac{1}{x} f'(1).}$$

We have now established that, if f is a logarithm, then

$$f(1) = 0 \quad \text{and} \quad f'(x) = \frac{1}{x} f'(1).$$

We can't have $f'(1) = 0$ for that would make f constant. (Explain.) The simplest alternative is to set $f'(1) = 1$. The essential properties then become

$$f(1) = 0 \quad \text{and} \quad f'(x) = \frac{1}{x}.$$

In this manner we are led to the function

$$L(x) = \int_1^x \frac{dt}{t}, \qquad x > 0.$$

It is the only function which takes on the value 0 at 1 and has the right derivative. It now remains for us to prove that L does transform multiplication into addition.

Theorem

If x and y are positive, then
$$L(xy) = L(x) + L(y).$$

PROOF
$$L(xy) = \int_1^{xy} \frac{dt}{t} = \int_1^x \frac{dt}{t} + \int_x^{xy} \frac{dt}{t}.$$

Since
$$L(x) = \int_1^x \frac{dt}{t}$$

we need only show that
$$L(y) = \int_x^{xy} \frac{dt}{t}.$$

At 1 both sides are 0:
$$L(1) = \int_1^1 \frac{dt}{t} = 0, \qquad \int_x^{x \cdot 1} \frac{dt}{t} = \int_x^x \frac{dt}{t} = 0$$

and for all $y > 0$ both sides have the same derivative:
$$\frac{d}{dy}[L(y)] = \frac{1}{y}, \qquad \frac{d}{dy}\left(\int_x^{xy} \frac{dt}{t}\right) = \frac{1}{xy} \cdot x = \frac{1}{y}.$$

property V, Section 4.9

By Theorem 3.11.4 both sides must be equal for all $y > 0$. □

5.2 The Logarithm Function, Part I

Definition

The function
$$L(x) = \int_1^x \frac{dt}{t}, \qquad x > 0$$

is called the (natural) logarithm.

Here are some obvious properties of L:

(1) $$L'(x) = \frac{1}{x} \quad \text{for all } x > 0.$$

Having a positive derivative on $(0, \infty)$, L increases on $(0, \infty)$.
(2) L is continuous on $(0, \infty)$ since it is there differentiable.

(3) $$L(x) \text{ is } \begin{cases} \text{negative,} & \text{if } 0 < x < 1 \\ 0, & \text{at } x = 1 \\ \text{positive,} & \text{if } x > 1 \end{cases}.$$

(4) For $x > 1$, $L(x)$ gives the area below the graph of

$$g(t) = \frac{1}{t}, \qquad 1 \leq t \leq x. \qquad\qquad \text{(Figure 5.2.1)}$$

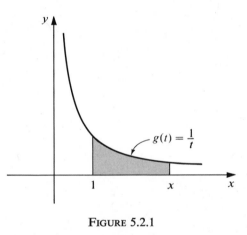

FIGURE 5.2.1

Not obvious at all but already proven is the following special property:
(5) For x and y positive,

$$L(xy) = L(x) + L(y).$$

We come now to something new.

Theorem 5.2.1

If x is positive and p/q is rational, then

$$L(x^{p/q}) = \frac{p}{q} L(x).$$

PROOF.   We show first that for all nonnegative integers n

$$L(x^n) = nL(x).$$

If n is 0 or 1, the result is trivial. Let's suppose that the result holds for $n = k$. Now

$$L(x^{k+1}) = L(x^k \cdot x) = L(x^k) + L(x),$$

the second equality holding because of L's special multiplicative property. By our inductive assumption

$$L(x^k) = kL(x).$$

Thus

$$L(x^{k+1}) = kL(x) + L(x) = (k+1)L(x),$$

which means that the result holds for $n = k + 1$. This shows that

$$L(x^n) = nL(x) \quad \text{for all nonnegative integers } n.$$

Since

$$0 = L(1) = L(x \cdot x^{-1}) = L(x) + L(x^{-1}),$$

we see that

$$L(x^{-1}) = -L(x).$$

If n is a positive integer, we have

$$L(x^{-n}) = -L(x^n) = -nL(x),$$

so that the formula holds for all negative integers.

Let q be a positive integer. Now

$$L(x) = L([x^{1/q}]^q) = qL(x^{1/q}).$$

Dividing by q, we obtain

$$L(x^{1/q}) = \frac{1}{q} L(x).$$

Let us now take p/q as an arbitrary rational number. We can assume that q is positive and absorb the sign in p. Since p is an integer, it follows that

$$L(x^{p/q}) = L([x^{1/q}]^p) = pL(x^{1/q}).$$

Since

$$L(x^{1/q}) = \frac{1}{q} L(x),$$

we can finally conclude that

$$L(x^{p/q}) = \frac{p}{q} L(x). \quad \square$$

The domain of L is $(0, \infty)$. What is the range of L?

Theorem

The range of L is $(-\infty, \infty)$.

PROOF. Since L is continuous on $(0, \infty)$, we know from the intermediate-value theorem that it "skips" no values. Thus, its range is an interval. To show that the interval is $(-\infty, \infty)$, we need only show that it is unbounded above and unbounded below. We can do this by taking M as an arbitrary positive number and showing that L takes on values greater than M and values less than $-M$.

Since

$$L(2) = \int_1^2 \frac{dt}{t}$$

is positive (explain), we know that some positive multiple of $L(2)$ must be greater than M; namely, we know that there exists a positive integer n such that

$$nL(2) > M.$$

Multiplying this equation by -1 we have

$$-nL(2) < -M.$$

Since

$$nL(2) = L(2^n) \quad \text{and} \quad -nL(2) = L(2^{-n}),$$

we have

$$L(2^n) > M \quad \text{and} \quad L(2^{-n}) < -M.$$

This proves the unboundedness. $\square$

The Number e

Since the range of L is $(-\infty, \infty)$ and L is an increasing function, we know that L takes on every value and it does so only once. In particular, there is one and only one number at which the function L takes on the value 1. *This unique number is denoted by the letter e.*[†]

With

$$\boxed{L(e) = 1}$$

it follows from Theorem 5.2.1 that

$$\boxed{L(e^{p/q}) = \frac{p}{q} \quad \text{for all rational numbers } \frac{p}{q}.}$$

Because of this relation, we call L the logarithm to the base e and sometimes write

$$L(x) = \log_e x.$$

---

† After Leonard Euler (1707–1783).

Usually we drop the subscript and write

$$L(x) = \log x.\dagger$$

The Graph of the Logarithm Function

You know that the logarithm function

$$\log x = \int_1^x \frac{dt}{t}$$

has domain $(0, \infty)$, range $(-\infty, \infty)$, and derivative

$$\frac{d}{dx}(\log x) = \frac{1}{x} > 0.$$

For small x the derivative is large (near 0 the curve is steep); for large x the derivative is small (far out the curve flattens out). At $x = 1$ the logarithm is 0 and its derivative $1/x$ is 1. (The graph crosses the x-axis at the point $(1, 0)$ and the tangent at that point is parallel to the line $y = x$.) The second derivative

$$\frac{d^2}{dx^2}(\log x) = -\frac{1}{x^2}$$

is negative throughout $(0, \infty)$. (The graph, pictured in Figure 5.2.2, is concave down throughout.)

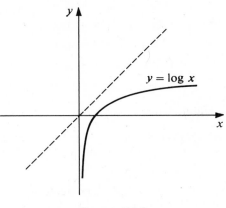

FIGURE 5.2.2

---

† Another common notation for $L(x)$ is $\ln x$. We will stick to $\log x$. Logarithms to bases other than e will be taken up later (they arise by other choices of $f'(1)$), but by far the most important logarithm in calculus is the logarithm to the base e. So much so, that when we speak of the logarithm of a number x and don't specify the base, you can be sure that we are talking about the *natural logarithm*

$$\log x = \int_1^x \frac{dt}{t}.$$

Problem.   Estimate log 2.

SOLUTION

$$\log 2 = \int_1^2 \frac{dt}{t}.$$

We can estimate the integral by upper and lower sums. Using the partition

$$P = \{1 = \tfrac{10}{10}, \tfrac{11}{10}, \tfrac{12}{10}, \tfrac{13}{10}, \tfrac{14}{10}, \tfrac{15}{10}, \tfrac{16}{10}, \tfrac{17}{10}, \tfrac{18}{10}, \tfrac{19}{10}, \tfrac{20}{10} = 2\},$$

we have

$$L_f(P) = \tfrac{1}{10}(\tfrac{10}{11} + \tfrac{10}{12} + \tfrac{10}{13} + \tfrac{10}{14} + \tfrac{10}{15} + \tfrac{10}{16} + \tfrac{10}{17} + \tfrac{10}{18} + \tfrac{10}{19} + \tfrac{10}{20})$$
$$= \tfrac{1}{11} + \tfrac{1}{12} + \tfrac{1}{13} + \tfrac{1}{14} + \tfrac{1}{15} + \tfrac{1}{16} + \tfrac{1}{17} + \tfrac{1}{18} + \tfrac{1}{19} + \tfrac{1}{20}$$

and

$$U_f(P) = \tfrac{1}{10}(\tfrac{10}{10} + \tfrac{10}{11} + \tfrac{10}{12} + \tfrac{10}{13} + \tfrac{10}{14} + \tfrac{10}{15} + \tfrac{10}{16} + \tfrac{10}{17} + \tfrac{10}{18} + \tfrac{10}{19})$$
$$= \tfrac{1}{10} + \tfrac{1}{11} + \tfrac{1}{12} + \tfrac{1}{13} + \tfrac{1}{14} + \tfrac{1}{15} + \tfrac{1}{16} + \tfrac{1}{17} + \tfrac{1}{18} + \tfrac{1}{19}.$$

With

$$0.100 = \tfrac{1}{10} = 0.100$$
$$0.090 < \tfrac{1}{11} < 0.091$$
$$0.083 < \tfrac{1}{12} < 0.084$$
$$0.076 < \tfrac{1}{13} < 0.077$$
$$0.071 < \tfrac{1}{14} < 0.072$$
$$0.066 < \tfrac{1}{15} < 0.067$$
$$0.062 < \tfrac{1}{16} < 0.063$$
$$0.058 < \tfrac{1}{17} < 0.059$$
$$0.055 < \tfrac{1}{18} < 0.056$$
$$0.052 < \tfrac{1}{19} < 0.053$$
$$0.050 = \tfrac{1}{20} = 0.050$$

you can see after a little addition that

$$0.663 < L_f(P) < \log 2 < U_f(P) < 0.722.$$

The average of these two estimates is

$$\tfrac{1}{2}(0.663 + 0.722) = 0.6925.$$

We are not far off. Three-place tables carry the estimate 0.693.   □

Table 5.2.1 gives the natural logarithms of the integers 2 through 10 rounded off to the nearest hundredth.†

---

† A more extended table appears at the end of the book, but throughout this chapter all numerical calculations will be based on Table 5.2.1.

TABLE 5.2.1

n	$\log n$
2	0.69
3	1.10
4	1.39
5	1.61
6	1.79
7	1.95
8	2.08
9	2.20
10	2.30

Problem.   Use Table 5.2.1 to estimate the following logarithms:

(a) log 0.2.   (b) log 0.25.   (c) log 2.4.   (d) log 90.

SOLUTION

(a) $\log 0.2 = \log \frac{1}{5} = -\log 5 \cong -1.61$.
(b) $\log 0.25 = \log (0.5)^2 = 2 \log \frac{1}{2} = -2 \log 2 \cong -0.138$.
(c) $\log 2.4 = \log [(3)(0.8)] = \log 3 + \log 8 - \log 10 \cong 0.88$.
(d) $\log 90 = \log [(9)(10)] = \log 9 + \log 10 \cong 4.50$. $\square$

Problem.   Use Table 5.2.1 to estimate e.

SOLUTION.   All we know about e so far is that $\log e = 1$.

$$\log 3 \quad \cong 1.10.$$
$$\therefore \log 27 = \log 3^3 = 3 \log 3 \cong 3.30.$$
$$\log 10 \cong 2.30.$$
$$\therefore \log 2.7 = \log \left(\tfrac{27}{10}\right) = \log 27 - \log 10 \cong 1.00.$$

Our calculations show that e is approximately 2.7. (Eight-place tables give the estimate $e \cong 2.71828182$. For most purposes we take $e \cong 2.72$.)

Exercises

Use Table 5.2.1 to estimate the following natural logarithms.

*1. log 20.   2. log 16.   *3. log 1.6.
*4. $\log 3^4$.   5. log 0.1.   *6. log 2.5.
*7. log 7.2.   8. $\log \sqrt{630}$.   *9. $\log \sqrt{2}$.

10. Interpret the equation
$$\log n = \log mn - \log m$$

in terms of area under the curve
$$y = \frac{1}{x}.$$

Draw a figure.

*11. How can you relate
$$\log x \qquad \text{for } 0 < x < 1$$

to area?

12. Estimate
$$\log 1.5 = \int_1^{1.5} \frac{dt}{t}$$

using the approximation
$$\tfrac{1}{2}[L_f(P) + U_f(P)] \quad \text{with} \quad P = \{1 = \tfrac{8}{8}, \tfrac{9}{8}, \tfrac{10}{8}, \tfrac{11}{8}, \tfrac{12}{8} = 1.5\}.$$

13. Taking $\log 5 \cong 1.61$, use differentials to estimate
 *(a) $\log 5.2$.   (b) $\log 4.8$.   *(c) $\log 5.5$.

14. Taking $\log 10 \cong 2.30$, use differentials to estimate
 (a) $\log 10.3$.   *(b) $\log 9.6$.   (c) $\log 11$.

Solve the following equations for x.

*15. $\log x = 2$.

16. $\log x = -1$.

17. $(2 - \log x) \log x = 0$.

*18. $\tfrac{1}{2} \log x = \log (2x - 1)$.

5.3   The Logarithm Function, Part II

Differentiation and Graphing

Problem.   Find the domain of f and find $f'(x)$, given that
$$f(x) = (\log x^2)^3.$$

SOLUTION.   Since the logarithm function is defined only for positive numbers, we must have $x^2 > 0$. The domain of f consists of all $x \neq 0$. Before differentiating, we make use of the special properties of the logarithm:
$$f(x) = (\log x^2)^3 = (2 \log x)^3 = 8(\log x)^3.$$

$$f'(x) = 24(\log x)^2 \frac{1}{x} = \frac{24(\log x)^2}{x}. \quad \square$$

Problem.   Find the domain of f and find $f'(x)$, if
$$f(x) = \log (x\sqrt{1 + x}).$$

SOLUTION.   For x to be in the domain of f, we must have

$$x\sqrt{1 + x} > 0$$

and consequently

$$x > 0.$$

The domain is the set of positive numbers. Before differentiating f, we make use of the special properties of the logarithm:

$$f(x) = \log(x\sqrt{1 + x}) = \log x + \tfrac{1}{2}\log(1 + x).$$

From this, we have

$$f'(x) = \frac{1}{x} + \frac{1}{2}\left(\frac{1}{1 + x}\right). \quad \square$$

Problem.   Sketch the graph of

$$f(x) = \log|x|.$$

SOLUTION.   This is an even function,

$$f(-x) = f(x),$$

defined for all numbers $x \neq 0$. The graph has two branches:

$$y = \log(-x), \quad x < 0 \qquad \text{and} \qquad y = \log x, \quad x > 0.$$

These branches are mirror images of one another. (Figure 5.3.1) $\square$

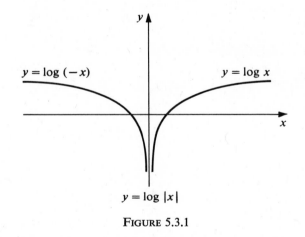

$y = \log(-x)$

$y = \log x$

$y = \log|x|$

FIGURE 5.3.1

Problem.   Show that

(5.3.1)

$$\frac{d}{dx}(\log|x|) = \frac{1}{x}$$

for all $x \neq 0$.

SOLUTION.   For $x > 0$,

$$\frac{d}{dx} (\log |x|) = \frac{d}{dx} (\log x) = \frac{1}{x}.$$

For $x < 0$,

$$\frac{d}{dx} (\log |x|) = \frac{d}{dx} [\log (-x)] = \frac{1}{-x} \frac{d}{dx} (-x) = \left(\frac{1}{-x}\right)(-1) = \frac{1}{x}. \quad \square$$

Problem.   Show that

(5.3.2)

$$\boxed{\frac{d}{dx} (\log |g(x)|) = \frac{g'(x)}{g(x)}}$$

provided that $g(x) \neq 0$.

SOLUTION.   Note that

$$\log |g(x)| = f(g(x)), \qquad \text{where} \quad f(x) = \log |x|.$$

By the chain rule

$$\frac{d}{dx} (\log |g(x)|) = f'(g(x))g'(x) = \frac{g'(x)}{g(x)}. \quad \square$$

Here are some examples:

$$\frac{d}{dx} (\log |1 - x^3|) = \frac{-3x^2}{1 - x^3} = \frac{3x^2}{x^3 - 1}.$$

$$\frac{d}{dx} \left(\log \left|\frac{x - 1}{x - 2}\right|\right) = \frac{d}{dx} (\log |x - 1|) - \frac{d}{dx} (|x - 2|)$$

$$= \frac{1}{x - 1} - \frac{1}{x - 2}. \quad \square$$

Problem.   Let

$$f(x) = \log \left(\frac{x^4}{x - 1}\right).$$

(1) Specify the domain of f. (2) On what intervals is f increasing? Decreasing? (3) Find the extreme values of f. (4) Determine the concavity of the graph and find the points of inflection. (5) Sketch the graph.

SOLUTION.   Since the logarithm function is defined only for positive numbers the domain of f consists of all numbers greater than 1.

Making use of the special properties of the logarithm we get

$$f(x) = \log x^4 - \log (x - 1) = 4 \log x - \log (x - 1).$$

This gives

$$f'(x) = \frac{4}{x} - \frac{1}{x-1} = \frac{3x-4}{x(x-1)}$$

and

$$f''(x) = -\frac{4}{x^2} + \frac{1}{(x-1)^2} = -\frac{(x-2)(3x-2)}{x^2(x-1)^2}.$$

Since the domain of f is $(1, \infty)$, we need concern ourselves only with $x > 1$. It is easy to see that

$$f'(x) \text{ is } \begin{cases} \text{negative,} & \text{if } 1 < x < \frac{4}{3} \\ 0, & \text{if } x = \frac{4}{3} \\ \text{positive,} & \text{if } x > \frac{4}{3} \end{cases}.$$

This means that f decreases on $(1, \frac{4}{3}]$ and increases on $[\frac{4}{3}, \infty)$. The value

$$f(\tfrac{4}{3}) = 4 \log 4 - 3 \log 3 \cong 2.26 \qquad \text{(Table 5.2.1)}$$

is a local and absolute minimum. There are no other extreme values.

Since

$$f''(x) \text{ is } \begin{cases} \text{positive,} & \text{if } 1 < x < 2 \\ 0, & \text{if } x = 2 \\ \text{negative,} & \text{if } x > 2 \end{cases},$$

the graph is concave up on $(1, 2]$ and concave down on $[2, \infty)$. The point

$$(2, f(2)) = (2, 4 \log 2) \cong (2, 2.76)$$

is the only point of inflection.

Before sketching the graph note that the derivative

$$f'(x) = \frac{4}{x} - \frac{1}{x-1}$$

is very large negative for x close to 1 and very close to 0 for x large. This means that the graph is very steep for x close to 1 and very flat for x large. The graph appears in Figure 5.3.2. $\square$

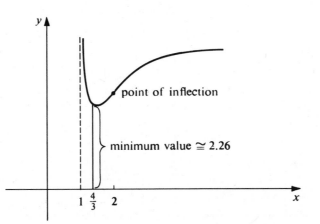

FIGURE 5.3.2

Integration

The integral counterpart of (5.3.1) is that

$$\int \frac{dx}{x} = \log |x| + C.$$

The integral counterpart of (5.3.2) is that

$$\int \frac{g'(x)}{g(x)}\, dx = \log |g(x)| + C,$$

provided of course that $g(x) \neq 0$. We can also derive this last formula by the usual u-substitution: set

$$u = g(x), \qquad du = g'(x)\, dx.$$

$$\int \frac{g'(x)}{g(x)}\, dx = \int \frac{du}{u} = \log |u| + C = \log |g(x)| + C. \quad \square$$

Problem.   Find

$$\int \frac{8x}{x^2 - 1}\, dx.$$

SOLUTION.   Set

$$u = x^2 - 1, \qquad du = 2x\, dx.$$

$$\int \frac{8x}{x^2 - 1}\, dx = 4 \int \frac{2x}{x^2 - 1}\, dx$$

$$= 4 \int \frac{du}{u} = 4 \log |u| + C$$

$$= 4 \log |x^2 - 1| + C. \quad \square$$

Problem.   Find

$$\int \frac{x^2}{1 - 4x^3}\, dx.$$

SOLUTION.   Set

$$u = 1 - 4x^3, \qquad du = -12x^2\, dx.$$

$$\int \frac{x^2}{1 - 4x^3}\, dx = -\frac{1}{12} \int \frac{-12x^2}{1 - 4x^3}\, dx$$

$$= -\frac{1}{12} \int \frac{du}{u} = -\frac{1}{12} \log |u| + C$$

$$= -\frac{1}{12} \log |1 - 4x^3| + C. \quad \square$$

Problem. Find

$$\int \frac{\log x}{x}\, dx.$$

SOLUTION. Set

$$u = \log x, \qquad du = \frac{1}{x}\, dx.$$

$$\int \frac{\log x}{x}\, dx = \int u\, du = \tfrac{1}{2}u^2 + C = \tfrac{1}{2}(\log x)^2 + C. \quad \square$$

Problem. Evaluate

$$\int_1^2 \frac{6x^2 - 2}{x^3 - x + 1}\, dx.$$

SOLUTION. Set

$$u = x^3 - x + 1, \qquad du = (3x^2 - 1)\, dx.$$

At $x = 1$, $u = 1$; at $x = 2$, $u = 7$.

$$\int_1^2 \frac{6x^2 - 2}{x^3 - x + 1}\, dx = 2 \int_1^2 \frac{3x^2 - 1}{x^3 - x + 1}\, dx$$

$$= 2 \int_1^7 \frac{du}{u} = 2 \left[\log |u| \right]_1^7.$$

$$= 2 (\log 7 - \log 1) = 2 \log 7. \quad \square$$

Logarithmic Differentiation

To differentiate a complicated product

$$g(x) = g_1(x)g_2(x) \cdots g_n(x)$$

you can first form

$$\log |g(x)| = \log |g_1(x)| + \log |g_2(x)| + \cdots + \log |g_n(x)|$$

and then differentiate

$$\frac{g'(x)}{g(x)} = \frac{g_1'(x)}{g_1(x)} + \frac{g_2'(x)}{g_2(x)} + \cdots + \frac{g_n'(x)}{g_n(x)}.$$

This process is called *logarithmic differentiation*.

Problem. Use logarithmic differentiation to find $g'(x)$ for

$$g(x) = x(x - 1)(x - 2)(x - 3).$$

SOLUTION

$$\log |g(x)| = \log |x| + \log |x - 1| + \log |x - 2| + \log |x - 3|.$$

$$\frac{g'(x)}{g(x)} = \frac{1}{x} + \frac{1}{x - 1} + \frac{1}{x - 2} + \frac{1}{x - 3},$$

$$g'(x) = x(x - 1)(x - 2)(x - 3) \left(\frac{1}{x} + \frac{1}{x - 1} + \frac{1}{x - 2} + \frac{1}{x - 3} \right). \quad \square$$

You can also use logarithmic differentiation when quotients are involved.

Problem. Use logarithmic differentiation to find $g'(x)$ for

$$g(x) = \frac{(x^2 + 1)^3(2x - 5)^2}{(x^2 + 5)^2}.$$

What is $g'(1)$?

SOLUTION

$$\log |g(x)| = 3 \log |x^2 + 1| + 2 \log |2x - 5| - 2 \log |x^2 + 5|$$

$$\frac{g'(x)}{g(x)} = \frac{6x}{x^2 + 1} + \frac{4}{2x - 5} - \frac{4x}{x^2 + 5}$$

$$g'(x) = \frac{(x^2 + 1)^3(2x - 5)^2}{(x^2 + 5)^2}\left(\frac{6x}{x^2 + 1} + \frac{4}{2x - 5} - \frac{4x}{x^2 + 5}\right)$$

$$g'(1) = \frac{2^3(-3)^2}{6^2}\left(\frac{6}{2} + \frac{4}{(-3)} - \frac{4}{6}\right) = 2. \quad \square$$

Exercises

Find the domain and compute the derivative.

*1. $f(x) = \log 4x.$

2. $f(x) = \log (2x + 1).$

*3. $f(x) = \log (x^3 + 1).$

4. $f(x) = \log [(x + 1)^3].$

*5. $f(x) = \log \sqrt{1 + x^2}.$

6. $f(x) = (\log x)^3.$

*7. $f(x) = \log |x^4 - 1|.$

8. $f(x) = \log (\log x).$

*9. $f(x) = \dfrac{1}{\log x}.$

10. $f(x) = \log \left|\dfrac{x + 2}{x^3 - 1}\right|.$

*11. $f(x) = x \log x.$

12. $f(x) = \log \sqrt[4]{x^2 + 1}.$

*13. $f(x) = \dfrac{\log (x + 1)}{x + 1}.$

14. $f(x) = \log \sqrt{\dfrac{1 - x}{2 - x}}.$

Work out the following integrals.

*15. $\displaystyle\int \frac{dx}{x + 1}.$

16. $\displaystyle\int \frac{dx}{3 - x}.$

*17. $\displaystyle\int \frac{x}{3 - x^2}\,dx.$

18. $\displaystyle\int \frac{x + 1}{x^2}\,dx.$

*19. $\displaystyle\int \frac{x}{(3 - x^2)^2}\,dx.$

20. $\displaystyle\int \frac{\log (x + a)}{x + a}\,dx.$

*21. $\displaystyle\int \left(\frac{1}{x - a} - \frac{1}{x - b}\right)dx.$

22. $\displaystyle\int \frac{x^2}{2x^3 - 1}\,dx.$

*23. $\displaystyle\int x\left(\frac{1}{x^2-a^2}-\frac{1}{x^2-b^2}\right)dx.$ 24. $\displaystyle\int \frac{2x-1}{x(x-1)}\,dx.$

*25. $\displaystyle\int \frac{\sqrt{x}}{1+x\sqrt{x}}\,dx.$ 26. $\displaystyle\int \frac{dx}{x\,(\log x)^2}.$

Evaluate.

*27. $\displaystyle\int_1^e \frac{dx}{x}.$ 28. $\displaystyle\int_1^{e^2} \frac{dx}{x}.$

*29. $\displaystyle\int_e^{e^2} \frac{dx}{x}.$ 30. $\displaystyle\int_3^4 \frac{dx}{1-x}.$

*31. $\displaystyle\int_4^5 \frac{x}{x^2-1}\,dx.$ 32. $\displaystyle\int_1^e \frac{\log x}{x}\,dx.$

*33. $\displaystyle\int_0^1 \frac{\log(x+1)}{x+1}\,dx.$ 34. $\displaystyle\int_0^1 \left(\frac{1}{x+1}-\frac{1}{x+2}\right)dx.$

Find the derivative by logarithmic differentiation.

*35. $g(x)=(x^2+1)^2(x-1)^5x^3.$ 36. $g(x)=(x^2+a^2)^2(x^3+b^3)^3.$

*37. $g(x)=\dfrac{x^4(x-1)}{x+2}.$ 38. $g(x)=\dfrac{(1+x)(2+x)x}{4+x}.$

*39. $g(x)=\sqrt{\dfrac{(x-1)(x-2)}{(x-3)(x-4)}}.$ 40. $g(x)=\dfrac{\sqrt{x^2+1}-x}{\sqrt{x^2+1}+x}.$

41. Find the area of that portion of the first quadrant which lies between
    *(a) $x+4y-5=0$ and $xy=1.$
    (b) $x+y-3=0$ and $xy=2.$

42. A particle moves along a line with acceleration

$$a(t)=-\frac{1}{(t+1)^2}\text{ feet per second per second.}$$

Find the distance traveled by the particle during the time interval $[0,4]$, given
that the initial velocity $v(0)$ is:
    *(a) 1 foot per second.   (b) 2 feet per second.

43. Find formulas for the nth derivatives:

    (a) $\dfrac{d^n}{dx^n}(\log x).$ (b) $\dfrac{d^n}{dx^n}[\log(1-x)].$

44. For each of the functions below, (i) find the domain, (ii) find the intervals where
    it is increasing and the intervals where it is decreasing, (iii) find the extreme

values, (iv) determine the concavity of the graph and find the points of inflection, and finally, (v) sketch the graph.

* (a) $f(x) = \log 2x$.   (b) $f(x) = x - \log x$.

* (c) $f(x) = x \log x$.   (d) $f(x) = \log(8x - x^2)$.

* (e) $f(x) = \log\left(\dfrac{x}{1 + x^2}\right)$.   (f) $f(x) = \log\left(\dfrac{x^3}{x - 1}\right)$.

5.4   The Exponential Function

Rational powers of e already have an established meaning: by $e^{p/q}$ we mean the qth root of e raised to the pth power. But what is meant by $e^{\sqrt{2}}$ or e^{π}?

Earlier we proved that each rational power $e^{p/q}$ had logarithm p/q:

$$(5.4.1) \qquad\qquad \log e^{p/q} = \frac{p}{q}.$$

Inspired by this relation, we make the following definition:

Definition

If z is irrational, then by e^z we mean the unique number which has logarithm z:

$$(5.4.2) \qquad\qquad \log e^z = z.$$

What is $e^{\sqrt{2}}$? It is the unique number which has logarithm $\sqrt{2}$. What is e^{π}? It is the unique number which has logarithm π. Note that e^x now has meaning for every value of x.

Definition

The function

$$E(x) = e^x \quad \text{for all real } x$$

is called the *exponential function*.

Some properties of the exponential function follow.

(1) Combining 5.4.1 and 5.4.2 we have

$$\boxed{\log e^x = x \quad \text{for all real } x.}$$

Writing

$$L(x) = \log x \quad \text{and} \quad E(x) = e^x$$

we have

$$L(E(x)) = x \quad \text{for all real } x.$$

This says that the *exponential function is the inverse of the logarithm function.*

(2) The graph of the exponential function appears in Figure 5.4.1. It can be obtained from the graph of the logarithm by reflection in the line $y = x$.

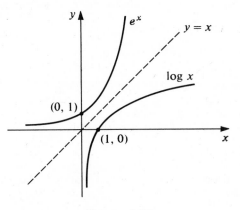

FIGURE 5.4.1

(3) Since the graph of the logarithm remains to the right of the y-axis, the graph of the exponential function remains above the x-axis; namely,

$$\boxed{e^x > 0 \quad \text{for all real } x.}$$

(4) Since the graph of the logarithm crosses the x-axis at $(1, 0)$, the graph of the exponential function crosses the y-axis at $(0, 1)$:

$$\log 1 = 0 \quad \text{gives} \quad e^0 = 1.$$

(5) That the exponential function and logarithm function are inverses can also be expressed by writing

$$\boxed{e^{\log x} = x \qquad \text{for all } x > 0.}$$

You can verify this equation directly by observing that both sides have the same logarithm:

$$\log (e^{\log x}) = \log x \log e = \log x. \quad \square$$

You know that for rational exponents

$$e^{(p/q) + (r/s)} = e^{p/q} \cdot e^{r/s}.$$

This fundamental property holds for all exponents, including the irrational ones.

Theorem 5.4.3

$$e^{x+y} = e^x \cdot e^y \quad \text{for all real } x \text{ and } y.$$

PROOF

$$\log (e^x \cdot e^y) = \log e^x + \log e^y = x + y = \log e^{x+y}.$$

Since the logarithm is one-to-one, we must have

$$e^{x+y} = e^x \cdot e^y. \quad \square$$

We leave it to you to show that

$$\boxed{e^{-y} = \frac{1}{e^y}} \quad \text{and that consequently} \quad \boxed{e^{x-y} = \frac{e^x}{e^y}.}$$

Here are some values of the exponential function rounded off to the nearest hundredth.†

TABLE 5.4.1

t	e^t	t	e^t
0.1	1.11	1.1	3.00
0.2	1.22	1.2	3.32
0.3	1.35	1.3	3.67
0.4	1.49	1.4	4.06
0.5	1.65	1.5	4.48
0.6	1.82	1.6	4.95
0.7	2.01	1.7	5.47
0.8	2.23	1.8	6.05
0.9	2.46	1.9	6.68
1.0	2.72	2.0	7.39

Problem. Use Table 5.4.1 to estimate the following powers of e: (a) $e^{-0.2}$. (b) $e^{2.4}$. (c) $e^{3.1}$.

SOLUTION. The idea, of course, is to use the laws of exponents.

(a) $e^{-0.2} = \dfrac{1}{e^{0.2}} \cong \dfrac{1}{1.22} \cong 0.82.$

(b) $e^{2.4} = e^{2+0.4} = (e^2)(e^{0.4}) \cong (7.39)(1.49) \cong 11.01.$

(c) $e^{3.1} = e^{1.7+1.4} = (e^{1.7})(e^{1.4}) \cong (5.47)(4.06) \cong 22.21. \quad \square$

---

† More extended tables appear at the end of the book, but throughout this chapter all numerical calculations will be based on Table 5.4.1.

We come now to one of the most important results in all of calculus. It is marvelously simple.

Theorem

The exponential function is its own derivative:

$$\frac{d}{dx}(e^x) = e^x.$$

PROOF. The logarithm function is differentiable and its derivative is never 0. It follows (Section 3.7) that its inverse, the exponential function, is also differentiable. Knowing this, we can show that

$$\frac{d}{dx}(e^x) = e^x$$

by differentiating the identity

$$\log e^x = x.$$

On the left-hand side, the chain rule gives

$$\frac{d}{dx}(\log e^x) = \frac{1}{e^x}\frac{d}{dx}(e^x).$$

On the right-hand side, the derivative is 1. Equating the derivatives, we get

$$\frac{1}{e^x}\frac{d}{dx}(e^x) = 1$$

and thus

$$\frac{d}{dx}(e^x) = e^x. \quad \square$$

You will frequently run across expressions of the form $e^{f(x)}$. If f is differentiable, the chain rule gives

$$\boxed{\frac{d}{dx}[e^{f(x)}] = e^{f(x)}f'(x).}^{\dagger}$$

---

† If you have trouble seeing this, write

$$e^{f(x)} = E(f(x)).$$

Then

$$\frac{d}{dx}[e^{f(x)}] = \frac{d}{dx}[E(f(x))] = E'(f(x))f'(x).$$

Since the exponential function is its own derivative,

$$E'(f(x)) = E(f(x)) = e^{f(x)}.$$

The result now follows. $\square$

Examples

$$\frac{d}{dx}(e^{kx}) = e^{kx} \cdot k = ke^{kx}.$$

$$\frac{d}{dx}(e^{\sqrt{x}}) = e^{\sqrt{x}} \frac{d}{dx}(\sqrt{x}) = e^{\sqrt{x}}\left(\frac{1}{2\sqrt{x}}\right) = \frac{1}{2\sqrt{x}}\, e^{\sqrt{x}}.$$

$$\frac{d}{dx}(e^{-x^2}) = e^{-x^2}\frac{d}{dx}(-x^2) = e^{-x^2}(-2x) = -2xe^{-x^2}. \quad \square$$

The relation

$$\frac{d}{dx}(e^x) = e^x \quad \text{and its corollary} \quad \frac{d}{dx}(e^{kx}) = ke^{kx}$$

have important applications to engineering, physics, chemistry, biology, and economics. We will discuss some of these applications in Section 5.6.

Problem.   Let

$$f(x) = xe^{-x} \quad \text{for all real } x.$$

(a) On what intervals is f increasing? Decreasing? (b) Find the extreme values of f. (c) Determine the concavity of the graph and find the points of inflection. (d) Sketch the graph.

SOLUTION.   We have

$$f(x) = xe^{-x},$$
$$f'(x) = xe^{-x}(-1) + e^{-x} = (1 - x)e^{-x},$$
$$f''(x) = (1 - x)e^{-x}(-1) - e^{-x} = (x - 2)e^{-x}.$$

Since

$$f'(x) \text{ is} \begin{cases} \text{positive,} & \text{for } x < 1 \\ 0, & \text{at } x = 1 \\ \text{negative,} & \text{for } x > 1 \end{cases},$$

f increases on $(-\infty, 1]$ and decreases on $[1, \infty)$.
   The number

$$f(1) = \frac{1}{e} \cong \frac{1}{2.72} \cong 0.37$$

is a local and absolute maximum. There are no other extreme values. Since

$$f''(x) \text{ is} \begin{cases} \text{negative,} & \text{for } x < 2 \\ 0, & \text{at } x = 2 \\ \text{positive,} & \text{for } x > 2 \end{cases},$$

the graph is concave down on $(-\infty, 2]$ and concave up on $[2, \infty)$. The point

$$(2, f(2)) = (2, 2e^{-2}) \cong \left(2, \frac{2}{(2.72)^2}\right) \cong (2, 0.27)$$

is the only point of inflection. A sketch of the graph is given in Figure 5.4.2. $\square$

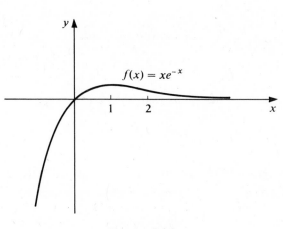

$$f(x) = xe^{-x}$$

FIGURE 5.4.2

Problem. Let

$$f(x) = xe^{x^2} \quad \text{for all } x \in (-1, 1).$$

(a) On what intervals is f increasing? Decreasing? (b) Find the extreme values of f. (c) Determine the concavity of the graph and find the points of inflection. (d) Sketch the graph.

SOLUTION.   We have

$$f(x) = xe^{x^2},$$
$$f'(x) = xe^{x^2}(2x) + e^{x^2} = (1 + 2x^2)e^{x^2},$$
$$f''(x) = (1 + 2x^2)e^{x^2}(2x) + 4xe^{x^2} = 2x(2x^2 + 3)e^{x^2}.$$

Since $f'(x) > 0$ for all $x \in (-1, 1)$, f increases on $(-1, 1)$ and there are no extreme values. Since

$$f''(x) \text{ is } \begin{cases} \text{negative,} & \text{for } x \in (-1, 0) \\ \quad\quad 0, & \text{for } x = 0 \\ \text{positive,} & \text{for } x \in (0, 1) \end{cases},$$

the graph is concave down on $(-1, 0]$ and concave up on $[0, 1)$. The point

$$(0, f(0)) = (0, 0)$$

is the only point of inflection. Since

$$f(-x) = -f(x),$$

f is an odd function and thus its graph, given in Figure 5.4.3, is symmetric with respect to the origin. $\square$

Problem.   Find

$$\int 6e^{3x} \, dx.$$

SOLUTION.   Set

$$u = e^{3x}, \qquad du = 3e^{3x} \, dx.$$

$$\int 6e^{3x} \, dx = 2 \int 3e^{3x} \, dx = 2 \int du = 2u + C = 2e^{3x} + C.$$

If you recognize at the very beginning that

$$3e^{3x} = \frac{d}{dx} (e^{3x}),$$

then you can dispense with the u-substitution and directly write

$$\int 6e^{3x} \, dx = 2 \int 3e^{3x} \, dx = 2e^{3x} + C. \quad \square$$

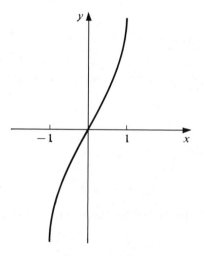

FIGURE 5.4.3

Problem.   Find

$$\int \frac{e^{\sqrt{x}}}{\sqrt{x}} \, dx.$$

SOLUTION.   Set

$$u = e^{\sqrt{x}}, \qquad du = e^{\sqrt{x}} \frac{1}{2\sqrt{x}} \, dx = \frac{1}{2} \left(\frac{e^{\sqrt{x}}}{\sqrt{x}} \right) dx.$$

$$\int \frac{e^{\sqrt{x}}}{\sqrt{x}} \, dx = 2 \int \frac{1}{2} \left(\frac{e^{\sqrt{x}}}{\sqrt{x}} \right) dx = 2 \int du = 2u + C = 2e^{\sqrt{x}} + C.$$

If you recognize immediately that

$$\frac{1}{2}\left(\frac{e^{\sqrt{x}}}{\sqrt{x}}\right) = \frac{d}{dx}(e^{\sqrt{x}}),$$

then you can dispense with the u-substitution and integrate directly:

$$\int \frac{e^{\sqrt{x}}}{\sqrt{x}}\,dx = 2\int \frac{1}{2}\left(\frac{e^{\sqrt{x}}}{\sqrt{x}}\right)dx = 2e^{\sqrt{x}} + C. \quad \square$$

REMARK.   The u-substitution simplifies many calculations but there is no reason to use it where you find that you can carry out the integration more quickly without it.

Problem.   Find

$$\int \frac{e^{3x}}{e^{3x} + 1}\,dx.$$

SOLUTION.   Set

$$u = e^{3x} + 1, \qquad du = e^{3x}\,dx.$$

$$\int \frac{e^{3x}}{e^{3x} + 1}\,dx = \int \frac{du}{u} = \log|u| + C = \log(e^{3x} + 1) + C. \quad \square$$

Problem.   Evaluate

$$\int_0^{\log 2} e^x\,dx.$$

SOLUTION

$$\int_0^{\log 2} e^x\,dx = \left[e^x\right]_0^{\log 2} = e^{\log 2} - e^0 = 2 - 1 = 1.$$

The u-substitution is obviously unnecessary here but if you prefer to use it you certainly can: set

$$u = e^x, \qquad du = e^x\,dx.$$
$$x = 0, \quad u = 1; \qquad x = \log 2, \quad u = 2.$$

$$\int_0^{\log 2} e^x\,dx = \int_1^2 du = \left[u\right]_1^2 = 2 - 1 = 1. \quad \square$$

Problem.   Evaluate

$$\int_0^1 e^x(e^x + 1)^{1/5}\,dx.$$

SOLUTION.   Set

$$u = e^x + 1, \qquad du = e^x\,dx.$$
$$x = 0, \quad u = 2; \qquad x = 1, \quad u = e + 1.$$

$$\int_0^1 e^x(e^x + 1)^{1/5}\, dx = \int_2^{e+1} u^{1/5}\, du$$

$$= \left[\tfrac{5}{6}u^{6/5} \right]_2^{e+1} = \tfrac{5}{6}[(e + 1)^{6/5} - 2^{6/5}] + C. \quad \square$$

Exercises

Differentiate the following functions.

*1. $y = e^{-2x}$.

2. $y = 3e^{2x+1}$.

*3. $y = e^{x^2-1}$.

4. $y = 2e^{-4x}$.

*5. $y = e^x \log x$.

6. $y = x^2 e^x$.

*7. $y = x^{-1}e^{-x}$.

8. $y = e^{\sqrt{x}+1}$.

*9. $y = \tfrac{1}{2}(e^x + e^{-x})$.

10. $y = \tfrac{1}{2}(e^x - e^{-x})$.

*11. $y = e^{\sqrt{x}} \log \sqrt{x}$.

12. $y = (1 - e^{4x})^2$.

*13. $y = e^{\sqrt{1-x^2}}$.

14. $y = x^2 e^x - x e^{x^2}$.

*15. $y = \dfrac{e^x - 1}{e^x + 1}$.

16. $y = \dfrac{e^{2x} - 1}{e^{2x} + 1}$.

Work out the following indefinite integrals.

*17. $\displaystyle\int e^{2x}\, dx$.

18. $\displaystyle\int e^{-2x}\, dx$.

*19. $\displaystyle\int e^{kx}\, dx$.

20. $\displaystyle\int e^{ax+b}\, dx$.

*21. $\displaystyle\int x e^{x^2}\, dx$.

22. $\displaystyle\int x e^{-x^2}\, dx$.

*23. $\displaystyle\int \frac{e^{1/x}}{x^2}\, dx$.

24. $\displaystyle\int \frac{e^{2\sqrt{x}}}{\sqrt{x}}\, dx$.

*25. $\displaystyle\int (e^{-x} - 1)^2\, dx$.

26. $\displaystyle\int \frac{4}{\sqrt{e^x}}\, dx$.

*27. $\displaystyle\int \frac{e^x}{e^x + 1}\, dx$.

28. $\displaystyle\int \frac{e^x}{\sqrt{e^x + 1}}\, dx$.

*29. $\displaystyle\int \frac{2e^x}{\sqrt[3]{e^x + 1}}\, dx$.

30. $\displaystyle\int \frac{e^{2x}}{2e^{2x} + 3}\, dx$.

Evaluate the following definite integrals.

*31. $\displaystyle\int_0^1 e^x\, dx$.

32. $\displaystyle\int_0^1 e^{-kx}\, dx$.

* 33. $\displaystyle\int_0^{\log \pi} e^{-6x}\, dx.$

34. $\displaystyle\int_0^1 xe^{-x^2}\, dx.$

* 35. $\displaystyle\int_0^1 \frac{e^x + e^{-x}}{2}\, dx.$

36. $\displaystyle\int_0^e \frac{3}{x + 2}\, dx.$

* 37. $\displaystyle\int_0^1 \frac{4 - e^x}{e^x}\, dx.$

38. $\displaystyle\int_0^1 \frac{e^x}{4 - e^x}\, dx.$

* 39. $\displaystyle\int_0^1 x(e^{x^2} + 2)\, dx.$

40. $\displaystyle\int_1^2 (2 - e^{-x})^2\, dx.$

Estimate by using Table 5.4.1.

* 41. $e^{-0.4}$.   42. $e^{2.6}$.   *43. $e^{2.8}$.   44. $e^{-2.1}$.

Estimate by using differentials.

* 45. $e^{2.03}$ $(e^2 \cong 7.39)$.   46. $e^{-0.15}$ $(e^0 = 1)$.
* 47. $e^{3.15}$ $(e^3 \cong 20.09)$.   48. e^{2a+h} $(e^a = A)$.

49. A particle moves along a coordinate line so that its position at time t is given by

$$x(t) = Ae^{ct} + Be^{-ct}.$$

Show that the acceleration of the particle is proportional to its position.

Sketch the region between the curves and compute the area.

50. $y = e^x$, $x \in [-1, 1]$; $y = e^{-x}$, $x \in [-1, 1]$.
* 51. $y = e^{ax}$, $x \in [-a, a]$; $y = e^{-ax}$, $x \in [-a, a]$.

52. We refer here to Figure 5.4.4 on page 240.

  (a) Find the points of tangency.
  (b) Find the area of region I.
  (c) Find the area of region II.

For each of the functions below, (i) find the domain, (ii) find the intervals where it is increasing and the intervals where it is decreasing, (iii) find the extreme values, (iv) determine the concavity and find the points of inflection, and finally, (v) sketch the graph.

* 53. $f(x) = e^{x^2}$.   54. $f(x) = e^{-x^2}$.
* 55. $f(x) = \frac{1}{2}(e^x + e^{-x})$.   56. $f(x) = \frac{1}{2}(e^x - e^{-x})$.
* 57. $f(x) = xe^x$.   58. $f(x) = (1 - x)e^x$.
* 59. $f(x) = e^{(1/x)^2}$.   60. $f(x) = x^2 e^{-x}$.

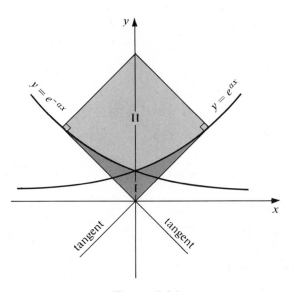

FIGURE 5.4.4

61. (*Optional*)   Prove that for all $x > 0$ and all positive integers n

$$e^x > 1 + x + \frac{x^2}{2!} + \frac{x^3}{3!} + \cdots + \frac{x^n}{n!},$$

where $n!$, read "n factorial," is shorthand for

$$n(n - 1)(n - 2) \cdots 2 \cdot 1.$$

HINT: $e^x = 1 + \displaystyle\int_0^x e^t \, dt > 1 + \int_0^x dt = 1 + x$

$$e^x = 1 + \int_0^x e^t \, dt > 1 + \int_0^x (1 + t) \, dt = 1 + x + \frac{x^2}{2}, \quad \text{etc.}$$

62. (*Optional*)   Prove that if n is a positive integer, then

$$e^x > x^n \quad \text{for all } x \text{ sufficiently large.}$$

HINT: Use Exercise 61.

5.5   Arbitrary Powers; Other Bases; Estimating e

The Function x^r

The elementary notion of exponent applies only to rational numbers. Expressions such as

$$10^{1/3}, \qquad 2^{-4/5}$$

make sense, but so far we have attached no meaning to expressions such as

$$10^{\sqrt{2}}, \qquad 2^{-\pi}.$$

The extension of our sense of exponent to allow for irrational exponents is conveniently done by making use of the logarithm and exponential functions. The heart of the matter is to observe that for $x > 0$ and p/q rational

$$x^{p/q} = e^{(p/q) \log x}.$$

(To verify this, take the log of both sides.) The right-hand side of this equation already makes sense for irrational numbers z; namely,

$$e^{z \log x}$$

already makes sense. To define x^z, we can simply set

$$x^z = e^{z \log x}.$$

Definition

If z is irrational and x is positive, we define the number x^z by setting

$$x^z = e^{z \log x}.$$

Having made this definition, we now have

$$x^r = e^{r \log x} \qquad \text{for all real numbers } r.$$

For example,

$$10^{\sqrt{2}} = e^{\sqrt{2} \log 10} \quad \text{and} \quad 2^{-\pi} = e^{-\pi \log 2}. \quad \square$$

With this extended sense of exponent the usual laws of exponents

$$x^{r+s} = x^r x^s, \qquad x^{r-s} = \frac{x^r}{x^s}, \qquad (x^r)^s = x^{rs}$$

still hold:

$$x^{r+s} = e^{(r+s) \log x} = e^{r \log x} \cdot e^{s \log x} = x^r x^s,$$

$$x^{r-s} = e^{(r-s) \log x} = e^{r \log x} \cdot e^{-s \log x} = \frac{e^{r \log x}}{e^{s \log x}} = \frac{x^r}{x^s},$$

$$(x^r)^s = e^{s \log x^r} = e^{rs \log x} = x^{rs}. \quad \square$$

The differentiation of arbitrary powers has a familiar look:

$$\frac{d}{dx}(x^r) = rx^{r-1}.$$

PROOF

$$\frac{d}{dx}(x^r) = \frac{d}{dx}(e^{r \log x}) = e^{r \log x}\frac{d}{dx}(r \log x) = x^r\frac{r}{x} = rx^{r-1}. \quad \square$$

For example,

$$\frac{d}{dx}(x^{\sqrt{2}}) = \sqrt{2}\, x^{\sqrt{2}-1}, \qquad \frac{d}{dx}(x^\pi) = \pi x^{\pi-1}. \quad \square$$

If g is a positive differentiable function, then

$$\frac{d}{dx}[g(x)]^r = r[g(x)]^{r-1}g'(x).$$

You can verify this by using the chain rule. (Exercise 12.) For example,

$$\frac{d}{dx}[(x^2 + 5)^{\sqrt{3}}] = \sqrt{3}(x^2 + 5)^{\sqrt{3}-1}(2x) = 2\sqrt{3}\, x(x^2 + 5)^{\sqrt{3}-1}. \quad \square$$

Problem.   Find

$$\int \frac{x^3}{(2x^4 + 1)^\pi}\, dx.$$

SOLUTION.   Set

$$u = 2x^4 + 1, \qquad du = 8x^3\, dx.$$

$$\int \frac{x^3}{(2x^4 + 1)^\pi}\, dx = \frac{1}{8}\int \frac{8x^3}{(2x^4 + 1)^\pi}\, dx$$

$$= \frac{1}{8}\int \frac{du}{u^\pi} = \frac{1}{8}\int u^{-\pi}\, du = \frac{1}{8}\left(\frac{u^{1-\pi}}{1-\pi}\right) + C$$

$$= \frac{(2x^4 + 1)^{1-\pi}}{8(1 - \pi)} + C. \quad \square$$

Problem.   Find

$$\frac{d}{dx}(x^x).$$

SOLUTION.   One way to do this is to observe that

$$x^x = e^{x \log x}$$

and then differentiate:

$$\frac{d}{dx}(x^x) = \frac{d}{dx}(e^{x \log x}) = e^{x \log x}\left(x \cdot \frac{1}{x} + \log x\right)$$

$$= x^x(1 + \log x).$$

Another way to do this problem is to set

$$g(x) = x^x$$

and use logarithmic differentiation:

$$\log g(x) = x \log x,$$

$$\frac{g'(x)}{g(x)} = x \cdot \frac{1}{x} + \log x = 1 + \log x,$$

$$g'(x) = g(x)(1 + \log x) = x^x(1 + \log x). \quad \square$$

The Function p^x

To form the function

$$f(x) = x^r$$

we take a positive variable x and raise it to a constant power r. To form the function

$$g(x) = p^x$$

we take a positive constant p and raise it to a variable power x.

The high status enjoyed by Euler's number e comes from the fact that

$$\frac{d}{dx}(e^x) = e^x.$$

For other bases the derivative is a little more complicated:

$$\boxed{\frac{d}{dx}(p^x) = p^x \log p.}$$

PROOF

$$\frac{d}{dx}(p^x) = \frac{d}{dx}\left[e^{x \log p}\right] = e^{x \log p} \log p = p^x \log p. \quad \square$$

The chain rule gives

$$\boxed{\frac{d}{dx}\left[p^{f(x)}\right] = p^{f(x)} \log p \, f'(x).}$$

Examples

$$\frac{d}{dx}(e^x) = e^x \quad \text{but} \quad \frac{d}{dx}(2^x) = 2^x \log 2.$$

$$\frac{d}{dx}(e^{3x^2}) = e^{3x^2}(6x) \quad \text{but} \quad \frac{d}{dx}(2^{3x^2}) = 2^{3x^2}(\log 2)(6x). \quad \square$$

Problem. Find

$$\int 2^x \, dx.$$

SOLUTION.   Set

$$u = 2^x, \qquad du = 2^x \log 2 \, dx.$$

$$\int 2^x \, dx = \frac{1}{\log 2} \int 2^x \log 2 \, dx$$

$$= \frac{1}{\log 2} \int du = \frac{1}{\log 2} u + C$$

$$= \frac{1}{\log 2} 2^x + C.$$

Without the *u*-substitution:

$$\int 2^x \, dx = \frac{1}{\log 2} \int 2^x \log 2 \, dx = \frac{1}{\log 2} 2^x + C. \quad \square$$

Problem. Evaluate

$$\int_1^2 3^{-x} \, dx.$$

SOLUTION.   Set

$$u = 3^{-x}, \qquad du = -3^{-x} \log 3 \, dx.$$

$$x = 1, \quad u = \tfrac{1}{3}; \qquad x = 2, \quad u = \tfrac{1}{9}.$$

$$\int_1^2 3^{-x} \, dx = -\frac{1}{\log 3} \int_1^2 -3^{-x} \log 3 \, dx$$

$$= -\frac{1}{\log 3} \int_{1/3}^{1/9} du = -\frac{1}{\log 3} \left(\frac{1}{9} - \frac{1}{3} \right) = \frac{2}{9 \log 3}.$$

Without the *u*-substitution:

$$\int_1^2 3^{-x} \, dx = -\frac{1}{\log 3} \int_1^2 -3^{-x} \log 3 \, dx$$

$$= -\frac{1}{\log 3} \left[3^{-x} \right]_1^2 = -\frac{1}{\log 3} (3^{-2} - 3^{-1})$$

$$= -\frac{1}{\log 3} \left(\frac{1}{9} - \frac{1}{3} \right) = \frac{2}{9 \log 3}. \quad \square$$

The Function $\log_p x$

If p is positive, then

$$\log p^t = t \log p.$$

If in addition p is different from 1, then $\log p \neq 0$, and therefore

$$\frac{\log p^t}{\log p} = t.$$

This means that the function

$$f(x) = \frac{\log x}{\log p}$$

satisfies the relation

$$f(p^t) = t.$$

In view of this we call

$$\frac{\log x}{\log p}$$

the log *of x to the base p* and write

$$\boxed{\log_p x = \frac{\log x}{\log p}.}$$

As examples, we have

$$\log_2 32 = \frac{\log 32}{\log 2} = \frac{\log 2^5}{\log 2} = \frac{5 \log 2}{\log 2} = 5$$

and

$$\log_{100} \left(\tfrac{1}{10}\right) = \frac{\log \left(\tfrac{1}{10}\right)}{\log 100} = \frac{\log 10^{-1}}{\log 10^2} = \frac{-\log 10}{2 \log 10} = -\frac{1}{2}. \quad \square$$

We can obtain the same results more quickly by using the fact that

$$\boxed{\log_p p^t = t.}$$

Thus

$$\log_2 32 = \log_2 2^5 = 5 \quad \text{and} \quad \log_{100} \left(\tfrac{1}{10}\right) = \log_{100} \left(100^{-1/2}\right) = -\tfrac{1}{2}. \quad \square$$

By differentiating

$$\log_p x = \frac{\log x}{\log p}$$

we obtain the formula

$$\frac{d}{dx}(\log_p x) = \frac{1}{x \log p} \cdot \quad \text{†}$$

When p is e, the factor $\log p$ in the denominator becomes 1, and the formula is simply

$$\frac{d}{dx}(\log_e x) = \frac{1}{x} \cdot$$

We view the logarithm to the base e

$$\log = \log_e$$

as the "natural logarithm" because it is the logarithm with the simplest derivative.

Estimating the Number e

We have defined the number e as the number which satisfies

$$1 = \int_1^e \frac{dt}{t} \cdot$$

It is time to discuss how to obtain a numerical estimate for e.

Since e is irrational (Exercise 28, Section 10.8), we cannot hope to express e as a

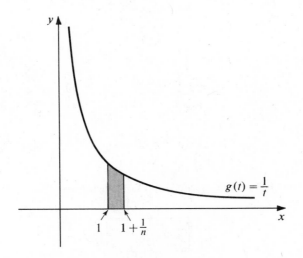

FIGURE 5.5.1

---

† The function $f(x) = \log_p x$ satisfies

$$f'(x) = \frac{1}{x \log p}, \qquad f'(1) = \frac{1}{\log p} \cdot$$

This means that in general

$$f'(x) = \frac{1}{x} f'(1).$$

We predicted this from general considerations in Section 5.1. (See Formula 5.1.1.)

terminating decimal. (Nor as a repeating decimal.) What we can do is describe a simple technique by which one can compute the value of e to any desired degree of accuracy.

Theorem 5.5.1

For each positive integer n,

$$\left(1 + \frac{1}{n}\right)^n \le e \le \left(1 + \frac{1}{n}\right)^{n+1}.$$

PROOF

$$\log\left(1 + \frac{1}{n}\right) = \int_1^{1+1/n} \frac{dt}{t} \le \int_1^{1+1/n} 1\ dt = \frac{1}{n}.$$

$$\text{since } \frac{1}{t} \le 1 \text{ for all } t \text{ in } \left[1, 1 + \frac{1}{n}\right]$$

From

$$\log\left(1 + \frac{1}{n}\right) \le \frac{1}{n}$$

we get

$$\left(1 + \frac{1}{n}\right) \le e^{1/n}$$

and thus

$$\left(1 + \frac{1}{n}\right)^n \le e.$$

Now

$$\log\left(1 + \frac{1}{n}\right) = \int_1^{1+1/n} \frac{dt}{t} \ge \int_1^{1+1/n} \frac{dt}{1 + 1/n} = \frac{1}{1 + 1/n} \cdot \frac{1}{n} = \frac{1}{n+1}.$$

$$\text{since } \frac{1}{t} \ge \frac{1}{1 + 1/n} \text{ for all } t \text{ in } \left[1, 1 + \frac{1}{n}\right]$$

From

$$\log\left(1 + \frac{1}{n}\right) \ge \frac{1}{n+1}$$

we get

$$\left(1 + \frac{1}{n}\right) \ge e^{1/(n+1)}$$

and thus

$$\left(1 + \frac{1}{n}\right)^{n+1} \ge e. \quad \square$$

It is easy to see how to estimate e by using Theorem 5.5.1. Since

$$(1 + \tfrac{1}{1})^1 = 2 \quad \text{and} \quad (1 + \tfrac{1}{1})^2 = 4,$$

we have

$$2 \le e \le 4.$$

Since
$$(1 + \tfrac{1}{2})^2 = \tfrac{9}{4} = 2.25 \quad \text{and} \quad (1 + \tfrac{1}{2})^3 = \tfrac{27}{8} = 3.375,$$
we have
$$2.25 \le e \le 3.375.$$
With a little more patience you can verify that
$$2.59 \le (1 + \tfrac{1}{10})^{10} \quad \text{and} \quad (1 + \tfrac{1}{10})^{11} \le 2.86$$
and thus arrive at
$$2.59 \le e \le 2.86.$$
Lengthier calculations show that
$$2.70 \le (1 + \tfrac{1}{100})^{100} \quad \text{and} \quad (1 + \tfrac{1}{100})^{101} \le 2.74$$
and thus
$$2.70 \le e \le 2.74.$$
Still lengthier calculations show that
$$2.71 \le (1 + \tfrac{1}{200})^{200} \quad \text{and} \quad (1 + \tfrac{1}{200})^{201} \le 2.73,$$
and thus
$$2.71 \le e \le 2.73.$$
A look at an eight-place table reveals the estimate
$$2.71828182.$$
For our purposes the estimate 2.72 is sufficient. □
   The inequality
$$\left(1 + \frac{1}{n}\right)^n \le e \le \left(1 + \frac{1}{n}\right)^{n+1}$$
is rather elegant but, as you saw, it does not provide a very efficient way of estimating e. (We have to go to $n = 200$ to show that $2.71 \le e \le 2.73$.) We shall have a much more efficient way of estimating e when we have infinite series at our disposal.

Exercises

1. Find.
   *(a) $\log_2 64$.
   (b) $\log_2 \tfrac{1}{64}$.
   (c) $\log_{64} \tfrac{1}{2}$.
   *(d) $\log_{10} 0.01$.
   (e) $\log_5 1$.
   *(f) $\log_5 0.2$.

2. Find.
   *(a) $\displaystyle\int 3^x \, dx.$

   (b) $\displaystyle\int x^3 \, dx.$

   *(c) $\displaystyle\int 2^{-x} \, dx.$

   *(d) $\displaystyle\int x10^{x^2} \, dx.$

   (e) $\displaystyle\int x10^{-x^2} \, dx.$

   *(f) $\displaystyle\int (2^x + 2^{-x}) \, dx.$

   *(g) $\displaystyle\int \frac{dx}{x \log 5}.$

   (h) $\displaystyle\int \frac{\log_5 x}{x} \, dx.$

3. Find $f'(e)$.

   *(a) $f(x) = \log_3 x$.   (b) $f(x) = x \log_3 x$.

   *(c) $f(x) = \log (\log x)$.   (d) $f(x) = \log_3 (\log_2 x)$.

4. Find by logarithmic differentiation.

   *(a) $\dfrac{d}{dx} [(x + 1)^x]$.   (b) $\dfrac{d}{dx} [(\log x)^x]$.

   *(c) $\dfrac{d}{dx} [(x^2 + 2)^{\log x}]$.   *(d) $\dfrac{d}{dx} \left[\left(\dfrac{1}{x}\right)^x \right]$.

   (e) $\dfrac{d}{dx} [(\log x)^{x^2 + 2}]$.   *(f) $\dfrac{d}{dx} [(\log x)^{\log x}]$.

5. Sketch figures in which you compare the following pairs of graphs.

   (a) $f(x) = e^x$ and $g(x) = 2^x$.   (b) $f(x) = e^x$ and $g(x) = 3^x$.

   (c) $f(x) = e^x$ and $g(x) = e^{-x}$.   (d) $f(x) = 2^x$ and $g(x) = 2^{-x}$.

   (e) $f(x) = 3^x$ and $g(x) = 3^{-x}$.   (f) $f(x) = \log x$ and $g(x) = \log_2 x$.

   (g) $f(x) = 2^x$ and $g(x) = \log_2 x$.   (h) $f(x) = 10^x$ and $g(x) = \log_{10} x$.

6. Show that if a, b, c are positive then

$$\log_a c = \log_a b \, \log_b c$$

provided that a and b are both different from 1.

7. Show that

   (a) $\log_a xy = \log_a x + \log_a y$,   (b) $\log_a \dfrac{x}{y} = \log_a x - \log_a y$,

   (c) $\log_a x^y = y \log_a x$.

8. Find those numbers x, if any, for which

   *(a) $10^x = e^x$.   (b) $\log_5 x = 0.04$.

   *(c) $\log_x 2 = \log_3 x$.   *(d) $\log_x 10 = \log_4 100$.

   *(e) $\log_2 x = \displaystyle\int_2^x \dfrac{dt}{t}$.   (f) $\log_x 10 = \log_2 (\tfrac{1}{10})$.

9. *(a) Estimate $\log a$ given that

$$e^{t_1} < a < e^{t_2}.$$

   *(b) Estimate e^b given that

$$\log x_1 < b < \log x_2.$$

10. For each of the functions below (i) specify the domain, (ii) find the intervals where the function is increasing and those where it is decreasing, (iii) find the extreme values.

   *(a) $f(x) = 10^{1 - x^2}$.   *(b) $f(x) = 10^{1/(1 - x^2)}$.

   *(c) $f(x) = 10^{\sqrt{1 - x^2}}$.   (d) $f(x) = \log_{10} \sqrt{1 - x^2}$.

11. Evaluate.

*(a) $\displaystyle\int_0^1 (1 + x)^{2\pi}\, dx.$ (b) $\displaystyle\int_1^2 2^{-x}\, dx.$

*(c) $\displaystyle\int_0^1 4^x\, dx.$ (d) $\displaystyle\int_0^1 x10^{1+x^2}\, dx.$

*(e) $\displaystyle\int_1^4 \frac{dx}{x \log 2}.$ (f) $\displaystyle\int_{10}^{100} \frac{dx}{x \log_{10} x}.$

*(g) $\displaystyle\int_0^2 p^{x/2}\, dx.$ (h) $\displaystyle\int_0^1 \frac{5p^{\sqrt{x+1}}}{\sqrt{x+1}}\, dx.$

12. Prove that if g is positive and differentiable then for each real exponent r

$$\frac{d}{dx}[g(x)]^r = r[g(x)]^{r-1}g'(x).$$

13. Use Table 5.2.1 to estimate the following logarithms:
    *(a) $\log_{10} 4.$ (b) $\log_{10} 7.$ *(c) $\log_{10} 12.$ (d) $\log_{10} 45.$
14. (*Optional*) Prove that
$$\lim_{h \to 0} (1 + h)^{1/h} = e.$$

HINT: Since at 1 the logarithm function has derivative 1,

$$\lim_{h \to 0} \frac{\log (1 + h) - \log 1}{h} = 1.$$

5.6   Exponential Growth and Decline

An exponential of the form
$$f(x) = Ce^{kx}$$
has the property that its derivative $f'(x)$ is proportional at each point to $f(x)$:
$$f'(x) = Cke^{kx} = kCe^{kx} = kf(x).$$
Moreover it is the oniy such function:

Theorem

If
$$f'(x) = kf(x) \quad \text{for all } x \text{ in some interval } I,$$
then f is of the form
$$f(x) = Ce^{kx} \quad \text{for all } x \text{ in } I.$$

PROOF

$$f'(x) = kf(x),$$
$$f'(x) - kf(x) = 0,$$
$$e^{-kx}f'(x) - ke^{-kx}f(x) = 0,$$

$$\frac{d}{dx}[e^{-kx}f(x)] = 0,$$
$$e^{-kx}f(x) = C,$$
$$f(x) = Ce^{kx}. \quad \square$$

The constant C is the value of f at 0:

$$f(0) = Ce^0 = C.$$

This is usually called the *initial value* of f.

Problem. A function everywhere defined has the property that at each point of its graph the slope is twice the y-coordinate. Given that the graph passes through the point $(0, \sqrt{2})$, find the function.

SOLUTION

$$f'(x) = 2f(x) \quad \text{for all real } x.$$

This means that f is of the form

$$f(x) = Ce^{2x}.$$

Since the graph passes through the point $(0, \sqrt{2})$, we know that

$$f(0) = \sqrt{2} = C.$$

The function must be

$$f(x) = \sqrt{2}e^{2x}. \quad \square$$

Problem. The number of bacteria present in a given culture increases at a rate proportional to the number present. When first observed, the culture contained n_0 bacteria, an hour later, n_1.

(a) Find the number present t hours after the observations began.
(b) How long did it take for the number of bacteria to double?

SOLUTION. (a) Let $n(t) = $ the number of bacteria present at time t. The basic equation is of the form

$$n'(t) = kn(t).$$

From this it follows that

$$n(t) = Ce^{kt}.$$

Since initially (at time 0) there were n_0 bacteria, we know that $C = n_0$. The equation is therefore

$$n(t) = n_0e^{kt}.$$

Since $n(1) = n_1$, we know that

$$n_1 = n_0 e^k,$$

$$\frac{n_1}{n_0} = e^k,$$

and therefore

$$\left(\frac{n_1}{n_0}\right)^t = e^{kt}.$$

The general law of growth is

$$n(t) = n_0 \left(\frac{n_1}{n_0}\right)^t.$$

(b) To find out how long it took for the number of bacteria to double we set

$$2n_0 = n_0 \left(\frac{n_1}{n_0}\right)^t$$

and solve for t:

$$2 = \left(\frac{n_1}{n_0}\right)^t,$$

$$\log 2 = t(\log n_1 - \log n_0),$$

$$t = \frac{\log 2}{\log n_1 - \log n_0} \text{ hours.} \quad \square$$

Problem. The rate of decay of radioactive material is proportional to the amount of such material present.

(a) Find the amount of material present t years later given that the initial amount is A_0 pounds and it takes 5 years for a third of the material to decay.
(b) How long does it take for half of the material to decay?

SOLUTION. (a) Let $A(t)$ be the amount of radioactive material present at time t. Since the rate of decay is proportional to the amount present, we know that

$$A'(t) = kA(t),$$

$$A(t) = Ce^{kt}.$$

Since A_0 is the initial amount, we have $C = A_0$. The general equation is thus of the form

(1) $$A(t) = A_0 e^{kt}.$$

At the end of 5 years, one-third of A_0 has decayed and therefore two-thirds of A_0 remains:

$$A(5) = \tfrac{2}{3}A_0.$$

We can use this relation to eliminate k from (1):

$$A(5) = A_0 e^{5k} = \tfrac{2}{3} A_0,$$

$$e^{5k} = \tfrac{2}{3},$$

$$e^k = (\tfrac{2}{3})^{1/5},$$

and consequently (1) becomes

$$A(t) = A_0 (\tfrac{2}{3})^{t/5}.$$

(b) Here we must find the value of t for which

$$A_0 (\tfrac{2}{3})^{t/5} = \tfrac{1}{2} A_0.$$

Dividing by A_0 we have

$$(\tfrac{2}{3})^{t/5} = \tfrac{1}{2}$$

so that

$$(\tfrac{2}{3})^t = (\tfrac{1}{2})^5,$$

$$t \log \tfrac{2}{3} = 5 \log \tfrac{1}{2},$$

$$t(\log 2 - \log 3) = -5 \log 2.$$

With $\log 2 \cong 0.69$ and $\log 3 \cong 1.10$, you'll find that $t \cong 8.41$ years. It takes about 8 years and 5 months for half of the material to decay. $\square$

Compound Interest

Consider money invested at interest rate r. If the accumulated interest is credited once a year, then the interest is said to be compounded annually; if twice a year, then semiannually; if four times a year, then quarterly. Many savings banks are now compounding interest daily.

The idea can be pursued further. Interest can be credited every hour, every second, every half-second, and so on. In the limiting case, interest is credited instantaneously. Economists call this *continuous compounding*.

The economists' formula for continuous compounding is a simple exponential:

$$\boxed{A(t) = A_0 e^{rt}.}$$

Here t is measured in years,

$$A(t) = \text{the principal in dollars at time } t,$$

$$A_0 = A(0) = \text{the initial investment},$$

$$r = \text{the annual interest rate}.$$

The rate r is called the *nominal* interest rate. Compounding makes the effective rate higher.

A derivation of the compound interest formula. We take h as a small time increment and observe that

$$A(t + h) - A(t) = \text{interest earned from time } t \text{ to time } t + h.$$

Had the principal remained $A(t)$ from time t to time $t + h$, the interest earned during this time period would have been

$$rhA(t).$$

Had the principal been $A(t + h)$ throughout the time interval, the interest earned would have been

$$rhA(t + h).$$

The actual interest earned must be somewhere in between:

$$rhA(t) \le A(t + h) - A(t) \le rhA(t + h).$$

Dividing by h, we get

$$rA(t) \le \frac{A(t + h) - A(t)}{h} \le rA(t + h).$$

If A varies continuously, then as h tends to zero, $rA(t + h)$ tends to $rA(t)$ and (by the pinching theorem) the difference quotient in the middle must also tend to $rA(t)$:

$$\lim_{h \to 0} \frac{A(t + h) - A(t)}{h} = rA(t).$$

This says that

$$A'(t) = rA(t),$$

from which it follows that

$$A(t) = Ce^{rt}.$$

With A_0 the initial investment, $C = A_0$, and

$$A(t) = A_0 e^{rt}. \quad \square$$

Problem. Find the amount of interest drawn by \$100 compounded continuously at 6% for 5 years.

SOLUTION

$$A(5) = 100e^{(0.06)5} = 100e^{0.30} \cong 135. \qquad \text{(Table 5.4.1)}$$

The interest drawn is approximately \$35. $\square$

Problem. A sum of money is drawing interest at the rate of 10% compounded continuously. What is the effective interest rate?

SOLUTION.   At the end of one year each dollar grows to

$$e^{0.10} \cong 1.11.$$

The interest accrued is approximately 11¢. The effective interest rate is about 11%. $\square$

Problem. How long does it take a sum of money to double at 5% compounded continuously?

SOLUTION.   In general

$$A(t) = A_0 e^{0.05t}.$$

We set

$$2A_0 = A_0 e^{0.05t}$$

and solve for t:

$$2 = e^{0.05t},$$
$$\log 2 = 0.05t = \tfrac{1}{20}t,$$
$$t = 20 \log 2 \cong 13.8. \qquad\qquad \text{(Table 5.2.1)}$$

It takes about 13 years, 9 months, and 18 days. ☐

Money today is better than money tomorrow because during that time we could be earning interest.

Problem. What is the present value of $1000 forty months from now? Assume continuous compounding at 6%.

SOLUTION.   We begin with

$$A(t) = A_0 e^{0.06t}$$

and solve for A_0:

$$A_0 = A(t)e^{-0.06t}.$$

Since forty months is $\tfrac{10}{3}$ years, $t = \tfrac{10}{3}$ and

$$A_0 = A(\tfrac{10}{3})e^{(-0.06)(10/3)} = 1000e^{-0.20} \cong 820.$$

The present value of $1000 forty months from now is about $820. ☐

Problem. A producer of brandy finds that the value of his inventory increases with time (the brandy improves with age) according to the formula

$$V(t) = V_0 e^{\sqrt{t}/2}. \qquad\qquad (t \text{ measured in years})$$

How long should he keep the brandy in storage? Neglect storage costs and assume continuous compounding at 8%.

SOLUTION.   Future dollars should be discounted by a factor of $e^{-0.08t}$. What we want to maximize is not $V(t)$ but the product

$$f(t) = V(t)e^{-0.08t} = V_0 e^{(\sqrt{t}/2)-0.08t}.$$

Differentiation gives

$$f'(t) = V_0 e^{(\sqrt{t}/2)-0.08t}\left(\frac{1}{4\sqrt{t}} - 0.08\right).$$

Setting $f'(t) = 0$, we find that

$$\frac{1}{4\sqrt{t}} = 0.08,$$

$$\sqrt{t} = \frac{1}{0.32},$$

$$t = \left(\frac{1}{0.32}\right)^2 \cong 9.76.$$

He should store it for about nine years and nine months. ☐

Exercises

1. A positive function everywhere defined has the property that at each point of its graph the slope is three times the *y*-coordinate. Find the function given that the graph passes through the point
   * (a) (0, 2).   (b) (1, 1).   * (c) (2, *e*).

2. Find the amount of interest drawn by $500 compounded continuously for 10 years
   * (a) at 5%.   (b) at 6%.   * (c) at 7%.

3. What is the present value of a $1000 bond that matures 5 years from now? Assume continuous compounding
   * (a) at 4%.   (b) at 6%.   * (c) at 8%.

4. How long does it take for a sum of money to double when compounded continuously
   * (a) at 4%?   (b) at 6%?   * (c) at 8%?

5. At what rate *r* of continuous compounding does a sum of money increase by a factor of *e*
   (a) in one year?   (b) in *n* years?

6. At what rate *r* of continuous compounding does a sum of money
   * (a) triple in 20 years?   (b) double in 10 years?

* 7. Water is run into a tank to dilute a saline solution. The volume *V* of the mixture is kept constant. Given that *s*, the amount of salt in the tank, varies with respect to *x*, the amount of water which has been run through, according to the formula

$$\frac{ds}{dx} = -\frac{s}{V},$$

find the amount of water which must be used to wash down 50% of the salt. Take *V* as 10,000 gallons.

8. (*Newton's law of cooling*) If the excess temperature of a body above the temperature of the surrounding air is *x* degrees, the time rate of decrease of *x* is proportional to *x*. If this excess temperature was at first 80 degrees, and after one minute is 70 degrees, what will it be after two minutes? In how many minutes will it decrease 20 degrees?

* 9. Atmospheric pressure *p* varies with altitude *h* according to the law

$$\frac{dp}{dh} = kp, \quad \text{where } k \text{ is a constant.}$$

Given that *p* is 15 pounds per square inch at sea level and 10 pounds per square inch at 10,000 feet, find *p* at
   (a) 5000 feet.   (b) 15,000 feet.

10. Given that

$$f'(t) = k[2 - f(t)] \quad \text{and} \quad f(0) = 0$$

show that the constant k satisfies the equation

$$k = \frac{1}{t} \log \left[\frac{2}{2 - f(t)} \right].$$

*11. In the inversion of raw sugar, the time rate of change varies as the amount of raw sugar remaining. If after 10 hours, 1000 pounds of raw sugar have been reduced to 800 pounds, how much raw sugar will remain after 20 hours?

12. The rate of decay of a radioactive substance is proportional to the amount of the substance present.
   (a) How long does it take for half of the substance to decay
      *(i) if it takes 4 years for a quarter of the substance to decay?
       (ii) if it takes 3 years for a quarter of the substance to decay?
   (b) A year ago there were 4 pounds of the substance. Now there are 3 pounds.
      *(i) How much was there 2 years ago?
       (ii) How much was there 10 years ago?
      *(iii) How much will remain 10 years from now?

13. Population tends to grow with time at a rate roughly proportional to the population present. According to the Bureau of the Census, the population of the United States in 1960 was approximately 179 million and in 1970, 205 million.
   (a) Use this information to estimate the population of 1940. (The actual figure was about 132 million.)
   (b) Predict the populations for 1980 and for the year 2000.
   (c) Estimate how long it takes for the population to double.

*14. A lumber company finds from experience that the value of its standing timber increases with time according to the formula

$$V(t) = V_0 (\tfrac{3}{2})^{\sqrt{t}}.$$

Here t is measured in years and V_0 is the value at planting time. How long should the company wait before cutting the timber? Neglect costs and assume continuous compounding at 5%.

5.7 Integration by Parts

We begin with the differentiation formula

$$f(x)g'(x) + f'(x)g(x) = (f \cdot g)'(x).$$

Integrating both sides, we get

$$\int f(x)g'(x)\, dx + \int f'(x)g(x)\, dx = \int (f \cdot g)'(x)\, dx.$$

Since

$$\int (f \cdot g)'(x)\, dx = f(x)g(x) + C,$$

we have

$$\int f(x)g'(x)\, dx + \int f'(x)g(x)\, dx = f(x)g(x) + C$$

and therefore

$$\int f(x)g'(x)\, dx = f(x)g(x) - \int f'(x)g(x)\, dx + C.$$

Since the computation of

$$\int f'(x)g(x)\, dx$$

will yield its own arbitrary constant, there is no reason to keep the constant C. We therefore drop it and write

$$\boxed{\int f(x)g'(x)\, dx = f(x)g(x) - \int f'(x)g(x)\, dx.}$$

This formula, called the formula for *integration by parts*, enables us to find

$$\int f(x)g'(x)\, dx$$

by computing

$$\int f'(x)g(x)\, dx$$

instead. It is of course of practical use only if the second integral is easier to compute than the first.

In practice we usually set

$$u = f(x), \qquad dv = g'(x)\, dx,$$
$$du = f'(x)\, dx, \qquad v = g(x).$$

The formula for integration by parts then becomes

$$\boxed{\int u\, dv = uv - \int v\, du.}$$

Problem. Find

$$\int xe^x\, dx.$$

SOLUTION. Set

$$u = x, \qquad dv = e^x\, dx,$$
$$du = dx, \qquad v = e^x.$$

$$\int xe^x\, dx = \int u\, dv = uv - \int v\, du = xe^x - \int e^x\, dx = xe^x - e^x + C. \quad \square$$

Problem. Find

$$\int \log x \, dx.$$

SOLUTION. Set

$$u = \log x, \qquad dv = dx,$$

$$du = \frac{1}{x} \, dx, \qquad v = x.$$

$$\int \log x \, dx = \int u \, dv = uv - \int v \, du$$

$$= x \log x - \int x \left(\frac{1}{x}\right) dx = x \log x - \int dx$$

$$= x \log x - x + C. \quad \square$$

Problem. Find

$$\int \frac{xe^x}{(x + 1)^2} \, dx.$$

SOLUTION.   The challenge here is to figure out how to choose u and dv. After several false starts we found that the following decomposition works out well:

$$u = xe^x, \qquad\qquad\qquad dv = \frac{1}{(x + 1)^2} \, dx,$$

$$du = (xe^x + e^x) \, dx = (x + 1)e^x \, dx, \qquad v = -\frac{1}{x + 1}.$$

$$\int \frac{xe^x}{(x + 1)^2} \, dx = \int u \, dv = uv - \int v \, du$$

$$= -\frac{xe^x}{x + 1} + \int e^x \, dx$$

$$= -\frac{xe^x}{x + 1} + e^x + C = \frac{e^x}{x + 1} + C. \quad \square$$

To compute some integrals you may have to integrate by parts more than once.

Problem.   Evaluate

$$\int_0^1 x^2 e^x \, dx.$$

SOLUTION.   First we compute the indefinite integral

$$\int x^2 e^x \, dx.$$

We set

$$u = x^2, \qquad dv = e^x \, dx,$$
$$du = 2x \, dx, \qquad v = e^x.$$

$$\int x^2 e^x \, dx = \int u \, dv = uv - \int v \, du = x^2 e^x - \int 2x e^x \, dx.$$

We now compute the integral on the right again by parts. This time we set

$$u = 2x, \qquad dv = e^x \, dx,$$
$$du = 2 \, dx, \qquad v = e^x.$$

$$\int 2x e^x \, dx = \int u \, dv = uv - \int v \, du = 2x e^x - \int 2e^x \, dx = 2x e^x - 2e^x + C.$$

This together with our earlier calculations gives

$$\int x^2 e^x \, dx = x^2 e^x - 2x e^x + 2e^x + C.$$

For

$$\int_0^1 x^2 e^x \, dx$$

we have

$$\left[x^2 e^x - 2x e^x + 2e^x \right]_0^1 = (e - 2e + 2e) - 2 = e - 2. \quad \square$$

Exercises

Compute the following integrals.

*1. $\int x e^{-x} \, dx.$

2. $\int x \log x \, dx.$

*3. $\int x \log (x + 1) \, dx.$

4. $\int x 2^x \, dx.$

*5. $\int x^2 \log x \, dx.$

6. $\int \log (-x) \, dx.$

*7. $\int \sqrt{x} \log x \, dx.$

8. $\int x^2 e^{-x^3} \, dx.$

*9. $\int x^3 e^{-x^2} \, dx.$

10. $\int x^2 (e^x - 1) \, dx.$

*11. $\int \dfrac{\log (x + 1)}{\sqrt{x + 1}} \, dx.$

12. $\int \dfrac{x^2}{\sqrt{1 - x}} \, dx.$

*13. $\displaystyle\int x\sqrt{x+1}\ dx.$

14. $\displaystyle\int x\log x^2\ dx.$

*15. $\displaystyle\int (\log x)^2\ dx.$

16. $\displaystyle\int x^2 e^{-x}\ dx.$

*17. $\displaystyle\int \frac{dx}{x\,(\log x)^3}.$

18. $\displaystyle\int \frac{(e^x + 2x)^2}{2}\ dx.$

*19. $\displaystyle\int_1^e x^n \log x\ dx.$

20. $\displaystyle\int_0^1 (2^x + x^2)^2\ dx.$

5.8   (Optional) The Equation $y'(x) + P(x)y(x) = Q(x)$

Differential equations (equations involving functions and their derivatives) arise so frequently in scientific problems that their study constitutes an important branch of mathematics.

The exponential function, being fixed under differentiation, that is, being endowed with the property

$$\frac{d}{dx}(e^x) = e^x,$$

plays an interesting role in the theory of differential equations.

In the case of linear differential equations

$$y^n(x) + P_1(x)y^{n-1}(x) + \cdots + P_{n-1}(x)y'(x) + P_n(x)y(x) = Q(x),$$

the exponential function plays a dominant role. In this section we consider the first-order case:

(1) $$y'(x) + P(x)y(x) = Q(x).$$

We assume that the coefficients P and Q are both continuous. To solve Equation (1) we begin by multiplying both sides by $e^{\int P(x)\,dx}$:

$$e^{\int P(x)\,dx}\,y'(x) + e^{\int P(x)\,dx}\,P(x)\,y(x) = e^{\int P(x)\,dx}\,Q(x).$$

Once we've done this, the left-hand side becomes

$$\frac{d}{dx}\left[e^{\int P(x)\,dx}y(x)\right]$$

and we can rewrite the equation as

$$\frac{d}{dx}\left[e^{\int P(x)\,dx}y(x)\right] = e^{\int P(x)\,dx}Q(x).$$

We can now integrate both sides and get

$$e^{\int P(x)\,dx}y(x) = \int e^{\int P(x)\,dx}Q(x)\ dx + C.$$

To get $y(x)$ by itself, we multiply the equation by $e^{-\int P(x)\,dx}$:

$$y(x) = e^{-\int P(x)\,dx}\left\{\int e^{\int P(x)\,dx}\,Q(x) + C\right\}.$$

This is called the *general solution* of Equation (1). Different choices of C give different *particular solutions*.

Example. To solve the differential equation

$$y'(x) + 2y(x) = e^x$$

we multiply through by

$$e^{\int 2\,dx} = e^{2x}. \quad \text{(we don't carry the constant here)}$$

This gives

$$e^{2x}y'(x) + e^{2x}2y(x) = e^{3x},$$

$$\frac{d}{dx}\left[e^{2x}y(x)\right] = e^{3x},$$

$$e^{2x}y(x) = \tfrac{1}{3}e^{3x} + C,$$

$$y(x) = \tfrac{1}{3}e^x + Ce^{-2x}.$$

This is the general solution. Different choices of C give different particular solutions. Setting $C = 0$, we get

$$y(x) = \tfrac{1}{3}e^x.$$

Setting $C = 1$, we get

$$y(x) = \tfrac{1}{3}e^x + e^{-2x}.$$

If we want the solution which takes on the value 1 at 0 then we set

$$y(0) = \tfrac{1}{3} + C = 1.$$

This forces

$$C = \tfrac{2}{3}$$

and

$$y(x) = \tfrac{1}{3}e^x + \tfrac{2}{3}e^{-2x}.$$

More generally, if we want the solution which takes on the value y_0 at x_0 then we set

$$y(x_0) = \tfrac{1}{3}e^{x_0} + Ce^{-2x_0} = y_0.$$

This forces

$$C = (y_0 - \tfrac{1}{3}e^{x_0})e^{2x_0}$$

and

$$y(x) = \tfrac{1}{3}e^x + (y_0 - \tfrac{1}{3}e^{x_0})e^{2(x_0-x)}. \quad \square$$

Example.   To solve the differential equation

$$xy'(x) - 2y(x) = 2$$

we first divide the equation by x so that the leading coefficient becomes 1:

$$y'(x) - \frac{2}{x}\,y(x) = \frac{2}{x}\,.$$

Then we multiply through by

$$e^{\int -2x^{-1}\,dx} = e^{-2\log x} = e^{\log x^{-2}} = x^{-2}.$$

This gives

$$x^{-2}y'(x) - 2x^{-3}y(x) = 2x^{-3},$$

$$\frac{d}{dx}\,[x^{-2}y(x)] = 2x^{-3},$$

$$x^{-2}y(x) = -x^{-2} + C,$$

$$y(x) = Cx^2 - 1.$$

This is the general solution.   ☐

Problem.   Water from a polluted reservoir (pollution level p_0 grams per gallon) is constantly being drawn off at the rate of n gallons an hour and replaced by less polluted water (pollution level p_1 grams per gallon). Given that the capacity of the reservoir is M gallons, how long will it take to reduce the pollution level to p grams per gallon?

SOLUTION.   Let $A(t) =$ total number of grams of pollution in the reservoir at time t. We want to find the time t at which

$$\frac{A(t)}{M} = p.$$

At each time t, pollutants leave the reservoir at the rate of

$$n\,\frac{A(t)}{M} \quad \text{grams per hour}$$

and enter the reservoir at the rate of

$$np_1 \quad \text{grams per hour.}$$

It follows that

$$A'(t) = -n\,\frac{A(t)}{M} + np_1$$

and

$$A'(t) + \frac{n}{M}\,A(t) = np_1.$$

To solve the equation for $A(t)$ we multiply through by

$$e^{\int (n/M)\,dt} = e^{nt/M}$$

This gives

$$e^{nt/M} A'(t) + e^{nt/M} \frac{n}{M} A(t) = np_1 e^{nt/M},$$

$$\frac{d}{dt} \left[e^{nt/M} A(t) \right] = np_1 e^{nt/M},$$

$$e^{nt/M} A(t) = Mp_1 e^{nt/M} + C,$$

$$A(t) = Mp_1 + Ce^{-nt/M},$$

$$\frac{A(t)}{M} = p_1 + \frac{C}{M} e^{-nt/M}.$$

The constant C is determined by the initial pollution level p_0:

$$p_0 = \frac{A(0)}{M} = p_1 + \frac{C}{M}$$

so that

$$C = M(p_0 - p_1).$$

We therefore have

$$\frac{A(t)}{M} = p_1 + (p_0 - p_1)e^{-nt/M}.$$

This equation gives the pollution level at each time t. To find the time at which the pollution level has been reduced to p, we set

$$p_1 + (p_0 - p_1)e^{-nt/M} = p$$

and solve for t:

$$e^{-nt/M} = \frac{p - p_1}{p_0 - p_1},$$

$$-\frac{n}{M} t = \log \frac{p - p_1}{p_0 - p_1},$$

$$t = \frac{M}{n} \log \frac{p_0 - p_1}{p - p_1}. \quad \square$$

Problem. In the problem just considered, how long would it take to reduce the pollution by a factor of 50% if the reservoir contained 50,000,000 gallons, the water were replaced at the rate of 25,000 gallons per hour, and the replacement water were completely pollution free?

SOLUTION.   In general

$$t = \frac{M}{n} \log \frac{p_0 - p_1}{p - p_1}.$$

Here

$$M = 50,000,000,$$
$$n = 25,000,$$
$$p_1 = 0,$$
$$p = \tfrac{1}{2}p_0,$$

so that

$$t = 2000 \log 2 \cong 2000(0.69) = 1380.$$

It would take about 1380 hours.   □

Exercises

Find the general solution.

*1. $y'(x) - 2y(x) = 1.$

*3. $2y'(x) + 5y(x) = 2.$

*5. $y'(x) - 2y(x) = 1 - 2x.$

*7. $xy'(x) - 3y(x) = -2nx.$

*9. $y'(x) + \dfrac{2}{x+1}\, y(x) = (x+1)^{5/2}.$

*11. $xy'(x) + y(x) = (1+x)e^x.$

2. $xy'(x) - 2y(x) = -x.$

4. $y'(x) - y(x) = -2e^{-x}.$

6. $xy'(x) - 2y(x) = -3x.$

8. $y'(x) + y(x) = 2 + 2x.$

10. $y'(x) + y(x) = \dfrac{1}{1+e^x}.$

12. $y'(x) - y(x) = e^x.$

Find the particular solution determined by the given side condition.

*13. $y'(x) + y(x) = x, \quad y(0) = 1.$

*15. $y'(x) + y(x) = \dfrac{1}{1+e^x}, \quad y(0) = e.$

16. $y'(x) + y(x) = \dfrac{1}{1+2e^x}, \quad y(0) = e.$

*17. $xy'(x) - 2y(x) = x^3e^x, \quad y(1) = 0.$

*19. $y'(x) + \dfrac{2}{x}\, y(x) = e^{-x}, \quad y(1) = -1.$

20. $y'(x) - \dfrac{2}{x+1}\, y(x) = (x+1)^3, \quad y(0) = 1.$

14. $y'(x) - y(x) = e^{2x}, \quad y(1) = 1.$

18. $x^2y'(x) + 2xy(x) = 1, \quad y(1) = 2.$

*21. A 200-gallon tank initially full of water develops a leak at the bottom. Given that 20% of the water leaks out in the first 5 minutes, find the amount of water left in the tank t minutes after the leak develops
   (a) if the water drains off at a rate that is proportional to the amount of water present;
   (b) if the water drains off at a rate that is proportional to the product of the time elapsed and the amount of water present.

22. An object falling in air is subject not only to gravitational force but also to air resistance. Find $v(t)$, the velocity of the object at time t, given that

$$v'(t) + Kv(t) = 32, \qquad K > 0,$$

and

$$v(0) = 0.$$

Show that $v(t)$ cannot exceed $32/K$. ($32/K$ is called the *terminal velocity*.)

*23. At a certain moment a 100-gallon mixing tank is full of brine containing 0.25 pound of salt per gallon. Find the amount of salt present t minutes later if the brine is being continuously drawn off at the rate of 3 gallons per minute and replaced by brine containing 0.2 pound of salt per gallon.

24. The current i in an electric circuit varies with time according to the formula

$$L\frac{di}{dt} + Ri = E.$$

Take E (the voltage), L (the inductance), and R (the resistance) as constants. Measure the current in amperes and t in seconds and suppose that initially the current is 0.

(a) Find a formula for the current at each subsequent time t.

(b) What upper limit does the current approach as t increases?

(c) In how many seconds will the current reach 90% of its upper limit?

5.9 Additional Exercises

*1. Find the minimum value of

$$y = ae^{kx} + be^{-kx}, \qquad a > 0, \quad b > 0.$$

2. Show that if

$$y = \tfrac{1}{2}a(e^{x/a} + e^{-x/a})$$

then

$$y'' = \frac{y}{a^2}.$$

*3. Find the points of inflection of the graph of

$$y = e^{-x^2}.$$

4. A rectangle has one side on the x-axis and the two upper vertices on the graph of

$$y = e^{-x^2}.$$

At what points should the upper vertices be so as to maximize the area of the rectangle?

*5. A telegraph cable consists of a core of copper wires with a covering made of nonconducting material. If x denotes the ratio of the radius of the core to the thickness of the covering, it is known that the speed of signaling is proportional to $x^2 \log (1/x)$. For what value of x is the speed a maximum?

6. A rectangle has two sides along the coordinate axes and one vertex at a point P which moves along the curve $y = e^x$ in such a way that y increases at the rate of $\frac{1}{2}$ unit per minute. How fast is the area of the rectangle increasing when $y = 3$?

*7. Find

$$\lim_{h \to 0} \frac{1}{h} (e^h - 1).$$

8. Show that

$$y = Ae^{\alpha x} + Be^{-\alpha x}$$

satisfies the differential equation

$$y'' - \alpha^2 y = 0.$$

9. Find a function y which satisfies the differential equation of Exercise 8 and the initial conditions below.

*(a) $y(0) = 0$, $y'(0) = \alpha$.   (b) $y(0) = \alpha$, $y'(0) = 0$.
*(c) $y(0) = 1$, $y'(0) = \alpha$.   (d) $y(0) = \alpha^2$, $y(1) = \alpha$.

10. Find the extreme values and points of inflection of

$$y = \frac{x}{\log x}.$$

*11. Find the area of the region bounded by the curve $xy = a^2$, the x-axis, and the vertical lines $x = a$, $x = 2a$.

12. A boat moving in still water is subject to a retardation proportional to its velocity. Show that the velocity t seconds after the power is shut off is given by the formula $v = ce^{-kt}$, where c is the velocity at the instant the power is shut off.

*13. At a certain instant a boat drifting in still water has a velocity of 4 miles per hour. One minute later the velocity is 2 miles per hour. Find the distance moved.

The Trigonometric and Hyperbolic Functions

6.1 Differentiating the Trigonometric Functions

An outline of what you are expected to remember about trigonometry appears in Chapter 1.

Just as the calculus of logarithms is simplified by the use of the base e, the calculus of the trigonometric functions is simplified by the use of radian measure. Throughout our discussion we will use radian measure and refer to degrees only in passing.

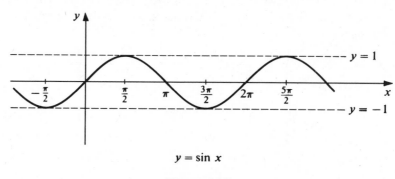

$$y = \sin x$$

FIGURE 6.1.1

We begin our discussion with the sine function, the graph of which appears in Figure 6.1.1.

To differentiate the sine function we will form the difference quotient

$$\frac{\sin (x + h) - \sin x}{h}$$

and take its limit as h tends to 0. Since

$$\sin (x + h) = \sin x \cos h + \cos x \sin h,$$

we obtain for $h \neq 0$

$$\frac{\sin (x + h) - \sin x}{h} = \cos x \frac{\sin h}{h} - \sin x \frac{1 - \cos h}{h}.$$

We are thus led to consider the behavior of

$$\frac{\sin h}{h} \quad \text{and} \quad \frac{1 - \cos h}{h}$$

as h tends to 0. In the exercises you are asked to show that

$$|\sin h| \leq |h|$$

and then that

$$\lim_{h \to 0} \sin h = 0 \quad \text{and} \quad \lim_{h \to 0} \cos h = 1.$$

Assuming these results, we can prove that

$$\lim_{h \to 0} \frac{\sin h}{h} = 1 \quad \text{and} \quad \lim_{h \to 0} \frac{1 - \cos h}{h} = 0.$$

PROOF.   First we prove that

$$\lim_{h \to 0} \frac{\sin h}{h} = 1.$$

For small $h > 0$ (see Figure 6.1.2 on page 270),

$$\text{area of triangle } OAP = \frac{1}{2} \sin h,$$

$$\text{area of sector} = \frac{1}{2} h,$$

$$\text{area of triangle } OAQ = \frac{1}{2} \tan h = \frac{1}{2} \frac{\sin h}{\cos h}.$$

Since

$$\text{triangle } OAP \subseteq \text{sector} \subseteq \text{triangle } OAQ$$

(and these are all proper containments), we must have

$$\frac{1}{2}\sin h < \frac{1}{2}h < \frac{1}{2}\frac{\sin h}{\cos h},$$

$$\sin h < h < \frac{\sin h}{\cos h},$$

$$\frac{\cos h}{\sin h} < \frac{1}{h} < \frac{1}{\sin h},$$

$$\cos h < \frac{\sin h}{h} < 1.$$

This last inequality was derived for $h > 0$, but, since

$$\cos(-h) = \cos h \quad \text{and} \quad \frac{\sin(-h)}{-h} = \frac{-\sin h}{-h} = \frac{\sin h}{h},$$

the inequality also holds for $h < 0$.
    With

$$\cos h < \frac{\sin h}{h} < 1$$

together with

$$\lim_{h \to 0} \cos h = 1,$$

we can conclude from the pinching theorem that

$$\lim_{h \to 0} \frac{\sin h}{h} = 1.$$

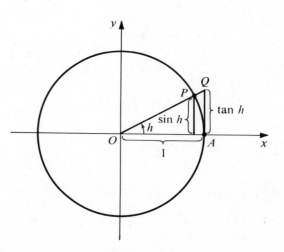

FIGURE 6.1.2

The proof that

$$\lim_{h \to 0} \frac{1 - \cos h}{h} = 0$$

is easier:

$$\left| \frac{1 - \cos h}{h} \right| \leq \left| \frac{1 - \cos h}{\sin h} \right| = \left| \frac{\sin h}{1 + \cos h} \right| \to \frac{0}{2} = 0.$$

since $|\sin h| < |h|$ since $1 - \cos^2 h = \sin^2 h$

The details are not hard to fill in.   □

From

$$\frac{\sin (x + h) - \sin x}{h} = \cos x \, \frac{\sin h}{h} - \sin x \, \frac{1 - \cos h}{h}$$

together with

$$\lim_{h \to 0} \frac{\sin h}{h} = 1 \quad \text{and} \quad \lim_{h \to 0} \frac{1 - \cos h}{h} = 0,$$

it follows that

$$\boxed{\frac{d}{dx} (\sin x) = \cos x.}$$

The other differentiation formulas are now easy to obtain. Since

$$\cos x = \sin \left(\frac{\pi}{2} - x \right),$$

the chain rule gives

$$\frac{d}{dx} (\cos x) = \frac{d}{dx} \left[\sin \left(\frac{\pi}{2} - x \right) \right] = -\cos \left(\frac{\pi}{2} - x \right) = -\sin x.$$

In short

$$\boxed{\frac{d}{dx} (\cos x) = -\sin x.}$$

Powers,

$$[\cos x]^2, \quad [\sin x]^2, \quad \text{etc.}$$

are usually written

$$\cos^2 x, \quad \sin^2 x, \quad \text{etc.}$$

Since

$$\tan x = \frac{\sin x}{\cos x},$$

we have

$$\frac{d}{dx}(\tan x) = \frac{\cos x \, \dfrac{d}{dx}(\sin x) - \sin x \, \dfrac{d}{dx}(\cos x)}{\cos^2 x}$$

$$= \frac{\cos^2 x + \sin^2 x}{\cos^2 x}$$

$$= \frac{1}{\cos^2 x}$$

$$= \sec^2 x$$

and thus the formula

$$\boxed{\frac{d}{dx}(\tan x) = \sec^2 x.}$$

The derivatives of the other trigonometric functions are as follows:

$$\boxed{\begin{aligned} \frac{d}{dx}(\cot x) &= -\operatorname{cosec}^2 x, \\[2mm] \frac{d}{dx}(\sec x) &= \sec x \tan x, \\[2mm] \frac{d}{dx}(\operatorname{cosec} x) &= -\operatorname{cosec} x \cot x. \end{aligned}}$$

The verification of these formulas is left as an exercise.   □

Problem.   Find

$$\frac{d}{dx}[(\cos ax)(\sin ax)].$$

SOLUTION

$$\frac{d}{dx}[(\cos ax)(\sin ax)] = \cos ax \, \frac{d}{dx}(\sin ax) + \sin ax \, \frac{d}{dx}(\cos ax)$$

$$= (\cos ax)(a \cos ax) + (\sin ax)(-a \sin ax)$$

$$= a(\cos^2 ax - \sin^2 ax)$$

$$= a \cos 2ax. \quad □$$

Problem.   Find

$$\frac{d}{dx}\left[\tan\left(\sqrt{1-x^2}\right)\right].$$

SOLUTION

$$\frac{d}{dx}\left[\tan\left(\sqrt{1-x^2}\right)\right] = \sec^2\left(\sqrt{1-x^2}\right)\cdot\frac{d}{dx}\left(\sqrt{1-x^2}\right)$$

$$= \sec^2\left(\sqrt{1-x^2}\right)\frac{-x}{\sqrt{1-x^2}}$$

$$= -\frac{x\sec^2\left(\sqrt{1-x^2}\right)}{\sqrt{1-x^2}}. \quad \square$$

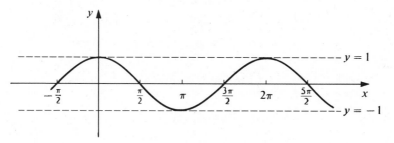

$$y = \cos x$$

FIGURE 6.1.3

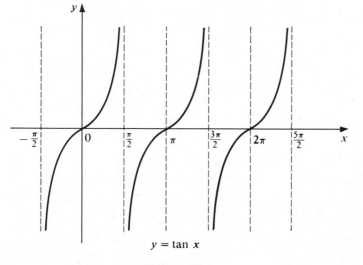

$$y = \tan x$$

FIGURE 6.1.4

Problem. Find

$$\frac{d}{dx} (\log |\cos x|).$$

SOLUTION

$$\frac{d}{dx} (\log |\cos x|) = \frac{1}{\cos x} \frac{d}{dx} (\cos x)$$

$$= \frac{1}{\cos x} (-\sin x)$$

$$= -\tan x. \quad \square$$

Problem. Find

$$\frac{d}{dx} (\sqrt{x} \sec \sqrt{x}).$$

SOLUTION

$$\frac{d}{dx} (\sqrt{x} \sec \sqrt{x}) = \sqrt{x} \frac{d}{dx} (\sec \sqrt{x}) + (\sec \sqrt{x}) \frac{d}{dx} (\sqrt{x})$$

$$= \sqrt{x} (\sec \sqrt{x})(\tan \sqrt{x}) \left(\frac{1}{2\sqrt{x}}\right) + (\sec \sqrt{x}) \left(\frac{1}{2\sqrt{x}}\right)$$

$$= \frac{\sec \sqrt{x}}{2\sqrt{x}} [(\sqrt{x} \tan \sqrt{x}) + 1]. \quad \square$$

Problem. Find

$$\frac{d}{dx} (x^{\sin x}).$$

SOLUTION

$$\frac{d}{dx} (x^{\sin x}) = \frac{d}{dx} (e^{\sin x \log x})$$

$$= e^{\sin x \log x} \frac{d}{dx} (\sin x \log x)$$

$$= x^{\sin x} \left[\sin x \frac{d}{dx} (\log x) + \log x \frac{d}{dx} (\sin x) \right]$$

$$= x^{\sin x} \left[\sin x \left(\frac{1}{x}\right) + \log x \cos x \right].$$

We can also set

$$g(x) = x^{\sin x}$$

and use logarithmic differentiation:

$$\log g(x) = \sin x \log x,$$

$$\frac{g'(x)}{g(x)} = \sin x \left(\frac{1}{x}\right) + \cos x \log x,$$

$$g'(x) = g(x)\left[\sin x \left(\frac{1}{x}\right) + \cos x \log x\right]$$

$$= x^{\sin x}\left[\sin x \left(\frac{1}{x}\right) + \cos x \log x\right]. \quad \square$$

Problem.   Given that

$$\log y = \sin (x + y)$$

find dy/dx by implicit differentiation.

SOLUTION

$$\frac{1}{y}\frac{dy}{dx} = \cos (x + y)\left[1 + \frac{dy}{dx}\right]$$

$$\frac{dy}{dx} = y \cos (x + y)\left[1 + \frac{dy}{dx}\right]$$

$$[1 - y \cos (x + y)]\frac{dy}{dx} = y \cos (x + y)$$

$$\frac{dy}{dx} = \frac{y \cos (x + y)}{1 - y \cos (x + y)}. \quad \square$$

Problem.   Use a differential to estimate sin 0.43.

SOLUTION.   From Table 6.1.1 on page 276,

$$\sin 0.40 \cong 0.389 \quad \text{and} \quad \cos 0.40 \cong 0.921.$$

What we need is an estimate for the increase of

$$f(x) = \sin x$$

from 0.40 to 0.43. Differentiation gives

$$f'(x) = \cos x$$

so that in general

$$df = f'(x)h = (\cos x)h.$$

With $x = 0.40$ and $h = 0.03$, df becomes

$$(\cos 0.40)(0.03) \cong (0.92)(0.03) \cong 0.028.$$

A change from 0.40 to 0.43 increases the sine by approximately 0.028:

$$\sin 0.43 \cong \sin 0.40 + 0.028 \cong 0.389 + 0.028 = 0.417. \quad \square$$

TABLE 6.1.1†

x	$\sin x$	$\cos x$	$\tan x$
0	0.000	1.000	0.000
0.1	0.100	0.995	0.100
0.2	0.199	0.980	0.203
0.3	0.296	0.955	0.309
0.4	0.389	0.921	0.423
0.5	0.479	0.878	0.546
0.6	0.565	0.825	0.684
0.7	0.644	0.765	0.842
0.8	0.717	0.697	1.030
0.9	0.783	0.622	1.260
1.0	0.841	0.540	1.557
1.1	0.891	0.454	1.965
1.2	0.932	0.362	2.572
1.3	0.964	0.267	3.602
1.4	0.985	0.170	5.798
1.5	0.997	0.071	14.101
1.57	1.000	0.001	1255.770

$\pi \cong 3.14159$

Problem.   Given that $1° \cong 0.0175$ radians, use a differential to estimate $\tan 46°$.

SOLUTION.   Table 6.1.2 gives

$$\tan 45° = 1, \qquad \sec 45° = \frac{1}{\cos 45°} = \sqrt{2}.$$

Note that

$$45° = \frac{\pi}{4} \text{ radians} \quad \text{and} \quad 46° \cong \frac{\pi}{4} + 0.0175 \text{ radians}.$$

TABLE 6.1.2

x	$\sin x$	$\cos x$	$\tan x$
0 [0°]	0	1	0
$\pi/6$ [30°]	1/2	$\sqrt{3}/2$	$1/\sqrt{3}$
$\pi/4$ [45°]	$1/\sqrt{2}$	$1/\sqrt{2}$	1
$\pi/3$ [60°]	$\sqrt{3}/2$	1/2	$\sqrt{3}$
$\pi/2$ [90°]	1	0	—
$2\pi/3$ [120°]	$\sqrt{3}/2$	$-1/2$	$-\sqrt{3}$
$3\pi/4$ [135°]	$1/\sqrt{2}$	$-1/\sqrt{2}$	-1
$5\pi/6$ [150°]	1/2	$-\sqrt{3}/2$	$-1/\sqrt{3}$
π [180°]	0	-1	0

$1° \cong 0.0175$
radians

$\sqrt{2} \cong 1.414$
$\sqrt{3} \cong 1.732$

† More extended tables appear at the end of the book.

For
$$f(x) = \tan x$$
the differential is
$$df = f'(x)h = (\sec^2 x)h.$$
It follows that
$$\tan 46° \cong \tan \left(\frac{\pi}{4} + 0.0175 \right) \cong \tan \frac{\pi}{4} + \left(\sec^2 \frac{\pi}{4} \right)(0.0175)$$
$$= 1 + 2(0.0175) = 1.035. \quad \square$$

Problem. Let
$$f(x) = \sin^2 x, \qquad x \in [0, \pi].$$

(a) On what intervals is f increasing? Decreasing? (b) Find the extreme values of f.
(c) Determine the concavity of the graph and find the points of inflection. (d) Sketch the graph.

SOLUTION.   Here
$$f(x) = \sin^2 x,$$
$$f'(x) = 2 \sin x \frac{d}{dx} (\sin x) = 2 \sin x \cos x = \sin 2x,$$
$$f''(x) = 2 \cos 2x.$$

Obviously f is positive between 0 and π and it is 0 at the endpoints. This makes 0 an endpoint minimum and also the absolute minimum.
Since
$$f'(x) \text{ is } \begin{cases} \text{positive,} & \text{for } 0 < x < \pi/2 \\ 0, & \text{at } x = \pi/2 \\ \text{negative,} & \text{for } \pi/2 < x < \pi \end{cases},$$
f increases on $[0, \pi/2]$ and decreases on $[\pi/2, \pi]$. The number
$$f(\pi/2) = 1$$
is a local and absolute maximum. There are no other extreme values. Since
$$f''(x) \text{ is } \begin{cases} \text{positive,} & \text{for } 0 < x < \pi/4 \\ 0, & \text{at } x = \pi/4 \\ \text{negative,} & \text{for } \pi/4 < x < 3\pi/4 \\ 0, & \text{at } x = 3\pi/4 \\ \text{positive} & \text{for } 3\pi/4 < x < \pi \end{cases},$$
the graph is concave up on $[0, \pi/4]$, concave down on $[\pi/4, 3\pi/4]$, and concave up again on $[3\pi/4, \pi]$. The points of inflection are
$$(\pi/4, f(\pi/4)) = (\pi/4, \tfrac{1}{2}) \quad \text{and} \quad (3\pi/4, f(3\pi/4)) = (3\pi/4, \tfrac{1}{2}).$$

The graph is given in Figure 6.1.5. $\square$

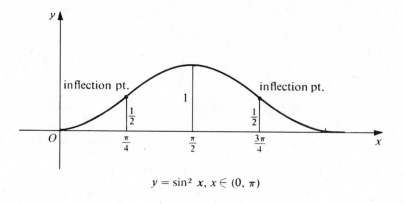

$$y = \sin^2 x, \, x \in (0, \pi)$$

FIGURE 6.1.5

Problem.   Sketch the graph of

$$y = e^{-x/4} \sin \tfrac{1}{2}\pi x, \qquad x \geq 0.$$

SOLUTION.   Since

$$-1 \leq \sin \tfrac{1}{2}\pi x \leq 1,$$

the graph will stay entirely between the curves

$$y = -e^{-x/4} \quad \text{and} \quad y = e^{-x/4}.$$

These are called *boundary curves*. The graph will touch the upper boundary when

$$\sin \tfrac{1}{2}\pi x = 1;$$

namely, at $x = 1, 5, 9$, etc. It will touch the lower boundary when

$$\sin \tfrac{1}{2}\pi x = -1;$$

namely, at $x = 3, 7, 11$, etc. It will cross the x-axis when

$$\sin \tfrac{1}{2}\pi x = 0;$$

namely, at $x = 0, 2, 4$, etc. The graph appears in Figure 6.1.6. The oscillations of the sine function have been *damped* by the boundary curves.   □

Problem.   Find

$$\lim_{x \to 0} \frac{\sin 4x}{x}.$$

SOLUTION.   You have seen that

$$\lim_{x \to 0} \frac{\sin x}{x} = 1.$$

It follows that

$$\lim_{x \to 0} \frac{\sin 4x}{4x} = 1$$

and

$$\lim_{x \to 0} \frac{\sin 4x}{x} = \lim_{x \to 0} 4 \left(\frac{\sin 4x}{4x} \right) = 4. \quad \square$$

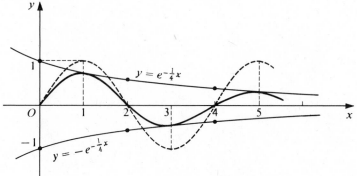

FIGURE 6.1.6

Our treatment of the trigonometric functions has been based entirely on radian measure. When degrees are used, the derivatives of the trigonometric functions contain the extra factor $\pi/180 \cong 0.0175$.

Problem. Find

$$\frac{d}{dx} (\sin x°).$$

SOLUTION. Since

$$x° = \frac{\pi}{180} x \text{ radians}$$

we have

$$\frac{d}{dx} (\sin x°) = \frac{d}{dx} \left(\sin \frac{\pi}{180} x \right) = \frac{\pi}{180} \cos \frac{\pi}{180} x = \frac{\pi}{180} \cos x°. \quad \square$$

This extra factor $\pi/180$ is a disadvantage, particularly in problems where it occurs repeatedly. It tends to discourage the use of degree measure in theoretical work.

Exercises

Differentiate.

*1. $y = \sin 3x$.   2. $y = \sin x \cos x$.

*3. $y = x \tan x$.   *4. $y = \sin x + \cos x$.

5. $y = \tan x^2$.   *6. $y = \sin \sqrt{x}$.

*7. $y = \cos \pi x \sin \pi x$.   8. $y = \sec x \tan x$.

*9. $y = x^2 \sin \pi x.$

*10. $y = \frac{1}{2} \sin^2 x.$

11. $y = \log |\sin x|.$

*12. $y = \tan^3 2x.$

*13. $y = e^x \sin x.$

14. $y = e^x \cos x.$

*15. $y = \sqrt{\cos x^2}.$

*16. $y = (\cos \pi x)^x.$

17. $y = \sec ax.$

*18. $y = e^{-kx} \sin kx.$

*19. $y = 10^x \tan \pi x.$

20. $y = x^{2\pi} \cos 2\pi x.$

*21. $y = \log \sqrt{\tan x}.$

*22. $y = e^x \sin \frac{1}{2}\pi x.$

23. $y = \dfrac{x - \sin x}{\tan x}.$

*24. $y = \tan^4 \frac{1}{4}\pi x.$

*25. $y = e^{\sin x}.$

26. $y = e^{\sin x} \cos x.$

*27. $y = \sec^2 x \tan^2 x.$

*28. $y = \sin nx \sin^n x.$

29. $y = \sin (x - a) \cos (x - a).$

*30. $y = \log \sqrt{\cos 2x}.$

*31. $\rho = \tan^2 \theta.$

32. $\rho = 2^\theta \cos \pi\theta.$

*33. $\rho = \operatorname{cosec}^2 3\theta.$

Find the second derivative.

*34. $y = \sin kx.$

35. $y = \cos kx.$

*36. $y = x \cos x.$

*37. $s = e^{2t} \cos t.$

38. $s = e^{-t} \sin t.$

*39. $s = e^{-t} \sin 2t.$

Find dy/dx by implicit differentiation.

*40. $y = \cos (x - y).$

41. $e^y = \sin (x + y).$

*42. $\cos y = \log (x + y).$

43. $\tan y = 3x^2 + \tan (x + y).$

*44. Find the angle at which the curves

$$y = \sin x \quad \text{and} \quad y = \cos x$$

intersect.

45. Find a formula for the nth derivative:

    (a) $y = \sin x.$ (b) $y = \cos x.$

Use differentials and Table 6.1.1 to estimate the following.

*46. $\cos 0.52.$ 47. $\sin 0.73.$ *48. $\cos 1.32.$

Use differentials and Table 6.1.2 to estimate the following.

*49. $\sin 62°.$ 50. $\cos 57°.$ *51. $\tan 33°.$

Find the limits which exist.

* 52. $\lim\limits_{x \to 0} \dfrac{\sin 3x}{x}$.

    53. $\lim\limits_{x \to 0} \dfrac{\sin x}{3x}$.

* 54. $\lim\limits_{x \to 0} \dfrac{\sin 3x}{2x}$.

* 55. $\lim\limits_{x \to 0} \dfrac{\sin x^2}{x}$.

    56. $\lim\limits_{x \to 0} \dfrac{\sin x^2}{x^2}$.

* 57. $\lim\limits_{x \to 0} \dfrac{\sin x}{x^2}$.

58. Using Figure 6.1.6 as a model, sketch the graph of
$$y = e^{-x/4} \cos \tfrac{1}{2}\pi x, \qquad x \geq 0.$$

For each of the functions below, (a) find the intervals where it is increasing and the intervals where it is decreasing, (b) find the extreme values, (c) determine the concavity of the graph and find the points of inflection, and finally, (d) sketch the graph.

* 59. $f(x) = x + \sin x, \quad 0 \leq x \leq 2\pi.$
   60. $f(x) = x - \sin x, \quad 0 \leq x \leq 2\pi.$
* 61. $f(x) = x \tan x, \quad -\pi/2 < x < \pi/2.$
   62. $f(x) = e^x \cos x, \quad 0 \leq x \leq 2\pi.$
* 63. $f(x) = e^x(\cos x + \sin x), \quad 0 \leq x \leq 2\pi.$
   64. $f(x) = \cos^2 x, \quad 0 \leq x \leq \pi.$

65. Prove that for all real x and y
    (a) $|\cos x - \cos y| \leq |x - y|.$ (b) $|\sin x - \sin y| \leq |x - y|.$
    HINT: Apply the mean-value theorem.

66. Show that
$$|\sin h| \leq |h|. \qquad \text{(begin with Figure 6.1.2)}$$

67. Show that
$$\lim_{h \to 0} \sin h = 0. \qquad \text{(use Exercise 66)}$$

68. Show that
$$\lim_{h \to 0} \cos h = 1. \qquad (\sin^2 \tfrac{1}{2}h = \tfrac{1}{2}[1 - \cos h])$$

6.2 Integrating the Trigonometric Functions

In the last section you saw that each trigonometric function is differentiable on its domain and that

$$\frac{d}{dx}(\sin x) = \cos x. \qquad\qquad \frac{d}{dx}(\cos x) = -\sin x.$$

$$\frac{d}{dx}(\tan x) = \sec^2 x. \qquad\qquad \frac{d}{dx}(\cot x) = -\csc^2 x.$$

$$\frac{d}{dx}(\sec x) = \sec x \tan x. \qquad\qquad \frac{d}{dx}(\csc x) = -\csc x \cot x.$$

Here we integrate the trigonometric functions. As we shall show in a moment, the following formulas are valid:

$$\text{(i)} \int \sin x \, dx = -\cos x + C.$$

$$\text{(ii)} \int \cos x \, dx = \sin x + C.$$

$$\text{(iii)} \int \tan x \, dx = \log |\sec x| + C.$$

$$\text{(iv)} \int \cot x \, dx = \log |\sin x| + C.$$

$$\text{(v)} \int \sec x \, dx = \log |\sec x + \tan x| + C.$$

$$\text{(vi)} \int \operatorname{cosec} x \, dx = \log |\operatorname{cosec} x - \cot x| + C.$$

Derivation of Formulas (i)–(vi)

$$\text{(i)} \int \sin x \, dx = -\int -\sin x \, dx = -\int \frac{d}{dx} (\cos x) \, dx = -\cos x + C. \quad \square$$

$$\text{(ii)} \int \cos x \, dx = \int \frac{d}{dx} (\sin x) \, dx = \sin x + C. \quad \square$$

The key to the remaining formulas is that

$$\int \frac{1}{f(x)} \frac{d}{dx} [f(x)] \, dx = \log |f(x)| + C.$$

$$\text{(iii)} \int \tan x \, dx = \int \frac{\sin x}{\cos x} \, dx = -\int \frac{1}{\cos x} \frac{d}{dx} (\cos x) \, dx = -\log |\cos x| + C$$

$$= \log \frac{1}{|\cos x|} + C = \log |\sec x| + C. \quad \square$$

$$\text{(iv)} \int \cot x \, dx = \int \frac{\cos x}{\sin x} \, dx = \int \frac{1}{\sin x} \frac{d}{dx} (\sin x) \, dx = \log |\sin x| + C. \quad \square$$

$$\text{(v)} \int \sec x \, dx = \int \sec x \, \frac{\sec x + \tan x}{\sec x + \tan x} \, dx = \int \frac{\sec x \tan x + \sec^2 x}{\sec x + \tan x} \, dx$$

$$= \int \frac{1}{\sec x + \tan x} \frac{d}{dx} (\sec x + \tan x) \, dx$$

$$= \log |\sec x + \tan x| + C. \quad \square$$

$$(vi) \int \operatorname{cosec} x \, dx = \int \operatorname{cosec} x \, \frac{\operatorname{cosec} x - \cot x}{\operatorname{cosec} x - \cot x} \, dx$$

$$= \int \frac{-\operatorname{cosec} x \cot x + \operatorname{cosec}^2 x}{\operatorname{cosec} x - \cot x} \, dx$$

$$= \int \frac{1}{\operatorname{cosec} x - \cot x} \frac{d}{dx} (\operatorname{cosec} x - \cot x) \, dx$$

$$= \log |\operatorname{cosec} x - \cot x| + C. \quad \square$$

Problem. Find

$$\int \sin x \cos x \, dx.$$

SOLUTION. Set

$$u = \sin x, \qquad du = \cos x \, dx.$$

$$\int \sin x \cos x \, dx = \int u \, du = \tfrac{1}{2}u^2 + C = \tfrac{1}{2} \sin^2 x + C.$$

Without the u-substitution:

$$\int \sin x \cos x \, dx = \int \sin x \frac{d}{dx} (\sin x) \, dx = \tfrac{1}{2} \sin^2 x + C. \quad \square$$

Problem. Find

$$\int \sec^3 x \tan x \, dx.$$

SOLUTION. Set

$$u = \sec x, \qquad du = \sec x \tan x \, dx.$$

$$\int \sec^3 x \tan x \, dx = \int \sec^2 x \, (\sec x \tan x) \, dx$$

$$= \int u^2 \, du = \tfrac{1}{3}u^3 + C$$

$$= \tfrac{1}{3} \sec^3 x + C. \quad \square$$

Problem. Find

$$\int x \cos \pi x^2 \, dx.$$

SOLUTION. Set

$$u = \pi x^2, \qquad du = 2\pi x \, dx.$$

$$\int x \cos \pi x^2 \, dx = \frac{1}{2\pi} \int (\cos \pi x^2)(2\pi x) \, dx$$

$$= \frac{1}{2\pi} \int \cos u \, du = \frac{1}{2\pi} \sin u + C$$

$$= \frac{1}{2\pi} \sin \pi x^2 + C. \quad \square$$

Problem. Find

$$\int \frac{\sec^2 x}{1 - \tan x} \, dx.$$

SOLUTION. Set

$$u = 1 - \tan x, \qquad du = -\sec^2 x \, dx.$$

$$\int \frac{\sec^2 x}{1 - \tan x} \, dx = - \int \frac{-\sec^2 x}{1 - \tan x} \, dx$$

$$= - \int \frac{du}{u} = -\log |u| + C$$

$$= -\log |1 - \tan x| + C. \quad \square$$

Problem. Find

$$\int x \sin x \, dx.$$

SOLUTION. What causes difficulty here is the factor x. We can eliminate the x by integrating by parts. Set

$$u = x, \qquad dv = \sin x \, dx,$$
$$du = dx, \qquad v = -\cos x.$$

$$\int x \sin x \, dx = \int u \, dv = uv - \int v \, du$$

$$= -x \cos x + \int \cos x \, dx = -x \cos x + \sin x + C. \quad \square$$

Problem. Find

$$\int e^x \cos x \, dx.$$

SOLUTION. Here we integrate by parts twice. First we set

$$u = e^x, \qquad dv = \cos x \, dx,$$
$$du = e^x \, dx, \qquad v = \sin x.$$

This gives

$$\int e^x \cos x \, dx = \int u \, dv = uv - \int v \, du$$

(1)
$$= e^x \sin x - \int e^x \sin x \, dx.$$

To find the integral on the right, we set

$$u = e^x, \qquad dv = \sin x \, dx,$$
$$du = e^x \, dx, \qquad v = -\cos x.$$

This gives

$$\int e^x \sin x \, dx = \int u \, dv = uv - \int v \, du$$

(2)
$$= -e^x \cos x + \int e^x \cos x \, dx.$$

Substituting (2) in (1), we get

$$\int e^x \cos x \, dx = e^x \sin x + e^x \cos x - \int e^x \cos x \, dx,$$

$$2 \int e^x \cos x \, dx = e^x(\sin x + \cos x),$$

$$\int e^x \cos x \, dx = \tfrac{1}{2}e^x(\sin x + \cos x).$$

Since this last integral is an indefinite integral we add an arbitrary constant C:

$$\int e^x \cos x \, dx = \tfrac{1}{2}e^x(\sin x + \cos x) + C. \quad \square$$

Exercises

Find the following indefinite integrals.

*1. $\displaystyle\int \cos (3x - 1) \, dx.$

2. $\displaystyle\int \cos \tfrac{1}{2}\pi x \, dx.$

*3. $\displaystyle\int 3 \sin^2 x \cos x \, dx.$

*4. $\displaystyle\int \sec 2x \tan 2x \, dx.$

5. $\displaystyle\int \cos^4 x \sin x \, dx.$

*6. $\displaystyle\int \sqrt{1 + \cos x} \, \sin x \, dx.$

*7. $\displaystyle\int e^{-\sin x} \cos x \, dx.$

8. $\displaystyle\int x \tan x^2 \, dx.$

*9. $\displaystyle\int x^{-1/2} \sin x^{1/2} \, dx.$

*10. $\displaystyle\int \frac{\cos x}{\sqrt{1 - \sin x}} \, dx.$

11. $\displaystyle\int \frac{\sin x}{\sqrt{1 + \cos x}} \, dx.$

*12. $\displaystyle\int \frac{\sin x}{1 + \cos x} \, dx.$

*13. $\displaystyle\int \cos^2 \pi x \sin \pi x \, dx.$

14. $\displaystyle\int \tan \pi x \, dx.$

*15. $\displaystyle\int \cot \pi x \, dx.$

*16. $\displaystyle\int \frac{\cos x}{1 + 2 \sin x} \, dx.$

17. $\displaystyle\int \frac{dx}{\cos^2 x}.$

*18. $\displaystyle\int \frac{\cos 2x}{2 - \sin 2x} \, dx.$

*19. $\displaystyle\int \sec^2 x \tan x \, dx.$

20. $\displaystyle\int (1 + \tan^2 x) \sec^2 x \, dx.$

*21. $\displaystyle\int \sqrt{1 + \tan x} \, \sec^2 x \, dx.$

*22. $\displaystyle\int \cos^2 \tfrac{1}{2} x \, dx.$

HINT: $\cos^2 \tfrac{1}{2} x = \tfrac{1}{2}(1 + \cos x).$

*23. $\displaystyle\int \cos^2 x \, dx.$

24. $\displaystyle\int \sin^2 x \, dx.$

*25. $\displaystyle\int \cos^2 \pi x \, dx.$

Find the following by integrating by parts.

*26. $\displaystyle\int x \cos x \, dx.$

27. $\displaystyle\int x \cos (x + \pi) \, dx.$

*28. $\displaystyle\int e^x \sin x \, dx.$

29. $\displaystyle\int \sec^3 x \, dx.$

6.3   The Inverse Trigonometric Functions

The Arc Sine

We begin with a number x in the interval $[-1, 1]$. Although there are infinitely many numbers at which the sine function takes on the value x, only one of these numbers lies in the interval $[-\tfrac{1}{2}\pi, \tfrac{1}{2}\pi]$. This unique number is written *arc sin x* and is called the *arc sine of x*. More simply,

$$
\boxed{\;\begin{aligned}
&\text{take } x \in [-1, 1] \quad \text{and} \quad y \in [-\tfrac{1}{2}\pi, \tfrac{1}{2}\pi]. \\
&\quad y = \text{arc sin } x \quad \text{iff} \quad \sin y = x.
\end{aligned}\;}
$$

The function

$$
y = \text{arc sin } x, \qquad x \in [-1, 1]
$$

is called the *inverse sine function*. It is, of course, *not* the inverse of the entire sine function but rather the inverse of

$$y = \sin x, \qquad x \in \left[-\tfrac{1}{2}\pi, \tfrac{1}{2}\pi\right].$$

The graphs are given in Figures 6.3.1 and 6.3.2.

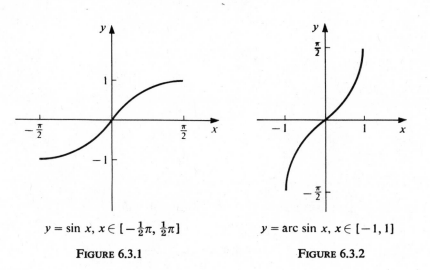

$$y = \sin x,\, x \in \left[-\tfrac{1}{2}\pi, \tfrac{1}{2}\pi\right] \qquad\qquad y = \text{arc sin } x,\, x \in [-1, 1]$$

FIGURE 6.3.1   FIGURE 6.3.2

Since the derivative of the sine function

$$\frac{d}{dx}(\sin x) = \cos x$$

does not take on the value 0 in the *open* interval $(-\tfrac{1}{2}\pi, \tfrac{1}{2}\pi)$, the arc sine function is differentiable on the *open* interval $(-1, 1)$. We can find its derivative

$$\frac{d}{dx}(\text{arc sin } x)$$

by differentiating the identity

$$\sin (\text{arc sin } x) = x.$$

So doing, we get

$$\frac{d}{dx}[\sin (\text{arc sin } x)] = 1,$$

$$\cos (\text{arc sin } x)\frac{d}{dx}(\text{arc sin } x) = 1,$$

$$\sqrt{1 - \sin^2 (\text{arc sin } x)}\,\frac{d}{dx}(\text{arc sin } x) = 1,$$

$$\sqrt{1 - x^2}\,\frac{d}{dx}(\text{arc sin } x) = 1,$$

and therefore

$$\frac{d}{dx} (\text{arc sin } x) = \frac{1}{\sqrt{1 - x^2}}$$

To find

$$\int \text{arc sin } x \, dx$$

we integrate by parts. We set

$$u = \text{arc sin } x, \qquad dv = dx,$$

$$du = \frac{dx}{\sqrt{1 - x^2}}, \qquad v = x.$$

This gives

$$\int \text{arc sin } x \, dx = \int u \, dv = uv - \int v \, du$$

$$= x \text{ arc sin } x - \int \frac{x}{\sqrt{1 - x^2}} \, dx.$$

As you can verify,

$$-\int \frac{x}{\sqrt{1 - x^2}} \, dx = \sqrt{1 - x^2} + C$$

so that

$$\int \text{arc sin } x \, dx = x \text{ arc sin } x + \sqrt{1 - x^2} + C.$$

The Arc Tangent

This time we begin with an arbitrary real number x. Although there are infinitely many numbers at which the tangent function takes on the value x, only one of these numbers lies in the interval $(-\frac{1}{2}\pi, \frac{1}{2}\pi)$. This unique number is written *arc tan x* and is called the *arc tangent of* x. More simply,

$$\text{take } x \in (-\infty, \infty) \quad \text{and} \quad y \in (-\tfrac{1}{2}\pi, \tfrac{1}{2}\pi).$$
$$y = \text{arc tan } x \quad \text{iff} \quad \text{tan } y = x.$$

The function

$$y = \text{arc tan } x, \qquad x \in (-\infty, \infty)$$

is called the *inverse tangent function.* It is, of course, *not* the inverse of the entire tangent function but rather the inverse of

$$y = \tan x, \qquad x \in (-\tfrac{1}{2}\pi, \tfrac{1}{2}\pi).$$

The graphs of these two functions are given in Figures 6.3.3 and 6.3.4.

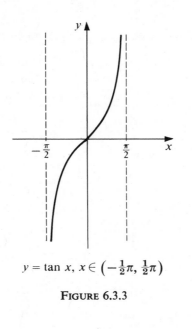

$$y = \tan x, \; x \in \left(-\tfrac{1}{2}\pi, \tfrac{1}{2}\pi\right)$$

FIGURE 6.3.3

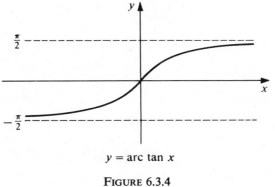

$$y = \text{arc tan } x$$

FIGURE 6.3.4

Since the derivative of the tangent function

$$\frac{d}{dx}(\tan x) = \sec^2 x = \frac{1}{\cos^2 x}$$

is never 0, the inverse tangent function is everywhere differentiable. We can find its derivative by differentiating the identity

$$\tan (\text{arc tan } x) = x.$$

This gives

$$\frac{d}{dx}\left[\tan\left(\text{arc tan } x\right)\right] = 1,$$

$$\sec^2\left(\text{arc tan } x\right)\frac{d}{dx}\left(\text{arc tan } x\right) = 1,$$

$$\left[1 + \tan^2\left(\text{arc tan } x\right)\right]\frac{d}{dx}\left(\text{arc tan } x\right) = 1,$$

$$\left(1 + x^2\right)\frac{d}{dx}\left(\text{arc tan } x\right) = 1,$$

so that

$$\boxed{\frac{d}{dx}\left(\text{arc tan } x\right) = \frac{1}{1 + x^2}.}$$

You can find

$$\int \text{arc tan } x\, dx$$

by integrating by parts. Set

$$u = \text{arc tan } x, \qquad dv = dx,$$

$$du = \frac{dx}{1 + x^2}, \qquad v = x,$$

and you'll see that

$$\boxed{\int \text{arc tan } x\, dx = x\,\text{arc tan } x - \tfrac{1}{2}\log\left(1 + x^2\right) + C.}$$

Problem.   Show that

$$\int \frac{dx}{a^2 + x^2} = \frac{1}{a}\,\text{arc tan }\frac{x}{a} + C.$$

SOLUTION.   We change variables so that the a^2 in the denominator becomes 1. We set

$$au = x, \qquad a\, du = dx.$$

$$\int \frac{dx}{a^2 + x^2} = \int \frac{a\, du}{a^2 + a^2 u^2} = \frac{1}{a}\int \frac{du}{1 + u^2}$$

$$= \frac{1}{a}\,\text{arc tan } u + C = \frac{1}{a}\,\text{arc tan }\frac{x}{a} + C. \quad \square$$

Problem. Find

$$\int_0^2 \frac{dx}{4 + x^2}.$$

SOLUTION.   By the last problem

$$\int \frac{dx}{4 + x^2} = \int \frac{dx}{2^2 + x^2} = \frac{1}{2} \text{ arc tan } \frac{x}{2} + C$$

so that

$$\int_0^2 \frac{dx}{4 + x^2} = \left[\frac{1}{2} \text{ arc tan } \frac{x}{2}\right]_0^2 = \frac{1}{2} \text{ arc tan } 1 = \frac{\pi}{8}. \quad \square$$

A Right Triangle Interpretation

In terms of a right triangle you can picture *arc sin x* as an angle which has x for its sine (Figure 6.3.5) and you can picture *arc tan x* as an angle which has x for its tangent. (Figure 6.3.6)

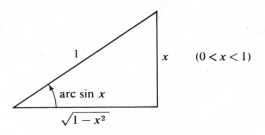

FIGURE 6.3.5

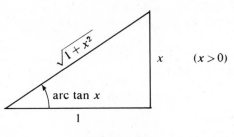

FIGURE 6.3.6

To obtain the formula for the derivative of the arc sine we used the identity

$$\cos (\text{arc sin } x) = \sqrt{1 - x^2}.$$

This is easy to read off from Figure 6.3.5. To differentiate the arc tangent we used the identity

$$\sec^2 (\text{arc tan } x) = 1 + x^2.$$

This is obvious from Figure 6.3.6.

Problem. Find: (a) tan (arc sin x) and (b) sin (arc tan x).

SOLUTION.   Figure 6.3.5 gives

$$\tan (\text{arc sin } x) = \frac{x}{\sqrt{1 - x^2}}$$

and Figure 6.3.6 gives

$$\sin (\text{arc tan } x) = \frac{x}{\sqrt{1 + x^2}}.$$

You can get the same results by using the proper trigonometric identities, but that requires some ingenuity and is definitely more work. □

A word of caution. While it is true that

$$\sin (\text{arc sin } x) = x$$

for all x in the domain of the arc sine, it is not true that

$$\text{arc sin } (\sin x) = x$$

for all x in the domain of the sine function. The relation

$$\text{arc sin } (\sin x) = x$$

holds only for $x \in [-\tfrac{1}{2}\pi, \tfrac{1}{2}\pi]$. For example,

$$\text{arc sin } (\sin \tfrac{1}{4}\pi) = \tfrac{1}{4}\pi$$

but

$$\text{arc sin } (\sin 2\pi) = \text{arc sin } 0 = 0.$$

The arc sine of the sine of 2π is that number in the interval $[-\tfrac{1}{2}\pi, \tfrac{1}{2}\pi]$ at which the sine takes on the same value as it does at 2π.

Similar remarks hold for the tangent and the arc tangent; namely, the relation

$$\tan (\text{arc tan } x) = x$$

is valid for all real x but the relation

$$\text{arc tan } (\tan x) = x$$

holds only for $x \in (-\tfrac{1}{2}\pi, \tfrac{1}{2}\pi)$.

There are four other trigonometric inverses:

the *arc cosine*, which is the inverse of

$$y = \cos x, \qquad x \in [0, \pi];$$

the *arc cotangent*, which is the inverse of

$$y = \cot x, \qquad x \in (0, \pi);$$

the *arc secant*, which is the inverse of

$$y = \sec x, \qquad x \in [0, \tfrac{1}{2}\pi) \cup (\tfrac{1}{2}\pi, \pi];$$

the *arc cosecant*, which is the inverse of

$$y = \operatorname{cosec} x, \qquad x \in [-\tfrac{1}{2}\pi, 0) \cup (0, \tfrac{1}{2}\pi] .$$

Of these four functions only the first two are much used. Figure 6.3.7 shows them all in terms of right triangles.

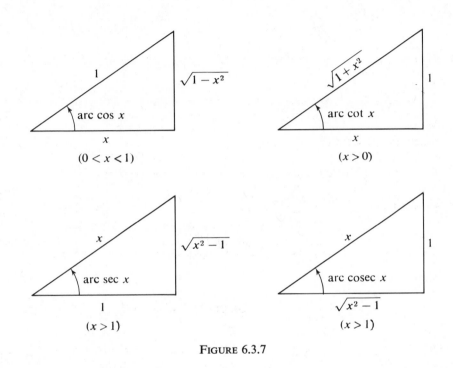

FIGURE 6.3.7

Exercises

Evaluate.

* 1. arc sin 1.   2. arc tan 1.   * 3. arc tan 0.
* 4. arc cos 0.   5. arc tan (-1).   * 6. arc sin $(1/\sqrt{2})$.
* 7. arc cos $(1/\sqrt{2})$.   8. arc sin $\tfrac{1}{2}$.   * 9. arc cos $\tfrac{1}{2}$.
* 10. arc sin $(-\tfrac{1}{2})$. 11. arc sin $(\sin 5\pi/4)$.   * 12. arc tan $[\sin (-3\pi)]$.

Differentiate.

* 13. $y = \operatorname{arc\,tan} (x + 1)$.   14. $y = \operatorname{arc\,tan} \sqrt{x}$.
* 15. $y = \operatorname{arc\,sin} x^2$.   16. $f(x) = e^x \operatorname{arc\,sin} x$.
* 17. $f(x) = x \operatorname{arc\,sin} 2x$.   18. $f(x) = e^{\operatorname{arc\,tan} x}$.
* 19. $u = (\operatorname{arc\,sin} x)^2$.   20. $v = \operatorname{arc\,tan} (e^x)$.

* 21. $y = \dfrac{\operatorname{arc\,tan} x}{x}$.   22. $y = \operatorname{arc\,tan} \dfrac{2}{x}$.

* 23. $f(x) = \sqrt{\arctan 2x}$.

24. $f(x) = \log(\arctan x)$.

* 25. $y = \arctan(\log x)$.

26. $y = \arctan(\sin x)$.

* 27. $\theta = \arcsin\left(\sqrt{1 - r^2}\right)$.

28. $\theta = \arcsin\left(\dfrac{r}{r + 1}\right)$.

* 29. $\theta = \arctan\left(\dfrac{c + r}{1 - cr}\right)$.

30. $\theta = \arctan\left(\dfrac{1}{1 + r^2}\right)$.

* 31. $f(x) = \sqrt{c^2 - x^2} + c \arcsin \dfrac{x}{c}$.†

32. $f(x) = \tfrac{1}{3} \arcsin(3x - 4x^2)$.

* 33. $y = \dfrac{x}{\sqrt{c^2 - x^2}} - \arcsin \dfrac{x}{c}$.†

34. $y = x\sqrt{c^2 - x^2} + c^2 \arcsin \dfrac{x}{c}$.†

35. Finish the proof that

$$\int \arctan x \, dx = x \arctan x - \tfrac{1}{2} \log(1 + x^2) + C.$$

36. Show that

$$\int \frac{dx}{\sqrt{a^2 - x^2}} = \arcsin\left(\frac{x}{a}\right) + C. \qquad\qquad (a > 0)$$

37. Show that

$$\int \frac{dx}{a^2 + (x + b)^2} = \frac{1}{a} \arctan\left(\frac{x + b}{a}\right) + C. \qquad\qquad (a \neq 0)$$

Evaluate.

* 38. $\displaystyle\int_0^1 \frac{dx}{1 + x^2}$.

39. $\displaystyle\int_{-1}^1 \frac{dx}{1 + x^2}$.

* 40. $\displaystyle\int_0^{1/\sqrt{2}} \frac{dx}{\sqrt{1 - x^2}}$.

* 41. $\displaystyle\int_0^1 \frac{dx}{\sqrt{4 - x^2}}$.

* 42. $\displaystyle\int_0^5 \frac{dx}{25 + x^2}$.

* 43. $\displaystyle\int_{-5}^5 \frac{dx}{25 + x^2}$.

44. (a) Use the relation

$$\sec(\text{arc sec } x) = x$$

and Figure 6.3.7 to show that

$$\boxed{\frac{d}{dx}(\text{arc sec } x) = \frac{1}{x\sqrt{x^2 - 1}}, \quad \text{for } x > 1.}$$

---

† Take $c > 0$.

Use a similar technique to figure out

*(b) $\dfrac{d}{dx}$ (arc cos x), for $0 < x < 1$,

*(c) $\dfrac{d}{dx}$ (arc cot x), for $x > 0$,

*(d) $\dfrac{d}{dx}$ (arc cosec x), for $x > 1$.

6.4 Additional Exercises

Find the area below the graph.

*1. $y = \dfrac{x}{x^2 + 1}$, $x \in [0, 1]$.

2. $y = \dfrac{1}{x^2 + 1}$, $x \in [0, 1]$.

*3. $y = \dfrac{1}{\sqrt{1 - x^2}}$, $x \in [0, \tfrac{1}{2}]$.

4. $y = \dfrac{x}{\sqrt{1 - x^2}}$, $x \in [0, \tfrac{1}{2}]$.

*5. $y = $ arc sin x, $x \in [0, \sqrt{3}/2]$.

6. $y = $ arc tan x, $x \in [0, 1]$.

*7. Find the maximum value of

$$y = a \sin x + b \cos x. \qquad\qquad (a > 0, \quad b > 0)$$

8. Find the maximum value of the slope of the curve in Exercise 7.

9. A particle moves along a coordinate line so that at time t it has coordinate $x(t)$, velocity $v(t)$, and acceleration $a(t)$. Find

  *(a) $x(1)$ given that $x(0) = 0$ and $v(t) = 1/(t^2 + 1)$.

  (b) $a(1)$ given that $v(t) = 1/(t^2 + 1)$.

  *(c) $v(0)$ given that $v(1) = 1$ and $a(t) = ($arc tan $t)/(t^2 + 1)$.

  (d) $x(1)$ given that $x(0) = 0$, $v(0) = 1$, and $a(t) = 1/(t^2 + 1)$.

  *(e) $a(0)$ given that $x(t) = $ arc tan $(1 + t)$.

10. A balloon, which leaves the ground 500 feet from an observer, rises vertically at the rate of 140 feet per minute. At what rate is the inclination of the observer's line of sight increasing when the balloon is 500 feet above the ground?

*11. The base of an isosceles triangle is 6 feet. If the altitude is 4 feet and increasing at the rate of 2 inches per minute, at what rate is the vertex angle changing?

*12. As a boy winds up the cord, his kite is moving horizontally at a height of 60 feet with a speed of 10 feet per minute. How fast is the inclination of the cord changing when its length is 100 feet?

*13. A revolving search light which is $\tfrac{1}{2}$ mile from shore makes 1 revolution per minute. How fast is the light traveling along the straight beach when at a distance of 1 mile from the nearest point of the shore?

*14. A tapestry 7 feet in height is hung on a wall so that its lower edge is 9 feet above an observer's eye. At what distance from the wall should he stand in order to obtain the most favorable view? HINT: The vertical angle subtended by the tapestry in the eye of the observer must be a maximum.

*15. Find the dimensions of the cylinder of maximum volume which can be inscribed in a sphere of radius 6 inches. (Use the central angle θ subtended by the radius of the base.)

16. Solve Exercise 15 if the curved surface of the cylinder is to be a maximum.

*17. A body of weight W is dragged along a horizontal plane by means of a force P whose line of action makes an angle θ with the plane. The magnitude of the force is given by the equation

$$P = \frac{mW}{m \sin \theta + \cos \theta},$$

where m denotes the coefficient of friction. For what value of θ is the pull a minimum?

*18. If a projectile is fired from O so as to strike an inclined plane which makes a constant angle α with the horizontal at O, the range is given by the formula

$$R = \frac{2v^2 \cos \theta \sin (\theta - \alpha)}{g \cos^2 \alpha},$$

where v and g are constants and θ is the angle of elevation. Calculate the value of θ which gives the maximum range.

19. (*Optional*) Set

$$F(x) = \begin{cases} \sin (1/x), & x \neq 0 \\ 0, & x = 0 \end{cases},$$

$$G(x) = \begin{cases} x \sin (1/x), & x \neq 0 \\ 0, & x = 0 \end{cases},$$

$$H(x) = \begin{cases} x^2 \sin (1/x), & x \neq 0 \\ 0, & x = 0 \end{cases}.$$

(a) Sketch a figure displaying the general nature of the graph of F.

(b) Sketch a figure displaying the general nature of the graph of G.

(c) Sketch a figure displaying the general nature of the graph of H.

(d) Which of these functions is continuous at 0?

(e) Which of these functions is differentiable at 0?

6.5 Simple Harmonic Motion

Sines and cosines play a central role in the study of all oscillations. Wherever there are waves to analyze (sound waves, radio waves, light waves, etc.) sines and cosines hold the center stage.

The simplest wave motion arises when an object moves back and forth along a coordinate line in such a way that its acceleration remains a constant negative multiple of its position:

$$a(t) = -kx(t).$$

Such motion is called *simple harmonic* motion.

With

$$a(t) = x''(t)$$

we have

$$x''(t) = -kx(t)$$

and thus

$$x''(t) + kx(t) = 0.$$

To emphasize that k is positive, we set $k = \beta^2$. The equation then takes the form

(6.5.1)

$$\boxed{x''(t) + \beta^2 x(t) = 0.}$$

It is easy to show that any function of the form

$$x(t) = C_2 \sin (\beta t + C_1)$$

satisfies the differential equation. Differentiation gives

$$x'(t) = C_2 \beta \cos (\beta t + C_1),$$
$$x''(t) = -C_2 \beta^2 \sin (\beta t + C_1) = -\beta^2 x(t)$$

so that

$$x''(t) + \beta^2 x(t) = 0.$$

What is not easy to see, but equally true, is that every solution of Equation (6.5.1) can be written in the form

(6.5.2)

$$x(t) = C_2 \sin (\beta t + C_1).$$

This we ask you to take on faith. (Or consult a text on differential equations.)

Let's study the function of (6.5.2). Observe first that

$$
\begin{aligned}
x(t + 2\pi/\beta) &= C_2 \sin \left[\beta(t + 2\pi/\beta) + C_1 \right] \\
&= C_2 \sin (\beta t + 2\pi + C_1) \\
&= C_2 \sin (\beta t + C_1 + 2\pi) \\
&= C_2 \sin (\beta t + C_1) \\
&= x(t).
\end{aligned}
$$

This says that the motion is *periodic* and the *period* is $2\pi/\beta$:

$$p = \frac{2\pi}{\beta}. \qquad \text{(Figure 6.5.1)}$$

If we measure t in seconds, then every $2\pi/\beta$ seconds the motion repeats itself. The reciprocal $\beta/2\pi$ gives the number of complete cycles per second. This number is called the *frequency*:

$$f = \frac{\beta}{2\pi}.$$

Since $\sin(\beta t + C_1)$ oscillates between -1 and 1, the product

$$x(t) = C_2 \sin(\beta t + C_1)$$

oscillates between $-|C_2|$ and $|C_2|$. The number $|C_2|$ is called the *amplitude* of the motion:

$$a = |C_2|. \hspace{3cm} \text{(Figure 6.5.1)}$$

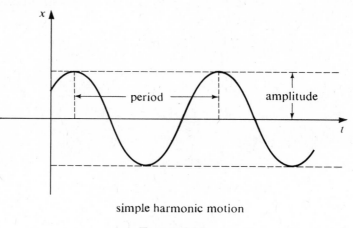

simple harmonic motion

FIGURE 6.5.1

Problem. Find an equation for the motion if the period is $2\pi/3$ and at time $t = 0$, $x = 1$ and $v = 3$.

SOLUTION. We begin with

$$x(t) = C_2 \sin(\beta t + C_1)$$

taking $C_2 > 0$ and $0 \le C_1 < 2\pi$. In general the period is $2\pi/\beta$ so that here

$$\frac{2\pi}{\beta} = \frac{2\pi}{3} \quad \text{and} \quad \beta = 3.$$

Consequently we have

$$x(t) = C_2 \sin(3t + C_1),$$

and by differentiation

$$v(t) = 3C_2 \cos(3t + C_1).$$

The conditions at $t = 0$ give

$$1 = x(0) = C_2 \sin C_1, \qquad 3 = v(0) = 3C_2 \cos C_1$$

and therefore

$$1 = C_2 \sin C_1, \qquad 1 = C_2 \cos C_1.$$

Adding the squares, we have

$$2 = C_2^2 \sin^2 C_1 + C_2^2 \cos^2 C_1 = C_2^2$$

and therefore

$$C_2 = \sqrt{2}.$$

To find C_1, we note that

$$1 = \sqrt{2} \sin C_1, \qquad 1 = \sqrt{2} \cos C_1.$$

These equations are satisfied by setting

$$C_1 = \frac{\pi}{4}.$$

The equation of motion can be written

$$x(t) = \sqrt{2} \sin\left(3t + \frac{\pi}{4}\right). \quad \square$$

Problem. An object in simple harmonic motion passes through the central point $x = 0$ at time 0 and every second thereafter. Find an equation for the motion if $v(0) = -4$.

SOLUTION

$$x(t) = C_2 \sin (\beta t + C_1).$$

On each complete cycle the object must pass through the central point twice—once going one way, once going the other way. The period is thus 2 seconds:

$$\frac{2\pi}{\beta} = 2 \quad \text{and} \quad \beta = \pi.$$

We now know that

$$x(t) = C_2 \sin (\pi t + C_1)$$

and consequently

$$v(t) = \pi C_2 \cos (\pi t + C_1).$$

The initial conditions can be written

$$0 = x(0) = C_2 \sin C_1, \qquad -4 = v(0) = \pi C_2 \cos C_1$$

so that

$$0 = C_2 \sin C_1, \qquad -\frac{4}{\pi} = C_2 \cos C_1$$

Taking $C_2 > 0$ and $0 \le C_1 < 2\pi$, we have

$$C_2 = \frac{4}{\pi}, \qquad C_1 = \pi.$$

The equation of motion can be written

$$x(t) = \frac{4}{\pi} \sin (\pi t + \pi). \quad \square$$

Problem.   A coil spring hangs naturally to a length l_0. When a mass m is attached to it, the spring stretches l_1 inches. The mass is later pulled down by hand an additional x_0 inches and released. What is the resulting motion?

SOLUTION.   Throughout we refer to Figure 6.5.2.

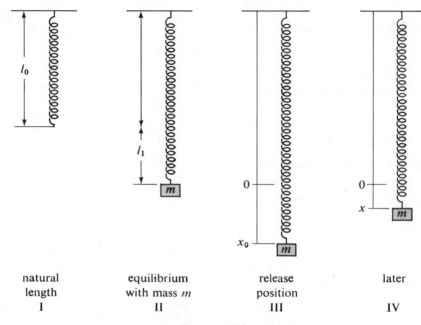

natural length	equilibrium with mass m	release position	later
I	II	III	IV

FIGURE 6.5.2

We begin by analyzing the forces acting on the mass at general position x (Stage IV). First there is the weight of the object:

$$F_1 = mg.$$

This is a downward force, and by the choice of our coordinate system, positive. Then there is the restoring force of the spring. This force, by Hooke's law, is proportional to the total displacement $l_1 + x$ and acts in the opposite direction:

$$F_2 = -k(l_1 + x) \qquad \text{with } k > 0.$$

If we neglect resistance, then these are the only forces acting on the object. Under these conditions the total force is

$$F = F_1 + F_2 = mg - k(l_1 + x),$$

which we rewrite as

(1) $$F = (mg - kl_1) - kx.$$

At stage II (Figure 6.5.2) there was equilibrium. The force of gravity, mg, plus the force of the spring, $-kl_1$, must have been 0:

$$mg - kl_1 = 0.$$

Equation (1) can therefore be simplified to

$$F = -kx.$$

Using Newton's

$$F = ma \qquad \text{(force} = \text{mass} \times \text{acceleration)}$$

we have

$$ma = -kx$$

and thus

$$a = -\frac{k}{m}x.$$

At any time t,

$$x''(t) = -\frac{k}{m}x(t).$$

Since $k/m > 0$ we can set $\beta = \sqrt{k/m}$ and write

$$x''(t) = -\beta^2 x(t).$$

The motion of the mass is simple harmonic motion. □

Exercises

*1. An object is in simple harmonic motion. Find an equation for the motion if the period is $\pi/4$ and at time $t = 0$, $x = 1$ and $v = 0$. What is the amplitude? What is the frequency?

2. An object is in simple harmonic motion. Find an equation for the motion if the frequency is $1/\pi$ and at time $t = 0$, $x = 0$ and $v = -2$. What is the amplitude? What is the period?

*3. An object is in simple harmonic motion with period p and amplitude a. What is its velocity at the central point?

4. An object is in simple harmonic motion with period p. Find the amplitude if at $x = x_0$, $v = \pm v_0$.

*5. An object in simple harmonic motion passes through the central point $x = 0$ at time $t = 0$ and every 3 seconds thereafter. Find the equation of motion if $v(0) = 5$.

6. Find the position at which an object in simple harmonic motion

$$x(t) = C_2 \sin (\beta t + C_1)$$

attains

(a) maximum speed.   (b) zero speed.

(c) maximum acceleration.   (d) zero acceleration.

*7. Where does the object of Exercise 6 take on half its maximum speed?

8. Show that harmonic motion

$$x(t) = C_2 \sin (\beta t + C_1)$$

can also be written in the form

$$x(t) = C_4 \cos (\beta t + C_3)$$

and also in the form

$$x(t) = C_5 \cos \beta t + C_6 \sin \beta t.$$

*9. What is $x(t)$ for the object of mass m discussed in the spring problem?

6.6   (Optional) The Hyperbolic Sine and Cosine

The *hyperbolic sine* and *cosine* are the functions

$$\sinh x = \tfrac{1}{2}(e^x - e^{-x}) \quad \text{and} \quad \cosh x = \tfrac{1}{2}(e^x + e^{-x}).$$

The reasons for these names will become apparent as we go on.
   Since

$$\frac{d}{dx} (\sinh x) = \frac{d}{dx} [\tfrac{1}{2}(e^x - e^{-x})] = \tfrac{1}{2}(e^x + e^{-x})$$

and

$$\frac{d}{dx} (\cosh x) = \frac{d}{dx} [\tfrac{1}{2}(e^x + e^{-x})] = \tfrac{1}{2}(e^x - e^{-x}),$$

we have

$$\frac{d}{dx} (\sinh x) = \cosh x \quad \text{and} \quad \frac{d}{dx} (\cosh x) = \sinh x.$$

In short, each of these functions is the derivative of the other.

The Graphs

We begin with the hyperbolic sine. Since

$$\sinh (-x) = \tfrac{1}{2}(e^{-x} - e^x) = -\tfrac{1}{2}(e^x - e^{-x}) = -\sinh x,$$

the hyperbolic tangent is an odd function. Its graph is therefore symmetric about the origin. Since

$$\frac{d}{dx} (\sinh x) = \cosh x = \tfrac{1}{2}(e^x + e^{-x}) > 0 \quad \text{for all real } x,$$

the hyperbolic sine increases everywhere. From

$$\frac{d^2}{dx^2} (\sinh x) = \frac{d}{dx} (\cosh x) = \sinh x = \tfrac{1}{2}(e^x - e^{-x})$$

you can see that

$$\frac{d^2}{dx^2} (\sinh x) \text{ is } \begin{cases} \text{negative,} & \text{for } x < 0 \\ 0, & \text{at } x = 0 \\ \text{positive,} & \text{for } x > 0 \end{cases}.$$

The graph is therefore concave down on $(-\infty, 0]$ and concave up on $[0, \infty)$. The point

$$(0, \sinh 0) = (0, 0)$$

is a point of inflection. The slope of the tangent at the origin is $\cosh 0 = 1$. A sketch of the graph appears in Figure 6.6.1. ☐

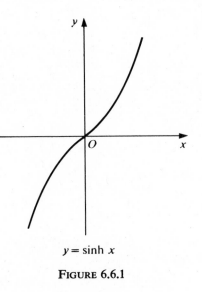

$$y = \sinh x$$

FIGURE 6.6.1

We turn now to the hyperbolic cosine. Since

$$\cosh (-x) = \tfrac{1}{2}(e^{-x} + e^x) = \tfrac{1}{2}(e^x + e^{-x}) = \cosh x,$$

the hyperbolic cosine is an even function. Its graph is therefore symmetric about the y-axis. Since

$$\frac{d}{dx} (\cosh x) = \sinh x,$$

you can see that

$$\frac{d}{dx} (\cosh x) \text{ is } \begin{cases} \text{negative,} & \text{for } x < 0 \\ 0, & \text{at } x = 0 \\ \text{positive,} & \text{for } x > 0 \end{cases}.$$

The function therefore decreases on $(-\infty, 0]$ and increases on $[0, \infty)$. The number

$$\cosh 0 = \tfrac{1}{2}(e^0 + e^{-0}) = \tfrac{1}{2}(1 + 1) = 1$$

is a minimum. There are no other extreme values. Since

$$\frac{d^2}{dx^2}(\cosh x) = \frac{d}{dx}(\sinh x) = \cosh x > 0 \quad \text{for all real } x,$$

the graph is everywhere concave up. For a picture of the graph see Figure 6.6.2.   □

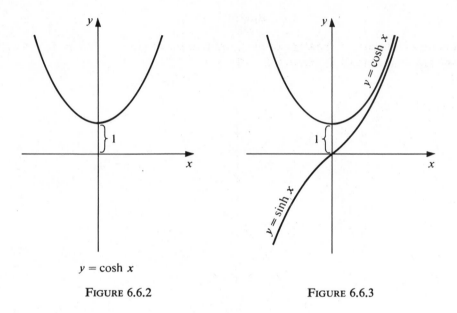

$$y = \cosh x$$

FIGURE 6.6.2   FIGURE 6.6.3

Figure 6.6.3 shows a comparison between the two graphs. The graph of $y = \cosh x$ remains above the graph of $y = \sinh x$,

$$\cosh x - \sinh x = \tfrac{1}{2}(e^x + e^{-x}) - \tfrac{1}{2}(e^x - e^{-x}) = e^{-x} > 0,$$

but the difference tends to 0 as x tends to $+\infty$.

Identities

These functions satisfy identities which are similar to those satisfied by the "circular" sine and cosine.

$$\cosh^2 t - \sinh^2 t = 1,$$
$$\sinh (t + s) = \sinh t \cosh s + \cosh t \sinh s,$$
$$\cosh (t + s) = \cosh t \cosh s + \sinh t \sinh s,$$
$$\sinh 2t = 2 \sinh t \cosh t,$$
$$\cosh 2t = \cosh^2 t + \sinh^2 t.$$

The verification of these identities is left to you as an exercise.

Relation to the Hyperbola $x^2 - y^2 = 1$

That

$$\cosh^2 t - \sinh^2 t = 1$$

means that points of the form

$$P(\cosh t, \sinh t)$$

all lie on the hyperbola

$$x^2 - y^2 = 1.$$

This in itself might justify calling these functions hyperbolic but, as you will see in a moment, there is actually a much deeper reason.

Let's return briefly to the unit circle $x^2 + y^2 = 1$ and the ordinary sine and cosine. If t lies between 0 and 2π, then $\frac{1}{2}t$ gives the area of the sector shaded in Figure 6.6.4.

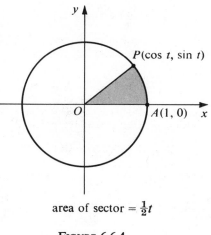

area of sector $= \frac{1}{2}t$

FIGURE 6.6.4

We have an entirely analogous situation in the case of the hyperbola

$$x^2 - y^2 = 1$$

and the hyperbolic sine and cosine.

If $t > 0$, then $\frac{1}{2}t$ gives the area of the hyperbolic sector shaded in Figure 6.6.5.

PROOF.   Let's call the area of the shaded sector $A(t)$. It is not hard to see that

$$A(t) = \frac{1}{2}\cosh t \sinh t - \int_1^{\cosh t} \sqrt{x^2 - 1}\, dx.$$

The first term

$$\frac{1}{2}\cosh t \sinh t$$

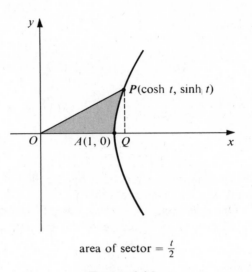

$$\text{area of sector} = \tfrac{t}{2}$$

FIGURE 6.6.5

gives the area of the triangle OPQ and the integral

$$\int_1^{\cosh t} \sqrt{x^2 - 1}\ dx$$

gives the area of the unshaded portion of the triangle. We wish to show that

$$A(t) = \tfrac{1}{2}t \qquad \text{for all } t \geq 0.$$

Differentiation of $A(t)$ gives

$$A'(t) = \frac{1}{2}\left[\cosh t\, \frac{d}{dt}\, (\sinh t) + \sinh t\, \frac{d}{dt}\, (\cosh t) \right] - \frac{d}{dt}\left(\int_1^{\cosh t} \sqrt{x^2 - 1}\ dx \right)$$

and therefore

(1) $$A'(t) = \tfrac{1}{2}(\cosh^2 t + \sinh^2 t) - \frac{d}{dt}\left(\int_1^{\cosh t} \sqrt{x^2 - 1}\ dx \right).$$

In Section 4.9 we showed that, in general,

$$\frac{d}{dt}\left(\int_a^{g(t)} f(x)\ dx \right) = f(g(t))g'(t).$$

In this case, then,

$$\frac{d}{dt}\left(\int_1^{\cosh t} \sqrt{x^2 - 1}\ dx \right) = \sqrt{\cosh^2 t - 1}\, \frac{d}{dt}\, (\cosh t)$$

$$= \sinh t \cdot \sinh t$$

$$= \sinh^2 t.$$

Substitution of this last expression into Equation (1) gives

$$A'(t) = \tfrac{1}{2}(\cosh^2 t + \sinh^2 t) - \sinh^2 t$$

$$= \tfrac{1}{2}(\cosh^2 t - \sinh^2 t)$$

$$= \tfrac{1}{2}.$$

It's not hard to see that $A(0) = 0$:

$$A(0) = \tfrac{1}{2} \cosh 0 \sinh 0 - \int_1^{\cosh 0} \sqrt{x^2 - 1}\, dx = 0.$$

The two conditions

$$A'(t) = \tfrac{1}{2} \quad \text{and} \quad A(0) = 0$$

together imply that

$$A(t) = \tfrac{1}{2}t. \quad \square$$

Exercises

Differentiate.

*1. $y = \sinh x^2$.

2. $y = \cosh (x + a)$.

*3. $y = \sqrt{\cosh ax}$.

4. $y = (\sinh ax)(\cosh ax)$.

*5. $y = \dfrac{\sinh x}{\cosh x - 1}$.

6. $y = \dfrac{\sinh x}{x}$.

*7. $y = a \sinh bx - b \cosh ax$.

8. $y = e^x(\cosh x + \sinh x)$.

*9. $y = \log |\sinh ax|$.

10. $y = \log |1 - \cosh ax|$.

Verify the following identities.

11. $\cosh^2 t - \sinh^2 t = 1$.

12. $\sinh (t + s) = \sinh t \cosh s + \cosh t \sinh s$.

13. $\cosh (t + s) = \cosh t \cosh s + \sinh t \sinh s$.

14. $\sinh 2t = 2 \sinh t \cosh t$.

15. $\cosh 2t = \cosh^2 t + \sinh^2 t$.

Find the extreme values.

*16. $y = 5 \cosh x + 4 \sinh x$.

17. $y = -5 \cosh x + 4 \sinh x$.

*18. $y = 4 \cosh x + 5 \sinh x$.

19. Show that for each positive integer n

$$(\cosh x + \sinh x)^n = \cosh nx + \sinh nx.$$

20. Verify that

$$y = A \cosh cx + B \sinh cx$$

satisfies the differential equation

$$y'' - c^2 y = 0.$$

* 21. Find the solution of the differential equation

$$y'' - 9y = 0$$

which satisfies

$$y(0) = 2 \quad \text{and} \quad y'(0) = 1.$$

22. Find the solution of the differential equation

$$4y'' - y = 0$$

which satisfies

$$y(0) = 1 \quad \text{and} \quad y'(0) = 2.$$

6.7   (Optional) Other Hyperbolic Functions

The hyperbolic tangent is defined by setting

$$\tanh x = \frac{\sinh x}{\cosh x} = \frac{e^x - e^{-x}}{e^x + e^{-x}}.$$

There is also a *hyperbolic cotangent*

$$\coth x = \frac{\cosh x}{\sinh x},$$

a *hyperbolic secant*

$$\operatorname{sech} x = \frac{1}{\cosh x},$$

and a *hyperbolic cosecant*

$$\operatorname{cosech} x = \frac{1}{\sinh x}.$$

The derivatives are as follows:

$$\frac{d}{dx}(\tanh x) = \operatorname{sech}^2 x,$$

$$\frac{d}{dx}(\coth x) = -\operatorname{cosech}^2 x,$$

$$\frac{d}{dx}(\operatorname{sech} x) = -\operatorname{sech} x \tanh x,$$

$$\frac{d}{dx}(\operatorname{cosech} x) = -\operatorname{cosech} x \coth x.$$

These formulas are easy to verify. For instance,

$$\frac{d}{dx}(\tanh x) = \frac{d}{dx}\left(\frac{\sinh x}{\cosh x}\right)$$

$$= \frac{\cosh x \frac{d}{dx}(\sinh x) - \sinh x \frac{d}{dx}(\cosh x)}{\cosh^2 x}$$

$$= \frac{\cosh^2 x - \sinh^2 x}{\cosh^2 x}$$

$$= \frac{1}{\cosh^2 x}$$

$$= \operatorname{sech}^2 x.$$

We leave it to you to verify the other formulas. □

Let's analyze the hyperbolic tangent a little further. Since

$$\tanh(-x) = \frac{\sinh(-x)}{\cosh(-x)} = \frac{-\sinh x}{\cosh x} = -\tanh x,$$

the hyperbolic sine is an odd function and therefore its graph is symmetric about the origin. Since

$$\frac{d}{dx}(\tanh x) = \operatorname{sech}^2 x > 0 \quad \text{for all real } x,$$

the function is everywhere increasing. From

$$\tanh x = \frac{e^x - e^{-x}}{e^x + e^{-x}} = 1 - \frac{2}{e^{2x} + 1}$$

you can see that tanh x always remains between -1 and 1. Moreover, it tends to 1 as x tends to $+\infty$ and it tends to -1 as x tends to $-\infty$. To check on the concavity of the graph, we take the second derivative:

$$\frac{d^2}{dx^2}(\tanh x) = \frac{d}{dx}(\operatorname{sech}^2 x) = 2\operatorname{sech} x \frac{d}{dx}(\operatorname{sech} x)$$

$$= 2\operatorname{sech} x (-\operatorname{sech} x \tanh x)$$

$$= -2\operatorname{sech}^2 x \tanh x.$$

Since

$$\tanh x = \frac{e^x - e^{-x}}{e^x + e^{-x}} \text{ is } \begin{cases} \text{negative,} & \text{for } x < 0 \\ 0, & \text{at } x = 0 \\ \text{positive,} & \text{for } x > 0 \end{cases},$$

it follows that

$$\frac{d^2}{dx^2}(\tanh x) \text{ is } \begin{cases} \text{positive,} & \text{for } x < 0 \\ 0, & \text{at } x = 0 \\ \text{negative,} & \text{for } x > 0 \end{cases}.$$

The graph is therefore concave up on $(-\infty, 0]$ and concave down on $[0, \infty)$. The point

$$(0, \tanh 0) = (0, 0)$$

is a point of inflection. At the origin the slope is

$$\text{sech}^2\, 0 = \frac{1}{\cosh^2 0} = 1.$$

For a picture of the graph see Figure 6.7.1.

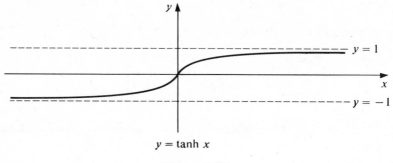

$$y = \tanh x$$

FIGURE 6.7.1

The Hyperbolic Inverses

The hyperbolic inverses that are important to us are the *inverse hyperbolic sine*, the *inverse hyperbolic cosine*, and the *inverse hyperbolic tangent*. These functions

$$y = \sinh^{-1} x, \qquad y = \cosh^{-1} x, \qquad y = \tanh^{-1} x$$

are the inverses of

$$y = \sinh x, \qquad y = \cosh x \quad (x \geq 0), \qquad y = \tanh x$$

respectively.

Theorem 6.7.1

(i) $\sinh^{-1} x = \log (x + \sqrt{x^2 + 1})$. $(x \text{ real})$

(ii) $\cosh^{-1} x = \log (x + \sqrt{x^2 - 1})$. $(x \geq 1)$

(iii) $\tanh^{-1} x = \frac{1}{2} \log \left(\frac{1 + x}{1 - x} \right)$. $(-1 < x < 1)$

PROOF. To show (i), we set

$$y = \sinh^{-1} x$$

and note that

$$\sinh y = x.$$

This gives in sequence

$$\tfrac{1}{2}(e^y - e^{-y}) = x,$$
$$e^y - e^{-y} = 2x,$$
$$e^{2y} - 2xe^y - 1 = 0.$$

This is a quadratic equation in e^y. From the general quadratic formula we get

$$e^y = \tfrac{1}{2}(2x \pm \sqrt{4x^2 + 4}) = x \pm \sqrt{x^2 + 1}.$$

Since $e^y > 0$, the minus sign on the right is impossible. Consequently we have

$$e^y = x + \sqrt{x^2 + 1}$$

and, taking logs,

$$y = \log (x + \sqrt{x^2 + 1}).$$

To show (ii), we set

$$y = \cosh^{-1} x, \qquad x \geq 1$$

and note that

$$\cosh y = x \quad \text{and} \quad y \geq 0.$$

This gives in sequence

$$\tfrac{1}{2}(e^y + e^{-y}) = x,$$
$$e^y + e^{-y} = 2x,$$
$$e^{2y} - 2xe^y + 1 = 0.$$

Again we have a quadratic in e^y. The general quadratic formula here gives

$$e^y = \tfrac{1}{2}(2x \pm \sqrt{4x^2 - 4}) = x \pm \sqrt{x^2 - 1}.$$

Since y is nonnegative,

$$e^y = x \pm \sqrt{x^2 - 1}$$

cannot be less than 1. This renders the negative sign impossible (check this out) and leaves as the only possibility

$$e^y = x + \sqrt{x^2 - 1}.$$

Taking logs, we get

$$y = \log (x + \sqrt{x^2 - 1}).$$

The proof of (iii) is left as an exercise. □

Exercises

Differentiate.

*1. $y = \tanh^2 x.$

2. $y = \tanh^2 3x.$

*3. $y = \log (\tanh x).$

4. $y = \tanh (\log x).$

* 5. $y = \sinh (\text{arc tan } e^{2x})$.

6. $y = \text{sech } (3x^2 + 1)$.

* 7. $y = \coth (\sqrt{x^2 + 1})$.

8. $y = \log (\text{sech } x)$.

* 9. $y = \dfrac{\text{sech } x}{1 + \cosh x}$.

10. $y = \dfrac{\cosh x}{1 + \text{sech } x}$.

11. Show that

$$\frac{d}{dx} (\coth x) = -\text{cosech}^2 x.$$

12. Show that

$$\frac{d}{dx} (\text{sech } x) = -\text{sech } x \tanh x.$$

13. Show that

$$\frac{d}{dx} (\text{cosech } x) = -\text{cosech } x \coth x.$$

14. Show that

$$\tanh (t + s) = \frac{\tanh t + \tanh s}{1 + \tanh t \tanh s}.$$

* 15. Given that

$$\tanh x_0 = \tfrac{4}{5}$$

find

(a) $\text{sech } x_0$.   HINT: $1 - \tanh^2 x = \text{sech}^2 x$.
Then find

(b) $\cosh x_0$.

(c) $\sinh x_0$.

(d) $\coth x_0$.

(e) $\text{cosech } x_0$.

16. Given that

$$\tanh t_0 = -\tfrac{5}{12},$$

evaluate the remaining hyperbolic functions at t_0.

17. Show that

$$\text{if}\quad x^2 \geq 1 \qquad \text{then}\quad x - \sqrt{x^2 - 1} \leq 1.$$

18. Show that

$$\tanh^{-1} x = \frac{1}{2} \log \left(\frac{1 + x}{1 - x} \right).$$

19. Show that

$$\boxed{\frac{d}{dx} (\sinh^{-1} x) = \frac{1}{\sqrt{x^2 + 1}}.}$$

20. Show that

$$\boxed{\frac{d}{dx} (\cosh^{-1} x) = \frac{1}{\sqrt{x^2 - 1}}.}$$

21. Show that

$$\frac{d}{dx}(\tanh^{-1} x) = \frac{1}{1 - x^2}.$$

22. Sketch the graphs of
    (a) $y = \sinh^{-1} x$.   (b) $y = \cosh^{-1} x$.   (c) $y = \tanh^{-1} x$.

23. Given that

$$\tan \phi = \sinh x$$

    show that

    (a) $\dfrac{d\phi}{dx} = \operatorname{sech} x$,   (b) $x = \log(\sec \phi + \tan \phi)$,   (c) $\dfrac{dx}{d\phi} = \sec \phi$.

The Technique of Integration

7.1 A Short Table of Integrals; Review

For review and ready reference we begin with a short table of integrals. A more extended table appears at the end of the book.

1. $\displaystyle\int \alpha\,dx = \alpha x + C.$

2. $\displaystyle\int x^n\,dx = \frac{1}{n+1}\,x^{n+1} + C, \qquad n \neq -1.$

3. $\displaystyle\int \frac{dx}{x} = \log |x| + C.$

4. $\displaystyle\int e^x\,dx = e^x + C.$

5. $\displaystyle\int p^x\,dx = \frac{p^x}{\log p} + C.$

6. $\displaystyle\int \log x\,dx = x \log x - x + C.$

7. $\displaystyle\int \cos x\,dx = \sin x + C.$

8. $\displaystyle\int \sin x\,dx = -\cos x + C.$

9. $\displaystyle\int \sec^2 x\,dx = \tan x + C.$

10. $\displaystyle\int \operatorname{cosec}^2 x\, dx = -\cot x + C.$

11. $\displaystyle\int \sec x \tan x\, dx = \sec x + C.$

12. $\displaystyle\int \operatorname{cosec} x \cot x\, dx = -\operatorname{cosec} x + C.$

13. $\displaystyle\int \tan x\, dx = \log |\sec x| + C.$

14. $\displaystyle\int \cot x\, dx = \log |\sin x| + C.$

15. $\displaystyle\int \sec x\, dx = \log |\sec x + \tan x| + C.$

16. $\displaystyle\int \operatorname{cosec} x\, dx = \log |\operatorname{cosec} x - \cot x| + C.$

17. $\displaystyle\int \frac{dx}{\sqrt{1 - x^2}} = \arcsin x + C.$

18. $\displaystyle\int \frac{dx}{1 + x^2} = \arctan x + C.$

19. $\displaystyle\int \arcsin x\, dx = x \arcsin x + \sqrt{1 - x^2} + C.$

20. $\displaystyle\int \arctan x\, dx = x \arctan x - \tfrac{1}{2} \log (1 + x^2) + C.$

21. $\displaystyle\int \frac{dx}{x^2 - 1} = \frac{1}{2} \log \left| \frac{x - 1}{x + 1} \right| + C.$

Only Formula 21 is new. To derive it observe that

$$\frac{1}{x^2 - 1} = \frac{1}{2}\left(\frac{1}{x - 1} - \frac{1}{x + 1} \right).$$

It follows then that

$$\int \frac{dx}{x^2 - 1} = \frac{1}{2} \int \frac{dx}{x - 1} - \frac{1}{2} \int \frac{dx}{x + 1}$$

$$= \frac{1}{2} \log |x - 1| - \frac{1}{2} \log |x + 1| + C$$

$$= \frac{1}{2} \log \left| \frac{x - 1}{x + 1} \right| + C. \quad \square$$

In the spirit of review we work out a few integrals.

Problem. Find

$$\int x \tan x^2 \, dx.$$

Solution.   Set

$$u = x^2, \qquad du = 2x \, dx.$$

$$\int x \tan x^2 \, dx = \tfrac{1}{2} \int 2x \tan x \, dx = \tfrac{1}{2} \int \tan u \, du$$

$$= \tfrac{1}{2} \log |\sec u| + C = \tfrac{1}{2} \log |\sec x^2| + C. \quad \square$$

Problem. Compute

$$\int_0^1 \frac{e^x}{e^x + 2} \, dx.$$

Solution.   First we figure out the indefinite integral

$$\int \frac{e^x}{e^x + 2} \, dx.$$

Set

$$u = e^x, \qquad du = e^x \, dx.$$

$$\int \frac{e^x}{e^x + 2} \, dx = \int \frac{du}{u + 2} = \log |u + 2| + C = \log |e^x + 2| + C.$$

Consequently

$$\int_0^1 \frac{e^x}{e^x + 2} \, dx = \Big[\log |e^x + 2| \Big]_0^1 = \log (e + 2) - \log 3 = \log \left(\frac{e + 2}{3} \right). \quad \square$$

The next two examples should reinforce the idea of integration by parts.

Problem. Find

$$\int \log (x^2 + 1) \, dx.$$

Solution.   Set

$$u = \log (x^2 + 1), \qquad dv = dx,$$

$$du = \frac{2x}{x^2 + 1} \, dx, \qquad v = x.$$

$$\int \log (x^2 + 1) \, dx = \int u \, dv = uv - \int v \, du$$

$$= x \log (x^2 + 1) - \int \frac{2x^2}{x^2 + 1} \, dx.$$

Since

$$\int \frac{2x^2}{x^2 + 1} \, dx = \int \frac{2(x^2 + 1)}{x^2 + 1} \, dx - \int \frac{2}{x^2 + 1} \, dx$$

$$= 2x - 2 \arctan x + C,$$

we have

$$\int \log (x^2 + 1) \, dx = x \log (x^2 + 1) - 2x + 2 \arctan x + C. \quad \square$$

Problem. Find

$$\int x \arctan x \, dx.$$

SOLUTION.   We first set

$$u = \arctan x, \qquad dv = x \, dx,$$

$$du = \frac{1}{x^2 + 1} \, dx, \qquad v = \tfrac{1}{2}x^2.$$

We could go on this way but since later we have to integrate $v \, du$ we may as well choose v so as to make $v \, du$ as simple as possible. By adding $\tfrac{1}{2}$ to our initial choice of v, we have

$$v = \tfrac{1}{2}x^2 + \tfrac{1}{2} = \tfrac{1}{2}(x^2 + 1)$$

and

$$v \, du = \tfrac{1}{2}.$$

The integration is now simple:

$$\int x \arctan x \, dx = \int u \, dv = uv - \int v \, du$$

$$= \tfrac{1}{2}(x^2 + 1) \arctan x - \int \tfrac{1}{2} \, dx$$

$$= \tfrac{1}{2}(x^2 + 1) \arctan x - \tfrac{1}{2}x + C. \quad \square$$

Exercises

Work out the following integrals.

*1. $\displaystyle\int e^{2-x} \, dx.$

2. $\displaystyle\int \frac{x}{\sqrt{1 - x^2}} \, dx.$

*3. $\displaystyle\int \cos \tfrac{2}{3}x \, dx.$

*4. $\displaystyle\int_0^c \frac{dx}{x^2 + c^2}.$

5. $\displaystyle\int_{-\pi/4}^{\pi/4} \frac{\sin x}{\cos^2 x} \, dx.$

*6. $\displaystyle\int_1^2 \frac{e^{1/x}}{x^2} \, dx.$

*7. $\displaystyle\int x \log (x + 1)\, dx.$

8. $\displaystyle\int \log 2x\, dx.$

*9. $\displaystyle\int \frac{dx}{5^x}.$

*10. $\displaystyle\int_{\pi/6}^{\pi/3} \operatorname{cosec} x\, dx.$

11. $\displaystyle\int_0^1 \sin \pi x\, dx.$

*12. $\displaystyle\int_{-\pi/4}^{\pi/4} \frac{dx}{\cos^2 x}.$

*13. $\displaystyle\int \sec^2 (1 - x)\, dx.$

14. $\displaystyle\int \frac{e^{\sqrt{x}}}{\sqrt{x}}\, dx.$

*15. $\displaystyle\int x\, 5^x\, dx.$

*16. $\displaystyle\int_0^1 \frac{x^3}{1 + x^4}\, dx.$

17. $\displaystyle\int_0^t \sec \pi x \tan \pi x\, dx.$

*18. $\displaystyle\int_{\pi/6}^{\pi/3} \cot x\, dx.$

*19. $\displaystyle\int e^{-x} \sin 2x\, dx.$

20. $\displaystyle\int \frac{x^3}{\sqrt{1 - x^4}}\, dx.$

*21. $\displaystyle\int \frac{x}{\sqrt{1 - x^4}}\, dx.$

*22. $\displaystyle\int x \arcsin 2x^2\, dx.$

23. $\displaystyle\int \frac{x}{(x + \alpha)^2 + a^2}\, dx.$

*24. $\displaystyle\int \frac{1 + \cos 2x}{\sin^2 2x}\, dx.$

*25. $\displaystyle\int \frac{\sec^2 \theta}{\sqrt{3 \tan \theta + 1}}\, d\theta.$

26. $\displaystyle\int \frac{\sin \phi}{3 - 2 \cos \phi}\, d\phi.$

*27. $\displaystyle\int \frac{e^x}{ae^x - b}\, dx.$

*28. $\displaystyle\int a^x e^x\, dx.$

29. $\displaystyle\int \frac{x}{x^2 + 1}\, dx.$

*30. $\displaystyle\int_0^{\pi/4} \frac{\sec^2 x \tan x}{\sqrt{2 + \sec^2 x}}\, dx.$

(*Optional*)   Work out the following integrals.

*31. $\displaystyle\int \sinh^2 x\, dx.$

32. $\displaystyle\int x \sinh x\, dx.$

*33. $\displaystyle\int \cos x \sinh x\, dx.$

*34. $\displaystyle\int \tanh^3 x\, dx.$

35. $\displaystyle\int \sinh^3 x\, dx.$

*36. $\displaystyle\int e^{ax} \cosh ax\, dx.$

7.2   Partial Fractions

A rational function is by definition the quotient of two polynomials. Thus, for example,

$$\frac{1}{x^2 - 4}, \quad \frac{2x^2 + 3}{x(x - 1)^2}, \quad \frac{-2x}{(x + 1)(x^2 + 1)}, \quad \frac{1}{x(x^2 + x + 1)},$$

$$\frac{3x^4 + x^3 + 20x^2 + 3x + 31}{(x + 1)(x^2 + 4)^2}, \quad \frac{x^5}{x^2 - 1}$$

are all rational functions, but

$$\frac{1}{\sqrt{x}}, \quad \log x, \quad \frac{|x - 2|}{x}$$

are not rational functions.

To integrate a rational function it is usually necessary to first rewrite it as a polynomial (which may be identically 0) plus fractions of the form

$$\frac{A}{(x - \alpha)^k} \quad \text{and} \quad \frac{Bx + C}{(x^2 + \beta x + \gamma)^k}$$

with the quadratic $x^2 + \beta x + \gamma$ irreducible (not factorable). Such fractions are called *partial fractions*.

It is shown in algebra that every rational function can be written in such a way. Here are some examples.

Example 1. (*The denominator splits into distinct linear factors.*) For

$$\frac{1}{x^2 - 4} = \frac{1}{(x - 2)(x + 2)},$$

we write

$$\frac{1}{x^2 - 4} = \frac{A}{x - 2} + \frac{B}{x + 2}.$$

Clearing fractions, we have

$$1 = A(x + 2) + B(x - 2),$$
$$1 = (A + B)x + 2A - 2B.$$

Since this last equation is an identity, we can equate coefficients:

$$A + B = 0,$$
$$2A - 2B = 1.$$

Solving the equations simultaneously, we obtain

$$A = \tfrac{1}{4}, \quad B = -\tfrac{1}{4}.$$

The desired decomposition is

$$\frac{1}{x^2 - 4} = \frac{1}{4(x - 2)} - \frac{1}{4(x + 2)}.$$

You can check this by carrying out the subtraction on the right. □

REMARK.   In general, each distinct linear factor $x - c$ in the denominator gives rise to a term of the form

$$\frac{A}{x - c}$$

in the decomposition.

Example 2.   (*The denominator has a repeated linear factor.*)   For

$$\frac{2x^2 + 3}{x(x - 1)^2},$$

we write

$$\frac{2x^2 + 3}{x(x - 1)^2} = \frac{A}{x} + \frac{B}{x - 1} + \frac{C}{(x - 1)^2}.$$

This leads to

$$2x^2 + 3 = A(x - 1)^2 + Bx(x - 1) + Cx,$$
$$2x^2 + 3 = (A + B)x^2 - (2A + B - C)x + A,$$

and thus to the equations

$$A + B = 2,$$
$$2A + B - C = 0,$$
$$A = 3.$$

These equations give

$$A = 3, \quad B = -1, \quad C = 5.$$

The decomposition is therefore

$$\frac{2x^2 + 3}{x(x - 1)^2} = \frac{3}{x} - \frac{1}{x - 1} + \frac{5}{(x - 1)^2}. \quad □$$

REMARK.   In general, each factor of the form $(x - c)^k$ in the denominator gives rise to an expression of the form

$$\frac{A_1}{x - c} + \frac{A_2}{(x - c)^2} + \cdots + \frac{A_k}{(x - c)^k}$$

in the decomposition.

Example 3.   (*The denominator has an irreducible quadratic factor.*)   For

$$\frac{-2x}{(x + 1)(x^2 + 1)},$$

we write

$$\frac{-2x}{(x + 1)(x^2 + 1)} = \frac{A}{x + 1} + \frac{Bx + C}{x^2 + 1}$$

and obtain

$$-2x = A(x^2 + 1) + (Bx + C)(x + 1),$$
$$-2x = (A + B)x^2 + (B + C)x + A + C.$$

This gives

$$A + B = 0,$$
$$B + C = -2,$$
$$A + C = 0.$$

Here the solutions are

$$A = 1, \quad B = -1, \quad C = -1$$

and we have the decomposition

$$\frac{-2x}{(x + 1)(x^2 + 1)} = \frac{1}{x + 1} - \frac{x + 1}{x^2 + 1}. \quad \square$$

Example 4. (*The denominator has an irreducible quadratic factor.*) For

$$\frac{1}{x(x^2 + x + 1)}$$

we write

$$\frac{1}{x(x^2 + x + 1)} = \frac{A}{x} + \frac{Bx + C}{x^2 + x + 1}$$

and obtain

$$1 = A(x^2 + x + 1) + (Bx + C)x,$$
$$1 = (A + B)x^2 + (A + C)x + A.$$

Here

$$A + B = 0,$$
$$A + C = 0,$$
$$A = 1.$$

We conclude that

$$A = 1, \quad B = -1, \quad C = -1,$$

and therefore

$$\frac{1}{x(x^2 + x + 1)} = \frac{1}{x} - \frac{x + 1}{x^2 + x + 1}. \quad \square$$

REMARK.   In general, each irreducible quadratic factor $x^2 + \beta x + \gamma$ in the denominator gives rise to a term of the form

$$\frac{Ax + B}{x^2 + \beta x + \gamma}$$

in the decomposition.

Example 5. (*The denominator has a repeated irreducible quadratic factor.*) For

$$\frac{3x^4 + x^3 + 20x^2 + 3x + 31}{(x + 1)(x^2 + 4)^2}$$

we write

$$\frac{3x^4 + x^3 + 20x^2 + 3x + 31}{(x + 1)(x^2 + 4)^2} = \frac{A}{x + 1} + \frac{Bx + C}{x^2 + 4} + \frac{Dx + E}{(x^2 + 4)^2}.$$

This gives

$$3x^4 + x^3 + 20x^2 + 3x + 31$$
$$= A(x^2 + 4)^2 + (Bx + C)(x + 1)(x^2 + 4) + (Dx + E)(x + 1).$$

We can rewrite the right-hand side as

$$(A + B)x^4 + (B + C)x^3 + (8A + 4B + C + D)x^2$$
$$+ (4B + 4C + D + E)x + (16A + 4C + E).$$

This leads to equations

$$A + B = 3,$$
$$B + C = 1,$$
$$8A + 4B + C + D = 20,$$
$$4B + 4C + D + E = 3,$$
$$16A + 4C + E = 31.$$

With a little patience you can see that

$$A = 2, \quad B = 1, \quad C = 0, \quad D = 0, \quad \text{and} \quad E = -1.$$

This gives the decomposition

$$\frac{3x^4 + x^3 + 20x^2 + 3x + 31}{(x + 1)(x^2 + 4)^2} = \frac{2}{x + 1} + \frac{x}{x^2 + 4} - \frac{1}{(x^2 + 4)^2}. \quad \square$$

REMARK.   In general each multiple irreducible quadratic factor $(x^2 + \beta x + \gamma)^k$ in the denominator gives rise to an expression of the form

$$\frac{A_1 x + B_1}{x^2 + \beta x + \gamma} + \frac{A_2 x + B_2}{(x^2 + \beta x + \gamma)^2} + \cdots + \frac{A_k x + B_k}{(x^2 + \beta x + \gamma)^k}$$

in the decomposition.

A polynomial appears in the decomposition when the degree of the numerator is greater than or equal to that of the denominator.

Example 6. (*The decomposition contains a polynomial.*) For

$$\frac{x^5 + 2}{x^2 - 1},$$

first carry out the suggested division:

$$
\begin{array}{r}
x^3 + x \\
x^2 - 1 \overline{)x^5 + 2} \\
\underline{x^5 - x^3 } \\
x^3 \\
\underline{x^3 - x } \\
x + 2.
\end{array}
$$

This gives

$$\frac{x^5 + 2}{x^2 - 1} = x^3 + x + \frac{x + 2}{x^2 - 1}.$$

Since the denominator of the fraction splits into linear factors, we write

$$\frac{x + 2}{x^2 - 1} = \frac{A}{x + 1} + \frac{B}{x - 1}.$$

This gives

$$x + 2 = A(x - 1) + B(x + 1),$$
$$x + 2 = (A + B)x - (A - B).$$

Here

$$A + B = 1,$$
$$A - B = -2,$$

so that

$$A = -\tfrac{1}{2}, \qquad B = \tfrac{3}{2}.$$

The decomposition takes the form

$$\frac{x^5 + 2}{x^2 - 1} = x^3 + x - \frac{1}{2(x + 1)} + \frac{3}{2(x - 1)}. \quad \square$$

We broke up the rational functions into partial fractions so that we could integrate them. Here are the integrations carried out.

Example 1

$$\int \frac{dx}{x^2 - 4} = \frac{1}{4} \int \left(\frac{1}{x - 2} - \frac{1}{x + 2} \right) dx$$

$$= \frac{1}{4} (\log |x - 2| - \log |x + 2|) + C$$

$$= \frac{1}{4} \log \left| \frac{x - 2}{x + 2} \right| + C. \quad \square$$

Example 2

$$\int \frac{2x^2 + 3}{x(x - 1)^2} \, dx = \int \left[\frac{3}{x} - \frac{1}{x - 1} + \frac{5}{(x - 1)^2} \right] dx$$

$$= 3 \log |x| - \log |x - 1| - \frac{5}{x - 1} + C. \quad \square$$

Example 3

$$\int \frac{-2x}{(x+1)(x^2+1)} \, dx = \int \left(\frac{1}{x+1} - \frac{x+1}{x^2+1} \right) dx.$$

Since

$$\int \frac{dx}{x+1} = \log |x+1| + C_1$$

and

$$\int \frac{x+1}{x^2+1} \, dx = \frac{1}{2} \int \frac{2x}{x^2+1} \, dx + \int \frac{dx}{x^2+1} = \tfrac{1}{2} \log (x^2+1) + \text{arc tan } x + C_2,$$

we have

$$\int \frac{-2x}{(x+1)(x^2+1)} \, dx = \log |x+1| - \tfrac{1}{2} \log (x^2+1) - \text{arc tan } x + C. \quad \square$$

Example 4

$$\int \frac{dx}{x(x^2+x+1)} = \int \left(\frac{1}{x} - \frac{x+1}{x^2+x+1} \right) dx = \log |x| - \int \frac{x+1}{x^2+x+1} \, dx.$$

To compute the remaining integral, note that

$$\int \frac{x+1}{x^2+x+1} \, dx = \frac{1}{2} \int \frac{2x+1}{x^2+x+1} \, dx + \frac{1}{2} \int \frac{dx}{x^2+x+1}.$$

The first integral is a logarithm:

$$\frac{1}{2} \int \frac{2x+1}{x^2+x+1} \, dx = \frac{1}{2} \log (x^2+x+1) + C_1.$$

To compute the second integral, we complete the square in the denominator:

$$\frac{1}{2} \int \frac{dx}{x^2+x+1} = \frac{1}{2} \int \frac{dx}{(x+\frac{1}{2})^2 + (\sqrt{3}/2)^2} = \frac{1}{\sqrt{3}} \text{arc tan} \left[\frac{2}{\sqrt{3}} \left(x + \frac{1}{2} \right) \right] + C_2.$$

Combining results, we have

$$\int \frac{dx}{x(x^2+x+1)} = \log |x| - \frac{1}{2} \log (x^2+x+1) - \frac{1}{\sqrt{3}} \text{arc tan} \left[\frac{2}{\sqrt{3}} \left(x + \frac{1}{2} \right) \right] + C. \quad \square$$

Example 5

$$\int \frac{3x^4+x^3+20x^2+3x+31}{(x+1)(x^2+4)^2} \, dx = \int \left[\frac{2}{x+1} + \frac{x}{x^2+4} - \frac{1}{(x^2+4)^2} \right] dx.$$

The first two fractions are easy to integrate:

$$\int \frac{2}{x + 1}\, dx = 2 \log |x + 1| + C_1,$$

$$\int \frac{x}{x^2 + 4}\, dx = \frac{1}{2} \int \frac{2x}{x^2 + 4}\, dx = \frac{1}{2} \log (x^2 + 4) + C_2.$$

To integrate the last fraction use the reduction formula

$$\int \frac{dx}{(x^2 + c^2)^n} = \frac{1}{2c^2(n - 1)}\left[\frac{x}{(x^2 + c^2)^{n-1}}\right] + \frac{2n - 3}{2c^2(n - 1)} \int \frac{dx}{(x^2 + c^2)^{n-1}}.$$

This gives

$$\int \frac{dx}{(x^2 + 4)^2} = \frac{1}{8}\left(\frac{x}{x^2 + 4}\right) + \frac{1}{8} \int \frac{dx}{x^2 + 4}$$

$$= \frac{1}{8}\left(\frac{x}{x^2 + 4}\right) + \frac{1}{16} \arctan\left(\frac{x}{2}\right) + C_3.$$

The integral we want is therefore

$$2 \log |x + 1| + \frac{1}{2} \log (x^2 + 4) - \frac{1}{8}\left(\frac{x}{x^2 + 4}\right) - \frac{1}{16} \arctan\left(\frac{x}{2}\right) + C. \quad \square$$

Example 6

$$\int \frac{x^5 + 2}{x^2 - 1}\, dx = \int \left[x^3 + x - \frac{1}{2(x + 1)} + \frac{3}{2(x - 1)}\right] dx$$

$$= \tfrac{1}{4}x^4 + \tfrac{1}{2}x^2 - \tfrac{1}{2} \log |x + 1| + \tfrac{3}{2} \log |x - 1| + C. \quad \square$$

Exercises

Work out the following integrals.

*1. $\displaystyle\int \frac{7}{(x - 2)(x + 5)}\, dx.$

2. $\displaystyle\int \frac{x}{(x + 1)(x + 2)(x + 3)}\, dx.$

*3. $\displaystyle\int \frac{x^2 + 1}{x(x^2 - 1)}\, dx.$

*4. $\displaystyle\int \frac{2x^2 + 3}{x^2(x - 1)}\, dx.$

5. $\displaystyle\int \frac{x^5}{x - 2}\, dx.$

*6. $\displaystyle\int \frac{x^5}{(x - 2)^2}\, dx.$

*7. $\displaystyle\int \frac{x + 3}{x^2 - 3x + 2}\, dx.$

8. $\displaystyle\int \frac{x^2 + 3}{x^2 - 3x + 2}\, dx.$

*9. $\displaystyle\int \frac{dx}{(x-1)^3}$.

*10. $\displaystyle\int \frac{x^2}{(x-1)^2(x+1)}\, dx$.

11. $\displaystyle\int \frac{x^3 + 4x^2 - 4x - 1}{(x^2 + 1)^2}\, dx$.

*12. $\displaystyle\int \frac{2x-1}{(x+1)^2(x-2)^2}\, dx$.

*13. $\displaystyle\int \frac{dx}{x^4 - 16}$.

14. $\displaystyle\int \frac{x}{x^3 - 1}\, dx$.

*15. $\displaystyle\int \frac{dx}{x^4 + 4}$.

*16. $\displaystyle\int_0^4 \frac{dx}{x^2 + 16}$.

17. $\displaystyle\int_0^4 \frac{dx}{(x^2 + 16)^2}$.

18. $\displaystyle\int_1^2 \frac{dx}{x^2 + 2x + 2}$.

19. Verify the reduction formula

$$\int \frac{dx}{(x^2 + c^2)^n} = \frac{1}{2c^2(n-1)}\left[\frac{x}{(x^2+c^2)^{n-1}}\right] + \frac{2n-3}{2c^2(n-1)}\int \frac{dx}{(x^2+c^2)^{n-1}}$$

by differentiation.

*20. (Optional) It is known that m parts of chemical A combine with n parts of chemical B to produce a compound C. Suppose that the rate at which C is produced varies directly with the product of the amounts of A and B present at that instant. Find the amount of C produced in t minutes from an initial mixing of A_0 pounds of A with B_0 pounds of B, given that

(a) $n = m$, $A_0 = B_0$, and A_0 pounds of C are produced in the first minute.

(b) $n = m$, $A_0 = \frac{1}{2}B_0$, and A_0 pounds of C are produced in the first minute.

(c) $n \neq m$, $A_0 = B_0$, and A_0 pounds of C are produced in the first minute.

HINT: Denote by $A(t)$, $B(t)$, and $C(t)$ the amounts of A, B, and C present at time t. Observe that

$$C'(t) = kA(t)B(t).$$

Note that

$$A_0 - A(t) = \frac{m}{m+n}\, C(t) \quad \text{and} \quad B_0 - B(t) = \frac{n}{m+n}\, C(t)$$

and thus

$$C'(t) = k\left[A_0 - \frac{m}{m+n}\, C(t)\right]\left[B_0 - \frac{n}{m+n}\, C(t)\right].$$

7.3   Powers and Products of Sines and Cosines

I.

$$\int \sin^m x \cos^n x\, dx \quad \text{with } m \text{ or } n \text{ odd.}$$

Suppose that n is odd. If $n = 1$, the integral is of the form

(1) $$\int \sin^m x \cos x \, dx = \frac{1}{m+1} \sin^{m+1} x + C.$$

If $n > 1$, write

$$\cos^n x = \cos^{n-1} x \cos x.$$

Since $n - 1$ is even, $\cos^{n-1} x$ can be expressed in powers of $\sin^2 x$ by substituting

$$\cos^2 x = 1 - \sin^2 x.$$

The integral then takes the form

$$\int (\text{sum of powers of } \sin x) \cdot \cos x \, dx,$$

which can be broken up into integrals of form (1).

Similarly if m is odd, write

$$\sin^m x = \sin^{m-1} x \sin x$$

and use the substitution

$$\sin^2 x = 1 - \cos^2 x.$$

Example

$$
\begin{aligned}
\int \sin^2 x \cos^5 x \, dx &= \int \sin^2 x \cos^4 x \cos x \, dx \\[6pt]
&= \int \sin^2 x \, (1 - \sin^2 x)^2 \cos x \, dx \\[6pt]
&= \int (\sin^2 x - 2 \sin^4 x + \sin^6 x) \cos x \, dx \\[6pt]
&= \int \sin^2 x \cos x \, dx - 2 \int \sin^4 x \cos x \, dx + \int \sin^6 x \cos x \, dx \\[6pt]
&= \tfrac{1}{3} \sin^3 x - \tfrac{2}{5} \sin^5 x + \tfrac{1}{7} \sin^7 x + C. \quad \square
\end{aligned}
$$

Example

$$
\begin{aligned}
\int \sin^5 x \, dx &= \int \sin^4 x \sin x \, dx \\[6pt]
&= \int (1 - \cos^2 x)^2 \sin x \, dx \\[6pt]
&= \int (1 - 2 \cos^2 x + \cos^4 x) \sin x \, dx \\[6pt]
&= \int \sin x \, dx - 2 \int \cos^2 x \sin x \, dx + \int \cos^4 x \sin x \, dx \\[6pt]
&= -\cos x + \tfrac{2}{3} \cos^3 x - \tfrac{1}{5} \cos^5 x + C. \quad \square
\end{aligned}
$$

II.
$$\int \sin^m x \cos^n x \, dx \quad \text{with } m \text{ and } n \text{ both even.}$$

Use the multiple angle formulas

$$\sin x \cos x = \tfrac{1}{2} \sin 2x,$$
$$\sin^2 x = \tfrac{1}{2} - \tfrac{1}{2} \cos 2x,$$
$$\cos^2 x = \tfrac{1}{2} + \tfrac{1}{2} \cos 2x.$$

Example

$$\int \cos^2 x \, dx = \int (\tfrac{1}{2} + \tfrac{1}{2} \cos 2x) \, dx$$

$$= \tfrac{1}{2} \int dx + \tfrac{1}{2} \int \cos 2x \, dx$$

$$= \tfrac{1}{2}x + \tfrac{1}{4} \sin 2x + C. \quad \square$$

Example

$$\int \sin^2 x \cos^2 x \, dx = \tfrac{1}{4} \int \sin^2 2x \, dx$$

$$= \tfrac{1}{4} \int (\tfrac{1}{2} - \tfrac{1}{2} \cos 4x) \, dx$$

$$= \tfrac{1}{8} \int dx - \tfrac{1}{8} \int \cos 4x \, dx$$

$$= \tfrac{1}{8}x - \tfrac{1}{32} \sin 4x + C. \quad \square$$

Example

$$\int \sin^4 x \cos^2 x \, dx = \int (\sin x \cos x)^2 \sin^2 x \, dx$$

$$= \int \tfrac{1}{4} \sin^2 2x \, (\tfrac{1}{2} - \tfrac{1}{2} \cos 2x) \, dx$$

$$= \tfrac{1}{8} \int \sin^2 2x \, dx - \tfrac{1}{8} \int \sin^2 2x \cos 2x \, dx$$

$$= \tfrac{1}{8} \int (\tfrac{1}{2} - \tfrac{1}{2} \cos 4x) \, dx - \tfrac{1}{8} \int \sin^2 2x \cos 2x \, dx$$

$$= \tfrac{1}{16}x - \tfrac{1}{64} \sin 4x - \tfrac{1}{48} \sin^3 2x + C. \quad \square$$

III.
$$\int \sin mx \cos nx \, dx, \qquad \int \sin mx \sin nx \, dx,$$
$$\int \cos mx \cos nx \, dx.$$

If $m = n$, there is no difficulty. Suppose that $m \neq n$. For the first integral use the identity

(1) $\sin mx \cos nx = \frac{1}{2} \sin [(m + n)x] + \frac{1}{2} \sin [(m - n)x]$.

You can confirm this identity by using the usual expansions of $\sin (\theta_1 + \theta_2)$ and $\sin (\theta_1 - \theta_2)$. (Section 1.2, part V.) Set

$$\theta_1 = mx \quad \text{and} \quad \theta_2 = nx$$

and you'll find that

$$\sin [(m + n)x] = \sin mx \cos nx + \cos mx \sin nx,$$
$$\sin [(m - n)x] = \sin mx \cos nx - \cos mx \sin nx.$$

Add these two equations, divide by 2, and you'll get identity (1).
   Integration of (1) gives

$$\int \sin mx \cos nx \, dx = \frac{1}{2} \int \sin [(m + n)x] \, dx + \frac{1}{2} \int \sin [(m - n)x] \, dx$$
$$= -\frac{\cos [(m + n)x]}{2(m + n)} - \frac{\cos [(m - n)x]}{2(m - n)} + C.$$

Similarly,

$$\int \sin mx \sin nx \, dx = -\frac{\sin [(m + n)x]}{2(m + n)} + \frac{\sin [(m - n)x]}{2(m - n)} + C,$$
$$\int \cos mx \cos nx \, dx = \frac{\sin [(m + n)x]}{2(m + n)} + \frac{\sin [(m - n)x]}{2(m - n)} + C.$$

Exercises

Work out the following integrals.

* 1. $\displaystyle\int \sin^3 x \, dx.$

2. $\displaystyle\int \sin^2 x \, dx.$

* 3. $\displaystyle\int \cos^4 x \sin^3 x \, dx.$

4. $\displaystyle\int_0^{\pi/2} \cos^3 x \, dx.$

* 5. $\displaystyle\int_0^\pi \sin^4 x \, dx.$

6. $\displaystyle\int \sin 2x \cos 3x \, dx.$

* 7. $\displaystyle\int_0^{\pi/2} \cos^4 x \, dx.$

8. $\displaystyle\int \cos 2x \sin 3x \, dx.$

*9. $\displaystyle\int \sin^3 x \cos^2 x\, dx.$

10. $\displaystyle\int \sin^3 x \cos^3 x\, dx.$

*11. $\displaystyle\int \cos 2x \cos 3x\, dx.$

12. $\displaystyle\int_0^{\pi/2} \cos^5 x\, dx.$

*13. $\displaystyle\int \sin^5 x \cos^2 x\, dx.$

14. $\displaystyle\int \sin^3 3x\, dx.$

*15. $\displaystyle\int \sin^2 x \cos^4 x\, dx.$

16. $\displaystyle\int (\sin^2 x + \cos x)^2\, dx.$

Evaluate the following integrals. They are important in applied mathematics. (For Exercises 19, 20, and 21 take $m \neq n$.)

*17. $\displaystyle\frac{1}{\pi} \int_0^{2\pi} \cos^2 nx\, dx.$

18. $\displaystyle\frac{1}{\pi} \int_0^{2\pi} \sin^2 nx\, dx.$

*19. $\displaystyle\int_0^{2\pi} \sin mx \cos nx\, dx.$

20. $\displaystyle\int_0^{2\pi} \sin mx \sin nx\, dx.$

*21. $\displaystyle\int_0^{2\pi} \cos mx \cos nx\, dx.$

7.4 Other Trigonometric Powers

I.
$$\int \tan^n x\, dx, \qquad \int \cot^n x\, dx.$$

For the first integral use

$$\tan^n x = \tan^{n-2} x \tan^2 x = (\tan^{n-2} x)(\sec^2 x - 1) = \tan^{n-2} x \sec^2 x - \tan^{n-2} x.$$

and for the second

$$\cot^n x = \cot^{n-2} x \cot^2 x = (\cot^{n-2} x)(\operatorname{cosec}^2 x - 1) = \cot^{n-2} x \operatorname{cosec}^2 x - \cot^{n-2} x.$$

Example

$$
\begin{aligned}
\int \tan^6 x\, dx &= \int (\tan^4 x \sec^2 x - \tan^4 x)\, dx \\
&= \int (\tan^4 x \sec^2 x - \tan^2 x \sec^2 x + \tan^2 x)\, dx \\
&= \int (\tan^4 x \sec^2 x - \tan^2 x \sec^2 x + \sec^2 x - 1)\, dx \\
&= \int \tan^4 x \sec^2 x\, dx - \int \tan^2 x \sec^2 x\, dx + \int \sec^2 x\, dx - \int dx \\
&= \tfrac{1}{5} \tan^5 x - \tfrac{1}{3} \tan^3 x + \tan x - x + C. \quad \square
\end{aligned}
$$

II.

$$\int \sec^n x \; dx, \qquad \int \operatorname{cosec}^n x \; dx.$$

For even powers use

$$\sec^n x = \sec^{n-2} x \sec^2 x = (\tan^2 x + 1)^{(n-2)/2} \sec^2 x$$

or

$$\operatorname{cosec}^n x = \operatorname{cosec}^{n-2} x \operatorname{cosec}^2 x = (\cot^2 x + 1)^{(n-2)/2} \operatorname{cosec}^2 x.$$

Example

$$\int \sec^4 x \; dx = \int \sec^2 x \sec^2 x \; dx = \int (\tan^2 x + 1) \sec^2 x \; dx$$

$$= \int \tan^2 x \sec^2 x \; dx + \int \sec^2 x \; dx$$

$$= \tfrac{1}{3} \tan^3 x + \tan x + C. \quad \square$$

The odd powers you can integrate by parts. For $\sec^n x$ set

$$u = \sec^{n-2} x, \qquad dv = \sec^2 x \; dx$$

and, when $\tan^2 x$ appears, use the identity

$$\tan^2 x = \sec^2 x - 1.$$

You can handle $\operatorname{cosec}^n x$ in a similar manner.

Example. For

$$\int \sec^3 x \; dx$$

set

$$u = \sec x, \qquad dv = \sec^2 x \; dx,$$
$$du = \sec x \tan x \; dx, \qquad v = \tan x.$$

$$\int \sec^3 x \; dx = \int \sec x \sec^2 x \; dx$$

$$= \int u \; dv = uv - \int v \; du$$

$$= \sec x \tan x - \int \tan^2 x \sec x \; dx$$

$$= \sec x \tan x - \int (\sec^2 x - 1) \sec x \; dx$$

$$= \sec x \tan x - \int \sec^3 x \; dx + \int \sec x \; dx$$

$$= \sec x \tan x - \int \sec^3 x \; dx + \log |\sec x + \tan x|.$$

$$2 \int \sec^3 x \, dx = \sec x \tan x + \log |\sec x + \tan x|,$$

$$\int \sec^3 x \, dx = \tfrac{1}{2} \sec x \tan x + \tfrac{1}{2} \log |\sec x + \tan x|.$$

Now add the arbitrary constant:

$$\int \sec^3 x \, dx = \tfrac{1}{2} \sec x \tan x + \tfrac{1}{2} \log |\sec x + \tan x| + C. \quad \square$$

III.
$$\int \tan^m x \sec^n x \, dx, \qquad \int \cot^m x \operatorname{cosec}^n x \, dx.$$

When n is even, write

$$\tan^m x \sec^n x = \tan^m x \sec^{n-2} x \sec^2 x$$

and express $\sec^{n-2} x$ entirely in terms of $\tan^2 x$ using

$$\sec^2 x = \tan^2 x + 1.$$

Example

$$\int \tan^5 x \sec^4 x \, dx = \int \tan^5 x \sec^2 x \sec^2 x \, dx$$

$$= \int \tan^5 x (\tan^2 x + 1) \sec^2 x \, dx$$

$$= \int \tan^7 x \sec^2 x \, dx + \int \tan^5 x \sec^2 x \, dx$$

$$= \tfrac{1}{8} \tan^8 x + \tfrac{1}{6} \tan^6 x + C. \quad \square$$

When n and m are both odd, write

$$\tan^m x \sec^n x = \tan^{m-1} x \sec^{n-1} x \sec x \tan x$$

and express $\tan^{m-1} x$ entirely in terms of $\sec^2 x$ using

$$\tan^2 x = \sec^2 x - 1.$$

Example

$$\int \tan^5 x \sec^3 x \, dx = \int \tan^4 x \sec^2 x \sec x \tan x \, dx$$

$$= \int (\sec^2 x - 1)^2 \sec^2 x \sec x \tan x \, dx$$

$$= \int (\sec^6 x - 2 \sec^4 x + \sec^2 x) \sec x \tan x \, dx$$

$$= \tfrac{1}{7} \sec^7 x - \tfrac{2}{5} \sec^5 x + \tfrac{1}{3} \sec^3 x + C. \quad \square$$

Finally, if n is odd and m is even, write the product entirely in terms of sec x and integrate by parts.

Example

$$\int \tan^2 x \sec x \, dx = \int (\sec^2 x - 1) \sec x \, dx$$

$$= \int (\sec^3 x - \sec x) \, dx$$

$$= \int \sec^3 x \, dx - \int \sec x \, dx.$$

We have computed each of these integrals by parts before:

$$\int \sec^3 x \, dx = \tfrac{1}{2} \sec x \tan x + \tfrac{1}{2} \log |\sec x + \tan x| + C$$

and

$$\int \sec x \, dx = \log |\sec x + \tan x| + C.$$

It follows that

$$\int \tan^2 x \sec x \, dx = \tfrac{1}{2} \sec x \tan x - \tfrac{1}{2} \log |\sec x + \tan x| + C. \quad \square$$

You can handle integrals of $\cot^m x \operatorname{cosec}^n x$ in a similar manner.

Exercises

*1. $\displaystyle\int \tan^3 x \, dx.$
      2. $\displaystyle\int \tan^2 (x + 1) \, dx.$

*3. $\displaystyle\int \sec^2 (x + 1) \, dx.$
      4. $\displaystyle\int \cot^3 x \, dx.$

*5. $\displaystyle\int \tan^2 x \sec^2 x \, dx.$
      6. $\displaystyle\int \cot^5 x \, dx.$

*7. $\displaystyle\int \tan^4 x \sec^4 x \, dx.$
      8. $\displaystyle\int \sec^4 4x \, dx.$

*9. $\displaystyle\int \tan^5 3x \, dx.$
      10. $\displaystyle\int \operatorname{cosec}^4 \tfrac{1}{4}x \, dx.$

*11. $\displaystyle\int \cot^3 2x \, dx.$
      12. $\displaystyle\int \cot^2 x \sec x \, dx.$

*13. $\displaystyle\int \operatorname{cosec}^3 x \, dx.$
      14. $\displaystyle\int \sec^5 x \, dx.$

7.5   Integrals Involving $\sqrt{a^2 \pm x^2}$ and $\sqrt{x^2 \pm a^2}$

Such integrals can generally be handled by trigonometric substitution.

$$
\begin{array}{lll}
\text{For} & \sqrt{a^2 - x^2} & \text{set} \quad a \sin u = x. \\
\text{For} & \sqrt{a^2 + x^2} & \text{set} \quad a \tan u = x. \\
\text{For} & \sqrt{x^2 - a^2} & \text{set} \quad a \sec u = x.
\end{array}
\qquad (\text{take } a > 0)
$$

Problem.   Find

$$
\int \frac{dx}{(a^2 - x^2)^{3/2}}.
$$

SOLUTION.   Set

$$
a \sin u = x, \qquad a \cos u \; du = dx
$$

and portray the substitution as in Figure 7.5.1.

$$
\int \frac{dx}{(a^2 - x^2)^{3/2}} = \int \frac{a \cos u}{(a^2 - a^2 \sin^2 u)^{3/2}} \, du = \frac{1}{a^2} \int \frac{\cos u}{\cos^3 u} \, du
$$

$$
= \frac{1}{a^2} \int \sec^2 u \; du = \frac{1}{a^2} \tan u + C.
$$

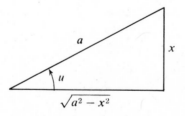

FIGURE 7.5.1

From Figure 7.5.1,

$$
\tan u = \frac{x}{\sqrt{a^2 - x^2}}
$$

so that

$$
\int \frac{dx}{(a^2 - x^2)^{3/2}} = \frac{x}{a^2 \sqrt{a^2 - x^2}} + C. \quad \square
$$

Problem.   Find

$$
\int \sqrt{a^2 + x^2} \; dx.
$$

SOLUTION. Set
$$a \tan u = x, \qquad a \sec^2 u \, du = dx$$
and portray the substitution as in Figure 7.5.2.

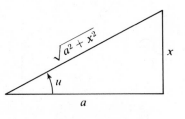

FIGURE 7.5.2

$$\int \sqrt{a^2 + x^2} \, dx = \int \sqrt{a^2 + a^2 \tan^2 u} \; a \sec^2 u \, du$$

$$= a^2 \int \sqrt{1 + \tan^2 u} \; \sec^2 u \, du$$

$$= a^2 \int \sec u \cdot \sec^2 u \, du$$

$$= a^2 \int \sec^3 u \, du$$

by Section 7.4
$$\;\downarrow\; \stackrel{}{=} \frac{a^2}{2} \left(\sec u \tan u + \log |\sec u + \tan u| \right) + C$$

from Figure 7.5.2
$$\;\downarrow\; \stackrel{}{=} \frac{a^2}{2} \left[\frac{\sqrt{a^2 + x^2}}{a} \left(\frac{x}{a} \right) + \log \left| \frac{\sqrt{a^2 + x^2}}{a} + \frac{x}{a} \right| \right] + C$$

$$= \tfrac{1}{2} x \sqrt{a^2 + x^2} + \tfrac{1}{2} a^2 \log (x + \sqrt{a^2 + x^2}) - \tfrac{1}{2} a^2 \log a + C.$$

We can absorb the constant
$$-\tfrac{1}{2} a^2 \log a$$
in C and write

$$\int \sqrt{a^2 + x^2} \, dx = \tfrac{1}{2} x \sqrt{a^2 + x^2} + \tfrac{1}{2} a^2 \log (x + \sqrt{a^2 + x^2}) + C.$$

This is one of the standard formulas. ☐

  Now a slight variation.

Problem. Find

$$\int \frac{dx}{x \sqrt{4x^2 + 9}} \, .$$

SOLUTION.   Set

$$3 \tan u = 2x, \qquad 3 \sec^2 u \, du = 2 \, dx$$

and portray the substitution as in Figure 7.5.3.

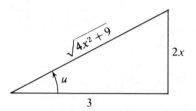

FIGURE 7.5.3

$$\int \frac{dx}{x\sqrt{4x^2 + 9}} = \int \frac{\frac{3}{2} \sec^2 u}{\frac{3}{2} \tan u \cdot 3 \sec u} \, du$$

$$= \frac{1}{3} \int \frac{\sec u}{\tan u} \, du$$

$$= \frac{1}{3} \int \operatorname{cosec} u \, du$$

$$= \frac{1}{3} \log |\operatorname{cosec} u - \cot u| + C.$$

From Figure 7.5.3,

$$\operatorname{cosec} u = \frac{\sqrt{4x^2 + 9}}{2x} \quad \text{and} \quad \cot u = \frac{3}{2x}$$

so that

$$\int \frac{dx}{x\sqrt{4x^2 + 9}} = \frac{1}{3} \log \left| \frac{\sqrt{4x^2 + 9} - 3}{2x} \right| + C. \quad \square$$

Exercises

Work out the following integrals.

*1. $\displaystyle\int \frac{dx}{\sqrt{a^2 - x^2}}$.

2. $\displaystyle\int \frac{dx}{(x^2 + 2)^{3/2}}$.

*3. $\displaystyle\int \frac{dx}{(5 - x^2)^{3/2}}$.

4. $\displaystyle\int \frac{x}{\sqrt{x^2 - 4}} \, dx$.

*5. $\displaystyle\int_1^2 \sqrt{x^2 - 1} \, dx$.

6. $\displaystyle\int \frac{x}{\sqrt{4 - x^2}} \, dx$.

*7. $\int \dfrac{x^2}{\sqrt{x^2 - 4}}\, dx.$

8. $\int \dfrac{x^2}{\sqrt{4 - x^2}}\, dx.$

*9. $\int \dfrac{x^2}{\sqrt{4 + x^2}}\, dx.$

10. $\int \dfrac{x}{(a^2 - x^2)^{3/2}}\, dx.$

*11. $\int \dfrac{x^2}{(a^2 - x^2)^{3/2}}\, dx.$

12. $\int \dfrac{x}{a^2 + x^2}\, dx.$

*13. $\int_0^2 x\sqrt{4 + x^2}\, dx.$

14. $\int x\sqrt{4 - x^2}\, dx.$

*15. $\int \dfrac{e^x}{\sqrt{9 - e^{2x}}}\, dx.$

16. $\int \dfrac{x^2}{(x^2 + 8)^{3/2}}\, dx.$

*17. $\int \dfrac{\sqrt{1 - x^2}}{x^4}\, dx.$

18. $\int \dfrac{\sqrt{x^2 - 1}}{x}\, dx.$

*19. $\int \dfrac{dx}{\sqrt{x^2 + a^2}}.$

20. $\int \dfrac{dx}{\sqrt{x^2 - a^2}}.$

*21. $\int \sqrt{a^2 - x^2}\, dx.$

22. $\int \sqrt{x^2 - a^2}\, dx.$

*23. $\int \dfrac{dx}{x\sqrt{a^2 - x^2}}.$

24. $\int \dfrac{dx}{x^2\sqrt{a^2 + x^2}}.$

*25. $\int \dfrac{dx}{x^2\sqrt{a^2 - x^2}}.$

26. $\int \dfrac{dx}{x^2\sqrt{x^2 - a^2}}.$

*27. $\int \dfrac{dx}{x\sqrt{a^2 + x^2}}.$

28. $\int \dfrac{dx}{x\sqrt{x^2 - a^2}}.$

7.6   Rational Expressions in sin *x* and cos *x*

There is a substitution which makes it possible to integrate all rational expressions of sin *x* and cos *x*. The result is significant enough to be cast as a theorem.

Theorem 7.6.1

If $f(x)$ is a rational expression in sin *x* and cos *x*, then the substitution

$$2 \text{ arc tan } u = x$$

transforms the integral

$$\int f(x)\, dx$$

into the integral of a rational function of *u*.

PROOF. Set

$$2 \text{ arc tan } u = x, \qquad \frac{2}{1 + u^2} \, du = dx$$

and observe that

$$u = \tan \tfrac{1}{2}x. \qquad \text{(see Figure 7.6.1)}$$

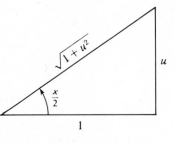

FIGURE 7.6.1

$$\int f(x) \, dx = 2 \int \frac{f(2 \text{ arc tan } u)}{1 + u^2} \, du.$$

Let's assume now that $f(x)$ is rational in $\cos x$ and $\sin x$. To prove the theorem, we need to show that $f(2 \text{ arc tan } u)$ is rational in u. This we can do by showing that $\sin (2 \text{ arc tan } u)$ and $\cos (2 \text{ arc tan } u)$ are both rational in u. This follows directly:

$$\sin (2 \text{ arc tan } u) = \sin x = 2 \sin \tfrac{1}{2}x \cos \tfrac{1}{2}x$$

$$= 2 \left(\frac{u}{\sqrt{1 + u^2}} \right) \left(\frac{1}{\sqrt{1 + u^2}} \right)$$

$$= 2 \left(\frac{u}{1 + u^2} \right).$$

$$\cos (2 \text{ arc tan } u) = \cos x = \cos^2 \tfrac{1}{2}x - \sin^2 \tfrac{1}{2}x$$

$$= \frac{1}{1 + u^2} - \frac{u^2}{1 + u^2}$$

$$= \frac{1 - u^2}{1 + u^2}. \quad \square$$

Note that in proving the theorem we showed that the substitution

$$\boxed{2 \text{ arc tan } u = x}$$

had the following consequences:

$$\sin (2 \arctan u) = \frac{2u}{1 + u^2}, \qquad \cos (2 \arctan u) = \frac{1 - u^2}{1 + u^2}.$$

Keep this in mind as you work the problems.

Problem. Find

$$\int \frac{dx}{1 + 2 \cos x}.$$

SOLUTION. Set

$$2 \arctan u = x, \qquad \frac{2}{1 + u^2} du = dx.$$

With $u = \tan \frac{1}{2}x$,

$$\int \frac{dx}{1 + 2 \cos x} = \int \frac{2}{1 + u^2} \left[\frac{1}{1 + 2 \cos (2 \arctan u)} \right] du$$

$$= \int \frac{2}{1 + u^2} \left[\frac{1}{1 + 2 \left(\dfrac{1 - u^2}{1 + u^2} \right)} \right] du$$

$$= -2 \int \frac{du}{u^2 - 3}$$

$$= \frac{1}{\sqrt{3}} \int \left(\frac{1}{u + \sqrt{3}} - \frac{1}{u - \sqrt{3}} \right) du$$

$$= \frac{1}{\sqrt{3}} (\log |u + \sqrt{3}| - \log |u - \sqrt{3}|) + C$$

$$= \frac{1}{\sqrt{3}} \log \left| \frac{u + \sqrt{3}}{u - \sqrt{3}} \right| + C$$

$$= \frac{1}{\sqrt{3}} \log \left| \frac{\tan \frac{1}{2}x + \sqrt{3}}{\tan \frac{1}{2}x - \sqrt{3}} \right| + C. \quad \square$$

Since the other trigonometric functions

$$\tan x, \quad \cot x, \quad \sec x, \quad \operatorname{cosec} x$$

are all rational combinations of

$$\sin x \quad \text{and} \quad \cos x$$

the substitution

$$2 \arctan u = x$$

also works for rational expressions involving these other functions.

Problem.   Find

$$\int \frac{\sec x}{2 \tan x + \sec x - 1}\, dx.$$

SOLUTION.   Set

$$2 \arctan u = x, \qquad \frac{2}{1 + u^2}\, du = dx.$$

With $u = \tan \tfrac{1}{2}x$,

$$\int \frac{\sec x}{2 \tan x + \sec x - 1}\, dx = \int \frac{\dfrac{1}{\cos x}}{2\, \dfrac{\sin x}{\cos x} + \dfrac{1}{\cos x} - 1}\, dx$$

$$= \int \frac{dx}{2 \sin x - \cos x + 1}$$

$$= \int \frac{\dfrac{2}{1 + u^2}}{2 \sin (2 \arctan u) - \cos (2 \arctan u) + 1}\, du$$

$$= \int \frac{\dfrac{2}{1 + u^2}}{2 \cdot \dfrac{2u}{1 + u^2} - \dfrac{1 - u^2}{1 + u^2} + 1}\, du$$

$$= \int \frac{du}{u^2 + 2u}$$

$$= \int \frac{du}{u(u + 2)}$$

$$= \frac{1}{2} \int \left(\frac{1}{u} - \frac{1}{u + 2} \right) du$$

$$= \frac{1}{2} \log \left| \frac{u}{u + 2} \right| + C$$

$$= \frac{1}{2} \log \left| \frac{\tan \tfrac{1}{2}x}{\tan \tfrac{1}{2}x + 2} \right| + C. \quad \square$$

Exercises

Work out the following integrals.

*1. $\displaystyle \int \frac{dx}{1 + \cos x}$.

2. $\displaystyle \int \frac{dx}{1 - \sin x}$.

*3. $\displaystyle \int \frac{dx}{1 - \cos x}$.

4. $\displaystyle \int_0^{\pi/2} \frac{dx}{3 + 2 \cos x}$.

* 5. $\displaystyle\int_0^{\pi/2} \frac{dx}{3 + \cos x}$.

6. $\displaystyle\int \frac{\sin x}{2 - \sin x}\, dx$.

* 7. $\displaystyle\int \frac{dx}{5 + 4 \cos x}$.

8. $\displaystyle\int \frac{dx}{1 + \tan x}$.

* 9. $\displaystyle\int \frac{dx}{5 \sec x - 3}$.

10. $\displaystyle\int \frac{\cos x}{1 - \cos x}\, dx$.

* 11. $\displaystyle\int \frac{1 - \cos x}{1 + \sin x}\, dx$.

12. $\displaystyle\int \frac{1 + \sin x}{1 + \cos x}\, dx$.

7.7　Some Rationalizing Substitutions

Problem.　Find

$$\int \frac{dx}{1 + \sqrt{x}} \, .$$

SOLUTION.　To rationalize the integrand we set

$$u^2 = x, \qquad 2u\, du = dx.$$

With $u = \sqrt{x}$,

$$\int \frac{dx}{1 + \sqrt{x}} = \int \frac{2u}{1 + u}\, du$$

$$= \int \left(2 - \frac{2}{1 + u}\right) du$$

$$= 2u - 2 \log |1 + u| + C$$

$$= 2\sqrt{x} - 2 \log |1 + \sqrt{x}| + C. \quad \square$$

Problem.　Find

$$\int \frac{x^{1/2}}{4(1 + x^{3/4})}\, dx.$$

SOLUTION.　Here we set

$$u^4 = x, \qquad 4u^3\, du = dx.$$

With $u = x^{1/4}$,

$$\int \frac{x^{1/2}}{4(1 + x^{3/4})}\, dx = \int \frac{(u^2)(4u^3)}{4(1 + u^3)}\, du$$

$$= \int \frac{u^5}{1 + u^3}\, du$$

$$= \int \left(u^2 - \frac{u^2}{1 + u^3}\right) du$$

$$= \tfrac{1}{3}u^3 - \tfrac{1}{3} \log |1 + u^3| + C$$

$$= \tfrac{1}{3}x^{3/4} - \tfrac{1}{3} \log |1 + x^{3/4}| + C. \quad \square$$

Problem. Find

$$\int \sqrt{1 - e^x} \, dx.$$

SOLUTION.   To rationalize the integrand we set

$$u^2 = 1 - e^x.$$

To find dx in terms of u and du we solve the equation for x:

$$1 - u^2 = e^x,$$

$$\log (1 - u^2) = x,$$

$$-\frac{2u}{1 - u^2} \, du = dx.$$

The rest is straightforward:

$$\int \sqrt{1 - e^x} \, dx = \int u \left(-\frac{2u}{1 - u^2} \right) du$$

$$= \int \frac{2u^2}{u^2 - 1} \, du$$

$$= \int \left(2 + \frac{1}{u - 1} - \frac{1}{u + 1} \right) du$$

$$= 2u + \log |u - 1| - \log |u + 1| + C$$

$$= 2u + \log \left| \frac{u - 1}{u + 1} \right| + C$$

$$= 2\sqrt{1 - e^x} + \log \left| \frac{\sqrt{1 - e^x} - 1}{\sqrt{1 + e^x} + 1} \right| + C. \quad \square$$

Exercises

Work out the following integrals.

* 1. $\displaystyle \int \frac{dx}{1 - \sqrt{x}}.$

2. $\displaystyle \int_0^3 x\sqrt{1 + x} \, dx.$

* 3. $\displaystyle \int \frac{\sqrt{x}}{1 + x} \, dx.$

4. $\displaystyle \int \frac{dx}{x(x^{1/3} - 1)}.$

* 5. $\displaystyle \int \sqrt{1 + e^x} \, dx.$

6. $\displaystyle \int x(1 + x)^{1/3} \, dx.$

* 7. $\displaystyle \int_2^3 \frac{x^3}{(1 + x^2)^3} \, dx.$

8. $\displaystyle \int x\sqrt{x + 1} \, dx.$

*9. $\displaystyle\int \frac{\sqrt{x}}{\sqrt{x}-1}\,dx.$

10. $\displaystyle\int \frac{x}{\sqrt{x}+1}\,dx.$

*11. $\displaystyle\int \frac{\sqrt{x-1}+1}{\sqrt{x-1}-1}\,dx.$

12. $\displaystyle\int \frac{1-e^x}{1+e^x}\,dx.$

*13. $\displaystyle\int \frac{dx}{\sqrt{1+e^x}}.$

14. $\displaystyle\int \frac{dx}{1+e^{-x}}.$

*15. $\displaystyle\int \frac{x}{\sqrt{x}+4}\,dx.$

16. $\displaystyle\int \frac{x+1}{x\sqrt{x}-2}\,dx.$

*17. $\displaystyle\int_0^1 2x^2(4x+1)^{-5/2}\,dx.$

18. $\displaystyle\int x^2\sqrt{x-1}\,dx.$

*19. $\displaystyle\int \frac{x}{(ax+b)^{3/2}}\,dx.$

20. $\displaystyle\int \frac{x}{\sqrt{ax+b}}\,dx.$

7.8 Approximate Integration

To evaluate a definite integral by the formula

$$\int_a^b f(x)\,dx = F(b) - F(a)$$

we must be able to find an antiderivative F and we must be able to evaluate it both at a and at b. When this is not possible, the method fails.

The method fails even for such simple-looking integrals as

$$\int_0^1 \sqrt{x}\,\sin x\,dx \quad\text{and}\quad \int_0^1 e^{-x^2}\,dx.$$

There are no *elementary functions* with derivatives $\sqrt{x}\,\sin x$ and e^{-x^2}.

Here we take up some simple numerical methods for estimating definite integrals—methods that you can use whether or not you can find an antiderivative. All the methods we describe involve only simple arithmetic and are ideally suited to the electronic computer.

We focus now on

$$\int_a^b f(x)\,dx.$$

As usual, we suppose that f is continuous on $[a, b]$ and, for pictorial convenience, assume that f is positive. We begin by subdividing $[a, b]$ into n nonoverlapping subintervals each of length $(b - a)/n$:

$$[a, b] = [x_0, x_1] \cup \cdots \cup [x_{i-1}, x_i] \cup \cdots \cup [x_{n-1}, x_n],$$

with

$$\Delta x_i = \frac{b - a}{n}.$$

The region Ω_i pictured in Figure 7.8.1 can be approximated in several ways.

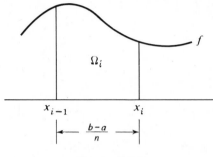

FIGURE 7.8.1

(1) By the left endpoint rectangle (Figure 7.8.2):

$$\text{area} = f(x_{i-1}) \, \Delta x_i = f(x_{i-1}) \left(\frac{b - a}{n} \right).$$

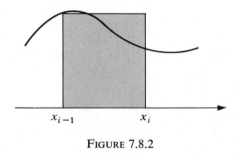

FIGURE 7.8.2

(2) By the right endpoint rectangle (Figure 7.8.3):

$$\text{area} = f(x_i) \, \Delta x_i = f(x_i) \left(\frac{b - a}{n} \right).$$

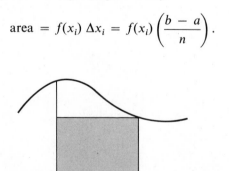

FIGURE 7.8.3

(3) By the midpoint rectangle (Figure 7.8.4):

$$\text{area} = f\left(\frac{x_i + x_{i-1}}{2}\right) \Delta x_i = f\left(\frac{x_i + x_{i-1}}{2}\right)\left(\frac{b - a}{n}\right).$$

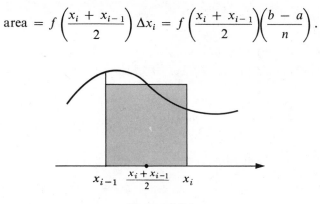

FIGURE 7.8.4

(4) By a trapezoid (Figure 7.8.5):

$$\text{area} = \tfrac{1}{2}[f(x_i) + f(x_{i-1})] \Delta x_i = \tfrac{1}{2}[f(x_i) + f(x_{i-1})]\left(\frac{b - a}{n}\right).$$

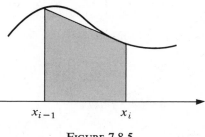

FIGURE 7.8.5

(5) By a parabolic region (Figure 7.8.6): take the parabola with vertical axis which passes through the three points indicated.†

$$\text{area} = \frac{1}{6}\left[f(x_{i-1}) + 4f\left(\frac{x_i + x_{i-1}}{2}\right) + f(x_i)\right] \Delta x_i$$

$$= \left[f(x_{i-1}) + 4f\left(\frac{x_i + x_{i-1}}{2}\right) + f(x_i)\right]\left(\frac{b - a}{6n}\right).$$

(You can later verify the validity of this formula by doing Exercises 16, 17, 18 of Section 8.3.)

The approximations to Ω_i just considered yield the following estimates for

$$\int_a^b f(x)\, dx.$$

---

† If the three points have the same ordinate, the parabola degenerates to a straight line.

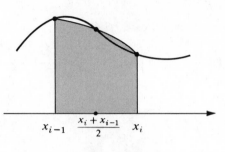

$$x_{i-1} \qquad \frac{x_i + x_{i-1}}{2} \qquad x_i$$

FIGURE 7.8.6

(1) The left endpoint estimate:

$$L_n = \frac{b - a}{n} \left[f(x_0) + f(x_1) + \cdots + f(x_{n-1}) \right].$$

(2) The right endpoint estimate:

$$R_n = \frac{b - a}{n} \left[f(x_1) + f(x_2) + \cdots + f(x_n) \right].$$

(3) The midpoint estimate:

$$M_n = \frac{b - a}{n} \left[f\!\left(\frac{x_1 + x_0}{2}\right) + \cdots + f\!\left(\frac{x_n + x_{n-1}}{2}\right) \right].$$

(4) The trapezoidal estimate (*trapezoidal rule*):

$$T_n = \frac{b - a}{2n} \left[f(x_0) + 2f(x_1) + \cdots + 2f(x_{n-1}) + f(x_n) \right].$$

(5) The parabolic estimate (*Simpson's rule*):

$$S_n = \frac{b - a}{6n} \left\{ f(x_0) + f(x_n) + 2[f(x_1) + \cdots + f(x_{n-1})] \right.$$
$$\left. + 4\left[f\!\left(\frac{x_1 + x_0}{2}\right) + \cdots + f\!\left(\frac{x_n + x_{n-1}}{2}\right) \right] \right\}.$$

As an example, we estimate

$$\log 2 = \int_1^2 \frac{dx}{x}$$

by each of these methods. Here we have

$$f(x) = \frac{1}{x}, \qquad [a, b] = [1, 2].$$

If we take $n = 5$, then each subinterval has length

$$\frac{b - a}{n} = \frac{2 - 1}{5} = \frac{1}{5}$$

and the partition points are

$$x_0 = \tfrac{5}{5}, \qquad x_1 = \tfrac{6}{5}, \qquad x_2 = \tfrac{7}{5}, \qquad x_3 = \tfrac{8}{5}, \qquad x_4 = \tfrac{9}{5}, \qquad x_5 = \tfrac{10}{5}.$$

This gives

$$L_5 = \tfrac{1}{5}(\tfrac{5}{5} + \tfrac{5}{6} + \tfrac{5}{7} + \tfrac{5}{8} + \tfrac{5}{9}) = (\tfrac{1}{5} + \tfrac{1}{6} + \tfrac{1}{7} + \tfrac{1}{8} + \tfrac{1}{9}),$$
$$R_5 = \tfrac{1}{5}(\tfrac{5}{6} + \tfrac{5}{7} + \tfrac{5}{8} + \tfrac{5}{9} + \tfrac{5}{10}) = (\tfrac{1}{6} + \tfrac{1}{7} + \tfrac{1}{8} + \tfrac{1}{9} + \tfrac{1}{10}),$$
$$M_5 = \tfrac{1}{5}(\tfrac{10}{11} + \tfrac{10}{13} + \tfrac{10}{15} + \tfrac{10}{17} + \tfrac{10}{19}) = 2(\tfrac{1}{11} + \tfrac{1}{13} + \tfrac{1}{15} + \tfrac{1}{17} + \tfrac{1}{19}),$$
$$T_5 = \tfrac{1}{10}(\tfrac{5}{5} + \tfrac{10}{6} + \tfrac{10}{7} + \tfrac{10}{8} + \tfrac{10}{9} + \tfrac{5}{10}) = (\tfrac{1}{10} + \tfrac{1}{6} + \tfrac{1}{7} + \tfrac{1}{8} + \tfrac{1}{9} + \tfrac{1}{20}),$$
$$S_5 = \tfrac{1}{30}[\tfrac{5}{5} + \tfrac{5}{10} + 2(\tfrac{5}{6} + \tfrac{5}{7} + \tfrac{5}{8} + \tfrac{5}{9}) + 4(\tfrac{10}{11} + \tfrac{10}{13} + \tfrac{10}{15} + \tfrac{10}{17} + \tfrac{10}{19})].$$

If you carry out the computations and round off to the nearest hundredth, you will have the following estimates:

$$L_5 \cong 0.75, \qquad R_5 \cong 0.65, \qquad M_5 \cong 0.69, \qquad T_5 \cong 0.70, \qquad S_5 \cong 0.69.$$

Since the integrand $1/x$ decreases throughout the interval $[1, 2]$ you can expect the left endpoint estimate, 0.75, to be too large and you can expect the right endpoint estimate, 0.65, to be too small. The other estimates should be better.

Table 1 at the end of the book gives log $2 \cong 0.693$. The estimate 0.69 is correct rounded off to the nearest hundredth.

Problem. Estimate

$$\int_0^2 \sqrt{4 + x^3}\, dx$$

by the trapezoidal rule. Take $n = 4$.

SOLUTION.   Each subinterval has length

$$\frac{b - a}{n} = \frac{2 - 0}{4} = \frac{1}{2}.$$

The partition points are

$$x_0 = 0, \qquad x_1 = \tfrac{1}{2}, \qquad x_2 = 1, \qquad x_3 = \tfrac{3}{2}, \qquad x_4 = 2.$$

Consequently

$$T_4 = \tfrac{1}{4}[f(0) + 2f(\tfrac{1}{2}) + 2f(1) + 2f(\tfrac{3}{2}) + f(2)].$$

With

$$f(0) = 2,$$
$$f(\tfrac{1}{2}) = \sqrt{4 + \tfrac{1}{8}} = \sqrt{4.125} \cong 2.031,$$
$$f(1) = \sqrt{5} \cong 2.236,$$
$$f(\tfrac{3}{2}) = \sqrt{4 + \tfrac{27}{8}} = \sqrt{7.375} \cong 2.716,$$
$$f(2) = \sqrt{12} \cong 3.464,$$

we have

$$T_4 \cong \tfrac{1}{4}(2 + 4.062 + 4.472 + 5.432 + 3.464) \cong 4.858. \quad \square$$

Problem. Estimate

$$\int_0^2 \sqrt{4 + x^3} \, dx$$

by Simpson's rule. Take $n = 2$.

SOLUTION.   There are two intervals each of length

$$\frac{b - a}{n} = \frac{2 - 0}{2} = 1.$$

We have

$$x_0 = 0, \qquad x_1 = 1, \qquad x_2 = 2.$$

$$\frac{x_0 + x_1}{2} = \frac{1}{2}, \qquad \frac{x_1 + x_2}{2} = \frac{3}{2}.$$

Simpson's rule yields

$$S_2 = \tfrac{1}{6}[f(0) + f(2) + 2f(1) + 4f(\tfrac{1}{2}) + 4f(\tfrac{3}{2})],$$

which, in the light of the square root estimates given in the last problem, gives

$$S_2 \cong \tfrac{1}{6}(2 + 3.464 + 4.472 + 8.124 + 10.864) \cong 4.821. \quad \square$$

Exercises

(In each of the numerical computations below round off your final answer to the nearest hundredth.)

  *1. Estimate

$$\int_0^{12} x^2 \, dx.$$

  (a) Use the left endpoint estimate, $n = 12$.
  (b) Use the right endpoint estimate, $n = 12$.
  (c) Use the midpoint estimate, $n = 6$.
  (d) Use the trapezoidal rule, $n = 12$.
  (e) Use Simpson's rule, $n = 6$.
  Check your results by performing the integration.
  2. Estimate

$$\int_0^1 \sin^2 \pi x \, dx.$$

  (a) Use the midpoint estimate, $n = 3$.
  (b) Use the trapezoidal rule, $n = 6$.
  (c) Use Simpson's rule, $n = 3$.
  Check your results by performing the integration.

* 3. Find the approximate value of π by estimating the integral

$$\frac{\pi}{4} = \text{arc tan } 1 = \int_0^1 \frac{dx}{1 + x^2}.$$

(a) Use the trapezoidal rule, $n = 4$.

(b) Use Simpson's rule, $n = 4$.

4. Estimate

$$\int_0^\pi \frac{\sin x}{\pi + x} dx.$$

(a) Use the trapezoidal rule, $n = 6$.

(b) Use Simpson's rule, $n = 3$.

* 5. Estimate

$$\int_0^4 \frac{dx}{\sqrt{4 + x^3}}.$$

(a) Use the trapezoidal rule, $n = 4$.

(b) Use Simpson's rule, $n = 2$.

7.9 Additional Exercises

Work out the following integrals and check your answers by differentiation.

1. $\displaystyle\int 10^{nx} \, dx.$

2. $\displaystyle\int e^x \tan (e^x) \, dx.$

3. $\displaystyle\int \sqrt{2x + 1} \, dx.$

4. $\displaystyle\int x\sqrt{2x + 1} \, dx.$

5. $\displaystyle\int \tan \left(\frac{\pi}{n} x\right) dx.$

6. $\displaystyle\int \frac{dx}{\sqrt{x + 1} - \sqrt{x}}.$

7. $\displaystyle\int \frac{dx}{a^2x^2 + b^2}.$

8. $\displaystyle\int \sin 2x \cos x \, dx.$

9. $\displaystyle\int \frac{e^{-\sqrt{x}}}{\sqrt{x}} dx.$

10. $\displaystyle\int \sec^3 (2x) \, dx.$

11. $\displaystyle\int \frac{dx}{\sqrt{1 - e^{2x}}}.$

12. $\displaystyle\int x2^x \, dx.$

13. $\displaystyle\int \sin^2 \left(\frac{\pi}{n} x\right) dx.$

14. $\displaystyle\int \frac{dx}{x^3 - 1}.$

15. $\displaystyle\int \frac{dx}{a\sqrt{x} + b}.$

16. $\displaystyle\int (1 - \sec x)^2 \, dx.$

17. $\displaystyle\int \frac{\sin 3x}{2 + \cos 3x} dx.$

18. $\displaystyle\int \frac{\sqrt{a - x}}{\sqrt{a + x}} dx.$

19. $\int \dfrac{\sqrt{a + x}}{\sqrt{a - x}}\, dx.$

20. $\int \dfrac{1 - \sin 2x}{1 + \sin 2x}\, dx.$

21. $\int \dfrac{\sqrt{a^2 - x^2}}{x^2}\, dx.$

22. $\int \log \sqrt{x + 1}\, dx.$

23. $\int \dfrac{x}{(x + 1)^2}\, dx.$

24. $\int \dfrac{dx}{\cos x - \sin x}.$

25. $\int a^{2x}\, dx.$

26. $\int \log (ax + b)\, dx.$

27. $\int x \log (ax + b)\, dx.$

28. $\int e^x \sin \pi x\, dx.$

29. $\int \dfrac{\sin \sqrt{x}}{\sqrt{x}}\, dx.$

30. $\int \log (x\sqrt{x})\, dx.$

31. $\int \dfrac{x^2}{1 + x^2}\, dx.$

32. $\int \sqrt{\dfrac{x^2}{9} - 1}\, dx.$

33. $\int \dfrac{-x^2}{\sqrt{1 - x^2}}\, dx.$

34. $\int (\tan x + \cot x)^2\, dx.$

35. $\int \dfrac{dx}{1 + 3 \sin x}.$

36. $\int x \log \sqrt{x^2 + 1}\, dx.$

37. $\int \dfrac{x^3}{\sqrt{1 + x^2}}\, dx.$

38. $\int x \tan^2 (\pi x)\, dx.$

39. $\int \dfrac{dx}{2 - \sqrt{x}}.$

40. $\int x \arctan (x - 3)\, dx.$

41. $\int \dfrac{2}{x(1 + x^2)}\, dx.$

42. $\int \sin 2x \cos 3x\, dx.$

43. $\int \dfrac{\cos^4 x}{\sin^2 x}\, dx.$

44. $\int \dfrac{x - 3}{x^2(x + 1)}\, dx.$

45. $\int \dfrac{\sqrt{x^2 + 4}}{x}\, dx.$

46. $\int \dfrac{dx}{2x^2 - 2x + 1}.$

47. $\int \dfrac{dx}{x\sqrt{9 - x^2}}.$

48. $\int \dfrac{dx}{e^x - 2e^{-x}}.$

49. $\int \dfrac{e^x}{\sqrt{e^x + 1}}\, dx.$

50. $\int \sin^5 \left(\dfrac{x}{2}\right)\, dx.$

51. $\int (\sin^2 x - \cos x)^2\, dx.$

52. $\int \dfrac{dx}{\sqrt{2x - x^2}}.$

53. $\displaystyle\int \log (1 - \sqrt{x})\, dx.$

54. $\displaystyle\int \frac{3}{\sqrt{2 - 3x - 4x^2}}\, dx.$

55. $\displaystyle\int \frac{\sin x}{\cos^2 x - 2 \cos x + 3}\, dx.$

56. $\displaystyle\int \left[\frac{x}{\sqrt{a^2 - x^2}} - \arcsin\left(\frac{x}{a}\right) \right] dx.$

(Optional)

57. $\displaystyle\int \sinh^2 x\, dx.$

58. $\displaystyle\int 2x \sinh x\, dx.$

59. $\displaystyle\int e^{-x} \cosh x\, dx.$

60. $\displaystyle\int \tanh^2 2x\, dx.$

Some Analytic Geometry

8.1 The Distance Between a Point and a Line; Translations

First we compute the distance between a line and the origin.

Theorem 8.1.1

The distance between the line $l: Ax + By + C = 0$ and the origin is given by the formula

$$d(0, l) = \frac{|C|}{\sqrt{A^2 + B^2}}.$$

PROOF. If the line passes through the origin, then $C = 0$ and the formula holds. Suppose that the line l does not pass through the origin and thus that $C \neq 0$. To find the desired distance, we will get the equation of the line l' which passes through the origin and is perpendicular to l. (See Figure 8.1.1.) From this equation we will obtain the point Q at which l and l' intersect. The distance we want is the distance between O and Q. Since l has equation

$$Ax + By + C = 0,$$

l' has equation

$$Bx - Ay = 0. \qquad\qquad \text{(check this out)}$$

By solving the two equations simultaneously, you will find that

$$Q = \left(-\frac{AC}{A^2 + B^2}, \; -\frac{BC}{A^2 + B^2} \right).$$

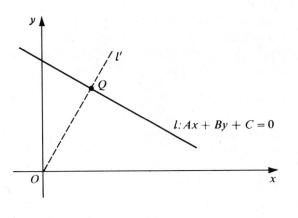

FIGURE 8.1.1

The distance between O and Q is therefore

$$\sqrt{A^2C^2/(A^2 + B^2)^2 + B^2C^2/(A^2 + B^2)^2}.$$

This expression can be simplified to

$$\frac{|C|}{\sqrt{A^2 + B^2}} \cdot \quad \square$$

To find the distance between the line $l: Ax + By + C = 0$ and an arbitrary point $P_0(x_0, y_0)$, we could proceed as we did in Theorem 8.1.1. The computations, however, would be cumbersome. They become much easier if we "translate" the coordinate system.

Translations

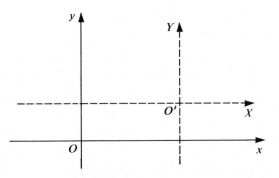

FIGURE 8.1.2

You already know how to impose a Cartesian coordinate system on the plane. Suppose that we have two such systems: the first possessing an origin O, an x-axis, and a y-axis; the second possessing an origin O', an X-axis and a Y-axis. If both systems are endowed with the same scale and if the corresponding axes are parallel and have the same orientation, then we say that the two coordinate systems differ by a *translation*. (See Figure 8.1.2.) Equivalently, we say that one system is obtained from the other by a *translation*.

Suppose now that we have in the plane two coordinate systems Oxy and $O'XY$ differing by a translation. A point P then possesses two pairs of coordinates: (x, y) and (X, Y). If the xy-coordinates of O' are (x_0, y_0), then clearly we have $x - x_0 = X$ and $y - y_0 = Y$, and therefore

$$x = X + x_0 \quad \text{and} \quad y = Y + y_0.$$

These formulas are useful in simplifying geometric arguments. We put them to use in the proof of the theorem below.

Theorem 8.1.2

The distance between the line l: $Ax + By + C = 0$ and the point $P_0(x_0, y_0)$ is given by the formula

$$d(P_0, l) = \frac{|Ax_0 + By_0 + C|}{\sqrt{A^2 + B^2}}.$$

PROOF. We introduce a new coordinate system $O'XY$ differing from Oxy by a translation. We choose O' to be the point $P_0(x_0, y_0)$. From our previous discussion, it follows that

$$x = X + x_0 \quad \text{and} \quad y = Y + y_0.$$

Our equation for l becomes, in the XY-system,

$$A(X + x_0) + B(Y + y_0) + C = 0,$$

from which we get

$$AX + BY + K = 0 \quad \text{with} \quad K = Ax_0 + By_0 + C.$$

The distance we want can now be expressed as the distance between the line l with equation $AX + BY + K = 0$ and the origin O'. By Theorem 8.1.1 this distance is

$$\frac{|K|}{\sqrt{A^2 + B^2}}.$$

Since $K = Ax_0 + By_0 + C$, we have

$$d(P_0, l) = \frac{|Ax_0 + By_0 + C|}{\sqrt{A^2 + B^2}}. \quad \square$$

Exercises

1. Find the distance between
    *(a) $P(-2, 1)$ and l: $3x - 2y = 0$.
     (b) $P(1, 2)$ and l: $3y - 2x = 1$.
    *(c) $P(3, 5)$ and l: $y = 2x$.
2. Which of the points $(0, 1)$, $(1, 0)$, and $(-1, 1)$ is the closest to l: $8x + 7y - 6 = 0$? Which is the farthest from l?
3. Compute the area of the triangle
    *(a) with vertices $(1, -2)$, $(-1, 3)$, and $(2, 4)$.
     (b) with vertices $(-1, 1)$, $(3, \sqrt{2})$, and $(\sqrt{2}, -1)$.
*4. A ray l is rotating clockwise about the point $Q(-b^2, 0)$ at the rate of one revolution per minute. How fast is the distance between l and the origin changing at the moment that l has slope $\frac{3}{4}$?
5. A ray l is rotating counterclockwise about the point $Q(-b^2, 0)$ at the rate of $\frac{1}{2}$ revolution per minute. What is the slope of l at the moment that l is receding most quickly from the point $P(b^2, -1)$?

8.2 The Conic Sections

If a "double right circular cone" is cut by a plane, the resulting intersection is called a *conic section* or, more briefly, a *conic*. In Figures 8.2.1, 8.2.2, and 8.2.3 we depict three important cases.

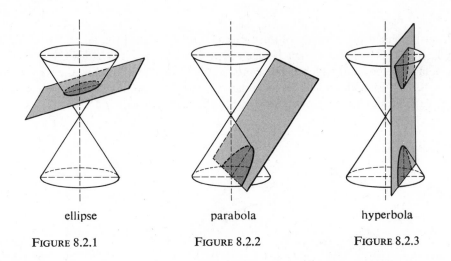

ellipse	parabola	hyperbola
FIGURE 8.2.1	FIGURE 8.2.2	FIGURE 8.2.3

By choosing a plane perpendicular to the axis of the cone, we can obtain a circle. Quite obviously, the other possibilities are: a point, a line, or a pair of lines.

The study of conic sections from this point of view goes back to Apollonius of Perga, a Greek of the third century B.C. He wrote eight books on the subject.

For our purposes, it is useful to dispense with cones and three-dimensional geometry and, instead, define ellipse, parabola, and hyperbola entirely in terms of plane geometry.

8.3   The Parabola

We begin with a line *l* and a point *F* not on *l*.

> The set of points *P* equidistant from *F* and *l* is called a *parabola*.   (Figure 8.3.1)

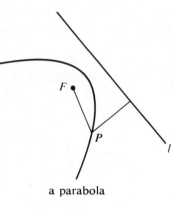

a parabola

FIGURE 8.3.1

The line *l* is called the *directrix* of the parabola and *F* is called the *focus*. (You will see why later on.) The line through *F* which is perpendicular to *l* is called the *axis* of the parabola. The point at which the axis intersects the parabola is called the *vertex*. (See Figure 8.3.2.) The equation of a parabola is particularly simple if we place the vertex at the origin and the focus along one of the coordinate axes.

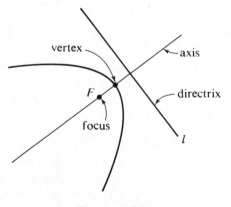

FIGURE 8.3.2

Suppose for a moment that the focus F is on the y-axis. Then F has coordinates of the form $(0, c)$, and, with the vertex at the origin, the directrix must have equation $y = -c$. (See Figure 8.3.3.)

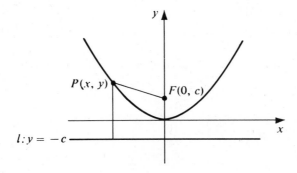

FIGURE 8.3.3

Any point $P(x, y)$ which lies on this parabola has the property that

$$d(P, F) = d(P, l).$$

Since

$$d(P, F) = \sqrt{x^2 + (y - c)^2} \quad \text{and} \quad d(P, l) = |y + c|,$$

we must have

$$\sqrt{x^2 + (y - c)^2} = |y + c|.$$

Squaring both sides gives

$$x^2 + (y - c)^2 = |y + c|^2 = (y + c)^2,$$
$$x^2 + y^2 - 2cy + c^2 = y^2 + 2cy + c^2.$$

This last equation simplifies to

$$x^2 = 4cy. \quad \square$$

You have just seen that the equation

$$\boxed{x^2 = 4cy} \qquad\qquad \text{(Figure 8.3.4)}$$

represents a parabola with vertex at the origin and focus at $(0, c)$.
By interchanging the roles of x and y it is obvious that the equation

$$\boxed{y^2 = 4cx} \qquad\qquad \text{(Figure 8.3.5)}$$

represents a parabola with vertex at the origin and focus at $(c, 0)$.

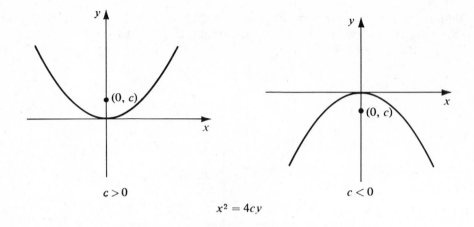

$$x^2 = 4cy$$

FIGURE 8.3.4

Every parabola with vertical axis is a translation of a parabola of the form

$$x^2 = 4cy$$

and every parabola with horizontal axis is a translation of a parabola of the form

$$y^2 = 4cx.$$

The equation

$$(x - x_0)^2 = 4c(y - y_0)$$

represents a parabola with vertex (x_0, y_0) and focus $(x_0, y_0 + c)$. The axis is vertical. The equation

$$(y - y_0)^2 = 4c(x - x_0)$$

represents a parabola with vertex (x_0, y_0) and focus $(x_0 + c, y_0)$. The axis is horizontal.

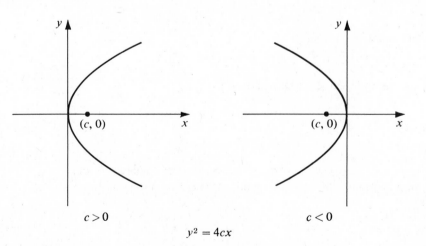

$$y^2 = 4cx$$

FIGURE 8.3.5

Problem. Identify the curve

$$(x - 1)^2 = 8(y + 3).$$

SOLUTION. We rewrite the equation as

$$(x - 1)^2 = 8[y - (-3)].$$

This is the parabola

$$x^2 = 8y \qquad\qquad\qquad (c = 2)$$

translated so that the vertex falls at $(1, -3)$. The axis is vertical and the focus is at $(1, -1)$. □

Problem. Identify the curve

$$(y - 1)^2 = 8(x + 3).$$

SOLUTION. We rewrite the equation as

$$(y - 1)^2 = 8[x - (-3)].$$

This is the parabola

$$y^2 = 8x \qquad\qquad\qquad (c = 2)$$

translated so that the vertex falls at $(-3, 1)$. The axis is horizontal and the focus is at $(-1, 1)$. □

Problem. Identify the curve

$$y = x^2 + 2x - 2.$$

SOLUTION. We first complete the square on the right by adding 3 to both sides of the equation:

$$y + 3 = x^2 + 2x + 1 = (x + 1)^2.$$

This gives

$$(x + 1)^2 = y + 3,$$
$$[x - (-1)]^2 = y - (-3).$$

This is the parabola

$$x^2 = y \qquad\qquad\qquad (c = \tfrac{1}{4})$$

translated so that the vertex falls at $(-1, -3)$. The axis is vertical and the focus is at $(-1, -\tfrac{11}{4})$. □

By the method of the last problem one can show that every quadratic

$$y = Ax^2 + Bx + C$$

represents a parabola with vertical axis. It looks like $\bigcup$ if A is positive and it looks like $\bigcap$ if A is negative.

Parabolic Reflectors

Parabolic curves play a prominent role in the design of searchlights and telescopes. To show how this comes about, we take a parabola and choose the coordinate system so that the equation of the parabola takes the form

$$x^2 = 4cy.$$

We can express y in terms of x by writing

$$y = \frac{x^2}{4c}.$$

Since

$$\frac{dy}{dx} = \frac{2x}{4c} = \frac{x}{2c},$$

the tangent at the point $P(x_0, y_0)$ has slope

$$m = \frac{x_0}{2c}$$

and the tangent line has equation

$$(y - y_0) = \frac{x_0}{2c}(x - x_0).$$

For the rest we refer to Figure 8.3.6. To find the coordinates of the point marked T in the figure, we set $x = 0$ and solve for y. Doing this, we find that T has coordinates

$$\left(0, \; y_0 - \frac{x_0^2}{2c}\right).$$

Since the point (x_0, y_0) is on the parabola,

$$x_0^2 = 4cy_0$$

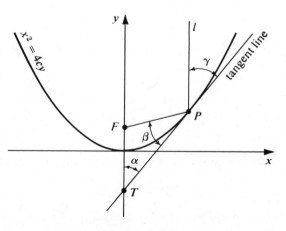

FIGURE 8.3.6

and we can rewrite the coordinates of T as

$$(0, -y_0).$$

Since the focus F has coordinates $(0, c)$,

$$d(F, T) = c + y_0.$$

Now

$$
\begin{aligned}
d(F, P) &= \sqrt{x_0^2 + (y_0 - c)^2} \\
&= \sqrt{4cy_0 + (y_0 - c)^2} \qquad\qquad (x_0^2 = 4cy_0) \\
&= \sqrt{4cy_0 + y_0^2 - 2cy_0 + c^2} \\
&= \sqrt{y_0^2 + 2cy_0 + c^2} \\
&= \sqrt{(y_0 + c)^2} \\
&= c + y_0.
\end{aligned}
$$

We have shown that

$$d(F, T) = d(F, P).$$

It follows that the triangle TFP is isosceles, and the base angles marked α and β in the figure are equal. With l parallel to the y-axis,

$$\alpha = \gamma$$

and thus

$$\gamma = \beta. \quad \square$$

From physics we know that when light is reflected, the angle of incidence equals the angle of reflection. We can take either γ or β as the angle of incidence; the other will then be the angle of reflection. The fact that $\gamma = \beta$ has important optical consequences. It means that *light from a source at the focus is reflected in a beam parallel to the axis*. It also means that *a beam of light parallel to the axis is reflected by the parabola through the focus*. The parabolic mirror of a searchlight uses the first principle, that of a reflecting telescope the second. (See Figure 8.3.7.)

parabolic mirrors

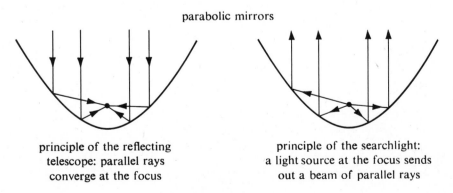

principle of the reflecting
telescope: parallel rays
converge at the focus

principle of the searchlight:
a light source at the focus sends
out a beam of parallel rays

FIGURE 8.3.7

Parabolic Trajectories

In the early part of the seventeenth century Galileo Galilei observed the motion of stones hurled from the tower of Pisa and noted that their trajectory was parabolic (in the shape of a parabola). By very simple calculus, together with some simplifying physical assumptions, we obtain results which agree with Galileo's observations.

Consider a projectile fired from a point (x_0, y_0) with initial velocity of v_0 feet per second at angle θ. (Figure 8.3.8)

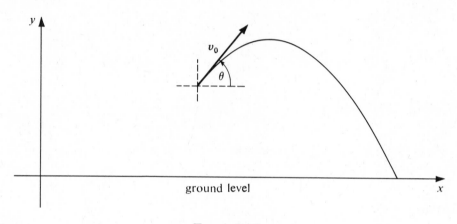

FIGURE 8.3.8

The horizontal component of v_0 is $v_0 \cos \theta$ and the vertical component of v_0 is $v_0 \sin \theta$. (Figure 8.3.9)

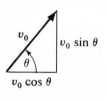

FIGURE 8.3.9

Let's neglect air resistance and the curvature of the earth. Under these circumstances there is no acceleration in the x-direction:

$$x''(t) = 0.$$

The only acceleration in the y-direction is due to gravity:

$$y''(t) = -32.$$

From the first equation,
$$x'(t) = C$$
and since $x'(0) = v_0 \cos \theta$,
$$x'(t) = v_0 \cos \theta.$$
Integrating again,
$$x(t) = (v_0 \cos \theta)t + C.$$
Since $x(0) = x_0$,

(1) $$x(t) = (v_0 \cos \theta)t + x_0.$$

From $y''(t) = -32$, it follows that
$$y'(t) = -32t + C.$$
Since $y'(0) = v_0 \sin \theta$,
$$y'(t) = -32t + v_0 \sin \theta.$$
Integrating again,
$$y(t) = -16t^2 + (v_0 \sin \theta)t + C$$
and since $y(0) = y_0$,

(2) $$y(t) = -16t^2 + (v_0 \sin \theta)t + y_0.$$

From (1)
$$t = \frac{1}{v_0 \cos \theta} [x(t) - x_0].$$

If you substitute this value of t in (2), you will find that
$$y(t) = -\frac{16 \sec^2 \theta}{v_0^2} [x(t) - x_0]^2 + \tan \theta [x(t) - x_0] + y_0.$$

The trajectory (the path followed by the projectile) is the curve

$$y = -\frac{16}{v_0^2} \sec^2 \theta [x - x_0]^2 + \tan \theta [x - x_0] + y_0.$$

This is a quadratic in x and therefore a parabola. $\square$

Exercises

1. Sketch the parabola and give an equation for it:
   * (a) vertex $(0, 0)$, focus $(2, 0)$.   * (b) vertex $(0, 0)$, focus $(-2, 0)$.
     (c) vertex $(0, 0)$, focus $(0, 2)$.   * (d) vertex $(1, 2)$, focus $(3, 2)$.
     (e) vertex $(1, 2)$, focus $(1, 3)$.   * (f) focus $(1, 1)$, directrix $y = -1$.
     (g) focus $(1, 1)$, directrix $x = 2$.   * (h) focus $(2, -2)$, directrix $x = 5$.
2. Find the vertex, the focus, the axis, and the directrix; then sketch the parabola.
   * (a) $y^2 = 2x$.   (b) $x^2 = -5y$.
   * (c) $y^2 = 2(x - 1)$.   (d) $2y = 4x^2 - 1$.

*(e) $(x + 2)^2 = 8y - 12$.   (f) $y - 3 = 2(x - 1)^2$.

*(g) $y = x^2 + x + 1$.   (h) $x = y^2 + y + 1$.

3. Find an equation for the indicated parabola:

*(a) focus $(1, 2)$, directrix $x + y + 1 = 0$.

(b) vertex $(2, 0)$, directrix $2x - y = 0$.

*(c) vertex $(2, 0)$, focus $(0, 2)$.

4. Show that every parabola has an equation of the form

$$(\alpha x + \beta y)^2 = \gamma x + \delta y + \epsilon \qquad \text{with} \quad \alpha^2 + \beta^2 \neq 0.$$

HINT: Take $l: Ax + By + C = 0$ as the directrix and $F(a, b)$ as the focus.

5. Show that not every equation of the form

$$(\alpha x + \beta y)^2 = \gamma x + \delta y + \epsilon \qquad \text{with} \quad \alpha^2 + \beta^2 \neq 0$$

represents a parabola.

6. Identify the graph of

$$(x - y)^2 = b \quad \text{for different values of } b.$$

*7. Find an equation for the parabola which has directrix $y = 1$, axis $x = 2$, and passes through the point $(5, 6)$.

8. Find an equation for the parabola which has horizontal axis, vertex $(-1, 1)$, and passes through the point $(-6, 13)$.

In Exercises 9 through 14 we measure distance in feet and time in seconds. We neglect air resistance and the curvature of the earth. We take O as the origin, the x-axis as ground level and consider a projectile fired from O at an angle θ with initial velocity v_0.

*9. Find an equation for the trajectory.

10. What is the maximum height attained by the projectile?

*11. Find the range of the projectile.

12. How many seconds after firing does the impact take place?

*13. How should θ be chosen so as to maximize the range?

14. How should θ be chosen so that the range becomes r?

*15. Find the vertex, the focus, and the directrix of the parabola

$$y = Ax^2 + Bx + C.$$

The following exercises will justify Simpson's rule.

16. Show that every parabola with axis parallel to the y-axis is the graph of a function of the form

$$g(x) = Ax^2 + Bx + C.$$

17. Show that there is one and only one parabola of the form

$$y = Ax^2 + Bx + C$$

which passes through three given noncollinear points with different abscissas.

18. Show that the function
$$g(x) = Ax^2 + Bx + C$$
satisfies the condition
$$\int_a^b g(x)\,dx = \frac{b-a}{6}\left[g(a) + 4g\left(\frac{a+b}{2}\right) + g(b)\right]$$
for every interval $[a, b]$.

8.4 The Ellipse

We begin with two points F_1 and F_2. Using the line segment $\overline{F_1F_2}$ as a base, we construct an isosceles triangle as in Figure 8.4.1.

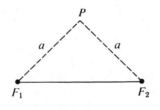

FIGURE 8.4.1

The point P has the property that $d(P, F_1) + d(P, F_2) = 2a$.

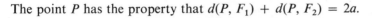

The set of all points P such that
$$d(P, F_1) + d(P, F_2) = 2a$$
is called an *ellipse*. F_1 and F_2 are called the *foci*.

To derive the most convenient equation for the ellipse, we place the foci along the x-axis at equal distances from the origin. (See Figure 8.4.2.) For some $c > 0$ we have
$$F_1 \text{ at } (-c, 0) \quad \text{and} \quad F_2 \text{ at } (c, 0).$$
For a point $P(x, y)$ to lie on the ellipse, it must satisfy the condition
$$d(P, F_1) + d(P, F_2) = 2a.$$
In terms of coordinates, we must have
$$\sqrt{(x + c)^2 + y^2} + \sqrt{(x - c)^2 + y^2} = 2a.$$
By transferring the second term to the right-hand side and squaring both sides, we obtain
$$(x + c)^2 + y^2 = 4a^2 + (x - c)^2 + y^2 - 4a\sqrt{(x - c)^2 + y^2}.$$

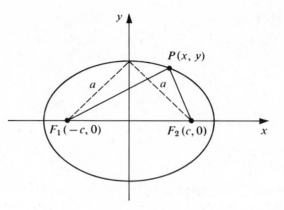

FIGURE 8.4.2

This reduces to

$$4a\sqrt{(x-c)^2 + y^2} = 4(a^2 - cx).$$

Canceling the factor 4 and squaring again, we obtain

$$a^2(x^2 - 2cx + c^2 + y^2) = a^4 - 2a^2cx + c^2x^2.$$

This in turn reduces to

$$(a^2 - c^2)x^2 + a^2y^2 = a^2(a^2 - c^2),$$

and thus to

$$\frac{x^2}{a^2} + \frac{y^2}{a^2 - c^2} = 1.$$

It is customary to set

$$b = \sqrt{a^2 - c^2}$$

and then write the equation as

$$\frac{x^2}{a^2} + \frac{y^2}{b^2} = 1.$$

The relations between a, b, and c are illustrated in Figure 8.4.3.

The line segment which joins $(-a, 0)$ to $(a, 0)$ is called the *major axis* and the line segment which joins $(0, -b)$ to $(0, b)$ is called the *minor axis*. The number $2a$ gives the length of the major axis and the number $2b$ gives the length of the minor axis.

If the foci are placed on the y-axis,

$$F_1 \text{ at } (0, -c) \quad \text{and} \quad F_2 \text{ at } (0, c),$$

the equation of the ellipse takes the form

$$\frac{x^2}{a^2 - c^2} + \frac{y^2}{a^2} = 1,$$

which we can rewrite as

$$\frac{x^2}{b^2} + \frac{y^2}{a^2} = 1.$$

The major axis is then vertical and the minor axis is horizontal.

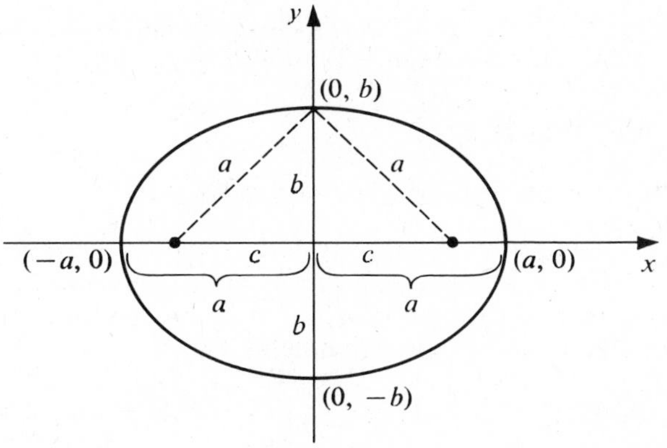

FIGURE 8.4.3

Example. The equation

$$16x^2 + 25y^2 = 400$$

can be rewritten as

$$\frac{x^2}{5^2} + \frac{y^2}{4^2} = 1$$

and thus as

$$\frac{x^2}{5^2} + \frac{y^2}{5^2 - 3^2} = 1.$$

This is the equation of an ellipse with foci at $(-3, 0)$ and $(3, 0)$. The major axis has length 10 and minor axis has length 8. □

Example. The ellipse

$$\frac{x^2}{4^2} + \frac{y^2}{5^2} = 1$$

has the same shape and size as the one of the previous example, but this time the foci are on the y-axis. The coordinates of the foci are $(0, -3)$ and $(0, 3)$. □

Example.   The equation

$$\frac{(x - 1)^2}{13^2} + \frac{(y + 4)^2}{12^2} = 1$$

can be rewritten as

$$\frac{(x - 1)^2}{13^2} + \frac{(y + 4)^2}{13^2 - 5^2} = 1.$$

This represents the ellipse

$$\frac{x^2}{13^2} + \frac{y^2}{13^2 - 5^2} = 1$$

translated 1 unit to the right and 4 units down. [The center, instead of being at the origin, is at the point $(1, -4)$.] The length of the major axis is 26 and the length of the minor axis is 24. The foci are at $(-4, -4)$ and $(6, -4)$. □

Example.   To identify the curve

$$4x^2 - 8x + y^2 + 4y - 8 = 0$$

we write

$$4(x^2 - 2x +) + (y^2 + 4y +) = 8.$$

By completing the squares within the parentheses, we get

$$4(x^2 - 2x + 1) + (y^2 + 4y + 4) = 16,$$

$$4(x - 1)^2 + (y + 2)^2 = 16,$$

$$\frac{(x - 1)^2}{4} + \frac{(y + 2)^2}{16} = 1,$$

$$\frac{(x - 1)^2}{2^2} + \frac{(y + 2)^2}{4^2} = 1.$$

This is the equation of an ellipse centered at $(1, -2)$. The major axis is vertical. The length of the major axis is 8 and the length of the minor axis is 4. As you can check, the foci are at $(1, -2 - 2\sqrt{3})$ and $(1, -2 + 2\sqrt{3})$. □

To sketch the ellipse

$$\frac{x^2}{a^2} + \frac{y^2}{a^2 - c^2} = 1,$$

you can loop a string of length $2(a + c)$ over tacks placed at the foci. A pencil placed in the loop so as to keep the string taut will describe the given ellipse. (Figure 8.4.4)

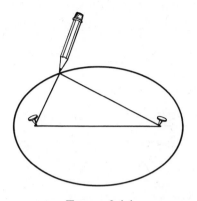

FIGURE 8.4.4

The Reflecting Property of the Ellipse

Like the parabola, the ellipse has an interesting reflecting property. To derive it, we consider the ellipse

$$\frac{x^2}{a^2} + \frac{y^2}{b^2} = 1.$$

Differentiation with respect to x gives

$$\frac{2x}{a^2} + \frac{2y}{b^2}\frac{dy}{dx} = 0$$

and thus

$$\frac{dy}{dx} = -\frac{b^2 x}{a^2 y}.$$

The slope at the point (x_0, y_0) is therefore

$$-\frac{b^2 x_0}{a^2 y_0}$$

and the tangent line has equation

$$y - y_0 = -\frac{b^2 x_0}{a^2 y_0}(x - x_0).$$

We can rewrite this last equation as

$$b^2 x_0 x + a^2 y_0 y - a^2 b^2 = 0.$$

We can now show the following:

At each point P of the ellipse, the focal radii $\overline{F_1 P}$ and $\overline{F_2 P}$ make equal angles with the tangent.

PROOF.   If P lies on the x-axis, the focal radii are coincident and there is nothing to show. To visualize the argument for P not on the x-axis, see Figure 8.4.5. To show

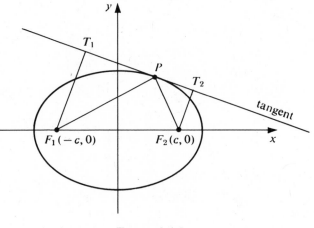

FIGURE 8.4.5

that $\overline{F_1P}$ and $\overline{F_2P}$ make equal angles with the tangent we need only show that the triangles PT_1F_1 and PT_2F_2 are similar. We can do this by showing that

$$\frac{d(T_1, F_1)}{d(F_1, P)} = \frac{d(T_2, F_2)}{d(F_2, P)}$$

or equivalently by showing that

$$\frac{|-b^2x_0c - a^2b^2|}{\sqrt{(x_0 + c)^2 + y_0^2}} = \frac{|b^2x_0c - a^2b^2|}{\sqrt{(x_0 - c)^2 + y_0^2}}.$$

The validity of this last equation can be seen by canceling the factor b^2 and then squaring. This gives

$$\frac{(x_0c + a^2)^2}{(x_0 + c)^2 + y_0^2} = \frac{(x_0c - a^2)^2}{(x_0 - c)^2 + y_0^2},$$

which can be simplified to

$$(a^2 - c^2)x_0^2 + a^2y_0^2 = a^2(a^2 - c^2)$$

and thus to

$$\frac{x_0^2}{a^2} + \frac{y_0^2}{b^2} = 1.$$

This last equation holds since the point $P(x_0, y_0)$ is on the ellipse.   □

The result we just proved has the following physical consequence:

An elliptical mirror takes light or sound originating at one focus and converges it at the other focus.

Exercises

For each of the following ellipses (a) find the foci, (b) find the length of the major axis, (c) find the length of the minor axis, and then (d) sketch the figure.

*1. $\dfrac{x^2}{9} + \dfrac{y^2}{4} = 1$. $\qquad\qquad$ 2. $\dfrac{x^2}{4} + \dfrac{y^2}{9} = 1$.

*3. $3x^2 + 2y^2 = 12$. $\qquad\qquad$ 4. $(x - 1)^2 + 4y^2 = 64$.

*5. $3x^2 + 4y^2 - 12 = 0$. $\qquad\qquad$ 6. $4x^2 + y^2 - 6y + 5 = 0$.

*7. $4(x - 1)^2 + y^2 = 64$. $\qquad\qquad$ 8. $16(x - 2)^2 + 25(y - 3)^2 = 400$.

Find an equation for the ellipse which satisfies the given conditions.

*9. Foci at $(-1, 0)$, $(1, 0)$; major axis 6.
 10. Foci at $(0, -1)$, $(0, 1)$; major axis 6.
*11. Foci at $(3, 1)$, $(9, 1)$; major axis 10.
 12. Foci at $(1, 3)$, $(1, 9)$; minor axis 8.
*13. Focus at $(1, 1)$; center at $(1, 3)$; major axis 10.
 14. Center at $(2, 1)$; vertices at $(2, 6)$ and $(1, 1)$.
 15. Major axis 10; vertices at $(3, 2)$ and $(3, -4)$.

The *eccentricity* of an ellipse is the ratio

$$\frac{d(F_1, F_2)}{\text{length of major axis}}.$$

For the ellipse in standard position

$$\frac{x^2}{a^2} + \frac{y^2}{a^2 - c^2} = 1$$

the eccentricity is the number

$$e = \frac{c}{a}.$$

*16. What happens to the ellipse if a remains constant but e tends to 0?
 17. What happens to the ellipse if a remains constant but e tends to 1?
*18. Find an equation for the ellipse in standard position which has major axis 10 and eccentricity $\frac{1}{2}$.

19. Find an equation for the ellipse in standard position which has minor axis $2\sqrt{3}$ and eccentricity $\frac{1}{2}$.

20. Like the parabola the ellipse can be defined in terms of one focus F and a directrix line l. Show that the ellipse

$$\frac{x^2}{a^2} + \frac{y^2}{a^2 - c^2} = 1$$

is the set of points $P(x, y)$ such that

$$d(P, F) = e\, d(P, l),$$

where

$$F \text{ is } (c, 0), \qquad e = \frac{c}{a}, \quad \text{and} \quad l: x = \frac{a^2}{c}.$$

Note that $0 < e < 1$.

8.5   The Hyperbola

We begin with two points F_1 and F_2 and a number $2a < d(F_1, F_2)$.

The set of all points P such that

$$|d(P, F_1) - d(P, F_2)| = 2a$$

is called a *hyperbola*. The points F_1 and F_2 are called the *foci* of the hyperbola.

The Equation of a Hyperbola

As in the case of the ellipse, to derive an equation for the hyperbola we take a system of coordinates in which F_1 is $(-c, 0)$ and F_2 is $(c, 0)$, with $c > 0$. Since $d(F_1, F_2) = 2c$, we must have $a < c$. (Figure 8.5.1)

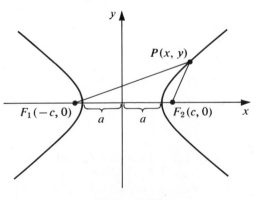

FIGURE 8.5.1

The condition

$$|d(P, F_1) - d(P, F_2)| = 2a$$

is equivalent to

$$d(P, F_1) - d(P, F_2) = \pm 2a,$$

the plus sign holding if $d(P, F_1) > d(P, F_2)$, the minus sign holding if $d(P, F_1) < d(P, F_2)$. In terms of coordinates, we get

$$\sqrt{(x + c)^2 + y^2} - \sqrt{(x - c)^2 + y^2} = \pm 2a.$$

Transferring the second term to the right and squaring both sides, we obtain

$$(x + c)^2 + y^2 = 4a^2 \pm 4a\sqrt{(x - c)^2 + y^2} + (x - c)^2 + y^2.$$

This equation reduces to

$$xc - a^2 = \pm a\sqrt{(x - c)^2 + y^2}.$$

Squaring once more, we obtain

$$x^2c^2 - 2a^2xc + a^4 = a^2(x^2 - 2xc + c^2 + y^2),$$

which reduces to

$$(c^2 - a^2)x^2 - a^2y^2 = a^2(c^2 - a^2),$$

and thus to

$$\boxed{\frac{x^2}{a^2} - \frac{y^2}{c^2 - a^2} = 1.}$$

In this setting it is customary to set

$$b = \sqrt{c^2 - a^2}$$

and write

$$\boxed{\frac{x^2}{a^2} - \frac{y^2}{b^2} = 1.}$$

The line determined by the foci intersects the hyperbola at two points, called the *vertices*. The line segment which joins the vertices is called the *transverse* axis. The number $2a$ gives the length of the transverse axis.

Excluded Region and Asymptotes

If

$$\frac{x^2}{a^2} - \frac{y^2}{b^2} = 1,$$

then $x^2/a^2 \geq 1$, so that

$$|x| \geq a.$$

It follows that no points of the hyperbola lie between the vertical lines $x = -a$ and $x = a$. The points (x, y) of the hyperbola that have $x \geq a$ constitute the *right-hand branch* of the hyperbola, and the others the *left-hand branch*.

Returning to the equation

$$\frac{x^2}{a^2} - \frac{y^2}{b^2} = 1,$$

it is obvious that

$$\frac{y^2}{b^2} < \frac{x^2}{a^2}$$

and therefore

$$|y| < \frac{b}{a} |x|.$$

For $x < 0$, this implies that

$$|y| < -\frac{b}{a} x$$

and thus

(1) $$\frac{b}{a} x < y < -\frac{b}{a} x.$$

For $x > 0$, we have

$$|y| < \frac{b}{a} x,$$

and thus

(2) $$-\frac{b}{a} x < y < \frac{b}{a} x.$$

From (1) and (2) together we see that all points of the hyperbola lie between the lines

$$\boxed{\; y = -\frac{b}{a} x \quad \text{and} \quad y = \frac{b}{a} x. \;}$$

These lines are called the *asymptotes* of the hyperbola.

The role of asymptotes in the theory of the hyperbola is best seen by considering the distance between a point of the hyperbola and one of the asymptotes. For convenience we choose $P(x_0, y_0)$ with positive coordinates and compute its distance from the asymptote $l: y = (b/a)x$. Rewriting the equation of the asymptote as

$$l: bx - ay = 0,$$

we see that

$$d(P, l) = \frac{|bx_0 - ay_0|}{\sqrt{a^2 + b^2}}.$$

From Equation (2)

$$bx_0 - ay_0 > 0$$

and thus

(3)
$$d(P, l) = \frac{bx_0 - ay_0}{\sqrt{a^2 + b^2}}.$$

If $P(x_0, y_0)$ lies on the hyperbola, we have

$$\frac{x_0^2}{a^2} - \frac{y_0^2}{b^2} = 1,$$

from which we get

$$b^2x_0^2 - a^2y_0^2 = a^2b^2,$$

and thus

$$bx_0 - ay_0 = \frac{a^2b^2}{bx_0 + ay_0}.$$

Substituting this into Equation (3), we get

$$d(P, l) = \frac{a^2b^2}{\sqrt{a^2 + b^2}}\left(\frac{1}{bx_0 + ay_0}\right),$$

and since $y_0 > 0$,

$$d(P, l) < \frac{a^2b^2}{\sqrt{a^2 + b^2}}\left(\frac{1}{bx_0}\right).$$

Since a and b are fixed, we can make $d(P, l)$ as small as we wish simply by choosing x_0 sufficiently large. By symmetry, a similar situation prevails in the other quadrants. In summary, we can state that *as a point moves away from a vertex along the branch of a hyperbola, it gets closer to one of the asymptotes and the distance between it and the asymptote tends to zero.* (See Figure 8.5.2.)

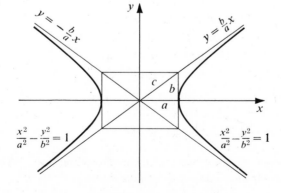

FIGURE 8.5.2

The Tangent to a Hyperbola

Proceeding as in the case of the ellipse, one can show that, if (x_0, y_0), $y_0 \neq 0$, is a point of the hyperbola

$$\frac{x^2}{a^2} - \frac{y^2}{b^2} = 1,$$

then the slope at (x_0, y_0) is given by the number

$$m = \frac{b^2 x_0}{a^2 y_0} \qquad \text{(Exercise 19)}$$

and the tangent line has equation

$$b^2 x_0 x - a^2 y_0 y - a^2 b^2 = 0. \qquad \text{(Exercise 20)}$$

Using this information, one can show that

> *at each point P of the hyperbola the tangent bisects the angle between the focal radii $\overline{F_1 P}$ and $\overline{F_2 P}$.*   (Exercise 21)

An Application to Range Finding

There is a simple application of the hyperbola to range-finding. If observers, located at two listening posts at a known distance apart, time the firing of a gun, the time difference multiplied by the velocity of sound gives the value of $2a$ and hence determines a hyperbola on which the gun must be located. A third listening post gives two more hyperbolas. The gun is found where the hyperbolas intersect.

Exercises

Find an equation for the indicated hyperbola.

* 1. Foci at $(-5, 0)$, $(5, 0)$; transverse axis 6.
  2. Foci at $(-13, 0)$, $(13, 0)$; transverse axis 10.
* 3. Foci at $(0, -13)$, $(0, 13)$; transverse axis 10.
  4. Foci at $(0, -13)$, $(0, 13)$; transverse axis 24.
* 5. Foci at $(-5, 1)$, $(5, 1)$; transverse axis 6.
  6. Foci at $(-3, 1)$, $(7, 1)$; transverse axis 6.
* 7. Foci at $(-1, -1)$, $(1, 1)$; transverse axis 2.

For each of the following hyperbolas find the length of the transverse axis, the vertices, the foci, and the asymptotes. Then sketch the figure.

* 8. $x^2 - y^2 = 1$.   9. $y^2 - x^2 = 1$.

*10. $\dfrac{x^2}{9} - \dfrac{y^2}{16} = 1.$ 11. $\dfrac{x^2}{16} - \dfrac{y^2}{9} = 1.$

*12. $\dfrac{y^2}{16} - \dfrac{x^2}{9} = 1.$ 13. $\dfrac{y^2}{9} - \dfrac{x^2}{16} = 1.$

*14. $\dfrac{(x-1)^2}{9} - \dfrac{(y-3)^2}{16} = 1.$ 15. $\dfrac{(x-1)^2}{16} - \dfrac{(y-3)^2}{9} = 1.$

*16. $-3x^2 + y^2 - 6x = 0.$ 17. $4x^2 - 8x - y^2 + 6y - 1 = 0.$

*18. Find the length of the transverse axis, the vertices, the foci, and the asymptotes of the hyperbola with equation $xy = 1$. HINT: Define new XY-coordinates by setting $x = X + Y$ and $y = X - Y$.

19. Show that if $y_0 \neq 0$ the slope of the hyperbola

$$\frac{x^2}{a^2} - \frac{y^2}{b^2} = 1$$

at the point (x_0, y_0) is given by

$$m = \frac{b^2 x_0}{a^2 y_0}.$$

20. Show that the tangent line at (x_0, y_0) has equation

$$b^2 x_0 x - a^2 y_0 y - a^2 b^2 = 0.$$

21. Show that the tangent to the hyperbola at the point P bisects the angle between the focal radii $\overline{F_1 P}$ and $\overline{F_2 P}$.

22. Like the parabola and the ellipse, the hyperbola can be defined in terms of one focus F and a directrix line l. Show that the hyperbola

$$\frac{x^2}{a^2} - \frac{y^2}{c^2 - a^2} = 1$$

is the set of all points $P(x, y)$ such that

$$d(P, F) = e\, d(P, l),$$

where

$$F \text{ is } (c, 0), \qquad e = \frac{c}{a}, \quad \text{and} \quad l: x = \frac{a^2}{c}.$$

Note that $e > 1$.

8.6   Polar Coordinates

Since you have probably seen polar coordinates before, a short review should suffice. As usual we will measure angles in radians.

In Figure 8.6.1 we have sketched a ray which begins at the origin and is at angle θ from the positive x-axis. The opposite ray is at an angle $\theta + \pi$.

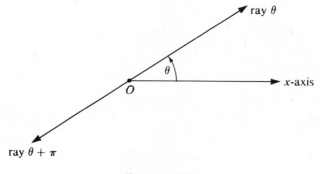

FIGURE 8.6.1

In Figure 8.6.2 we have marked off some points along the same rays and have labeled the points with polar coordinates.

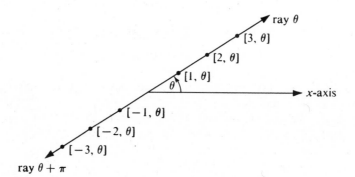

FIGURE 8.6.2

In general, the point $[r, \theta]$ is at a distance $|r|$ from the origin

$$\begin{cases} \text{along the ray } \theta, & \text{if } r > 0 \\ \text{along the ray } \theta + \pi, & \text{if } r < 0 \end{cases}.$$

Polar coordinates are not unique. Many pairs $[r, \theta]$ can represent the same point.

1. If $r = 0$, it does not matter how we choose θ. The resulting point is still the origin:

$$O = [0, \theta] \quad \text{for all } \theta.$$

2. Geometrically there is no distinction between angles that differ by an integral multiple of 2π. Consequently,

$$[r, \theta] = [r, \theta + 2n\pi] \quad \text{for all integers } n.$$

(Figure 8.6.3)

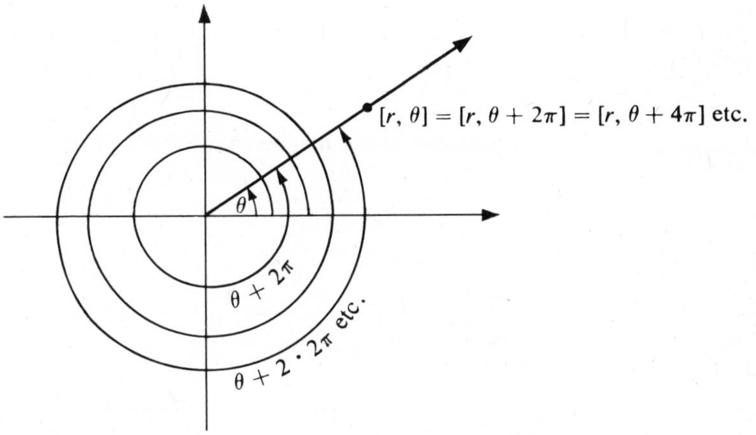

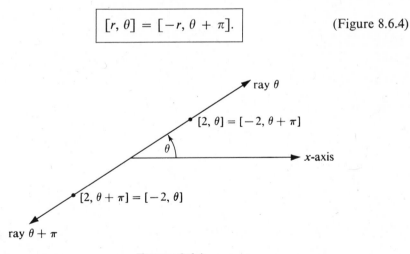

$[r, \theta] = [r, \theta + 2\pi] = [r, \theta + 4\pi]$ etc.

$\theta + 2\pi$

$\theta + 2 \cdot 2\pi$ etc.

FIGURE 8.6.3

3. Finally, we observe that

$$[r, \theta] = [-r, \theta + \pi].$$

(Figure 8.6.4)

ray θ

$[2, \theta] = [-2, \theta + \pi]$

θ

x-axis

$[2, \theta + \pi] = [-2, \theta]$

ray $\theta + \pi$

FIGURE 8.6.4

Symmetry with respect to each of the coordinate axes and with respect to the origin is illustrated in Figures 8.6.5, 8.6.6, and 8.6.7. The coordinates marked are, of course, not the only ones possible.

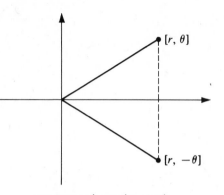

symmetry about the x-axis

FIGURE 8.6.5

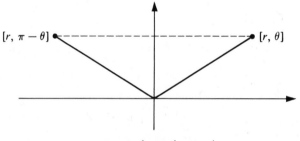

symmetry about the y-axis

FIGURE 8.6.6

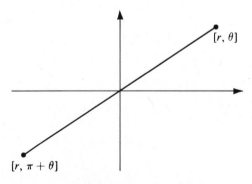

symmetry about the origin

FIGURE 8.6.7

Problem. Test the lemniscate
$$r^2 = \cos 2\theta$$
for symmetry.

SOLUTION. Since
$$\cos [2(-\theta)] = \cos (-2\theta) = \cos 2\theta,$$
you can see that if $[r, \theta]$ is on the curve then so is $[r, -\theta]$. This says that the curve is symmetric about the x-axis.
    Since
$$\cos [2(\pi - \theta)] = \cos (2\pi - 2\theta) = \cos (-2\theta) = \cos 2\theta,$$
you can see that if $[r, \theta]$ is on the curve then so is $[r, \pi - \theta]$. The curve is therefore symmetric about the y-axis.
    Being symmetric about both axes, the curve must also be symmetric about the origin. You can also verify this directly by noting that
$$\cos [2(\pi + \theta)] = \cos (2\pi + 2\theta) = \cos 2\theta,$$
so that, if $[r, \theta]$ lies on the curve, then so does $[r, \pi + \theta]$. A sketch of the lemniscate appears in Figure 8.6.8. □

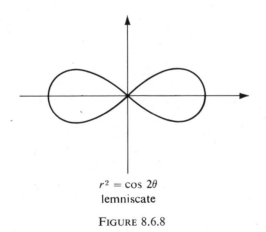

$r^2 = \cos 2\theta$
lemniscate

FIGURE 8.6.8

    In applications it is often useful to change from one coordinate system to another. Keep in mind that in changing between polar coordinates $[r, \theta]$ and rectangular coordinates (x, y) the following relations hold:

$$x = r \cos \theta, \qquad y = r \sin \theta.$$

We also have

$$x^2 + y^2 = r^2.$$

Problem. Find the rectangular coordinates of the point

$$[-2, \tfrac{1}{3}\pi].$$

Solution.   The relations

$$x = r \cos \theta, \qquad y = r \sin \theta$$

give here

$$x = -2 \cos \tfrac{1}{3}\pi = -2(\tfrac{1}{2}) = -1$$

and

$$y = -2 \sin \tfrac{1}{3}\pi = -2(\tfrac{1}{2}\sqrt{3}) = -\sqrt{3}. \quad \square$$

Problem. Find all possible polar coordinates for the point with rectangular coordinates $(-2, 2\sqrt{3})$.

Solution.   Here

$$-2 = r \cos \theta, \qquad 2\sqrt{3} = r \sin \theta$$

so that

$$r^2 = r^2 \cos^2 \theta + r^2 \sin^2 \theta = (-2)^2 + (2\sqrt{3})^2 = 4 + 12 = 16.$$

If we take $r = 4$, then we have

$$-2 = 4 \cos \theta, \qquad 2\sqrt{3} = 4 \sin \theta$$

and thus

$$-\tfrac{1}{2} = \cos \theta, \qquad \tfrac{1}{2}\sqrt{3} = \sin \theta.$$

This means that we can take

$$\theta = \tfrac{2}{3}\pi \qquad\qquad \text{(Figure 8.6.9)}$$

or more generally

$$\theta = \tfrac{2}{3}\pi + 2n\pi.$$

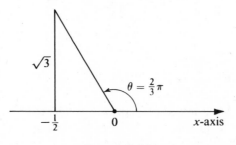

FIGURE 8.6.9

If we take $r = -4$, then we can take

$$\theta = \tfrac{2}{3}\pi + \pi = \tfrac{5}{3}\pi$$

or more generally

$$\theta = \tfrac{5}{3}\pi + 2n\pi. \quad \square$$

Let's specify some simple sets in polar coordinates.

(1) The circle of radius a

$$x^2 + y^2 = a^2$$

becomes in polar coordinates

$$r = a.$$

The interior of the circle is given by

$$0 \le r < a$$

and the exterior by

$$r > a. \quad \square$$

(2) The line through the origin with inclination α is given by the equation

$$\theta = \alpha. \quad \square$$

(3) The vertical line $x = a$ becomes in polar coordinates

$$r \cos \theta = a. \quad \square$$

(4) The horizontal line $y = b$ becomes

$$r \sin \theta = b. \quad \square$$

(5) The line $Ax + By + C = 0$ becomes

$$r(A \cos \theta + B \sin \theta) + C = 0. \quad \square$$

Problem.   Find an equation for the equilateral hyperbola

$$x^2 - y^2 = a^2$$

in terms of polar coordinates.

SOLUTION

$$r^2 \cos^2 \theta - r^2 \sin^2 \theta = a^2,$$
$$r^2(\cos^2 \theta - \sin^2 \theta) = a^2,$$
$$r^2 \cos 2\theta = a^2. \quad \square$$

Problem.   Show that the equation

$$r = 2a \cos \theta$$

represents a circle.

SOLUTION.   Multiplication by r gives

$$r^2 = 2ar \cos \theta,$$
$$x^2 + y^2 = 2ax,$$
$$x^2 - 2ax + y^2 = 0,$$
$$x^2 - 2ax + a^2 + y^2 = a^2,$$
$$(x - a)^2 + y^2 = a^2.$$

This is a circle of radius a centered at the point with rectangular coordinates $(a, 0)$. $\quad \square$

Problem.   Sketch the cardioid

$$r = a(1 + \cos \theta).$$ (take $a > 0$)

SOLUTION.   We will concern ourselves only with θ from $-\pi$ to π. Outside that interval the curve repeats itself.

Since $\cos(-\theta) = \cos \theta$, if a point $[r, \theta]$ lies on the curve then so does $[r, -\theta]$. This tells us that the cardioid is symmetric about the x-axis. We will need to study the curve carefully only in the upper half plane, in this case only at points where θ is between 0 and π. We can sketch in the rest by symmetry.

TABLE 8.6.1

θ	r
0	$2a$
$\frac{1}{6}\pi$	$a(1 + \frac{1}{2}\sqrt{3})$
$\frac{1}{3}\pi$	$\frac{3}{2}a$
$\frac{1}{2}\pi$	a
$\frac{2}{3}\pi$	$\frac{1}{2}a$
π	0

When θ is 0, $\cos \theta$ is 1, and r is $2a$. As θ increases from 0 to π, $\cos \theta$ decreases from 1 to -1, and r decreases from $2a$ to 0.

With this general information, together with the few points given in Table 8.6.1, you can see that the cardioid must look as in Figure 8.6.10.   □

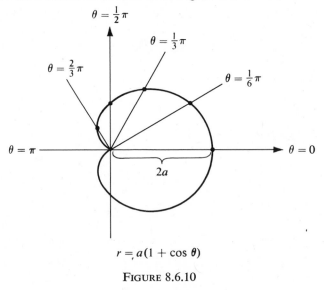

$$r = a(1 + \cos \theta)$$

FIGURE 8.6.10

Problem.   Sketch the graph of

$$r = a \sin 3\theta.$$ (take $a > 0$)

SOLUTION.   As θ increases from 0 to $\frac{1}{6}\pi$, r increases from 0 to a:

$$a \sin 3(\tfrac{1}{6}\pi) = a \sin \tfrac{1}{2}\pi = a.$$

As θ increases from $\frac{1}{6}\pi$ to $\frac{1}{3}\pi$, r decreases back to 0:

$$a \sin 3(\tfrac{1}{3}\pi) = a \sin \pi = 0. \qquad\qquad \text{(Figure 8.6.11)}$$

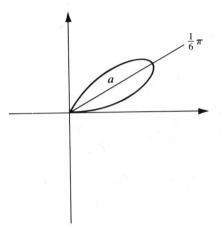

FIGURE 8.6.11

As θ increases from $\frac{1}{3}\pi$ to $\frac{1}{3}\pi + \frac{1}{6}\pi = \frac{1}{2}\pi$, r decreases from 0 to $-a$:

$$a \sin 3(\tfrac{1}{2}\pi) = -a.$$

As θ increases from $\frac{1}{2}\pi$ to $\frac{1}{2}\pi + \frac{1}{6}\pi = \frac{2}{3}\pi$, r increases back to 0:

$$a \sin 3(\tfrac{2}{3}\pi) = a \sin 2\pi = 0. \qquad\qquad \text{(Figure 8.6.12)}$$

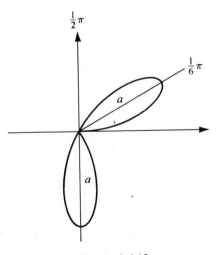

FIGURE 8.6.12

As θ increases from $\frac{2}{3}\pi$ to $\frac{2}{3}\pi + \frac{1}{6}\pi = \frac{5}{6}\pi$, r increases from 0 to a:

$$a \sin 3(\tfrac{5}{6}\pi) = a \sin \tfrac{5}{2}\pi = a \sin \tfrac{1}{2}\pi = a.$$

As θ increases from $\frac{5}{6}\pi$ to π, r decreases back to 0:

$$a \sin 3\pi = a \sin \pi = 0. \hspace{2cm} \text{(Figure 8.6.13)}$$

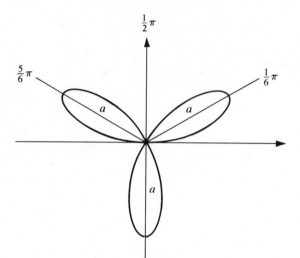

FIGURE 8.6.13

From here on, the curve repeats itself. The value $\theta + \pi$ gives rise to the same point on the curve as did θ:

$$[a \sin 3(\theta + \pi), \theta + \pi] = [-a \sin 3\theta, \theta + \pi] = [a \sin 3\theta, \theta].$$

The last equality follows from the fact that in general

$$[-r, \theta + \pi] = [r, \theta]. \quad \square$$

The fact that a single point has many pairs of polar coordinates can cause complications. In particular, it means that a point $[r_1, \theta_1]$ can lie on a curve

$$r = f(\theta)$$

although its coordinates r_1 and θ_1 do not satisfy the equation. For example, the coordinates of $[2, \pi]$ do not satisfy the equation

$$r^2 = 4 \cos \theta.$$

$$(r^2 = 2^2 = 4 \quad \text{but} \quad 4 \cos \theta = 4 \cos \pi = -4.)$$

Nevertheless the point $[2, \pi]$ does lie on the curve

$$r^2 = 4 \cos \theta.$$

It does lie on the curve because

$$[2, \pi] = [-2, 0] \qquad \text{(check this out)}$$

and the coordinates of $[-2, 0]$ do satisfy the equation:

$$r^2 = (-2)^2 = 4, \qquad 4 \cos \theta = 4 \cos 0 = 4. \quad \square$$

The difficulties are compounded when we deal with two or more curves. Here is an example.

Problem. Find the points where the cardioids

$$r = a(1 - \cos \theta) \quad \text{and} \quad r = a(1 + \cos \theta)$$

intersect.

SOLUTION. We begin by solving the two equations simultaneously. First we rewrite the equations as

$$\frac{r}{a} = 1 - \cos \theta, \qquad \frac{r}{a} = 1 + \cos \theta.$$

Adding the equations, we get

$$2\frac{r}{a} = 2$$

and therefore

$$r = a.$$

This means that

$$\cos \theta = 0 \quad \text{and} \quad \theta = \tfrac{1}{2}\pi + n\pi.$$

As you can check, the points

$$[a, \tfrac{1}{2}\pi + n\pi]$$

all lie on both curves. Not all of these points are distinct, however:

$$\text{for } n \text{ even, } [a, \tfrac{1}{2}\pi + n\pi] = [a, \tfrac{1}{2}\pi]$$
$$\text{for } n \text{ odd, } [a, \tfrac{1}{2}\pi + n\pi] = [a, \tfrac{3}{2}\pi].$$

In short, by solving the two equations simultaneously we have arrived at two common points:

$$[a, \tfrac{1}{2}\pi] = (0, a) \quad \text{and} \quad [a, \tfrac{3}{2}\pi] = (0, -a).$$

There is however a third point at which the curves intersect and that is the origin O. (Figure 8.6.14)

The origin clearly lies on both curves:

$$\text{for } r = a(1 - \cos \theta) \quad \text{take } \theta = 0, 2\pi, \text{ etc.}$$
$$\text{for } r = a(1 + \cos \theta) \quad \text{take } \theta = \pi, 3\pi, \text{ etc.}$$

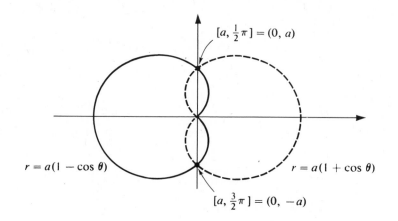

$$[a, \tfrac{1}{2}\pi] = (0, a)$$

$$r = a(1 - \cos \theta)$$

$$r = a(1 + \cos \theta)$$

$$[a, \tfrac{3}{2}\pi] = (0, -a)$$

FIGURE 8.6.14

The reason that the origin does not appear when we solve the two equations simultaneously is that the curves do not pass through the origin "simultaneously"; that is, they do not pass through the origin for the same values of θ.†   □

Tangent lines are taken up in the next section. Area problems are studied in Chapter 9.

Exercises

Plot the following points.

1. $[1, \tfrac{1}{3}\pi]$.   2. $[1, \tfrac{1}{2}\pi]$.   3. $[-1, \tfrac{1}{3}\pi]$.   4. $[-1, -\tfrac{1}{3}\pi]$.

5. $[4, \tfrac{5}{4}\pi]$.   6. $[-2, 0]$.   7. $[-\tfrac{1}{2}, \pi]$.   8. $[\tfrac{1}{3}, \tfrac{2}{3}\pi]$.

Find the rectangular coordinates of each of the following points.

*9. $[2, 3\pi]$.   10. $[4, \tfrac{1}{6}\pi]$.   *11. $[-3, -\tfrac{1}{3}\pi]$.   12. $[-1, \tfrac{1}{4}\pi]$.

*13. $[-1, -\pi]$.   14. $[2, 0]$.   *15. $[3, \tfrac{1}{2}\pi]$.   16. $[3, -\tfrac{1}{2}\pi]$.

The following points are given in rectangular coordinates. Find all possible polar coordinates for each point.

*17. $(0, 1)$.   18. $(1, 0)$.   *19. $(-3, 0)$.

20. $(4, 4)$.   *21. $(2, -2)$.   22. $(3, -3\sqrt{3})$.

*23. $(4\sqrt{3}, 4)$.   24. $(\sqrt{3}, -1)$.

---

† Think of each of the equations

$$r = a(1 - \cos \theta) \quad \text{and} \quad r = a(1 + \cos \theta)$$

as giving the position of an object at time θ. At the points we found by solving the two equations simultaneously, the objects will collide. (They both arrive there at the same time.) At the origin the situation is different. Both objects will pass through the origin, but no collision takes place because the objects pass through the origin at different times.

Test the following curves for symmetry.

* 25. $r = 2 + \cos \theta$.
                                      26. $r^2 = \cos 2\theta$.

* 27. $r^2 = \sin \theta$.
                                      28. $r(\sin \theta + \cos \theta) = 1$.

* 29. $r^2 \sin 2\theta = 1$.
                                      30. $r = \cos 2\theta$.

Express the following equations in terms of polar coordinates.

* 31. $2xy = 1$.
                                      32. $y = mx$.

* 33. $x^2 + y^2 = 4$.
                                      34. $x^2 + (y - b)^2 = b^2$.

* 35. $x^2 + y^2 + ax = a\sqrt{x^2 + y^2}$.
         36. $(x^2 + y^2)^2 = a^2(x^2 - y^2)$.

Identify the following curves. Change to rectangular coordinates.

* 37. $r \sin \theta = 4$.
                                      38. $r \cos \theta = 4$.

* 39. $\theta = \frac{1}{3}\pi$.
                                      40. $\theta^2 = \frac{1}{9}\pi^2$.

* 41. $r = \dfrac{2}{1 - \cos \theta}$.
                           42. $r = 4 \sin (\theta + \pi)$.

Sketch the following curves.

* 43. $r = \cos \theta$.
                                      44. $r = \sin \theta$.

* 45. $r = \theta, \quad \theta \geq 0$.
                                46. $r = 2\theta, \quad \theta \geq 0$.

* 47. $r = e^\theta, \quad \theta \geq 0$.
                              48. $r = e^{2\theta}, \quad \theta \geq 0$.

Determine whether the given point lies on the given curve.

* 49. $r^2 \cos \theta = 1; \quad [1, \pi]$.
                  50. $r^2 = \cos 2\theta; \quad [1, \frac{1}{4}\pi]$.

* 51. $r = \sin \frac{1}{3}\theta; \quad [\frac{1}{2}, \frac{1}{2}\pi]$.
          52. $r^2 = \sin 3\theta; \quad [1, -\frac{5}{6}\pi]$.

* 53. Show that the point $[2, \pi]$ lies both on

$$r^2 = 4 \cos \theta \quad \text{and} \quad r = 3 + \cos \theta.$$

Find the points at which the curves intersect. Express your answers in rectangular coordinates.

* 54. $r = 2 \sin \theta, \quad r = 2 \cos \theta$.

55. $r = \dfrac{1}{1 - \cos \theta}, \quad r \sin \theta = b$.

* 56. $r^2 = \sin \theta, \quad r = 2 - \sin \theta$.

(*Optional*)   Sketch the following curves.

* 57. $r = a \cos 2\theta$.
                                      58. $r = a \sin 2\theta$.

* 59. $r = a \cos 3\theta$.
                                      60. $r^2 = \sin \theta$.

61. (*Optional*)   Let $e > 0$. Show that the polar equation

$$r = \frac{ed}{1 - e \cos \theta}$$

represents

$$\begin{cases} \text{an ellipse} & \text{if } e < 1 \\ \text{a parabola} & \text{if } e = 1 \\ \text{a hyperbola} & \text{if } e > 1 \end{cases}.$$

8.7   Curves Given Parametrically

We begin with a pair of functions x and y differentiable on some interval I. At the endpoints we require only continuity.

For each t in I we can interpret $(x(t), y(t))$ as the point with x-coordinate $x(t)$ and y-coordinate $y(t)$. As t then ranges over I, the point $(x(t), y(t))$ traces out a curve in the plane. The curve so traced out is said to be given *parametrically* by the functions x and y.

Problem.   An object moves so that at time t it has coordinates

$$x(t) = t + 1, \qquad y(t) = 2t - 5.$$

Find the path generated by the motion.

SOLUTION.   Here we can express $y(t)$ in terms of $x(t)$:

$$y(t) = 2[x(t) - 1] - 5 = 2x(t) - 7.$$

The path generated is the line

$$y = 2x - 7. \quad \square$$

Problem.   Find the curve generated by

$$x(t) = a \cos t, \qquad y(t) = a \sin t.$$

SOLUTION.   Here

$$[x(t)]^2 + [y(t)]^2 = a^2 \cos^2 t + a^2 \sin^2 t = a^2$$

so that the curve is the circle

$$x^2 + y^2 = a^2. \quad \square$$

Problem.   Find the curve generated by

$$x(t) = a \cos t, \qquad y(t) = b \sin t.$$

SOLUTION.   Here

$$\frac{[x(t)]^2}{a^2} + \frac{[y(t)]^2}{b^2} = \cos^2 t + \sin^2 t = 1$$

so that the curve is the ellipse

$$\frac{x^2}{a^2} + \frac{y^2}{b^2} = 1. \quad \square$$

A single curve can be parametrized in many different ways. As an example take the ellipse

$$\frac{x^2}{a^2} + \frac{y^2}{b^2} = 1$$

and think of t as time measured in seconds. A particle whose position is given by the equations

$$x(t) = a \cos t, \qquad y(t) = b \sin t, \qquad t \in [0, 2\pi]$$

traverses the ellipse in a counterclockwise manner. It begins at the point

$$(x(0), y(0)) = (a, 0)$$

and makes a full circuit in 2π seconds. If the equations of motion are

$$x(t) = a \cos 2\pi t, \qquad y(t) = -b \sin 2\pi t, \qquad t \in [0, 1]$$

the particle still travels the same ellipse, but in a different manner. Once again it starts at

$$(x(0), y(0)) = (a, 0)$$

but this time it moves clockwise and makes the full circuit in only one second. If the equations of motion are

$$x(t) = a \sin 4\pi t, \qquad y(t) = b \cos 4\pi t, \qquad t \in [0, \infty)$$

the motion begins at

$$(x(0), y(0)) = (0, b)$$

and goes on in perpetuity. The motion is clockwise, a complete circuit taking place every half second. $\quad \square$

Problem. Two particles start at the same instant, the first along the linear path

$$x_1(t) = \tfrac{16}{3} - \tfrac{8}{3}t, \qquad y_1(t) = 4t - 5, \qquad t \geq 0$$

and the second along the elliptical path

$$x_2(t) = 2 \sin \tfrac{1}{2}\pi t, \qquad y_2(t) = -3 \cos \tfrac{1}{2}\pi t, \qquad t \geq 0.$$

(i) At what points, if any, do the paths intersect?
(ii) At what points, if any, do the particles collide?

SOLUTION.   To see where the paths intersect, we first express them both as equations in x and y. The linear path can be written as

$$3x + 2y - 6 = 0, \qquad x \leq \tfrac{16}{3}$$

and the elliptical path as

$$\frac{x^2}{4} + \frac{y^2}{9} = 1.$$

Solving the two equations simultaneously, we get

$$x = 2, \quad y = 0 \quad \text{and} \quad x = 0, \quad y = 3.$$

This means that the paths intersect at the points

$$(2, 0) \quad \text{and} \quad (0, 3).$$

This answers part (i). Now for part (ii). The first particle passes through (2, 0) only when

$$x_1(t) = \tfrac{16}{3} - \tfrac{8}{3}t = 2 \quad \text{and} \quad y_1(t) = 4t - 5 = 0.$$

As you can check, this happens only when $t = \tfrac{5}{4}$. When $t = \tfrac{5}{4}$, the second particle is elsewhere. Hence no collision takes place at (2, 0). There is however a collision at (0, 3) because both particles get there at exactly the same time, $t = 2$:

$$x_1(2) = 0 = x_2(2), \qquad y_1(2) = 3 = y_2(2). \quad \square$$

Tangents

The problem of finding tangent lines to a parametrized curve

$$C: x(t), y(t), \qquad t \in I$$

is complicated by the fact that such a curve can intersect itself. This means that at a given point a curve can have

(i) one tangent,
(ii) two or more tangents, or
(iii) no tangent at all.

We illustrate these possibilities in Figure 8.7.1.

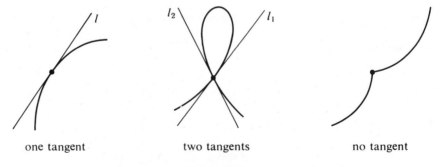

one tangent   two tangents   no tangent

FIGURE 8.7.1

To make sure that at least one tangent line exists at each point we make the additional assumption that

$$x' \text{ and } y' \text{ are continuous} \quad \text{and} \quad [x'(t)]^2 + [y'(t)]^2 \neq 0.$$

We now choose a point (x_0, y_0) of the curve C and a time t_0 at which

$$x(t_0) = x_0 \quad \text{and} \quad y(t_0) = y_0.$$

What we want is the slope of the curve as it passes through the point (x_0, y_0) at time t_0.† To find this slope, we assume that $x'(t_0) \neq 0$. With $x'(t_0) \neq 0$, we can be sure that for h sufficiently small

$$x(t_0 + h) - x(t_0) \neq 0. \tag{why?}$$

We can therefore form the quotient

$$\frac{y(t_0 + h) - y(t_0)}{x(t_0 + h) - x(t_0)}.$$

This quotient is the slope of the secant line pictured in Figure 8.7.2. The limit of this

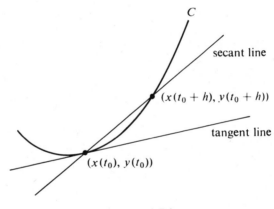

FIGURE 8.7.2

quotient as h tends to zero is the slope we want. Since

$$\frac{y(t_0 + h) - y(t_0)}{x(t_0 + h) - x(t_0)} = \frac{(1/h)[y(t_0 + h) - y(t_0)]}{(1/h)[x(t_0 + h) - x(t_0)]} \to \frac{y'(t_0)}{x'(t_0)},$$

we have

$$\boxed{m = \frac{y'(t_0)}{x'(t_0)}.}$$

As an equation for the tangent line, we can write

$$y - y(t_0) = \frac{y'(t_0)}{x'(t_0)} [x - x(t_0)],$$

---

† It could pass through the point (x_0, y_0) at other times also.

which simplifies to

$$y'(t_0)[x - x(t_0)] - x'(t_0)[y - y(t_0)] = 0,$$

and thus to

(8.7.1)

$$y'(t_0)[x - x_0] - x'(t_0)[y - y_0] = 0.$$

We derived this equation under the assumption that $x'(t_0) \neq 0$. If $x'(t_0) = 0$, Equation 8.7.1 still makes sense. It is simply

$$y'(t_0)[x - x_0] = 0,$$

which, since $y'(t_0) \neq 0,$† can be simplified to

$$x = x_0.$$

In this instance the line is vertical and we say that the curve has a *vertical tangent*.

Problem. Find the tangent(s) to the curve

$$x(t) = t^3, \qquad y(t) = 1 - t$$

at the point $(8, -1)$.

SOLUTION. Since the curve passes through the point only when $t = 2$, there is only one tangent line at $(8, -1)$. With

$$x(t) = t^3, \qquad y(t) = 1 - t$$

we have

$$x'(t) = 3t^2, \qquad y'(t) = -1$$

and therefore

$$x'(2) = 12, \qquad y'(2) = -1.$$

The tangent line has equation

$$(-1)[x - 8] - (12)[y - (-1)] = 0.$$

This reduces to

$$x + 12y + 4 = 0. \quad \square$$

Problem. Find the points of the curve

$$x(t) = 3 - 4 \sin t, \qquad y(t) = 4 + 3 \cos t$$

at which there is (i) a horizontal tangent, (ii) a vertical tangent.

SOLUTION. Since

$$x'(t) = -4 \cos t \quad \text{and} \quad y'(t) = -3 \sin t$$

---

† We are assuming that $[x'(t)]^2 + [y'(t)]^2$ is never 0. Since $x'(t_0) = 0$, $y'(t_0) \neq 0$.

the derivatives are never 0 simultaneously. To find the points at which there is a horizontal tangent, we set

$$y'(t) = 0.$$

This gives

$$t = n\pi.$$

Horizontal tangents occur therefore at points of the form

$$(x(n\pi), y(n\pi)).$$

Since

$$x(n\pi) = 3 - 4 \sin n\pi = 3$$

and

$$y(n\pi) = 4 + 3 \cos n\pi = \begin{cases} 7, & n \text{ even} \\ 1, & n \text{ odd} \end{cases},$$

there is a horizontal tangent only at (3, 7) and (3, 1).

To find the vertical tangents, we set $x'(t) = 0$. This gives

$$t = \tfrac{1}{2}\pi + n\pi.$$

Vertical tangents occur therefore at points of the form

$$(x(\tfrac{1}{2}\pi + n\pi), y(\tfrac{1}{2}\pi + n\pi)).$$

Since

$$x(\tfrac{1}{2}\pi + n\pi) = 3 - 4 \sin (\tfrac{1}{2}\pi + n\pi) = \begin{cases} -1, & n \text{ even} \\ 7, & n \text{ odd} \end{cases}$$

and

$$y(\tfrac{1}{2}\pi + n\pi) = 4 + 3 \cos (\tfrac{1}{2}\pi + n\pi) = 4,$$

there are vertical tangents at $(-1, 4)$ and $(7, 4)$. □

Problem.   Find the tangent(s) to the curve

$$x(t) = t^2 - 2t + 1, \qquad y(t) = t^4 - 4t^2 + 4$$

at the point (1, 4).

SOLUTION.   The curve passes through the point (1, 4) when $t = 0$ and when $t = 2$. (Check this out.) Differentiation gives

$$x'(t) = 2t - 2, \qquad y'(t) = 4t^3 - 8t.$$

At $t = 0$,

$$x'(t) \neq 0 \quad \text{and} \quad y'(t) = 0,$$

so that the tangent line is horizontal.

At $t = 2$,

$$x'(t) = 2 \quad \text{and} \quad y'(t) = 16,$$

and therefore the tangent line is given by

$$y - 4 = 8(x - 1). \quad \square$$

We can apply these ideas to a curve given in polar coordinates by first using the relations

$$x = r \cos \theta, \qquad y = r \sin \theta.$$

The polar curve

$$r = f(\theta)$$

can then be written

$$x(\theta) = f(\theta) \cos \theta, \qquad y(\theta) = f(\theta) \sin \theta.$$

Problem.   Find the slope of the curve

$$r = a\theta \quad \text{at } \theta = \tfrac{1}{2}\pi.$$

SOLUTION

$$x(\theta) = r \cos \theta = a\theta \cos \theta,$$

$$y(\theta) = r \sin \theta = a\theta \sin \theta.$$

Now we differentiate:

$$x'(\theta) = -a\theta \sin \theta + a \cos \theta,$$

$$y'(\theta) = a\theta \cos \theta + a \sin \theta.$$

$$x'(\tfrac{1}{2}\pi) = -\tfrac{1}{2}\pi a, \qquad y'(\tfrac{1}{2}\pi) = a,$$

$$\text{slope} = \frac{y'(\tfrac{1}{2}\pi)}{x'(\tfrac{1}{2}\pi)} = -\frac{2}{\pi}. \quad \square$$

Problem.   Find the points of the cardioid

$$r = 1 - \cos \theta \qquad\qquad \text{(Figure 8.6.10)}$$

at which there is a vertical tangent.

SOLUTION.   Since the cosine function has period 2π, we need only concern ourselves with θ in $[0, 2\pi)$. Parametrically we have

$$x(\theta) = (1 - \cos \theta) \cos \theta, \qquad y(\theta) = (1 - \cos \theta) \sin \theta.$$

Differentiating and simplifying, we obtain

$$x'(\theta) = \sin \theta (2 \cos \theta - 1), \qquad y'(\theta) = (1 - \cos \theta)(1 + 2 \cos \theta).$$

The only numbers in the interval $[0, 2\pi)$ at which x' is zero and y' is not are $\tfrac{1}{3}\pi$, π, and $\tfrac{5}{3}\pi$. We have a vertical tangent at each of the following points:

$$[\tfrac{1}{2}, \tfrac{1}{3}\pi], \quad [2, \pi], \quad [\tfrac{1}{2}, \tfrac{5}{3}\pi].$$

In rectangular coordinates, these points are:

$$(\tfrac{1}{4}, \tfrac{1}{4}\sqrt{3}), \quad (-2, 0), \quad (\tfrac{1}{4}, -\tfrac{1}{4}\sqrt{3}). \quad \square$$

This section has been introductory. We return to curves and their parametrizations in Chapter 13.

Exercises

Express the curve as an equation in x and y.

* 1. $x(t) = t^2$, $y(t) = 2t + 1$.
  2. $x(t) = t^2$, $y(t) = t^4 + 1$.
* 3. $x(t) = at + b$, $y(t) = t^3$.
  4. $x(t) = 2 \cos t$, $y(t) = 3 \sin t$.
* 5. $x(t) = 2t + 1$, $y(t) = \log(t^2 + 1)$.
  6. $x(t) = \cos^2 t$, $y(t) = \sin t$.

Find the slope at the point indicated.

* 7. $x(t) = t$, $y(t) = t^3 - 1$; $t = 1$.
  8. $x(t) = t^2$, $y(t) = t + 5$; $t = 2$.
* 9. $x(t) = 2t$, $y(t) = \cos \pi t$; $t = 0$.
  10. $x(t) = 2t - 1$, $y(t) = t^4$; $t = 1$.
* 11. $x(t) = t^2$, $y(t) = (2 - t)^2$; $t = \frac{1}{2}$.
  12. $x(t) = 1/t$, $y(t) = 1/(t + 1)$; $t = 1$.
* 13. $x(t) = \cos t$, $y(t) = 1 + \sin t$; $t = \frac{1}{4}\pi$.
  14. $r = 4 - 2 \sin \theta$, $\theta = \frac{1}{3}\pi$.
* 15. $r = 4 \cos 2\theta$, $\theta = \frac{1}{2}\pi$.

  16. $r = \dfrac{4}{5 - \cos \theta}$, $\theta = \frac{1}{2}\pi$.

* 17. $r = \dfrac{5}{4 - \cos \theta}$, $\theta = \frac{1}{6}\pi$.

  18. $r = \dfrac{\sin \theta - \cos \theta}{\sin \theta + \cos \theta}$, $\theta = 0$.

* 19. $x(t) = \log(t - 2)$, $y(t) = \frac{1}{3}t$; $t = 3$.
  20. $x(t) = e^t$, $y(t) = 3e^{-t}$; $t = 0$.

21. Let $P = [r_1, \theta_1]$ be a point on the polar curve

$$r = f(\theta). \qquad\qquad \text{(Figure 8.7.3)}$$

Show that if $f'(\theta_1) = 0$, the tangent line at P is perpendicular to the line segment $\overline{OP}$.

Find the points (x, y) at which the curve has (a) a horizontal tangent, (b) a vertical tangent. Then sketch the curve.

* 22. $x(t) = 3t - t^3$, $y(t) = t + 1$.   23. $x(t) = t^2 - 2t$, $y(t) = t^3 - 12t$.
* 24. $x(t) = \sin 2t$, $y(t) = \sin t$.   25. $x(t) = \cos^4 t$, $y(t) = \sin^4 t$.

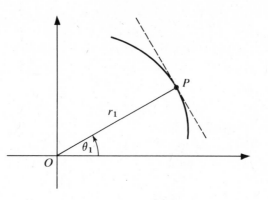

FIGURE 8.7.3

* 26. Two particles start at the same instant, the first along the elliptical path

$$x_1(t) = 2 - 3 \cos \pi t, \qquad y_1(t) = 3 + 7 \sin \pi t, \qquad t \geq 0$$

and the second along the parabolic path

$$x_2(t) = 3t + 2, \qquad y_2(t) = -\tfrac{7}{15}(3t + 1)^2 + \tfrac{157}{15}, \qquad t \geq 0.$$

(a) At what points, if any, do these paths intersect?
(b) At what points, if any, will the particles collide?

8.8   Rotations; Eliminating the *xy*-Term

We begin with a rectangular coordinate system Oxy. If we rotate this system α radians about the origin we obtain a new coordinate system OXY. See Figure 8.8.1.

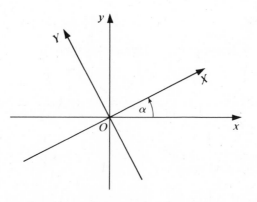

FIGURE 8.8.1

A point P will now have two pairs of rectangular coordinates:

$$(x, y) \text{ with respect to the } Oxy \text{ system}$$

and

$$(X, Y) \text{ with respect to the } OXY \text{ system.}$$

Here we investigate the relation between (x, y) and (X, Y). With P as in Figure 8.8.2,

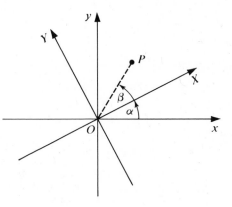

FIGURE 8.8.2

$$x = r \cos (\alpha + \beta), \qquad y = r \sin (\alpha + \beta)$$

and

$$X = r \cos \beta, \qquad Y = r \sin \beta.$$

Since

$$\cos (\alpha + \beta) = \cos \alpha \cos \beta - \sin \alpha \sin \beta,$$
$$\sin (\alpha + \beta) = \sin \alpha \cos \beta + \cos \alpha \sin \beta,$$

we have

$$x = r \cos (\alpha + \beta) = (\cos \alpha) \, r \cos \beta - (\sin \alpha) \, r \sin \beta,$$
$$y = r \sin (\alpha + \beta) = (\sin \alpha) \, r \cos \beta + (\cos \alpha) \, r \sin \beta,$$

and therefore

$$\boxed{x = (\cos \alpha)X - (\sin \alpha)Y \quad \text{and} \quad y = (\sin \alpha)X + (\cos \alpha)Y.}$$

These formulas give the algebraic consequences of a rotation of α radians.

Eliminating the *xy*-Term

Rotations of the coordinate system enable us to simplify equations of the second degree by eliminating the *xy*-term; that is, if in the Oxy system, S has an equation of the form

(1) $ax^2 + bxy + cy^2 + dx + ey + f = 0, \qquad \text{with } b \neq 0,$

then there exists a coordinate system OXY, differing from Oxy by a rotation, such that in the OXY system S has an equation of the form

$$AX^2 + CY^2 + DX + EY + F = 0.$$

To see this, substitute

$$x = (\cos \alpha)X - (\sin \alpha)Y, \qquad y = (\sin \alpha)X + (\cos \alpha)Y$$

in Equation (1). This will give you a second-degree equation in X and Y in which the coefficient of XY is

$$-2a \cos \alpha \sin \alpha + b(\cos^2 \alpha - \sin^2 \alpha) + 2c \cos \alpha \sin \alpha.$$

This coefficient can be simplified to

$$(c - a) \sin 2\alpha + b \cos 2\alpha.$$

It is zero provided that

$$(a - c) \sin 2\alpha = b \cos 2\alpha.$$

The case $a = c$ yields

$$\cos 2\alpha = 0,$$

$$2\alpha = \tfrac{1}{2}\pi + n\pi, \qquad\qquad (n \text{ an arbitrary integer})$$

$$\alpha = \tfrac{1}{4}\pi + \tfrac{1}{2}n\pi.$$

The case $a = c$ can therefore be satisfied by setting

$$\alpha = \tfrac{1}{4}\pi. \qquad\qquad\qquad (\text{choose } n = 0)$$

The case $a \neq c$ yields

$$\tan 2\alpha = \frac{b}{a - c},$$

$$2\alpha = \arctan\left(\frac{b}{a - c}\right) + n\pi,$$

$$\alpha = \frac{1}{2} \arctan\left(\frac{b}{a - c}\right) + \tfrac{1}{2}n\pi.$$

The case $a \neq c$ can therefore be satisfied by setting

$$\alpha = \frac{1}{2} \arctan\left(\frac{b}{a - c}\right). \qquad\qquad (\text{choose } n = 0) \quad \square$$

Example. In the case of

$$xy - 2 = 0,$$

we have $a = c$ and thus can choose $\alpha = \tfrac{1}{4}\pi$. Setting

$$x = (\cos \tfrac{1}{4}\pi)X - (\sin \tfrac{1}{4}\pi)Y = \tfrac{1}{2}\sqrt{2}(X - Y),$$

$$y = (\sin \tfrac{1}{4}\pi)X + (\cos \tfrac{1}{4}\pi)Y = \tfrac{1}{2}\sqrt{2}(X + Y),$$

we find that $xy - 2 = 0$ becomes

$$\tfrac{1}{2}(X^2 - Y^2) - 2 = 0,$$

which simplifies to

$$\frac{X^2}{4} - \frac{Y^2}{4} = 1.$$

This is the equation of a hyperbola in standard position in the OXY system. The graph is shown in Figure 8.8.3. □

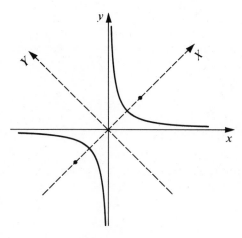

FIGURE 8.8.3

Example.   In the case of

$$11x^2 + 4\sqrt{3}xy + 7y^2 - 1 = 0,$$

we have $a = 11$, $b = 4\sqrt{3}$, and $c = 7$. Thus we can choose

$$\alpha = \frac{1}{2} \text{ arc tan}\left(\frac{b}{a - c}\right) = \frac{1}{2} \text{ arc tan } \sqrt{3} = \frac{1}{6}\pi.$$

Setting

$$x = (\cos \tfrac{1}{6}\pi)X - (\sin \tfrac{1}{6}\pi)Y = \tfrac{1}{2}(\sqrt{3}X - Y),$$
$$y = (\sin \tfrac{1}{6}\pi)X + (\cos \tfrac{1}{6}\pi)Y = \tfrac{1}{2}(X + \sqrt{3}Y),$$

we find after simplification that our initial equation becomes

$$13X^2 + 5Y^2 - 1 = 0,$$

or

$$\frac{X^2}{(1/\sqrt{13})^2} + \frac{Y^2}{(1/\sqrt{5})^2} = 1.$$

This is the equation of an ellipse. Its graph is pictured in Figure 8.8.4. □

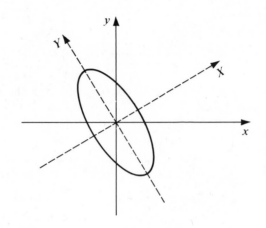

FIGURE 8.8.4

Exercises

For each of the equations below, (a) find a rotation $\alpha \in (-\frac{1}{4}\pi, \frac{1}{4}\pi]$ that eliminates the xy-term; (b) rewrite the equation in terms of the new coordinate system; (c) sketch the graph displaying both coordinate systems.

*1. $xy = 1$.

2. $xy - y + x = 1$.

*3. $11x^2 + 10\sqrt{3}xy + y^2 - 4 = 0$.

4. $52x^2 - 72xy + 73y^2 - 100 = 0$.

*5. $x^2 - 2xy + y^2 + x + y = 0$.

6. $3x^2 + 2\sqrt{3}xy + y^2 - 2x + 2\sqrt{3}y = 0$.

*7. $x^2 + 2\sqrt{3}xy + 3y^2 + 2\sqrt{3}x - 2y = 0$.

8. $2x^2 + 4\sqrt{3}xy + 6y^2 + (8 - \sqrt{3})x + (8\sqrt{3} + 1)y + 8 = 0$.

For each of the equations below, find a rotation $\alpha \in (-\frac{1}{4}\pi, \frac{1}{4}\pi]$ that eliminates the xy-term. Then find $\cos \alpha$ and $\sin \alpha$.

*9. $x^2 + xy + Kx + Ly + M = 0$.

10. $5x^2 + 24xy + 12y^2 + Kx + Ly + M = 0$.

Supplement to Section 8.8

(Optional) The Second-Degree Equation

It is possible to draw general conclusions about the graph of a second-degree equation

$$ax^2 + bxy + cy^2 + dx + ey + f = 0 \qquad a, b, c \text{ not all } 0,$$

just from its *discriminant*

$$\Delta = b^2 - 4ac.$$

There are three cases.

CASE 1.   If $\Delta < 0$, the graph is an ellipse, a circle, a point, or empty.

CASE 2.   If $\Delta > 0$, the graph is a hyperbola or a pair of intersecting lines.

CASE 3.   If $\Delta = 0$, the graph is a parabola, a line, a pair of lines, or empty.

Below we outline how these assertions may be verified. A useful first step is to rotate the coordinate system so that the equation takes the form

(1) $$AX^2 + CY^2 + DX + EY + F = 0.$$

An elementary but time-consuming computation shows that the discriminant is unchanged by a rotation, so that in this instance we have

$$\Delta = b^2 - 4ac = -4AC.$$

Moreover, A and C cannot both be zero. If $\Delta < 0$, then $AC > 0$ and we can rewrite (1) as

$$\frac{X^2}{C} + \frac{D}{AC}X + \frac{Y^2}{A} + \frac{EY}{AC} + \frac{F}{AC} = 0.$$

By completing the squares, we obtain an equation of the form

$$\frac{(X - \alpha)^2}{(\sqrt{|C|})^2} + \frac{(Y - \beta)^2}{(\sqrt{|A|})^2} = K.$$

If $K > 0$, we have an ellipse or a circle. If $K = 0$, we have the point (α, β). If $K < 0$, the set is empty.

If $\Delta > 0$, then $AC < 0$. Proceeding as before, we obtain an equation of the form

$$\frac{(X - \alpha)^2}{(\sqrt{|C|})^2} - \frac{(Y - \beta)^2}{(\sqrt{|A|})^2} = K.$$

If $K \neq 0$, we have a hyperbola. If $K = 0$, the equation becomes

$$\left(\frac{X - \alpha}{\sqrt{|C|}} - \frac{Y - \beta}{\sqrt{|A|}}\right)\left(\frac{X - \alpha}{\sqrt{|C|}} + \frac{Y - \beta}{\sqrt{|A|}}\right) = 0,$$

so that we have a pair of lines intersecting at the point (α, β).

If $\Delta = 0$, then $AC = 0$, so that either $A = 0$ or $C = 0$. Since A and C are not both zero, there is no loss in generality in assuming that $A \neq 0$ and $C = 0$. In this case Equation (1) reduces to

$$AX^2 + DX + EY + F = 0.$$

Dividing by A, we obtain an equation of the form

$$X^2 + HX + IY + J = 0.$$

Completion of the square gives us

$$(X - \alpha)^2 = \beta Y + K.$$

If $\beta \neq 0$, we have a parabola. If $\beta = 0$ and $K = 0$, we have a line. If $\beta = 0$ and $K > 0$, we have a pair of parallel lines. If $\beta = 0$ and $K < 0$, the set is empty. $\square$

Applications of the Definite Integral

9.1 The Average of a Continuous Function

Take a function f continuous on an interval $[a, b]$. What number should we call the *average* of f on $[a, b]$?

We could define the average as

$$\tfrac{1}{2}[f(a) + f(b)],$$

but that would take into consideration only what happens at the endpoints. We want a notion of average that takes into consideration what happens throughout the interval $[a, b]$.

To see how to proceed, let's look at averages in a more familiar setting. Think of a student taking tests. During the first marking period he takes N_1 tests, during the next marking period N_2 tests, and so on, until the last marking period, when he takes N_n tests.

Now let N be the total number of tests taken by the student:

$$N = N_1 + \cdots + N_n.$$

If we give equal weight to each of the tests, then the student's average has the following two properties:

(i) lowest of N grades $\leq$ average on N tests $\leq$ highest of N grades;

(ii) $\left(\begin{array}{c} \text{average on} \\ N \text{ tests} \end{array}\right) \cdot N = \left(\begin{array}{c} \text{average on} \\ \text{first } N_1 \text{ tests} \end{array}\right) \cdot N_1 + \cdots + \left(\begin{array}{c} \text{average on} \\ \text{final } N_n \text{ tests} \end{array}\right) \cdot N_n.$

Property (i) is obvious. Property (ii) holds since each side of the equation represents the total number of points scored.

Let's return now to the function f. In view of (i) and (ii), we make two requirements of the number which we call the average of f:

Requirement 1

Minimum value of $f \leq$ average of $f \leq$ maximum value of f.

Requirement 2

If $P = \{x_0, x_1, \ldots, x_n\}$ is a partition of $[a, b]$, then

$$\begin{pmatrix} \text{average of } f \\ \text{on } [a, b] \end{pmatrix} (b - a) = \begin{pmatrix} \text{average of } f \\ \text{on } [x_0, x_1] \end{pmatrix} \Delta x_1 + \cdots + \begin{pmatrix} \text{average of } f \\ \text{on } [x_{n-1}, x_n] \end{pmatrix} \Delta x_n.$$

To meet Requirements 1 and 2, there is only one way to define average, and that is to set

$$\boxed{\text{average of } f \text{ on } [a, b] = \frac{1}{b - a} \int_a^b f(x)\, dx.}$$

PROOF.   Take an arbitrary partition

$$P = \{x_0, x_1, \ldots, x_n\} \quad \text{of} \quad [a, b].$$

For each subinterval $[x_{i-1}, x_i]$, Requirement 1 gives

$$m_i \leq \text{average of } f \text{ on } [x_{i-1}, x_i] \leq M_i,$$

where, as usual,

$$m_i = \text{minimum value of } f \text{ on } [x_{i-1}, x_i]$$

and

$$M_i = \text{maximum value of } f \text{ on } [x_{i-1}, x_i].$$

Multiplying by Δx_i, we have

$$m_i\, \Delta x_i \leq \begin{pmatrix} \text{average of } f \\ \text{on } [x_{i-1}, x_i] \end{pmatrix} \Delta x_i \leq M_i\, \Delta x_i.$$

Summing from $i = 1$ to $i = n$, we get

$$L_f(P) \leq \begin{pmatrix} \text{average of } f \\ \text{on } [x_0, x_1] \end{pmatrix} \Delta x_1 + \cdots + \begin{pmatrix} \text{average of } f \\ \text{on } [x_{n-1}, x_n] \end{pmatrix} \Delta x_n \leq U_f(P).\dagger$$

---

† $L_f(P)$ and $U_f(P)$ are the lower and upper sums discussed in Section 4.2.

Requirement 2 now gives

$$L_f(P) \leq \left(\begin{array}{c}\text{average of } f \\ \text{on } [a, b]\end{array}\right)(b - a) \leq U_f(P).$$

Since P was chosen arbitrarily, this inequality must hold for all partitions P.

There is only one number with this property and that is the integral of f on $[a, b]$. (See Section 4.2.) Thus we must have

$$\left(\begin{array}{c}\text{average of } f \\ \text{on } [a, b]\end{array}\right)(b - a) = \int_a^b f(x)\, dx$$

and

$$\text{average of } f \text{ on } [a, b] = \frac{1}{b - a} \int_a^b f(x)\, dx. \quad \Box$$

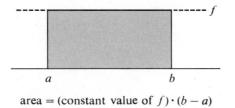

area = (constant value of f) · $(b - a)$

FIGURE 9.1.1

This sense of average provides a powerful and intuitive way of viewing the definite integral. Think for a moment in terms of area. If f is constant and positive on $[a, b]$, then Ω, the region below the graph, is a rectangle. Its area is given simply by the formula

area of Ω = (constant value of f on $[a, b]$) · $(b - a)$. (Figure 9.1.1)

If f is now allowed to vary continuously on $[a, b]$, then we have

$$\text{area of } \Omega = \int_a^b f(x)\, dx.$$

In terms of average values, the formula becomes

area of Ω = (average value of f on $[a, b]$) · $(b - a)$.   (Figure 9.1.2)

Think now in terms of motion. If an object moves along a line with constant speed $|v|$ during the time interval $[a, b]$, then

the distance traveled = (the constant value of $|v|$ on $[a, b]$) · $(b - a)$.

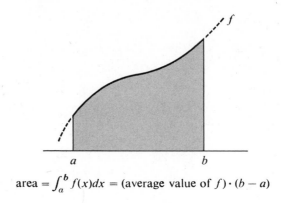

$$\text{area} = \int_a^b f(x)dx = \text{(average value of } f) \cdot (b - a)$$

FIGURE 9.1.2

If the speed $|v|$ varies, then we have

$$\text{distance traveled} = \int_a^b |v(t)|\ dt$$

and the formula reads

distance traveled = (average value of $|v|$ on $[a, b]$) $\cdot$ $(b - a)$.

In general we have the identity

$$\int_a^b f(x)\ dx = \textit{(average value of } f \textit{ on } [a, b]) \cdot (b - a).$$

Theorem.   *The First Mean-Value Theorem for Integrals*

If f is continuous on $[a, b]$, then there is a number c in (a, b) such that

$$\int_a^b f(x)\ dx = f(c)(b - a).$$

In other words, there is a number c in (a, b) at which f takes on its average value.

PROOF.   Observe that

min value of f on $[a, b] \leq$ average value of f on $[a, b] \leq$ max value of f on $[a, b]$

and apply the intermediate-value theorem.   □

To see this theorem geometrically, look at Figure 9.1.3. Here we are assuming that f is positive. The theorem says that somewhere in (a, b) there exists a number c such that the rectangle displayed has the same area as the region under the curve.

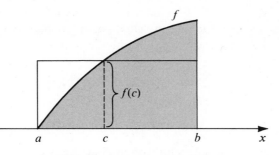

$f(c)$ is the average value of f:
the area of the rectangle = the area under the curve

FIGURE 9.1.3

Exercises

Find the average value on $[0, a]$.

*1. $f(x) = \alpha x + \beta$.

2. $f(x) = x^2$.

*3. $f(x) = x^3$.

4. $f(x) = e^x$.

Find the average value.

*5. $f(x) = \dfrac{1}{x}, \quad x \in [1, a]$.

6. $f(x) = \dfrac{a}{a^2 + x^2}, \quad x \in [0, a]$.

*7. $f(x) = \sqrt{a^2 - x^2}, \quad x \in [-a, a]$.

8. $f(x) = \tan \frac{1}{4}\pi x, \quad x \in [0, a]$.

*9. What are the averages of the sine and cosine on an interval of length 2π?

10. Find the average value of $\sin^2 x$ between $x = 0$ and $x = \pi$. (This average is frequently used in the theory of alternating currents.)

11. Given that f is continuous on $[a, b]$, compare

$$f(b)(b - a) \quad \text{and} \quad \int_a^b f(x)\, dx$$

if (a) f is constant on $[a, b]$; (b) f is increasing on $[a, b]$; (c) f is decreasing on $[a, b]$.

12. In Chapter 3, we viewed $[f(b) - f(a)]/(b - a)$ as the average rate of change of f on $[a, b]$ and $f'(x)$ as the rate of change at x. If our new sense of average is to be consistent with the old one, we must have

$$\frac{f(b) - f(a)}{b - a} = \text{average of } f' \text{ on } [a, b].$$

Prove that this is the case.

13. Show that the average slope of the logarithm curve from $x = a$ to $x = b$ is

$$\frac{1}{b - a} \log\left(\frac{a}{b}\right).$$

*14. A stone falls from rest in a vacuum for x seconds. (a) Compare its terminal velocity to its average velocity; (b) compare its average velocity during the first $x/2$ seconds to its average velocity during the next $x/2$ seconds.

15. Show that

$$\text{the average of } f \text{ on } [a, b]$$

need not be

$$\sqrt{\text{the average of } f^2 \text{ on } [a, b]}.$$

*16. Find the average vertical and horizontal dimensions of the ellipse

$$\frac{x^2}{a^2} + \frac{y^2}{b^2} = 1.$$

17. Consider the circle

$$x^2 + y^2 = r^2.$$

Find the average of y in the first quadrant
(a) when y is expressed as a function of x.
(b) when y is expressed as a function of the polar angle θ.

18. Prove that two distinct continuous functions cannot have the same average on every interval.

19. (*Optional*) Let f be continuous on $[a, b]$. Let $a < c < b$. Prove that

$$f(c) = \lim_{h \downarrow 0} (\text{average of } f \text{ on } [c - h, c + h]).$$

9.2   More on Area

We begin with two functions f and g both continuous on an interval $[a, b]$. If f and g are both positive and if $f(x) \geq g(x)$ for all x in $[a, b]$ (see Figure 9.2.1), then the area between the graphs is given by the integral

$$\int_a^b [f(x) - g(x)] \, dx.$$

This you already know.

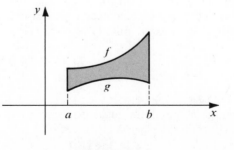

FIGURE 9.2.1

What we want now is a formula which applies also to the region depicted in Figure 9.2.2. Here neither f nor g remains positive. Moreover, neither function remains greater than the other.

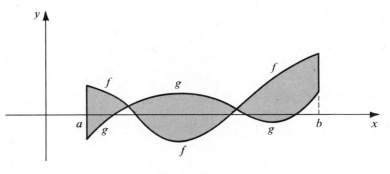

FIGURE 9.2.2

One way to proceed is to raise the entire region a fixed number of units so that both functions become positive. (See Figure 9.2.3.) The new boundaries are now of the form

$$F(x) = f(x) + C \quad \text{and} \quad G(x) = g(x) + C.$$

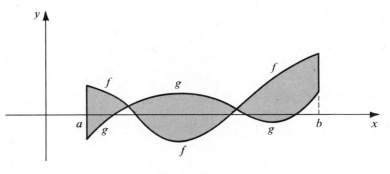

FIGURE 9.2.3

The area of the first part is

$$\int_a^s [F(x) - G(x)] \, dx = \int_a^s [f(x) - g(x)] \, dx = \int_a^s |f(x) - g(x)| \, dx;$$

the area of the second part is

$$\int_s^t [G(x) - F(x)] \, dx = \int_s^t [g(x) - f(x)] \, dx = \int_s^t |f(x) - g(x)| \, dx;$$

the area of the third part is

$$\int_t^b [F(x) - G(x)] \, dx = \int_t^b [f(x) - g(x)] \, dx = \int_t^b |f(x) - g(x)| \, dx.$$

The total area is therefore the sum

$$\int_a^s |f(x) - g(x)| \, dx + \int_s^t |f(x) - g(x)| \, dx + \int_t^b |f(x) - g(x)| \, dx.$$

By the additivity of the integral, this sum is simply

$$\int_a^b |f(x) - g(x)| \, dx.$$

In short,

$$\boxed{\text{the area between the graphs} = \int_a^b |f(x) - g(x)| \, dx.}$$

The number $|f(x) - g(x)|$ gives the vertical separation between the two boundaries:

$$\boxed{\text{the area} = (\text{the average vertical separation}) \cdot (b - a).}$$

Problem.   Find the area between

$$y = \cos x \quad \text{and} \quad y = \sin x, \qquad x \in [0, 2\pi].$$

SOLUTION.   The region in question is displayed in Figure 9.2.4. The area is given by the integral

$$\int_0^{2\pi} |\cos x - \sin x| \, dx.$$

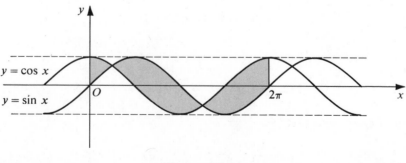

FIGURE 9.2.4

The easiest way to evaluate this integral is to note that

$$\cos x - \sin x \text{ is } \begin{cases} \geq 0, & 0 \leq x \leq \frac{1}{4}\pi \\ \leq 0, & \frac{1}{4}\pi \leq x \leq \frac{5}{4}\pi \\ \geq 0, & \frac{5}{4}\pi \leq x \leq 2\pi \end{cases}.$$

This means that

$$\text{area} = \int_0^{\pi/4} (\cos x - \sin x)\, dx + \int_{\pi/4}^{5\pi/4} (\sin x - \cos x)\, dx + \int_{5\pi/4}^{2\pi} (\cos x - \sin x)\, dx$$

$$= \Big[\sin x + \cos x \Big]_0^{\pi/4} + \Big[-\cos x - \sin x \Big]_{\pi/4}^{5\pi/4} + \Big[\sin x + \cos x \Big]_{5\pi/4}^{2\pi}$$

$$= (\sqrt{2} - 1) + 2\sqrt{2} + (1 + \sqrt{2}) = 4\sqrt{2}. \quad \square$$

In Figures 9.2.5 and 9.2.6 we display regions in which the boundary curves are not functions of x but functions of y instead. In such cases the area is given by the integral

$$\int_c^d |f(y) - g(y)|\, dy.$$

This is simply

$$(\text{the average horizontal separation}) \cdot (d - c).$$

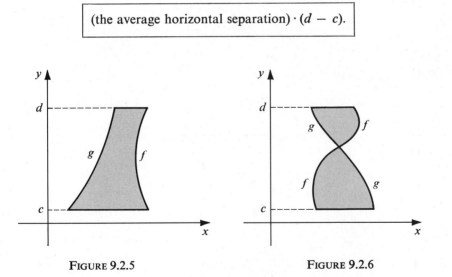

FIGURE 9.2.5   FIGURE 9.2.6

Problem. Find the area between the two parabolas

$$x = y^2 \quad \text{and} \quad x = -2y^2 + 3.$$

SOLUTION. The region is displayed in Figure 9.2.7. The two parabolas intersect at $y = -1$ and $y = 1$. For each y between -1 and 1 the horizontal separation between the two curves is

$$|(-2y^2 + 3) - y^2| = |-3y^2 + 3| = -3y^2 + 3.$$

We can find the area of the region by integrating from $y = -1$ to $y = 1$:

$$A = \int_{-1}^{1} (-3y^2 + 3)\, dy = 4. \quad \square$$

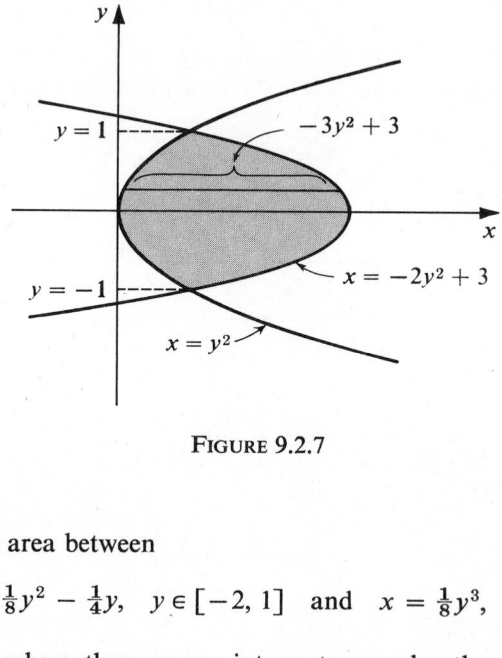

FIGURE 9.2.7

Problem. Find the area between

$$x = \tfrac{1}{4}y^3 + \tfrac{1}{8}y^2 - \tfrac{1}{4}y, \quad y \in [-2, 1] \quad \text{and} \quad x = \tfrac{1}{8}y^3, \quad y \in [-2, 1].$$

SOLUTION. To see where these curves intersect, we solve the two equations simultaneously.

$$\tfrac{1}{4}y^3 + \tfrac{1}{8}y^2 - \tfrac{1}{4}y = \tfrac{1}{8}y^3,$$
$$\tfrac{1}{8}y^3 + \tfrac{1}{8}y^2 - \tfrac{1}{4}y = 0,$$
$$y^3 + y^2 - 2y = 0,$$
$$y(y^2 + y - 2) = 0,$$
$$y(y - 1)(y + 2) = 0.$$

The two curves intersect at

$$y = -2, \quad y = 0, \quad y = 1.$$

The region between the two curves is displayed in Figure 9.2.8.

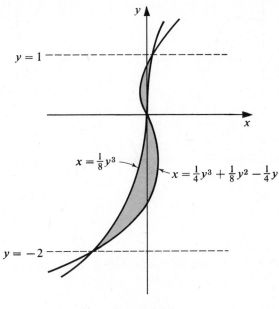

$$y = 1$$

$$x = \tfrac{1}{8}y^3$$

$$x = \tfrac{1}{4}y^3 + \tfrac{1}{8}y^2 - \tfrac{1}{4}y$$

$$y = -2$$

FIGURE 9.2.8

To find the area, we integrate the horizontal separation

$$|(\tfrac{1}{4}y^3 + \tfrac{1}{8}y^2 - \tfrac{1}{4}y) - (\tfrac{1}{8}y^3)| = |\tfrac{1}{8}y^3 + \tfrac{1}{8}y^2 - \tfrac{1}{4}y|$$

from $y = -2$ to $y = 1$. Note that

$$\tfrac{1}{8}y^3 + \tfrac{1}{8}y^2 - \tfrac{1}{4}y = \tfrac{1}{8}(y^3 + y^2 - 2y) = \tfrac{1}{8}y(y-1)(y+2)$$

is positive between $y = -2$ and $y = 0$, and is negative between $y = 0$ and $y = 1$.
It follows that

$$A = \int_{-2}^{1} |\tfrac{1}{8}y^3 + \tfrac{1}{8}y^2 - \tfrac{1}{4}y|\, dy$$

$$= \int_{-2}^{0} (\tfrac{1}{8}y^3 + \tfrac{1}{8}y^2 - \tfrac{1}{4}y)\, dy + \int_{0}^{1} -(\tfrac{1}{8}y^3 + \tfrac{1}{8}y^2 - \tfrac{1}{4}y)\, dy$$

$$= \tfrac{1}{3} + \tfrac{5}{96} = \tfrac{37}{96}. \quad \square$$

You can sketch a curve

$$x = f(y)$$

by first drawing

$$y = f(x) \qquad \qquad \text{(Figure 9.2.9)}$$

and then reflecting what you have in the line $y = x$. (Figure 9.2.10)

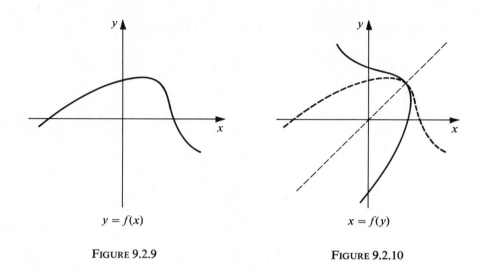

$$y = f(x)$$

FIGURE 9.2.9

$$x = f(y)$$

FIGURE 9.2.10

Exercises

Sketch the region that is bounded by the following curves and find the area.

* 1. $y = x^2$, $y = 2x + 3$.
  2. $y^2 - 27x = 0$, $x + y = 0$.
* 3. $x - y^2 + 3 = 0$, $x - 2y = 0$.
  4. $y^2 = 2x$, $x - y = 4$.
* 5. $x^3 - 10y^2 = 0$, $x - y = 0$.
  6. $a^2 x = a^2 y - y^2$, $4x - y = 0$.
* 7. $3y - 3x + x^2 = 0$, $x - 2y = 0$.
  8. $x + y^2 - 4 = 0$, $x + y = 2$.
* 9. $x + y - y^3 = 0$, $x - y + y^2 = 0$.
  10. $xy = 9$, $\sqrt{x} + \sqrt{y} = 4$.

* 11. Find the area enclosed by the ellipse

$$\frac{x^2}{a^2} + \frac{y^2}{b^2} = 1.$$

12. Find the area of the region that lies between the right-hand branch of the hyperbola

$$\frac{x^2}{a^2} - \frac{y^2}{b^2} = 1$$

and the vertical line

$$x = 2a.$$

* 13. Find the area between

$$y = \sin 2x, \quad x \in [0, \pi] \quad \text{and} \quad y = \sin x, \quad x \in [0, \pi].$$

14. Find the area between

$$y = \sin^2 x, \quad x \in [0, 2\pi] \quad \text{and} \quad y = \cos^2 x, \quad x \in [0, 2\pi].$$

*15. Find the area between

$$x = y^3 + 4y^2 - 4y + 1, \quad y \in [-1, 2] \quad \text{and} \quad x = 4y^2 + 1, \quad y \in [-1, 2].$$

16. Find the area between

$$x = y^3 - 3y^2 + 2y + 2, \quad y \in [0, 3] \quad \text{and} \quad x = 2y^2 - 4y + 2, \quad y \in [0, 3].$$

9.3 Volume: Parallel Cross Sections

We begin with a solid through which we pass a coordinate line l. (Figure 9.3.1) As in the figure we suppose that the solid lies entirely between $x = a$ and $x = b$. By $A(x)$ we mean the area of the cross section that has coordinate x.

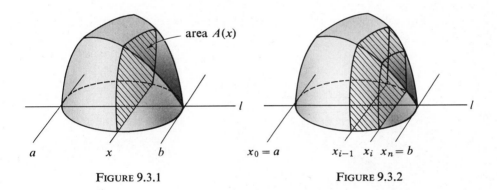

FIGURE 9.3.1   **FIGURE 9.3.2**

If the cross-sectional area, $A(x)$, varies continuously with x, then we can find the volume of the solid by integrating $A(x)$ from a to b:

(9.3.1)
$$V = \int_a^b A(x)\, dx.$$

PROOF.   Let S be the solid in question and let $P = \{x_0, x_1, \ldots, x_n\}$ be a partition of $[a, b]$. For each subinterval $[x_{i-1}, x_i]$ let m_i and M_i be respectively the minimum and maximum values of A on $[x_{i-1}, x_i]$. Let S_i be that portion of the solid (Figure 9.3.2) that corresponds to the interval $[x_{i-1}, x_i]$ and let V_i be its volume. S_i contains a solid of cross-sectional area m_i and thickness Δx_i, and is contained in a solid of cross-sectional area M_i and thickness Δx_i. This suggests that

$$m_i \Delta x_i \leq V_i \leq M_i \Delta x_i.$$

Summing these inequalities from $i = 1$ to $i = n$, we get

(1)
$$L_A(P) \le V_1 + \cdots + V_n \le U_A(P).$$

Since

$$S = S_1 \cup S_2 \cup \cdots \cup S_n$$

and the S_i do not overlap, V, the total volume of S, must satisfy

$$V = V_1 + \cdots + V_n.$$

Inequality (1) now implies that

$$L_A(P) \le V \le U_A(P).$$

Since this last inequality must hold for every partition P, we must have

$$V = \int_a^b A(x)\, dx. \quad \square$$

REMARK.   For solids of constant cross-sectional area (such as those of Figure 9.3.3),

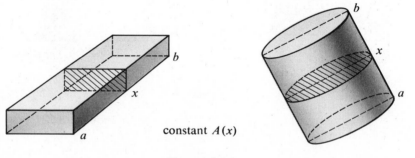

constant $A(x)$

FIGURE 9.3.3

the volume is simply the cross-sectional area times the thickness:

$$V = (\text{constant cross-sectional area}) \cdot (b - a).$$

When the cross-sectional area varies we have the formula

$$V = \int_a^b A(x)\, dx,$$

which says:

$$\boxed{V = (\text{average cross-sectional area}) \cdot (b - a).}$$

Problem.   Find the volume of a cone that has base radius r and height h.

SOLUTION.   As a basis for the coordinate system we choose the line that passes through the vertex of the cone and the center of the base. (See Figure 9.3.4.) The cross section

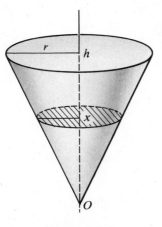

FIGURE 9.3.4

with coordinate x is a circular disk. Its radius $r(x)$ satisfies the equation

$$\frac{r(x)}{x} = \frac{r}{h}.$$

This you can see by similar triangles. Since

$$A(x) = \pi[r(x)]^2 \quad \text{and} \quad r(x) = \frac{r}{h}\,x,$$

we have

$$A(x) = \frac{\pi r^2}{h^2} \cdot x^2.$$

The volume of the cone is thus

$$\int_0^h \frac{\pi r^2}{h^2}\, x^2\, dx = \frac{\pi r^2}{h^2}\left[\frac{x^3}{3}\right]_0^h = \frac{1}{3}\,\pi r^2 h. \quad \square$$

The next result is more general and contains the cone as a special instance.

Solids of Revolution

Suppose that f is nonnegative and continuous on $[a, b]$. (See Figure 9.3.5.) If we revolve the region below the graph of f about the x-axis, we obtain a solid. The volume of this solid is given by the formula

(9.3.2)

$$V = \int_a^b \pi[f(x)]^2\, dx.$$

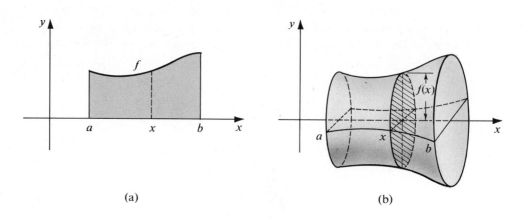

FIGURE 9.3.5

PROOF.   The cross section with coordinate x is a circular disk of radius $f(x)$. The cross-sectional area is thus

$$\pi[f(x)]^2.$$

The result follows from Formula 9.3.1.   ☐

As special instances of such solids of revolution we have the cone and the sphere.

Formula (9.3.2) Applied to the Cone

To generate a cone of base radius r and height h, we can take

$$f(x) = \frac{r}{h}\, x, \qquad 0 \le x \le h$$

and revolve the region below the graph about the x-axis. Formula (9.3.2) then gives

$$V = \pi \int_0^h \frac{r^2}{h^2}\, x^2 \, dx = \tfrac{1}{3}\pi r^2 h. \quad ☐$$

Formula (9.3.2) Applied to the Sphere

A sphere of radius r can be obtained by revolving the region below the graph of

$$f(x) = \sqrt{r^2 - x^2}, \qquad -r \le x \le r$$

about the x-axis. Consequently, we have

$$\text{volume of sphere} = \pi \int_{-r}^{r} (r^2 - x^2)\, dx = \pi \left[r^2 x - \tfrac{1}{3}x^3 \right]_{-r}^{r} = \tfrac{4}{3}\pi r^3.$$

Incidentally, this result was obtained by Archimedes in the third century B.C.   ☐

Problem.   Find the volume of a triangular pyramid with base area B and height h.

SOLUTION.   As in Figure 9.3.6 we choose the coordinate line so that it passes through the apex and is perpendicular to the base. To the apex we assign coordinate 0, and to the base point coordinate h. The cross section at distance x from the apex ($0 \le x \le h$) is a triangle similar to the base triangle. The corresponding sides of these

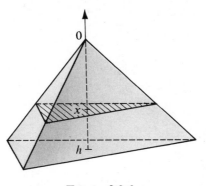

FIGURE 9.3.6

triangles are proportional, the factor of proportionality being x/h. This means that the areas are also proportional, but now the factor of proportionality is $(x/h)^2$. Thus we have

$$A(x) = \frac{Bx^2}{h^2},$$

and

$$V = \int_0^h A(x)\,dx = \frac{B}{h^2} \int_0^h x^2\,dx = \frac{1}{3}\,Bh.$$

This formula too was discovered in ancient times. It was known to Eudoxos of Cnidos (408–355 B.C.).   □

Problem.   The base of a solid is the region bounded by the ellipse

$$\frac{x^2}{a^2} + \frac{y^2}{b^2} = 1.$$

Find the volume of the solid given that the cross sections perpendicular to the x-axis are equilateral triangles.

SOLUTION.   Take x as in Figure 9.3.7. The cross section with coordinate x is an equilateral triangle with side $\overline{PQ}$. The equation of the ellipse can be rewritten as

$$y^2 = \frac{b^2}{a^2}\,(a^2 - x^2).$$

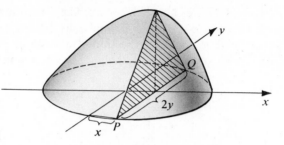

$$\text{FIGURE 9.3.7}$$

Since

$$\text{length of } \overline{PQ} = 2y = \frac{2b}{a} \sqrt{a^2 - x^2},$$

the equilateral triangle has area

$$A(x) = \frac{\sqrt{3}b^2}{a^2} (a^2 - x^2).†$$

To find the volume we integrate $A(x)$ from $-a$ to a:

$$V = \int_{-a}^{a} A(x)\, dx = \frac{\sqrt{3}b^2}{a^2} \int_{-a}^{a} (a^2 - x^2)\, dx$$

$$= \frac{\sqrt{3}b^2}{a^2} \left[a^2 x - \frac{x^3}{3} \right]_{-a}^{a}$$

$$= \frac{\sqrt{3}b^2}{a^2} \left(\frac{4}{3} a^3 \right) = \frac{4}{3} \sqrt{3}\, ab^2. \quad \square$$

Problem.   The base of a solid is the region between the parabolas

$$x = y^2 \quad \text{and} \quad x = -2y^2 + 3.$$

Find the volume of the solid if the cross sections perpendicular to the x-axis are squares.

SOLUTION.   The two parabolas intersect at $x = 1$. (See Figure 9.3.8.) From $x = 0$ to $x = 1$, the x cross section has area

$$A(x) = (2y)^2 = 4y^2 = 4x.$$

The volume from $x = 0$ to $x = 1$ is

$$V_1 = \int_0^1 4x\, dx = 4 \left[\tfrac{1}{2} x^2 \right]_0^1 = 2.$$

---

† In general, the area of an equilateral triangle is $\tfrac{1}{4}\sqrt{3} s^2$, where s is the length of a side.

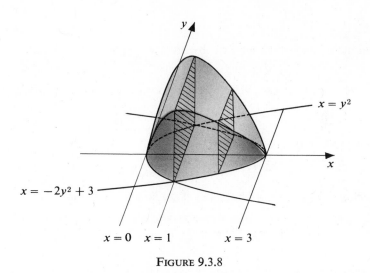

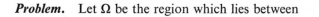

FIGURE 9.3.8

From $x = 1$ to $x = 3$, the x cross section has area

$$A(x) = (2y)^2 = 4y^2 = 2(3 - x).$$

(Here we are measuring the span across the second parabola $x = -2y^2 + 3$.)
The volume from $x = 1$ to $x = 3$ is

$$V_2 = \int_1^3 2(3 - x)\, dx = 2\left[-\tfrac{1}{2}(3 - x)^2\right]_1^3 = 4.$$

The total volume is

$$V = V_1 + V_2 = 6. \quad \square$$

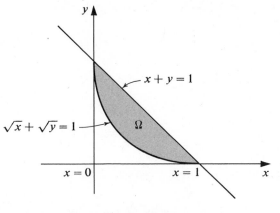

FIGURE 9.3.9

Problem.   Let Ω be the region which lies between

$$\sqrt{x} + \sqrt{y} = 1 \quad \text{and} \quad x + y = 1.$$

Find the volume of the solid generated by revolving Ω about the x-axis.

SOLUTION.   We refer to Figure 9.3.9. In terms of x, the upper boundary becomes

$$y = 1 - x, \qquad 0 \le x \le 1$$

and the lower boundary

$$y = (1 - \sqrt{x})^2, \qquad 0 \le x \le 1.$$

The solid enclosed by revolving the upper boundary has volume

$$V_1 = \int_0^1 \pi y^2 \, dx = \pi \int_0^1 (1 - x)^2 \, dx = \tfrac{1}{3}\pi.$$

The solid enclosed by revolving the lower boundary has volume

$$V_2 = \int_0^1 \pi y^2 \, dx$$

$$= \pi \int_0^1 (1 - \sqrt{x})^4 \, dx$$

$$= \pi \int_0^1 (1 - 4x^{1/2} + 6x - 4x^{3/2} + x^2) \, dx$$

$$= \tfrac{1}{15}\pi.$$

The volume of the solid generated by Ω is the difference

$$V = V_1 - V_2 = \tfrac{1}{3}\pi - \tfrac{1}{15}\pi = \tfrac{4}{15}\pi. \quad \square$$

Problem.   Find the volume if the region OAB of Figure 9.3.10 is revolved about AB.

SOLUTION.   The curve OB given by

$$y = x^2, \qquad 0 \le x \le 2$$

is also given by

$$x = \sqrt{y}, \qquad 0 \le y \le 4. \qquad \text{(Figure 9.3.11)}$$

The horizontal cross section at height y is a circular disk of radius $2 - \sqrt{y}$. The area of this disk is

$$A(y) = \pi(2 - \sqrt{y})^2 = \pi(4 - 4\sqrt{y} + y).$$

We can find the volume we want by integrating $A(y)$ from $y = 0$ to $y = 4$:

$$V = \int_0^4 A(y) \, dy = \int_0^4 \pi(4 - 4\sqrt{y} + y) \, dy = \tfrac{8}{3}\pi. \quad \square$$

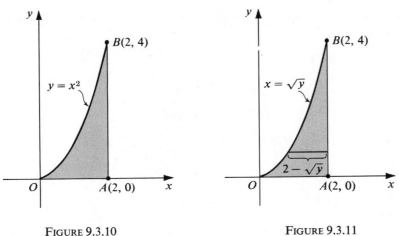

FIGURE 9.3.10   FIGURE 9.3.11

Exercises

Sketch the graph and find the volume generated by revolving the region below it about the x-axis.

*1. $f(x) = x^2$, $x \in [0, 1]$.

2. $f(x) = x^3$, $x \in [1, 2]$.

*3. $f(x) = e^x$, $x \in [0, 1]$.

4. $f(x) = \sin x$, $x \in [0, \pi]$.

*5. $f(x) = x^{2/3}$, $x \in [0, 2]$.

6. $f(x) = \sec \frac{1}{2}\pi x$, $x \in [-\frac{1}{2}, \frac{1}{2}]$.

*7. $f(x) = 1/x$, $x \in [1, 2]$.

8. $f(x) = \log x$, $x \in [1, e]$.

Sketch the graphs and find the volume generated by revolving the region between them about the x-axis.

*9. $f(x) = \sqrt{x}$, $x \in [0, 1]$; $g(x) = x^2$, $x \in [0, 1]$.

10. $f(x) = \sqrt{1 - x^2}$, $g(x) = \frac{1}{2}$.

*11. $f(x) = \sin x$, $x \in [0, \pi]$; $g(x) = x$, $x \in [0, \pi]$.

12. Find the volume of the solid generated by revolving the ellipse

$$\frac{x^2}{a^2} + \frac{y^2}{b^2} = 1$$

about the x-axis.

*13. Find the volume of the solid generated by revolving the equilateral triangle with vertices

$$(0, 0), \quad (a, 0), \quad (\tfrac{1}{2}a, \tfrac{1}{2}\sqrt{3}a)$$

about the x-axis.

*14. Derive a formula for the volume of the frustum of a cone in terms of its height h, the lower base radius R, and the upper base radius r. (See Figure 9.3.12.)

15. The base of a solid is the circle

$$x^2 + y^2 = r^2.$$

Find the volume of the solid given that the cross sections perpendicular to the x-axis are

(a) squares.   (b) equilateral triangles.

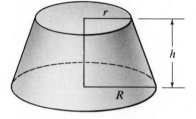

FIGURE 9.3.12

*16. The base of a solid is the region bounded by the ellipse

$$\frac{x^2}{a^2} + \frac{y^2}{b^2} = 1.$$

Find the volume of the solid, given that the cross sections perpendicular to the x-axis are (a) isosceles right triangles each with hypotenuse on the xy-plane, (b) squares, (c) triangles of height 2.

*17. A hemispheric basin of radius r feet is being used to store water. To what percent of capacity is it filled when the water is (a) $\frac{1}{2}r$ feet deep, (b) $\frac{1}{3}r$ feet deep?

18. A sphere of radius r is cut by two parallel planes: in the one case, a units above the equator; in the other case, b units above the equator. Assume that $a < b$ and find the volume of that portion of the sphere that lies between the two planes.

*19. The base of the solid is the region between the parabolas

$$x = y^2 \quad \text{and} \quad x = -2y^2 + 3.$$

Find the volume of the solid, if the cross sections perpendicular to the x-axis are, (a) rectangles of height h, (b) equilateral triangles, (c) isosceles right triangles each with hypotenuse on the xy-plane.

20. See Figure 9.3.13. Find the volume when the region

 *(a) OAB is revolved about the x-axis.

 *(b) OAB is revolved about AB.

 *(c) OAB is revolved about CA.

 *(d) OAB is revolved about the y-axis.

 (e) OAC is revolved about the y-axis.

 (f) OAC is revolved about CA.

 (g) OAC is revolved about AB.

 (h) OAC is revolved about the x-axis.

*21. Find the volume of the intersection of two cylinders of radius r if the axes meet at right angles.

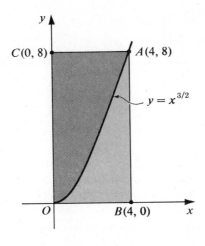

$C(0, 8)$

$A(4, 8)$

$y = x^{3/2}$

O

$B(4, 0)$

x

FIGURE 9.3.13

9.4   Volume: The Shell Method

First we have to go a little further into the theory of integration.

If f and g are both continuous on $[a, b]$, then so is their product. The definite integral of the product

$$\int_a^b f(x)g(x)\, dx$$

is then the only number I which satisfies the inequality

$$L_{fg}(P) \le I \le U_{fg}(P) \quad \text{for all partitions } P \text{ of } [a, b].$$

This you already know. For some applications (including the one we're after here) we need a slightly different way of getting to

$$\int_a^b f(x)g(x)\, dx.$$

Take two functions f and g which are continuous and nonnegative on $[a, b]$. Let $P = \{x_0, x_1, \ldots, x_n\}$ be a partition of $[a, b]$. For each subinterval $[x_{i-1}, x_i]$ let

$$M_i(f) = \text{max value of } f \text{ on } [x_{i-1}, x_i]$$
$$M_i(g) = \text{max value of } g \text{ on } [x_{i-1}, x_i]$$
$$m_i(f) = \text{min value of } f \text{ on } [x_{i-1}, x_i]$$
$$m_i(g) = \text{min value of } g \text{ on } [x_{i-1}, x_i].$$

Now form the sums

$$U_{f,g}^\star(P) = M_1(f)M_1(g)\,\Delta x_1 + M_2(f)M_2(g)\,\Delta x_2 + \cdots + M_n(f)M_n(g)\,\Delta x_n,$$
$$L_{f,g}^\star(P) = m_1(f)\,m_1(g)\,\Delta x_1 + m_2(f)\,m_2(g)\,\Delta x_2 + \cdots + m_n(f)\,m_n(g)\,\Delta x_n.$$

Although $U_{f,g}^\star(P)$ and $L_{f,g}^\star(P)$ are not the usual upper and lower sums,† in this setting they do play a similar role.

Theorem 9.4.1 (Bliss)‡

If f and g are continuous and nonnegative on $[a, b]$, then the definite integral

$$\int_a^b f(x)g(x)\, dx$$

is the only number I which satisfies the inequality

$$L_{f,g}^\star(P) \leq I \leq U_{f,g}^\star(P) \quad \text{for all partitions } P \text{ of } [a, b].$$

This result is a little too technical to be proved here. You can, however, find a proof in the Appendix at the end of the book. □

To describe the shell method of computing volumes, we begin with a solid cylinder of radius R and height h and from it we cut out a cylindrical core of radius r. (Figure 9.4.1)

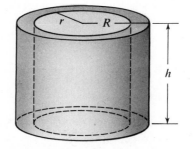

FIGURE 9.4.1

Since the original cylinder had volume $\pi R^2 h$ and the piece removed had volume $\pi r^2 h$, the cylindrical shell that remains has volume

(9.4.2) $\pi R^2 h - \pi r^2 h.$

We'll use this shortly.

---

† For example, in the case of $U_{f,g}^\star(P)$ the coefficient of Δx_i is the product

$$M_i(f)M_i(g) = \left(\begin{array}{c}\text{max value of } f \\ \text{on } [x_{i-1}, x_i]\end{array}\right) \cdot \left(\begin{array}{c}\text{max value of } g \\ \text{on } [x_{i-1}, x_i]\end{array}\right).$$

In the case of the usual upper sum $U_{fg}(P)$ the coefficient of Δx_i is

$$M_i(fg) = \text{max value of } fg \text{ on } [x_{i-1}, x_i].$$

These are in general not the same. (See Exercise 17.)

‡ This is a simplified version of a theorem first proved by the American mathematician G. A. Bliss (1876–1951).

Consider now a function f which is nonnegative on an interval $[a, b]$. For convenience assume that $a \geq 0$. If the region below the graph is revolved about the y-axis, then a solid S is generated. See Figure 9.4.2.

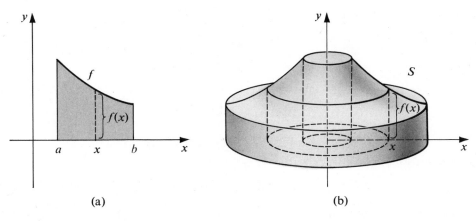

(a)   (b)

FIGURE 9.4.2

The volume of this solid is given by the formula

(9.4.3)
$$V = \int_a^b 2\pi x f(x) \, dx.$$

PROOF. We take a partition $P = \{x_0, x_1, \ldots, x_n\}$ of $[a, b]$ and concentrate on what's happening on the ith interval $[x_{i-1}, x_i]$.

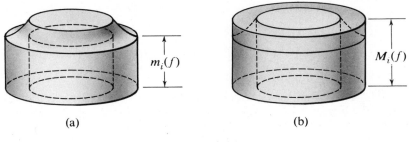

(a)   (b)

FIGURE 9.4.3

Let S_i be the solid generated by the ith interval and let V_i be its volume. S_i contains a cylindrical shell of height $m_i(f)$, outer radius x_i, and inner radius x_{i-1}. [part (a) of Figure 9.4.3] On the other hand, S_i is itself contained in a cylindrical shell of height $M_i(f)$, outer radius x_i, and inner radius x_{i-1}. [part (b) of Figure 9.4.3] It follows now (from 9.4.2) that

$$\pi x_i^2 m_i(f) - \pi x_{i-1}^2 m_i(f) \le V_i \le \pi x_i^2 M_i(f) - \pi x_{i-1}^2 M_i(f)$$

and therefore that

$$\pi m_i(f)[x_i^2 - x_{i-1}^2] \le V_i \le \pi M_i(f)[x_i^2 - x_{i-1}^2],$$
$$\pi m_i(f)[x_i + x_{i-1}][x_i - x_{i-1}] \le V_i \le \pi M_i(f)[x_i + x_{i-1}][x_i - x_{i-1}].$$

Note that

$$2x_{i-1} \le x_i + x_{i-1} \le 2x_i$$

and as usual set

$$x_i - x_{i-1} = \Delta x_i.$$

This gives

(1) $$m_i(f)[2\pi x_{i-1}] \, \Delta x_i \le V_i \le M_i(f)[2\pi x_i] \, \Delta x_i.$$

Since for the function $g(x) = 2\pi x$

$$2\pi x_{i-1} = m_i(g) \quad \text{and} \quad 2\pi x_i = M_i(g)$$

you can rewrite (1) to read

$$m_i(f)m_i(g) \, \Delta x_i \le V_i \le M_i(f)M_i(g) \, \Delta x_i.$$

If you sum these inequalities from $i = 1$ to $i = n$, you'll find that the total volume

$$V = V_1 + V_2 + \cdots + V_n$$

must satisfy the inequality

$$L_{f,g}^{*}(P) \le V \le U_{f,g}^{*}(P).$$

Since the partition P was chosen arbitrarily, the inequality must hold for all partitions P. It follows from Theorem 9.4.1 that

$$V = \int_a^b f(x)g(x) \, dx$$

and since $g(x) = 2\pi x$, that

$$V = \int_a^b f(x)[2\pi x] \, dx = \int_a^b 2\pi x f(x) \, dx. \quad \square$$

REMARK.   The product $2\pi x f(x)$ gives the lateral area of a cylinder of radius x and height $f(x)$. (See Figure 9.4.2.) You can therefore interpret the shell method formula

$$V = \int_a^b 2\pi x f(x) \, dx$$

as saying:

$$\boxed{\text{volume} = (\text{average lateral area}) \cdot (b - a).}$$

Problem.   Use the shell method to find a formula for the volume of a cone.

SOLUTION.   A cone of base radius r and height h can be obtained by revolving the region below the graph of

$$f(x) = -\frac{h}{r} x + h, \qquad 0 \le x \le r$$

about the y-axis. (You can convince yourself of this by drawing your own diagram.) The shell method then gives

$$V = \int_0^r 2\pi x \left(-\frac{h}{r} x + h \right) dx.$$

A short computation yields the familiar formula

$$V = \tfrac{1}{3}\pi r^2 h. \quad \square$$

Problem.   Use the shell method to find the volume of the ring generated by revolving the circular disk

$$(x - c)^2 + y^2 \le r^2, \qquad c > r$$

about the y-axis. (See Figure 9.4.4.)

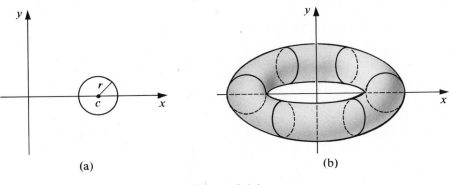

(a)   (b)

FIGURE 9.4.4

SOLUTION.   The upper half of the ring is generated by the region below the graph of

$$f(x) = \sqrt{r^2 - (x - c)^2}, \qquad c - r \le x \le c + r.$$

If we let V equal the volume of the entire ring, then

$$V = 2 \int_{c-r}^{c+r} 2\pi x \sqrt{r^2 - (x - c)^2} \; dx = 2\pi \int_{c-r}^{c+r} 2x\sqrt{r^2 - (x - c)^2} \; dx.$$

Since

$$\int_{c-r}^{c+r} 2x\sqrt{r^2 - (x - c)^2} \; dx$$

$$= \int_{c-r}^{c+r} 2(x - c)\sqrt{r^2 - (x - c)^2} \; dx + 2c \int_{c-r}^{c+r} \sqrt{r^2 - (x - c)^2} \; dx$$

and the first integral on the right is zero, we have

$$V = (2\pi c)\left(2\int_{c-r}^{c+r}\sqrt{r^2-(x-c)^2}\,dx\right) = (2\pi c)(\text{area of circle of radius } r)$$

$$= (2\pi c)(xr^2) = 2\pi^2 cr^2. \quad \square$$

Problem. Let Ω be the region which lies between

$$\sqrt{x} + \sqrt{y} = 1 \quad \text{and} \quad x + y = 1.$$

Find the volume of the solid generated by revolving Ω about the y-axis.

SOLUTION. We refer to Figure 9.3.9. In terms of x, the upper boundary becomes

$$y = 1 - x, \qquad 0 \le x \le 1$$

and the lower boundary

$$y = (1 - \sqrt{x})^2, \qquad 0 \le x \le 1.$$

The solid enclosed by revolving the upper boundary has volume

$$V_1 = \int_0^1 2\pi xy\,dx = \int_0^1 2\pi x(1-x)\,dx = \tfrac{1}{3}\pi.$$

The solid enclosed by revolving the lower boundary has volume

$$V_2 = \int_0^1 2\pi xy\,dx = \int_0^1 2\pi x(1-\sqrt{x})^2\,dx = \tfrac{1}{15}\pi.$$

The volume of the solid generated by Ω is the difference

$$V = V_1 - V_2 = \tfrac{1}{3}\pi - \tfrac{1}{15}\pi = \tfrac{4}{15}\pi. \quad \square$$

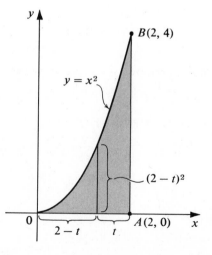

FIGURE 9.4.5

Problem.   Use the shell method to find the volume when the region OAB of Figure 9.3.10 is revolved about AB.

SOLUTION.   Introduce a new variable t as in Figure 9.4.5. The height of the curve t units from AB is $(2 - t)^2$. Since t ranges from $t = 0$ to $t = 2$,

$$V = \int_0^2 2\pi t (2 - t)^2 \, dt = 2\pi \int_0^2 (4t - 4t^2 + t^3) \, dt = \tfrac{8}{3}\pi. \quad \square$$

Exercises

Use the shell method to find the volume when the region below the graph is revolved about the y-axis.

*1. $f(x) = x^2, \quad x \in [0, 1]$.

2. $f(x) = x^3, \quad x \in [1, 2]$.

*3. $f(x) = e^x, \quad x \in [0, 1]$.

4. $f(x) = \sin x, \quad x \in [0, \pi]$.

*5. $f(x) = x^{2/3}, \quad x \in [0, 2]$.

6. $f(x) = \sin x^2, \quad x \in [0, \tfrac{1}{2}\sqrt{\pi}]$.

*7. $f(x) = 1/x, \quad x \in [1, 2]$.

8. $f(x) = e^{x^2}, \quad x \in [0, 1]$.

*9. $f(x) = \sqrt{x}, \quad x \in [0, 1]$.

10. $f(x) = \log x, \quad x \in [1, e]$.

Use the shell method to find the volume when the region between the graphs is revolved about the y-axis.

*11. $f(x) = \sqrt{x}, \quad x \in [0, 1]; \quad g(x) = x^2, \quad x \in [0, 1]$.

12. $f(x) = \sqrt{1 - x^2}, \quad g(x) = \tfrac{1}{2}$.

*13. $f(x) = \sin x, \quad x \in [0, \pi]; \quad g(x) = x, \quad x \in [0, \pi]$.

14. Use the shell method to find the volume enclosed when the ellipse

$$\frac{x^2}{a^2} + \frac{y^2}{b^2} = 1$$

is revolved about the y-axis.

*15. Use the shell method to find the volume enclosed when the equilateral triangle with vertices

$$(0, 0), \ (a, 0), \ (\tfrac{1}{2}a, \tfrac{1}{2}\sqrt{3}a)$$

is revolved about the y-axis.

16. The region between the x-axis and the graph of

$$y = \sin x, \qquad x \in [2n\pi, 2(n + 1)\pi]$$

is revolved about the y-axis. Find the volume of the solid generated.

17. Show that if f and g are continuous and nonnegative on $[x_{i-1}, x_i]$ then

$$M_i(fg) \leq M_i(f)M_i(g).$$

Show that this inequality does not necessarily hold if f and g are allowed to take on negative values.

9.5   Area in Polar Coordinates

Here we develop a technique for calculating the area of a region the boundary of which is given in polar coordinates.

As a start, we suppose that α and β are two real numbers with

$$\alpha < \beta \le \alpha + 2\pi.$$

We take ρ as a function continuous and nonnegative on $[\alpha, \beta]$ and ask for the area of the region Γ which is bounded by the rays

$$\theta = \alpha, \qquad \theta = \beta$$

and the curve

$$r = \rho(\theta).$$

Such a region is portrayed in Figure 9.5.1.

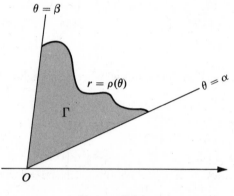

FIGURE 9.5.1

It's not hard to show that the area of Γ is given by the formula

(9.5.1)
$$A = \int_{\alpha}^{\beta} \tfrac{1}{2}[\rho(\theta)]^2 \, d\theta.$$

PROOF.   As usual we begin by partitioning the interval.   We take $P = \{\theta_0, \theta_1, \ldots, \theta_n\}$ as a partition of $[\alpha, \beta]$ and direct our attention to what happens between θ_{i-1} and θ_i.   We set

$$r_i = \text{min value of } \rho \text{ on } [\theta_{i-1}, \theta_i],$$
$$R_i = \text{max value of } \rho \text{ on } [\theta_{i-1}, \theta_i].$$

That portion of Γ which lies between θ_{i-1} and θ_i contains a circular sector of radius r_i and central angle $\Delta\theta_i = \theta_i - \theta_{i-1}$ and is itself contained in a circular sector of radius R_i and central angle $\Delta\theta_i = \theta_i - \theta_{i-1}$. (See Figure 9.5.2.) Its area A_i must therefore satisfy the inequality

$$\tfrac{1}{2}r_i^2 \, \Delta\theta_i \le A_i \le \tfrac{1}{2}R_i^2 \, \Delta\theta_i.\dagger$$

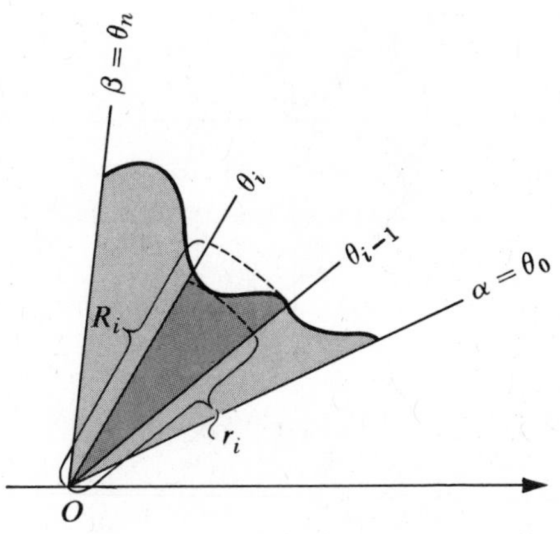

FIGURE 9.5.2

By summing these inequalities from $i = 1$ to $i = n$, you can see that the total area A must satisfy the inequality

(1) $$L_f(P) \le A \le U_f(P)$$

with

$$f(\theta) = \tfrac{1}{2}[\rho(\theta)]^2.$$

Since (1) must hold for every partition P of $[a, b]$, we must have

$$A = \int_\alpha^\beta f(\theta)\, d\theta = \int_\alpha^\beta \tfrac{1}{2}[\rho(\theta)]^2 \, d\theta. \quad \square$$

REMARK.   When ρ is constant, Γ is a circular sector and

the area of Γ = (constant value of $\tfrac{1}{2}\rho^2$ on $[\alpha, \beta]$) $\cdot (\beta - \alpha)$.

When ρ varies continuously,

the area of Γ = (average value of $\tfrac{1}{2}\rho^2$ on $[\alpha, \beta]$) $\cdot (\beta - \alpha)$.

$\dagger$ The area of a circular sector = $\tfrac{1}{2}r^2\theta$ where r is the radius and θ is the central angle.

Problem. Calculate the area of the region Γ pictured in Figure 9.5.3.

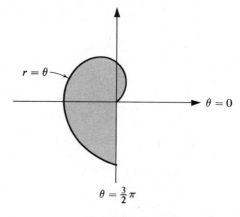

$$\theta = \tfrac{3}{2}\pi$$

FIGURE 9.5.3

SOLUTION

$$\text{area of } \Gamma = \int_0^{3\pi/2} \tfrac{1}{2}\theta^2 \, d\theta = \left[\tfrac{1}{6}\theta^3\right]_0^{3\pi/2} = \tfrac{9}{16}\pi^3. \quad \square$$

Problem. Calculate the area enclosed by the cardioid

$$r = 1 - \cos\theta. \qquad\qquad\text{(Figure 8.6.14)}$$

SOLUTION

$$\text{area} = \int_0^{2\pi} \tfrac{1}{2}(1 - \cos\theta)^2 \, d\theta$$

$$= \tfrac{1}{2}\int_0^{2\pi} (1 - 2\cos\theta + \cos^2\theta) \, d\theta$$

$$= \tfrac{1}{2}\left[\int_0^{2\pi} (1 - 2\cos\theta) \, d\theta + \int_0^{2\pi} \cos^2\theta \, d\theta\right].$$

Since

$$\int_0^{2\pi} (1 - 2\cos\theta) \, d\theta = \left[\theta - 2\sin\theta\right]_0^{2\pi} = 2\pi$$

and

$$\int_0^{2\pi} \cos^2\theta \, d\theta = \int_0^{2\pi} \tfrac{1}{2}(1 + \cos 2\theta) \, d\theta = \left[\tfrac{1}{2}\theta + \tfrac{1}{4}\sin 2\theta\right]_0^{2\pi} = \pi,$$

we have

$$\text{area} = \tfrac{3}{2}\pi. \quad \square$$

A slightly more complicated type of region is pictured in Figure 9.5.4. We can calculate the area of such a region Ω by taking the area up to $r = \rho_2(\theta)$ and subtracting from it the area up to $r = \rho_1(\theta)$. This gives the formula

$$\text{area of } \Omega = \int_\alpha^\beta \tfrac{1}{2}([\rho_2(\theta)]^2 - [\rho_1(\theta)]^2)\, d\theta.$$

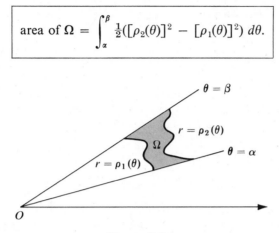

FIGURE 9.5.4

Exercises

Calculate the area enclosed by the following curves. Take $a > 0$.

*1. $r = a\cos\theta$, $-\tfrac{1}{2}\pi \le \theta \le \tfrac{1}{2}\pi$.

2. $r = a\cos 3\theta$, $-\tfrac{1}{6}\pi \le \theta \le \tfrac{1}{6}\pi$.

*3. $r = a\sqrt{\cos 2\theta}$, $-\tfrac{1}{4}\pi \le \theta \le \tfrac{1}{4}\pi$.

4. $r = a(1 + \cos 3\theta)$, $-\tfrac{1}{3}\pi \le \theta \le \tfrac{1}{3}\pi$.

*5. $r^2 = a^2 \sin^2\theta$.

6. $r^2 = a^2 \sin^2 2\theta$.

Calculate the area of the region bounded by the following.

*7. $r = \tan 2\theta$ and the rays $\theta = 0$, $\theta = \tfrac{1}{8}\pi$.

8. $r = \cos\theta$, $r = \sin\theta$, and the rays $\theta = 0$, $\theta = \tfrac{1}{4}\pi$.

*9. $r = 2\cos\theta$, $r = \cos\theta$, and the rays $\theta = 0$, $\theta = \tfrac{1}{4}\pi$.

10. $r = 1 + \cos\theta$, $r = \cos\theta$, and the rays $\theta = 0$, $\theta = \tfrac{1}{2}\pi$.

*11. $r = a(4\cos\theta - \sec\theta)$ and the rays $\theta = 0$, $\theta = \tfrac{1}{4}\pi$.

12. The parabola $r = \tfrac{1}{2}\sec^2 \tfrac{1}{2}\theta$ and the vertical line through the origin.

Find the area of the region bounded by the graphs of the following.

*13. $r = e^\theta$, $0 \le \theta \le \pi$; $r = 0$, $0 \le \theta \le \pi$; the rays $\theta = 0$ and $\theta = \pi$.

14. $r = e^\theta$, $2\pi \le \theta \le 3\pi$; $r = 0$, $0 \le \theta \le \pi$; the rays $\theta = 0$ and $\theta = \pi$.

*15. $r = e^\theta$, $0 \le \theta \le \pi$; $r = e^{\theta/2}$, $0 \le \theta \le \pi$; the rays $\theta = 2\pi$ and $\theta = 3\pi$.

16. $r = e^\theta$, $0 \le \theta \le \pi$; $r = e^\theta$, $2\pi \le \theta \le 3\pi$; the rays $\theta = 0$ and $\theta = \pi$.

9.6 The Least Upper Bound Axiom; Arc Length

The Least Upper Bound Axiom

We begin with a set S of real numbers. As we mentioned in Chapter 1, a number M is called an *upper bound* for S iff

$$x \leq M \qquad \text{for all} \quad x \in S.$$

Obviously, not all sets of real numbers have upper bounds. Those that do are said to be *bounded above*.

It is obvious that every set that has a largest element has an upper bound; if b is the largest element of S, then

$$x \leq b \qquad \text{for all} \quad x \in S$$

and therefore b is an upper bound for S. The converse is false: for example, the sets

$$(-\infty, 0) \quad \text{and} \quad \left\{ \frac{1}{2}, \frac{2}{3}, \frac{3}{4}, \ldots, \frac{n}{n+1}, \ldots \right\}$$

both have upper bounds (2 for instance) but neither has a largest element.

Let's return to the set $(-\infty, 0)$. While $(-\infty, 0)$ does not have a largest element, the set of its upper bounds, $[0, \infty)$, does have a least element, namely 0. A similar remark can be made about the second set. While the set of quotients

$$\frac{n}{n+1} = 1 - \frac{1}{n+1}$$

does not have a greatest element, the set of its upper bounds, $[1, \infty)$, does have a least element, namely 1. Along these lines there is a key *assumption* that we make about the real number system. It is called the *least upper bound axiom*.

The Least Upper Bound Axiom

Every nonempty set of real numbers that has an upper bound has a *least* upper bound.

The least upper bound of a set has a property that deserves particular attention. The idea is this: the fact that M is the least upper bound of the set S does not tell us that M is in S (it need not be) but it does tell us that we can approximate M as closely as we wish by members of S.

Theorem 9.6.1

If M is the least upper bound of the set S and ϵ is positive, then there is a number s in S such that

$$M - \epsilon < s \leq M.$$

PROOF.   Since all numbers in S are less than or equal to M, the right-hand side of the inequality causes no difficulty. All we have to show therefore is that there exists some number s in S such that

$$M - \epsilon < s.$$

Suppose on the contrary that no such number exists. We then have

$$x \leq M - \epsilon \qquad \text{for all} \quad x \in S$$

and so that $M - \epsilon$ becomes an upper bound for S. But this cannot happen, for it makes $M - \epsilon$ an upper bound that is less than M, and M is by assumption the *least* upper bound. □

Similar observations can be made about lower bounds. In the first place, a number m is called a *lower bound* for S iff

$$m \leq x \qquad \text{for all} \quad x \in S.$$

Sets that have lower bounds are said to be *bounded below*. Not all sets have lower bounds, but those that do, have *greatest lower bounds*. This need not be taken as an axiom. Using the least upper bound axiom we can prove it as a theorem.

Theorem

Every nonempty set of real numbers that has a lower bound has a *greatest* lower bound.

PROOF.   Suppose that S is nonempty and that it has a lower bound x. Then

$$x \leq s \qquad \text{for all } s \in S.$$

It follows that

$$-s \leq -x \qquad \text{for all } s \in S;$$

that is,

$$\{-s : s \in S\} \quad \text{has an upper bound } -x.$$

From the least upper bound axiom we conclude that $\{-s : s \in S\}$ has a least upper bound. Call it x_0. Since

$$-s \leq x_0 \qquad \text{for all } s \in S,$$

we see that

$$-x_0 \leq s \qquad \text{for all } s \in S,$$

and thus $-x_0$ is a lower bound for S. We now assert that $-x_0$ is the greatest lower bound of the set S. To see this, note that if there existed x_1 satisfying

$$-x_0 < x_1 \leq s \qquad \text{for all } s \in S,$$

then we would have

$$-s \leq -x_1 < x_0 \qquad \text{for all } s \in S,$$

and thus x_0 would not be the *least* upper bound of $\{-s : s \in S\}$. □

As in the case of the least upper bound, the greatest lower bound of a set need not be in the set but it can be approximated as closely as we wish by members of the set. In short, we have the following theorem, the proof of which is left as an exercise.

Theorem 9.6.2

If m is the greatest lower bound of the set S and ϵ is positive, then there is a number s in S such that

$$m \leq s < m + \epsilon.$$

Arc Length Defined

We come now to the notion of arc length. In Figure 9.6.1 we have sketched a curve C which we assume is parametrized by a pair of continuously differentiable functions†:

$$C: \quad x(t),\, y(t), \qquad t \in [a, b].$$

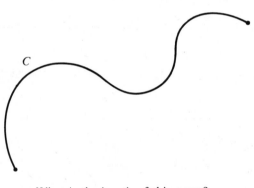

What is the length of this curve?

FIGURE 9.6.1

What we want to do is assign a length to the curve C. Here our experience in Chapter 4 can be used as a model.

To decide what should be meant by the area of a region Ω, we approximated Ω by the union of a finite number to rectangles. To decide what should be meant by the length of the curve C, we will approximate C by the union of a finite number of line segments.

Each point t in $[a, b]$ gives rise to a point $P = P(x(t), y(t))$ that lies on the curve C. By choosing a finite number of points in $[a, b]$,

$$a = t_0 < t_1 < \cdots < t_{i-1} < t_i < \cdots < t_{n-1} < t_n = b,$$

we obtain a finite number of points of C,

$$P_0, P_1, \ldots, P_{i-1}, P_i, \ldots, P_{n-1}, P_n.$$

† By this we mean functions that have continuous first derivatives.

We now join these points consecutively by line segments and call the resulting path

$$L = \overline{P_0 P_1} \cup \cdots \cup \overline{P_{i-1} P_i} \cup \cdots \cup \overline{P_{n-1} P_n},$$

a *polygonal path* inscribed in the curve C. (See Figure 9.6.2.)

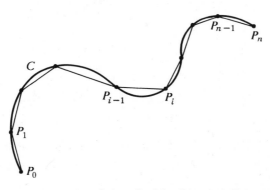

a polygonal path inscribed in the curve C

FIGURE 9.6.2

The length of such a polygonal path is the sum of the distances between consecutive vertices:

$$\text{length of } L = d(P_0, P_1) + \cdots + d(P_{i-1}, P_i) + \cdots + d(P_{n-1}, P_n).$$

The polygon L serves as an approximation to the curve C, but obviously a better approximation can be obtained by adding more vertices to L. At this point let's ask ourselves exactly what it is we require of the number that we shall call the length of C. Certainly we require that

$$l(L) \le \text{the length of } C \qquad \text{for each } L \text{ inscribed in } C.$$

But that is not enough. There is another requirement that seems reasonable. If we can choose L to approximate C as closely as we wish, then we should be able to choose L so that $l(L)$ approximates the length of C as closely as we wish. Namely, for each positive ϵ there should exist a polygonal path L such that

$$(\text{length of } C) - \epsilon < l(L) \le \text{length of } C.$$

Theorem 9.6.1 tells us that we can achieve this result by defining the length of C as the least upper bound of all the $l(L)$. This is in fact what we do:

Definition

$$\text{length of } C = \left\{ \begin{array}{l} \text{the least upper bound of the} \\ \text{set of all lengths of polygonal} \\ \text{paths inscribed in } C. \end{array} \right]$$

Arc-Length Formulas

Now that we have explained what we mean by the length of a curve C it is time to describe a practical way to compute it. The basic result is easy to state. If C is parametrized by a pair of continuously differentiable functions

$$C: \quad x(t), y(t), t \in [a, b]$$

then

(9.6.3)

$$\text{length of } C = \int_a^b \sqrt{[x'(t)]^2 + [y'(t)]^2} \, dt.$$

Obviously this is not something to be taken on faith. It has to be proven. We will do so, but not until Chapter 13. Right now we assume that the result is true and carry out some computations.

Example. If C is a circle of radius r, say

$$C: \quad x(t) = r \cos t, \quad y(t) = r \sin t, \qquad t \in [0, 2\pi]$$

the formula yields

$$l(C) = \int_0^{2\pi} \sqrt{r^2 \sin^2 t + r^2 \cos^2 t} \, dt = \int_0^{2\pi} r \, dt = 2\pi r,$$

which is reassuring. □

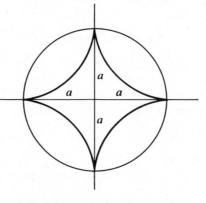

FIGURE 9.6.3

Example. In the case of the astroid

$$C: \quad x(t) = a \cos^3 t, \quad y(t) = a \sin^3 t$$

shown in Figure 9.6.3, we have

$$[x'(t)]^2 + [y'(t)]^2 = 9a^2 \sin^2 t \cos^2 t,$$

so that

$$l(C) = 4 \int_0^{\pi/2} 3a \sin t \cos t \, dt = 12a \left[\tfrac{1}{2} \sin^2 t \right]_0^{\pi/2} = 6a.$$

This is a little less than the circumference of the circumscribed circle. $\square$

Suppose now that the curve C is actually the graph of a continuously differentiable function

$$y = f(x), \qquad x \in [a, b].$$

As a parametrization for C we can take

$$x(t) = t, \quad y(t) = f(t), \qquad t \in [a, b].$$

Since

$$x'(t) = 1 \quad \text{and} \quad y'(t) = f'(t),$$

the arc length formula reduces to

$$l(C) = \int_a^b \sqrt{1 + [f'(t)]^2} \, dt.$$

Replacing t by x we can write

(9.6.4) $$\boxed{\text{length of the graph of } f = \int_a^b \sqrt{1 + [f'(x)]^2} \, dx.}$$

Example. For

$$f(x) = x^2, \qquad x \in [0, 1]$$

the graph is a parabolic arc from $(0, 0)$ to $(1, 1)$. The length of this arc is given by

$$\int_0^1 \sqrt{1 + [f'(x)]^2} \, dx = \int_0^1 \sqrt{1 + 4x^2} \, dx = 2 \int_0^1 \sqrt{x^2 + (\tfrac{1}{2})^2} \, dx$$

$$= \left[x\sqrt{x^2 + (\tfrac{1}{2})^2} + (\tfrac{1}{2})^2 \log |x + \sqrt{x^2 + (\tfrac{1}{2})^2}| \right]_0^1$$

$$= \tfrac{1}{2}\sqrt{5} + \tfrac{1}{4} \log (2 + \sqrt{5}). \quad \square$$

Curves given in polar coordinates can also be measured in a similar manner. If C is given by a polar equation

$$r = \rho(\theta), \qquad \alpha \le \theta \le \beta$$

then we have the parametric representation

$$x(\theta) = \rho(\theta) \cos \theta, \quad y(\theta) = \rho(\theta) \sin \theta, \qquad \alpha \le \theta \le \beta.$$

With
$$x'(\theta) = -\rho(\theta) \sin \theta + \rho'(\theta) \cos \theta \quad \text{and} \quad y'(\theta) = \rho(\theta) \cos \theta + \rho'(\theta) \sin \theta,$$
you'll find, after simplification, that
$$[x'(\theta)]^2 + [y'(\theta)]^2 = [\rho(\theta)]^2 + [\rho'(\theta)]^2.$$
Consequently, we have the formula

$$l(C) = \int_\alpha^\beta \sqrt{[\rho(\theta)]^2 + [\rho'(\theta)]^2} \, d\theta.$$

Example.   As a polar equation for the circle of radius ρ we have simply
$$r = \rho.$$
Since in this case ρ is a constant, its derivative is zero.   The circumference of the circle is therefore
$$\int_0^{2\pi} \sqrt{\rho^2} \, d\theta = \int_0^{2\pi} \rho \, d\theta = 2\pi\rho. \quad \square$$

Example.   In the case of the cardioid
$$r = 1 - \cos \theta, \qquad\qquad\qquad \text{(Figure 8.6.14)}$$
we have
$$\rho(\theta) = 1 - \cos \theta.$$
Here
$$[\rho(\theta)]^2 + [\rho'(\theta)]^2 = 1 - 2 \cos \theta + \cos^2 \theta + \sin^2 \theta = 2(1 - \cos \theta).$$
In view of the identity
$$\tfrac{1}{2}(1 - \cos \theta) = \sin^2 \tfrac{1}{2}\theta,$$
we have
$$[\rho(\theta)]^2 + [\rho'(\theta)]^2 = 4 \sin^2 \tfrac{1}{2}\theta.$$
The length of the cardioid is therefore
$$\int_0^{2\pi} \sqrt{[\rho(\theta)]^2 + [\rho'(\theta)]^2} \, d\theta = \int_0^{2\pi} 2 \sin \tfrac{1}{2}\theta \, d\theta = 4 \int_0^{2\pi} \tfrac{1}{2} \sin \tfrac{1}{2}\theta \, d\theta$$
$$= 4 \left[-\cos \tfrac{1}{2}\theta \right]_0^{2\pi} = 8. \quad \square$$

Exercises

Find the least upper bound for each of the following sets.

*1. $(0, 2)$.   2. $[0, 2]$.

* 3. $\{x: x^2 < 4\}$.   (the set of all x such that $x^2 < 4$)

4. $\{x: x^2 < 2\}$.   *5. $\{x: x^2 \le 2\}$.

6. $\{0.9, 0.99, 0.999, \ldots\}$.   *7. $\{x: \log x < 1\}$.

Find the greatest lower bound for each of the following sets.

*8. $(0, 2)$.   9. $[0, 2]$.

10. $\{x: \log x > 0\}$.   *11. $\{2\frac{1}{2}, 2\frac{1}{3}, 2\frac{1}{4}, \ldots\}$.

Find the length of the graph of f.

*12. $f(x) = mx + b, \quad x \in [c, d]$.

13. $f(x) = x^{3/2}, \quad x \in [0, 44]$.

*14. $f(x) = \frac{1}{3}\sqrt{x}\,(x - 3), \quad x \in [0, 3]$.

15. $f(x) = 2e^{x/2}, \quad x \in [0, \log 8]$.

*16. $f(x) = \log(\sec x), \quad x \in [-\frac{1}{4}\pi, \frac{1}{4}\pi]$.

17. $f(x) = \frac{1}{2}(e^x + e^{-x}), \quad x \in [0, \log 2]$.

*18. $f(x) = e^x, \quad x \in [0, \frac{1}{2}\log 3]$.

19. $f(x) = \frac{1}{3}(x^2 + 2)^{3/2}, \quad x \in [0, 1]$.

*20. $f(x) = \frac{2}{3}(x - 1)^{3/2}, \quad x \in [1, 2]$.

21. $f(x) = \frac{1}{2}x^2, \quad x \in [0, 1]$.

*22. $f(x) = \frac{1}{2}x\sqrt{3 - x^2} + \frac{3}{2} \arcsin (\frac{1}{3}\sqrt{3}\,x), \quad x \in [0, 1]$.

Find the length of the curve.

23. $x(t) = at^2, \quad y(t) = 2at, \quad t \in [0, b]$.

*24. $x(t) = t - 1, \quad y(t) = \frac{1}{2}t^2, \quad t \in [0, 1]$.

25. $x(t) = t^2, \quad y(t) = t^3, \quad t \in [0, 1]$.

*26. $x(t) = e^t \sin t, \quad y(t) = e^t \cos t, \quad t \in [0, \pi]$.

27. $x(t) = \cos t + t \sin t, \quad y(t) = \sin t - t \cos t, \quad t \in [0, \pi]$.

*28. $r = e^\theta, \quad 0 \le \theta \le 4\pi$. (logarithmic spiral)

29. $r = ae^\theta, \quad -2\pi \le \theta \le 2\pi$.

*30. $r = e^{2\theta}, \quad 0 \le \theta \le 2\pi$.

31. $r = a\theta, \quad 0 \le \theta \le 2\pi$. (spiral of Archimedes)

32. Prove Theorem 9.6.2.

33. At time t a particle has position

$$x(t) = 1 - \cos t, \qquad y(t) = t - \sin t.$$

Find the total distance traveled between $t = 0$ and $t = 2\pi$.

*34. At time t a particle has position

$$x(t) = 1 + \arctan t, \qquad y(t) = 1 - \log \sqrt{1 + t^2}.$$

Find the total distance traveled between $t = 0$ and $t = 1$.

*35. (*Optional*)   Let f be a positive nonconstant function with the property that for every interval $[a, b]$ the length of the graph equals the area of the region under the graph. Find f, given that $f(0) = 1$.

36. (*Optional*)   Show that a homogeneous, flexible, inelastic rope hanging from two fixed points assumes the shape of a catenary:

$$f(x) = a \cosh\left(\frac{x}{a}\right) = \frac{a}{2}\,(e^{x/a} + e^{-x/a}).$$

HINT: Refer to Figure 9.6.4. That part of the rope which corresponds to the interval $[0, x]$ is subject to the following forces:

(1) its weight, which is proportional to its length;
(2) a horizontal pull at 0, $p(0)$;
(3) a tangential pull at x, $p(x)$.

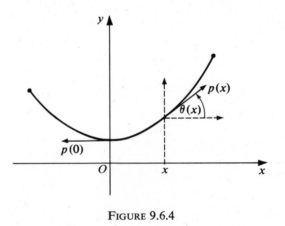

FIGURE 9.6.4

Balancing the vertical forces, we have

$$k \int_0^x \sqrt{1 + [f'(t)]^2}\; dt = p(x) \sin \theta. \qquad \text{(weight is the vertical pull at } x\text{)}$$

Balancing the horizontal forces, we have

$$p(0) = p(x) \cos \theta.$$

$$\text{(pull at 0 is the horizontal pull at } x\text{)}$$

Note also that

$$f'(x) = \tan \theta.$$

9.7   Area of a Surface of Revolution

Let f be a nonnegative function with derivative f' continuous on $[a, b]$. By revolving the graph of f from $x = a$ to $x = b$ about the x-axis, we obtain a surface of revolution. (See Figure 9.7.1.) The area of this surface is given by the formula

(9.7.1)
$$S = \int_a^b 2\pi f(x)\sqrt{[f'(x)]^2 + 1}\ dx.$$

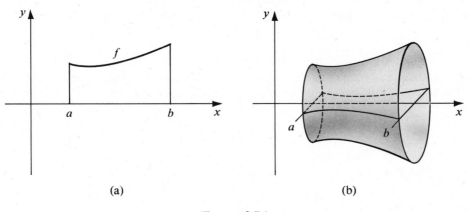

(a)   (b)

FIGURE 9.7.1

PROOF. We begin with a partition $P = \{x_0, x_1, \ldots, x_n\}$ of $[a, b]$ and focus our attention on that portion of the surface which is generated by the ith subinterval $[x_{i-1}, x_i]$. Call the area of this portion S_i. (Figure 9.7.2) The arc which generates

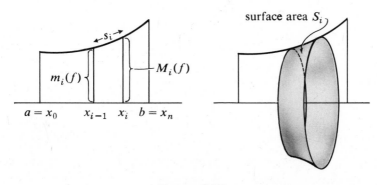

FIGURE 9.7.2

this portion has length

$$s_i = \int_{x_{i-1}}^{x_i} \sqrt{[f'(x)]^2 + 1}\ dx.$$

Its minimum distance from the x-axis is

$$m_i(f) = \text{min value of } f \text{ on } [x_{i-1}, x_i]$$

and its maximum distance from the x-axis is

$$M_i(f) = \text{max value of } f \text{ on } [x_{i-1}, x_i].$$

If we replace this arc by an arc of the same length but closer to the x-axis, then we can expect the surface area to be diminished. If, on the other hand, we replace this arc by an arc of the same length but further away from the x-axis, then we can expect the surface area to be increased. This reasoning leads to the inequality

(1) $$2\pi m_i(f)s_i \le S_i \le 2\pi M_i(f)s_i.$$

The lower bound for S_i is the surface area generated by a horizontal line segment of length s_i at a distance $m_i(f)$ from the x-axis. The upper bound is the surface area generated by a horizontal line segment of length s_i at a distance $M_i(f)$ from the x-axis. (Figure 9.7.3)

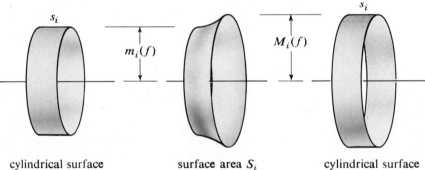

cylindrical surface   surface area S_i cylindrical surface
surface area $= 2\pi m_i(f)s_i$ surface area $= 2\pi M_i(f)s_i$

FIGURE 9.7.3

If we set

$$g(x) = \sqrt{[f'(x)]^2 + 1},$$

then we have

$$s_i = \int_{x_{i-1}}^{x_i} g(x)\ dx$$

and consequently the inequality

$$m_i(g)\,\Delta x_i \le s_i \le M_i(g)\,\Delta x_i.$$ (Property IV, Section 4.9)

Combining this with (1), we have

$$2\pi m_i(f)m_i(g)\,\Delta x_i \le S_i \le 2\pi M_i(f)M_i(g)\,\Delta x_i.$$

Summing these inequalities from $i = 1$ to $i = n$, you can see that S, the total surface area generated by the graph, must satisfy the inequality

$$L^*_{2\pi f, g}(P) \leq S \leq U^*_{2\pi f, g}(P). \qquad \text{(Section 9.4)}$$

Since P was chosen arbitrarily, this must hold for all partitions P. It follows from Theorem 9.4.1 that

$$S = \int_a^b 2\pi f(x)g(x)\, dx = \int_a^b 2\pi f(x)\sqrt{[f'(x)]^2 + 1}\, dx. \quad \square$$

Problem. Find a formula for the slant surface area of a cone.

SOLUTION. A cone of base radius r and height h can be generated by revolving the graph of

$$f(x) = \frac{r}{h}\, x, \qquad x \in [0, h]$$

about the x-axis. (See Figure 9.7.4.) For such a cone we have

$$\text{slant surface area} = \int_0^h 2\pi f(x)\sqrt{[f'(x)]^2 + 1}\, dx$$

$$= 2\pi \int_0^h \frac{r}{h}\, x \sqrt{\frac{r^2}{h^2} + 1}\, dx$$

$$= \frac{2\pi r \sqrt{r^2 + h^2}}{h^2} \int_0^h x\, dx$$

$$= \pi r \sqrt{r^2 + h^2}. \quad \square$$

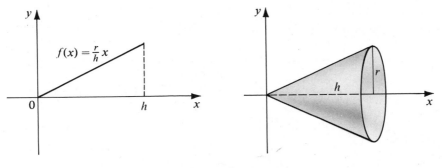

FIGURE 9.7.4

Problem. Find the area of the surface generated by revolving the graph of

$$f(x) = \sin x, \qquad x \in [0, \tfrac{1}{2}\pi]$$

about the x-axis.

SOLUTION.   In this case, we have

$$f(x) = \sin x, \qquad f'(x) = \cos x$$

and thus

$$\text{surface area} = \int_0^{\pi/2} 2\pi \sin x \sqrt{\cos^2 x + 1} \, dx.$$

To find

$$\int 2\pi \sin x \sqrt{\cos^2 x + 1} \, dx,$$

we set

$$u = \cos x, \qquad du = -\sin x \, dx.$$

This gives

$$\int 2\pi \sin x \sqrt{\cos^2 x + 1} \, dx = -2\pi \int \sqrt{u^2 + 1} \, du$$

$$= -\pi [u\sqrt{u^2 + 1} + \log(\sqrt{u^2 + 1} + u)] + C$$

$$= -\pi [\cos x \sqrt{\cos^2 x + 1} + \log(\sqrt{\cos^2 x + 1} + \cos x)] + C$$

and, as you can check,

$$\int_0^{\pi/2} 2\pi \sin x \sqrt{\cos^2 x + 1} \, dx = \pi [\sqrt{2} + \log(\sqrt{2} + 1)]. \quad \square$$

Suppose now that the graph of a nonnegative function f is given parametrically:

$$\text{graph of } f: x(t), \, y(t), \qquad t \in [c, d].$$

If x' and y' are continuous and x' remains positive, then the surface area generated by revolving the graph about the x-axis is given by the formula

(9.7.2)

$$S = \int_c^d 2\pi y(t) \sqrt{[x'(t)]^2 + [y'(t)]^2} \, dt.$$

PROOF.   Set

$$a = x(c), \qquad b = x(d).$$

Note that

$$f(x(t)) = y(t) \quad \text{and} \quad f'(x(t))x'(t) = y'(t).$$

By 9.7.1,

$$S = \int_a^b 2\pi f(u) \sqrt{[f'(u)]^2 + 1} \, du.$$

The substitution
$$u = x(t), \qquad du = x'(t) \, dt$$
gives
$$S = \int_c^d 2\pi f(x(t)) \sqrt{[f'(x(t))]^2 + 1} \; x'(t) \, dt$$
$$= \int_c^d 2\pi y(t) \sqrt{\left[\frac{y'(t)}{x'(t)}\right]^2 + 1} \; x'(t) \, dt$$
$$= \int_c^d 2\pi y(t) \sqrt{[x'(t)]^2 + [y'(t)]^2} \, dt. \quad \square$$

Problem.   Derive the formula for the surface area of a sphere.

SOLUTION.   We can generate a sphere of radius r by revolving the arc
$$x(t) = r \cos t, \quad y(t) = r \sin t, \qquad t \in [0, \pi]$$
about the x-axis. Formula 9.7.2 then gives
$$S = \int_0^\pi 2\pi y(t) \sqrt{[x'(t)]^2 + [y'(t)]^2} \, dt = 2\pi \int_0^\pi r \sin t \sqrt{r^2(\sin^2 t + \cos^2 t)} \, dt$$
$$= 2\pi r^2 \int_0^\pi \sin t \, dt = 2\pi r^2 \Big[-\cos t \Big]_0^\pi = 4\pi r^2. \quad \square$$

Problem.   Find the surface area when the curve
$$y^2 - 2 \log y = 4x \qquad \text{from} \quad y = 1 \text{ to } y = 2$$
is revolved about the x-axis.

SOLUTION.   We can represent the curve parametrically by setting
$$x(t) = \tfrac{1}{4}(t^2 - 2 \log t), \quad y(t) = t, \qquad t \in [1, 2].$$
Here
$$x'(t) = \tfrac{1}{2}(t - t^{-1}), \qquad y'(t) = 1$$
and
$$\sqrt{[x'(t)]^2 + [y'(t)]^2} = \tfrac{1}{2}(t + t^{-1}).$$
It follows that
$$S = \int_1^2 2\pi y(t) \sqrt{[x'(t)]^2 + [y'(t)]^2} \, dt$$
$$= \int_1^2 2\pi t [\tfrac{1}{2}(t + t^{-1})] \, dt$$
$$= \int_1^2 \pi(t^2 + 1) \, dt$$
$$= \pi \left[\tfrac{1}{3}t^3 + t \right]_1^2 = \tfrac{10}{3}\pi. \quad \square$$

Exercises

Find the surface area when the curve is revolved about the x-axis.

*1. $f(x) = r, \quad x \in [0, 1]$.
                     2. $f(x) = \sqrt{x}, \quad x \in [1, 2]$.

*3. $y = \cos x, \quad x \in [-\frac{1}{2}\pi, \frac{1}{2}\pi]$.
               4. $f(x) = 2\sqrt{1 - x}, \quad x \in [-1, 0]$.

*5. $f(x) = \frac{1}{3}x^3, \quad x \in [0, 2]$.
                 6. $f(x) = |x - 3|, \quad x \in [0, 6]$.

*7. $4y = x^3, \quad x \in [0, 1]$.
                    8. $y^2 = 9x, \quad x \in [0, 4]$.

*9. $x(t) = 2t, \quad y(t) = 3t, \quad t \in [0, 1]$.

10. $x(t) = t^2, \quad y(t) = 2t, \quad t \in [0, \sqrt{3}]$.

*11. $x^2 + y^2 = r^2, \quad x \in [-\frac{1}{2}r, \frac{1}{2}r]$.

12. $y = \frac{1}{2}(e^x + e^{-x}), \quad x \in [0, 1]$.

*13. $x(t) = a \cos^3 t, \quad y(t) = a \sin^3 t, \quad t \in [0, \frac{1}{2}\pi]$.

14. $x(t) = e^t \sin t, \quad y(t) = e^t \cos t, \quad t \in [0, \frac{1}{2}\pi]$.

*15. $r = \rho(\theta), \quad \theta \in [\theta_1, \theta_2]$.

16. $r = e^\theta, \quad \theta \in [0, \frac{1}{2}\pi]$.

*17. Find a formula for the lateral surface area of the frustrum of a cone. (Figure 9.3.12)

18. Find the surface area when the curve

$$2x = y\sqrt{y^2 - 1} + \log |y - \sqrt{y^2 - 1}| \qquad \text{from } y = 2 \text{ to } y = 5$$

is revolved about the x-axis.

*19. Find the area of the ellipsoid obtained by revolving the ellipse

$$\frac{x^2}{a^2} + \frac{y^2}{b^2} = 1$$

(a) about its major axis; (b) about its minor axis.

20. Find the surface area when the curve

$$6a^2xy = y^4 + 3a^4 \qquad \text{from } y = a \text{ to } y = 3a$$

is revolved about the x-axis.

21. Where in the proof of 9.7.2 did we use the hypothesis that x' remained positive?

9.8 (Optional) The Notion of Work

If an object moves a distance s in a fixed direction subject to a constant force f applied in the direction of the motion, then the work done by f is defined by the physicist by the equation

$$W = f \cdot s. \qquad \qquad \text{(work} = \text{force} \cdot \text{distance)}$$

If the force is measured in pounds and the distance is measured in feet, then the units of work are called foot-pounds. For example, if a constant force of 500 pounds is applied to pushing a car 60 feet, the work done by the pushing force is said to be 30,000 foot-pounds.

If we coordinatize the line along which the motion takes place, assigning coordinate a to the initial point and coordinate b to the terminal point $(b > a)$, then we have

$$\text{work} = (\text{constant force}) \cdot (b - a).$$

In other words,

$$\text{work} = (\text{constant value of } f \text{ on } [a, b]) \cdot (b - a).$$

If f, the force applied varies continuously, then we replace the constant value of f by its average value, and define

$$\text{work} = (\text{average value of } f \text{ on } [a, b]) \cdot (b - a).$$

Thus, in the case of a variable force, we have

(9.8.1)
$$\text{work} = \int_a^b f(x)\, dx.$$
(Figure 9.8.1)

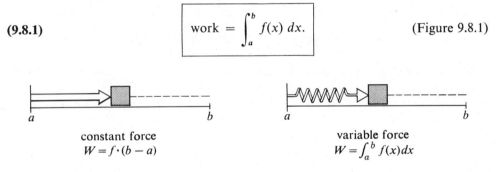

constant force
$$W = f \cdot (b - a)$$

variable force
$$W = \int_a^b f(x)\,dx$$

FIGURE 9.8.1

Stretching a Spring

A simple instance of a variable force is offered by the behavior of a steel spring. According to Hooke's law (Robert Hooke, 1635–1703), a spring stretched x units beyond its natural length exerts a force which is proportional to x:

$$f(x) = kx.$$

The constant k depends on the spring and on the units used.

Problem. A certain spring exerts a force of $\frac{1}{2}$ pound when stretched $\frac{1}{3}$ foot beyond its natural length. What is the work done in stretching the spring $\frac{1}{10}$ foot beyond its natural length? What is the work done in stretching it an additional $\frac{1}{10}$ foot?

SOLUTION. We know that the force function is of the form

$$f(x) = kx.$$

Since $f(\frac{1}{3}) = \frac{1}{2}$, we have

$$k \cdot \tfrac{1}{3} = \tfrac{1}{2},$$

so that $k = \frac{3}{2}$. Thus, for this particular spring, the force function is

$$f(x) = \tfrac{3}{2}x.$$

The work done in the initial stretching is

$$W_1 = \int_0^{1/10} \tfrac{3}{2}x \, dx = \tfrac{3}{2}\left[\tfrac{1}{2}x^2\right]_0^{1/10} = \tfrac{3}{400} \text{ foot-pounds.}$$

The work done in the next stretching is

$$W_2 = \int_{1/10}^{2/10} \tfrac{3}{2}x \, dx = \tfrac{3}{2}\left[\tfrac{1}{2}x^2\right]_{1/10}^{2/10} = \tfrac{9}{400} \text{ foot-pounds.} \quad \square$$

Emptying Out a Tank

To lift an object, one must counteract the force of gravity. Consequently, the work done in lifting an object is given by the simple formula

$$\text{work} = (\text{weight of the object}) \cdot (\text{distance lifted}).$$

Problem.   A tank containing a liquid that weighs σ pounds per cubic foot is being pumped out from above. (See Figure 9.8.2.) What is the work done in lowering the level of the liquid from a feet below the top of the tank to b feet below the top of the tank?

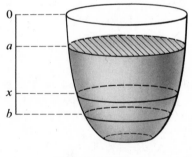

FIGURE 9.8.2

SOLUTION.   For each $x \in [a, b]$, we let

$$A(x) = \text{cross-sectional area } x \text{ feet below the top of the tank,}$$
$$s(x) = \text{distance that the } x\text{-level must be lifted.}$$

We let $P = \{x_0, x_1, \ldots, x_n\}$ be an arbitrary partition of $[a, b]$ and focus our attention on the ith subinterval $[x_{i-1}, x_i]$. (Figure 9.8.3) Now set

$$m_i(A) = \text{min value of } A \text{ on } [x_{i-1}, x_i],$$
$$M_i(A) = \text{max value of } A \text{ on } [x_{i-1}, x_i],$$
$$m_i(s) = \text{min value of } s \text{ on } [x_{i-1}, x_i],$$
$$M_i(s) = \text{max value of } s \text{ on } [x_{i-1}, x_i].$$

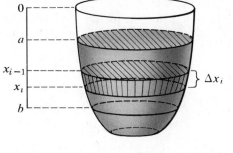

FIGURE 9.8.3

V_i, the volume of the liquid between x_{i-1} and x_i, satisfies the inequality

$$m_i(A) \, \Delta x_i \le V_i \le M_i(A) \, \Delta x_i. \qquad \text{(explain)}$$

Its weight, w_i, therefore satisfies the inequality

$$\sigma m_i(A) \, \Delta x_i \le w_i \le \sigma M_i(A) \, \Delta x_i.$$

W_i, the work necessary to lift this weight to the top of the tank, must in turn satisfy the inequality

$$\sigma m_i(s) m_i(A) \, \Delta x_i \le W_i \le \sigma M_i(s) M_i(A) \, \Delta x_i. \qquad \text{(explain)}$$

Setting

$$f(x) = \sigma s(x),$$

we have with obvious notation

$$m_i(f) m_i(A) \, \Delta x_i \le W_i \le M_i(f) M_i(A) \, \Delta x_i.$$

If you now sum this last inequality from $i = 1$ to $i = n$, then you'll see that the total work done

$$W = W_1 + W_2 + \cdots + W_n$$

must satisfy the inequality

$$L^{\star}_{f,A}(P) \le W \le U^{\star}_{f,A}(P). \qquad \text{(Section 9.4)}$$

Since this must hold for all partitions P, you can conclude from Theorem 9.4.1 that

$$W = \int_a^b f(x) A(x) \, dx.$$

Since $f(x) = \sigma s(x)$,

$$\boxed{W = \int_a^b \sigma s(x) A(x) \, dx.} \qquad \square$$

Problem. A hemispherical water tank of radius 10 feet is being pumped out. Find the work done in lowering the water level from 2 feet below the top of the tank to 4 feet below the top of the tank, (a) given that the pump is placed right on top of the tank, (b) given that the pump is placed 3 feet above the tank.

SOLUTION. As the weight of water take 62.5 pounds per cubic foot. It's not hard to see that the cross section x feet from the top of the tank is a disk of radius $\sqrt{100 - x^2}$. Its area is therefore

$$A(x) = \pi(100 - x^2).$$

For part (a)

$$s(x) = x$$

and therefore

$$W = \int_2^4 62.5\pi x(100 - x^2)\, dx = 33,750\pi \text{ foot-pounds.}$$

For part (b)

$$s(x) = x + 3$$

and

$$W = \int_2^4 62.5\pi(x + 3)(100 - x^2)\, dx = 67,750\pi \text{ foot-pounds.} \quad \square$$

Exercises

*1. A weight of 2 pounds stretches a certain spring 4 feet beyond its natural length. Find the work done in stretching the spring (a) 1 foot beyond its natural length, (b) $1\frac{1}{2}$ feet beyond its natural length.

2. A certain spring has natural length l. Given that W is the work done in stretching the spring from l feet to $l + a$ feet, find the work done in stretching the spring (a) from l feet to $l + 2a$ feet, (b) from l feet to $l + na$ feet, (c) from $l + a$ feet to $l + 2a$ feet, (d) from $l + a$ feet to $l + na$ feet.

*3. Find the natural length of a heavy metal spring, given that the work done in stretching it from a length of 2 feet to a length of 3 feet is one-half the work done in stretching it from a length of 3 feet to a length of 4 feet.

4. A vertical cylindrical tank of radius 2 feet and height 6 feet is full of water. Find the work done in pumping out the water (a) to the top of the tank, (b) to a level 5 feet above the top of the tank. (Assume that water weighs 62.5 pounds per cubic foot.)

*5. A horizontal cylindrical tank of radius 3 feet and length 8 feet is half full of oil weighing 60 pounds per cubic foot. What is the work done in pumping out the oil (a) to the top of the tank, (b) to a level 4 feet above the top of the tank?

6. A conical container (vertex down) of radius r feet and height h feet is full of a liquid weighing σ pounds per cubic foot. Find the work done in pumping out the top $h/2$ feet of liquid (a) to the top of the tank, (b) to a level k feet above the top of the tank.

*7. A rope of length l feet that weighs σ pounds per foot is lying on the ground. What is the work done in lifting the rope so that it hangs from a beam (a) l feet high, (b) $2l$ feet high?

8. An 800-pound steel beam hangs from a 50-foot cable which weighs 6 pounds per foot. Find the work done in winding 20 feet of the cable about a steel drum.

9.9   (Optional) Fluid Pressure

Here we take up the study of fluid pressure and discuss how to calculate the force exerted by a fluid on a vertical wall. By pressure we mean *force per unit area*.

When a horizontal surface is submerged in a liquid the force against it is the total weight of the fluid above it. If the horizontal surface has area A, if the depth is h, and the density of the fluid is σ, then the force is given by the formula

$$f = \sigma h A.$$

In this discussion we'll measure area in square feet, depth in feet, and density in pounds per cubic foot. The force is then expressed in pounds.

In Figure 9.9.1 we have represented a vertical wall. By $w(x)$ we mean the width of the wall at depth x. To determine the force exerted on the wall by the fluid from

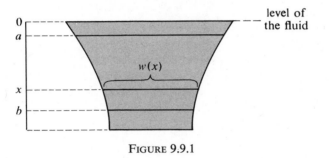

FIGURE 9.9.1

depth a to depth b, we let $P = \{x_0, x_1, \ldots, x_n\}$ be a partition of $[a, b]$ and focus our attention on that portion of the wall which corresponds to the ith subinterval $[x_{i-1}, x_i]$. (See Figure 9.9.2.) With pressure the same in all directions (this the physicists tell us is correct), we can estimate the force against this ith strip. If we take A_i as the area of this strip, then f_i, the force against this strip, satisfies the inequality

(1) $$\sigma x_{i-1} A_i \le f_i \le \sigma x_i A_i.$$ (explain)

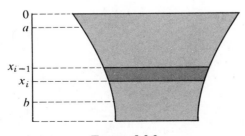

FIGURE 9.9.2

Now set

$$m_i(w) = \text{min value of } w \text{ on } [x_{i-1}, x_i],$$
$$M_i(w) = \text{max value of } w \text{ on } [x_{i-1}, x_i],$$

and note that

$$m_i(w) \, \Delta x_i \le A_i \le M_i(w) \, \Delta x_i.$$

Combining this with (1), we have

$$\sigma x_{i-1} m_i(w) \, \Delta x_i \le f_i \le \sigma x_i M_i(w) \, \Delta x_i.$$

By setting

$$k(x) = \sigma x,$$

we have, with obvious notation,

$$m_i(k) m_i(w) \, \Delta x_i \le f_i \le M_i(k) M_i(w) \, \Delta x_i.$$

Summing from $i = 1$ to $i = n$, you can see that the total force

$$f = f_1 + f_2 + \cdots + f_n$$

must satisfy the inequality

$$L_{k,w}^{\star}(P) \le f \le U_{k,w}^{\star}(P) \qquad\qquad \text{(Section 9.4)}$$

and from the arbitrariness of P conclude (from Theorem 9.4.1) that

$$f = \int_a^b k(x) w(x) \, dx.$$

Since we chose

$$k(x) = \sigma x,$$

we have

$$\boxed{f = \int_a^b \sigma x w(x) \, dx.} \qquad \square$$

Problem.   A circular water main (Figure 9.9.3) 6 feet in diameter is half full of water. Find the total force on the gate that closes the main.

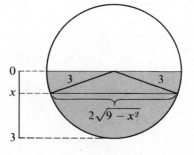

FIGURE 9.9.3

SOLUTION. Here

$$w(x) = 2\sqrt{9 - x^2} \quad \text{and} \quad \sigma = 62.5 \text{ pounds per cubic foot.}$$

The force exerted against the main is

$$f = \int_0^3 (62.5)x(2\sqrt{9 - x^2}) \, dx = 62.5 \int_0^3 2x\sqrt{9 - x^2} \, dx = 1125 \text{ pounds.} \quad \square$$

Problem. The trapezoidal gate of a dam is shown in Figure 9.9.4. Find the force on the gate when the surface of the water is 4 feet above the top of the gate.

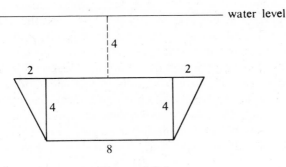

FIGURE 9.9.4

SOLUTION. First we find the width of the gate x feet below the water level. By similar triangles (see Figure 9.9.5)

$$t = \tfrac{1}{2}(8 - x)$$

so that

$$w(x) = 8 + 2t = 16 - x.$$

The force against the gate is

$$f = \int_4^8 62.5x(16 - x) \, dx = 14{,}666\tfrac{2}{3} \text{ pounds.} \quad \square$$

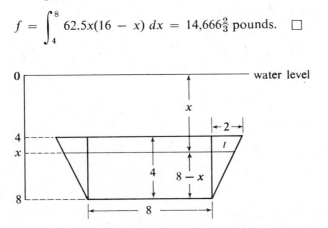

FIGURE 9.9.5

Exercises

* 1. Each end of a horizontal oil tank is an ellipse with horizontal axis 12 feet long and vertical axis 6 feet long. Calculate the force on one end when the tank is half full of oil weighing 60 pounds per cubic foot.

2. The vertical end of a vat is a segment of a parabola (vertex down) 8 feet across the top and 16 feet deep. Calculate the force on this end when the vat is full of liquid weighing 70 pounds per cubic foot.

* 3. The vertical end of a water trough is an isosceles right triangle with the 90° angle at the bottom. Calculate the force on this triangle when the trough is full of water given that the legs of the triangle are 8 feet long.

4. The vertical end of a water trough is an isosceles triangle 5 feet across the top and 5 feet deep. Calculate the force on the end when the trough is full of water.

* 5. A horizontal cylindrical tank of diameter 8 feet is half full of oil weighing 60 pounds per cubic foot. Calculate the force on one end.

6. Calculate the force on one end if the tank of Exercise 5 is full.

* 7. A rectangular gate in a vertical dam is 10 feet wide and 6 feet high. Find (a) the force when the level of the water is 8 feet above the top of the gate; (b) how much higher must the water rise to double the force found in (a)?

8. A vertical cylindrical tank of diameter 30 feet and height 50 feet is full of oil weighing 60 pounds per cubic foot. Find the force on the curved surface.

9.10   (Optional) Revenue Streams

In Section 5.6 we talked about continuous compounding and you saw that the present value of A dollars t years from now is given by the formula

$$P.V. = Ae^{-rt},$$

where r is the rate of continuous compounding.

Suppose now that revenue flows continually at a constant rate of R dollars per year for n years. What is the present value of such a revenue stream?

To answer this, we let $P = \{t_0, t_1, \ldots, t_n\}$ be a partition of the time interval $[0, n]$ and focus on what happens during a typical time subinterval $[t_{i-1}, t_i]$.

The revenue that flows during that time interval is the product $R \, \Delta t_i$. To get its present value, $(P.V.)_i$, we must discount, by more than $e^{-rt_{i-1}}$ but by less than e^{-rt_i}. (Explain.) This says that

$$R \, \Delta t_i e^{-rt_i} \le (P.V.)_i \le R \, \Delta t_i e^{-rt_{i-1}},$$

which we can rewrite as

(1) $$Re^{-rt_i} \, \Delta t_i \le (P.V.)_i \le Re^{-rt_{i-1}} \, \Delta t_i.$$

Setting

$$f(t) = Re^{-rt}$$

we have in our usual notation

(2) $$m_i(f) \, \Delta t_i \le (P.V.)_i \le M_i(f) \, \Delta t_i.$$

The present value of the entire stream of revenues is the sum

$$P.V. = (P.V.)_1 + (P.V.)_2 + \cdots + (P.V.)_n.$$

By summing (2) from $i = 1$ to $i = n$, you can see that

$$L_f(P) \le P.V. \le U_f(P),$$

and, since P was chosen arbitrarily, that

$$P.V. = \int_0^n f(t)\, dt.$$

Since $f(t) = Re^{-rt}$,

(9.10.1)
$$P.V. = \int_0^n Re^{-rt}\, dt. \qquad \square$$

Problem 1. What is the present value of an 8-year stream of revenues expected to flow continually at the rate of \$100 per year? Assume continuous compounding at 5%.

SOLUTION

$$P.V. = \int_0^8 100e^{-0.05t}\, dt$$

$$= 2000(1 - e^{-0.40})$$

$$\cong 2000(1 - 0.670) \qquad \text{(Table 2 at the end of the book)}$$

$$= 666.$$

The present value is about \$666. $\square$

Revenues usually do not flow at a constant rate R but rather at a time-dependent rate $R(t)$. In general, $R(t)$ tends to increase when business is good and tends to decrease when business is poor.

"Growth companies" are dubbed so because of their continually increasing $R(t)$. The "cyclical companies" owe their name to a fluctuating $R(t)$; $R(t)$ increases for a while, then decreases for a while, and then the cycle is supposedly repeated.

If we postulate a varying $R(t)$ then the present value of an n-year stream is given by the formula

(9.10.2)
$$P.V. = \int_0^n R(t)e^{-rt}\, dt.$$

Here, as before, r is the rate of continuous compounding used to discount future dollars.

You should be able to convince yourself of the validity of this formula on your own. A proof is outlined in the exercises.

Problem 2. What is the present value of an 8-year stream of revenues which is expected to flow at the rate of

$$R(t) = (1 + 0.04t)100$$

dollars per year? As in Problem 1, assume continuous compounding at 5%.

SOLUTION

$$P.V. = \int_0^8 (1 + 0.04t)100e^{-0.05t}\, dt$$

$$= \int_0^8 100e^{-0.05t}\, dt + \int_0^8 4te^{-0.05t}\, dt.$$

From Problem 1,

$$\int_0^8 100e^{-0.05t}\, dt \cong 666.$$

To evaluate the second integral, we first find

$$\int 4te^{-0.05t}\, dt$$

using integration by parts. We set

$$u = 4t, \qquad dv = e^{-0.05t}\, dt,$$
$$du = 4\, dt, \qquad v = -20e^{-0.05t}.$$

$$\int 4te^{-0.05t}\, dt = \int u\, dv = uv - \int v\, du$$

$$= -80te^{-0.05t} + \int 80e^{-0.05t}\, dt$$

$$= -80te^{-0.05t} - 1600e^{-0.05t} + C$$

$$= -80(t + 20)e^{-0.05t} + C.$$

$$\int_0^8 4te^{-0.05t}\, dt = \left[-80(t + 20)e^{-0.05t} \right]_0^8$$

$$= -80[28e^{-0.4} - 20]$$

$$\cong -80[28(0.670) - 20]$$

$$\cong 99.20.$$

The present value is about \$765.20. □

Exercises

Let S be a continuous revenue stream at the rate of $1000 per year.

* 1. What is the present value of the first 4 years of revenue? Assume continuous compounding (a) at 4%; (b) at 8%.
   2. What is the present value of the fifth year with continuous compounding (a) at 4%? (b) at 8%?

Let S be a continuous revenue stream at the rate of $(1000 + 60t)$ per year.

   3. What is the present value of the first 2 years of revenue? Assume continuous compounding (a) at 5%; (b) at 10%.
* 4. What is the present value of the third year with continuous compounding (a) at 5%? (b) at 10%?

   5. Prove the validity of (9.10.2). As we did for (9.10.1), partition the interval $[0, n]$. Take the ith subinterval and argue that

$$m_i(R)m_i(f)\, \Delta t_i \le (P.V.)_i \le M_i(R)M_i(f)\, \Delta t_i,$$

where

$$f(t) = e^{-rt}.$$

9.11   Additional Exercises

Find the areas bounded by the following curves. In each case draw the figure.

* 1. $y^2 = 2x, \quad x^2 = 3y.$ 2. $y^2 = 6x, \quad x^2 + 4y = 0.$
* 3. $y = 4 - x^2, \quad y = x + 2.$ 4. $y = 4 - x^2, \quad x + y + 2 = 0.$
* 5. $y^2 = x, \quad y = x^3.$ 6. $x + y = 2y^2, \quad y = x^3.$
* 7. $4y = x^2 - x^4, \quad x + y + 1 = 0.$ 8. $r^2 = 4 \sin 2\theta.$
* 9. $r = 2 + \sin 3\theta.$ 10. $r = a(\sin 2\theta + \cos 2\theta).$
* 11. $r = \sec \theta + \tan \theta$ and the rays $\theta = 0, \theta = \frac{1}{4}\pi.$
* 12. $xy = a^2$, the x-axis, $x = a$, and $x = 2a.$

* 13. Find the area bounded by the parabola $y = 6 + 4x - x^2$ and the chord joining $(-2, -6)$ to $(4, 6)$.
* 14. Find the area bounded by the semicubical parabola $y^3 = x^2$ and the chord joining $(-1, 1)$ to $(8, 4)$.

The square with one vertex at the origin and opposite vertex at $(1, 1)$ is divided into two parts by each of the following curves. In each instance, find the ratio of the larger area to the smaller area.

* 15. $y^2 = x^3.$ 16. $y = x^3.$
* 17. $\sqrt{x} + \sqrt{y} = 1.$ 18. $y = x^n, \quad n > 1.$
* 19. $y = \sin \frac{1}{2}\pi x.$ 20. $y = xe^{x-1}.$

The region bounded by the following curves is revolved about the x-axis. Find the volume of the solid generated.

*21. $ay^2 = x^3$, $y = 0$, $x = a$.

  22. $y^2 = (2 - x)^3$, $y = 0$, $x = 0$, $x = 1$.

*23. $\sqrt{x} + \sqrt{y} = \sqrt{a}$, $x = 0$, $y = 0$.

  24. $x^{2/3} + y^{2/3} = a^{2/3}$.

*25. $(x/a)^2 + (y/b)^{2/3} = 1$.

  26. $a^2 y^2 = x^2(a^2 - x^2)$.

The region bounded by the following curves is revolved about the y-axis. Find the volume of the solid generated.

*27. $y = x^3$, $y = 0$, $x = 2$.   28. $y^2 = 9 - x$, $x = 0$.

*29. $(x/a)^2 + (y/b)^{2/3} = 1$.   30. $x^2 = 16 - y$, $y = 0$.

Find the length of each curve.

*31. $y^2 = x^3$ from $x = 0$ to $x = \frac{5}{9}$.   (upper branch)

  32. $9y^2 = 4(1 + x^2)^3$ from $x = 0$ to $x = 3$.   (upper branch)

*33. $ay^2 = x^3$ from $x = 0$ to $x = 5a$.   (upper branch)

  34. $y = \log(1 - x^2)$ from $x = 0$ to $x = \frac{1}{2}$.

*35. $x^{2/3} + y^{2/3} = a^{2/3}$.

  36. $y = \log(\operatorname{cosec} x)$ from $x = \frac{1}{6}\pi$ to $x = \frac{1}{2}\pi$.

*37. $(x/a)^{2/3} + (y/b)^{2/3} = 1$ in the first quadrant.

Find the area of the surface generated by revolving the indicated curve about the x-axis.

*38. $y^2 = 2px$ from $x = 0$ to $x = 4p$.

  39. $y^2 = 4x$ from $x = 0$ to $x = 24$.

*40. $6a^2 xy = x^4 + 3a^4$ from $x = a$ to $x = 2a$.

  41. $3x^2 + 4y^2 = 3a^2$.

  42. (*Optional*)   A load of weight W is lifted from the bottom of a shaft h feet deep. Find the work done if the rope used to hoist the load weighs σ pounds per foot.

*43. (*Optional*)   A wedge is cut from a cylinder of radius r by two planes, one perpendicular to the axis of the cylinder and the other passing through a diameter of the section made by the first plane and inclined to this plane at an angle θ. Find the volume of the wedge.

Sequences and Series

10.1 Sequences of Real Numbers

So far our attention has been fixed on functions defined on an interval or on the union of intervals. Here we study functions that are defined on the set of positive integers.

Definition

A function that takes on real values and is defined on the set of positive integers is called a *sequence of real numbers* or, more briefly, a *sequence*.

Thus the functions defined for each positive integer by

$$a(n) = n - 1, \qquad b(n) = \frac{n^2 - 1}{n}, \qquad c(n) = \sqrt{\log n}, \qquad d(n) = \frac{e^n}{n}$$

are all sequences of real numbers. The notions developed for functions in general carry over to sequences. For example, a sequence is said to be bounded (bounded above, bounded below) iff its range is bounded (bounded above, bounded below). If a and b are sequences, then their linear combination $\alpha a + \beta b$ and their product ab, defined by setting

$$(\alpha a + \beta b)(n) = \alpha a(n) + \beta b(n) \quad \text{and} \quad (ab)(n) = a(n) \cdot b(n),$$

are also sequences. If the sequence b does not take on the value 0, then the reciprocal $1/b$ is a sequence and so is the quotient a/b:

$$\frac{1}{b}(n) = \frac{1}{b(n)} \quad \text{and} \quad \frac{a}{b}(n) = \frac{a(n)}{b(n)}.$$

The number $a(n)$, called *the nth term of the sequence a*, is usually abbreviated as a_n. The sequence itself can then be denoted by

$$\{a_1, a_2, a_3, \ldots\}$$

or even by

$$\{a_n\}.$$

For instance, the sequence of reciprocals

$$a(n) = \frac{1}{n} \quad \text{for each positive integer } n$$

can be defined by simply setting

$$a_n = \frac{1}{n}$$

or by writing

$$\{1, \tfrac{1}{2}, \tfrac{1}{3}, \ldots\},$$

or even by writing

$$\left\{\frac{1}{n}\right\}.$$

In our abbreviated notation, we have

$$\alpha\{a_n\} + \beta\{b_n\} = \{\alpha a_n + \beta b_n\}, \qquad \{a_n\}\{b_n\} = \{a_n b_n\}$$

and, provided that none of the b_n are zero,

$$\frac{1}{\{b_n\}} = \left\{\frac{1}{b_n}\right\} \quad \text{and} \quad \frac{\{a_n\}}{\{b_n\}} = \left\{\frac{a_n}{b_n}\right\}.$$

We illustrate the notation by some trivial assertions.

(1) $$5\{1, \tfrac{1}{2}, \tfrac{1}{4}, \tfrac{1}{8}, \ldots\} = \{5, \tfrac{5}{2}, \tfrac{5}{4}, \tfrac{5}{8}, \ldots\}$$

or, equivalently,

$$5\left\{\frac{1}{2^{n-1}}\right\} = \left\{\frac{5}{2^{n-1}}\right\}.$$

(2) $$\{1, 2, 3, \ldots\} + \{1, \tfrac{1}{2}, \tfrac{1}{3}, \ldots\} = \{2, 2\tfrac{1}{2}, 3\tfrac{1}{3}, \ldots\}$$

or, equivalently,

$$\{n\} + \left\{\frac{1}{n}\right\} = \left\{n + \frac{1}{n}\right\}.$$

(3) $$\frac{\{n\}}{\{n+1\}} = \left\{\frac{n}{n+1}\right\}.$$

(4) The sequence $\left\{\dfrac{1}{n}\right\}$ is bounded below by 0 and bounded above by 1:

$$0 \le \frac{1}{n} \le 1 \quad \text{for all } n.$$

(5) The sequence $\{2^n\}$ is bounded below by 2:

$$2 \leq 2^n \quad \text{for all } n.$$

It is not bounded above: there is no fixed number M which satisfies

$$2^n \leq M \quad \text{for all } n.$$

(6) The sequence

$$a_n = (-1)^n 2^n$$

can be written

$$\{-2, 4, -8, 16, -32, 64, \ldots\}.$$

It is unbounded in both senses. □

Definition

A sequence $\{a_n\}$ is said to be

  (i) *increasing*   iff $a_n < a_{n+1}$ for each positive integer n,
 (ii) *nondecreasing* iff $a_n \leq a_{n+1}$ for each positive integer n,
(iii) *decreasing*   iff $a_n > a_{n+1}$ for each positive integer n,
(iv) *nonincreasing*   iff $a_n \geq a_{n+1}$ for each positive integer n.

If any of these four properties holds, the sequence is said to be *monotonic*.

The sequences

$$\{1, \tfrac{1}{2}, \tfrac{1}{3}, \ldots\}, \quad \{1, 1, \tfrac{1}{2}, \tfrac{1}{2}, \tfrac{1}{3}, \tfrac{1}{3}, \ldots\}, \quad \{2, 4, 8, 16, \ldots\}, \quad \{2, 2, 4, 4, 8, 8, 16, 16, \ldots\}$$

are all monotonic, but

$$\{1, \tfrac{1}{2}, 1, \tfrac{1}{3}, 1, \tfrac{1}{4}, \ldots\}$$

is not.

The next examples are less trivial.

Example. The sequence

$$a_n = \frac{n}{n+1}$$

is increasing. It is bounded below by $\tfrac{1}{2}$ and above by 1.

PROOF. Since

$$a_n = \frac{n}{n+1} = \frac{1}{1 + 1/n} < \frac{1}{1 + 1/(n+1)} = \frac{n+1}{(n+1)+1} = a_{n+1},$$

we have

$$a_n < a_{n+1}.$$

This tells us that the sequence is increasing. The bounds are obvious. □

Example.   The sequence $a_n = 2^n/n!$ is nonincreasing.

PROOF.   Since

$$a_{n+1} = \frac{2^{n+1}}{(n+1)!} = \frac{2}{n+1} \cdot \frac{2^n}{n!} = \frac{2}{n+1} a_n,$$

we see that

$$a_n > a_{n+1} \qquad \text{for all } n \geq 2.$$

Since $a_1 = a_2 = 2$, we have

$$a_n \geq a_{n+1} \qquad \text{for all } n. \quad \square$$

Example.   The sequence $\{|c|^n\}$ is bounded if $|c| \leq 1$ and unbounded if $|c| > 1$.

PROOF.   The first statement is obvious. For the second we suppose that $|c| > 1$. To show the unboundedness, we take an arbitrary positive number M and show that there exists a positive integer k for which

$$|c|^k \geq M.$$

A suitable k is one that satisfies

$$k \geq \frac{\log M}{\log |c|},$$

for then we have

$$k \log |c| \geq \log M, \qquad \log |c|^k \geq \log M,$$

and consequently

$$|c|^k \geq M. \quad \square$$

Since sequences are defined on the set of positive integers and not on an interval, they are not directly susceptible to the methods of calculus. Fortunately, we can often get around this difficulty by dealing initially, not with the sequence itself, but with a function which is defined on all of $[1, \infty)$ and agrees with the given sequence on the integers.

We illustrate this idea in the next two examples. In the first example, we study the sequence

$$a_n = \frac{e^n}{n}$$

by directing our attention first to the function

$$f(x) = \frac{e^x}{x}.$$

Example.   The sequence

$$a_n = \frac{e^n}{n}$$

is an increasing sequence that is unbounded above.

PROOF.   We begin by setting

$$f(x) = \frac{e^x}{x}.$$

It is not hard to see that

$$f'(x) = \frac{xe^x - e^x}{x^2} = \frac{e^x(x - 1)}{x^2}.$$

Since

$$f'(x) \text{ is } \begin{cases} 0, & \text{at } x = 1 \\ \text{positive}, & \text{for } x > 1 \end{cases},$$

the function f increases on $[1, \infty)$. Obviously then the sequence

$$a_n = \frac{e^n}{n}$$

is itself increasing. To see the unboundedness of the sequence, observe that

$$e^n > \frac{n^p}{p!} \quad \text{for each positive integer } p.$$

(See Exercise 61, Section 5.4). Setting $p = 2$, we have

$$e^n > \frac{n^2}{2!}$$

and thus

$$\frac{e^n}{n} > \frac{n}{2}. \quad \square$$

Example.   The sequence

$$a_n = (n + 2)^{1/(n + 2)}$$

is a decreasing sequence.

PROOF.   We could compare a_n with a_{n+1} directly, but it is easier to consider the function

$$f(x) = x^{1/x}$$

instead. Since

$$f(x) = e^{(1/x) \log x},$$

we have

$$f'(x) = e^{(1/x) \log x} \frac{d}{dx} \left(\frac{1}{x} \log x \right) = x^{1/x} \left(\frac{1 - \log x}{x^2} \right).$$

For $x > e$, $f'(x) < 0$. This tells us that f decreases on $[e, \infty)$. Since $3 > e$, f decreases on $[3, \infty)$. It follows that $\{a_n\}$ is a decreasing sequence. $\square$

Exercises

Discuss the boundedness and monotonicity of the sequence whose nth term is given by the following.

*1. $\dfrac{2}{n}$.

2. $\dfrac{(-1)^n}{n}$.

*3. $\sqrt{n}$.

4. $\dfrac{n + (-1)^n}{n}$.

*5. $\dfrac{n}{2^n}$.

6. $\dfrac{n-1}{n}$.

*7. $\dfrac{n^2}{n+1}$.

8. $\dfrac{n+1}{n^2}$.

*9. $\dfrac{4n}{\sqrt{4n^2+1}}$.

10. $\dfrac{4^n}{2^n + 100}$.

*11. $\dfrac{2^n}{4^n+1}$.

12. $\dfrac{10^{10}\sqrt{n}}{n+1}$.

*13. $(-1)^n\sqrt{n}$.

14. $\dfrac{n+1}{n}$.

*15. $\log\left(\dfrac{2n}{n+1}\right)$.

16. $\left(-\tfrac{1}{2}\right)^n$.

*17. $\dfrac{(n+1)^2}{n^2}$.

18. $\sqrt{4 - \dfrac{1}{n}}$.

*19. $\dfrac{2^n - 1}{2^n}$.

20. $(0.9)^n$.

*21. $\dfrac{1}{n} - \dfrac{1}{n+1}$.

22. $\dfrac{\log(n+2)}{n+2}$.

*23. $\log\left(\dfrac{n+1}{n}\right)$.

24. $\dfrac{\sqrt{n+1}}{\sqrt{n}}$.

*25. $(-1)^n \dfrac{n^2}{3^n}$.

26. $\dfrac{1 - (\tfrac{1}{2})^n}{(\tfrac{1}{2})^n}$.

*27. $\cos n\pi$.

28. Show that linear combinations and products of bounded sequences are bounded.
29. Show that the sequence

$$a_n = (c^n + d^n)^{1/n} \qquad \text{with} \quad 0 < c < d$$

is bounded and monotonic.

10.2   The Limit of a Sequence

The gist of

$$\lim_{x \to c} f(x) = l$$

is that we can make $f(x)$ as close as we wish to the number l simply by requiring that x be sufficiently close to c. The gist of

$$\lim_{n \to \infty} a_n = l$$

(read "the limit of a_n as n tends to infinity is l") is that we can make a_n as close as we wish to the number l simply by requiring that n be sufficiently large.

Definition of Limit

$$\lim_{n \to \infty} a_n = l \quad \text{iff} \quad \begin{cases} \text{for each } \epsilon > 0 \text{ there exists an integer } k > 0 \text{ such that,} \\ \quad\quad \text{if } n \geq k, \quad\quad \text{then } |a_n - l| < \epsilon. \end{cases}$$

Example

$$\lim_{n \to \infty} \frac{1}{n} = 0.$$

PROOF. We take $\epsilon > 0$ and note that if $k > 1/\epsilon$, then $1/k < \epsilon$. Consequently, for $n \geq k$, we have

$$\frac{1}{n} \leq \frac{1}{k} < \epsilon \quad \text{and thus} \quad \left| \frac{1}{n} - 0 \right| < \epsilon. \quad \square$$

Example

$$\lim_{n \to \infty} \frac{2n - 1}{n} = 2.$$

PROOF. Let $\epsilon > 0$. We must show that there exists an integer k such that

$$\left| \frac{2n - 1}{n} - 2 \right| < \epsilon \quad\quad \text{for all } n \geq k.$$

Since

$$\left| \frac{2n - 1}{n} - 2 \right| = \left| \frac{2n - 1 - 2n}{n} \right| = \left| -\frac{1}{n} \right| = \frac{1}{n},$$

again we need only choose $k > 1/\epsilon$. $\quad \square$

Example. For the decimal fractions

$$b_n = 0.\overbrace{33 \ldots 3}^{n}$$

we have

$$\lim_{n \to \infty} b_n = \frac{1}{3}.$$

PROOF. Let $\epsilon > 0$. We have

$$\left| b_n - \frac{1}{3} \right| = \left| 0.\overbrace{33 \ldots 3}^{n} - \frac{1}{3} \right| = \left| \frac{0.\overbrace{99 \ldots 9}^{n} - 1}{3} \right| = \frac{1}{3} \cdot \frac{1}{10^n} < \frac{1}{10^n}.$$

By choosing k such that $1/10^k < \epsilon$, we ensure that

$$\left| b_n - \frac{1}{3} \right| < \epsilon \quad\quad \text{for all } n \geq k. \quad \square$$

Limit Theorems

The limit process for sequences is so similar to the limit process you have already studied that you will probably find that you can prove most of the limit theorems on your own. In any case, it is a good idea to try to come up with your own proofs and refer to the given proofs only if necessary.

Uniqueness Theorem

If
$$\lim_{n \to \infty} a_n = l \quad \text{and} \quad \lim_{n \to \infty} a_n = m,$$
then
$$l = m.$$

PROOF.   If $l \neq m$, then
$$\tfrac{1}{2}|l - m| > 0.$$
From
$$\lim_{n \to \infty} a_n = l \quad \text{and} \quad \lim_{n \to \infty} a_n = m,$$
we conclude that there exists k_1 such that
$$\text{if} \quad n \geq k_1, \quad \text{then} \quad |a_n - l| < \tfrac{1}{2}|l - m|$$
and there exists k_2 such that
$$\text{if} \quad n \geq k_2, \quad \text{then} \quad |a_n - m| < \tfrac{1}{2}|l - m|.$$
For $n \geq \max\{k_1, k_2\}$ we have
$$|a_n - l| + |a_n - m| < |l - m|.$$
But by the triangle inequality we have
$$|a_n - l| + |a_n - m| = |l - a_n| + |a_n - m| \geq |l - m|.$$
Combining the last two statements, we obtain
$$|l - m| < |l - m|.$$
The hypothesis $l \neq m$ has led to an absurdity. We conclude that $l = m$.   □

Definition

A sequence that has a limit is said to be *convergent*. A sequence that has no limit is said to be *divergent*.

Instead of writing

$$\lim_{n \to \infty} a_n = l,$$

we shall use the simpler notation

$$a_n \to l. \qquad\qquad \text{(read ``}a_n \text{ converges to } l\text{'')}$$

Theorem 10.2.1

A convergent sequence is bounded.

PROOF. Suppose that

$$a_n \to l$$

and take any positive number: 1, for instance. Using 1 as ϵ, you can see that there must exist k such that

$$|a_n - l| < 1 \qquad \text{for all} \quad n \geq k.$$

This means that

$$|a_n| < 1 + |l| \qquad \text{for all} \quad n \geq k,$$

and consequently

$$|a_n| \leq \max \{|a_1|, |a_2|, \ldots, |a_{k-1}|, 1 + |l|\} \qquad \text{for all} \quad n.$$

This proves that $\{a_n\}$ is bounded. □

Since convergent sequences are bounded, unbounded sequences are divergent. The following sequences,

$$\left\{ \frac{n}{2} \right\}, \quad \{\sqrt{\log n}\}, \quad \left\{ \frac{n}{\cos (n + 1)} \right\}, \quad \left\{ \frac{e^n}{n} \right\},$$

being unbounded, are all divergent. Boundedness does not imply convergence. As a counterexample, take the oscillating sequence

$$\{1, 0, 1, 0, \ldots\}.$$

This sequence is certainly bounded (above by 1 and below by 0), but obviously it does not converge.

Boundedness together with monotonicity does imply convergence.

Theorem 10.2.2

A bounded, nondecreasing sequence converges to the least upper bound of its range; a bounded, nonincreasing sequence converges to the greatest lower bound of its range.

PROOF.   Suppose that $\{a_n\}$ is bounded and nondecreasing. Let ϵ be an arbitrary positive number. If
$$l = \text{lub } \{a_n : n \text{ a positive integer}\},$$
then there must exist a_k such that
$$l - \epsilon < a_k.$$
(Otherwise $l - \epsilon$ would be an upper bound less than l.) By the monotonicity, we see that
$$\text{if } n \geq k, \quad \text{then } a_k \leq a_n.$$
Thus we have
$$l - \epsilon < a_n \leq l \quad \text{for all } n \geq k.$$
It follows that $|a_n - l| < \epsilon$ and
$$a_n \to l.$$
The nonincreasing case can be handled in a similar manner.   □

Example.   Take the sequence
$$\{(3^n + 4^n)^{1/n}\}.$$
Since
$$3 = (3^n)^{1/n} < (3^n + 4^n)^{1/n} < (2 \cdot 4^n)^{1/n} = 2^{1/n} \cdot 4 \leq 8,$$
the sequence is bounded. Note that
$$(3^n + 4^n)^{(n+1)/n} = (3^n + 4^n)^{1/n}(3^n + 4^n) = (3^n + 4^n)^{1/n}3^n + (3^n + 4^n)^{1/n}4^n$$
$$> 3 \cdot 3^n + 4 \cdot 4^n = 3^{n+1} + 4^{n+1}.$$
Taking the $(n + 1)$st root of both extremes, we get
$$(3^n + 4^n)^{1/n} > (3^{n+1} + 4^{n+1})^{1/n+1}.$$
This tells us that the sequence is a decreasing one. Being bounded, it must be convergent. (Later you will be asked to show that the limit is 4.)   □

Theorem

If $a_n \to l$ and $b_n \to m$, then

(i) $a_n + b_n \to l + m,$
(ii) $\alpha a_n \to \alpha l \quad$ for each real $\alpha,$
(iii) $a_n b_n \to lm.$

If, in addition, $m \neq 0$ and b_n is never 0, then

(iv) $\dfrac{1}{b_n} \to \dfrac{1}{m} \quad$ and

(v) $\dfrac{a_n}{b_n} \to \dfrac{l}{m}.$

PROOF. Parts (i) and (ii) are left as exercises. To prove (iii), we set $\epsilon > 0$. For each n,

$$|a_n b_n - lm| = |(a_n b_n - a_n m) + (a_n m - lm)|$$
$$\leq |a_n|\,|b_n - m| + |m|\,|a_n - l|.$$

Since $\{a_n\}$ is convergent, $\{a_n\}$ is bounded; that is, there exists $M > 0$ such that

$$|a_n| \leq M \quad \text{for all } n.$$

Since $|m| < |m| + 1$, we have

(1) $$|a_n b_n - lm| \leq M|b_n - m| + (|m| + 1)|a_n - l|.$$

Since $b_n \to m$, we know that there exists k_1 such that

$$\text{if} \quad n \geq k_1, \qquad \text{then} \quad |b_n - m| < \frac{\epsilon}{2M}.$$

Since $a_n \to l$, we know that there exists k_2 such that

$$\text{if} \quad n \geq k_2, \qquad \text{then} \quad |a_n - l| < \frac{\epsilon}{2(|m| + 1)}.$$

For $n \geq \max\{k_1, k_2\}$ both conditions hold, and consequently

$$M|b_n - m| + (|m| + 1)|a_n - l| < \frac{\epsilon}{2} + \frac{\epsilon}{2} = \epsilon.$$

In view of (1), we can conclude that

$$\text{if} \quad n \geq \max\{k_1, k_2\}, \qquad \text{then} \quad |a_n b_n - lm| < \epsilon.$$

This proves that

$$a_n b_n \to lm.$$

To prove (iv), once again we set $\epsilon > 0$. In the first place

$$\left|\frac{1}{b_n} - \frac{1}{m}\right| = \left|\frac{m - b_n}{b_n m}\right| = \frac{|b_n - m|}{|b_n|\,|m|}.$$

Since $b_n \to m$ and $|m|/2 > 0$, there exists k_1 such that

$$\text{if} \quad n \geq k_1, \qquad \text{then} \quad |b_n - m| < \frac{|m|}{2}.$$

This tells us that for $n \geq k_1$ we have

$$|b_n| > \frac{|m|}{2} \quad \text{and thus} \quad \frac{1}{|b_n|} < \frac{2}{|m|}.$$

Thus for $n \geq k_1$ we have

(2) $$\left|\frac{1}{b_n} - \frac{1}{m}\right| \leq \frac{2}{|m|^2}|b_n - m|.$$

Since $b_n \to m$, there exists k_2 such that

$$\text{if} \quad n \geq k_2, \qquad \text{then} \quad |b_n - m| < \frac{\epsilon|m|^2}{2}.$$

Thus for $n \geq k_2$ we have

$$\frac{2}{|m|^2} |b_n - m| < \epsilon.$$

In view of (2), we can be sure that

$$\text{if} \quad n \geq \max\{k_1, k_2\}, \qquad \text{then} \quad \left| \frac{1}{b_n} - \frac{1}{m} \right| < \epsilon.$$

This proves that

$$\frac{1}{b_n} \to \frac{1}{m}.$$

The proof of (v) is now easy:

$$\frac{a_n}{b_n} = a_n \cdot \frac{1}{b_n} \to l \cdot \frac{1}{m} = \frac{l}{m}. \quad \square$$

We are now in a position to handle any rational sequence

$$a_n = \frac{\alpha_k n^k + \alpha_{k-1} n^{k-1} + \cdots + \alpha_0}{\beta_j n^j + \beta_{j-1} n^{j-1} + \cdots + \beta_0}.$$

To determine the behavior of such a sequence we need only divide both numerator and denominator by the highest power of n that occurs.

Example

$$\frac{3n^4 - 2n^2 + 1}{n^5 - 3n^3} = \frac{\dfrac{3}{n} - \dfrac{2}{n^3} + \dfrac{1}{n^5}}{1 - \dfrac{3}{n^2}} \to \frac{0}{1} = 0. \quad \square$$

Example

$$\frac{1 - 4n^7}{n^7 + 12n} = \frac{\dfrac{1}{n^7} - 4}{1 + \dfrac{12}{n^6}} \to \frac{-4}{1} = -4. \quad \square$$

Example

$$\frac{n^4 - 3n^2 + n + 2}{n^3 + 7n} = \frac{1 - \dfrac{3}{n^2} + \dfrac{1}{n^3} + \dfrac{2}{n^4}}{\dfrac{1}{n} + \dfrac{7}{n^3}}.$$

Since the numerator tends to 1 and the denominator tends to 0, the sequence is unbounded. It can therefore not converge. $\square$

Theorem

$$a_n \to l \quad \text{iff} \quad a_n - l \to 0 \quad \text{iff} \quad |a_n - l| \to 0.$$

The proof is left as an exercise. $\square$

The Pinching Theorem for Sequences

Suppose that

$$a_n \leq b_n \leq c_n \quad \text{for all } n \text{ sufficiently large.}$$

If

$$a_n \to l \quad \text{and} \quad c_n \to l,$$

then

$$b_n \to l.$$

Once again the proof is left as an exercise. $\square$

As an immediate and obvious consequence of the pinching theorem we have the following corollary.

Corollary

Suppose that

$$|b_n| \leq |a_n| \quad \text{for all } n \text{ sufficiently large.}$$

If $|a_n| \to 0,$ then $|b_n| \to 0.$

Example

$$\frac{\cos \pi n}{n} \to 0$$

since

$$\left| \frac{\cos \pi n}{n} \right| \leq \frac{1}{n} \quad \text{and} \quad \frac{1}{n} \to 0. \quad \square$$

Example

$$\sqrt{4 + \left(\frac{1}{n}\right)^2} \to 2$$

since

$$2 \leq \sqrt{4 + \left(\frac{1}{n}\right)^2} \leq \sqrt{4 + 4\left(\frac{1}{n}\right) + \left(\frac{1}{n}\right)^2} = 2 + \frac{1}{n}$$

and

$$2 + \frac{1}{n} \to 2. \quad \square$$

Example

$$\lim_{n \to \infty} \left(1 + \frac{1}{n}\right)^n = e.$$

PROOF.   You have already seen that

$$\left(1 + \frac{1}{n}\right)^n \le e \le \left(1 + \frac{1}{n}\right)^{n+1}.$$

From this it follows that

$$\left|\left(1 + \frac{1}{n}\right)^n - e\right| \le \left|\left(1 + \frac{1}{n}\right)^n - \left(1 + \frac{1}{n}\right)^{n+1}\right|$$

$$= \left(1 + \frac{1}{n}\right)^n\left|1 - \left(1 + \frac{1}{n}\right)\right| = \left(1 + \frac{1}{n}\right)^n\left(\frac{1}{n}\right) \le \frac{e}{n}.$$

In short, we have the estimate

$$\left|\left(1 + \frac{1}{n}\right)^n - e\right| \le \frac{e}{n}.$$

Since

$$\frac{e}{n} \to 0,$$

we have

$$\left|\left(1 + \frac{1}{n}\right)^n - e\right| \to 0 \quad \text{and therefore} \quad \left(1 + \frac{1}{n}\right)^n \to e. \quad \square$$

The sequences

$$\left\{\cos\frac{\pi}{n}\right\}, \quad \left\{\log\left(\frac{n}{n+1}\right)\right\}, \quad \{e^{1/n}\}, \quad \left\{\tan\left(\sqrt{\frac{\pi^2 n^2 - 8}{16n^2}}\right)\right\}$$

are all of the form

$$\{f(c_n)\}$$

with f a continuous function. Such sequences are easy to deal with. The basic idea is this: when a continuous function is applied to a convergent sequence, the result is itself a convergent sequence. More precisely, we have the following theorem.

Theorem

Suppose that

$$c_n \to c$$

and that, for each n, c_n is in the domain of f. If f is continuous at c, then

$$f(c_n) \to f(c).$$

PROOF. We assume that f is continuous at c and take $\epsilon > 0$. From the continuity of f at c we know that there exists $\delta > 0$ such that

$$\text{if} \quad |x - c| < \delta, \qquad \text{then} \quad |f(x) - f(c)| < \epsilon.$$

Since $c_n \to c$, we know that there exists a positive integer k such that

$$\text{if} \quad n \geq k, \qquad \text{then} \quad |c_n - c| < \delta.$$

It follows therefore that

$$\text{if} \quad n \geq k, \qquad \text{then} \quad |f(c_n) - f(c)| < \epsilon. \quad \Box$$

Examples

(1) Since

$$\frac{\pi}{n} \to 0$$

and the cosine function is continuous at 0,

$$\cos \frac{\pi}{n} \to \cos 0 = 1. \quad \Box$$

(2) Since

$$\frac{n}{n + 1} = \frac{1}{1 + 1/n} \to 1$$

and the logarithm function is continuous at 1,

$$\log \left(\frac{n}{n + 1} \right) \to \log 1 = 0. \quad \Box$$

(3) Since

$$\frac{1}{n} \to 0$$

and the exponential function is continuous at 0,

$$e^{1/n} \to e^0 = 1. \quad \Box$$

(4) Since

$$\frac{\pi^2 n^2 - 8}{16n^2} = \frac{\pi^2 - 8/n^2}{16} \to \frac{\pi^2}{16}$$

and the function $f(x) = \tan \sqrt{x}$ is continuous at $\pi^2/16$,

$$\tan \left(\sqrt{\frac{\pi^2 n^2 - 8}{16n^2}} \right) \to \tan \left(\sqrt{\frac{\pi^2}{16}} \right) = \tan \frac{\pi}{4} = 1. \quad \Box$$

(5) Since

$$\frac{2n + 1}{n} + \left(5 - \frac{1}{n^2} \right) \to 7$$

and the square-root function is continuous at 7, we have

$$\sqrt{\frac{2n + 1}{n} + \left(5 - \frac{1}{n^2}\right)} \to \sqrt{7}. \quad \square$$

(6) For each $x > 0$,

$$x^{1/n} \to 1.$$

To see this, we note that

$$\frac{1}{n} \log x \to 0.$$

Since the exponential function is continuous at 0,

$$x^{1/n} = e^{(1/n)\log x} \to e^0 = 1. \quad \square$$

(7) Since the absolute-value function is everywhere continuous,

$$a_n \to l \quad \text{implies} \quad |a_n| \to |l|. \quad \square$$

Exercises

State whether or not the indicated sequence converges and, if it does, find the limit.

*1. $\dfrac{2}{n}$.

2. $\dfrac{(-1)^n}{n}$.

*3. $\sqrt{n}$.

4. $\dfrac{n - 1}{n}$.

*5. $\dfrac{n + (-1)^n}{n}$.

6. $\sin\dfrac{\pi}{2n}$.

*7. $\dfrac{n^2}{n + 1}$.

8. $\dfrac{n + 1}{n^2}$.

*9. $\dfrac{4n}{\sqrt{n^2 + 1}}$.

10. $\dfrac{4^n}{2^n + 10^6}$.

*11. $\dfrac{2^n}{4^n + 1}$.

12. $\dfrac{10^{10}\sqrt{n}}{n + 1}$.

*13. $(-1)^n\sqrt{n}$.

14. 2^n.

*15. $\log\left(\dfrac{2n}{n + 1}\right)$.

16. $(-\frac{1}{2})^n$.

*17. $\dfrac{n^4 - 1}{n^4 + n - 6}$.

18. $\dfrac{n^5}{17n^4 + 12}$.

*19. $\dfrac{(n + 1)^2}{n^2}$.

20. $\sqrt{4 - \dfrac{1}{n}}$.

*21. $\dfrac{2^n - 1}{2^n}$.

22. $\dfrac{n^2}{\sqrt{2n^4 + 1}}$.

*23. $\dfrac{1}{n} - \dfrac{1}{n + 1}$.

24. $\cos n\pi$.

*25. $(0.9)^n$.

26. $(0.9)^{-n}$.

*27. $\log n - \log (n + 1)$.

*28. $\dfrac{\sqrt{n + 1}}{2\sqrt{n}}$.

29. $\dfrac{\sqrt{n}\sin(e^n\pi)}{n + 1}$.

*30. $\left(1 + \dfrac{2}{n}\right)^n$.

31. Prove that a bounded nonincreasing sequence converges to the greatest lower bound of its range.

32. Show that

$$\left(1 + \frac{1}{n}\right)^{n+1} \to e.$$

33. Prove the pinching theorem for sequences.

10.3   Some Important Limits

The following limits occur so frequently that it is actually worth remembering them:

(10.3.1) $x^{1/n} \to 1$ for each $x > 0.$

(10.3.2) $x^n \to 0$ for $|x| < 1.$

(10.3.3) $\dfrac{x^n}{n!} \to 0$ for each real $x.$

(10.3.4) $\dfrac{1}{n^\alpha} \to 0$ for each $\alpha > 0.$

(10.3.5) $\dfrac{\log n}{n} \to 0.$

(10.3.6) $n^{1/n} \to 1.$

(10.3.7) $\dfrac{n^\alpha}{e^n} \to 0$ for each $\alpha > 0.$

(10.3.8) $\left(1 + \dfrac{x}{n}\right)^n \to e^x$ for each real $x.$

PROOF OF (10.3.1). Given in the last section.   □

PROOF OF (10.3.2). Suppose that $|x| < 1$. Let $\epsilon > 0$. By (10.3.1),

$$\epsilon^{1/n} \to 1.$$

Thus there exists an integer $k > 0$ such that

$$|x| < \epsilon^{1/k}. \hspace{3cm} \text{(explain)}$$

This means that

$$|x^k| = |x|^k < \epsilon,$$

and thus

$$\text{if}\ \ n \geq k, \ \ \ \ \text{then}\ \ |x^n| < \epsilon. \ \ \square$$

PROOF OF (10.3.3). Choose an integer k such that

$$k > |x|.$$

For $n > k$,

$$n! = \underbrace{[n(n-1)\cdots(k+1)]}_{n-k \text{ terms}}k! > (k^{n-k})(k!) = (k^n)(k^{-k})(k!).$$

Consequently

$$\frac{|x^n|}{n!} = \frac{|x|^n}{n!} < \frac{|x|^n}{(k^n)(k^{-k})(k!)} = \left(\frac{|x|}{k}\right)^n \frac{k^k}{k!}.$$

Since $|x|/k < 1$,

$$\left(\frac{|x|}{k}\right)^n \to 0 \quad \text{and therefore} \quad \left(\frac{|x|}{k}\right)^n \frac{k^k}{k!} \to 0.$$

It follows that

$$\frac{|x^n|}{n!} \to 0. \quad \square$$

PROOF OF (10.3.4).   Since $\alpha > 0$, there exists an odd positive integer p such that

$$\frac{1}{p} < \alpha.$$

Obviously

$$0 < \frac{1}{n^\alpha} = \left(\frac{1}{n}\right)^\alpha \le \left(\frac{1}{n}\right)^{1/p}.$$

Since

$$\frac{1}{n} \to 0 \quad \text{and} \quad f(x) = x^{1/p} \text{ is continuous at } 0,$$

we have

$$\left(\frac{1}{n}\right)^{1/p} \to 0 \quad \text{and thus} \quad \frac{1}{n^\alpha} \to 0. \quad \square$$

PROOF OF (10.3.5)

$$\frac{\log n}{n} = \frac{1}{n}\int_1^n \frac{dt}{t} \le \frac{1}{n}\int_1^n \frac{dt}{\sqrt{t}} = \frac{2}{n}\left[\sqrt{t}\,\right]_1^n = \frac{2}{n}(\sqrt{n}-1) = 2\left(\frac{1}{\sqrt{n}} - \frac{1}{n}\right) \to 0. \quad \square$$

PROOF OF (10.3.6).   We have

$$n^{1/n} = e^{(1/n)\log n}.$$

Since

$$(1/n)\log n \to 0$$

and the exponential function is continuous at 0,

$$n^{1/n} \to e^0 = 1. \quad \square$$

PROOF OF (10.3.7).   Let p be a positive integer greater than α. Obviously

$$\frac{n^\alpha}{e^n} \le \frac{n^p}{e^n}.$$

Since

$$e^n > \frac{n^{p+1}}{(p+1)!},\qquad\text{(Exercise 61, Section 5.4)}$$

we have

$$\frac{n^p}{e^n} < \frac{(p+1)!}{n}.$$

Thus

$$\frac{n^\alpha}{e^n} < \frac{(p+1)!}{n}.$$

Since $(p+1)!$ is fixed,

$$\frac{(p+1)!}{n} \to 0,\quad\text{and consequently}\quad \frac{n^\alpha}{e^n} \to 0.\ \ \square$$

PROOF OF (10.3.8).   For $x = 0$, the result is obvious. For $x \ne 0$,

$$\log\left(1 + \frac{x}{n}\right)^n = n\log\left(1 + \frac{x}{n}\right) = x\left[\frac{\log(1 + x/n) - \log 1}{x/n}\right].$$

The crux here is to recognize the bracketed expression as a difference quotient for the logarithm function. Once we see this, we can easily obtain

$$\lim_{n\to\infty}\left[\frac{\log(1 + x/n) - \log 1}{x/n}\right] = \lim_{h\to 0}\left[\frac{\log(1 + h) - \log 1}{h}\right] = 1.$$

This tells us that

$$\log\left(1 + \frac{x}{n}\right)^n \to x,$$

and consequently

$$\left(1 + \frac{x}{n}\right)^n = e^{\log(1 + x/n)^n} \to e^x.\ \ \square$$

Exercises

State whether or not the indicated sequence converges and, if it does, find the limit.

*1. $2^{2/n}$.

*2. $e^{-\alpha/n}$.

* 3. $\left(\dfrac{2}{n}\right)^n$.

4. $\dfrac{n}{2^n}$.

5. $\dfrac{\log(n+1)}{n}$.

6. $\log\left(\dfrac{n+1}{n}\right)$.

*7. $\dfrac{\log_{10} n}{n}$.

* 8. $n^{\alpha/n},\ \alpha > 0$.

*9. $\dfrac{3^n}{4^n}$.

10. $\dfrac{3^{n+1}}{4^{n-1}}$.

*11. $n^{1/(n+2)}$.

12. $(n+2)^{1/n}$.

*13. $\dfrac{n^2}{3^n}$.

14. $(n+2)^{1/(n+2)}$.

*15. $\displaystyle\int_0^n e^{-x}\,dx$.

16. $\displaystyle\int_{-n}^0 e^{2x}\,dx$.

*17. $\displaystyle\int_0^n e^{-nx}\,dx$.

18. $\displaystyle\int_{-1/n}^{1/n} \sin x^3\,dx$.

*19. $\displaystyle\int_{-1+1/n}^{1-1/n} \dfrac{dx}{\sqrt{1-x^2}}$.

20. $\displaystyle\int_n^{n+1} e^{-x^2}\,dx$.

*21. $\displaystyle\int_{1/n}^1 \dfrac{dx}{\sqrt{x}}$.

*22. $\left(1-\dfrac{1}{n}\right)^n$.

23. $n^2 \sin n\pi$.

*24. $n^2 \sin \dfrac{\pi}{n}$.

25. Show that

$$\lim_{n\to\infty} (\sqrt{n+1} - \sqrt{n}) = 0.$$

26. Show that

$$\text{if} \quad 0 < c < d, \quad \text{then} \quad (c^n + d^n)^{1/n} \to d.$$

*27. Find

$$\lim_{n\to\infty} 2n \sin (\pi/n).$$

What is the geometric significance of this limit?

HINT: Think of a regular polygon of n sides inscribed in the unit circle.

Find the following limits.

*28. $\displaystyle\lim_{n\to\infty} \dfrac{1+2+\cdots+n}{n^2}$.   HINT: $1 + 2 + \cdots + n = \dfrac{n(n+1)}{2}$.

29. $\displaystyle\lim_{n\to\infty} \dfrac{1^2+2^2+\cdots+n^2}{(1+n)(2+n)}$.   HINT: $1^2 + 2^2 + \cdots + n^2 = \dfrac{n(n+1)(n+2)}{6}$.

*30. $\displaystyle\lim_{n\to\infty} \dfrac{1^3+2^3+\cdots+n^3}{2n^4+n-1}$.   HINT: $1^3 + 2^3 + \cdots + n^3 = \dfrac{n^2(n+1)^2}{4}$.

31. A sequence $\{a_n\}$ is said to be a *Cauchy sequence* [after the French mathematician Augustin-Louis (baron de) Cauchy, 1789–1857] iff

> for each $\epsilon > 0$ there exists a positive integer k such that
> $$|a_n - a_m| < \epsilon \quad \text{for all} \quad m, n \geq k.$$

Show that

$$
\boxed{\text{every convergent sequence is a Cauchy sequence.}} \; †
$$

32. Let

$$
m_n = \frac{1}{n}(a_1 + a_2 + \cdots + a_n)
$$

be the sequence of arithmetic means. (a) Prove that

$$
\text{if} \quad \{a_n\} \text{ is increasing,} \qquad \text{then} \quad \{m_n\} \text{ is increasing.}
$$

(b) (*Optional*) Prove that

$$
\text{if} \quad a_n \to 0, \qquad \text{then} \quad m_n \to 0.
$$

HINT: Choose an integer $j > 0$ such that if

$$
n \geq j, \qquad \text{then} \quad a_n < \frac{\epsilon}{2}.
$$

Then for $n \geq j$,

$$
|m_n| < \frac{|a_1 + a_2 + \cdots + a_j|}{n} + \frac{\epsilon}{2}\left(\frac{n - j}{n}\right).
$$

(*Optional*) State whether or not the indicated sequence converges and, if it does, find the limit.

* 33. $\left(\dfrac{n + 1}{n + 2}\right)^n$.

34. $\left(1 + \dfrac{1}{n}\right)^{n^2}$.

* 35. $\left(1 + \dfrac{1}{n^2}\right)^n$.

36. $2^{n^2} \cdot n^{-n}$.

* 37. $\left(t + \dfrac{x}{n}\right)^n$, $\quad t > 0, \quad x > 0$.

38. $\displaystyle\int_{-n}^{n} \frac{dx}{1 + x^2}$.

10.4   Some Comments on Notation

On Indexing Sequences

In indexing sequences it is not necessary to begin with the index 1. For example, the sequence

$$
\{1, \tfrac{1}{2}, \tfrac{1}{4}, \ldots\}
$$

indicated by

$$
\{a_1, a_2, a_3, \ldots\} \qquad \text{with } a_n = (\tfrac{1}{2})^{n-1}
$$

---

† It is also true that every Cauchy sequence is convergent, but this is much more difficult to prove.

can also be indicated by

$$\{b_0, b_1, b_2, \ldots\} \qquad \text{with } b_n = (\tfrac{1}{2})^n,$$

or more generally by

$$\{c_p, c_{p+1}, c_{p+2}, \ldots\} \qquad \text{with } c_n = (\tfrac{1}{2})^{n-p}.$$

A Review of Σ-Notation for Finite Sums

The symbol Σ is the capital Greek letter "sigma." We use the expression

(1)
$$\sum_{k=0}^{n} a_k$$

(read "the sum of the a sub k from k equals 0 to n") to indicate the sum

$$a_0 + a_1 + \cdots + a_n.$$

More generally, if $n \geq m$, we use

(2)
$$\sum_{k=m}^{n} a_k$$

to indicate the sum

$$a_m + a_{m+1} + \cdots + a_n.$$

In (1) and (2) the letter "k" is being used as a "dummy" variable. That is, it can be replaced by any letter not already engaged. For instance,

$$\sum_{i=3}^{7} a_i, \qquad \sum_{j=3}^{7} a_j, \qquad \sum_{k=3}^{7} a_k$$

can all be used to indicate the sum

$$a_3 + a_4 + a_5 + a_6 + a_7.$$

Translating

$$(a_0 + \cdots + a_n) + (b_0 + \cdots + b_n) = (a_0 + b_0) + \cdots + (a_n + b_n),$$
$$\alpha(a_0 + \cdots + a_n) = \alpha a_0 + \cdots + \alpha a_n,$$
$$(a_0 + \cdots + a_m) + (a_{m+1} + \cdots + a_n) = a_0 + \cdots + a_n,$$

into the Σ-notation, we have

$$\sum_{k=0}^{n} a_k + \sum_{k=0}^{n} b_k = \sum_{k=0}^{n} (a_k + b_k), \qquad \alpha \sum_{k=0}^{n} a_k = \sum_{k=0}^{n} \alpha a_k,$$

$$\sum_{k=0}^{m} a_k + \sum_{k=m+1}^{n} a_k = \sum_{k=0}^{n} a_k.$$

At times it is convenient to change indices. In doing this, note that

$$\sum_{k=j}^{n} a_k = \sum_{k=0}^{n-j} a_{k+j}.$$

Both are abbreviations for $a_j + a_{j+1} + \cdots + a_n.$

10.5   Infinite Series

To place our discussion on a solid footing, we begin by establishing the basic terminology. Our starting point is an arbitrary real sequence

$$\{a_0, a_1, a_2, \ldots\}.$$

Definition

(1) The *infinite series*

$$\sum_{k=0}^{\infty} a_k$$

is the sequence

$$\{a_0, a_0 + a_1, a_0 + a_1 + a_2, \ldots\}.$$

(2) The number a_k is called the kth *term of the series*, and the number

$$s_n = a_0 + a_1 + \cdots + a_n = \sum_{k=0}^{n} a_k$$

is called the *nth partial sum.*

(3) To say that the series $\sum_{k=0}^{\infty} a_k$ *converges* to the number l is to say that the sequence $\{s_n\}$ converges to the number l. To say that the series $\sum_{k=0}^{\infty} a_k$ *diverges* is to say that the sequence $\{s_n\}$ diverges.

(4) If the series $\sum_{k=0}^{\infty} a_k$ converges to the number l, then l is called *the sum of the series* and we write

$$\sum_{k=0}^{\infty} a_k = l \quad \text{or} \quad a_0 + a_1 + a_2 + \cdots = l.$$

Example.   The series

$$\sum_{k=0}^{\infty} 2^k$$

diverges. To see this, note that the nth partial sum is given by

$$s_n = 1 + 2 + \cdots + 2^n.$$

Since the sequence $\{s_n\}$ is unbounded, it cannot converge. (Theorem 10.2.1)   ☐

Example

$$\sum_{k=0}^{\infty} \frac{1}{(k + 1)(k + 2)} = 1.$$

To show this, we must show that

$$s_n = \sum_{k=0}^{n} \frac{1}{(k+1)(k+2)} \to 1.$$

Since

$$\frac{1}{(k+1)(k+2)} = \frac{1}{k+1} - \frac{1}{k+2},$$

it follows that

$$s_n = \frac{1}{1\cdot 2} + \frac{1}{2\cdot 3} + \cdots + \frac{1}{(n+1)(n+2)}$$

$$= \left(\frac{1}{1} - \frac{1}{2}\right) + \left(\frac{1}{2} - \frac{1}{3}\right) + \cdots + \left(\frac{1}{n+1} - \frac{1}{n+2}\right).$$

The sum on the right telescopes to give

$$s_n = 1 - \frac{1}{n+2}.$$

It is now obvious that $s_n \to 1$. $\square$

Example.   The series

$$\sum_{k=1}^{\infty} (-1)^k$$

diverges. Here

$$s_n = -1 \quad \text{for } n \text{ odd} \qquad \text{and} \qquad s_n = 0 \quad \text{for } n \text{ even}.$$

Consequently $\{s_n\}$ does not converge. $\square$

The next example has many applications.

Example.   *The Geometric Series*

(i)   If $|x| < 1$,   then $\displaystyle\sum_{k=0}^{\infty} x^k = \frac{1}{1-x}$.

(ii)   If $|x| \geq 1$,   then $\displaystyle\sum_{k=0}^{\infty} x^k$ diverges.

PROOF.   The nth partial sum of the geometric series

$$\sum_{k=0}^{\infty} x^k$$

is the number

(1) $s_n = 1 + x + \cdots + x^n.$

Multiplication by x gives

$$xs_n = x + x^2 + \cdots + x^{n+1}.$$

We now subtract the second equation from the first one and obtain

$$(1 - x)s_n = 1 - x^{n+1}.$$

For $x \neq 1$ this gives

(2)
$$s_n = \frac{1 - x^{n+1}}{1 - x}.$$

If $|x| < 1$, then $x^{n+1} \to 0$ and thus

$$s_n \to \frac{1}{1 - x}.$$

This proves (i). For $x = 1$, we use (1) and see that

$$s_n = n + 1.$$

Consequently the series diverges. For $|x| \geq 1$, $x \neq 1$, we use (2). Since in this instance $\{x^{n+1}\}$ diverges, $\{s_n\}$ diverges. $\square$

As a special case of the geometric series, we have

$$\sum_{k=0}^{\infty} \frac{1}{2^k} = \frac{1}{1 - \frac{1}{2}} = 2.$$

By beginning the summation at $k = 1$, instead of at $k = 0$, we get

$$\boxed{\sum_{k=1}^{\infty} \frac{1}{2^k} = 1.}$$

The partial sums of this series

$$s_1 = \frac{1}{2},$$
$$s_2 = \frac{1}{2} + \frac{1}{4} = \frac{3}{4},$$
$$s_3 = \frac{1}{2} + \frac{1}{4} + \frac{1}{8} = \frac{7}{8},$$
$$s_4 = \frac{1}{2} + \frac{1}{4} + \frac{1}{8} + \frac{1}{16} = \frac{15}{16},$$
$$s_5 = \frac{1}{2} + \frac{1}{4} + \frac{1}{8} + \frac{1}{16} + \frac{1}{32} = \frac{31}{32},$$

etc.

are illustrated in Figure 10.5.1. Each new partial sum lies halfway between the previous one and the number 1.

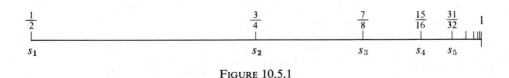

FIGURE 10.5.1

The geometric series plays an important role in our decimal system. To see how this comes about, take the geometric series and consider the case $x = \frac{1}{10}$. Here

$$\sum_{k=0}^{\infty} \frac{1}{10^k} = \frac{1}{1 - \frac{1}{10}} = \frac{10}{9} \quad \text{and} \quad \sum_{k=1}^{\infty} \frac{1}{10^k} = \frac{1}{9}.$$

The nth partial sum in this last instance is

$$s_n = \frac{1}{10} + \frac{1}{10^2} + \cdots + \frac{1}{10^n}.$$

Take now the series

$$\sum_{k=1}^{\infty} \frac{a_k}{10^k} \quad \text{with} \quad a_k \in \{0, 1, 2, \ldots, 9\}.$$

The partial sums here are

$$t_n = \frac{a_1}{10} + \frac{a_2}{10^2} + \cdots + \frac{a_n}{10^n} \le 9\left(\frac{1}{10} + \frac{1}{10^2} + \cdots + \frac{1}{10^n}\right) = 9s_n.$$

Since $\{s_n\}$ converges, $\{s_n\}$ is bounded. This tells us that $\{t_n\}$ is bounded. Since $\{t_n\}$ is nondecreasing, $\{t_n\}$ is itself convergent; that is, the series

$$\sum_{k=1}^{\infty} \frac{a_k}{10^k}$$

converges. The sum of this series is what we mean by the decimal fraction

$$0.a_1 a_2 a_3 \ldots . \quad \square$$

The geometric series arises naturally in many settings. Here is a simple example.

Problem. At what time between 2 and 3 o'clock is the minute hand directly over the hour hand?

SOLUTION. The hour hand travels $\frac{1}{12}$ as fast as the minute hand. At 2 o'clock the minute hand points to 12 and the hour hand points to 2. By the time the minute hand reaches 2, the hour hand points to $2 + \frac{1}{6}$. By the time the minute hand reaches $2 + \frac{1}{6}$, the hour hand points to

$$2 + \frac{1}{6} + \frac{1}{6 \cdot 12}.$$

By the time the minute hand reaches

$$2 + \frac{1}{6} + \frac{1}{6 \cdot 12},$$

the hour hand points to

$$2 + \frac{1}{6} + \frac{1}{6 \cdot 12} + \frac{1}{6 \cdot 12^2}$$

and so on. In general, by the time the minute hand reaches

$$2 + \frac{1}{6} + \frac{1}{6 \cdot 12} + \cdots + \frac{1}{6 \cdot 12^{n-1}} = 2 + \frac{1}{6} \sum_{k=0}^{n-1} \frac{1}{12^n},$$

the hour hand points to

$$2 + \frac{1}{6} + \frac{1}{6 \cdot 12} + \cdots + \frac{1}{6 \cdot 12^{n-1}} + \frac{1}{6 \cdot 12^n} = 2 + \frac{1}{6} \sum_{k=0}^{n} \frac{1}{12^k}.$$

The two hands coincide when both of them point to the limiting value

$$2 + \frac{1}{6} \sum_{k=0}^{\infty} \frac{1}{12^k} = 2 + \frac{1}{6}\left(\frac{1}{1 - \frac{1}{12}}\right) = 2 + \frac{2}{11}.$$

This happens at

$$2 + \tfrac{2}{11} \text{ o'clock.\dagger} \quad \square$$

We will return to the geometric series later. Right now we turn our attention to some results about series in general. First we prove something that is essentially obvious:

If

$$\sum_{k=0}^{\infty} a_k = l \quad \text{and} \quad \sum_{k=0}^{\infty} b_k = m,$$

then

(i)

$$\sum_{k=0}^{\infty} (a_k + b_k) = l + m,$$

and

(ii)

$$\sum_{k=0}^{\infty} \alpha a_k = \alpha l \quad \text{for each real } \alpha.$$

PROOF. Let

$$s_n = \sum_{k=0}^{n} a_k, \qquad t_n = \sum_{k=0}^{n} b_k,$$

$$u_n = \sum_{k=0}^{n} (a_k + b_k), \qquad v_n = \sum_{k=0}^{n} \alpha a_k.$$

Note that

$$u_n = s_n + t_n \quad \text{and} \quad v_n = \alpha s_n.$$

Since $s_n \to l$ and $t_n \to m$, we have

$$u_n \to l + m \quad \text{and} \quad v_n \to \alpha l. \quad \square$$

---

† This particular problem can also be solved without infinite series. See if you can figure out how.

Theorem 10.5.1

If $\sum_{k=0}^{\infty} a_k$ converges,   then $a_k \to 0$.

PROOF.   For each n, set

$$s_n = \sum_{k=0}^{n} a_k.$$

To say that $\sum_{k=0}^{\infty} a_k$ converges to some number l is to say that

$$s_n \to l.$$

Obviously we must also have

$$s_{n-1} \to l.$$

Since $a_n = s_n - s_{n-1}$, we have

$$a_n \to l - l = 0. \quad \square$$

    Caution.   The converse of Theorem 10.5.1 is false. There are divergent series $\sum_{k=0}^{\infty} a_k$ for which $a_k \to 0$.

    Example.   In the case of

$$\sum_{k=1}^{\infty} \frac{1}{\sqrt{k}}$$

we have

$$a_k = \frac{1}{\sqrt{k}} \to 0$$

but

$$s_n = 1 + \frac{1}{\sqrt{2}} + \frac{1}{\sqrt{3}} + \cdots + \frac{1}{\sqrt{n}} > \frac{n}{\sqrt{n}} = \sqrt{n}.$$

Being unbounded, $\{s_n\}$ cannot converge. $\square$

Exercises

Each of the series below converges. Find its sum.

*1. $\displaystyle\sum_{k=3}^{\infty} \frac{1}{(k+1)(k+2)}$.

2. $\displaystyle\sum_{k=0}^{\infty} \frac{1}{(k+3)(k+4)}$.

*3. $\displaystyle\sum_{k=0}^{\infty} \frac{1}{(k+2)(2k+2)}$.

4. $\displaystyle\sum_{k=0}^{\infty} \frac{3}{10^k}$.

*5. $\displaystyle\sum_{k=0}^{\infty} \frac{12}{100^k}$.

6. $\displaystyle\sum_{k=0}^{\infty} \frac{67}{1000^k}$.

*7. $\displaystyle\sum_{k=0}^{\infty} \frac{(-1)^k}{5^k}$.

8. $\displaystyle\sum_{k=0}^{\infty} \left(\frac{3}{4}\right)^k$.

*9. $\displaystyle\sum_{k=0}^{\infty} \frac{3^k + 4^k}{5^k}$.

10. $\displaystyle\sum_{k=0}^{\infty} \frac{1 - 2^k}{3^k}$.

*11. $\displaystyle\sum_{k=0}^{\infty} \left(\frac{25}{10^k} - \frac{6}{100^k}\right)$.

12. $\displaystyle\sum_{k=3}^{\infty} \frac{1}{2^{k-1}}$.

*13. $\displaystyle\sum_{k=0}^{\infty} \frac{1}{2^{k+3}}$.

14. $\displaystyle\sum_{k=0}^{\infty} \frac{2^{k+3}}{3^k}$.

In the following, write each decimal fraction as an infinite series and express the sum as the quotient of two integers.

*15. $0.777\ldots$.

16. $0.999\ldots$.

*17. $0.2424\ldots$.

18. $0.8989\ldots$.

*19. $0.\overset{\frown}{112}\overset{\frown}{112}\overset{\frown}{112}\ldots$.

20. $0.\overset{\frown}{315}\overset{\frown}{315}\overset{\frown}{315}\ldots$.

*21. $0.62\overset{\frown}{45}\overset{\frown}{45}\ldots$.

22. $0.112\overset{\frown}{019}\overset{\frown}{019}\ldots$.

23. Using series, show that every repeating decimal represents a rational number (the quotient of two integers).

24. Show that

$$\sum_{k=0}^{\infty} a_k = l \quad \text{iff} \quad \sum_{k=j+1}^{\infty} a_k = l - (a_0 + \cdots + a_j).$$

Use the geometric series to prove the following.

25. $\displaystyle\frac{1}{1 + x} = \sum_{k=0}^{\infty} (-1)^k x^k \quad \text{for} \quad |x| < 1.$

26. $\displaystyle\frac{1}{1 + x^2} = \sum_{k=0}^{\infty} (-1)^k x^{2k} \quad \text{for} \quad |x| < 1.$

27. Find a series expansion for

*(a) $\displaystyle\frac{x}{1 + x}$ valid for $|x| < 1$.

(b) $\displaystyle\frac{x}{1 + x^2}$ valid for $|x| < 1$.

*(c) $\displaystyle\frac{1}{1 - x^2}$ valid for $|x| < 1$.

(d) $\displaystyle\frac{1}{1 + 4x^2}$ valid for $|x| < \frac{1}{2}$.

28. At what time between 4 and 5 o'clock is the minute hand directly above the hour hand? Express your answer as a geometric series.

*29. How much money must a man deposit at 5% interest (compounded annually) if he wishes to enable his descendants to withdraw $100 a year in perpetuity? Express your answer as a geometric series.

30. A ball dropped from a height of h feet is known to rebound to a height of σh feet where σ is a positive constant less than 1. Find the total distance traveled by the ball if it is dropped initially from a height of h_0 feet.

10.6 Series with Nonnegative Terms

The easiest series to handle are those with nonnegative terms. To determine whether or not such series converge, the following simple theorem is fundamental.

Theorem 10.6.1

A series with nonnegative terms converges iff the sequence of partial sums is bounded.

PROOF. If the sequence of partial sums is not bounded, it cannot converge (Theorem 10.2.1). Suppose that the sequence of partial sums is bounded. Since the terms are nonnegative, the sequence of partial sums is nondecreasing. By Theorem 10.2.2 it must converge. □

To decide upon the boundedness or unboundedness of a sequence of partial sums, it is sometimes useful to compare it to a sequence of integrals.

The Integral Test

If the function f is continuous, decreasing, and positive on the interval $[1, \infty)$, then the series

$$\sum_{k=1}^{\infty} f(k) \quad \text{converges} \quad \text{iff} \quad \text{the sequence} \left\{ \int_{1}^{n} f(x) \, dx \right\} \text{converges.}$$

PROOF. To visualize the proof, see Figure 10.6.1. Since f decreases on the interval $[1, n]$,

$$f(2) + \cdots + f(n) \quad \text{is a lower sum for } f \text{ on } [1, n]$$

and

$$f(1) + \cdots + f(n - 1) \quad \text{is an upper sum for } f \text{ on } [1, n].$$

Consequently

$$f(2) + \cdots + f(n) \leq \int_{1}^{n} f(x) \, dx \quad \text{and} \quad \int_{1}^{n} f(x) \, dx \leq f(1) + \cdots + f(n - 1).$$

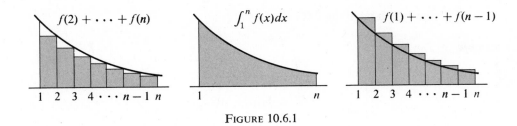

$f(2) + \cdots + f(n)$ $\int_1^n f(x)dx$ $f(1) + \cdots + f(n-1)$

FIGURE 10.6.1

If the sequence

$$\left\{ \int_1^n f(x) \, dx \right\}$$

converges, it is bounded. By the first inequality the series itself must be bounded and therefore convergent. Suppose now that the sequence

$$\left\{ \int_1^n f(x) \, dx \right\}$$

diverges. Since f is positive,

$$\int_1^n f(x) \, dx < \int_1^{n+1} f(x) \, dx.$$

This tells us that the sequence of integrals is monotonic. Being divergent and monotonic, it must be unbounded. By the second inequality, the series itself must be unbounded and therefore divergent. $\square$

Applying the Integral Test

Example. *The Harmonic Series*

$$\boxed{\sum_{k=1}^{\infty} \frac{1}{k} \quad \text{diverges.}}$$

PROOF.   Set $f(x) = 1/x$ and note that

$$f(k) = \frac{1}{k} \, .$$

Here

$$\int_1^n f(x) \, dx = \int_1^n \frac{dx}{x} = \log n.$$

Since the sequence $\{\log n\}$ diverges, the series diverges. $\square$

More generally, we have the following result:

Example. *The p-series*

$$\sum_{k=1}^{\infty} \frac{1}{k^p} \quad \text{converges} \quad \text{iff} \quad p > 1.$$

PROOF. Set $f(x) = 1/x^p$ and note that

$$f(k) = \frac{1}{k^p}.$$

For $p \neq 1$,

$$\int_1^n f(x)\,dx = \int_1^n \frac{dx}{x^p} = \left[\frac{x^{1-p}}{1-p}\right]_1^n = \frac{1}{1-p}(n^{1-p} - 1).$$

If $p > 1$, then $\{n^{1-p}\}$ converges and thus the sequence of integrals converges. If $p < 1$, then $\{n^{1-p}\}$ diverges and the sequence of integrals diverges. The case $p = 1$ gives the harmonic series which, as you saw in the last example, diverges. □

Example. To show that

$$\sum_{k=1}^{\infty} \frac{1}{k \log (k+1)} \quad \text{diverges,}$$

we set

$$f(x) = \frac{1}{x \log (x+1)}.$$

Since f is continuous, decreasing, and positive, we can use the integral test:

$$\int_1^n \frac{dx}{x \log (x+1)} > \int_1^n \frac{dx}{(x+1) \log (x+1)}$$

$$= \left[\log (\log (x+1))\right]_1^n$$

$$= \log (\log (n+1)) - \log (\log 2).$$

Since $\{\log (\log (n+1))\}$ diverges, the sequence of integrals diverges. □

REMARK ON NOTATION. You have seen that for each $j \geq 0$

$$\sum_{k=0}^{\infty} a_k \quad \text{converges} \quad \text{iff} \quad \sum_{k=j+1}^{\infty} a_k \quad \text{converges.}$$

(Exercise 24, Section 10.5.) This tells you that, in determining whether or not a series converges, it is immaterial where we begin the indexing. Where such detailed indexing is unnecessary, we can omit it and speak simply of the series

$$\sum a_k.$$

For instance, it makes sense to say that

$$\sum \frac{1}{k^2} \quad \text{converges} \quad \text{and} \quad \sum \frac{1}{k} \quad \text{diverges.}$$

Much of the work on series with nonnegative terms is based on comparison with series of known behavior. The basic comparison theorem is extremely simple.

Basic Comparison Theorem

Let $\sum a_k$ be a series with nonnegative terms.

(i) $\sum a_k$ converges if there exists a convergent series $\sum c_k$ with nonnegative terms such that

$$a_k \le c_k \quad \text{for all } k \text{ sufficiently large;}$$

(ii) $\sum a_k$ diverges if there exists a divergent series $\sum d_k$ with nonnegative terms such that

$$d_k \le a_k \quad \text{for all } k \text{ sufficiently large.}$$

PROOF. The proof is just a matter of noting that in the first instance

$$\sum a_k \quad \text{and} \quad \sum c_k \quad \text{are both bounded,}$$

and in the second instance

$$\sum a_k \quad \text{and} \quad \sum d_k \quad \text{are both unbounded.}$$

The details are left to you. □

Comparison with the geometric series

$$\sum q^k$$

and with the p-series

$$\sum \frac{1}{k^p}$$

leads to two important tests for convergence: the root test and the ratio test.

The Root Test

Let $\sum a_k$ be a series with nonnegative terms and suppose that

$$(a_k)^{1/k} \to \rho.$$

If $\rho < 1$, $\sum a_k$ converges. If $\rho > 1$, $\sum a_k$ diverges. If $\rho = 1$, the test is inconclusive.

PROOF. We suppose first that $\rho < 1$ and choose $\epsilon > 0$ so small that

$$\rho + \epsilon < 1.$$

Since $(a_k)^{1/k} \to \rho$, we have

$$(a_k)^{1/k} < \rho + \epsilon \quad \text{for all } k \text{ sufficiently large.}$$

Thus

$$a_k < (\rho + \epsilon)^k \quad \text{for all } k \text{ sufficiently large.}$$

Since

$$\sum (\rho + \epsilon)^k \quad \text{converges}$$

(a geometric series with $0 < \rho + \epsilon < 1$), we know by the comparison theorem that

$$\sum a_k \quad \text{converges.}$$

We suppose now that $\rho > 1$ and choose $\epsilon > 0$ so small that

$$\rho - \epsilon > 1.$$

Since $(a_k)^{1/k} \to \rho$, we have

$$\rho - \epsilon < (a_k)^{1/k} \quad \text{for all } k \text{ sufficiently large.}$$

Thus

$$(\rho - \epsilon)^k < a_k \quad \text{for all } k \text{ sufficiently large.}$$

Since

$$\sum (\rho - \epsilon)^k \quad \text{diverges}$$

(a geometric series with $\rho - \epsilon > 1$), the comparison theorem tells us that

$$\sum a_k \quad \text{diverges.}$$

To see the inconclusiveness of the root test in the case of $\rho = 1$, we need only observe that the series $\sum (1/k^2)$ converges and the series $\sum (1/k)$ diverges but in both instances

$$(a_k)^{1/k} \to 1. \quad \square$$

Applying the Root Test

Example. In the case of

$$\sum \frac{1}{(\log k)^k},$$

we have

$$(a_k)^{1/k} = \frac{1}{\log k} \to 0.$$

The series converges. $\quad \square$

Example. In the case of

$$\sum \frac{2^k}{k^3},$$

we have

$$(a_k)^{1/k} = 2 \left(\frac{1}{k} \right)^{3/k} = 2 \left[\left(\frac{1}{k} \right)^{1/k} \right]^3 \to 2 \cdot 1^3 = 2.$$

The series diverges. $\quad \square$

Example. In the case of

$$\sum \left(1 - \frac{1}{k}\right)^k,$$

we have

$$(a_k)^{1/k} = 1 - \frac{1}{k} \to 1.$$

Here the root test is inconclusive. Since $a_k = (1 - 1/k)^k$ tends to $1/e$ and not to 0, we know by Theorem 10.5.1 that actually the series diverges. $\square$

The Ratio Test

Let $\sum a_k$ be a series with positive terms and suppose that

$$\frac{a_{k+1}}{a_k} \to \lambda.$$

If $\lambda < 1$, $\sum a_k$ converges. If $\lambda > 1$, $\sum a_k$ diverges. If $\lambda = 1$, the test is inconclusive.

PROOF. We suppose first that $\lambda < 1$ and choose $\epsilon > 0$ so small that

$$\lambda + \epsilon < 1.$$

Since

$$\frac{a_{k+1}}{a_k} \to \lambda,$$

we know that there exists $m > 0$ such that

$$\text{if} \quad k \geq m, \quad \text{then} \quad \frac{a_{k+1}}{a_k} < \lambda + \epsilon.$$

This gives us

$$a_{m+1} < (\lambda + \epsilon)a_m, \qquad a_{m+2} < (\lambda + \epsilon)a_{m+1} < (\lambda + \epsilon)^2 a_m,$$

and more generally,

$$a_{m+j} < (\lambda + \epsilon)^j a_m.$$

In other words, for $k > m$ we have

$$a_k < (\lambda + \epsilon)^{k-m} a_m = \frac{a_m}{(\lambda + \epsilon)^m} (\lambda + \epsilon)^k.$$

Since $\lambda + \epsilon < 1$,

$$\sum \frac{a_m}{(\lambda + \epsilon)^m} (\lambda + \epsilon)^k = \frac{a_m}{(\lambda + \epsilon)^m} \sum (\lambda + \epsilon)^k \quad \text{converges.}$$

By the comparison theorem,

$$\sum a_k \quad \text{converges.}$$

The proof of the rest of the theorem is left to the exercises. $\square$

Applying the Ratio Test

Example. The convergence of the series

$$\sum \frac{1}{k!}$$

is easy to verify by the ratio test:

$$\frac{a_{k+1}}{a_k} = \frac{1}{(k+1)!} \cdot \frac{k!}{1} = \frac{1}{k+1} \to 0. \quad \square$$

Example. In the case of

$$\sum \frac{k}{10^k},$$

we have

$$\frac{a_{k+1}}{a_k} = \frac{k+1}{10^{k+1}} \cdot \frac{10^k}{k} = \frac{1}{10} \frac{k+1}{k} \to \frac{1}{10}.$$

The series therefore converges. $\square$

Example. In the case of

$$\sum \frac{k^k}{k!},$$

we have

$$\frac{a_{k+1}}{a_k} = \frac{(k+1)^{k+1}}{(k+1)!} \cdot \frac{k!}{k^k} = \left(\frac{k+1}{k}\right)^k = \left(1 + \frac{1}{k}\right)^k \to e.$$

Since $e > 1$, the series diverges. $\square$

Example. For the series

$$\sum \frac{1}{2k+1}$$

the ratio test is inconclusive:

$$\frac{a_{k+1}}{a_k} = \frac{1}{2(k+1)+1} \cdot \frac{2k+1}{1} = \frac{2k+1}{2k+3} = \frac{2+1/k}{2+3/k} \to 1.$$

The integral test shows that the series diverges. $\square$

Exercises

In the following, indicate whether the series converges or diverges.

*1. $\sum \dfrac{k}{k^3+1}$.

2. $\sum \dfrac{1}{3k+2}$.

*3. $\sum \dfrac{1}{k^k}$.

4. $\sum \dfrac{e^k}{k!}$.

*5. $\sum \dfrac{1}{k2^k}$.

6. $\sum \left(\dfrac{k}{2k+1}\right)^k$.

*7. $\displaystyle\sum \frac{1}{(2k+1)^2}$.

8. $\displaystyle\sum \frac{\log k}{k}$.

*9. $\displaystyle\sum \frac{1}{\sqrt{k+1}}$.

10. $\displaystyle\sum \frac{1}{\log k \cdot \log (k+1)}$.

*11. $\displaystyle\sum \frac{2^{k-1}}{(k+1)(k-2)}$.

12. $\displaystyle\sum \frac{1}{k^2+1}$.

*13. $\displaystyle\sum \frac{1}{1+2\log k}$.

14. $\displaystyle\sum \left(\frac{1}{2}\right)^{-k}$.

*15. $\displaystyle\sum k\left(\frac{2}{3}\right)^{k}$.

16. $\displaystyle\sum \frac{k}{(k+1)(2k+1)(4k+1)}$.

*17. $\displaystyle\sum \frac{k!}{10^k}$.

18. $\displaystyle\sum \frac{k!}{10^{4k}}$.

*19. $\displaystyle\sum \frac{k^2}{e^k}$.

20. $\displaystyle\sum \frac{\sqrt{k}}{k^2+1}$.

*21. $\displaystyle\sum \frac{2^k k!}{k^k}$.

22. $\displaystyle\sum \frac{k!}{(k+2)!}$.

*23. $\displaystyle\sum \frac{1}{k}\left(\frac{1}{\log k}\right)^{3/2}$.

24. $\displaystyle\sum \frac{1}{k}\left(\frac{1}{\log k}\right)^{1/2}$.

*25. $\displaystyle\sum \frac{1}{\sqrt{k^3-1}}$.

26. $\displaystyle\sum \left(\frac{k}{k+100}\right)^{k}$.

*27. $\displaystyle\sum \frac{(k!)^2}{(2k)!}$.

28. $\displaystyle\sum k^{-(1+1/k)}$.

*29. $\displaystyle\sum \frac{11}{1+100^{-k}}$.

30. $\displaystyle\sum \frac{\log k}{e^k}$.

*31. $\displaystyle\sum \frac{1}{1+\sqrt{k}}$.

32. $\displaystyle\sum \frac{\log k}{k^2}$.

33. Complete the proof of the ratio test.
   (a) Prove that if $\lambda > 1$, then $\sum a_k$ diverges.
   (b) Prove that if $\lambda = 1$, the ratio test is inconclusive.
      HINT: Consider $\sum (1/k)$ and $\sum (1/k^2)$.

34. Let $\sum a_k$ be a series with nonnegative terms and let $\{c_k\}$ be a sequence of positive numbers which converges to some positive number c. Show that

$$\sum a_k \quad \text{converges} \quad \text{iff} \quad \sum c_k a_k \quad \text{converges}.$$

   HINT: There exists an integer n such that

$$\tfrac{1}{2}c < c_k < \tfrac{3}{2}c \quad \text{for } k \geq n.$$

10.7 Absolute Convergence; Alternating Series

In this section we consider series which have both positive and negative terms.

Absolute Convergence

Let $\sum a_k$ be a series with both positive and negative terms. One way to show that $\sum a_k$ converges is to show that $\sum |a_k|$ converges.

Theorem

If $\sum |a_k|$ converges,   then $\sum a_k$ converges.

Proof.   For each k,

$$-|a_k| \le a_k \le |a_k|$$

and therefore

$$0 \le a_k + |a_k| \le 2|a_k|.$$

If $\sum |a_k|$ converges, then $\sum 2|a_k| = 2\sum |a_k|$ converges, and therefore, by the comparison test,

$$\sum (a_k + |a_k|)   \text{ converges.}$$

Since

$$a_k = (a_k + |a_k|) - |a_k|,$$

we can conclude that

$$\sum a_k   \text{ converges.}   \square$$

A series $\sum a_k$ for which $\sum |a_k|$ converges is called *absolutely convergent*. The theorem we just proved says that

---

if $\sum a_k$ converges absolutely,   then $\sum a_k$ converges.

As you will see a little later, the converse is false: There are series that converge but do not converge absolutely.

Here are some examples that should help you get the feeling of these ideas.

Example

$$1 - \frac{1}{2^2} + \frac{1}{3^2} - \frac{1}{4^2} + \frac{1}{5^2} - \frac{1}{6^2} + \cdots.$$

If we replace each of the terms by its absolute value, we obtain the series

$$1 + \frac{1}{2^2} + \frac{1}{3^2} + \frac{1}{4^2} + \frac{1}{5^2} + \frac{1}{6^2} + \cdots.$$

This is a p-series with $p = 2$. It is therefore convergent. This means that the initial series is absolutely convergent and thus convergent. $\square$

Example

$$1 - \frac{1}{2} - \frac{1}{2^2} + \frac{1}{2^3} - \frac{1}{2^4} - \frac{1}{2^5} + \frac{1}{2^6} - \frac{1}{2^7} - \frac{1}{2^8} + \cdots.$$

If we replace each of the terms by its absolute value, we obtain the series

$$1 + \frac{1}{2} + \frac{1}{2^2} + \frac{1}{2^3} + \frac{1}{2^4} + \frac{1}{2^5} + \frac{1}{2^6} + \frac{1}{2^7} + \frac{1}{2^8} + \cdots.$$

This is a convergent geometric series. The initial series is therefore absolutely convergent and thus convergent. □

Example

$$1 - \frac{1}{2} + \frac{1}{3} - \frac{1}{4} + \frac{1}{5} - \frac{1}{6} + \cdots.$$

If we replace each of the terms by its absolute value, we obtain the series

$$1 + \frac{1}{2} + \frac{1}{3} + \frac{1}{4} + \frac{1}{5} + \frac{1}{6} + \cdots.$$

This is the harmonic series. Since the harmonic series diverges, the initial series is not absolutely convergent. It is however, convergent. (See the next theorem.) □

Alternating Series

A series in which successive terms have opposite signs is called an *alternating series*. As examples of alternating series we have

$$1 - \frac{1}{2} + \frac{1}{3} - \frac{1}{4} + \frac{1}{5} - \frac{1}{6} + \cdots,$$

$$1 - \frac{1}{\sqrt{2}} + \frac{1}{\sqrt{3}} - \frac{1}{\sqrt{4}} + \frac{1}{\sqrt{5}} - \frac{1}{\sqrt{6}} + \cdots,$$

$$1 - \frac{2}{\log 2} + \frac{3}{\log 3} - \frac{4}{\log 4} + \frac{5}{\log 5} - \frac{6}{\log 6} + \cdots.$$

The series

$$1 - \frac{1}{2} - \frac{1}{3} + \frac{1}{4} - \frac{1}{5} - \frac{1}{6} + \cdots$$

is not an alternating series because there are consecutive terms with the same sign.

Theorem on Alternating Series

Let $\{a_k\}$ be a decreasing sequence of positive numbers.

$$\text{If} \quad a_k \to 0, \quad \text{then} \quad \sum_{k=0}^{\infty} (-1)^k a_k \quad \text{converges.}$$

PROOF. First we look at the even partial sums, s_{2m}. Since

$$s_{2m} = (a_0 - a_1) + (a_2 - a_3) + \cdots + (a_{2m-2} - a_{2m-1}) + a_{2m}$$

is the sum of positive numbers, the even partial sums are all positive. Since

$$s_{2m+2} = s_{2m} - (a_{2m+1} - a_{2m+2}) \quad \text{and} \quad a_{2m+1} - a_{2m+2} > 0,$$

we have

$$s_{2m+2} < s_{2m}.$$

This means that the sequence of even partial sums is decreasing. Being bounded below by 0, it is convergent; say,

$$s_{2m} \to l.$$

Now

$$s_{2m+1} = s_{2m} - a_{2m+1}.$$

Since $a_{2m+1} \to 0$, we also have

$$s_{2m+1} \to l.$$

With both the even and the odd partial sums tending to l, it is obvious that all the partial sums must tend to l. □

The theorem we just proved makes it clear that each of the following series converges:

$$1 - \frac{1}{2} + \frac{1}{3} - \frac{1}{4} + \frac{1}{5} - \frac{1}{6} + \cdots,$$

$$1 - \frac{1}{\sqrt{2}} + \frac{1}{\sqrt{3}} - \frac{1}{\sqrt{4}} + \frac{1}{\sqrt{5}} - \frac{1}{\sqrt{6}} + \cdots,$$

$$1 - \frac{1}{2!} + \frac{1}{3!} - \frac{1}{4!} + \frac{1}{5!} - \frac{1}{6!} + \cdots.$$

Of these only the last one is absolutely convergent.

Exercises

Test the following series for (a) convergence, (b) absolute convergence.

*1. $1 + (-1) + 1 + \cdots + (-1)^k + \cdots$.

2. $\dfrac{1}{4} - \dfrac{1}{6} + \dfrac{1}{8} - \dfrac{1}{10} + \cdots + \dfrac{(-1)^k}{2k} + \cdots$.

*3. $\dfrac{1}{2} - \dfrac{2}{3} + \dfrac{3}{4} - \dfrac{4}{5} + \cdots + (-1)^k \dfrac{k}{k+1} + \cdots$.

4. $\dfrac{1}{2 \log 2} - \dfrac{1}{3 \log 3} + \dfrac{1}{4 \log 4} - \dfrac{1}{5 \log 5} + \cdots + (-1)^k \dfrac{1}{k \log k} + \cdots$.

*5. $\sum (-1)^k \dfrac{\log k}{k}$.

6. $\sum (-1)^k \dfrac{k}{\log k}$.

*7. $\sum \left(\dfrac{1}{k} - \dfrac{1}{k!} \right)$.

8. $\sum \dfrac{k^3}{2^k}$.

*9. $\sum (-1)^k \dfrac{1}{2k+1}$.

10. $\sum (-1)^k \dfrac{(k!)^2}{(2k)!}$.

*11. $\sum \dfrac{k!}{(-2)^k}$.

12. $\sum \sin \left(\dfrac{k\pi}{4} \right)$.

*13. $\sum \sin \left(\dfrac{\pi}{4k^2} \right)$.

14. $\sum \dfrac{(-1)^k}{\sqrt{k} + \sqrt{k+1}}$.

*15. $\sum (-1)^k \dfrac{k}{k^2+1}$.

16. $\sum \dfrac{(-1)^k}{\sqrt{k(k+1)}}$.

*17. $\sum (-1)^k \dfrac{k}{2^k}$.

18. $\sum \left(\dfrac{1}{\sqrt{k}} - \dfrac{1}{\sqrt{k+1}} \right)$.

*19. $\dfrac{1}{2} - \dfrac{1}{3} - \dfrac{1}{4} + \dfrac{1}{5} - \dfrac{1}{6} - \dfrac{1}{7} + \cdots + \dfrac{1}{3k+2} - \dfrac{1}{3k+3} - \dfrac{1}{3k+4} + \cdots$.

20. $\dfrac{2 \cdot 3}{4 \cdot 5} - \dfrac{5 \cdot 6}{7 \cdot 8} + \dfrac{8 \cdot 9}{10 \cdot 11} - \dfrac{11 \cdot 12}{13 \cdot 14} + \cdots + (-1)^k \dfrac{(3k+2)(3k+3)}{(3k+4)(3k+5)} + \cdots$.

21. (*Important*) *An Estimate for Alternating Series.* If $\{a_k\}$ is a decreasing sequence of positive numbers, then the alternating series

$$\sum_{k=0}^{\infty} (-1)^k a_k = a_0 - a_1 + a_2 - a_3 + a_4 - a_5 + a_6 - \cdots$$

converges to some number l. Show that this number l lies between consecutive partial sums s_n, s_{n+1} and that

$$|s_n - l| < a_{n+1}.$$

22. (*Optional*) In Section 10 of this chapter we prove that if

$$|x| < |x_1| \quad \text{and} \quad \sum a_k x_1^k \text{ converges}$$

then

$$\sum a_k x^k \quad \text{converges absolutely}.$$

See if you can show this now.

10.8   Taylor Polynomials and Taylor Series

Taylor Polynomials

We begin with a function f that is differentiable at 0. The linear function that best approximates f at 0 is the function

$$P_1(x) = f(0) + f'(0)x.$$

At 0 it has the same value as f and also the same derivative:

$$P_1(0) = f(0), \qquad P_1'(0) = f'(0).$$

If f has two derivatives at 0, then we can get a better approximation by using the quadratic polynomial

$$P_2(x) = f(0) + f'(0)x + f''(0)\frac{x^2}{2!}.$$

Here, as you can check,

$$P_2(0) = f(0), \qquad P_2'(0) = f'(0), \quad \text{and} \quad P_2''(0) = f''(0).$$

If f has three derivatives at 0, then we can use the cubic polynomial

$$P_3(x) = f(0) + f'(0)x + \frac{f''(0)}{2!}x^2 + \frac{f'''(0)}{3!}x^3,$$

and thereby take the third derivative into account. Here

$$P_3(0) = f(0), \quad P_3'(0) = f'(0), \quad P_3''(0) = f''(0), \quad \text{and} \quad P_3'''(0) = f'''(0).$$

More generally, for a function f with n derivatives at 0, we can use the polynomial

$$P_n(x) = f(0) + f'(0)x + \frac{f''(0)}{2!}x^2 + \cdots + \frac{f^{(n)}(0)}{n!}x^n.$$

This is the polynomial of degree n whose first n derivatives agree with those of f at 0:

$$P_n(0) = f(0), \quad P_n'(0) = f'(0), \quad P_n''(0) = f''(0), \ldots, P_n^{(n)}(0) = f^{(n)}(0).$$

The polynomials $P_1(x), P_2(x), \ldots, P_n(x)$ that we have just described are called *Taylor polynomials*, after the English mathematician Brook Taylor (1685–1731). The groundwork for all that we discuss here was laid by Taylor in 1712.

The fact that the Taylor polynomials

$$P_k(x) = f(0) + f'(0)x + \frac{f''(0)}{2!}x^2 + \cdots + \frac{f^{(k)}(0)}{k!}x^k$$

approximate $f(x)$ is of little practical use unless we know something about the accuracy of the approximations. The key to this question of accuracy is Taylor's theorem.

Taylor's Theorem

Let I be an interval containing the number 0. If f has n continuous derivatives on I, then for each x in I

$$f(x) = f(0) + f'(0)x + \frac{f''(0)}{2!}x^2 + \cdots + \frac{f^{(n-1)}(0)}{(n-1)!}x^{n-1} + R_n,$$

where the remainder R_n is given by the formula

$$R_n = \frac{1}{(n-1)!}\int_0^x f^{(n)}(t)(x-t)^{n-1}\,dt.$$

PROOF. Integration by parts (see Exercise 27 at the end of the section) gives

$$f'(0)x = \int_0^x f'(t)\,dt - \int_0^x f''(t)(x - t)\,dt,$$

$$\frac{f''(0)}{2!}x^2 = \int_0^x f''(t)(x - t)\,dt - \frac{1}{2!}\int_0^x f'''(t)(x - t)^2\,dt,$$

$$\frac{f'''(0)}{3!}x^3 = \frac{1}{2!}\int_0^x f'''(t)(x - t)^2\,dt - \frac{1}{3!}\int_0^x f^{(iv)}(t)(x - t)^3\,dt,$$

$$\vdots$$

$$\frac{f^{(n-1)}(0)}{(n-1)!}x^{n-1} = \frac{1}{(n-2)!}\int_0^x f^{(n-1)}(t)(x - t)^{(n-2)}\,dt - \frac{1}{(n-1)!}\int_0^x f^{(n)}(t)(x - t)^{n-1}\,dt.$$

We now add these equations. The sum on the left is simply

$$f'(0)x + \frac{f''(0)}{2!}x^2 + \frac{f'''(0)}{3!}x^3 + \cdots + \frac{f^{(n-1)}(0)}{(n-1)!}x^{n-1}.$$

The sum on the right telescopes to

$$\int_0^x f'(t)\,dt - \frac{1}{(n-1)!}\int_0^x f^{(n)}(t)(x - t)^{n-1}\,dt = f(x) - f(0) - R_n.$$

The result is now obvious. □

Taylor's formula can be written

$$f(x) = P_{n-1}(x) + R_n,$$

where

$$R_n = \frac{1}{(n-1)!}\int_0^x f^{(n)}(t)(x - t)^{n-1}\,dt.$$

For most purposes it is convenient to write the remainder term, R_n, in the manner of Lagrange†:

(10.8.1)

$$\boxed{R_n = \frac{f^{(n)}(c)}{n!}x^n,}$$

where c is some number between 0 and x. (So as not to obscure the main ideas by too many technical arguments we have placed the proof of Lagrange's form of the remainder in the supplement to this section. It's not that the argument is difficult, it's just that it is distracting.)

---

† Joseph Louis Lagrange, French mathematician (1736–1813).

Taylor Series

If we adopt the conventions that $0! = 1$ and that the zeroth derivative of f is f itself, then we can write Taylor's formula in the form

$$f(x) = \sum_{k=0}^{n-1} \frac{f^{(k)}(0)}{k!} x^k + R_n.$$

Of special interest is the case in which the function has derivatives of all orders. In this case Taylor's formula is valid for all positive integers n. If it happens that $R_n \to 0$, then

$$f(x) = \lim_{n \to \infty} \sum_{k=0}^{n-1} \frac{f^{(k)}(0)}{k!} x^k$$

and we have the Taylor series expansion

$$f(x) = \sum_{k=0}^{\infty} \frac{f^{(k)}(0)}{k!} x^k.$$

It is time for some examples.

Example.   For all real x,

(10.8.2)
$$e^x = \sum_{k=0}^{\infty} \frac{x^k}{k!} = 1 + x + \frac{x^2}{2!} + \frac{x^3}{3!} + \cdots.$$

PROOF.   Let x be a fixed real number. Setting

$$f(t) = e^t,$$

we find that

$$f^{(k)}(t) = e^t \quad \text{for all } k.$$

This tells us that for all k

$$f^{(k)}(0) = e^0 = 1.$$

The coefficients in (10.8.2) are thus the Taylor coefficients. To verify that the Taylor series does indeed converge to e^x, we must verify that

$$R_n \to 0.$$

Let M be the maximum value of the exponential function on the interval joining 0 and x. For all t in that interval

$$|f^{(n)}(t)| = e^t \leq M.$$

Using Lagrange's form of the remainder, we have

$$|R_n| = \frac{|e^c|}{n!} |x|^n \leq M \frac{|x|^n}{n!}.$$

Since

$$\frac{|x|^n}{n!} \to 0, \qquad\qquad [\text{by } (10.3.3)]$$

we have

$$R_n \to 0. \quad \square$$

Example. For all real x,

(10.8.3)

$$\sin x = \sum_{k=0}^{\infty} \frac{(-1)^k}{(2k+1)!} x^{2k+1} = x - \frac{x^3}{3!} + \frac{x^5}{5!} - \frac{x^7}{7!} + \cdots$$

and

(10.8.4)

$$\cos x = \sum_{k=0}^{\infty} \frac{(-1)^k}{(2k)!} x^{2k} = 1 - \frac{x^2}{2!} + \frac{x^4}{4!} - \frac{x^6}{6!} + \cdots.$$

PROOF.   We focus on the sine function and leave the cosine as an exercise. We take x as a fixed real number and let

$$f(t) = \sin t.$$

Differentiating, we find

$$f'(t) = \cos t, \qquad f''(t) = -\sin t,$$
$$f'''(t) = -\cos t, \qquad f^{(\mathrm{iv})}(t) = \sin t.$$

More generally,

$$f^{(2k)}(t) = (-1)^k \sin t, \qquad f^{(2k+1)}(t) = (-1)^k \cos t.$$

Since

$$f^{(2k)}(0) = 0 \quad \text{and} \quad f^{(2k+1)}(0) = (-1)^k,$$

we see that the coefficients in (10.8.3) are actually the Taylor coefficients. Since for all t, $|f^{(n)}(t)| \le 1$, Lagrange's form of the remainder gives

$$|R_n| = \frac{|f^{(n)}(c)|}{n!} |x|^n \le \frac{|x|^n}{n!}.$$

Since $|x|^n/n! \to 0$, we have

$$R_n \to 0.$$

This tells us that the series does converge to $\sin x$. $\square$

Some Numerical Calculations

The fact that the Taylor polynomial

$$P_{n-1}(x) = f(0) + f'(0)x + \frac{f''(0)}{2!} x^2 + \cdots + \frac{f^{(n-1)}(x)}{(n-1)!} x^{n-1}$$

approximates $f(x)$ to within R_n enables us to calculate certain functional values as accurately as we wish. Below we give some sample calculations. For ready reference we list in Tables 10.8.1 and 10.8.2 some values of $n!$ and of its reciprocal $1/n!$.

TABLE 10.8.2

$\dfrac{1}{n!}$
$0.50000 = \dfrac{1}{2!} = 0.50000$
$0.16666 < \dfrac{1}{3!} < 0.16667$
$0.04166 < \dfrac{1}{4!} < 0.04167$
$0.00833 < \dfrac{1}{5!} < 0.00834$
$0.00138 < \dfrac{1}{6!} < 0.00139$
$0.00019 < \dfrac{1}{7!} < 0.00020$

TABLE 10.8.1

$n!$
$2! = 2$
$3! = 6$
$4! = 24$
$5! = 120$
$6! = 720$
$7! = 5{,}040$
$8! = 40{,}320$

Problem.   Find a numerical estimate for e that is accurate within 0.001.

SOLUTION.   You are already familiar with the series expansion

$$e^x = 1 + x + \frac{x^2}{2!} + \frac{x^3}{3!} + \cdots + \frac{x^{n-1}}{(n-1)!} + \cdots .$$

Taking $x = 1$, we have

$$e = 1 + 1 + \frac{1}{2!} + \frac{1}{3!} + \cdots + \frac{1}{(n-1)!} + \cdots .$$

To estimate e within 0.001, we need only consider the expansion to the point where

$$|R_n| < 0.001.$$

Since

$$|R_7| = \frac{e^c}{7!} < \frac{e}{7!} < 3(0.0002) = 0.0006,$$

we see that

$$P_6(1) = 1 + 1 + \frac{1}{2!} + \frac{1}{3!} + \frac{1}{4!} + \frac{1}{5!} + \frac{1}{6!} = \frac{1957}{720}$$

differs from e by less than 0.0006. To turn this into a decimal estimate for e, we note that

$$2.7180 < \frac{1957}{720} < 2.7181.$$

Within the prescribed limits of accuracy we can take $e \cong 2.718$.   □

Problem. Find a numerical estimate for $e^{0.2}$ that is accurate within 0.001.

SOLUTION. Here we take

$$1 + 0.2 + \frac{(0.2)^2}{2!} + \frac{(0.2)^3}{3!} + \cdots + \frac{(0.2)^{n-1}}{(n-1)!} + \cdots$$

to the point where

$$|R_n| < 0.001.$$

Since

$$|R_4| = |e^c| \frac{|0.2|^4}{4!} \leq \frac{e^{0.2}16}{240{,}000} < \frac{3 \cdot 16}{240{,}000} < 0.001,$$

we can conclude that

$$1 + 0.2 + \frac{(0.2)^2}{2!} + \frac{(0.2)^3}{3!} = \frac{7326}{6000} = 1.221$$

differs from $e^{0.2}$ by less than 0.001.   □

Problem. Find a numerical estimate for $\sin 0.5$ that is accurate within 0.001.

SOLUTION. Recall the series expansion

$$\sin x = x - \frac{x^3}{3!} + \frac{x^5}{5!} - \frac{x^7}{7!} + \cdots$$

and the remainder estimate

$$|R_n| \leq \frac{|x|^n}{n!}.$$

Since at $x = 0.5$, we have

$$|R_5| \leq \frac{(0.5)^5}{5!} < \frac{1}{32}(0.00834) < 0.0003$$

you can see that

$$P_4(0.5) \overset{\text{the coefficient of } x^4 \text{ is } 0}{=} P_3(0.5) = 0.5 - \frac{(0.5)^3}{3!} = \frac{23}{48}$$

differs from $\sin 0.5$ by less than 0.0003. For a decimal estimate note that

$$0.4791 < \tfrac{23}{48} < 0.4792.$$

Within the prescribed limits of accuracy we can take $\sin 0.5 \cong 0.479$.   □

Exercises

Find the Taylor polynomial $P_5(x)$ for each of the following functions.

*1. e^{-x}. 2. e^{2x}. *3. arc tan x. 4. arc sin x.
*5. $(1 + x)^{-1}$. 6. tan x. *7. x^2 tan x. 8. e^x sin x.

Find the Taylor polynomial $P_4(x)$ for each of the following functions.

*9. sec x. 10. $x - \cos x$. *11. $\log (1 + x)$. 12. $\sqrt{1 + x}$.
*13. $\log \cos x$. 14. $\sqrt{\cos x}$. *15. e^x tan x. 16. $\sqrt{\sec x}$.

Use Taylor polynomials to find a numerical estimate that is accurate within 0.01.

*17. $\sqrt{e}$. 18. sin 0.3. *19. cos 1. 20. cos 0.5.
*21. arc sin 0.9. 22. sin 1. *23. $e^{2.1}$. 24. $e^{-1.4}$.

25. Show that a polynomial

$$P(x) = a_0 + a_1 x + \cdots + a_n x^n$$

   is its own Taylor series.

26. Show that

$$\cos x = \sum_{k=0}^{\infty} \frac{(-1)^k}{(2k)!} x^{2k} \quad \text{for all real } x.$$

27. Verify the identity

$$\frac{f^{(k)}(0)}{k!} x^k = \frac{1}{(k - 1)!} \int_0^x f^{(k)}(t)(x - t)^{(k-1)} \, dt - \frac{1}{k!} \int_0^x f^{(k+1)}(t)(x - t)^k \, dt$$

   by computing the second integral by parts.

28. (*Optional*)   Show that e is irrational by following these steps.
    (1) Take the expansion

$$e = \sum_{k=0}^{\infty} \frac{1}{k!}$$

   and show that the qth partial sum

$$s_q = \sum_{k=0}^{q} \frac{1}{k!}$$

   satisfies the inequality

$$0 < q! \, (e - s_q) < \frac{1}{q}.$$

   (2) Show that $q! \, s_q$ is an integer and argue that, if e were of the form p/q, then we would have $q! \, (e - s_q)$ as a positive integer less than 1.

Here is a more general version of Taylor's theorem:

(10.8.5)

Let I be an interval containing the number a. If g has n continuous derivatives on I, then for each $x \in I$

$$g(x) = g(a) + g'(a)(x - a) + \cdots + \frac{g^{(n-1)}(a)}{(n-1)!}(x - a)^{n-1} + R_n,$$

where

$$R_n = \frac{1}{(n-1)!} \int_a^x g^{(n)}(s)(x - s)^{n-1}\, ds.$$

In this more general setting the Lagrange form of the remainder takes the form

$$R_n = \frac{g^{(n)}(c)}{n!}(x - a)^n \quad \text{with } c \text{ between } a \text{ and } x.$$

If $R_n \to 0$,

$$g(x) = g(a) + g'(a)(x - a) + \cdots + \frac{g^{(n)}(a)}{n!}(x - a)^n + \cdots.$$

This is called the Taylor series expansion of $g(x)$ in powers of $(x - a)$.

29. Verify the expansion

$$e^x = e^a\left[1 + (x - a) + \frac{(x - a)^2}{2!} + \frac{(x - a)^3}{3!} + \cdots\right].$$

Expand the following in powers of $(x - a)$.

* 30. $\sin x$.   31. $\cos x$.   * 32. $\log x$.

(*Optional*) Find the Taylor polynomial.

* 33. $P_5(x)$ for $\sinh x$.   * 34. $P_4(x)$ for $\cosh x$.

35. (*Optional*) Prove the more general version of Taylor's theorem by setting

$$f(x - a) = g(x)$$

and noting that

$$f'(x - a) = g'(x), \ldots, f^{(n)}(x - a) = g^{(n)}(x)$$

and that

$$f(0) = g(a),\ f'(0) = g'(a), \ldots, f^{(n)}(0) = g^{(n)}(a).$$

Supplement to Section 10.8

Before proving the validity of Lagrange's form of the remainder, we prove a simple extension of the first mean-value theorem for integrals.

Theorem 10.8.6

If f and g are continuous on $[a, b]$ and g keeps a constant sign on $[a, b]$, then

$$\int_a^b f(t)g(t) \, dt = f(c) \int_a^b g(t) \, dt$$

for some number c in the interval $[a, b]$.

PROOF.   Set

$$m = \text{minimum value of } f \text{ on } [a, b],$$
$$M = \text{maximum value of } f \text{ on } [a, b].$$

For each number t in $[a, b]$

$$m \le f(t) \le M$$

and, if g remains nonnegative,

$$mg(t) \le f(t)g(t) \le Mg(t).$$

(If g remains nonpositive, the inequality is reversed and the rest of the argument must be adjusted accordingly.)

By integrating the last inequality from a to b, we get

$$\int_a^b mg(t) \, dt \le \int_a^b f(t)g(t) \, dt \le \int_a^b Mg(t) \, dt$$

and therefore

$$m \int_a^b g(t) \, dt \le \int_a^b f(t)g(t) \, dt \le M \int_a^b g(t) \, dt.$$

If

$$\int_a^b g(t) \, dt = 0, \qquad \text{then} \qquad \int_a^b f(t)g(t) = 0$$

and the result holds. If

$$\int_a^b g(t) \, dt \ne 0, \qquad \text{then} \quad m \le \frac{\displaystyle\int_a^b f(t)g(t) \, dt}{\displaystyle\int_a^b g(t) \, dt} \le M$$

and since f, being continuous, skips no values, there must exist c in $[a, b]$ such that

$$f(c) = \frac{\displaystyle\int_a^b f(t)g(t)\,dt}{\displaystyle\int_a^b g(t)\,dt}.$$

Obviously then

$$\int_a^b f(t)g(t) = f(c)\int_a^b g(t)\,dt. \quad \square$$

Lagrange's form of the remainder is now easy to obtain. For $x > 0$,

$$R_n = \frac{1}{(n-1)!}\int_0^x f^{(n)}(t)(x-t)^{n-1}\,dt = \frac{f^{(n)}(c)}{(n-1)!}\int_0^x (x-t)^{n-1}\,dt$$

here is where we use the — *(arrow)*
last theorem

$$= \frac{f^{(n)}(c)}{(n-1)!}\left[-\frac{(x-t)^n}{n}\right]_0^x$$

$$= \frac{f^{(n)}(c)}{n!}\,x^n.$$

For $x < 0$,

$$R_n = \frac{1}{(n-1)!}\int_0^x f^{(n)}(t)(x-t)^{n-1}\,dt = \frac{1}{(n-1)!}\int_x^0 f^{(n)}(t)[-(x-t)^{n-1}]\,dt$$

$$= \frac{f^{(n)}(c)}{(n-1)!}\int_x^0 [-(x-t)^{n-1}]\,dt$$

$$= \frac{f^{(n)}(c)}{(n-1)!}\left[\frac{(x-t)^n}{n}\right]_x^0$$

$$= \frac{f^{(n)}(c)}{n!}\,x^n. \quad \square$$

10.9   The Logarithm and the Arc Tangent; Computing π

Taylor series expansions for $\log(1+x)$ and arc tan x can be derived by the method of the last section. It is, however, easier to follow the method described below.

A Series for log $(1 + x)$

In proving the convergence of the geometric series we noted that for $x \neq 1$,

$$\frac{1-x^n}{1-x} = 1 + x + \cdots + x^{n-1}.$$

Since

$$\frac{1 - x^n}{1 - x} = \frac{1}{1 - x} - \frac{x^n}{1 - x},$$

we can rewrite the identity as

$$\frac{1}{1 - x} = 1 + x + x^2 + \cdots + x^{n-1} + \frac{x^n}{1 - x}.$$

Setting $x = -t$, we have

$$\frac{1}{1 + t} = 1 + (-t) + (-t)^2 + \cdots + (-t)^{n-1} + \frac{(-t)^n}{1 + t}.$$

It follows that

$$\log(1 + x) = \int_0^x \frac{dt}{1 + t} = x - \frac{x^2}{2} + \frac{x^3}{3} - \cdots + (-1)^{n-1} \frac{x^n}{n} + T_n(x),$$

where

$$T_n(x) = (-1)^n \int_0^x \frac{t^n}{1 + t} \, dt.$$

Below we show that

$$T_n(x) \to 0 \quad \text{if} \quad -1 < x \leq 1.$$

For $0 \leq x \leq 1$, we have

$$|T_n(x)| = \int_0^x \frac{t^n}{1 + t} \, dt \leq \int_0^x t^n \, dt = \frac{x^{n+1}}{n + 1} \to 0.$$

For $-1 < x < 0$, we have

$$|T_n(x)| = \left| \int_0^x \frac{t^n}{1 + t} \, dt \right| \leq \int_x^0 \frac{|t|^n}{|1 + t|} \, dt = \int_x^0 \frac{(-t)^n}{1 + t} \, dt$$

$$\leq \int_x^0 \frac{(-t)^n}{1 + x} \, dt = \frac{1}{1 + x} \left[\frac{(-x)^{n+1}}{n + 1} \right] \to 0.$$

You have now seen that

$$T_n(x) \to 0 \quad \text{for} \quad -1 < x \leq 1.$$

It follows that

(10.9.1) $$\log(1 + x) = x - \frac{x^2}{2} + \frac{x^3}{3} - \frac{x^4}{4} + \cdots \qquad \text{for} \quad -1 < x \leq 1.$$

There is no hope of extending the result to $x = -1$ since $\log 0$ is not defined.
Using $x = 1$, we get

$$\log 2 = 1 - \tfrac{1}{2} + \tfrac{1}{3} - \tfrac{1}{4} + \cdots.$$

This intrigued some of the early enthusiasts of the calculus.

Problem. Use series (10.9.1) to find a numerical estimate for log 1.3 that is accurate within 0.01.

SOLUTION

$$\log 1.3 = \log(1 + 0.3) = 0.3 - \tfrac{1}{2}(0.3)^2 + \tfrac{1}{3}(0.3)^3 - \tfrac{1}{4}(0.3)^4 + \cdots.$$

Since this is an alternating series and the terms form a decreasing sequence, we know that log 1.3 lies between consecutive odd and even partial sums.

The first term less than 0.01 is

$$\tfrac{1}{3}(0.3)^3 = \tfrac{1}{3}(0.027) = 0.009.$$

The relation

$$0.3 - \tfrac{1}{2}(0.3)^2 < \log 1.3 < 0.3 - \tfrac{1}{2}(0.3)^2 + \tfrac{1}{3}(0.3)^3$$

gives

$$0.255 < \log 1.3 < 0.264.$$

Within the prescribed limits of accuracy we can take

$$\log 1.3 \cong 0.26. \quad \square$$

A Series for arc tan x

The development here is so similar to the one we just used for log $(1 + x)$ that we can leave out some of the details. We begin with

$$\frac{1}{1 + t^2} = \frac{1}{1 - (-t^2)} = 1 + (-t^2) + (-t^2)^2 + \cdots + (-t^2)^{n-1} + \frac{(-t^2)^n}{1 + t^2}.$$

Since

$$\text{arc tan } x = \int_0^x \frac{dt}{1 + t^2},$$

we have

$$\text{arc tan } x = x + (-1)\frac{x^3}{3} + (-1)^2\frac{x^5}{5} + \cdots + (-1)^{n-1}\frac{x^{2n-1}}{2n - 1} + S_n(x),$$

where

$$S_n(x) = (-1)^n \int_0^x \frac{t^{2n}}{1 + t^2}\, dt.$$

If $x \in [-1, 1]$, then $S_n(x) \to 0$. (See Exercise 8.) It follows that

(10.9.2)
$$\boxed{\text{arc tan } x = x - \frac{x^3}{3} + \frac{x^5}{5} - \frac{x^7}{7} + \cdots \qquad \text{for} \quad -1 \le x \le 1.}$$

This series was known to the Scottish mathematician James Gregory in 1671, but the discovery was not published until 1712.

Since
$$\text{arc tan } 1 = \tfrac{1}{4}\pi,$$
we have
$$\tfrac{1}{4}\pi = 1 - \tfrac{1}{3} + \tfrac{1}{5} - \tfrac{1}{7} + \tfrac{1}{9} - \cdots.$$

Leibniz seems to have found this formula in 1673. It is an elegant formula for π, but it converges too slowly to be efficient for computational purposes.

Computing π

A more convenient way to compute π is to use the identity

(10.9.3) $\tfrac{1}{4}\pi = 4 \text{ arc tan } \tfrac{1}{5} - \text{arc tan } \tfrac{1}{239}.$

This identity was discovered in 1706 by John Machin, a Scotsman. It can be verified by repeated application of the addition formula
$$\tan (A + B) = \frac{\tan A + \tan B}{1 - \tan A \tan B}.$$
(See Exercises 10, 11, 12.)

The arc tangent series (10.9.2) gives
$$\text{arc tan } \tfrac{1}{5} = \tfrac{1}{5} - \tfrac{1}{3}(\tfrac{1}{5})^3 + \tfrac{1}{5}(\tfrac{1}{5})^5 - \tfrac{1}{7}(\tfrac{1}{5})^7 + \cdots$$
and
$$\text{arc tan } \tfrac{1}{239} = \tfrac{1}{239} - \tfrac{1}{3}(\tfrac{1}{239})^3 + \tfrac{1}{5}(\tfrac{1}{239})^5 - \tfrac{1}{7}(\tfrac{1}{239})^7 + \cdots.$$

Since these are alternating series and the terms in each case form a decreasing sequence, we know that
$$\tfrac{1}{5} - \tfrac{1}{3}(\tfrac{1}{5})^3 \leq \text{arc tan } \tfrac{1}{5} \leq \tfrac{1}{5} - \tfrac{1}{3}(\tfrac{1}{5})^3 + \tfrac{1}{5}(\tfrac{1}{5})^5$$
and
$$\tfrac{1}{239} - \tfrac{1}{3}(\tfrac{1}{239})^3 \leq \text{arc tan } \tfrac{1}{239} \leq \tfrac{1}{239}.$$

With these inequalities together with (10.9.3) and a little arithmetic, you can verify that
$$3.14 < \pi < 3.147.$$

By using six terms of the series for arc tan $\tfrac{1}{5}$ and still only two of the series for arc tan $\tfrac{1}{239}$, one can show that
$$3.14159262 < \pi < 3.14159267.$$

Greater accuracy can of course be obtained by taking more terms into account. We shall not pursue this any further but instead simply give the value of π to twenty decimal places:
$$\pi = 3.14159\ 26535\ 89793\ 23846\ldots.$$

Exercises

Use series to obtain a numerical estimate that is accurate within 0.01.

*1. log 1.1.   2. log 1.2.   *3. log 0.8.

  4. log 1.4.   *5. arc tan 0.4.   6. arc tan 0.5.

*7. Let $a > 0$. Find a series expansion for $\log (a + x)$ which is valid for

$$-a < x \leq a.$$

8. Prove that

$$\text{if} \quad x \in [-1, 1], \quad \text{then} \quad (-1)^n \int_0^x \frac{t^{2n}}{1 + t^2} \, dt \to 0.$$

*9. Evaluate

$$\sum_{k=1}^{\infty} \frac{1}{k(k + 1)(k + 2)}.$$

HINT:

$$\frac{1}{k(k + 1)(k + 2)} = \frac{1}{2}\left(\frac{1}{k} - \frac{1}{k + 1}\right) - \frac{1}{2}\left(\frac{1}{k + 1} - \frac{1}{k + 2}\right).$$

Verify that

$$\tfrac{1}{4}\pi = 4 \arctan \tfrac{1}{5} - \arctan \tfrac{1}{239}$$

by the three-step computation of Exercises 10 through 12.

10. First, compute $\tan (2 \arctan \tfrac{1}{5})$.

*11. Then, compute $\tan (4 \arctan \tfrac{1}{5})$.

12. Finally, compute $\tan (4 \arctan \tfrac{1}{5} - \arctan \tfrac{1}{239})$.

10.10 Power Series, Part I

You have seen how certain functions can be expanded in Taylor series

$$\sum_{k=0}^{\infty} \frac{f^{(k)}(0)}{k!} x^k.$$

Such series are special instances of series of the form

$$\sum_{k=0}^{\infty} a_k x^k.$$

The latter are called *power series*.†

Power series have some interesting properties and they raise some interesting questions. A natural question to ask is: What power series are actually Taylor series? We will be able to give a complete answer to this question a little later.

As before, when it is immaterial where we begin the summation, we will dispense with detailed indexing. For example, to indicate a power series we will often simply write

$$\sum a_k x^k.$$

---

† Power series can also be written in the form

$$\sum_{k=0}^{\infty} a_k(x - c)^k.$$

A translation converts this into

$$\sum_{k=0}^{\infty} a_k x^k.$$

We launch our discussion with a definition.

Definition

A power series $\sum a_k x^k$ is said to converge

(i) at x_1 iff $\sum a_k x_1^k$ converges.
(ii) on the set S iff $\sum a_k x^k$ converges for each $x \in S$.

For determining at what numbers a power series converges, the following theorem is fundamental.

Theorem 10.10.1

If $\sum a_k x^k$ converges at x_1, $x_1 \neq 0$, then it converges absolutely at all x with $|x| < |x_1|$. If $\sum a_k x^k$ diverges at x_1, then it diverges at all x with $|x| > |x_1|$.

PROOF.   If $\sum a_k x_1^k$ converges, then $a_k x_1^k \to 0$. In particular, for k sufficiently large,

$$|a_k x_1^k| \leq 1$$

and thus

$$|a_k x^k| = |a_k x_1^k| \left|\frac{x}{x_1}\right|^k \leq \left|\frac{x}{x_1}\right|^k.$$

For x satisfying $|x| < |x_1|$, we have

$$\left|\frac{x}{x_1}\right| < 1.$$

The convergence of

$$\sum |a_k x^k|$$

follows by comparison with the geometric series. This proves the first statement.
   Suppose now that

$$\sum a_k x_1^k \quad \text{diverges.}$$

By the previous argument, there cannot exist x with $|x_1| < |x|$ such that

$$\sum a_k x^k \quad \text{converges.}$$

The existence of such an x would imply the absolute convergence of

$$\sum a_k x_1^k.$$

This proves the second statement.   □

From the theorem we just proved you can see that for a power series there are exactly three possibilities:

(i) *The series may converge only at 0.* This is what happens with

$$\sum k^k x^k.$$

(The kth term, $k^k x^k$, tends to 0 only if $x = 0$.)

(ii) *It may converge everywhere absolutely.* This is what happens with the exponential series

$$\sum \frac{x^k}{k!}.$$

(iii) The only remaining possibility is that *there exists a positive number r such that the series converges absolutely for $|x| < r$ and diverges for $|x| > r$.*

This is what happens with the geometric series

$$\sum x^k.$$

In this instance, there is absolute convergence for $|x| < 1$ and divergence for $|x| > 1$. Associated with each case there is a *radius of convergence:*

In case (i), we say that the radius of convergence is 0.
In case (ii), we say that the radius of convergence is ∞.
In case (iii), we say that the radius of convergence is r.

The three cases are pictured in Figure 10.10.1.

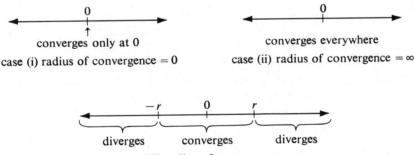

case (i) radius of convergence $= 0$ case (ii) radius of convergence $= \infty$

case (iii) radius of convergence $= r$

FIGURE 10.10.1

In general the behavior of a power series at $-r$ and at r is not predictable. The series

$$\sum x^k, \quad \sum \frac{1}{k} x^k, \quad \sum \frac{(-1)^k}{k} x^k, \quad \sum \left(\frac{1}{k}\right)^2 x^k$$

all have radius of convergence 1, but the intervals of convergence are respectively

$$(-1, 1), \quad [-1, 1), \quad (-1, 1], \quad [-1, 1].$$

Problem.   Verify that

(1)
$$\sum \frac{(-1)^k}{k} x^k$$

has interval of convergence $(-1, 1]$.

SOLUTION.   First we show that the radius of convergence is 1. We do this by examining the series

(2)
$$\sum \left| \frac{(-1)^k}{k} x^k \right| = \sum \frac{1}{k} |x|^k.$$

We set

$$c_k = \frac{1}{k} |x|^k$$

and note that

$$\frac{c_{k+1}}{c_k} = \frac{k}{k+1} \frac{|x|^{k+1}}{|x|^k} = \frac{k}{k+1} |x| \to |x|.$$

By the ratio test, series (2) converges for $|x| < 1$ and diverges for $|x| > 1$.† This shows that series (1) converges absolutely for $|x| < 1$ and diverges for $|x| > 1$. The radius of convergence is therefore 1.

Now we test the endpoints $x = -1$ and $x = 1$. At $x = -1$

$$\sum \frac{(-1)^k}{k} x^k \qquad \text{becomes} \qquad \sum \frac{(-1)^k}{k} (-1)^k = \sum \frac{1}{k}.$$

This is the harmonic series which, as you know, diverges. At $x = 1$,

$$\sum \frac{(-1)^k}{k} x^k \qquad \text{becomes} \qquad \sum \frac{(-1)^k}{k}.$$

This is a convergent alternating series.

We have shown that series (1) converges absolutely for $|x| < 1$, diverges at -1 and converges at 1. The interval of convergence is $(-1, 1]$. $\square$

Problem.   Verify that

(1)
$$\sum \frac{1}{k^2} x^k$$

has interval of convergence $[-1, 1]$.

SOLUTION.   First we examine the series

(2)
$$\sum \left| \frac{1}{k^2} x^k \right| = \sum \frac{1}{k^2} |x|^k.$$

---

† We could also have used the root test:

$$c_k^{1/k} = \left| \frac{1}{k} \right|^{1/k} |x| = \frac{1}{k^{1/k}} |x| \to |x|.$$

We set

$$c_k = \frac{1}{k^2}\,|x|^k$$

and note that

$$\frac{c_{k+1}}{c_k} = \frac{k^2}{(k+1)^2}\,\frac{|x|^{k+1}}{|x|^k} = \left(\frac{k}{k+1}\right)^2 |x| \to |x|.$$

By the ratio test, (2) converges for $|x| < 1$ and diverges for $|x| > 1$.† This shows that (1) converges absolutely for $|x| < 1$ and diverges for $|x| > 1$. The radius of convergence is therefore 1.

Now for the endpoints. At $x = -1$

$$\sum \frac{1}{k^2}\,x^k \qquad \text{becomes} \qquad \sum \frac{1}{k^2}\,(-1)^k,$$

which expanded looks like this:

$$-1 + \tfrac{1}{4} - \tfrac{1}{9} + \tfrac{1}{16} - \cdots.$$

This is a convergent alternating series. At $x = 1$,

$$\sum \frac{1}{k^2}\,x^k \qquad \text{becomes} \qquad \sum \frac{1}{k^2}.$$

This is a convergent p-series. The interval of convergence is therefore the entire closed interval $[-1, 1]$. $\square$

Problem. Find the interval of convergence of

(1) $$\sum \frac{k}{6^k}\,x^k.$$

SOLUTION. We begin by examining the series

(2) $$\sum \left| \frac{k}{6^k}\,x^k \right| = \sum \frac{k}{6^k}\,|x|^k.$$

We set

$$c_k = \frac{k}{6^k}\,|x|^k$$

and apply the root test. (The ratio test will also work.) Since

$$c^{1/k} = \tfrac{1}{6}k^{1/k}|x| \to \tfrac{1}{6}|x|,$$

you can see that (2) converges

for $\tfrac{1}{6}|x| < 1$ (for $|x| < 6$)

and diverges

for $\tfrac{1}{6}|x| > 1$.   (for $|x| > 6$)

---

† Once again we could have used the root test:

$$c_k^{1/k} = \frac{1}{k^{2/k}}\,|x| \to |x|.$$

This shows that (1) converges absolutely for $|x| < 6$ and diverges for $|x| > 6$. The radius of convergence is 6.

It is easy to see that (1) diverges both at -6 and at 6. The interval of convergence is therefore $(-6, 6)$. □

Problem. Find the interval of convergence of

(1) $$\sum \frac{(2k)!}{(3k)!} x^k.$$

SOLUTION.   We begin by examining

(2) $$\sum \left| \frac{(2k)!}{(3k)!} x^k \right| = \sum \frac{(2k)!}{(3k)!} |x|^k.$$

Set

$$c_k = \frac{(2k)!}{(3k)!} |x|^k.$$

When factorials are involved it is generally easier to use the ratio test. Note that

$$\frac{c_{k+1}}{c_k} = \frac{[2(k + 1)]! \, (3k)! \, |x|^{k+1}}{[3(k + 1)]! \, (2k)! \, |x|^k} = \frac{(2k+2)(2k+1)}{(3k + 3)(3k + 2)(3k + 1)} |x|.$$

As $k \to \infty$, this ratio tends to 0 no matter what x is. By the ratio test, (2) converges for all x and therefore (1) converges absolutely for all x. The radius of convergence is ∞ and the interval of convergence is $(-\infty, \infty)$. □

Problem. Find the interval of convergence of

$$\sum (\tfrac{1}{2}k)^k x^k.$$

SOLUTION.   Since

$$(\tfrac{1}{2}k)^k x^k \to 0 \qquad \text{only if} \quad x = 0,$$

the series converges only at $x = 0$. □

Exercises

Find the interval of convergence.

* 1. $\sum k x^k$.

2. $\sum \dfrac{1}{k} x^k$.

* 3. $\sum \dfrac{x^k}{(2k)!}$.

4. $\sum \dfrac{2^k}{k^2} x^k$.

* 5. $\sum (-k)^{2k} x^{2k}$.

6. $\sum \dfrac{(-1)^k}{\sqrt{k}} x^k$.

* 7. $\sum \dfrac{1}{k 2^k} x^k$.

8. $\sum \dfrac{1}{k^2 2^k} x^k$.

* 9. $\sum \dfrac{1}{k(k + 1)} x^k$.

10. $\sum \frac{k^2}{1 + k^2} x^k.$ *11. $\sum \frac{2^k}{\sqrt{k}} x^k.$ 12. $\sum \frac{1}{\log k} x^k.$

*13. $\sum \frac{k - 1}{k} x^k.$ 14. $\sum k a^k x^k.$ *15. $\sum \frac{k}{10^k} x^k.$

16. $\sum \frac{(-1)^k}{k^k} x^k.$ *17. $\sum \frac{(-1)^k a^k}{k^2} x^k.$

18. $\sum \frac{x^k}{(\log k)^k}.$ *19. $\sum \frac{2^{1/k} \pi^k}{k(k + 1)(k + 2)} x^k.$

20. $\sum (-1)^k (\frac{2}{3})^k x^k.$ *21. $\sum \frac{\log k}{2^k} x^k.$

10.11 Power Series, Part II

In the interior of its interval of convergence a power series represents an infinitely differentiable function. As a first step toward proving this, we have the following theorem.

Theorem 10.11.1

If

$$\sum a_k x^k \quad \text{converges on } (-c, c),$$

then

$$\sum \frac{d}{dx} (a_k x^k) = \sum k a_k x^{k-1} \quad \text{also converges on } (-c, c).$$

PROOF.   Let's assume that $\sum a_k x^k$ converges on $(-c, c)$.   By Theorem 10.10.1 we know that it converges there absolutely.

We now take $t \in (-c, c)$ and choose $\epsilon > 0$ such that

$$|t| < |t| + \epsilon < |c|.$$

Since $|t| + \epsilon$ is within the interval of convergence,

$$\sum |a_k(|t| + \epsilon)^k| \quad \text{converges.}$$

In Exercise 13 you are asked to show that for all k sufficiently large

$$|kt^{k-1}| \leq (|t| + \epsilon)^k.$$

Using this inequality, you can see that

$$|ka_k t^{k-1}| \leq |a_k(|t| + \epsilon)^k|.$$

Since $\sum |a_k(|t| + \epsilon)^k|$ converges, it follows that

$$\sum |ka_k t^{k-1}| \quad \text{converges.} \quad \square$$

Repeated application of this theorem shows that the series

$$\sum \frac{d^2}{dx^2}(a_k x^k) = \sum k(k-1)a_k x^{k-2},$$

$$\sum \frac{d^3}{dx^3}(a_k x^k) = \sum k(k-1)(k-2)a_k x^{k-3},$$

etc.,

also converge on $(-c, c)$. For example, since the geometric series

$$\sum x^k = 1 + x + x^2 + x^3 + x^4 + \cdots + x^n + \cdots$$

converges on $(-1, 1)$, you know now that

$$\sum \frac{d}{dx}(x^k) = \sum kx^{k-1} = 1 + 2x + 3x^2 + 4x^3 + \cdots + nx^{n-1} + \cdots$$

$$\sum \frac{d^2}{dx^2}(x^k) = \sum k(k-1)x^{k-2} = 2 + 6x + 12x^2 + \cdots + n(n-1)x^{n-2} + \cdots, \text{etc.,}$$

also converge on $(-1, 1)$. $\square$

Suppose now that

$$\sum a_k x^k \quad \text{converges on } (-c, c).$$

Then, as you just saw,

$$\sum ka_k x^{k-1} \quad \text{also converges on } (-c, c).$$

Using the first series, we can define a function f on $(-c, c)$ by setting

$$f(x) = \sum a_k x^k \quad \text{for all } x \text{ in } (-c, c).$$

Using the second series, we can also define a function g by setting

$$g(x) = \sum ka_k x^{k-1} \quad \text{for all } x \text{ in } (-c, c).$$

The crucial point is that

$$f'(x) = g(x).$$

For emphasis we cast this as a theorem.

The Differentiability of Power-Series Theorem

If

$$f(x) = \sum_{k=0}^{\infty} a_k x^k \quad \text{for all } x \text{ in } (-c, c),$$

then f is differentiable on $(-c, c)$ and

$$f'(x) = \sum_{k=1}^{\infty} ka_k x^{k-1} \quad \text{for all } x \text{ in } (-c, c).$$

By applying this theorem to f', you will see that f' is itself differentiable. This in turn implies that f'' is differentiable, and so on. In short, f has derivatives of all orders.

We can summarize this way:

> In the interior of its interval of convergence a power series defines an infinitely differentiable function, the derivatives of which can be obtained by differentiating term by term.

The term-by-term differentiation refers to the fact that

$$\frac{d}{dx}\left(\sum a_k x^k\right) = \sum \frac{d}{dx}(a_k x^k).$$

For a proof of the differentiability theorem we refer to the supplement at the end of the section. Right now we give some examples.

Example. You already know that

$$\frac{d}{dx}(e^x) = e^x.$$

You can see this directly by differentiating the exponential series:

$$\frac{d}{dx}(e^x) = \frac{d}{dx}\left(\sum_{k=0}^{\infty} \frac{x^k}{k!}\right) = \sum_{k=0}^{\infty} \frac{d}{dx}\left(\frac{x^k}{k!}\right) = \sum_{k=1}^{\infty} \frac{x^{k-1}}{(k-1)!} = \sum_{k=0}^{\infty} \frac{x^k}{k!} = e^x. \quad \square$$

Example. You know that

$$\sin x = x - \frac{x^3}{3!} + \frac{x^5}{5!} - \frac{x^7}{7!} + \frac{x^9}{9!} - \cdots$$

and

$$\cos x = 1 - \frac{x^2}{2!} + \frac{x^4}{4!} - \frac{x^6}{6!} + \frac{x^8}{8!} - \cdots.$$

Relations

$$\frac{d}{dx}(\sin x) = \cos x, \qquad \frac{d}{dx}(\cos x) = -\sin x$$

can be easily verified by differentiating the series term by term:

$$\frac{d}{dx}(\sin x) = 1 - \frac{3x^2}{3!} + \frac{5x^4}{5!} - \frac{7x^6}{7!} + \frac{9x^8}{9!} - \cdots$$

$$= 1 - \frac{x^2}{2!} + \frac{x^4}{4!} - \frac{x^6}{6!} + \frac{x^8}{8!} - \cdots = \cos x,$$

$$\frac{d}{dx}(\cos x) = -\frac{2x}{2!} + \frac{4x^3}{4!} - \frac{6x^5}{6!} + \frac{8x^7}{8!} - \cdots = -x + \frac{x^3}{3!} - \frac{x^5}{5!} + \frac{x^7}{7!} - \cdots$$

$$= -\left(x - \frac{x^3}{3!} + \frac{x^5}{5!} - \frac{x^7}{7!} + \cdots\right) = -\sin x. \quad \square$$

Example.   We can sum the series

$$\sum_{n=1}^{\infty} \frac{x^n}{n}, \qquad x \in (-1, 1)$$

by setting

$$g(x) = \sum_{n=1}^{\infty} \frac{x^n}{n}, \qquad x \in (-1, 1)$$

and noting that

$$g'(x) = \sum_{n=1}^{\infty} \frac{n x^{n-1}}{n} = \sum_{n=1}^{\infty} x^{n-1} = \sum_{n=0}^{\infty} x^n = \frac{1}{1-x}.$$

Since

$$g(0) = 0 \quad \text{and} \quad g'(x) = \frac{1}{1-x},$$

we must have

$$g(x) = -\log |1 - x|. \quad \square$$

Power series can also be integrated term by term.

Term-by-Term Integration Theorem

If

$$f(x) = \sum a_k x^k \quad \text{converges on } (-c, c),$$

then

$$g(x) = \sum \frac{a_k x^{k+1}}{k+1} \quad \text{also converges on } (-c, c)$$

and

$$\int f(x) \, dx = g(x) + C.$$

PROOF.   If $\sum a_k x^k$ converges on $(-c, c)$, then $\sum |a_k x^k|$ converges on $(-c, c)$. Since

$$\left| \frac{a_k x^k}{k+1} \right| \leq |a_k x^k|,$$

we know by comparison that

$$\sum \left| \frac{a_k x^k}{k+1} \right| \quad \text{also converges on } (-c, c).$$

It follows that

$$x \sum \frac{a_k x^k}{k+1} = \sum \frac{a_k x^{k+1}}{k+1} \quad \text{converges on } (-c, c).$$

With

$$f(x) = \sum a_k x^k \quad \text{and} \quad g(x) = \sum \frac{a_k x^{k+1}}{k+1},$$

we know from the differentiability theorem that

$$g'(x) = f(x)$$

and therefore

$$\int f(x)\, dx = g(x) + C. \quad \square$$

The idea of term-by-term integration can also be expressed by writing

$$\int \left(\sum a_k x^k \right) dx = \left(\sum \frac{a_k x^{k+1}}{k+1} \right) + C.$$

For definite integrals, we have

$$\int_a^b \left(\sum a_k x^k \right) dx = \sum \left(\int_a^b \frac{a_k x^k}{k+1}\, dx \right),$$

so long, of course, as $[a, b]$ is contained in the interval of convergence.

Here are some examples of term-by-term integration.

Example. By integrating

$$\frac{1}{1+x} = \frac{1}{1-(-x)} = \sum_{k=0}^{\infty} (-1)^k x^k, \qquad |x| < 1,$$

we get

$$\log(1+x) = \int \left(\sum_{k=0}^{\infty} (-1)^k x^k \right) dx = \left(\sum_{k=0}^{\infty} \frac{(-1)^k x^{k+1}}{k+1} \right) + C, \qquad |x| < 1.$$

The constant C is 0 as you can tell from the fact that the series on the right and $\log(1+x)$ are both 0 at $x = 0$. $\quad \square$

Example. The arc tangent series can also be obtained by term-by-term integration:

$$\frac{1}{1+x^2} = \frac{1}{1-(-x^2)} = \sum_{k=0}^{\infty} (-1)^k x^{2k}$$

and therefore

$$\text{arc tan } x = \int \left(\sum_{k=0}^{\infty} (-1)^k x^{2k} \right) dx = \left(\sum_{k=0}^{\infty} \frac{(-1)^k x^{2k+1}}{2k+1} \right) + C.$$

The constant C is once again 0, this time because the series on the right and arc tan x are both 0 at $x = 0$. $\quad \square$

Term-by-term integration can be a useful technique even if at the beginning there are no series in sight. Suppose that you are trying to evaluate an integral

$$\int_a^b f(x)\, dx$$

and that you cannot find an antiderivative. If you know that $f(x)$ has a convergent power series expansion, then you can evaluate the integral in question by first forming the series and then integrating term by term.

Example. We can apply these ideas to

$$\int_0^1 e^{-x^2}\, dx.$$

Since

$$e^y = 1 + y + \frac{y^2}{2!} + \frac{y^3}{3!} + \frac{y^4}{4!} + \cdots,$$

we have

$$e^{-x^2} = 1 - x^2 + \frac{x^4}{2!} - \frac{x^6}{3!} + \frac{x^8}{4!} - \cdots.$$

Term-by-term integration gives

$$\int_0^1 e^{-x^2}\, dx = \left[x - \frac{x^3}{3} + \frac{x^5}{5 \cdot 2!} - \frac{x^7}{7 \cdot 3!} + \frac{x^9}{9 \cdot 4!} - \cdots \right]_0^1$$

$$= 1 - \frac{1}{3} + \frac{1}{5 \cdot 2!} - \frac{1}{7 \cdot 3!} + \frac{1}{9 \cdot 4!} - \cdots.$$

This last series is an alternating series of the type we have discussed. It is not hard to see that

$$1 - \frac{1}{3} + \frac{1}{5 \cdot 2!} - \frac{1}{7 \cdot 3!} \leq \int_0^1 e^{-x^2}\, dx \leq 1 - \frac{1}{3} + \frac{1}{5 \cdot 2!} - \frac{1}{7 \cdot 3!} + \frac{1}{9 \cdot 4!}$$

and therefore

$$0.74 < \int_0^1 e^{-x^2}\, dx < 0.75.$$

Greater accuracy can of course be obtained by using more terms.   □

It is time to relate Taylor series

$$\sum_{k=0}^{\infty} \frac{f^{(k)}(0)}{k!} x^k$$

to power series in general. The relation is very simple:

| On its interval of convergence a power series is the Taylor series of its sum. |

To see this, all you have to do is differentiate

$$f(x) = a_0 + a_1x + a_2x^2 + \cdots + a_kx^k + \cdots$$

term by term. By doing this you will find that

$$f^{(k)}(0) = k! \, a_k \quad \text{and therefore} \quad a_k = \frac{f^{(k)}(0)}{k!}. \quad \square$$

Exercises

*1. Find a series for $\sec^2 x$ by differentiating the series for $\tan x$.

2. Find a series for $\log(\cos x)$ by integrating the series for $\tan x$.

*3. Find a series for $\log(1 - x)$ by integration.

Estimate within 0.01.

*4. $\displaystyle\int_0^1 e^{-x^3} \, dx.$ 5. $\displaystyle\int_0^1 \sin x^2 \, dx.$

*6. $\displaystyle\int_0^1 \sin \sqrt{x} \, dx.$ *7. $\displaystyle\int_0^1 \arctan x^2 \, dx.$

8. $\displaystyle\int_0^1 x^4 e^{-x^2} \, dx.$ *9. $\displaystyle\int_1^2 \frac{1 - \cos x}{x} \, dx.$

10. $\displaystyle\int_0^{1/3} e^x \log(1 + x) \, dx.$ *11. $\displaystyle\int_0^1 x \sin \sqrt{x} \, dx.$

12. Show that, if $\sum a_k x^k$ and $\sum b_k x^k$ both converge to the same sum on some interval, then

$$a_k = b_k \quad \text{for each } k.$$

13. Show that if $\epsilon > 0$, then

$$|kt^{k-1}| < (|t| + \epsilon)^k \quad \text{for all } k \text{ sufficiently large.}$$

HINT: $k^{1/k}|t|^{(k-1)/k} = k^{1/k}|t|^{1-(1/k)} \to |t|$.

Supplement to Section 10.11

Proof of the Differentiability of Power Series

We begin by setting

$$f(x) = \sum_{k=0}^{\infty} a_k x^k \quad \text{and} \quad g(x) = \sum_{k=1}^{\infty} k a_k x^{k-1}.$$

Taking x from $(-c, c)$, we want to show that

$$\lim_{h \to 0} \frac{f(x + h) - f(x)}{h} = g(x).$$

With $x + h$ in $(-c, c)$, $h \neq 0$, we have

$$\left| g(x) - \frac{f(x + h) - f(x)}{h} \right| = \left| \sum_{k=1}^{\infty} k a_k x^{k-1} - \sum_{k=0}^{\infty} \frac{a_k(x + h)^k - a_k x^k}{h} \right|$$

$$= \left| \sum_{k=1}^{\infty} k a_k x^{k-1} - \sum_{k=1}^{\infty} a_k \left[\frac{(x + h)^k - x^k}{h} \right] \right|.$$

By the mean-value theorem we know that there exists t_k between x and $x + h$ such that

$$\frac{(x + h)^k - x^k}{h} = k t_k^{k-1}.$$

Using this result, we have

$$\left| g(x) - \frac{f(x + h) - f(x)}{h} \right| = \left| \sum_{k=1}^{\infty} k a_k x^{k-1} - \sum_{k=1}^{\infty} k a_k t_k^{k-1} \right|$$

$$= \left| \sum_{k=1}^{\infty} k a_k (x^{k-1} - t_k^{k-1}) \right| = \left| \sum_{k=2}^{\infty} k a_k (x^{k-1} - t_k^{k-1}) \right|.$$

Again by the mean-value theorem we know that there exists p_{k-1} between x and t_k such that

$$\frac{x^{k-1} - t_k^{k-1}}{x - t_k} = (k - 1)(p_{k-1})^{k-2}.$$

Now

$$|x^{k-1} - t_k^{k-1}| = |x - t_k| \, |(k - 1)(p_{k-1})^{k-2}|.$$

Since $|x - t_k| < |h|$ and

$$|p_{k-1}| \le |\alpha|, \qquad \text{where} \quad |\alpha| = \max \{|x|, |x + h|\},$$

we have

$$|x^{k-1} - t_k^{k-1}| \le |h| \, |(k - 1)\alpha^{k-2}|.$$

Thus

$$\left| g(x) - \frac{f(x + h) - f(x)}{h} \right| \le |h| \sum_{k=2}^{\infty} |k(k - 1)a_k \alpha^{k-2}|.$$

Since the series on the right converges, we have

$$\lim_{h \to 0} \left(|h| \sum_{k=2}^{\infty} |k(k - 1)a_k \alpha^{k-2}| \right) = 0,$$

so that

$$\lim_{h \to 0} \left| g(x) - \frac{f(x + h) - f(x)}{h} \right| = 0$$

and

$$\lim_{h \to 0} \frac{f(x + h) - f(x)}{h} = g(x). \quad \square$$

10.12   Problems on the Binomial Series

Through a collection of problems we invite you to derive on your own the basic properties of one of the most celebrated series of all—the *binomial series*.

Beginning with the binomial $1 + x$ (2 terms) and a real number α, form the function

$$f(x) = (1 + x)^{\alpha}.$$

Problem 1.   Show that

$$\frac{f^{(k)}(0)}{k!} = \frac{\alpha[\alpha - 1][\alpha - 2] \cdots [\alpha - (k - 1)]}{k!}.$$

The number you just obtained is the coefficient of x^k in the expansion of $(1 + x)^{\alpha}$. It is called the kth *binomial coefficient* and it is usually denoted by $\binom{\alpha}{k}$:

$$\binom{\alpha}{k} = \frac{\alpha[\alpha - 1][\alpha - 2] \cdots [\alpha - (k - 1)]}{k!}. \qquad †$$

Problem 2.   Show that the binomial series

$$\Sigma \binom{\alpha}{k} x^k$$

has radius of convergence 1. HINT: Use the ratio test.

From Problem 2 you know that the binomial series converges on the open interval $(-1, 1)$ and defines there an infinitely differentiable function. The next thing to show is that this function (the one defined by the series) is actually $(1 + x)^{\alpha}$. To do this, you first need some other results.

Problem 3.   Verify the identity

$$(k + 1) \binom{\alpha}{k + 1} + k \binom{\alpha}{k} = \alpha \binom{\alpha}{k}.$$

Problem 4.   Use the identity of Problem 3 to show that the sum of the binomial series

$$\phi(x) = \sum_{k=0}^{\infty} \binom{\alpha}{k} x^k, \qquad x \in (-1, 1)$$

satisfies the differential equation

$$(1 + x)\phi'(x) = \alpha\phi(x).$$

You are now in a position to prove the main result.

---

† By special definition $\binom{\alpha}{0} = 1$.

Problem 5. Show that for all x in $(-1, 1)$

$$(1 + x)^\alpha = \sum_{k=0}^{\infty} \binom{\alpha}{k} x^k.$$

To get a better feeling for what the series looks like it is a good idea to expand it a few terms:

(10.12.1) $(1 + x)^\alpha = 1 + \alpha x + \dfrac{\alpha(\alpha - 1)}{2!} x^2 + \dfrac{\alpha(\alpha - 1)(\alpha - 2)}{3!} x^3 + \cdots.$

Problem 6. Using (10.12.1) as a model expand each of the following in powers of x up to x^4.

*(a) $\sqrt{1 + x}$.   (b) $\dfrac{1}{\sqrt{1 + x}}$.   *(c) $\sqrt{1 + x^2}$.   (d) $\dfrac{1}{\sqrt{1 - x^2}}$.

Problem 7. Use series to obtain a numerical estimate that is accurate within 0.01.

*(a) $\sqrt{98}$. HINT: $\sqrt{98} = (100 - 2)^{1/2} = 10(1 - \frac{1}{50})^{1/2}$.
(b) $\sqrt[3]{9}$.   *(c) $\sqrt[4]{620}$.   (d) $\sqrt[5]{36}$.   *(e) $17^{-1/4}$.

10.13 Additional Exercises

State whether or not the indicated sequence converges, and if it does, find the limit.

*1. $3^{1/n}$.

2. $n2^{1/n}$.

*3. $\cos n\pi \sin n\pi$.

4. $\dfrac{(n + 1)(n + 2)}{(n + 3)(n + 4)}$.

*5. $\left(\dfrac{n}{1 + n}\right)^{1/n}$.

6. $\dfrac{n^2 + 5n + 1}{n^3 + 1}$.

*7. $\cos \dfrac{\pi}{n} \sin \dfrac{\pi}{n}$.

8. $\dfrac{n^\pi}{1 - n^\pi}$.

*9. $\left(2 + \dfrac{1}{n}\right)^n$.

10. $\dfrac{\pi}{n} \cos \dfrac{\pi}{n}$.

*11. $\dfrac{\log [n(n + 1)]}{n}$.

12. $\left[\log \left(1 + \dfrac{1}{n}\right)\right]^n$.

*13. $\dfrac{\pi}{n} \log \dfrac{n}{\pi}$.

14. $\dfrac{\pi}{n} e^{\pi/n}$.

*15. $\dfrac{n}{\pi} \sin n\pi$.

Sum the following series.

*16. $\displaystyle\sum_{k=0}^{\infty} \left(\dfrac{1}{4}\right)^k$.

17. $\displaystyle\sum_{k=0}^{\infty} \left(\dfrac{3}{4}\right)^{k+1}$.

*18. $\displaystyle\sum_{k=0}^{\infty} (-1)^k \left(\dfrac{1}{2}\right)^k$.

19. $\displaystyle\sum_{k=0}^{\infty} \frac{(\log 2)^k}{k!}$.

*20. $\displaystyle\sum_{k=1}^{\infty} \left(\frac{1}{k} - \frac{1}{k+1} \right)$.

21. $\displaystyle\sum_{k=2}^{\infty} \left(\frac{1}{k^2} - \frac{1}{(k+1)^2} \right)$.

Test the following series for (a) convergence, (b) absolute convergence.

*22. $1 + \dfrac{1}{3} + \dfrac{1}{5} + \cdots + \dfrac{1}{2n+1} + \cdots$.

23. $\dfrac{1}{1\cdot 3} + \dfrac{1}{3\cdot 5} + \dfrac{1}{5\cdot 7} + \cdots + \dfrac{1}{(2n+1)(2n+3)} + \cdots$.

*24. $\dfrac{1}{2\cdot 3} - \dfrac{1}{3\cdot 4} + \dfrac{1}{4\cdot 5} - \cdots + \dfrac{(-1)^{n+1}}{(n+1)(n+2)} + \cdots$.

25. $\dfrac{1}{2 \log 2} + \dfrac{1}{3 \log 3} + \dfrac{1}{4 \log 4} + \cdots + \dfrac{1}{n \log n} + \cdots$.

*26. $1 - \dfrac{1}{3} + \dfrac{1}{5} - \cdots + (-1)^n \dfrac{1}{2n+1} + \cdots$.

27. $100 - \dfrac{100^2}{2!} + \dfrac{100^3}{3!} - \cdots + (-1)^{n+1} \dfrac{100^n}{n!} + \cdots$.

*28. $1 - \dfrac{2}{3} + \dfrac{3}{3^2} - \cdots + (-1)^{n-1} \dfrac{n}{3^{n-1}} + \cdots$.

29. $\dfrac{3}{4} + 2\left(\dfrac{3}{4}\right)^2 + 3\left(\dfrac{3}{4}\right)^3 + \cdots + n\left(\dfrac{3}{4}\right)^n + \cdots$.

*30. $\dfrac{1}{\sqrt{2\cdot 3}} - \dfrac{1}{\sqrt{3\cdot 4}} + \dfrac{1}{\sqrt{4\cdot 5}} - \cdots + \dfrac{(-1)^{n-1}}{\sqrt{(n+1)(n+2)}} + \cdots$.

31. $\dfrac{1}{5} - \dfrac{1}{\sqrt{5}} + \dfrac{1}{\sqrt[3]{5}} - \cdots + \dfrac{(-1)^{n-1}}{\sqrt[n]{5}} + \cdots$.

Find the interval of convergence.

*32. $\displaystyle\sum \frac{x^{2k+1}}{2k+1}$.

33. $\displaystyle\sum \frac{k!}{2} x^k$.

*34. $\displaystyle\sum \frac{k!}{2} x^{k!}$.

35. $\displaystyle\sum \frac{x^k}{2^{k!}}$.

Expand in powers of x.

*36. a^x.

37. $e^{\sin x}$.

*38. $\log (1 + x^2)$.

39. Verify the expansion

$$\log \left(\frac{1+x}{1-x} \right) = 2\left[x + \frac{x^3}{3} + \frac{x^5}{5} + \cdots \right].$$

Use the first three nonzero terms of the series to find a fractional approximation for $\log 2$.   HINT: Use $x = \frac{1}{3}$.

40. Deduce a series expansion for

$$\frac{1}{(1 - x)^2}$$

from the geometric series.

Estimate within 0.01.

* 41. $\displaystyle\int_0^{1/2} \frac{dx}{1 + x^4}$.   42. $e^{2/3}$.   * 43. $\sqrt[3]{60}$.   44. $\displaystyle\int_0^1 x \sin x^4 \, dx$.

* 45. (*Optional*) Find a series expansion for $\tanh^{-1} x$ by integrating the series of its derivative.

L'Hospital's Rule; Improper Integrals

11.1 Limits as $x \to \pm\infty$

Now that you are familiar with convergence for sequences, limits as $x \to \infty$ should present no difficulties. To say that

$$\lim_{x \to \infty} f(x) = l$$

[read "the limit of $f(x)$ as x tends to infinity is l"] is to say that we can make the difference between $f(x)$ and l as small as we wish simply by requiring that x be sufficiently large. More precisely,

$$\lim_{x \to \infty} f(x) = l \quad \text{iff} \quad \begin{cases} \text{for each } \epsilon > 0 \text{ there exists } K > 0 \text{ such that,} \\ \quad \text{if } x \geq K, \quad \text{then } |f(x) - l| < \epsilon. \end{cases}$$

Here are some simple examples:

$$\lim_{x \to \infty} \frac{1}{x} = 0, \qquad \lim_{x \to \infty} \frac{x - 1}{2x + 5} = \lim_{x \to \infty} \frac{1 - 1/x}{2 + 5/x} = \frac{1}{2}, \qquad \lim_{x \to \infty} e^{1/x} = 1.$$

It is easy to interpret the idea geometrically. To say that $\lim_{x \to \infty} f(x) = A$ (see Figure 11.1.1) is to say that, as x becomes arbitrarily large, the graph of f approaches the horizontal line $y = A$. The line is called a *horizontal asymptote*.

It is obvious that limits as $x \to \infty$ are unique (if they exist) and that the usual rules for computation hold. For example, if

$$\lim_{x \to \infty} f(x) = l \quad \text{and} \quad \lim_{x \to \infty} g(x) = m,$$

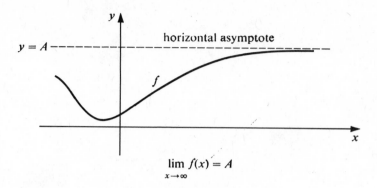

$$\lim_{x \to \infty} f(x) = A$$

FIGURE 11.1.1

then

(i)
$$\lim_{x \to \infty} \alpha f(x) + \beta g(x) = \alpha l + \beta m;$$

(ii)
$$\lim_{x \to \infty} f(x)g(x) = lm;$$

(iii)
$$\lim_{x \to \infty} \frac{f(x)}{g(x)} = \frac{l}{m} \qquad \text{if } m \neq 0.$$

Moreover, the pinching theorem is valid; namely, if

$$h(x) \leq f(x) \leq g(x) \qquad \text{for all } x \text{ sufficiently large}$$

and

$$\lim_{x \to \infty} h(x) = l = \lim_{x \to \infty} g(x),$$

then

$$\lim_{x \to \infty} f(x) = l.$$

We illustrate these ideas by carrying out some computations.

Problem. Find

$$\lim_{x \to \infty} \frac{\sin^2 x}{x^2}.$$

SOLUTION.   For $x \neq 0$,

$$0 \leq \frac{\sin^2 x}{x^2} \leq \frac{1}{x^2}. \qquad\qquad \text{(since } |\sin x| \leq 1\text{)}$$

Since obviously

$$\lim_{x \to \infty} \frac{1}{x^2} = 0,$$

we have

$$\lim_{x \to \infty} \frac{\sin^2 x}{x^2} = 0. \quad \square$$

Problem.   Find

$$\lim_{x \to \infty} \frac{x(e^{-x} + 1)}{2x - 1}.$$

SOLUTION.   For x different from 0 and $\frac{1}{2}$,

$$\frac{x(e^{-x} + 1)}{2x - 1} = \frac{e^{-x} + 1}{2 - 1/x}.$$

Since

$$\lim_{x \to \infty} (e^{-x} + 1) = 1 \quad \text{and} \quad \lim_{x \to \infty} (2 - 1/x) = 2,$$

we have

$$\lim_{x \to \infty} \frac{x(e^{-x} + 1)}{2x - 1} = \frac{1}{2}. \quad \square$$

Problem.   Find

$$\lim_{x \to \infty} \frac{x + \log x}{x \log x}.$$

SOLUTION.   For x different from 0 and 1,

$$\frac{x + \log x}{x \log x} = \frac{x}{x \log x} + \frac{\log x}{x \log x} = \frac{1}{\log x} + \frac{1}{x}.$$

Since obviously

$$\lim_{x \to \infty} \frac{1}{\log x} = 0 \quad \text{and} \quad \lim_{x \to \infty} \frac{1}{x} = 0,$$

we have

$$\lim_{x \to \infty} \frac{x + \log x}{x \log x} = 0. \quad \square$$

There is a simple relation between limits at ∞ and one-sided limits at 0. Setting $x = 1/t$, it is easy to see that

$$\boxed{\lim_{x \to \infty} f(x) = l \quad \text{iff} \quad \lim_{t \downarrow 0} f(1/t) = l.}$$

We'll use this idea in the next section.

Limits as $x \to -\infty$ are defined by setting

$$\boxed{\lim_{x \to -\infty} f(x) = l \quad \text{iff} \quad \begin{cases} \text{for each } \epsilon > 0 \text{ there exist } K < 0 \text{ such that,} \\ \text{if } x \le K, \quad \text{then } |f(x) - l| < \epsilon. \end{cases}}$$

The behavior of such limits is entirely analogous to the case $x \to \infty$. As simple examples, we have

$$\lim_{x \to -\infty} e^x = 0, \qquad \lim_{x \to -\infty} \frac{3 - x}{2 + x} = \lim_{x \to -\infty} \frac{3/x - 1}{2/x + 1} = -1.$$

In the case of

$$h(x) = \begin{cases} -1, & x < 0 \\ 1, & x > 0 \end{cases},$$

we have

$$\lim_{x \to \infty} h(x) = 1 \quad \text{but} \quad \lim_{x \to -\infty} h(x) = -1. \quad \square$$

Exercises

Find the following limits.

*1. $\lim_{x \to \infty} \dfrac{3}{x}$.

2. $\lim_{x \to \infty} \dfrac{x}{x^2 + 1}$.

*3. $\lim_{x \to \infty} \dfrac{x - 1}{x + 1}$.

*4. $\lim_{x \to \infty} \dfrac{2 \sin x}{x}$.

5. $\lim_{x \to \infty} \log\left(\dfrac{2x}{x + 1}\right)$.

*6. $\lim_{x \to \infty} 10^{1/x}$.

*7. $\lim_{x \to \infty} \dfrac{x^2 e^{-x}}{2 - x^2}$.

8. $\lim_{x \to \infty} \dfrac{(x - a)(x - b)}{(ax + b)^2}$.

*9. $\lim_{x \to \infty} \dfrac{(ax - b)^2}{(a - x)(b - x)}$.

*10. $\lim_{x \to \infty} \sin \dfrac{1}{x}$.

11. $\lim_{x \to \infty} \left(1 - \sin \dfrac{1}{x}\right)^2$.

*12. $\lim_{x \to \infty} x \sin \dfrac{1}{x}$.

*13. $\lim_{x \to \infty} \displaystyle\int_0^x \dfrac{dt}{1 + t^2}$.

14. $\lim_{x \to \infty} \displaystyle\int_e^x \dfrac{dt}{t(\log t)^2}$.

*15. $\lim_{x \to \infty} \left(x - x \cos \dfrac{1}{x}\right)$.

16. $\lim_{x \to \infty} (\sqrt{x + 1} - \sqrt{x})$.

*17. $\lim_{x \to \infty} (\sqrt{x^2 + x} - \sqrt{x^2 + 1})$.

18. $\lim_{x \to -\infty} \dfrac{\sqrt{x^2 + 1}}{2x}$.

19. (a) Prove that if f is increasing and bounded on $[a, \infty)$, then f has a limit as x tends to ∞.

(b) Prove that if f is decreasing and bounded on $[a, \infty)$, then f has a limit as x tends to ∞.

20. Prove that if f is continuous on $[a, \infty)$ and has a finite limit as x tends to ∞, then f is bounded on $[a, \infty)$.

11.2 L'Hospital's Rule $(\frac{0}{0})$

Here we are concerned with limits of quotients

$$\frac{f(x)}{g(x)}$$

where numerator and denominator both tend to 0. For convenience we will dispense with the formality of writing

$$\lim_{x \to c} f(x) = 0, \qquad \lim_{x \to \infty} f(x) = 0, \qquad \text{etc.}$$

and write instead

$$\text{as} \quad x \to c, \qquad f(x) \to 0,$$
$$\text{as} \quad x \to \infty, \qquad f(x) \to 0, \qquad \text{etc.}$$

L'Hospital's Rule $(\frac{0}{0})$

Suppose that

as $\ x \downarrow c, \ x \uparrow c, \ x \to c, \ x \to \infty, \text{ or } x \to -\infty, \ \ f(x) \to 0 \text{ and } g(x) \to 0.$

If $\ \dfrac{f'(x)}{g'(x)} \to l, \qquad$ then $\quad \dfrac{f(x)}{g(x)} \to l.$

We will prove the validity of L'Hospital's rule a little later. First we demonstrate its usefulness.

Problem. Find

$$\lim_{x \to \pi/2} \frac{\cos x}{\pi - 2x}.$$

SOLUTION. As $x \to \pi/2$, both numerator and denominator tend to 0 and it is not at all obvious what happens to the quotient

$$\frac{\cos x}{\pi - 2x}.$$

To see whether we can apply L'Hospital's rule, we will examine the quotient of the derivatives. Since

$$\frac{\dfrac{d}{dx}(\cos x)}{\dfrac{d}{dx}(\pi - 2x)} = \frac{-\sin x}{-2} = \frac{\sin x}{2}$$

and

$$\frac{\sin x}{2} \to \frac{1}{2} \qquad \text{as} \quad x \to \frac{\pi}{2},$$

we have

$$\frac{\dfrac{d}{dx}(\cos x)}{\dfrac{d}{dx}(\pi - 2x)} \to \frac{1}{2}.$$

It follows from L'Hospital's rule that

$$\frac{\cos x}{\pi - 2x} \to \frac{1}{2}.$$

We can express this all on one line:

$$\lim_{x \to \pi/2} \frac{\cos x}{\pi - 2x} = \lim_{x \to \pi/2} \frac{\dfrac{d}{dx}(\cos x)}{\dfrac{d}{dx}(\pi - 2x)} = \lim_{x \to \pi/2} \frac{\sin x}{2} = \frac{1}{2}. \quad \square$$

Problem. Find

$$\lim_{x \downarrow 0} \frac{x}{\sin \sqrt{x}}.$$

SOLUTION.   As $x \downarrow 0$, both numerator and denominator tend to 0. Since

$$\frac{\dfrac{d}{dx}(x)}{\dfrac{d}{dx}(\sin \sqrt{x})} = \frac{2\sqrt{x}}{\cos \sqrt{x}} \to 0 \qquad \text{as} \quad x \downarrow 0,$$

it follows from L'Hospital's rule that

$$\frac{x}{\sin \sqrt{x}} \to 0 \qquad \text{as} \quad x \downarrow 0.$$

For short we write

$$\lim_{x \downarrow 0} \frac{x}{\sin \sqrt{x}} = \lim_{x \downarrow 0} \frac{\dfrac{d}{dx}(x)}{\dfrac{d}{dx}(\sin \sqrt{x})} = \lim_{x \downarrow 0} \frac{2\sqrt{x}}{\cos \sqrt{x}} = 0. \quad \square$$

Sometimes it is necessary to differentiate numerator and denominator more than once. The next problem gives such an instance.

Problem. Find

$$\lim_{x \to 0} \frac{1 - \cos 2x}{x^2}.$$

SOLUTION.   As $x \to 0$, both numerator and denominator tend to 0. Trying to apply L'Hospital's rule, we form the quotient

$$\frac{\dfrac{d}{dx}(1 - \cos 2x)}{\dfrac{d}{dx}(x^2)} = \frac{2 \sin 2x}{2x} = \frac{\sin 2x}{x}.$$

Since numerator and denominator still tend to 0, we differentiate again:

$$\frac{\dfrac{d^2}{dx^2}(1 - \cos 2x)}{\dfrac{d^2}{dx^2}(x^2)} = \frac{\dfrac{d}{dx}(\sin 2x)}{\dfrac{d}{dx}(x)} = \frac{2 \cos 2x}{1}.$$

Since this last quotient tends to 2, we can conclude that

$$\frac{\sin 2x}{x} \to 2 \quad \text{and therefore} \quad \frac{1 - \cos 2x}{x^2} \to 2.$$

For short we can write

$$\lim_{x \to 0} \frac{1 - \cos 2x}{x^2} = \lim_{x \to 0} \frac{\sin 2x}{x} = \lim_{x \to 0} \frac{2 \cos 2x}{1} = 2. \quad \square$$

Problem.   Find

$$\lim_{x \downarrow 0} (1 + x)^{1/x}.$$

SOLUTION.   As the expression stands, it is not amenable to L'Hospital's rule. To make it so, we take its logarithm. This gives

$$\lim_{x \downarrow 0} \log (1 + x)^{1/x} = \lim_{x \downarrow 0} \frac{\log (1 + x)}{x} \underset{\text{differentiating}}{\overset{}{=}} \lim_{x \downarrow 0} \frac{1}{1 + x} = 1.$$

Since $\log (1 + x)^{1/x}$ tends to 1, $(1 + x)^{1/x}$ must itself tend to e. $\square$

To prove the validity of L'Hospital's rule, we need first a generalization of the mean-value theorem.

The Cauchy Mean-Value Theorem

Suppose that f and g are differentiable on (a, b) and continuous on $[a, b]$. If g' is never 0 in (a, b), then there is a number c in (a, b) such that

$$\frac{f'(c)}{g'(c)} = \frac{f(b) - f(a)}{g(b) - g(a)}.$$

PROOF.   We can prove this by applying the mean-value theorem to the function

$$G(x) = [g(b) - g(a)][f(x) - f(a)] - [g(x) - g(a)][f(b) - f(a)].$$

Since

$$G(a) = 0 \quad \text{and} \quad G(b) = 0,$$

there exists (by the mean-value theorem) a number c in (a, b) such that

$$G'(c) = 0.$$

Since in general

$$G'(x) = [g(b) - g(a)]f'(x) - g'(x)[f(b) - f(a)],$$

we must have

$$[g(b) - g(a)]f'(c) - g'(c)[f(b) - f(a)] = 0,$$

and therefore

$$[g(b) - g(a)]f'(c) = g'(c)[f(b) - f(a)].$$

Since g' is never 0 in (a, b),

$$g'(c) \neq 0 \quad \text{and} \quad g(b) - g(a) \neq 0.$$

$\uparrow$ ——————— explain

We can therefore divide by these numbers and obtain

$$\frac{f'(c)}{g'(c)} = \frac{f(b) - f(a)}{g(b) - g(a)}. \quad \square$$

Now we prove L'Hospital's rule for the case $x \downarrow c$. We assume that, as $x \downarrow c$,

$$f(x) \to 0, \qquad g(x) \to 0, \quad \text{and} \quad \frac{f'(x)}{g'(x)} \to l$$

and show that

$$\frac{f(x)}{g(x)} \to l.$$

PROOF.   The fact that

$$\frac{f'(x)}{g'(x)} \to l \quad \text{as} \quad x \downarrow c$$

assures us that both f' and g' exist on a set of the form $(c, c + h]$ and g' is not zero there. By setting $f(c) = 0$ and $g(c) = 0$, we ensure that f and g are both continuous on $[c, c + h]$. We can now apply the Cauchy mean-value theorem and conclude that there exists a number c_h between c and $c + h$ such that

$$\frac{f'(c_h)}{g'(c_h)} = \frac{f(c + h) - f(c)}{g(c + h) - g(c)} = \frac{f(c + h)}{g(c + h)}.$$

The result is now obvious. $\square$

Here is a proof of L'Hospital's rule for the case $x \to \infty$.

PROOF.   The key here is to set $x = 1/t$:

$$\lim_{x \to \infty} \frac{f'(x)}{g'(x)} = \lim_{t \downarrow 0} \frac{f'(1/t)}{g'(1/t)}$$

$$= \lim_{t \downarrow 0} \frac{-t^{-2}f'(1/t)}{-t^{-2}g'(1/t)}$$

by L'Hospital's rule for the case $t \downarrow 0$ $= \lim_{t \downarrow 0} \frac{f(1/t)}{g(1/t)}$

$$= \lim_{x \to \infty} \frac{f(x)}{g(x)} . \quad \square$$

CAUTION.   L'Hospital's rule does not apply when the numerator or the denominator has a finite nonzero limit. For example,

$$\lim_{x \to 0} \frac{x}{x + \cos x} = \frac{0}{1} = 0.$$

A blind application of L'Hospital's rule would lead to

$$\lim_{x \to 0} \frac{x}{x + \cos x} = \lim_{x \to 0} \frac{1}{1 - \sin x} = 1.$$

This is wrong. $\square$

Exercises

Calculate the following limits.

*1. $\displaystyle \lim_{x \downarrow 0} \frac{\sin x}{\sqrt{x}}$.

2. $\displaystyle \lim_{x \to 1} \frac{\log x}{1 - x}$.

*3. $\displaystyle \lim_{x \to 0} \frac{e^x - 1}{\log (1 + x)}$.

*4. $\displaystyle \lim_{x \to a} \frac{x - a}{x^n - a^n}$.

5. $\displaystyle \lim_{x \to \pi/2} \frac{\cos x}{\sin 2x}$.

*6. $\displaystyle \lim_{x \to 0} \frac{e^x - 1}{x(1 + x)}$.

*7. $\displaystyle \lim_{x \to 0} \frac{1 - \cos x}{3x}$.

8. $\displaystyle \lim_{x \to 1} \frac{x^{1/2} - x^{1/4}}{x - 1}$.

*9. $\displaystyle \lim_{x \to 0} \frac{e^x - e^{-x}}{\sin x}$.

*10. $\displaystyle \lim_{x \to 0} \frac{a^x - (a + 1)^x}{x}$.

11. $\displaystyle \lim_{x \to 0} \frac{\tan \pi x}{e^x - 1}$.

*12. $\displaystyle \lim_{x \to 0} \frac{x + \sin \pi x}{x - \sin \pi x}$.

*13. $\displaystyle \lim_{x \to 0} \frac{\sqrt{a + x} - \sqrt{a - x}}{x}$.

14. $\displaystyle \lim_{x \to 0} \frac{1 + x - e^x}{x(e^x - 1)}$.

*15. $\lim\limits_{x \to 0} \dfrac{x - \tan x}{x - \sin x}$.

*16. $\lim\limits_{x \to 0} \dfrac{\log (\sec x)}{x^2}$.

17. $\lim\limits_{x \uparrow 1} \dfrac{\sqrt{1 - x^2}}{\sqrt{1 - x^3}}$.

*18. $\lim\limits_{x \to 0} \left(\dfrac{1}{\sin x} - \dfrac{1}{x} \right)$.

*19. $\lim\limits_{x \to 1} \left(\dfrac{1}{\log x} - \dfrac{x}{x - 1} \right)$.

20. $\lim\limits_{x \to \infty} \left(1 + \dfrac{1}{x} \right)^x$.

*21. $\lim\limits_{x \to 0} (e^x + x)^{1/x}$.

*22. $\lim\limits_{x \to \pi/4} \dfrac{\sec^2 x - 2 \tan x}{1 + \cos 4x}$.

23. $\lim\limits_{x \downarrow 0} \dfrac{\sqrt{x}}{\sqrt{x} + \sin \sqrt{x}}$.

*24. $\lim\limits_{x \to 0} \dfrac{x e^{nx} - x}{1 - \cos nx}$.

*25. $\lim\limits_{x \to 0} \dfrac{x - \arcsin x}{\sin^3 x}$.

26. $\lim\limits_{x \to \pi/2} \dfrac{\log (\sin x)}{(\pi - 2x)^2}$.

27. Find the fallacy:

$$\lim_{x \to 0} \frac{2 + x + \sin x}{x^3 + x - \cos x} = \lim_{x \to 0} \frac{1 + \cos x}{3x^2 + 1 + \sin x} = \lim_{x \to 0} \frac{-\sin x}{6x + \cos x} = \frac{0}{1} = 0.$$

28. (*Optional*)   Calculate

$$\lim_{x \to 0} \frac{\sinh x - \sin x}{\sin^3 x}.$$

11.3   Infinite Limits; L'Hospital's Rule ($\frac{\infty}{\infty}$)

It is convenient at times to allow infinite limits. Here is the basic idea. We write

$$\lim_{x \to c} f(x) = \infty$$

iff we can make $f(x)$ as large as we wish simply by requiring that x be sufficiently close to c. More formally, we have

$$\lim_{x \to c} f(x) = \infty \qquad \text{iff} \qquad \begin{cases} \text{for each } M > 0 \text{ there exists } \delta > 0 \text{ such that,} \\ \text{if } \ 0 < |x - c| < \delta, \qquad \text{then} \quad f(x) \geq M. \end{cases}$$

Here are some examples.

Examples

$$\lim_{x \to c} \frac{1}{|x - c|} = \infty, \qquad \lim_{x \to c} \frac{1}{(\sqrt{x} - \sqrt{c})^2} = \infty, \qquad \lim_{x \to c} \frac{1}{\sin^2 (x - c)} = \infty. \ \ \square$$

One-sided infinite limits

$$\lim_{x \downarrow c} f(x) = \infty \quad \text{and} \quad \lim_{x \uparrow c} f(x) = \infty$$

are defined in a similar manner. We won't bother to spell out the details. The idea must be obvious. It must also be obvious to you that all these notions can be extended to allow for a limit of $-\infty$.

Geometrically limits of the form

$$\lim_{x \to c} f(x) = \pm\infty, \qquad \lim_{x \downarrow c} f(x) = \pm\infty, \qquad \lim_{x \uparrow c} f(x) = \pm\infty$$

signal the presence of a *vertical asymptote*. We give some pictorial examples in Figures 11.3.1–11.3.4.

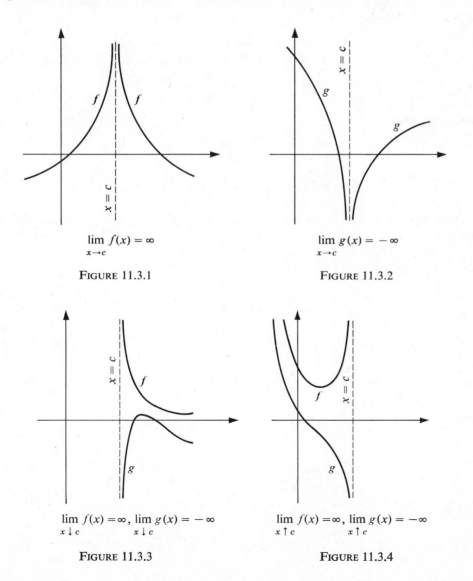

$$\lim_{x \to c} f(x) = \infty$$

FIGURE 11.3.1

$$\lim_{x \to c} g(x) = -\infty$$

FIGURE 11.3.2

$$\lim_{x \downarrow c} f(x) = \infty, \; \lim_{x \downarrow c} g(x) = -\infty$$

FIGURE 11.3.3

$$\lim_{x \uparrow c} f(x) = \infty, \; \lim_{x \uparrow c} g(x) = -\infty$$

FIGURE 11.3.4

There still remain limits of the form

$$\lim_{x \to \pm \infty} f(x) = \pm \infty$$

to discuss. There are four possibilities here but we will spell out details only for one of them:

$$\lim_{x \to \infty} f(x) = \infty.$$

We leave the others to your imagination. By writing

$$\lim_{x \to \infty} f(x) = \infty$$

we mean that we can make $f(x)$ as large as we wish simply by requiring that x be sufficiently large:

$$\lim_{x \to \infty} f(x) = \infty \quad \text{iff} \quad \begin{cases} \text{for each } M > 0 \text{ there exists } K > 0 \text{ such that,} \\ \quad \text{if } x \geq K, \quad \text{then } f(x) \geq M. \end{cases}$$

As simple examples you can take

$$\lim_{x \to \infty} x = \infty, \qquad \lim_{x \to \infty} e^x = \infty,$$

$$\lim_{x \to \infty} \log x = \infty, \qquad \lim_{x \to \infty} \left(x + \frac{1}{x} \right) = \infty.$$

To illustrate these ideas further we calculate some limits which are not quite so obvious.

Problem. Find

$$\lim_{x \to \infty} x(2 + \sin x).$$

SOLUTION. Since

$$-1 \leq \sin x,$$

we have

$$1 \leq 2 + \sin x$$

and thus for $x > 0$

$$x \leq x(2 + \sin x).$$

Since

$$\lim_{x \to \infty} x = \infty,$$

it is clear that

$$\lim_{x \to \infty} x(2 + \sin x) = \infty. \quad \square$$

Problem. Find

$$\lim_{x \to \infty} \frac{1 - x^2}{2x + 1}.$$

SOLUTION

$$\lim_{x \to \infty} \frac{1 - x^2}{2x + 1} = \lim_{x \to \infty} \frac{1/x^2 - 1}{2/x + 1/x^2} = -\infty. \quad \square$$

Problem. Find

$$\lim_{x \to \infty} \log\left(\frac{x}{x^2 + 1}\right).$$

SOLUTION. Note first that

$$\lim_{x \to \infty} \frac{x}{x^2 + 1} = \lim_{x \to \infty} \frac{1}{x + 1/x} = 0.$$

It follows that

$$\lim_{x \to \infty} \log\left(\frac{x}{x^2 + 1}\right) = -\infty. \quad \square$$

Problem. Find

$$\lim_{x \to \infty} x \sin \frac{1}{x}.$$

SOLUTION

$$\lim_{x \to \infty} x \sin \frac{1}{x} = \lim_{x \to \infty} \frac{\sin 1/x}{1/x} = \lim_{t \downarrow 0} \frac{\sin t}{t} = 1. \quad \square$$

Problem. Find

$$\lim_{x \to 0} \left(\frac{1}{x} - \frac{1}{x^2}\right).$$

SOLUTION

$$\lim_{x \to 0} \left(\frac{1}{x} - \frac{1}{x^2}\right) = \lim_{x \to 0} \frac{x - 1}{x^2} = -\infty. \quad \square$$

We come now to limits of quotients

$$\frac{f(x)}{g(x)}$$

where numerator and denominator both tend to ∞. We take l as a fixed real number.

L'Hospital's Rule ($\frac{\infty}{\infty}$)

Suppose that

as $\quad x \uparrow c, \; x \downarrow c, \; x \to c, \; x \to \infty, \; \text{or} \; x \to -\infty, \quad f(x) \to \infty \; \text{and} \; g(x) \to \infty.$

If $\quad \dfrac{f'(x)}{g'(x)} \to l, \quad$ then $\quad \dfrac{f(x)}{g(x)} \to l.$

While the proof of L'Hospital's rule in this setting is a little more complicated than it was in the ($\frac{0}{0}$) case,† the application of the rule is much the same.

Problem. Calculate

$$\lim_{x \to \infty} \frac{\log x}{x}.$$

SOLUTION. Both numerator and denominator tend to ∞ with x. L'Hospital's rule gives

$$\lim_{x \to \infty} \frac{\log x}{x} = \lim_{x \to \infty} \frac{\dfrac{d}{dx}(\log x)}{\dfrac{d}{dx}(x)} = \lim_{x \to \infty} \frac{1}{x} = 0. \quad \square$$

Problem. Calculate

$$\lim_{x \to \infty} \frac{x^n}{e^x}.$$

SOLUTION. Here we differentiate n times:

$$\lim_{x \to \infty} \frac{x^n}{e^x} = \lim_{x \to \infty} \frac{nx^{n-1}}{e^x} = \lim_{x \to \infty} \frac{n(n-1)x^{n-2}}{e^x} = \cdots = \lim_{x \to \infty} \frac{n!}{e^x} = 0. \quad \square$$

Problem. Calculate

$$\lim_{x \downarrow 0} x^x.$$

SOLUTION. We begin by taking the logarithm of x^x:

$$\lim_{x \downarrow 0} \log x^x = \lim_{x \downarrow 0} x \log x = \lim_{t \to \infty} \frac{\log 1/t}{t} = \lim_{t \to \infty} \left(-\frac{\log t}{t} \right) = 0.$$

Since $\log x^x$ tends to 0, x^x must tend to 1. $\quad \square$

We repeat: L'Hospital's rule does not apply when the numerator or denominator has a finite nonzero limit. While

$$\lim_{x \downarrow 0} \frac{1 + x}{\sin x} = \infty,$$

a misapplication of L'Hospital's rule can lead to

$$\lim_{x \downarrow 0} \frac{1 + x}{\sin x} = \lim_{x \downarrow 0} \frac{1}{-\cos x} = -1. \quad \square$$

---

† We omit the proof.

Exercises

Calculate the following limits.

*1. $\lim\limits_{x \uparrow \pi/2} \tan x$.

2. $\lim\limits_{x \downarrow 0} \log \dfrac{1}{x}$.

*3. $\lim\limits_{x \to \infty} \dfrac{x^3 - 1}{2 - x}$.

*4. $\lim\limits_{x \to \infty} \dfrac{x^3}{1 - x^3}$.

5. $\lim\limits_{x \to \infty} x^2 \sin \dfrac{1}{x}$.

*6. $\lim\limits_{x \to \infty} \dfrac{\log x}{\sqrt{x}}$.

*7. $\lim\limits_{x \uparrow \pi/2} \dfrac{\tan 5x}{\tan x}$.

8. $\lim\limits_{x \to 0} x \log |\sin x|$.

*9. $\lim\limits_{x \downarrow 0} x^{2x}$.

*10. $\lim\limits_{x \to 0} x \, (\log |x|)^2$.

11. $\lim\limits_{x \to \infty} x \sin \dfrac{\pi}{x}$.

*12. $\lim\limits_{x \downarrow 0} \dfrac{\log x}{\cot x}$.

*13. $\lim\limits_{x \to \infty} \dfrac{\sqrt{1 + x^2}}{x}$.

14. $\lim\limits_{x \to \infty} \log \left(\dfrac{x^2 - 1}{x^2 + 1} \right)^3$.

*15. $\lim\limits_{x \downarrow 0} x^{\sin x}$.

*16. $\lim\limits_{x \to 0} |\sin x|^x$.

17. $\lim\limits_{x \to 0} \left[\dfrac{1}{\log (1 + x)} - \dfrac{1}{x} \right]$.

*18. $\lim\limits_{x \to 1} \left[\dfrac{x}{x - 1} - \dfrac{1}{\log x} \right]$.

*19. $\lim\limits_{x \to 1} x^{1/(x-1)}$.

20. $\lim\limits_{x \to \infty} \left(\cos \dfrac{1}{x} \right)^x$.

*21. $\lim\limits_{x \to \infty} (x^2 + a^2)^{(1/x)^2}$.

*22. $\lim\limits_{x \to 0} \left[\dfrac{1}{\sin^2 x} - \dfrac{1}{x^2} \right]$.

23. $\lim\limits_{x \to \pi/2} |\sec x|^{\cos x}$.

*24. $\lim\limits_{x \to \infty} \dfrac{1}{x} \displaystyle\int_0^x e^{t^2} \, dt$.

*25. $\lim\limits_{x \uparrow 3} \dfrac{3x^2 - 4}{x - 3}$.

26. $\lim\limits_{x \to \infty} (\sqrt{x^2 + 2x} - x)$.

27. Find the fallacy:

$$\lim_{x \downarrow 0} \frac{x^2}{\sin x} = \lim_{x \downarrow 0} \frac{2x}{\cos x} = \lim_{x \downarrow 0} \frac{2}{-\sin x} = -\infty.$$

28. Show that the Taylor series

$$\sum_{n=0}^{\infty} \frac{f^{(n)}(0)}{n!} x^n$$

of the function

$$f(x) = \left\{ \begin{matrix} e^{-1/x^2}, & x \neq 0 \\ 0, & x = 0 \end{matrix} \right]$$

is identically zero and hence does not represent the function.

11.4   Improper Integrals

We begin with a function f continuous on an unbounded interval $[a, \infty)$. For each number $b > a$ we can form the integral

$$\int_a^b f(x)\, dx.$$

If, as b tends to ∞, this integral tends to a finite limit l,

$$\lim_{b \to \infty} \int_a^b f(x)\, dx = l,$$

then we write

$$\int_a^\infty f(x)\, dx = l$$

and call

$$\int_a^\infty f(x)\, dx$$

a *convergent improper integral*. Otherwise we say that

$$\int_a^\infty f(x)\, dx$$

is a *divergent improper integral*. Improper integrals of the form

$$\int_{-\infty}^b f(x)\, dx$$

are introduced in a similar manner.

Examples

(1) $\displaystyle \int_1^\infty e^{-x}\, dx = \frac{1}{e}\,.$ (2) $\displaystyle \int_1^\infty \frac{dx}{x}$ diverges.

(3) $\displaystyle \int_1^\infty \frac{dx}{x^2} = 1.$ (4) $\displaystyle \int_1^\infty \cos \pi x\, dx$ diverges.

VERIFICATION

(1) $\displaystyle \int_1^\infty e^{-x}\, dx = \lim_{b \to \infty} \int_1^b e^{-x}\, dx = \lim_{b \to \infty} \left[-e^{-x} \right]_1^b = \lim_{b \to \infty} \left(\frac{1}{e} - \frac{1}{e^b} \right) = \frac{1}{e}\,.$

(2) $\displaystyle \int_1^\infty \frac{dx}{x} = \lim_{b \to \infty} \int_1^b \frac{dx}{x} = \lim_{b \to \infty} \log b = \infty.$

(3) $\displaystyle \int_1^\infty \frac{dx}{x^2} = \lim_{b \to \infty} \int_1^b \frac{dx}{x^2} = \lim_{b \to \infty} \left[-\frac{1}{x} \right]_1^b = \lim_{b \to \infty} \left(1 - \frac{1}{b} \right) = 1.$

(4) Note first that

$$\int_1^b \cos \pi x \; dx = \left[\frac{1}{\pi} \sin \pi x\right]_1^b = \frac{1}{\pi} \sin \pi b.$$

As b tends to ∞, $\sin \pi b$ oscillates between -1 and 1. The integral

$$\int_1^b \cos \pi x \; dx$$

oscillates between $-1/\pi$ and $1/\pi$. It does not converge. □

The usual formulas for area and volume are extended to the unbounded case by means of improper integrals.

Example. If Ω is the region below the graph of

$$f(x) = \frac{1}{x^p}, \qquad x \geq 1 \qquad\qquad \text{(Figure 11.4.1)}$$

then we have

$$\text{area of } \Omega = \left\{\begin{array}{ll} \dfrac{1}{p-1}, & \text{if } p > 1 \\[2mm] \infty, & \text{if } p \leq 1 \end{array}\right].$$

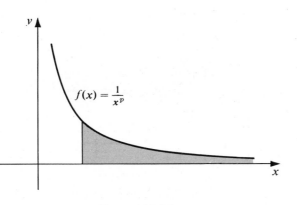

$$f(x) = \frac{1}{x^p}$$

FIGURE 11.4.1

This comes about from setting

$$\text{area of } \Omega = \lim_{b \to \infty} \int_1^b \frac{dx}{x^p} = \int_1^\infty \frac{dx}{x^p}.$$

For $p \neq 1$, we have

$$\int_1^\infty \frac{dx}{x^p} = \lim_{b \to \infty} \int_1^b \frac{dx}{x^p} = \lim_{b \to \infty} \frac{1}{1-p}(b^{1-p} - 1) = \left\{\begin{array}{ll} \dfrac{1}{p-1}, & \text{if } p > 1 \\[2mm] \infty, & \text{if } p < 1 \end{array}\right].$$

For $p = 1$, we have

$$\int_1^\infty \frac{dx}{x^p} = \int_1^\infty \frac{dx}{x} = \infty,$$

as you have seen already. □

Example. From the last example you know that the region below the graph of

$$f(x) = \frac{1}{x}, \qquad x \geq 1,$$

has infinite area. Suppose that this region (with infinite area) is revolved about the x-axis. (See Figure 11.4.2.) What is the volume of the resulting solid? It may surprise you somewhat but the volume is not infinite. In fact, it is π:

$$V = \pi \int_1^\infty \frac{dx}{x^2} = \pi \cdot 1 = \pi.$$

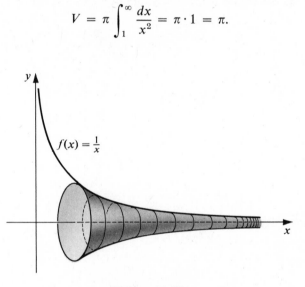

FIGURE 11.4.2

This result should not be disquieting. You have seen this kind of thing before:

$$\sum_{n=1}^\infty \frac{1}{n} \quad \text{diverges} \qquad \text{but} \qquad \sum_{n=1}^\infty \frac{1}{n^2} \quad \text{converges.} \quad \square$$

Suppose now that f is continuous on $(-\infty, \infty)$. The *improper integral*

$$\int_{-\infty}^\infty f(x)\, dx$$

is said to *converge* iff both

$$\int_{-\infty}^0 f(x)\, dx \quad \text{and} \quad \int_0^\infty f(x)\, dx$$

converge. If

$$\int_{-\infty}^{0} f(x)\,dx = l \quad \text{and} \quad \int_{0}^{\infty} f(x)\,dx = m,$$

then we set

$$\int_{-\infty}^{\infty} f(x)\,dx = l + m.$$

Improper integrals can also arise on bounded intervals. Suppose that f is continuous on the half-open interval $[a, b)$ but unbounded there. See Figure 11.4.3. If

$$\lim_{\epsilon \downarrow 0} \int_{a}^{b-\epsilon} f(x)\,dx = l \qquad (l \text{ finite}),$$

then we write

$$\int_{a}^{b-} f(x)\,dx = l$$

and call

$$\int_{a}^{b-} f(x)\,dx$$

a *convergent improper integral.* Otherwise we say that

$$\int_{a}^{b-} f(x)\,dx$$

is a *divergent improper integral.*

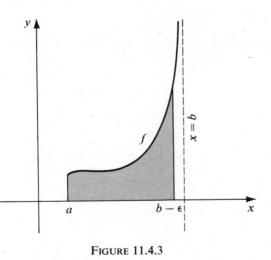

FIGURE 11.4.3

Functions continuous but unbounded on intervals of the form $(a, b]$ lead to limits of the form

$$\lim_{\epsilon \downarrow 0} \int_{a+\epsilon}^{b} f(x)\,dx$$

and thus to improper integrals

$$\int_{a+}^{b} f(x) \, dx.$$

Examples

(1) $\int_{0}^{1-} (1 - x)^{-2/3} \, dx = 3.$ (2) $\int_{0+}^{1} \frac{dx}{x}$ diverges.

VERIFICATION

(1) $\int_{0}^{1-} (1 - x)^{-2/3} \, dx = \lim_{\epsilon \downarrow 0} \int_{0}^{1-\epsilon} (1 - x)^{-2/3} \, dx$

$$= \lim_{\epsilon \downarrow 0} \left[-3(1 - x)^{1/3} \right]_{0}^{1-\epsilon}$$

$$= \lim_{\epsilon \downarrow 0} 3(1 - \epsilon^{1/3})$$

$$= 3.$$

(2) $\int_{0+}^{1} \frac{dx}{x} - \lim_{\epsilon \downarrow 0} \int_{\epsilon}^{1} \frac{dx}{x} = \lim_{\epsilon \downarrow 0} (-\log \epsilon) = \infty.$ $\square$

Exercises

Evaluate the improper integrals that converge.

*1. $\int_{1}^{\infty} \frac{dx}{x^2}.$ 2. $\int_{0}^{\infty} e^{px} \, dx, \quad p > 0.$

*3. $\int_{0}^{\infty} e^{-px} \, dx, \quad p > 0.$ *4. $\int_{0}^{\infty} \frac{dx}{1 + x^2}.$

5. $\int_{0}^{\infty} \frac{dx}{4 + x^2}.$ *6. $\int_{0+}^{1} \frac{dx}{\sqrt{x}}.$

*7. $\int_{0+}^{8} \frac{dx}{x^{2/3}}.$ 8. $\int_{0+}^{1} \frac{dx}{x^2}.$

*9. $\int_{0}^{1-} \frac{dx}{\sqrt{1 - x^2}}.$ *10. $\int_{0}^{1-} \frac{dx}{\sqrt{1 - x}}.$

11. $\int_{0}^{2-} \frac{x}{\sqrt{4 - x^2}} \, dx.$ *12. $\int_{0}^{a-} \frac{dx}{\sqrt{a^2 - x^2}}.$

*13. $\int_{e}^{\infty} \frac{\log x}{x} \, dx.$ 14. $\int_{e}^{\infty} \frac{dx}{x \log x}.$

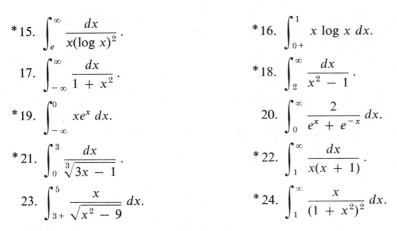

*15. $\displaystyle\int_{e}^{\infty} \frac{dx}{x(\log x)^2}$.

*16. $\displaystyle\int_{0+}^{1} x \log x \, dx.$

17. $\displaystyle\int_{-\infty}^{\infty} \frac{dx}{1 + x^2}$.

*18. $\displaystyle\int_{2}^{\infty} \frac{dx}{x^2 - 1}$.

*19. $\displaystyle\int_{-\infty}^{0} xe^x \, dx.$

20. $\displaystyle\int_{0}^{\infty} \frac{2}{e^x + e^{-x}} \, dx.$

*21. $\displaystyle\int_{0}^{3} \frac{dx}{\sqrt[3]{3x - 1}}$.

*22. $\displaystyle\int_{1}^{\infty} \frac{dx}{x(x + 1)}$.

23. $\displaystyle\int_{3+}^{5} \frac{x}{\sqrt{x^2 - 9}} \, dx.$

*24. $\displaystyle\int_{1}^{\infty} \frac{x}{(1 + x^2)^2} \, dx.$

*25. Let Ω be the region under the curve
$$y = e^{-x}, \qquad x \geq 0.$$
(a) Sketch Ω. (b) Find the area of Ω. (c) Find the volume of the solid obtained by revolving Ω about the x-axis. (d) Find the volume obtained by revolving Ω about the y-axis. (e) Find the surface area of the solid in part (c).

26. Find the area of the region bounded above by $xy = 1$, below by $y(x^2 + 1) = x$, and to the left by $x = 1$.

*27. Let Ω be the region below the curve
$$y = x^{-1/4}, \qquad 0 < x \leq 1.$$
(a) Sketch Ω. (b) Find the area of Ω. (c) Find the volume of the solid obtained by revolving Ω about the x-axis. (d) Find the volume of the solid obtained by revolving Ω about the y-axis.

28. (*A comparison test*)   Let f and g be continuous functions satisfying
$$0 \leq f(x) \leq g(x) \qquad \text{for all} \quad x \in [a, \infty).$$
Show that

(a) if $\displaystyle\int_{a}^{\infty} g(x) \, dx$ converges,   then $\displaystyle\int_{a}^{\infty} f(x) \, dx$ converges.

(b) if $\displaystyle\int_{a}^{\infty} f(x) \, dx$ diverges,   then $\displaystyle\int_{a}^{\infty} g(x) \, dx$ diverges.

Use the comparison test of Exercise 28 to determine which of the following integrals converge.

*29. $\displaystyle\int_{1}^{\infty} \frac{dx}{\sqrt{1 + x^3}}$.

30. $\displaystyle\int_{1}^{\infty} e^{-x^2} \, dx.$

*31. $\displaystyle\int_{0}^{\infty} (1 + x^5)^{-1/6} \, dx.$

*32. $\displaystyle\int_{\pi}^{\infty} \frac{\sin^2 2x}{x^2} \, dx.$

33. $\displaystyle\int_{1}^{\infty} \frac{\log x}{x^2} \, dx.$

*34. $\displaystyle\int_{e}^{\infty} \frac{dx}{\sqrt{x + 1} \log x}$.

The present value of a perpetual stream of revenues which flows continually at the rate of $R(t)$ dollars per year is given by the formula

$$P.V. = \int_0^\infty R(t)e^{-rt}\, dt,$$

where r is the rate of continuous compounding.

*35. What is the present value of such a stream if $R(t)$ is constantly $100 per year and r is 5%?

36. What is the present value if

$$R(t) = 100 + 60t$$

dollars per year and r is 10%?

Vectors

12

12.1 Cartesian Space Coordinates

To introduce a Cartesian coordinate system in space we begin with a plane Cartesian coordinate system O-xy. Through the point O, which we continue to call the origin, we pass a third line, perpendicular to the other two. This third line we call the z-axis. We coordinatize it using the same scale as before, assigning z-coordinate 0 to the origin O.

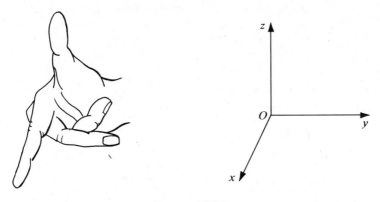

FIGURE 12.1.1

For later convenience we orient the z-axis so that O-xyz forms a "right-handed" system. That is, if the index finger of the right hand points along the positive x-axis and the middle finger along the positive y-axis, then the thumb will point along the positive z-axis. See Figure 12.1.1.

A point $P(x_0, y_0, z_0)$ is shown in Figure 12.1.2. A plane drawn through P perpendicular to the x-axis will intersect the x-axis at $x = x_0$; a plane drawn through P

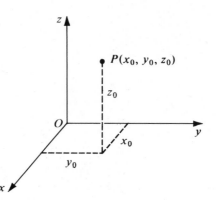

FIGURE 12.1.2

perpendicular to the y-axis will intersect the y-axis at $y = y_0$; a plane drawn through P perpendicular to the z-axis will intersect the z-axis at $z = z_0$.

The Distance Formula

The distance $d(P_1, P_2)$ between two points $P_1(x_1, y_1, z_1)$ and $P_2(x_2, y_2, z_2)$ can be found by applying the Pythagorean theorem twice. With Q and R as in Figure 12.1.3, P_1P_2R and P_1RQ are both right triangles. From the first triangle

$$[d(P_1, P_2)]^2 = [d(P_1, R)]^2 + [d(R, P_2)]^2,$$

and from the second triangle

$$[d(P_1, R)]^2 = [d(Q, R)]^2 + [d(P_1, Q)]^2.$$

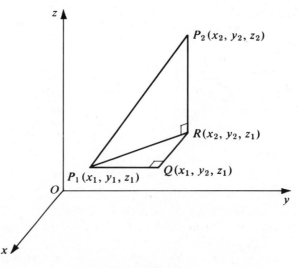

FIGURE 12.1.3

Combining equations,

$$[d(P_1, P_2)]^2 = [d(Q, R)]^2 + [d(P_1, Q)]^2 + [d(R, P_2)]^2$$
$$= (x_2 - x_1)^2 + (y_2 - y_1)^2 + (z_2 - z_1)^2.$$

Taking square roots, we have the distance formula:

$$\boxed{d(P_1, P_2) = \sqrt{(x_2 - x_1)^2 + (y_2 - y_1)^2 + (z_2 - z_1)^2}.}$$

Symmetry

You are already familiar with two kinds of symmetry: symmetry with respect to a line and symmetry with respect to a point. In space we can also speak of symmetry with respect to a plane. The idea is obvious.

Exercises

1. Sketch a right-handed coordinate system and plot the following points.

   (a) $A(1, 0, 0)$.   (b) $B(0, 1, 0)$.   (c) $C(0, 0, 1)$.
   (d) $D(1, 1, 1)$.   (e) $E(-1, -1, -1)$.   (f) $F(1, -1, 1)$.
   (g) $G(-1, 1, 2)$.   (h) $H(1, -2, 1)$.   (i) $I(-2, 1, -1)$.

2. Find the distance between the given points.

   * (a) $P(1, 0, 0)$ and $Q(0, 1, 0)$.
      (b) $P(-1, 1, 2)$ and $Q(1, -2, 1)$.
   * (c) $P(4, 2, 5)$ and $Q(1, 4, -1)$.
      (d) $P(-1, -\sqrt{3}, -3)$ and $Q(3, \sqrt{3}, 0)$.
   * (e) $P(x_1, x_2, 0)$ and $Q(y_1, y_2, 0)$.

* 3. Find an equation for the sphere of radius r centered at
      (a) the origin.   (b) $Q(a_1, a_2, a_3)$.

4. Find an equation for the plane through $Q(a_1, a_2, a_3)$ that is parallel to
   * (a) the xy-plane.   (b) the xz-plane.   * (c) the yz-plane.

* 5. Find the distance between an arbitrary point $P(a, b, c)$ and each of the coordinate axes. (Draw figures.)

6. What are the coordinates of the point halfway between $P(a_1, a_2, a_3)$ and $Q(b_1, b_2, b_3)$?

* 7. What are the coordinates of the point two-thirds of the way from $P(a_1, a_2, a_3)$ to $Q(b_1, b_2, b_3)$?

8. Let $P(x_0, y_0, z_0)$ be as in Figure 12.1.2. Find the coordinates of Q given that the line segment $\overline{PQ}$ is symmetric
   * (a) about the xy-plane.   (b) about the xz-plane.
      (c) about the yz-plane.   * (d) about the origin.
   Draw the figure in each case.

12.2 Displacements and Forces

Our purpose here is to motivate the notion of vector. A quick reading of this section will do.

A displacement along a coordinate line can be specified by a real number and depicted by an arrow. For a displacement of a_1 units we can use the number a_1 and an arrow that begins at some number x and ends at $x + a_1$. If $a_1 > 0$ (and we follow convention), the arrow will point to the right and if $a_1 < 0$, the arrow will point to the left. (See Figure 12.2.1.)

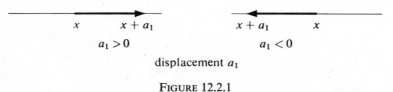

$$x \qquad x + a_1 \qquad\qquad x + a_1 \qquad x$$
$$a_1 > 0 \qquad\qquad\qquad a_1 < 0$$

displacement a_1

FIGURE 12.2.1

Displacements in the xy-plane are more interesting because instead of having two possible directions there are now an infinite number of directions.

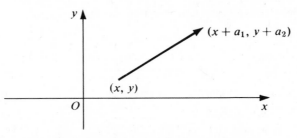

a displacement (a_1, a_2) in the xy-plane

FIGURE 12.2.2

The most general displacements take place in space. Here ordered triples of real numbers come into play. A displacement of a_1 units in x-coordinate, a_2 units in

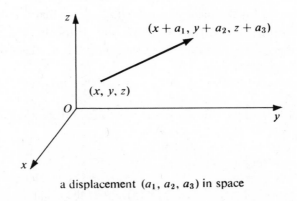

a displacement (a_1, a_2, a_3) in space

FIGURE 12.2.3

y-coordinate, and a_3 units in z-coordinate can be indicated by an arrow that begins at some point (x, y, z) and ends at $(x + a_1, y + a_2, z + a_3)$.

We can follow one displacement by another. A displacement (a_1, a_2, a_3) followed by a displacement (b_1, b_2, b_3) results in a total displacement $(a_1 + b_1, a_2 + b_2, a_3 + b_3)$. We can express this by writing

$$(a_1, a_2, a_3) + (b_1, b_2, b_3) = (a_1 + b_1, a_2 + b_2, a_3 + b_3).$$

We can picture the first displacement by an arrow that begins at some point $P(x, y, z)$ and ends at $Q(x + a_1, y + a_2, z + a_3)$, the second displacement by an arrow that begins at $Q(x + a_1, y + a_2, z + a_3)$ and ends at $R(x + a_1 + b_1, y + a_2 + b_2, z + a_3 + b_3)$ and the resultant displacement by an arrow that begins at $P(x, y, z)$ and ends at $R(x + a_1 + b_1, y + a_2 + b_2, z + a_3 + b_3)$. The three arrows then form a triangular pattern that is easy to remember. See Figure 12.2.4.

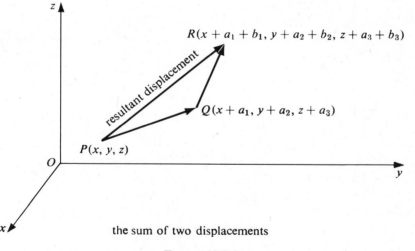

the sum of two displacements

FIGURE 12.2.4

From a displacement (a_1, a_2, a_3) and a real number α, we can form a new displacement $(\alpha a_1, \alpha a_2, \alpha a_3)$. We call this new displacement α times the initial one and write

$$\alpha(a_1, a_2, a_3) = (\alpha a_1, \alpha a_2, \alpha a_3).$$

The effect of multiplying a displacement by a number α is to change its magnitude by a factor of $|\alpha|$, keeping the same direction if $\alpha > 0$ and reversing it if $\alpha < 0$. We illustrate this idea in Figure 12.2.5. The displacement

$$2(a_1, a_2, a_3) = (2a_1, 2a_2, 2a_3)$$

is twice as long as

$$(a_1, a_2, a_3)$$

and has the same direction; the displacement

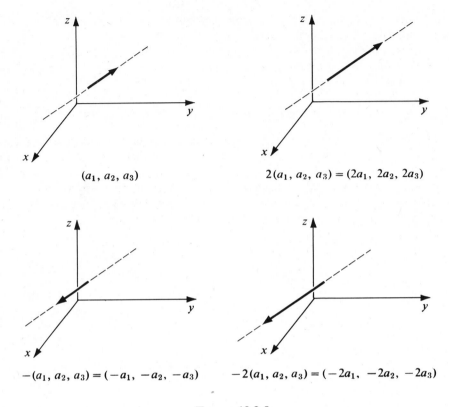

(a_1, a_2, a_3) $2(a_1, a_2, a_3) = (2a_1, \ 2a_2, \ 2a_3)$

$-(a_1, a_2, a_3) = (-a_1, \ -a_2, \ -a_3)$ $-2(a_1, a_2, a_3) = (-2a_1, \ -2a_2, \ -2a_3)$

FIGURE 12.2.5

$$-(a_1, a_2, a_3) = (-a_1, \ -a_2, \ -a_3)$$

has the same length as

$$(a_1, a_2, a_3)$$

but the opposite direction; the displacement

$$-2(a_1, a_2, a_3) = (-2a_1, \ -2a_2, \ -2a_3)$$

is twice as long as

$$(a_1, a_2, a_3)$$

and has the opposite direction.

This same pattern

$$(a_1, a_2, a_3) + (b_1, b_2, b_3) = (a_1 + b_1, a_2 + b_2, a_3 + b_3)$$
$$\alpha(a_1, a_2, a_3) = (\alpha a_1, \alpha a_2, \alpha a_3)$$

arises naturally in other settings; for example, in the analysis of forces. To specify a force completely, we need to know its components along the x, y, and z axes. If these are respectively a_1, a_2, a_3, then the force is totally specified by the ordered triple (a_1, a_2, a_3). See Figure 12.2.6.

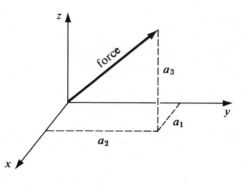

FIGURE 12.2.6

If two forces
$$F_1 = (a_1, a_2, a_3) \quad \text{and} \quad F_2 = (b_1, b_2, b_3)$$
are applied simultaneously at the same point, the effect is the same as that produced by the single force
$$F_3 = (a_1 + b_1, a_2 + b_2, a_3 + b_3).$$
We call F_3 the *resultant* or *total force* and write
$$F_1 + F_2 = F_3.$$
For the ordered triples,
$$(a_1, a_2, a_3) + (b_1, b_2, b_3) = (a_1 + b_1, a_2 + b_2, a_3 + b_3).$$
Pictorially we have the usual force diagram, Figure 12.2.7. It is a parallelogram with the sides representing
$$F_1 = (a_1, a_2, a_3) \quad \text{and} \quad F_2 = (b_1, b_2, b_3)$$

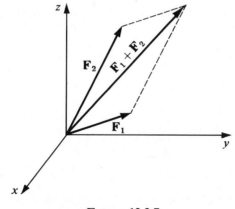

FIGURE 12.2.7

and the diagonal representing

$$F_3 = F_1 + F_2 = (a_1 + b_1, a_2 + b_2, a_3 + b_3).$$

For any force $F = (a_1, a_2, a_3)$ and a real number α, the force αF is defined by the equation

$$\alpha F = (\alpha a_1, \alpha a_2, \alpha a_3).$$

Thus once again we have

$$\alpha(a_1, a_2, a_3) = (\alpha a_1, \alpha a_2, \alpha a_3).$$

12.3   Vectors

The algebra of number triples which we introduced in the last section to discuss displacements and forces is so prodigiously rich in applications that it has found a firm place in the world of science and engineering and has generated much mathematics. It is to this mathematics that we now turn.

Definition

Ordered triples of real numbers subject to addition

$$(a_1, a_2, a_3) + (b_1, b_2, b_3) = (a_1 + b_1, a_2 + b_2, a_3 + b_3)$$

and multiplication by scalars (real numbers)

$$\alpha(a_1, a_2, a_3) = (\alpha a_1, \alpha a_2, \alpha a_3)$$

are called *vectors*.

Two vectors are called *equal* iff they are identical; namely

$$(a_1, a_2, a_3) = (b_1, b_2, b_3) \quad \text{iff} \quad a_1 = b_1, a_2 = b_2, a_3 = b_3.$$

Geometrically we can depict vectors by arrows. To depict the vector (a_1, a_2, a_3) we can choose any initial point $Q = Q(x, y, z)$ and use the arrow

$$\overrightarrow{QR} \quad \text{with} \quad R = R(x + a_1, y + a_2, z + a_3).$$

Most of the time we will choose the origin as the initial point and use the arrow

$$\overrightarrow{OP} \quad \text{with} \quad P = P(a_1, a_2, a_3).\dagger \qquad\qquad \text{(Figure 12.3.1)}$$

We will use boldface letters to denote vectors. Thus for

$$\mathbf{a} = (a_1, a_2, a_3) \quad \text{and} \quad \mathbf{b} = (b_1, b_2, b_3)$$

we have

$$\mathbf{a} + \mathbf{b} = (a_1 + b_1, a_2 + b_2, a_3 + b_3),$$

---

† The ordered triple $(0, 0, 0)$ will not be represented by an arrow but simply by the origin.

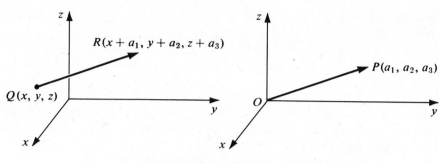

representations of (a_1, a_2, a_3)

FIGURE 12.3.1

and if α is a scalar (a real number)

$$\alpha \mathbf{a} = (\alpha a_1, \alpha a_2, \alpha a_3).$$

For the zero vector $(0, 0, 0)$ we will use the symbol $\mathbf{0}$. Obviously

$$0\mathbf{a} = \mathbf{0} \quad \text{for all vectors } \mathbf{a}.$$

By the vector $-\mathbf{b}$ we mean $(-1)\mathbf{b}$; that is,

$$-(b_1, b_2, b_3) = (-b_1, -b_2, -b_3).$$

By $\mathbf{a} - \mathbf{b}$ we mean $\mathbf{a} + (-\mathbf{b})$; that is,

$$(a_1, a_2, a_3) - (b_1, b_2, b_3) = (a_1 - b_1, a_2 - b_2, a_3 - b_3).$$

Problem. Given

$$\mathbf{a} = (1, -1, 2), \quad \mathbf{b} = (2, 3, -1), \quad \mathbf{c} = (8, 7, 1)$$

find

(a) $\mathbf{a} - \mathbf{b}$.　　　　　　　　　　　(b) $2\mathbf{a} + \mathbf{b}$.

(c) $3\mathbf{a} - 7\mathbf{b}$.　　　　　　　　　　(d) $2\mathbf{a} + 3\mathbf{b} - \mathbf{c}$.

SOLUTION

(a) $\mathbf{a} - \mathbf{b} = (1, -1, 2) - (2, 3, -1)$
$\quad\quad\quad = (1 - 2, -1 - 3, 2 + 1)$
$\quad\quad\quad = (-1, -4, 3).$

(b) $2\mathbf{a} + \mathbf{b} = 2(1, -1, 2) + (2, 3, -1)$
$\quad\quad\quad\quad = (2, -2, 4) + (2, 3, -1)$
$\quad\quad\quad\quad = (4, 1, 3).$

(c) $3\mathbf{a} - 7\mathbf{b} = 3(1, -1, 2) - 7(2, 3, -1)$
$\quad\quad\quad\quad = (3, -3, 6) - (14, 21, -7)$
$\quad\quad\quad\quad = (-11, -24, 13).$

(d) $2\mathbf{a} + 3\mathbf{b} - \mathbf{c} = 2(1, -1, 2) + 3(2, 3, -1) - (8, 7, 1)$

$$= (2, -2, 4) + (6, 9, -3) - (8, 7, 1)$$

$$= (0, 0, 0)$$

$$= \mathbf{0}. \quad \square$$

The addition of vectors can be visualized as the "addition" of arrows by the parallelogram law (as in the case of forces). If you picture $\mathbf{a}$, $\mathbf{b}$, and $\mathbf{a} + \mathbf{b}$ all as arrows emanating from the same point, say the origin, then $\mathbf{a} + \mathbf{b}$ acts as the diagonal of the parallelogram generated by $\mathbf{a}$ and $\mathbf{b}$. (See Figure 12.3.2.)

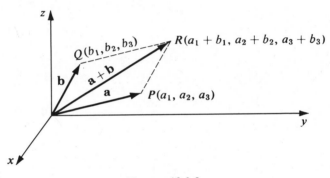

FIGURE 12.3.2

The addition of vectors can also be visualized as the "tail-to-head" addition of arrows (as in the case of displacements). If, instead of starting $\mathbf{b}$ at the origin, you start $\mathbf{b}$ at the tip of $\mathbf{a}$, then $\mathbf{a} + \mathbf{b}$ goes from the tail of $\mathbf{a}$ to the tip of $\mathbf{b}$. (See Figure 12.3.3.) These two pictorial representations of vector addition lead to the same result.

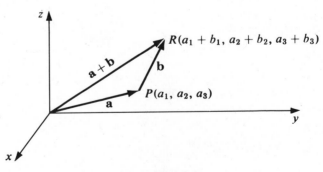

FIGURE 12.3.3

Definition 12.3.1

Two nonzero vectors **a** and **b** are said to be *parallel* provided that

$$\mathbf{a} = \alpha \mathbf{b} \quad \text{for some real number} \quad \alpha.$$

If $\alpha > 0$, **a** and **b** are said to have the *same direction*; if $\alpha < 0$, they are said to have *opposite directions*.

In the case of

$$\mathbf{a} = (2, -2, 6), \quad \mathbf{b} = (1, -1, 3), \quad \mathbf{c} = (-1, 1, -3)$$

we have

$$\mathbf{a} = 2\mathbf{b} \quad \text{and} \quad \mathbf{a} = -2\mathbf{c}.$$

This tells us that **a** and **b** are parallel and have the same direction but **a** and **c**, though parallel, have opposite directions. (See Figure 12.3.4.)

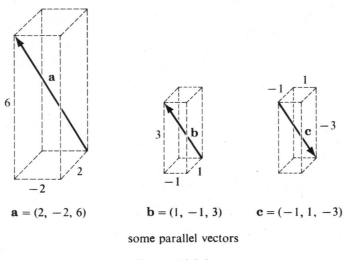

$$\mathbf{a} = (2, -2, 6) \qquad\qquad \mathbf{b} = (1, -1, 3) \qquad \mathbf{c} = (-1, 1, -3)$$

some parallel vectors

FIGURE 12.3.4

Definition 12.3.1 did not include the zero vector **0**. By special convention, **0** is called *parallel* to *every vector*. We will come back to this idea later.

Definition 12.3.2

The norm of a vector $\mathbf{a} = (a_1, a_2, a_3)$ is the number

$$\|\mathbf{a}\| = \sqrt{a_1^2 + a_2^2 + a_3^2}.$$

The norm of $\mathbf{a}$ is also called the *length* or *magnitude* of $\mathbf{a}$. The reason for this is simple: if we represent the vector $\mathbf{a}$ by an arrow, say

$$\overrightarrow{QR} \quad \text{with} \quad Q = Q(x, y, z) \quad \text{and} \quad R = R(x + a_1, y + a_2, z + a_3),$$

then $\|\mathbf{a}\|$ gives the length of $\overrightarrow{QR}$.

The norm properties of vectors are very similar to the absolute value properties of real numbers. In particular, we have

---

(1) $\|\mathbf{a}\| \geq 0$ and $\|\mathbf{a}\| = 0$ iff $\mathbf{a} = \mathbf{0}$.

(2) $\|\alpha\mathbf{a}\| = |\alpha| \, \|\mathbf{a}\|$.

(3) $\|\mathbf{a} + \mathbf{b}\| \leq \|\mathbf{a}\| + \|\mathbf{b}\|$.   (the triangle inequality)

Property (1) is obvious. Property (2) is easy to verify:

$$\|\alpha\mathbf{a}\| = \sqrt{(\alpha a_1)^2 + (\alpha a_2)^2 + (\alpha a_3)^2} = |\alpha| \sqrt{a_1^2 + a_2^2 + a_3^2} = |\alpha| \, \|\mathbf{a}\|.$$

We prove Property (3), the triangle inequality, in the next section, where we have "dot products" at our disposal. A proof at this time would be laborious.

You can visualize the triangle inequality as saying that the length of a side of a triangle cannot exceed the sum of the lengths of the other two sides. (See Figure 12.3.5.)

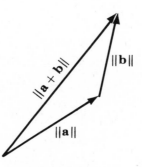

FIGURE 12.3.5

Problem.   Given

$$\mathbf{a} = (1, -2, 3) \quad \text{and} \quad \mathbf{b} = (-4, 1, 0),$$

find

(a) $\|\mathbf{a}\|$.   (b) $\|\mathbf{b}\|$.   (c) $\|\mathbf{a} + \mathbf{b}\|$.

(d) $\|\mathbf{a} - \mathbf{b}\|$.   (e) $\|-7\mathbf{a}\|$.   (f) $\|2\mathbf{a} - 3\mathbf{b}\|$.

SOLUTION

(a) $\|\mathbf{a}\| = \sqrt{1^2 + (-2)^2 + (3)^2} = \sqrt{1 + 4 + 9} = \sqrt{14}.$

(b) $\|\mathbf{b}\| = \sqrt{(-4)^2 + 1^2 + 0^2} = \sqrt{16 + 1 + 0} = \sqrt{17}.$

(c) $\|\mathbf{a} + \mathbf{b}\| = \|(-3, -1, 3)\| = \sqrt{(-3)^2 + (-1)^2 + (3)^2} = \sqrt{9 + 1 + 9}$
$= \sqrt{19}.$

(d) $\|\mathbf{a} - \mathbf{b}\| = \|(5, -3, 3)\| = \sqrt{(5)^2 + (-3)^2 + (3)^2} = \sqrt{25 + 9 + 9} = \sqrt{43}.$

(e) $\|-7\mathbf{a}\| = |-7|\,\|\mathbf{a}\| = 7\sqrt{14}.$

(f) $\|2\mathbf{a} - 3\mathbf{b}\| = \|2(1, -2, 3) - 3(-4, 1, 0)\| = \|(14, -7, 6)\|$
$= \sqrt{(14)^2 + (-7)^2 + (6)^2} = \sqrt{196 + 49 + 36} = \sqrt{281}. \quad \square$

To multiply a nonzero vector by a nonzero scalar α is to change its length by a factor of $|\alpha|$,

$$\|\alpha\mathbf{a}\| = |\alpha|\,\|\mathbf{a}\|,$$

keeping the same direction if $\alpha > 0$ and reversing the direction if $\alpha < 0$. The vector obtained from $\mathbf{a}$ simply by reversing its direction is the vector $(-1)\mathbf{a} = -\mathbf{a}$. (See Figure 12.3.6.)

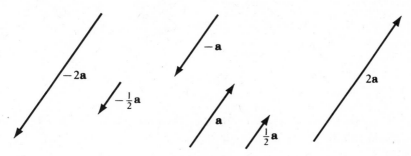

FIGURE 12.3.6

Problem. Let $\mathbf{a}$ be a nonzero vector. Find the unit vector (the vector of norm 1) which has the same direction as $\mathbf{a}$.

SOLUTION. If $\mathbf{b}$ is the required vector then

$$\|\mathbf{b}\| = 1 \quad \text{and} \quad \mathbf{b} = \alpha\mathbf{a} \qquad \text{for some } \alpha > 0.$$

It follows that

$$1 = \|\mathbf{b}\| = \|\alpha\mathbf{a}\| = |\alpha|\,\|\mathbf{a}\| = \alpha\|\mathbf{a}\|.$$

This says that

$$\alpha = \frac{1}{\|\mathbf{a}\|}$$

and consequently

$$\mathbf{b} = \frac{\mathbf{a}}{\|\mathbf{a}\|}. \quad \square$$

While

$$\frac{\mathbf{a}}{\|\mathbf{a}\|}$$

is the unit vector in the direction of **a**,

$$-\frac{\mathbf{a}}{\|\mathbf{a}\|}$$

is the unit vector in the opposite direction.

Since

$$\mathbf{a} - \mathbf{b} = \mathbf{a} + (-\mathbf{b})$$

we can draw the vector $\mathbf{a} - \mathbf{b}$ by drawing $-\mathbf{b}$ and adding it to the vector **a**. (Figure 12.3.7)

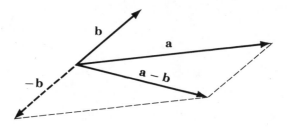

FIGURE 12.3.7

We can obtain the same result more easily by noting that $\mathbf{a} - \mathbf{b}$ is the vector that we must add to **b** to obtain **a**. See Figure 12.3.8.

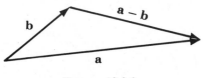

FIGURE 12.3.8

We single out for special attention the vectors

$$\mathbf{i} = (1, 0, 0), \qquad \mathbf{j} = (0, 1, 0), \qquad \mathbf{k} = (0, 0, 1).$$

These vectors all have norm 1 and, if pictured as emanating from the origin, lie along the positive coordinate axes. They are called *unit coordinate vectors*. (See Figure 12.3.9.)

Every vector can be expressed as a linear combination of the unit coordinate vectors:

$$\boxed{\text{for} \quad \mathbf{a} = (a_1, a_2, a_3) \quad \text{we have} \quad \mathbf{a} = a_1\mathbf{i} + a_2\mathbf{j} + a_3\mathbf{k}.}$$

Proof

$$(a_1, a_2, a_3) = (a_1, 0, 0) + (0, a_2, 0) + (0, 0, a_3)$$
$$= a_1(1, 0, 0) + a_2(0, 1, 0) + a_3(0, 0, 1)$$
$$= a_1\mathbf{i} + a_2\mathbf{j} + a_3\mathbf{k}. \quad \square$$

If

$$\mathbf{a} = a_1\mathbf{i} + a_2\mathbf{j} + a_3\mathbf{k},$$

the numbers a_1, a_2, a_3 are called the *scalar components* of $\mathbf{a}$ or more simply the *components* of $\mathbf{a}$. The vectors $a_1\mathbf{i}$, $a_2\mathbf{j}$, $a_3\mathbf{k}$ are called the *vector components* of $\mathbf{a}$.

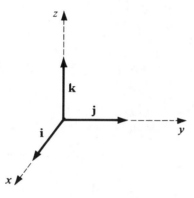

FIGURE 12.3.9

Problem. Given that

$$\mathbf{a} = 3\mathbf{i} - \mathbf{j} + \mathbf{k} \quad \text{and} \quad \mathbf{b} = 2\mathbf{i} + 3\mathbf{j} - \mathbf{k},$$

(1) express $2\mathbf{a} - \mathbf{b}$ as a linear combination of $\mathbf{i}$, $\mathbf{j}$, $\mathbf{k}$.

(2) compute $\|2\mathbf{a} - \mathbf{b}\|$.

(3) find the unit vector $\mathbf{c}$ in the direction of $2\mathbf{a} - \mathbf{b}$.

Solution

(1)
$$2\mathbf{a} - \mathbf{b} = 2(3\mathbf{i} - \mathbf{j} + \mathbf{k}) - (2\mathbf{i} + 3\mathbf{j} - \mathbf{k})$$
$$= 6\mathbf{i} - 2\mathbf{j} + 2\mathbf{k} - 2\mathbf{i} - 3\mathbf{j} + \mathbf{k}$$
$$= 4\mathbf{i} - 5\mathbf{j} + 3\mathbf{k}.$$

(2)
$$\|2\mathbf{a} - \mathbf{b}\| = \|4\mathbf{i} - 5\mathbf{j} + 3\mathbf{k}\|$$
$$= \sqrt{16 + 25 + 9}$$
$$= \sqrt{50}$$
$$= 5\sqrt{2}.$$

(3)
$$\mathbf{c} = \frac{2\mathbf{a} - \mathbf{b}}{\|2\mathbf{a} - \mathbf{b}\|} = \frac{1}{5\sqrt{2}}(2\mathbf{a} - \mathbf{b}) = \frac{\sqrt{2}}{10}(2\mathbf{a} - \mathbf{b}). \quad \square$$

Exercises

Simplify the following linear combinations.

*1. $(2\mathbf{i} - \mathbf{j} + \mathbf{k}) + (\mathbf{i} - 3\mathbf{j} + 5\mathbf{k})$.   2. $(6\mathbf{j} - \mathbf{k}) + (3\mathbf{i} - \mathbf{j} + 2\mathbf{k})$.

*3. $2(\mathbf{j} + \mathbf{k}) - 3(\mathbf{i} + \mathbf{j} - 2\mathbf{k})$.   4. $2(\mathbf{i} - \mathbf{j}) + 6(2\mathbf{i} + \mathbf{j} - 2\mathbf{k})$.

Compute the norm of the following vectors.

*5. $3\mathbf{i} + 4\mathbf{j}$.   6. $\mathbf{i} - \mathbf{j}$.

*7. $2\mathbf{i} + \mathbf{j} - 2\mathbf{k}$.   8. $6\mathbf{i} + 2\mathbf{j} - \mathbf{k}$.

*9. $\frac{1}{2}(\mathbf{i} + 4\mathbf{j}) - (\frac{3}{2}\mathbf{i} + \mathbf{k})$.   10. $(\mathbf{i} - \mathbf{j}) + 2(\mathbf{j} - \mathbf{i}) + (\mathbf{k} - \mathbf{j})$.

*11. Let

$$\mathbf{a} = \mathbf{i} - \mathbf{j} + 2\mathbf{k}, \qquad \mathbf{b} = 2\mathbf{i} - \mathbf{j} + 2\mathbf{k},$$
$$\mathbf{c} = 3\mathbf{i} - 3\mathbf{j} + 6\mathbf{k}, \qquad \mathbf{d} = -2\mathbf{i} + 2\mathbf{j} - 4\mathbf{k}.$$

(a) Which vectors are parallel?

(b) Which vectors have the same direction?

(c) Which vectors have opposite directions?

*12. Label the vectors.

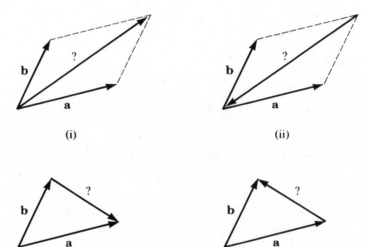

(i)   (ii)

(iii)   (iv)

13. (*Important*)   Prove the following version of the triangle inequality:

$$\big|\, \|\mathbf{a}\| - \|\mathbf{b}\| \,\big| \le \|\mathbf{a} - \mathbf{b}\|.$$

HINT: Apply the previously stated triangle inequality to

$$\mathbf{a} = (\mathbf{a} - \mathbf{b}) + \mathbf{b}.$$

*14. Given that

$$\mathbf{a} = (1, 1, 1), \qquad \mathbf{b} = (-1, 3, 2), \qquad \mathbf{c} = (-3, 0, 1), \qquad \mathbf{d} = (4, -1, 1),$$

(a) express $\mathbf{a} + 2\mathbf{b} + 3\mathbf{c} + 4\mathbf{d}$ as a linear combination of $\mathbf{i}, \mathbf{j}, \mathbf{k}$.
(b) find scalars A, B, C such that

$$\mathbf{d} = A\mathbf{a} + B\mathbf{b} + C\mathbf{c}.$$

15. Find α given that

$$3\mathbf{i} + \mathbf{j} - \mathbf{k} \quad \text{and} \quad \alpha\mathbf{i} - 4\mathbf{j} + 4\mathbf{k}$$

are parallel.
*16. Find α given that

$$3\mathbf{i} + \mathbf{j} \quad \text{and} \quad \alpha\mathbf{j} - \mathbf{k}$$

have the same length.
17. Find the unit vector in the direction of $\mathbf{i} - 2\mathbf{j} + 2\mathbf{k}$.
18. Find α given that

$$\|\alpha\mathbf{i} + (\alpha - 1)\mathbf{j} + (\alpha + 1)\mathbf{k}\| = 2.$$

*19. Find the vector of norm 2 in the direction of $\mathbf{i} + 2\mathbf{j} - \mathbf{k}$.
20. Find the vectors of norm 2 parallel to $3\mathbf{j} + 2\mathbf{k}$.

12.4   The Dot Product

We come now to a notion that is widely used in geometry and in physics.

Definition 12.4.1

For vectors $\mathbf{a} = (a_1, a_2, a_3)$ and $\mathbf{b} = (b_1, b_2, b_3)$ we define

$$\mathbf{a} \cdot \mathbf{b} = a_1 b_1 + a_2 b_2 + a_3 b_3.$$

The number $\mathbf{a} \cdot \mathbf{b}$, read "$\mathbf{a}$ dot $\mathbf{b}$", is called the *dot product* of $\mathbf{a}$ and $\mathbf{b}$.

Examples. For

$$\mathbf{a} = 2\mathbf{i} - \mathbf{j} + 3\mathbf{k}, \qquad \mathbf{b} = -3\mathbf{i} + \mathbf{j} + 4\mathbf{k}, \qquad \mathbf{c} = \mathbf{i} - \mathbf{j}$$

we have

$$\mathbf{a} \cdot \mathbf{b} = (2)(-3) + (-1)(1) + (3)(4) = -6 - 1 + 12 = 5,$$
$$\mathbf{a} \cdot \mathbf{c} = (2)(1) + (-1)(-1) + (3)(0) = 2 + 1 = 3,$$
$$\mathbf{b} \cdot \mathbf{c} = (-3)(1) + (1)(-1) + (4)(0) = -3 - 1 = -4. \quad \square$$

The dot product is also called the *scalar product* or *inner product*. We'll generally call it the dot product and speak of "dotting" $\mathbf{a}$ with $\mathbf{b}$. If we dot a vector with itself, we obtain the square of the norm of the vector:

$$\mathbf{a} \cdot \mathbf{a} = \|\mathbf{a}\|^2.$$

$$[\mathbf{a} \cdot \mathbf{a} = a_1 a_1 + a_2 a_2 + a_3 a_3 = a_1^2 + a_2^2 + a_3^2 = \|\mathbf{a}\|^2.]$$

Consequently

$$\mathbf{a} \cdot \mathbf{a} \geq 0 \qquad \text{and} \qquad \mathbf{a} \cdot \mathbf{a} = 0 \quad \text{iff} \quad \mathbf{a} = \mathbf{0}.$$

For a geometric interpretation of $\mathbf{a} \cdot \mathbf{b}$ we form the triangle with vertices $P(x, y, z)$, $Q(x + a_1, y + a_2, z + a_3)$, $R(x + b_1, y + b_2, z + b_3)$. See Figure 12.4.1. Since the sides of the triangle have lengths $\|\mathbf{a}\|$, $\|\mathbf{b}\|$, and $\|\mathbf{a} - \mathbf{b}\|$, the law of cosines gives

$$\|\mathbf{a} - \mathbf{b}\|^2 = \|\mathbf{a}\|^2 + \|\mathbf{b}\|^2 - 2\|\mathbf{a}\| \|\mathbf{b}\| \cos \theta.$$

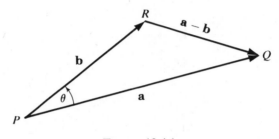

FIGURE 12.4.1

Consequently

$$
\begin{aligned}
2\|\mathbf{a}\| \|\mathbf{b}\| \cos \theta &= \|\mathbf{a}\|^2 + \|\mathbf{b}\|^2 - \|\mathbf{a} - \mathbf{b}\|^2 \\
&= a_1^2 + a_2^2 + a_3^2 + b_1^2 + b_2^2 + b_3^2 - (a_1 - b_1)^2 - (a_2 - b_2)^2 - (a_3 - b_3)^2 \\
&= 2(a_1 b_1 + a_2 b_2 + a_3 b_3) \\
&= 2(\mathbf{a} \cdot \mathbf{b}),
\end{aligned}
$$

so that

(12.4.2)

$$\boxed{\mathbf{a} \cdot \mathbf{b} = \|\mathbf{a}\| \|\mathbf{b}\| \cos \theta.}$$

The number θ (which in the figure is the angle between $\overrightarrow{PQ}$ and $\overrightarrow{PR}$) we take to satisfy $0 < \theta < \pi$ and call it the *angle between* $\mathbf{a}$ and $\mathbf{b}$. It can be obtained from the relation

$$\boxed{\cos \theta = \frac{\mathbf{a} \cdot \mathbf{b}}{\|\mathbf{a}\| \|\mathbf{b}\|}}$$

unless either $\mathbf{a}$ or $\mathbf{b}$ is the zero vector, in which case the angle is not defined.

If $\mathbf{a} \cdot \mathbf{b} = 0$, which happens if $\theta = \frac{1}{2}\pi$ or if one of the vectors is the zero vector, we call $\mathbf{a}$ and $\mathbf{b}$ perpendicular (Figure 12.4.2).† To indicate that $\mathbf{a}$ and $\mathbf{b}$ are perpendicular, we will sometimes write $\mathbf{a} \perp \mathbf{b}$. Obviously the coordinate vectors $\mathbf{i}, \mathbf{j}, \mathbf{k}$ are mutually perpendicular.

---

† This makes the zero vector both parallel and perpendicular to every vector. There is, however, no contradiction since we never apply the notion of angle to the zero vector.

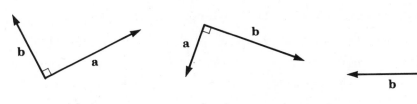

$$\mathbf{a} \cdot \mathbf{b} = 0$$

FIGURE 12.4.2

Problem.   Find the angle between

$$\mathbf{a} = 2\mathbf{i} + \mathbf{j} + \mathbf{k} \quad \text{and} \quad \mathbf{b} = \mathbf{i} + \mathbf{j} - 3\mathbf{k}.$$

SOLUTION

$$\mathbf{a} \cdot \mathbf{b} = (2)(1) + (1)(1) + (1)(-3) = 2 + 1 - 3 = 0.$$

Since the dot product is 0, the two vectors must be perpendicular. The angle between them is $\frac{1}{2}\pi$ radians.   □

Problem.   Find the angle between

$$\mathbf{a} = 2\mathbf{i} + 3\mathbf{j} + 2\mathbf{k} \quad \text{and} \quad \mathbf{b} = \mathbf{i} + 2\mathbf{j} - \mathbf{k}.$$

SOLUTION

$$\mathbf{a} \cdot \mathbf{b} = (2)(1) + (3)(2) + (2)(-1) = 6,$$
$$\|\mathbf{a}\| = \sqrt{2^2 + 3^2 + 2^2} = \sqrt{17},$$
$$\|\mathbf{b}\| = \sqrt{1^2 + 2^2 + (-1)^2} = \sqrt{6}.$$

In general

$$\cos \theta = \frac{\mathbf{a} \cdot \mathbf{b}}{\|\mathbf{a}\| \, \|\mathbf{b}\|}$$

so that here

$$\cos \theta = \frac{6}{\sqrt{17}\sqrt{6}} = \sqrt{\frac{6}{17}} \cong \sqrt{0.353} \cong 0.594.$$

From Table 4 at the end of the book you can see that θ is about 0.93 radians.   □

Problem.   Find α given that

$$\mathbf{i} + 2\mathbf{j} \quad \text{and} \quad 3\mathbf{i} - \alpha\mathbf{j} + \mathbf{k}$$

are perpendicular.

SOLUTION.   Since the two vectors are perpendicular, their dot product

$$(\mathbf{i} + 2\mathbf{j}) \cdot (3\mathbf{i} - \alpha\mathbf{j} + \mathbf{k}) = (1)(3) + (2)(-\alpha) + (0)(1) = 3 - 2\alpha$$

is zero and $\alpha = \frac{3}{2}$.   □

Note that the dot product has the following properties:

$$\mathbf{a} \cdot \mathbf{b} = \mathbf{b} \cdot \mathbf{a} \qquad \text{(the dot product is commutative)}$$

$$\alpha \mathbf{a} \cdot \beta \mathbf{b} = \alpha \beta (\mathbf{a} \cdot \mathbf{b}) \qquad \text{(scalars can be factored out)}$$

$$\mathbf{a} \cdot (\mathbf{b} + \mathbf{c}) = \mathbf{a} \cdot \mathbf{b} + \mathbf{a} \cdot \mathbf{c} \qquad \text{(the dot distributes over addition)}$$

$$|\mathbf{a} \cdot \mathbf{b}| \leq \|\mathbf{a}\| \, \|\mathbf{b}\|. \qquad \text{(Schwarz's inequality)}$$

The first three properties follow easily from the definition. Schwarz's inequality is obvious from the relation

$$\mathbf{a} \cdot \mathbf{b} = \|\mathbf{a}\| \, \|\mathbf{b}\| \cos \theta$$

together with the observation that

$$|\cos \theta| \leq 1. \quad \square$$

Problem.   Show that

$$(\mathbf{a} + \mathbf{b}) \cdot \mathbf{c} = \mathbf{a} \cdot \mathbf{c} + \mathbf{b} \cdot \mathbf{c}.$$

SOLUTION

$$(\mathbf{a} + \mathbf{b}) \cdot \mathbf{c} = \mathbf{c} \cdot (\mathbf{a} + \mathbf{b}) \qquad \text{(by commutativity)}$$

$$= \mathbf{c} \cdot \mathbf{a} + \mathbf{c} \cdot \mathbf{b} \qquad \text{(by distributivity)}$$

$$= \mathbf{a} \cdot \mathbf{c} + \mathbf{b} \cdot \mathbf{c}. \quad \square \qquad \text{(by commutativity)}$$

We are now in a position to verify the triangle inequality

$$\|\mathbf{a} + \mathbf{b}\| \leq \|\mathbf{a}\| + \|\mathbf{b}\|.$$

VERIFICATION

$$\|\mathbf{a} + \mathbf{b}\|^2 = (\mathbf{a} + \mathbf{b}) \cdot (\mathbf{a} + \mathbf{b})$$

$$= \mathbf{a} \cdot \mathbf{a} + \mathbf{b} \cdot \mathbf{a} + \mathbf{a} \cdot \mathbf{b} + \mathbf{b} \cdot \mathbf{b}$$

$$= \|\mathbf{a}\|^2 + 2(\mathbf{a} \cdot \mathbf{b}) + \|\mathbf{b}\|^2$$

$$\leq \|\mathbf{a}\|^2 + 2\|\mathbf{a}\| \, \|\mathbf{b}\| + \|\mathbf{b}\|^2$$

$$= (\|\mathbf{a}\| + \|\mathbf{b}\|)^2$$

so that

$$\|\mathbf{a} + \mathbf{b}\| \leq \|\mathbf{a}\| + \|\mathbf{b}\|. \quad \square$$

We go back now to the formula

$$\mathbf{a} \cdot \mathbf{b} = \|\mathbf{a}\| \, \|\mathbf{b}\| \cos \theta.$$

If **b** is a nonzero vector, the number

$$\frac{\mathbf{a} \cdot \mathbf{b}}{\|\mathbf{b}\|} = \|\mathbf{a}\| \cos \theta \qquad \text{(see Figure 12.4.3)}$$

measures the "advance" of **a** in the direction of **b**. It is called the (*scalar*) *component* of **a** in the direction of **b**.

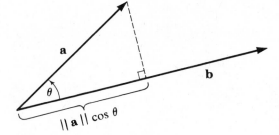

FIGURE 12.4.3

In abbreviated form we write

$$\text{comp}_b \, \mathbf{a} = \frac{\mathbf{a} \cdot \mathbf{b}}{\|\mathbf{b}\|} = \|\mathbf{a}\| \, \cos \theta.$$

Note that

$$\text{comp}_b \, \mathbf{a} \quad \text{is} \quad \begin{cases} \text{positive,} & \text{if } 0 \le \theta < \tfrac{1}{2}\pi \\ \text{zero,} & \text{if } \theta = \tfrac{1}{2}\pi \\ \text{negative,} & \text{if } \tfrac{1}{2}\pi < \theta \le \pi \end{cases}. \qquad \text{(see Figure 12.4.4)}$$

Problem.   Given

$$\mathbf{a} = \mathbf{j} - 2\mathbf{i} + \mathbf{k}, \qquad \mathbf{b} = 4\mathbf{i} - 3\mathbf{j} + \mathbf{k}, \qquad \mathbf{c} = 3\mathbf{i} - 4\mathbf{j} + 12\mathbf{k},$$

compute $\text{comp}_b \, \mathbf{a}$ and $\text{comp}_c \, \mathbf{a}$.

SOLUTION.   We begin by rewriting $\mathbf{a}$ as

$$\mathbf{a} = -2\mathbf{i} + \mathbf{j} + \mathbf{k}.$$

$$\text{comp}_b \, \mathbf{a} = \frac{\mathbf{a} \cdot \mathbf{b}}{\|\mathbf{b}\|} = \frac{(-2)(4) + (1)(-3) + (1)(1)}{\sqrt{(4)^2 + (-3)^2 + (1)^2}} = -\frac{10}{\sqrt{26}} = -\frac{5}{14}\sqrt{26}.$$

$$\text{comp}_c \, \mathbf{a} = \frac{\mathbf{a} \cdot \mathbf{c}}{\|\mathbf{c}\|} = \frac{(-2)(3) + (1)(-4) + (1)(12)}{\sqrt{(3)^2 + (-4)^2 + (12)^2}} = \frac{2}{13}. \quad \square$$

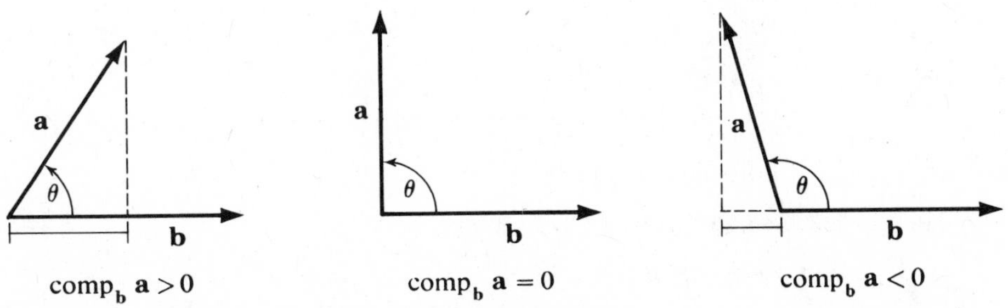

comp$_b$ a > 0 comp$_b$ a $= 0$ comp$_b$ a < 0

FIGURE 12.4.4

For each vector

$$\mathbf{a} = a_1\mathbf{i} + a_2\mathbf{j} + a_3\mathbf{k}$$

we have

$$\operatorname{comp}_i \mathbf{a} = \frac{\mathbf{a} \cdot \mathbf{i}}{\|\mathbf{i}\|} = \frac{a_1}{1} = a_1,$$

$$\operatorname{comp}_j \mathbf{a} = \frac{\mathbf{a} \cdot \mathbf{j}}{\|\mathbf{j}\|} = \frac{a_2}{1} = a_2,$$

$$\operatorname{comp}_k \mathbf{a} = \frac{\mathbf{a} \cdot \mathbf{k}}{\|\mathbf{k}\|} = \frac{a_3}{1} = a_3.$$

These results agree with our previous use of the word "component."†

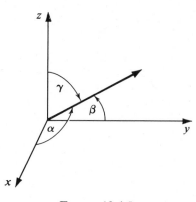

FIGURE 12.4.5

In Figure 12.4.5 we have portrayed a nonzero vector $\mathbf{a}$. The angles α, β, γ that the vector makes with the coordinate vectors are called the *direction angles* of $\mathbf{a}$ and their cosines are called the direction cosines. Since

$$\cos \alpha = \frac{\mathbf{a} \cdot \mathbf{i}}{\|\mathbf{a}\| \, \|\mathbf{i}\|} = \frac{a_1}{\|\mathbf{a}\|}, \qquad \cos \beta = \frac{\mathbf{a} \cdot \mathbf{j}}{\|\mathbf{a}\| \, \|\mathbf{j}\|} = \frac{a_2}{\|\mathbf{a}\|}, \qquad \cos \gamma = \frac{\mathbf{a} \cdot \mathbf{k}}{\|\mathbf{a}\| \, \|\mathbf{k}\|} = \frac{a_3}{\|\mathbf{a}\|},$$

you can see that

$$\boxed{\mathbf{a} = \|\mathbf{a}\|(\cos \alpha \, \mathbf{i} + \cos \beta \, \mathbf{j} + \cos \gamma \, \mathbf{k}).}$$

Taking the norm of both sides we have

$$\|\mathbf{a}\| = \|\mathbf{a}\| \, \|\cos \alpha \, \mathbf{i} + \cos \beta \, \mathbf{j} + \cos \gamma \, \mathbf{k}\|,$$

$$1 = \|\cos \alpha \, \mathbf{i} + \cos \beta \, \mathbf{j} + \cos \gamma \, \mathbf{k}\| = \sqrt{\cos^2 \alpha + \cos^2 \beta + \cos^2 \gamma},$$

---

† At the end of Section 12.3.

and therefore

$$\cos^2 \alpha + \cos^2 \beta + \cos^2 \gamma = 1.$$

The sum of the squares of the direction cosines is always 1.

If $\mathbf{u}$ is a unit vector, the relation

$$\mathbf{u} = \|\mathbf{u}\|(\cos \alpha \, \mathbf{i} + \cos \beta \, \mathbf{j} + \cos \gamma \, \mathbf{k})$$

gives

$$\mathbf{u} = \cos \alpha \, \mathbf{j} + \cos \beta \, \mathbf{j} + \cos \gamma \, \mathbf{k}.$$

For a unit vector, the scalar components are the direction cosines.

Problem. Find the direction cosines of

$$\mathbf{a} = \mathbf{i} + 2\mathbf{j} - 2\mathbf{k}.$$

What are the direction angles?

SOLUTION. Here

$$\|\mathbf{a}\| = \sqrt{1^2 + 2^2 + (-2)^2} = 3$$

so that

$$\cos \alpha = \frac{a_1}{\|\mathbf{a}\|} = \tfrac{1}{3}, \qquad \cos \beta = \frac{a_2}{\|\mathbf{a}\|} = \tfrac{2}{3}, \qquad \cos \gamma = \frac{a_3}{\|\mathbf{a}\|} = -\tfrac{2}{3}.$$

$\alpha = \text{arc cos } \tfrac{1}{3} \cong 1.23 \text{ radians,}$ (about 71 degrees)

$\beta = \text{arc cos } \tfrac{2}{3} \cong 0.84 \text{ radians,}$ (about 48 degrees)

$\gamma = \text{arc cos } (-\tfrac{2}{3}) \cong 1.57 + \text{arc cos } \tfrac{2}{3} \cong 2.31 \text{ radians.}$ (about 132 degrees)

Problem. Find the vector which has norm 3 and direction angles

$$\alpha = \tfrac{1}{4}\pi, \qquad \beta = \tfrac{2}{3}\pi, \qquad \gamma = \tfrac{1}{3}\pi.$$

SOLUTION

$$\mathbf{a} = \|\mathbf{a}\|(\cos \alpha \, \mathbf{i} + \cos \beta \, \mathbf{j} + \cos \gamma \, \mathbf{k})$$
$$= 3(\tfrac{1}{2}\sqrt{2}\,\mathbf{i} - \tfrac{1}{2}\mathbf{j} + \tfrac{1}{2}\mathbf{k})$$
$$= \tfrac{3}{2}\sqrt{2}\,\mathbf{i} - \tfrac{3}{2}\mathbf{j} + \tfrac{3}{2}\mathbf{k}. \quad \square$$

Exercises

For the first four exercises take

$$\mathbf{a} = 2\mathbf{i} + \mathbf{j}, \qquad \mathbf{b} = 3\mathbf{i} - \mathbf{j} + 2\mathbf{k}, \qquad \mathbf{c} = 4\mathbf{i} + 3\mathbf{k}.$$

*1. Compute the three dot products, $\mathbf{a} \cdot \mathbf{b}$, $\mathbf{a} \cdot \mathbf{c}$, $\mathbf{b} \cdot \mathbf{c}$.

2. Find the cosines of the angles between these vectors.

* 3. Find the component of **a** in the direction of **b** and in the direction of **c**.

4. Find the component of **b** in the direction of **a** and in the direction of **c**.

* 5. Find the vector which has norm 2 and direction angles $\frac{1}{4}\pi, \frac{1}{4}\pi, \frac{1}{2}\pi$.

6. Find the direction angles of $\mathbf{i} - \mathbf{j} + \sqrt{2}\mathbf{k}$.

* 7. What can you conclude about **a** and **b** given that

  (a) $\|\mathbf{a}\|^2 + \|\mathbf{b}\|^2 = \|\mathbf{a} + \mathbf{b}\|^2$?

  (b) $\|\mathbf{a}\|^2 + \|\mathbf{b}\|^2 = \|\mathbf{a} - \mathbf{b}\|^2$?

  HINT: Draw figures.

* 8. (*Important*) Show that $\mathbf{a} \cdot \mathbf{b} = \mathbf{a} \cdot \mathbf{c}$ does not necessarily imply that $\mathbf{b} = \mathbf{c}$. Draw a figure illustrating this fact for nonzero vectors.

9. Verify that

$$4(\mathbf{a} \cdot \mathbf{b}) = \|\mathbf{a} + \mathbf{b}\|^2 - \|\mathbf{a} - \mathbf{b}\|^2.$$

10. Show that

$$\mathbf{a} \perp \mathbf{b} \quad \text{iff} \quad \|\mathbf{a} + \mathbf{b}\| = \|\mathbf{a} - \mathbf{b}\|.$$

11. What are the direction angles of $-\mathbf{a}$ if the direction angles of **a** are α, β, γ?

12. Under what conditions does

$$|\mathbf{a} \cdot \mathbf{b}| = \|\mathbf{a}\| \, \|\mathbf{b}\|?$$

13. If a force **F** is applied throughout a displacement **r** then the work done is given by the formula

$$W = (\text{comp}_{\mathbf{r}} \, \mathbf{F})\|\mathbf{r}\|.$$

* (a) Express the work W as a dot product.

  (b) What is the work done if $\mathbf{F} \perp \mathbf{r}$?

* (c) Two forces of the same magnitude $\mathbf{F}_1$ and $\mathbf{F}_2$ are applied throughout a displacement **r** at angles θ_1 and θ_2 respectively. Compare the work done by $\mathbf{F}_1$ to that done by $\mathbf{F}_2$ if

  (i) $\theta_1 = -\theta_2$.   (ii) $\theta_1 = \frac{1}{3}\pi, \quad \theta_2 = \frac{1}{6}\pi$.

  (d) What is the work done by **F** if the object moves completely around a triangle? Justify your answer.

* 14. Find the numbers x for which

$$x\mathbf{i} + 11\mathbf{j} - 3\mathbf{k} \perp 2x\mathbf{i} - x\mathbf{j} - 5\mathbf{k}.$$

15. Find the numbers x for which the angle between

$$\mathbf{c} = x\mathbf{i} + \mathbf{j} + \mathbf{k} \quad \text{and} \quad \mathbf{d} = \mathbf{i} + x\mathbf{j} + \mathbf{k}$$

  is $\frac{1}{3}\pi$.

* 16. Find the angle between the diagonal of a cube and one of its edges.

17. Find the angle between the diagonal of a cube and the diagonal of one of its faces.

18. Show that $\frac{1}{4}\pi, \frac{1}{6}\pi, \frac{2}{3}\pi$ cannot be the direction angles of a vector.

19. Let **b** be a nonzero vector. Express the vector **a** as the sum of a vector parallel to **b** and a vector perpendicular to **b**.

    HINT: Set $\mathbf{a} = s\mathbf{b} + (\mathbf{a} - s\mathbf{b})$ and choose s so that $\mathbf{a} - s\mathbf{b} \perp \mathbf{b}$.

12.5   Lines

Vectors to us are simply ordered triples of real numbers and, as you have seen, they can be given many interpretations. When vectors are used to indicate position in three-dimensional space, they are called *position vectors*. In general we picture position vectors as emanating from the origin. In this section we will use position vectors to characterize lines.

We begin with the idea that two distinct points determine a line. In Figure 12.5.1 we have marked two points and the line *l* that they determine. To obtain a vector characterization of *l*, we choose the vectors $\mathbf{r}_0$ and **d** as in Figure 12.5.2. (Since we

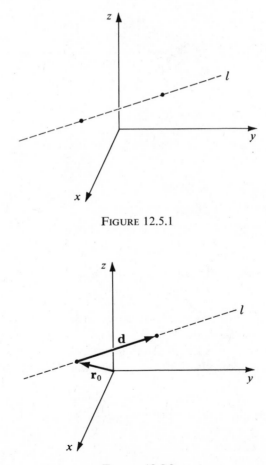

FIGURE 12.5.1

FIGURE 12.5.2

began with two distinct points, the vector $\mathbf{d}$ is nonzero. Keep this in mind.) In Figure 12.5.3 we have drawn an additional vector $\mathbf{r}$. What we want is a necessary and sufficient condition on the vector $\mathbf{r}$ for the tip of $\mathbf{r}$ to fall on the line l. This is not hard to find: The tip of $\mathbf{r}$ will fall on l iff

$$\mathbf{r} - \mathbf{r}_0 \quad \text{and} \quad \mathbf{d} \quad \text{are parallel;}$$

this in turn will happen iff

$$\mathbf{r} - \mathbf{r}_0 = t\mathbf{d} \quad \text{for some real } t$$

or, equivalently, iff

$$\mathbf{r} = \mathbf{r}_0 + t\mathbf{d} \quad \text{for some real } t.$$

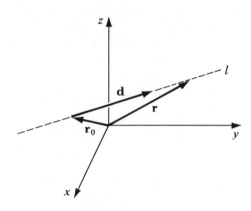

FIGURE 12.5.3

In view of this we call the equation

(12.5.1)
$$\boxed{\mathbf{r}(t) = \mathbf{r}_0 + t\mathbf{d}, \qquad t \text{ real}}$$

a *vector parametrization* for the line l. By varying t, we vary the vector $\mathbf{r}(t)$, but its tip remains on the line l. As t ranges over the set of real numbers, the tip of $\mathbf{r}(t)$ traces out the line l.

The roles of $\mathbf{r}_0$ and $\mathbf{d}$ in this matter deserve further attention. The vector $\mathbf{r}_0 = x_0\mathbf{i} + y_0\mathbf{j} + z_0\mathbf{k}$ places the point $P(x_0, y_0, z_0)$ on the line. The vector $\mathbf{d}$, which as we remarked is nonzero, gives the direction of the line. It is called a *direction vector* for l, and its scalar components d_1, d_2, d_3 are *direction numbers* for the line.

Problem.   Find a vector parametrization for the line that passes through the point $P(1, -1, 2)$ and is parallel to the vector $2\mathbf{i} - 3\mathbf{j} + \mathbf{k}$.

SOLUTION.   Here we can take

$$\mathbf{r}_0 = \mathbf{i} - \mathbf{j} + 2\mathbf{k} \quad \text{and} \quad \mathbf{d} = 2\mathbf{i} - 3\mathbf{j} + \mathbf{k}.$$

As a vector parametrization for the line we have

$$\mathbf{r}(t) = (\mathbf{i} - \mathbf{j} + 2\mathbf{k}) + t(2\mathbf{i} - 3\mathbf{j} + \mathbf{k}),$$

which we can rewrite as

$$\mathbf{r}(t) = (1 + 2t)\mathbf{i} - (1 + 3t)\mathbf{j} + (2 + t)\mathbf{k}. \quad \square$$

Problem. Given that the line

$$\mathbf{r}(t) = \mathbf{r}_0 + t\mathbf{d}$$

passes through the origin, what can we conclude about $\mathbf{r}_0$ and $\mathbf{d}$?

SOLUTION. At first glance it may appear that $\mathbf{r}_0$ has to be $\mathbf{0}$ but this is not true. From the fact that the line passes through the origin we cannot conclude that $\mathbf{r}_0$ is $\mathbf{0}$ but only that there exists a real number t_0 such that

$$\mathbf{r}_0 + t_0\mathbf{d} = \mathbf{0}.$$

This tells us that

$$\mathbf{r}_0 = -t_0\mathbf{d}$$

and thus that $\mathbf{r}_0$ is a scalar multiple of $\mathbf{d}$. $\quad \square$

There are other ways to represent lines. The line

$$l: \mathbf{r}(t) = \mathbf{r}_0 + t\mathbf{d}$$

which passes through $P(x_0, y_0, z_0)$ and has direction numbers d_1, d_2, d_3 can also be represented parametrically by the three scalar functions

(12.5.2) $\boxed{\quad x(t) = x_0 + d_1t, \qquad y(t) = y_0 + d_2t, \qquad z(t) = z_0 + d_3t. \quad}$

These are just the scalar components of $\mathbf{r}(t)$.

Problem. Find a scalar parametrization for the line which passes through $P(-1, 4, 2)$ and has direction numbers $(1, 2, 3)$.

SOLUTION

$$x(t) = -1 + t, \qquad y(t) = 4 + 2t, \qquad z(t) = 2 + 3t. \quad \square$$

Problem. Find the direction numbers of the line

$$x(t) = 3 - t, \qquad y(t) = 2 + 4t, \qquad z(t) = 1 - 5t.$$

What other direction numbers can be used for the same line?

SOLUTION. The direction numbers are $-1, 4, -5$. All multiples

$$-k, 4k, -5k \qquad \text{with} \quad k \neq 0$$

can be used as direction numbers for the same line. $\quad \square$

If none of the direction numbers is zero, then you can solve each of the equations in (12.5.2) for t:

$$t = \frac{x(t) - x_0}{d_1} = \frac{y(t) - y_0}{d_2} = \frac{z(t) - z_0}{d_3}.$$

Elimination of the parameter t gives

(12.5.3)
$$\boxed{\frac{x - x_0}{d_1} = \frac{y - y_0}{d_2} = \frac{z - z_0}{d_3}.}$$

These are called the *symmetric* equations for the line.

Problem. Find a set of symmetric equations for the line which passes through the points $P(x_0, y_0, z_0)$ and $Q(x_1, y_1, z_1)$. Under what conditions are the symmetric equations valid?

SOLUTION. As direction numbers we can take the triple

$$x_1 - x_0, y_1 - y_0, z_1 - z_0.$$

We can then write the symmetric equations as

$$\frac{x - x_0}{x_1 - x_0} = \frac{y - y_0}{y_1 - y_0} = \frac{z - z_0}{z_1 - z_0}$$

or equivalently as

$$\frac{x - x_1}{x_1 - x_0} = \frac{y - y_1}{y_1 - y_0} = \frac{z - z_1}{z_1 - z_0}.$$

These equations are valid provided that $x_1 \neq x_0$, $y_1 \neq y_0$, and $z_1 \neq z_0$.   □

Two distinct lines

$$l_1 \colon \mathbf{r}(t) = \mathbf{r}_0 + t\mathbf{d},$$
$$l_2 \colon \mathbf{R}(u) = \mathbf{R}_0 + u\mathbf{D}$$

intersect iff there are numbers t and u at which

$$\mathbf{r}(t) = \mathbf{R}(u).$$

Problem. Find the point at which the lines

$$l_1 \colon \mathbf{r}(t) = (\mathbf{i} - 6\mathbf{j} + 2\mathbf{k}) + t(\mathbf{i} + 2\mathbf{j} + \mathbf{k}),$$
$$l_2 \colon \mathbf{R}(u) = (4\mathbf{j} + \mathbf{k}) + u(2\mathbf{i} + \mathbf{j} + 2\mathbf{k}),$$

intersect.

SOLUTION.   We set
$$\mathbf{r}(t) = \mathbf{R}(u)$$
and solve for t and u:

$$(\mathbf{i} - 6\mathbf{j} + 2\mathbf{k}) + t(\mathbf{i} + 2\mathbf{j} + \mathbf{k}) = (4\mathbf{j} + \mathbf{k}) + u(2\mathbf{i} + \mathbf{j} + 2\mathbf{k}),$$
$$\mathbf{i} - 6\mathbf{j} + 2\mathbf{k} + t\mathbf{i} + 2t\mathbf{j} + t\mathbf{k} = 4\mathbf{j} + \mathbf{k} + 2u\mathbf{i} + u\mathbf{j} + 2u\mathbf{k},$$
$$(1 + t - 2u)\mathbf{i} + (-10 + 2t - u)\mathbf{j} + (1 + t - 2u)\mathbf{k} = 0.$$

This tells us that
$$1 + t - 2u = 0,$$
$$-10 + 2t - u = 0,$$
$$1 + t - 2u = 0.$$

Note that the first and third equations are the same. Solving the first two equations simultaneously, we obtain
$$t = 7, \qquad u = 4.$$
As you can verify,
$$\mathbf{r}(7) = 8\mathbf{i} + 8\mathbf{j} + 9\mathbf{k} = \mathbf{R}(4).$$

The two lines intersect at the tip of this vector, which is the point $P(8, 8, 9)$. □

In the setting of plane geometry we can think of two lines as parallel iff they do not intersect. This point of view is not satisfactory in three-dimensional space where there are many lines which fail to intersect which yet we would hesitate to call parallel. See Figure 12.5.4. We can get around this difficulty by using direction vectors; namely, we will call two lines *parallel* iff their direction vectors are parallel.

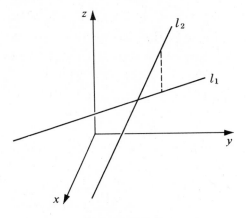

FIGURE 12.5.4

If two lines intersect, we can compute the angle between them by finding the angle between their direction vectors. Depending upon our choice of direction vectors ($\mathbf{d}$ or $-\mathbf{d}$, for example) there are two such angles, θ and $\pi - \theta$. We choose the smaller of the two, the one with nonnegative cosine.

Problem. Find the angle between

$$l_1: \mathbf{r}(t) = (\mathbf{i} - 6\mathbf{j} + 2\mathbf{k}) + t(\mathbf{i} + 2\mathbf{j} + \mathbf{k}) \quad \text{and} \quad l_2: \mathbf{R}(u) = (4\mathbf{j} + \mathbf{k}) + u(2\mathbf{i} + \mathbf{j} + 2\mathbf{k}).$$

SOLUTION.   As direction vectors we have

$$\mathbf{d} = \mathbf{i} + 2\mathbf{j} + \mathbf{k} \quad \text{and} \quad \mathbf{D} = 2\mathbf{i} + \mathbf{j} + 2\mathbf{k}.$$

As you can check,

$$\cos \theta = \frac{|\mathbf{d} \cdot \mathbf{D}|}{\|\mathbf{d}\| \, \|\mathbf{D}\|} = \frac{\sqrt{6}}{3} \cong 0.816$$

so that

$$\theta \cong 0.62 \text{ radians.} \quad \square$$

If two parallel vectors are both depicted by arrows emanating from the origin, then the two arrows will both fall on the same line. Because of this, parallel vectors are often called *collinear vectors*.

Exercises

*1. Given that

$$l: \mathbf{r}(t) = (\mathbf{i} + 2\mathbf{j}) + t(6\mathbf{i} + \mathbf{j} - 5\mathbf{k}),$$

determine which of the following points lie on l:

$$P(1, 2, 0), \quad Q(-5, 1, 5), \quad R(-4, 2, 5).$$

*2. Determine which of the following lines are parallel:

$$l_1: \mathbf{r}_1(t) = (\mathbf{i} + 2\mathbf{k}) + t(\mathbf{i} - 2\mathbf{j} + 3\mathbf{k}),$$
$$l_2: \mathbf{r}_2(u) = (\mathbf{i} + 2\mathbf{k}) + u(\mathbf{i} + 2\mathbf{j} - 3\mathbf{k}),$$
$$l_3: \mathbf{r}_3(v) = (6\mathbf{i} - \mathbf{j}) - v(2\mathbf{i} - 4\mathbf{j} + 6\mathbf{k}),$$
$$l_4: \mathbf{r}_4(w) = (\tfrac{1}{2} + \tfrac{1}{2}w)\mathbf{i} - w\mathbf{j} + (1 + \tfrac{3}{2}w)\mathbf{k}.$$

*3. The vector $\mathbf{d}$ serves as a direction vector for the line

$$l: \mathbf{r}(t) = \mathbf{r}_0 + t\mathbf{d}.$$

What other vectors could be used as direction vectors for this same line?

4. Find a vector parametrization for the line that
   *(a) contains $P(3, 1, 0)$ and is parallel to the line $\mathbf{r}(t) = (\mathbf{i} - \mathbf{j}) + t\mathbf{k}$.
   (b) contains $P(1, -1, 2)$ and is parallel to the line $\mathbf{r}(t) = t(3\mathbf{i} - \mathbf{j} + \mathbf{k})$.
   (c) contains the origin and $Q(x_1, y_1, z_1)$.
   *(d) contains $P(x_0, y_0, z_0)$ and $Q(x_1, y_1, z_1)$.

5. Find a set of scalar parametric equations for the line that
   *(a) contains $P(1, 0, 3)$ and $Q(2, -1, 4)$.
   (b) contains $P(x_0, y_0, z_0)$ and $Q(x_1, y_1, z_1)$.
   *(c) contains $P(2, -2, 3)$ and is perpendicular to the xz-plane.

6. Give a vector parametrization for the line that contains $P(-1, 2, -3)$ and is parallel to
$$\frac{x+1}{2} = \frac{y-3}{1} = \frac{z}{4}.$$

*7. Write a set of symmetric equations for the line that contains the origin and the point $P(x_0, y_0, z_0)$.

8. Where does the line
$$\frac{x-x_0}{d_1} = \frac{y-y_0}{d_2} = \frac{z-z_0}{d_3}$$

intersect the xy-plane?

9. Find the point where l_1 and l_2 intersect and give the angle of intersection.

(a) $l_1: \mathbf{r}_1(t) = \mathbf{i} + t\mathbf{j}$,
$\quad l_2: \mathbf{r}_2(u) = \mathbf{j} + u(\mathbf{i} + \mathbf{j})$.

*(b) $l_1: \mathbf{r}_1(t) = (\mathbf{i} - 4\sqrt{3}\mathbf{j}) + t(\mathbf{i} + \sqrt{3}\mathbf{j})$,
$\quad l_2: \mathbf{r}_2(u) = (4\mathbf{i} + 3\sqrt{3}\mathbf{j}) + u(\mathbf{i} - \sqrt{3}\mathbf{j})$.

(c) $l_1: \mathbf{r}_1(t) = -6\mathbf{k} + t(\mathbf{i} - \mathbf{j} + 2\mathbf{k})$,
$\quad l_2: \mathbf{r}_2(u) = \mathbf{i} + \mathbf{j} - u(\mathbf{i} - 3\mathbf{j} - 2\mathbf{k})$.

*(d) $l_1: x_1(t) = 3 + t, \quad y_1(t) = 1 - t, \quad z_1(t) = 5 + 2t$.
$\quad l_2: x_2(u) = 1, \quad y_2(u) = 4 + u, \quad z_2(u) = 2 + u$.

*10. What can you conclude about the lines
$$\frac{x-x_0}{d_1} = \frac{y-y_0}{d_2} = \frac{z-z_0}{d_3}, \qquad \frac{x-x_0}{D_1} = \frac{y-y_0}{D_2} = \frac{z-z_0}{D_3}$$

from knowing that
$$d_1D_1 + d_2D_2 + d_3D_3 = 0?$$

*11. What can you conclude about the lines
$$\frac{x-x_0}{d_1} = \frac{y-y_0}{d_2} = \frac{z-z_0}{d_3}, \qquad \frac{x-x_1}{D_1} = \frac{y-y_1}{D_2} = \frac{z-z_1}{D_3}$$

from knowing that
$$\frac{d_1}{D_1} = \frac{d_2}{D_2} = \frac{d_3}{D_3}?$$

12. Let P_0, P_1 be two distinct points and let $\mathbf{r}_0, \mathbf{r}_1$ be the position vectors that they determine:
$$\mathbf{r}_0 = \overrightarrow{OP_0}, \qquad \mathbf{r}_1 = \overrightarrow{OP_1}.$$

As t ranges over the set of real numbers,
$$\mathbf{r}(t) = \mathbf{r}_0 + t(\mathbf{r}_1 - \mathbf{r}_0)$$

traces out the line determined by P_0 and P_1. Restrict t so that $\mathbf{r}(t)$ traces out only the line segment P_0P_1.

*13. Given the line

$$l: \mathbf{r}(t) = \mathbf{r}_0 + t\mathbf{d},$$

(a) find the scalar t_0 for which the vector $\mathbf{r}(t_0)$ is perpendicular to l.

(b) find the parametrizations

$$\mathbf{R}(t) = \mathbf{R}_0 + t\mathbf{D}$$

for l in which $\mathbf{R}_0 \perp l$ and $\|\mathbf{D}\| = 1$.

*14. Use Exercise 13 to find the distance between the line

$$l: \mathbf{r}(t) = \mathbf{r}_0 + t\mathbf{d}$$

and the origin given that

(a) $\mathbf{r}_0 = \mathbf{j}$ and $\mathbf{d} = \mathbf{i} - \mathbf{j}$.

(b) $\mathbf{r}_0 = \sqrt{3}\mathbf{i}$ and $\mathbf{d} = \frac{1}{3}\sqrt{3}(\mathbf{i} + \mathbf{j} + \mathbf{k})$.

12.6   Planes

We derived the equation of a line in space by using a point from it and a nonzero vector *parallel* to it, called a *direction vector*. We can derive the equation of a plane in space by using a point from it and a nonzero vector *perpendicular* to it, called a *normal vector*.

Let the point be $P(x_0, y_0, z_0)$ and let the normal vector be

$$\mathbf{N} = A\mathbf{i} + B\mathbf{j} + C\mathbf{k}.$$

For convenience we depict $\mathbf{N}$ as an arrow starting at $P(x_0, y_0, z_0)$. See Figure 12.6.1. The point $Q(x, y, z)$ of the figure will lie on the given plane iff the normal

$$\mathbf{N} = A\mathbf{i} + B\mathbf{j} + C\mathbf{k}$$

is perpendicular to the vector

$$\overrightarrow{PQ} = (x - x_0)\mathbf{i} + (y - y_0)\mathbf{j} + (z - z_0)\mathbf{k}.$$

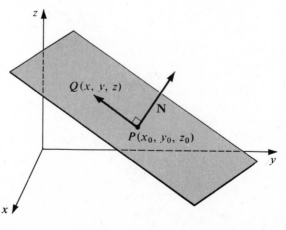

FIGURE 12.6.1

This happens iff

$$\mathbf{N} \cdot \overrightarrow{PQ} = 0$$

and consequently iff

(12.6.1)
$$\boxed{A(x - x_0) + B(y - y_0) + C(z - z_0) = 0.}$$

This is the usual xyz-equation for the plane that contains $P(x_0, y_0, z_0)$ and has normal $\mathbf{N} = A\mathbf{i} + B\mathbf{j} + C\mathbf{k}$. Note that the A, B, C of (12.6.1) are the components of the normal vector and hence direction numbers for the normal line.

Problem. Write an equation for the plane that contains $P(1, 0, 2)$ and has normal $\mathbf{N} = 3\mathbf{i} - 2\mathbf{j} + \mathbf{k}$.

SOLUTION. The general equation

$$A(x - x_0) + B(y - y_0) + C(z - z_0) = 0$$

becomes

$$3(x - 1) + (-2)(y - 0) + (1)(z - 2) = 0,$$

which we can simplify to
$$3x - 2y + z - 5 = 0. \quad \square$$

Problem. Find an equation for the plane that contains $P(1, 3, -2)$ and is perpendicular to the vector $\mathbf{i} - \mathbf{j}$.

SOLUTION. We can take $\mathbf{i} - \mathbf{j}$ as the normal:

$$\mathbf{N} = \mathbf{i} - \mathbf{j}.$$

The equation becomes

$$1(x - 1) + (-1)(y - 3) + 0(z + 2) = 0.$$

This simplifies to
$$x - y + 2 = 0.$$

This last equation looks very much like the equation of a line. If our domain of discourse were the xy-plane, then

$$x - y + 2 = 0$$

would represent a line. In this case, however, our domain of discourse is three-space. Here the equation
$$x - y + 2 = 0$$

represents the set of all points $Q(x, y, z)$, where $x - y + 2 = 0$ and z is unrestricted. This set forms a vertical plane which intersects the xy-plane in the line just mentioned. $\quad \square$

We can write the equation of a plane entirely in vector notation. With
$$\mathbf{N} = A\mathbf{i} + B\mathbf{j} + C\mathbf{k}$$
and
$$\mathbf{r}_0 = x_0\mathbf{i} + y_0\mathbf{j} + z_0\mathbf{k}, \qquad \mathbf{r} = x\mathbf{i} + y\mathbf{j} + z\mathbf{k},$$
Equation (12.6.1) becomes

(12.6.2)
$$\mathbf{N} \cdot (\mathbf{r} - \mathbf{r}_0) = 0.$$

This is the plane with normal $\mathbf{N}$ determined by the position vector $\mathbf{r}_0$. (See Figure 12.6.2.)

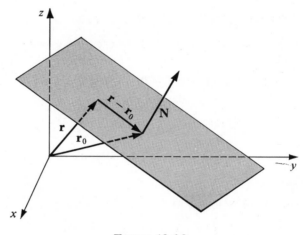

FIGURE 12.6.2

Three vectors $\mathbf{a}$, $\mathbf{b}$, $\mathbf{c}$ are said to be *coplanar* iff there are scalars s, t, u not all zero such that
$$s\mathbf{a} + t\mathbf{b} + u\mathbf{c} = 0.$$

Problem. Show that if $\mathbf{a}$, $\mathbf{b}$, $\mathbf{c}$ are coplanar then the points
$$P(a_1, a_2, a_3), \qquad Q(b_1, b_2, b_3), \qquad R(c_1, c_2, c_3)$$
all lie on the same plane through the origin.

SOLUTION. If $\mathbf{a}$, $\mathbf{b}$, $\mathbf{c}$ are coplanar then we can write
$$s\mathbf{a} + t\mathbf{b} + u\mathbf{c} = 0 \quad \text{with } s, t, u \text{ not all } 0.$$
If $s \neq 0$, then
$$\mathbf{a} = -\frac{t}{s}\mathbf{b} - \frac{u}{s}\mathbf{c}.$$

The plane through the origin that contains Q and R has a vector equation of the form

$$\mathbf{N} \cdot \mathbf{r} = 0. \qquad\qquad \text{(we take } \mathbf{r}_0 = \mathbf{0})$$

Here $\mathbf{N}$ is a normal vector and $\mathbf{r}$ is the position vector of any point of the plane. Since Q and R lie on the plane

$$\mathbf{N} \cdot \mathbf{b} = 0 \quad \text{and} \quad \mathbf{N} \cdot \mathbf{c} = 0.$$

Consequently

$$\mathbf{N} \cdot \mathbf{a} = \mathbf{N} \cdot \left(-\frac{t}{s}\mathbf{b} - \frac{u}{s}\mathbf{c} \right)$$

$$= -\frac{t}{s}(\mathbf{N} \cdot \mathbf{b}) - \frac{u}{s}(\mathbf{N} \cdot \mathbf{c})$$

$$= 0.$$

This means that $P(a_1, a_2, a_3)$ lies on this same plane. The cases $t \neq 0$ and $u \neq 0$ can be handled in a similar manner. $\square$

Problem. Show that every equation of the form

$$ax + by + cz + d = 0 \qquad \text{with} \qquad \sqrt{a^2 + b^2 + c^2} \neq 0$$

represents a plane in space.

SOLUTION. Let x_0, y_0, z_0 be a triple that satisfies the given equation. In other words, suppose that

$$ax_0 + by_0 + cz_0 + d = 0.$$

Equation

$$ax + by + cz + d = 0$$

can then be rewritten as

$$(ax + by + cz + d) - (ax_0 + by_0 + cz_0 + d) = 0$$

and therefore as

$$a(x - x_0) + b(y - y_0) + c(z - z_0) = 0.$$

This equation (and hence the initial equation) represents the plane which contains the point $P(x_0, y_0, z_0)$ and is perpendicular to the vector

$$\mathbf{N} = a\mathbf{i} + b\mathbf{j} + c\mathbf{k}. \quad \square$$

If $\mathbf{N}$ is normal to a given plane, then all other normals to that plane are parallel to $\mathbf{N}$ and hence scalar multiples of $\mathbf{N}$. In particular there are only two normals of length 1:

$$\frac{\mathbf{N}}{\|\mathbf{N}\|} \quad \text{and} \quad -\frac{\mathbf{N}}{\|\mathbf{N}\|}.$$

These are called the *unit normals.*

Problem. Find the unit normals for the plane

$$3x - 4y + 12z + 8 = 0.$$

SOLUTION. We can take

$$\mathbf{N} = 3\mathbf{i} - 4\mathbf{j} + 12\mathbf{k}.$$

Since

$$\|\mathbf{N}\| = \sqrt{(3)^2 + (-4)^2 + (12)^2} = \sqrt{169} = 13,$$

the unit normals are

$$\frac{\mathbf{N}}{\|\mathbf{N}\|} = \frac{1}{13}(3\mathbf{i} - 4\mathbf{j} + 12\mathbf{k})$$

and

$$-\frac{\mathbf{N}}{\|\mathbf{N}\|} = -\frac{1}{13}(3\mathbf{i} - 4\mathbf{j} + 12\mathbf{k}). \quad \square$$

Two planes are called *parallel* iff their normals are parallel. If two planes p_1, p_2 intersect, we can find the angle between them by finding the angle between their normals, $\mathbf{N}_1$, $\mathbf{N}_2$. (See Figure 12.6.3.) Depending on our choice of normals, there are two such angles, each the supplement of the other. We will choose the acute angle, the one with the nonnegative cosine:

$$\cos \theta = \frac{|\mathbf{N}_1 \cdot \mathbf{N}_2|}{\|\mathbf{N}_1\| \, \|\mathbf{N}_2\|}.$$

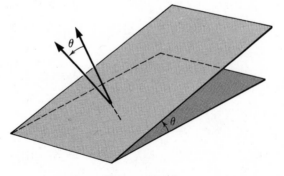

FIGURE 12.6.3

Problem. Let

$$p_1 : 2(x - 1) - 3y + 5(z - 2) = 0,$$
$$p_2 : -4x + 6y + 10z + 24 = 0,$$
$$p_3 : 4x - 6y - 10z + 1 = 0,$$
$$p_4 : 2x - 3y + 5z - 12 = 0.$$

(a) Indicate which planes are identical.
(b) Indicate which planes are distinct but parallel.
(c) Find the angle between p_1 and p_2.

Solution

(a) p_1 and p_4 are identical, as you can verify by simplifying the equation of p_1.
(b) p_2 and p_3 are distinct but parallel. The planes are distinct since $P(0, 0, \frac{1}{10})$ lies
    on p_3 but not on p_2. They are parallel since the normals

$$-4\mathbf{i} + 6\mathbf{j} + 10\mathbf{k} \quad \text{and} \quad 4\mathbf{i} - 6\mathbf{j} + 10\mathbf{k}$$

are parallel.
(c) As normals we can take

$$\mathbf{N}_1 = 2\mathbf{i} - 3\mathbf{j} + 5\mathbf{k} \quad \text{and} \quad \mathbf{N}_2 = -4\mathbf{i} + 6\mathbf{j} + 10\mathbf{k}.$$

$$\cos \theta = \frac{|\mathbf{N}_1 \cdot \mathbf{N}_2|}{\|\mathbf{N}_1\| \, \|\mathbf{N}_2\|} = \frac{|-8 - 18 + 50|}{\sqrt{4 + 9 + 25}\sqrt{16 + 36 + 100}} = \frac{6}{19} \cong 0.316.$$

The angle between the planes is about 1.25 radians. □

Problem. Show that the distance from the point $P_1(x_1, y_1, z_1)$ to the plane
$Ax + By + Cz + D = 0$ is

$$\frac{|Ax_1 + By_1 + Cz_1 + D|}{\sqrt{A^2 + B^2 + C^2}}.$$

Solution. As in Figure 12.6.4 we drop a perpendicular from P_1 to the plane in
question. This perpendicular will intersect the plane at some point P_2. The distance
we want is the length of the vector $\overrightarrow{P_2 P_1}$.

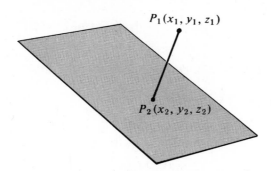

$P_1(x_1, y_1, z_1)$

$P_2(x_2, y_2, z_2)$

Figure 12.6.4

Since

$$\overrightarrow{P_2 P_1} = (x_1 - x_2)\mathbf{i} + (y_1 - y_2)\mathbf{j} + (z_1 - z_2)\mathbf{k}$$

is parallel to the normal

$$\mathbf{N} = A\mathbf{i} + B\mathbf{j} + C\mathbf{k},$$

we know that

$$\overrightarrow{P_2 P_1} = t\mathbf{N} \quad \text{for some scalar } t.$$

To find t, we write this vector equation in terms of its components:

$$x_1 - x_2 = tA, \qquad y_1 - y_2 = tB, \qquad z_1 - z_2 = tC.$$

Multiplying the first equation by A, the second by B, and the third by C, we get

$$A(x_1 - x_2) = tA^2, \qquad B(y_1 - y_2) = tB^2, \qquad C(z_1 - z_2) = tC^2.$$

Adding these equations and noting that

$$Ax_2 + By_2 + Cz_2 + D = 0, \qquad (P_2 \text{ lies on the plane})$$

we have

$$Ax_1 + By_1 + Cz_1 + D = t(A^2 + B^2 + C^2)$$

so that

$$t = \frac{Ax_1 + By_1 + Cz_1 + D}{A^2 + B^2 + C^2}.$$

Since

$$\|\mathbf{N}\| = \sqrt{A^2 + B^2 + C^2},$$

you can see that

$$\|\overrightarrow{P_2P_1}\| = \|t\mathbf{N}\| = |t| \, \|\mathbf{N}\| = \frac{|Ax_1 + By_1 + Cz_1 + D|}{\sqrt{A^2 + B^2 + C^2}}. \qquad \square$$

Exercises

* 1. Which of the points
$$P(3, 2, 1), \; Q(2, 3, 1), \; R(1, 4, 1)$$
lie on the plane
$$3(x - 1) + 4y - 5(z + 2) = 0?$$

* 2. Which of the points
$$P(2, 1, -2), \; Q(2, 0, 0), \; R(4, 1, -1), \; S(0, -1, -3)$$
lie on the plane
$$\mathbf{N} \cdot (\mathbf{r} - \mathbf{r}_0) = 0$$
if
$$\mathbf{N} = \mathbf{i} - 3\mathbf{j} + \mathbf{k} \quad \text{and} \quad \mathbf{r}_0 = 4\mathbf{i} + \mathbf{j} - \mathbf{k}?$$

* 3. Write an equation for the plane that contains the point $P(2, 3, 4)$ and is perpendicular to $\mathbf{i} - 4\mathbf{j} + 3\mathbf{k}$.

4. Write an equation for the plane that contains the point $P(1, -2, 3)$ and is perpendicular to $\mathbf{j} + 2\mathbf{k}$.

* 5. Write an equation for the plane that is parallel to the plane
$$3x - 2y + 5z - 2 = 0$$
and contains the point $(2, 1, 1)$.

6. Write an equation for the plane that contains $P_0(x_0, y_0, z_0)$ and is perpendicular to $\overrightarrow{OP_0}$.

7. Show that the plane

$$\frac{x}{a} + \frac{y}{b} + \frac{z}{c} = 1$$

intersects the coordinate axes at $x = a$, $y = b$, $z = c$. This is called the *intercept form* of the equation of a plane.

8. Write the equation of the plane

$$4x + 5y - 6z = 60$$

in intercept form.

*9. Where does the plane

$$3x - y + 4z + 2 = 0$$

intersect the coordinate axes?

*10. Find the unit normals for the plane

$$2x - 3y + 7z - 3 = 0.$$

Find the angle between the planes:

*11. $5(x - 1) - 3(y + 2) + 2z = 0,$
       $x + 3(y - 1) + 2(z + 4) = 0.$

12. $2x - y + 3z = 5,$
      $5x + 5y - z = 1.$

*13. $x - y + z - 1 = 0,$
       $2x + y + 3z + 5 = 0.$

Determine whether or not the following sets of vectors are coplanar.

14. $\mathbf{i}, \mathbf{i} - 2\mathbf{j}, 3\mathbf{j} + \mathbf{k}$.

*15. $4\mathbf{j} - \mathbf{k}, 3\mathbf{i} + \mathbf{j} + 2\mathbf{k}, \mathbf{0}$.

16. $\mathbf{j} - \mathbf{k}, 3\mathbf{i} - \mathbf{j} + 2\mathbf{k}, 3\mathbf{i} - 2\mathbf{j} + 3\mathbf{k}$.

*17. $\mathbf{i} + \mathbf{j} + \mathbf{k}, 2\mathbf{i} - \mathbf{j}, 3\mathbf{i} - \mathbf{j} - \mathbf{k}$.

18. Find the distance from the point $P(2, -1, 3)$ to the plane

$$2x + 4y - z + 1 = 0.$$

*19. Find the distance between two parallel planes

$$Ax + By + Cz + D_1 = 0 \quad \text{and} \quad Ax + By + Cz + D_2 = 0.$$

20. Show that two nonparallel lines

$$\mathbf{r}(t) = \mathbf{a} + t\mathbf{b}$$

and

$$\mathbf{R}(u) = \mathbf{A} + u\mathbf{B}$$

intersect iff the vectors $\mathbf{a} - \mathbf{A}$, $\mathbf{b}$, and $\mathbf{B}$ are coplanar.

12.7   The Cross Product

Here we explore a notion that should be of special value to those of you who have an interest in physics, particularly in mechanics and in electromagnetism. In mechanics cross products come up in connection with angular momentum, torque, and other phenomena of turning. In electromagnetism cross products play an important role. In particular they give us a convenient formulation of the law of forces between moving charges.

For vectors

$$\mathbf{a} = a_1\mathbf{i} + a_2\mathbf{j} + a_3\mathbf{k} \quad \text{and} \quad \mathbf{b} = b_1\mathbf{i} + b_2\mathbf{j} + b_3\mathbf{k},$$

we define the *cross product* $\mathbf{a} \times \mathbf{b}$ by

$$\mathbf{a} \times \mathbf{b} = (a_2b_3 - a_3b_2)\mathbf{i} + (a_3b_1 - a_1b_3)\mathbf{j} + (a_1b_2 - a_2b_1)\mathbf{k}.$$

While the dot product $\mathbf{a} \cdot \mathbf{b}$ is a scalar and is often called the scalar product of $\mathbf{a}$ and $\mathbf{b}$, the cross product $\mathbf{a} \times \mathbf{b}$ is a vector and it is often called the *vector product* of $\mathbf{a}$ and $\mathbf{b}$.

Examples.   For

$$\mathbf{a} = \mathbf{i} - 2\mathbf{j} + 3\mathbf{k} \quad \text{and} \quad \mathbf{b} = 2\mathbf{i} + \mathbf{j} - \mathbf{k}$$

we have

$$\begin{aligned}
\mathbf{a} \times \mathbf{b} &= (a_2b_3 - a_3b_2)\mathbf{i} + (a_3b_1 - a_1b_3)\mathbf{j} + (a_1b_2 - a_2b_1)\mathbf{k} \\
&= [(-2)(-1) - (3)(1)]\mathbf{i} + [(3)(2) - (1)(-1)]\mathbf{j} + [(1)(1) - (-2)(2)]\mathbf{k} \\
&= -\mathbf{i} + 7\mathbf{j} + 5\mathbf{k}. \quad \square
\end{aligned}$$

For

$$\mathbf{a} = \mathbf{i} - \mathbf{j} \quad \text{and} \quad \mathbf{b} = \mathbf{i} + \mathbf{k},$$

we have

$$\begin{aligned}
\mathbf{a} \times \mathbf{b} &= (a_2b_3 - a_3b_2)\mathbf{i} + (a_3b_1 - a_1b_3)\mathbf{j} + (a_1b_2 - a_2b_1)\mathbf{k} \\
&= [(-1)(1) - (0)(0)]\mathbf{i} + [(0)(1) - (1)(1)]\mathbf{j} + [(1)(0) - (-1)(1)]\mathbf{k} \\
&= -\mathbf{i} - \mathbf{j} + \mathbf{k}. \quad \square
\end{aligned}$$

If you are familiar with determinants, you will find it easy to remember the defining formula by noting that

$$\mathbf{a} \times \mathbf{b} = \begin{vmatrix} \mathbf{i} & \mathbf{j} & \mathbf{k} \\ a_1 & a_2 & a_3 \\ b_1 & b_2 & b_3 \end{vmatrix} = \begin{vmatrix} a_2 & a_3 \\ b_2 & b_3 \end{vmatrix}\mathbf{i} - \begin{vmatrix} a_1 & a_3 \\ b_1 & b_3 \end{vmatrix}\mathbf{j} + \begin{vmatrix} a_1 & a_2 \\ b_1 & b_2 \end{vmatrix}\mathbf{k}.$$

Straightforward calculations that you can carry out yourself show that the cross product has the following properties:

$$\mathbf{b} \times \mathbf{a} = -(\mathbf{a} \times \mathbf{b}).$$ (anticommutative)

$$\mathbf{a} \times \mathbf{a} = \mathbf{0}.$$ (self-annihilating)

$$\alpha \mathbf{a} \times \beta \mathbf{b} = \alpha\beta(\mathbf{a} \times \mathbf{b}).$$ (scalars can be factored out)

$$\mathbf{a} \times (\mathbf{b} + \mathbf{c}) = (\mathbf{a} \times \mathbf{b}) + (\mathbf{a} \times \mathbf{c}).$$ ($\times$ distributes over $+$)

$$\mathbf{a} \times \mathbf{b} \perp \mathbf{a}, \qquad \mathbf{a} \times \mathbf{b} \perp \mathbf{b}.$$

$$\|\mathbf{a} \times \mathbf{b}\|^2 = \|\mathbf{a}\|^2 \|\mathbf{b}\|^2 - (\mathbf{a} \cdot \mathbf{b})^2.$$

We focus on the last two properties. The first of these tells us that $\mathbf{a} \times \mathbf{b}$ is perpendicular both to $\mathbf{a}$ and to $\mathbf{b}$. The second one gives us the length of $\mathbf{a} \times \mathbf{b}$. From it we see that if $\mathbf{a} = \mathbf{0}$ or $\mathbf{b} = \mathbf{0}$, then $\mathbf{a} \times \mathbf{b} = \mathbf{0}$. If, on the other hand, neither $\mathbf{a}$ nor $\mathbf{b}$ is $\mathbf{0}$, then we have

$$\mathbf{a} \cdot \mathbf{b} = \|\mathbf{a}\| \|\mathbf{b}\| \cos \theta,$$

so that

$$\begin{aligned}
\|\mathbf{a} \times \mathbf{b}\|^2 &= \|\mathbf{a}\|^2 \|\mathbf{b}\|^2 - (\mathbf{a} \cdot \mathbf{b})^2 \\
&= \|\mathbf{a}\|^2 \|\mathbf{b}\|^2 - \|\mathbf{a}\|^2 \|\mathbf{b}\|^2 \cos^2 \theta \\
&= \|\mathbf{a}\|^2 \|\mathbf{b}\|^2 (1 - \cos^2 \theta) \\
&= \|\mathbf{a}\|^2 \|\mathbf{b}\|^2 \sin^2 \theta,
\end{aligned}$$

and, taking square roots

$$\|\mathbf{a} \times \mathbf{b}\| = \|\mathbf{a}\| \|\mathbf{b}\| \sin \theta.$$

This means (see Figure 12.7.1) that the length of $\mathbf{a} \times \mathbf{b}$ is the area of the parallelogram generated by $\mathbf{a}$ and $\mathbf{b}$.

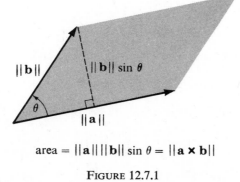

$$\text{area} = \|\mathbf{a}\| \|\mathbf{b}\| \sin \theta = \|\mathbf{a} \times \mathbf{b}\|$$

Figure 12.7.1

From the formula

$$\|\mathbf{a} \times \mathbf{b}\| = \|\mathbf{a}\| \|\mathbf{b}\| \sin \theta,$$

it follows that

$$\mathbf{a} \times \mathbf{b} = \mathbf{0} \quad \text{iff} \quad \mathbf{a} \text{ and } \mathbf{b} \text{ are parallel.}$$

We know that the cross product of $\mathbf{a}$ and $\mathbf{b}$ is perpendicular to $\mathbf{a}$, perpendicular to $\mathbf{b}$, and that it has length $\|\mathbf{a}\| \|\mathbf{b}\| \sin \theta$. Does this determine it completely? The answer is no. In general there are two such vectors, each the negative of the other. It can be argued that the vector we want, $\mathbf{a} \times \mathbf{b}$, is the one which makes

$$\mathbf{a}, \mathbf{b}, \mathbf{a} \times \mathbf{b}$$

a right-handed system. (See Figure 12.7.2.)

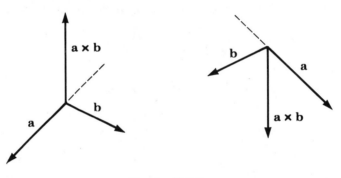

FIGURE 12.7.2

As you can verify,

$$\mathbf{i} \times \mathbf{j} = \mathbf{k}, \quad \mathbf{j} \times \mathbf{k} = \mathbf{i}, \quad \mathbf{k} \times \mathbf{i} = \mathbf{j}. \qquad \text{(Figure 12.7.3)}$$

An easy way to remember these products is to arrange $\mathbf{i}, \mathbf{j}, \mathbf{k}$ in cyclic order,

$$\mathbf{i} \to \mathbf{j} \to \mathbf{k},$$

and note that

$$[\text{each unit vector}] \times [\text{the next one}] = [\text{the third one}].$$

From this observation and properties

$$\mathbf{b} \times \mathbf{a} = -(\mathbf{a} \times \mathbf{b}),$$
$$\mathbf{a} \times \mathbf{a} = \mathbf{0},$$
$$\alpha\mathbf{a} \times \beta\mathbf{b} = \alpha\beta(\mathbf{a} \times \mathbf{b}),$$
$$\mathbf{a} \times (\mathbf{b} + \mathbf{c}) = (\mathbf{a} \times \mathbf{b}) + (\mathbf{a} \times \mathbf{c}),$$

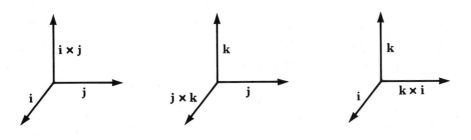

FIGURE 12.7.3

we can compute cross products without having to appeal to the somewhat complicated defining formula.

Examples

$$\mathbf{j} \times \mathbf{i} = -(\mathbf{i} \times \mathbf{j}) = -\mathbf{k}, \quad \mathbf{k} \times \mathbf{j} = -(\mathbf{j} \times \mathbf{k}) = -\mathbf{i},$$

$$\mathbf{i} \times \mathbf{k} = -(\mathbf{k} \times \mathbf{i}) = -\mathbf{j}.$$

$$\mathbf{i} \times \mathbf{i} = \mathbf{j} \times \mathbf{j} = \mathbf{k} \times \mathbf{k} = \mathbf{0}.$$

$$2\mathbf{i} \times 3\mathbf{k} = 6(\mathbf{i} \times \mathbf{k}) = -6(\mathbf{k} \times \mathbf{i}) = -6\mathbf{j}.$$

$$\mathbf{k} \times (-3\mathbf{j} + 2\mathbf{k}) = -3(\mathbf{k} \times \mathbf{j}) + 2(\mathbf{k} \times \mathbf{k}) = 3(\mathbf{j} \times \mathbf{k}) + \mathbf{0} = 3\mathbf{i}.$$

$$\begin{aligned}
(\mathbf{i} + \mathbf{j}) \times (-2\mathbf{j} + \mathbf{k}) &= [(\mathbf{i} + \mathbf{j}) \times (-2\mathbf{j})] + [(\mathbf{i} + \mathbf{j}) \times \mathbf{k}] \\
&= [2\mathbf{j} \times (\mathbf{i} + \mathbf{j})] - (\mathbf{k} \times (\mathbf{i} + \mathbf{j})] \\
&= 2(\mathbf{j} \times \mathbf{i}) + 2(\mathbf{j} \times \mathbf{j}) - (\mathbf{k} \times \mathbf{i}) - (\mathbf{k} \times \mathbf{j}) \\
&= -2(\mathbf{i} \times \mathbf{j}) + \mathbf{0} - \mathbf{j} + (\mathbf{j} \times \mathbf{k}) \\
&= -2\mathbf{k} - \mathbf{j} + \mathbf{i} = \mathbf{i} - \mathbf{j} - 2\mathbf{k}. \quad \square
\end{aligned}$$

Problem. Show that

$$(\mathbf{a} + \mathbf{b}) \times \mathbf{c} = (\mathbf{a} \times \mathbf{c}) + (\mathbf{b} \times \mathbf{c}).$$

SOLUTION

$$\begin{aligned}
(\mathbf{a} + \mathbf{b}) \times \mathbf{c} &= -[\mathbf{c} \times (\mathbf{a} + \mathbf{b})] \\
&= -[(\mathbf{c} \times \mathbf{a}) + (\mathbf{c} \times \mathbf{b})] \\
&= -(\mathbf{c} \times \mathbf{a}) - (\mathbf{c} \times \mathbf{b}) \\
&= (\mathbf{a} \times \mathbf{c}) + (\mathbf{b} \times \mathbf{c}). \quad \square
\end{aligned}$$

We use this last result in the next problem.

Problem. Compute $\mathbf{a} \times \mathbf{b}$ given that

$$\mathbf{a} = 2\mathbf{i} - \mathbf{k}, \quad \mathbf{b} = \mathbf{i} - 3\mathbf{j} + 2\mathbf{k}.$$

SOLUTION

$$\mathbf{a} \times \mathbf{b} = (2\mathbf{i} - \mathbf{k}) \times (\mathbf{i} - 3\mathbf{j} + 2\mathbf{k})$$
$$= [2\mathbf{i} \times (\mathbf{i} - 3\mathbf{j} + 2\mathbf{k})] - [\mathbf{k} \times (\mathbf{i} - 3\mathbf{j} + 2\mathbf{k})]$$
$$= 2(\mathbf{i} \times \mathbf{i}) - 6(\mathbf{i} \times \mathbf{j}) + 4(\mathbf{i} \times \mathbf{k}) - (\mathbf{k} \times \mathbf{i}) + 3(\mathbf{k} \times \mathbf{j}) - 2(\mathbf{k} \times \mathbf{k})$$
$$= 0 - 6\mathbf{k} - 4\mathbf{j} - \mathbf{j} - 3\mathbf{i} - 0$$
$$= -3\mathbf{i} - 5\mathbf{j} - 6\mathbf{k}. \quad \square$$

Problem. Find the volume of the parallelepiped with edges **a**, **b**, **c** pictured in Figure 12.7.4.

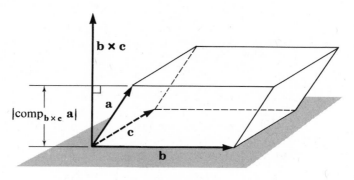

FIGURE 12.7.4

SOLUTION.   As a base we take the parallelogram with edges **b** and **c**. The area of the base is $\|\mathbf{b} \times \mathbf{c}\|$. Since $\mathbf{b} \times \mathbf{c}$ is perpendicular to the base, the height is

$$|\text{comp}_{\mathbf{b} \times \mathbf{c}}\, \mathbf{a}| = \frac{|\mathbf{a} \cdot (\mathbf{b} \times \mathbf{c})|}{\|\mathbf{b} \times \mathbf{c}\|} .$$

Consequently

$$V = (\text{area of base}) (\text{height}) = |\mathbf{a} \cdot (\mathbf{b} \times \mathbf{c})|. \quad \square$$

The number $\mathbf{a} \cdot (\mathbf{b} \times \mathbf{c})$ that arises in the last problem is called the *triple scalar product* of the vectors **a**, **b**, **c**. In terms of coordinates we have

$$\mathbf{a} \cdot (\mathbf{b} \times \mathbf{c}) = \mathbf{a} \cdot [(b_2 c_3 - b_3 c_2)\mathbf{i} + (b_3 c_1 - b_1 c_3)\mathbf{j} + (b_1 c_2 - b_2 c_1)\mathbf{k}]$$

and thus

(12.7.1) $\mathbf{a} \cdot (\mathbf{b} \times \mathbf{c}) = a_1(b_2 c_3 - b_3 c_2) + a_2(b_3 c_1 - b_1 c_3) + a_3(b_1 c_2 - b_2 c_1).$

If you are familiar with determinants, you have probably recognized that we have

$$\mathbf{a} \cdot (\mathbf{b} \times \mathbf{c}) = \begin{vmatrix} a_1 & a_2 & a_3 \\ b_1 & b_2 & b_3 \\ c_1 & c_2 & c_3 \end{vmatrix} .$$

Problem.   Given that

$$\mathbf{a} \cdot (\mathbf{b} \times \mathbf{c}) = 0$$

what can you conclude about $\mathbf{a}$, $\mathbf{b}$, and $\mathbf{c}$?

SOLUTION.   If $\mathbf{a} \cdot (\mathbf{b} \times \mathbf{c}) = 0$, then the parallelepiped with edges $\mathbf{a}$, $\mathbf{b}$, $\mathbf{c}$ has volume 0. This can only happen if $\mathbf{a}$, $\mathbf{b}$, and $\mathbf{c}$ are coplanar.   □

From the last problem you know that

(12.7.2) $\mathbf{a} \cdot (\mathbf{b} \times \mathbf{c}) = 0$ iff $\mathbf{a}, \mathbf{b}, \mathbf{c}$ are coplanar.

We can use this to find a convenient equation for the plane that passes through three noncollinear points P_1, P_2, P_3. A point P will lie on this plane iff the arrows

$$\overrightarrow{P_1P}, \ \overrightarrow{P_1P_2}, \ \overrightarrow{P_1P_3}$$

are coplanar. By (12.7.2) this will happen only if

(12.7.3) $\overrightarrow{P_1P} \cdot (\overrightarrow{P_1P_2} \times \overrightarrow{P_1P_3}) = 0.$

Problem.   Find an equation in x, y, z for the plane that passes through $P_1(0, 1, 1)$, $P_2(1, 0, 1)$, $P_3(1, 1, 0)$.

SOLUTION.   The point $P = P(x, y, z)$ will lie on this plane only if

$$\overrightarrow{P_1P} \cdot (\overrightarrow{P_1P_2} \times \overrightarrow{P_1P_3}) = 0.$$

Here

$$\overrightarrow{P_1P} = x\mathbf{i} + (y - 1)\mathbf{j} + (z - 1)\mathbf{k}, \qquad \overrightarrow{P_1P_2} = \mathbf{i} - \mathbf{j}, \qquad \overrightarrow{P_1P_3} = \mathbf{i} - \mathbf{k}.$$

A calculation that you can carry out using (12.7.1) shows that

$$x + y + z - 2 = 0. \quad \square$$

Exercises

Compute.

*1. $(\mathbf{i} + \mathbf{j}) \times (\mathbf{i} - \mathbf{j})$.

*3. $(\mathbf{i} - \mathbf{j}) \times (\mathbf{j} - \mathbf{k})$.

*5. $(2\mathbf{j} - \mathbf{k}) \times (\mathbf{i} - 3\mathbf{j})$.

*7. $\mathbf{j} \cdot (\mathbf{i} \times \mathbf{k})$.

*9. $(\mathbf{i} \times \mathbf{j}) \times \mathbf{k}$.

2. $(\mathbf{i} - \mathbf{j}) \times (\mathbf{j} - \mathbf{i})$.

4. $\mathbf{j} \times (2\mathbf{i} - \mathbf{k})$.

6. $\mathbf{i} \cdot (\mathbf{j} \times \mathbf{k})$.

8. $(\mathbf{j} \times \mathbf{i}) \cdot (\mathbf{i} \times \mathbf{k})$.

10. $(\mathbf{j} \times \mathbf{k}) \times \mathbf{i}$.

Compute.

*11. $(\mathbf{i} + \mathbf{j} + \mathbf{k}) \times (2\mathbf{i} + \mathbf{k})$.

12. $(2\mathbf{i} - \mathbf{k}) \times (\mathbf{i} - 2\mathbf{j} + 2\mathbf{k})$.

*13. $[2\mathbf{i} + \mathbf{j}] \cdot [(\mathbf{i} - 3\mathbf{j} + \mathbf{k}) \times (4\mathbf{i} + \mathbf{k})]$.

14. $[(-2\mathbf{i} + \mathbf{j} - 3\mathbf{k}) \times \mathbf{i}] \times [\mathbf{i} + \mathbf{j}]$.

*15. $[(\mathbf{i} - \mathbf{j}) \times (\mathbf{j} - \mathbf{k})] \times [\mathbf{i} + 5\mathbf{k}]$.

16. $[\mathbf{i} - \mathbf{j}] \times [(\mathbf{j} - \mathbf{k}) \times (\mathbf{j} + 5\mathbf{k})]$.

17. Show that
$$\mathbf{b} \times \mathbf{a} = -(\mathbf{a} \times \mathbf{b})$$
and deduce from it that
$$\mathbf{a} \times \mathbf{a} = \mathbf{0}.$$

18. Show that
$$\mathbf{a} \times (\mathbf{b} + \mathbf{c}) = (\mathbf{a} \times \mathbf{b}) + (\mathbf{a} \times \mathbf{c}).$$

19. Show that
$$\mathbf{a} \times \mathbf{b} \perp \mathbf{a} \quad \text{and} \quad \mathbf{a} \times \mathbf{b} \perp \mathbf{b}.$$

20. Show that
$$\|\mathbf{a} \times \mathbf{b}\|^2 = \|\mathbf{a}\|^2 \|\mathbf{b}\|^2 - (\mathbf{a} \cdot \mathbf{b})^2.$$

21. Is $\times$ associative? Namely, is it always true that
$$\mathbf{a} \times (\mathbf{b} \times \mathbf{c}) = (\mathbf{a} \times \mathbf{b}) \times \mathbf{c}?$$

*22. Use a cross product to find the area of triangle PQR.

    (a) $P(0, 1, 0)$, $Q(-1, 1, 2)$, $R(2, 1, -1)$.

    (b) $P(1, 2, 3)$, $Q(-1, 3, 2)$, $R(3, -1, 2)$.

*23. Find the volume of the parallelepiped with edges

    (a) $\mathbf{i} + \mathbf{j}$, $2\mathbf{i} - \mathbf{k}$, $3\mathbf{j} + \mathbf{k}$.

    (b) $\mathbf{i} - 3\mathbf{j} + \mathbf{k}$, $2\mathbf{j} - \mathbf{k}$, $\mathbf{i} + \mathbf{j} - 2\mathbf{k}$.

24. Express
$$(\mathbf{a} + \mathbf{b}) \times (\mathbf{a} - \mathbf{b})$$
as a scalar multiple of $\mathbf{a} \times \mathbf{b}$.

*25. Find an equation in x, y, z for the plane that passes through the points

    (a) $P_1(1, 0, 1)$, $P_2(2, 1, 0)$, $P_3(1, 1, 1)$.

    (b) $P_1(1, 1, 1)$, $P_2(2, -2, -1)$, $P_3(0, 2, 1)$.

26. Let **a**, **b**, **c** be distinct nonzero vectors. Show that

$$\mathbf{a} \times \mathbf{b} = \mathbf{a} \times \mathbf{c} \quad \text{iff} \quad \mathbf{a} \text{ and } \mathbf{b} - \mathbf{c} \text{ are parallel.}$$

27. Sketch a figure depicting all those vectors **c** that satisfy

$$\mathbf{a} \times \mathbf{b} = \mathbf{a} \times \mathbf{c}.$$

28. Some authors write

$$\mathbf{a} \times \mathbf{b} = \|\mathbf{a}\| \|\mathbf{b}\| \sin \theta \, \mathbf{u},$$

where θ is the angle between **a** and **b** and **u** is a unit vector. Find the vector u which makes the equation correct.

12.8   Some Geometry by Vector Methods

Try your hand at proving the following theorems by vector methods. Follow the hints if you like, but you may find it more interesting to disregard them and come up with proofs that are entirely your own.

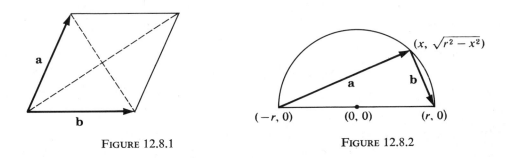

FIGURE 12.8.1   FIGURE 12.8.2

Proposition 1

The diagonals of a parallelogram are perpendicular iff the parallelogram is a rhombus.

HINT FOR PROOF. With **a** and **b** as in Figure 12.8.1 the diagonals are $\mathbf{a} + \mathbf{b}$ and $\mathbf{a} - \mathbf{b}$. Show that

$$(\mathbf{a} + \mathbf{b}) \cdot (\mathbf{a} - \mathbf{b}) = 0 \quad \text{iff} \quad \|\mathbf{a}\| = \|\mathbf{b}\|. \quad \square$$

Proposition 2

Every angle inscribed in a semicircle is a right angle.

HINT FOR PROOF. Take **a** and **b** as in Figure 12.8.2 and show that

$$\mathbf{a} \cdot \mathbf{b} = 0. \quad \square$$

Proposition 3

In a parallelogram the sum of the squares of the lengths of the diagonals equals the sum of the squares of the lengths of the sides.

HINT FOR PROOF. With **a** and **b** as in Figure 12.8.1 the diagonals are $\mathbf{a} + \mathbf{b}$ and $\mathbf{a} - \mathbf{b}$. Show that

$$\|\mathbf{a} + \mathbf{b}\|^2 + \|\mathbf{a} - \mathbf{b}\|^2 = 2\|\mathbf{a}\|^2 + 2\|\mathbf{b}\|^2. \quad \square$$

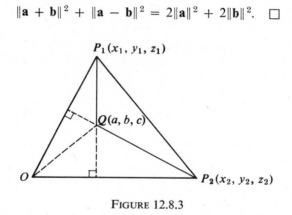

FIGURE 12.8.3

Proposition 4

The three altitudes of a triangle meet at one point.

HINT FOR PROOF. As in Figure 12.8.3 assume that the altitudes drawn from P_1 and P_2 intersect at Q. Use the fact that

$$\overrightarrow{P_1Q} \perp \overrightarrow{OP_2} \quad \text{and} \quad \overrightarrow{P_2Q} \perp \overrightarrow{OP_1}$$

to show that

$$\overrightarrow{OQ} \perp \overrightarrow{P_1P_2}. \quad \square$$

Proposition 5

The three medians of a triangle meet at one point.

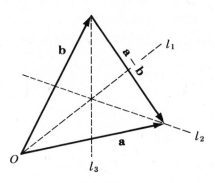

FIGURE 12.8.4

HINT FOR SOLUTION.　With l_1, l_2, l_3 as in Figure 12.8.4,

$$l_1: \mathbf{r}_1(t) = t(\mathbf{a} + \mathbf{b}),$$
$$l_2: \mathbf{r}_2(u) = \tfrac{1}{2}\mathbf{b} + u(\mathbf{a} - \tfrac{1}{2}\mathbf{b}),$$
$$l_3: \mathbf{r}_3(v) = \tfrac{1}{2}\mathbf{a} + v(\mathbf{b} - \tfrac{1}{2}\mathbf{a}).$$

Show that l_1 intersects both l_2 and l_3 at the same point.　□

Proposition 6　(The law of sines)

If a triangle has sides $\mathbf{a}$, $\mathbf{b}$, $\mathbf{c}$ and opposite angles A, B, C, then

$$\frac{\sin A}{\|\mathbf{a}\|} = \frac{\sin B}{\|\mathbf{b}\|} = \frac{\sin C}{\|\mathbf{c}\|}.$$

HINT FOR PROOF.　With $\mathbf{a}$, $\mathbf{b}$, $\mathbf{c}$ as in Figure 12.8.5

$$\mathbf{a} + \mathbf{b} + \mathbf{c} = \mathbf{0}.$$

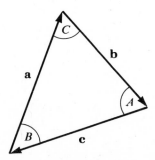

FIGURE 12.8.5

Observe that

$$\mathbf{a} \times [\mathbf{a} + \mathbf{b} + \mathbf{c}] = \mathbf{0} \quad \text{and} \quad \mathbf{b} \times [\mathbf{a} + \mathbf{b} + \mathbf{c}] = \mathbf{0}. \quad \square$$

Proposition 7

If two distinct planes have a point in common, then they have a line in common.

HINT FOR PROOF. If the point $P(a_1, a_2, a_3)$ lies on both planes

$$\mathbf{n} \cdot (\mathbf{r} - \mathbf{r}_0) = 0 \quad \text{and} \quad \mathbf{N} \cdot (\mathbf{R} - \mathbf{R}_0) = 0,$$

then the vector

$$\mathbf{a} = a_1\mathbf{i} + a_2\mathbf{j} + a_3\mathbf{k}$$

satisfies both conditions

$$\mathbf{n} \cdot (\mathbf{a} - \mathbf{r}_0) = 0 \quad \text{and} \quad \mathbf{N} \cdot (\mathbf{a} - \mathbf{R}_0) = 0.$$

Consider the line

$$\mathbf{r}(t) = \mathbf{a} + t(\mathbf{n} \times \mathbf{N}). \quad \square$$

Vector Calculus

13.1 Vector Functions

If f_1, f_2, f_3 are real-valued functions defined on some interval I, then for each $t \in I$ we can form the vector

$$\mathbf{f}(t) = f_1(t)\mathbf{i} + f_2(t)\mathbf{j} + f_3(t)\mathbf{k}$$

and thereby create a *vector-valued function* $\mathbf{f}$. For short we will call such a function a *vector function*.

Example. Taking

$$f_1(t) = x_0 + d_1 t, \qquad f_2(t) = y_0 + d_2 t, \qquad f_3(t) = z_0 + d_3 t$$

we can form the vector function

$$\mathbf{f}(t) = (x_0 + d_1 t)\mathbf{i} + (y_0 + d_2 t)\mathbf{j} + (z_0 + d_3 t)\mathbf{k}.$$

If d_1, d_2, d_3 are not all 0, then $\mathbf{f}(t)$ traces out the line that passes through the point $P(x_0, y_0, z_0)$ and has direction numbers d_1, d_2, d_3. If d_1, d_2, d_3 are all 0, then we have the constant function

$$\mathbf{f}(t) = x_0\mathbf{i} + y_0\mathbf{j} + z_0\mathbf{k}. \quad \square$$

Example. From

$$f_1(t) = \cos t, \qquad f_2(t) = \sin t, \qquad f_3(t) = 0$$

we can form the function

$$\mathbf{f}(t) = \cos t \, \mathbf{i} + \sin t \, \mathbf{j}.$$

For each t

$$\|\mathbf{f}(t)\| = \sqrt{\cos^2 t + \sin^2 t} = 1.$$

As t ranges over the set of real numbers, the position vector $\mathbf{f}(t)$ remains in the xy-plane and its tip traces out the unit circle in a counterclockwise manner, effecting a complete revolution each time that t increases by 2π. (Figure 13.1.1) □

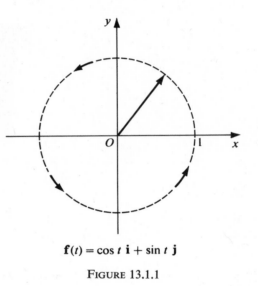

$$\mathbf{f}(t) = \cos t\, \mathbf{i} + \sin t\, \mathbf{j}$$

FIGURE 13.1.1

The ideas of limit, continuity, derivative, and integral are easily extended to vector functions. Suppose that

$$\mathbf{f}(t) = f_1(t)\mathbf{i} + f_2(t)\mathbf{j} + f_3(t)\mathbf{k}, \qquad t \in I.$$

We say that $\mathbf{f}$ has a *limit* at t_0 provided that each of the component functions f_1, f_2, f_3 has a limit at t_0; namely, for $\mathbf{L} = l_1\mathbf{i} + l_2\mathbf{j} + l_3\mathbf{k}$, we set

$$\lim_{t \to t_0} \mathbf{f}(t) = \mathbf{L} \quad \text{iff} \quad \lim_{t \to t_0} f_1(t) = l_1, \quad \lim_{t \to t_0} f_2(t) = l_2, \quad \lim_{t \to t_0} f_3(t) = l_3.$$

We say that $\mathbf{f}$ is *continuous* at t_0 provided that each of the component functions is continuous at t_0; that is, provided that

$$\lim_{t \to t_0} f_1(t) = f_1(t_0), \quad \lim_{t \to t_0} f_2(t) = f_2(t_0), \quad \lim_{t \to t_0} f_3(t) = f_3(t_0).$$

More briefly

$$
\mathbf{f} \text{ is } \textit{continuous} \text{ at } t_0 \quad \text{iff} \quad \lim_{t \to t_0} \mathbf{f}(t) = \mathbf{f}(t_0).
$$

Problem. Find

$$
\lim_{t \to 0} \mathbf{f}(t)
$$

if

$$
\mathbf{f}(t) = \cos (t + \pi)\, \mathbf{i} + 2\, \frac{\sin t}{t}\, \mathbf{j} + e^{-t^2}\, \mathbf{k}.
$$

SOLUTION

$$
\lim_{t \to 0} \mathbf{f}(t) = \lim_{t \to 0} \left[\cos (t + \pi)\, \mathbf{i} + 2\, \frac{\sin t}{t}\, \mathbf{j} + e^{-t^2}\, \mathbf{k} \right]
$$

$$
= \left[\lim_{t \to 0} \cos (t + \pi) \right] \mathbf{i} + \left[\lim_{t \to 0} 2\, \frac{\sin t}{t} \right] \mathbf{j} + \left[\lim_{t \to 0} e^{-t^2} \right] \mathbf{k}
$$

$$
= -\mathbf{i} + 2\mathbf{j} + \mathbf{k}. \quad \square
$$

We say that $\mathbf{f}$ is *differentiable* at t provided that each of the component functions is differentiable at t. If this happens, then we write

$$
\mathbf{f}'(t) = f_1'(t)\mathbf{i} + f_2'(t)\mathbf{j} + f_3'(t)\mathbf{k}
$$

and call $\mathbf{f}'(t)$ the *derivative* of $\mathbf{f}$ at t. The derivative $\mathbf{f}'(t)$ can also be written as the limit of a *vector difference quotient*:

$$
\mathbf{f}'(t) = \lim_{h \to 0} \frac{1}{h} \left[\mathbf{f}(t + h) - \mathbf{f}(t) \right].
$$

Here is a proof:

$$
\mathbf{f}'(t) = f_1'(t)\mathbf{i} + f_2'(t)\mathbf{j} + f_3'(t)\mathbf{k}
$$

$$
= \left[\lim_{h \to 0} \frac{f_1(t + h) - f_1(t)}{h} \right] \mathbf{i} + \left[\lim_{h \to 0} \frac{f_2(t + h) - f_2(t)}{h} \right] \mathbf{j} + \left[\lim_{h \to 0} \frac{f_3(t + h) - f_3(t)}{h} \right] \mathbf{k}
$$

$$
= \lim_{h \to 0} \left[\frac{f_1(t + h) - f_1(t)}{h}\, \mathbf{i} + \frac{f_2(t + h) - f_2(t)}{h}\, \mathbf{j} + \frac{f_3(t + h) - f_3(t)}{h}\, \mathbf{k} \right]
$$

$$
= \lim_{h \to 0} \frac{1}{h} \left[\mathbf{f}(t + h) - \mathbf{f}(t) \right]. \quad \square
$$

Interpretations of the vector derivative and applications of vector differentiation are postponed until Sections 13.3 and 13.4. Here we limit ourselves to computations.

Problem. Given that

$$\mathbf{f}(t) = t\mathbf{i} + \sqrt{t + 1}\,\mathbf{j} - e^t\mathbf{k}$$

find:

(1) the domain of **f**.   (2) **f**(0).   (3) **f**′(t).

(4) **f**′(0).   (5) ‖**f**(t)‖.   (6) **f**(t) · **f**′(t).

SOLUTION

(1) For a number to be in the domain of **f** it is necessary only that it be in the domain of each of its components. The domain of **f** is thus $[-1, \infty)$.

(2) $\mathbf{f}(0) = 0\,\mathbf{i} + \sqrt{0 + 1}\,\mathbf{j} - e^0\mathbf{k}$

    $= \mathbf{j} - \mathbf{k}$.

(3) $\mathbf{f}(t) = t\mathbf{i} + \sqrt{t + 1}\,\mathbf{j} - e^t\mathbf{k}$

    $\mathbf{f}'(t) = \mathbf{i} + \dfrac{1}{2\sqrt{t + 1}}\,\mathbf{j} - e^t\mathbf{k}$.

(4) $\mathbf{f}'(0) = \mathbf{i} + \dfrac{1}{2\sqrt{0 + 1}}\,\mathbf{j} - e^0\mathbf{k} = \mathbf{i} + \dfrac{1}{2}\mathbf{j} - \mathbf{k}$.

(5) $\|\mathbf{f}(t)\| = \sqrt{t^2 + (\sqrt{t + 1})^2 + (e^t)^2} = \sqrt{t^2 + t + 1 + e^{2t}}$.

(6) $\mathbf{f}(t) \cdot \mathbf{f}'(t) = (t\mathbf{i} + \sqrt{t + 1}\,\mathbf{j} - e^t\mathbf{k}) \cdot \left(\mathbf{i} + \dfrac{1}{2\sqrt{t + 1}}\,\mathbf{j} - e^t\mathbf{k}\right)$

    $= (t)(1) + (\sqrt{t + 1})\left(\dfrac{1}{2\sqrt{t + 1}}\right) + (-e^t)(-e^t)$

    $= t + \frac{1}{2} + e^{2t}$. □

In differentiating vector functions we need not stop with the first derivative. If **f**′ is itself differentiable, we can form the second derivative **f**″ and so on.

Problem. Find **f**″(t) if

$$\mathbf{f}(t) = t \sin t\, \mathbf{i} + e^{-t}\mathbf{j} + t\mathbf{k}.$$

SOLUTION

$$\mathbf{f}'(t) = (t \cos t + \sin t)\mathbf{i} - e^{-t}\mathbf{j} + \mathbf{k},$$
$$\mathbf{f}''(t) = (-t \sin t + \cos t + \cos t)\mathbf{i} + e^{-t}\mathbf{j}$$
$$= (2 \cos t - t \sin t)\mathbf{i} + e^{-t}\mathbf{j}. □$$

Integration can also be handled component by component; namely we set

$$\int_a^b \mathbf{f}(t)\, dt = \left(\int_a^b f_1(t)\, dt\right)\mathbf{i} + \left(\int_a^b f_2(t)\, dt\right)\mathbf{j} + \left(\int_a^b f_3(t)\, dt\right)\mathbf{k}.$$

Problem.   Find

$$\int_0^1 \mathbf{f}(t)\, dt$$

for

$$\mathbf{f}(t) = t\mathbf{i} + \sqrt{t + 1}\,\mathbf{j} - e^t\,\mathbf{k}.$$

SOLUTION

$$\int_0^1 f(t)\, dt = \left(\int_0^1 t\, dt\right)\mathbf{i} + \left(\int_0^1 \sqrt{t + 1}\, dt\right)\mathbf{j} + \left(\int_0^1 (-e^t)\, dt\right)\mathbf{k}$$

$$= \left[\tfrac{1}{2}t^2\right]_0^1 \mathbf{i} + \left[\tfrac{2}{3}(t + 1)^{3/2}\right]_0^1 \mathbf{j} + \left[-e^t\right]_0^1 \mathbf{k}$$

$$= \tfrac{1}{2}\mathbf{i} + \tfrac{2}{3}(2\sqrt{2} - 1)\mathbf{j} + (1 - e)\mathbf{k}. \quad \square$$

Problem.   Find $\mathbf{f}(t)$ given that

$$\mathbf{f}'(t) = 2\cos t\,\mathbf{i} - t\sin t^2\,\mathbf{j} + 2t\mathbf{k} \quad \text{and} \quad \mathbf{f}(0) = \mathbf{i} + 3\mathbf{k}.$$

SOLUTION.   By integrating $\mathbf{f}'(t)$ we get

$$\mathbf{f}(t) = (2\sin t + c_1)\mathbf{i} + (\tfrac{1}{2}\cos t^2 + c_2)\mathbf{j} + (t^2 + c_3)\mathbf{k},$$

where c_1, c_2, c_3 are constants to be determined. Since

$$\mathbf{i} + 3\mathbf{k} = \mathbf{f}(0) = c_1\mathbf{i} + (\tfrac{1}{2} + c_2)\mathbf{j} + c_3\mathbf{k},$$

you can see that

$$c_1 = 1, \qquad c_2 = -\tfrac{1}{2}, \qquad c_3 = 3.$$

Thus

$$\mathbf{f}(t) = (2\sin t + 1)\mathbf{i} + (\tfrac{1}{2}\cos t^2 - \tfrac{1}{2})\mathbf{j} + (t^2 + 3)\mathbf{k}. \quad \square$$

In passing, we point out that

$$\int_a^b [\mathbf{f}(t) + \mathbf{g}(t)]\, dt = \int_a^b \mathbf{f}(t)\, dt + \int_a^b \mathbf{g}(t)\, dt$$

and

$$\int_a^b [\alpha\mathbf{f}(t)]\, dt = \alpha \int_a^b \mathbf{f}(t)\, dt \quad \text{for every scalar } \alpha.$$

Not so obvious is that

(1) $$\int_a^b [\mathbf{c} \cdot \mathbf{f}(t)]\, dt = \mathbf{c} \cdot \int_a^b \mathbf{f}(t)\, dt \quad \text{for every vector } \mathbf{c}$$

and

$$(2) \qquad \left\| \int_a^b \mathbf{f}(t)\, dt \right\| \le \int_a^b \|\mathbf{f}(t)\|\, dt.$$

The proof of (1) we leave as an exercise. Here is a proof of (2): Set

$$\mathbf{c} = \int_a^b \mathbf{f}(t)\, dt.$$

Note that

$$\|\mathbf{c}\|^2 = \mathbf{c} \cdot \mathbf{c} = \mathbf{c} \cdot \int_a^b \mathbf{f}(t)\, dt = \int_a^b [\mathbf{c} \cdot \mathbf{f}(t)]\, dt \le \int_a^b \|\mathbf{c}\|\, \|\mathbf{f}(t)\|\, dt = \|\mathbf{c}\| \int_a^b \|\mathbf{f}(t)\|\, dt.$$

by property (1)   by Schwarz's inequality

If $\mathbf{c} \ne \mathbf{0}$, then we can divide by $\|\mathbf{c}\|$ and thereby get

$$\|\mathbf{c}\| \le \int_a^b \|\mathbf{f}(t)\|\, dt.$$

If $\mathbf{c} = \mathbf{0}$, the result is obvious in the first place. $\square$

Exercises

Differentiate.

*1. $\mathbf{f}(t) = (1 + 2t)\mathbf{i} + (3 - t)\mathbf{j} + (2 + 3t)\mathbf{k}$.

2. $\mathbf{f}(t) = 2\mathbf{i} - \cos t\, \mathbf{k}$.

*3. $\mathbf{f}(t) = \sqrt{1 - t}\,\mathbf{i} + \sqrt{1 + t}\,\mathbf{j} + \dfrac{1 - t}{1 + t}\mathbf{k}$.

4. $\mathbf{f}(t) = \dfrac{1 - t^2}{1 + t^2}\mathbf{i} + \dfrac{2t}{1 + t^2}\mathbf{j}$.

*5. $\mathbf{f}(t) = \sin t\, \mathbf{i} + \cos t\, \mathbf{j} + \tan t\, \mathbf{k}$.

6. $\mathbf{f}(t) = e^t\mathbf{i} + \log t\, \mathbf{j} + \arctan t\, \mathbf{k}$.

*7. $\mathbf{f}(t) = \arcsin t\, \mathbf{i} + \log(1 + 2t)\mathbf{j} + t^2\mathbf{k}$.

Compute.

8. $\displaystyle\int_1^2 \mathbf{f}(t)\, dt$ for $\mathbf{f}(t) = \mathbf{i} + 2t\mathbf{j}$.

*9. $\displaystyle\int_0^\pi \mathbf{r}(t)\, dt$ for $\mathbf{r}(t) = \sin t\, \mathbf{i} + \cos t\, \mathbf{j} + t\, \mathbf{k}$.

10. $\displaystyle\int_0^1 \mathbf{g}(t)\, dt$ for $\mathbf{g}(t) = e^t\mathbf{i} + \sqrt{2}\,\mathbf{j} + e^{-t}\mathbf{k}$.

* 11. $\displaystyle\int_0^1 \mathbf{h}(t)\,dt$ for $\mathbf{h}(t) = e^{-t}[t^2\mathbf{i} + \sqrt{2t}\,\mathbf{j} + \mathbf{k}]$.

12. Show that if $\mathbf{f}'(t) = \mathbf{0}$ for all t in some interval I, then $\mathbf{f}$ is constant on I. (That is, there exists a vector $\mathbf{c}$ such that $\mathbf{f}(t) = \mathbf{c}$ for all $t \in I$.)

13. Show that if $\mathbf{f}'(t) = \mathbf{g}'(t)$ for all t in some interval I, then $\mathbf{f}$ and $\mathbf{g}$ differ by a constant on I.

Find $\mathbf{f}(t)$ given that

14. $\mathbf{f}'(t) = \mathbf{i} + t^2\mathbf{j}$ and $\mathbf{f}(0) = \mathbf{j} - \mathbf{k}$.

* 15. $\mathbf{f}'(t) = t\mathbf{i} + \dfrac{t}{\sqrt{1 + t^2}}\mathbf{j} + te^t\mathbf{k}$ and $\mathbf{f}(0) = \mathbf{i} + 2\mathbf{j} + 3\mathbf{k}$.

* 16. $\mathbf{f}'(t) = 2\,\mathbf{f}(t)$ and $\mathbf{f}(0) = \mathbf{i} - \mathbf{k}$.

17. (*Important*) Show that

$$\lim_{t \to t_0} \mathbf{f}(t) = \mathbf{L} \quad \text{iff} \quad \lim_{t \to t_0} \|\mathbf{f}(t) - \mathbf{L}\| = 0.$$

18. Show that

$$\int_a^b [\mathbf{c} \cdot \mathbf{f}(t)]\,dt = \mathbf{c} \cdot \int_a^b \mathbf{f}(t)\,dt \quad \text{for every vector } \mathbf{c}.$$

13.2   Differentiation Formulas

Vector functions with a common domain can be combined in many ways to form new functions. From $\mathbf{f}$ and $\mathbf{g}$ we can form the sum $\mathbf{f} + \mathbf{g}$:

$$(\mathbf{f} + \mathbf{g})(t) = \mathbf{f}(t) + \mathbf{g}(t).$$

We can form scalar multiples $\alpha\mathbf{f}$ and thus linear combinations $\alpha\mathbf{f} + \beta\mathbf{g}$:

$$(\alpha\mathbf{f})(t) = \alpha\mathbf{f}(t), \qquad (\alpha\mathbf{f} + \beta\mathbf{g})(t) = \alpha\mathbf{f}(t) + \beta\mathbf{g}(t).$$

We can form the dot product $\mathbf{f} \cdot \mathbf{g}$:

$$(\mathbf{f} \cdot \mathbf{g})(t) = \mathbf{f}(t) \cdot \mathbf{g}(t).$$

We can also form the cross product $\mathbf{f} \times \mathbf{g}$:

$$(\mathbf{f} \times \mathbf{g})(t) = \mathbf{f}(t) \times \mathbf{g}(t).$$

There are two ways to bring scalar functions (real-valued functions) into play. If the scalar function u has the same domain as $\mathbf{f}$, we can form the product $u\mathbf{f}$:

$$(u\mathbf{f})(t) = u(t)\mathbf{f}(t).$$

If for each t in some interval, $u(t)$ is in the domain of $\mathbf{f}$, then we can form the composition $\mathbf{f} \circ u$:

$$(\mathbf{f} \circ u)(t) = \mathbf{f}(u(t)).$$

It is not hard to show that if the original functions are continuous, then so are the newly constructed ones. More interestingly, if the original functions are differentiable, then so are the new ones. The derivatives can be computed using the following rules:

(1) $(\mathbf{f} + \mathbf{g})'(t) = \mathbf{f}'(t) + \mathbf{g}'(t)$.

(2) $(\alpha\mathbf{f})'(t) = \alpha\mathbf{f}'(t)$.

(3) $(u\mathbf{f})'(t) = u(t)\mathbf{f}'(t) + u'(t)\mathbf{f}(t)$.

(4) $(\mathbf{f} \cdot \mathbf{g})'(t) = [\mathbf{f}(t) \cdot \mathbf{g}'(t)] + [\mathbf{f}'(t) \cdot \mathbf{g}(t)]$.

(5) $(\mathbf{f} \times \mathbf{g})'(t) = [\mathbf{f}(t) \times \mathbf{g}'(t)] + [\mathbf{f}'(t) \times \mathbf{g}(t)]$.

(6) $(\mathbf{f} \circ u)'(t) = \mathbf{f}'(u(t))u'(t) = u'(t)\mathbf{f}'(u(t))$.   (chain rule)

The validity of these rules we show later. First a few comments. Rules (1) and (2) look exactly like those we used for differentiating ordinary functions. Rule (2) is just a special case of Rule (3): set $u(t) = \alpha$ for all t. Rules (3), (4), and (5) are all "product" rules and should remind you of the rule for differentiating the product of ordinary functions. Keep in mind that $\times$ is not commutative and therefore the order in Rule (5) is important. In Rule (6) we first wrote the scalar part $u'(t)$ on the right so that the formula would look like the chain rule for ordinary functions. In general $\mathbf{a}\alpha$ means the same as $\alpha\mathbf{a}$.

Problem.   Given the vector functions

$$\mathbf{f}(t) = \cos t \, \mathbf{i} + \sin t \, \mathbf{j}, \qquad \mathbf{g}(t) = \mathbf{i} + \arctan t \, \mathbf{j} - t^2 \, \mathbf{k}$$

and the scalar function

$$u(t) = \log (t + 1)$$

find

(a) $(\mathbf{f} + \mathbf{g})'(t)$.   (b) $(u\mathbf{f})'(t)$.

(c) $(\mathbf{f} \cdot \mathbf{g})'(t)$.   (d) $(\mathbf{f} \times \mathbf{g})'(t)$.

(e) $(\mathbf{f} \circ u)'(t)$.

SOLUTION

(a) $(\mathbf{f} + \mathbf{g})'(t) = \mathbf{f}'(t) + \mathbf{g}'(t)$

$$= (-\sin t \, \mathbf{i} + \cos t \, \mathbf{j}) + \left(\frac{1}{1 + t^2} \mathbf{j} - 2t \, \mathbf{k} \right)$$

$$= -\sin t \, \mathbf{i} + \left(\cos t + \frac{1}{1 + t^2} \right) \mathbf{j} - 2t \, \mathbf{k}.$$

We can achieve the same result by adding first and then differentiating:

$$(\mathbf{f} + \mathbf{g})(t) = \mathbf{f}(t) + \mathbf{g}(t)$$
$$= (\cos t\,\mathbf{i} + \sin t\,\mathbf{j}) + (\mathbf{i} + \arctan t\,\mathbf{j} - t^2\mathbf{k})$$
$$= (1 + \cos t)\mathbf{i} + (\sin t + \arctan t)\mathbf{j} - t^2\mathbf{k}.$$

Thus

$$(\mathbf{f} + \mathbf{g})'(t) = -\sin t\,\mathbf{i} + \left(\cos t + \frac{1}{1 + t^2}\right)\mathbf{j} - 2t\,\mathbf{k}.$$

(b) $(u\mathbf{f})'(t) = u(t)\mathbf{f}'(t) + u'(t)\mathbf{f}(t)$

$$= \log(t + 1)[-\sin t\,\mathbf{i} + \cos t\,\mathbf{j}] + \frac{1}{t + 1}[\cos t\,\mathbf{i} - \sin t\,\mathbf{j}]$$

$$= \left[\frac{\cos t}{t + 1} - \log(t + 1)\sin t\right]\mathbf{i} + \left[\log(t + 1)\cos t + \frac{\sin t}{t + 1}\right]\mathbf{j}.$$

Another method:

$$(u\mathbf{f})(t) = u(t)\mathbf{f}(t)$$
$$= \log(t + 1)[\cos t\,\mathbf{i} + \sin t\,\mathbf{j}]$$
$$= \log(t + 1)\cos t\,\mathbf{i} + \log(t + 1)\sin t\,\mathbf{j}.$$

$$(u\mathbf{f})'(t) = \left[\log(t + 1)(-\sin t) + \frac{\cos t}{t + 1}\right]\mathbf{i} + \left[\log(t + 1)\cos t + \frac{\sin t}{t + 1}\right]\mathbf{j}$$

$$= \left[\frac{\cos t}{t + 1} - \log(t + 1)\sin t\right]\mathbf{i} + \left[\log(t + 1)\cos t + \frac{\sin t}{t + 1}\right]\mathbf{j}.$$

(c) $(\mathbf{f} \cdot \mathbf{g})'(t) = \mathbf{f}(t) \cdot \mathbf{g}'(t) + \mathbf{f}'(t) \cdot \mathbf{g}(t)$

$$= (\cos t\,\mathbf{i} + \sin t\,\mathbf{j}) \cdot \left(\frac{1}{1 + t^2}\mathbf{j} - 2t\,\mathbf{k}\right)$$

$$+ (-\sin t\,\mathbf{i} + \cos t\,\mathbf{j}) \cdot (\mathbf{i} + \arctan t\,\mathbf{j} - t^2\mathbf{k})$$

$$= \frac{\sin t}{1 + t^2} - \sin t + \cos t \arctan t.$$

Another method:

$$(\mathbf{f} \cdot \mathbf{g})(t) = \mathbf{f}(t) \cdot \mathbf{g}(t)$$
$$= (\cos t\,\mathbf{i} + \sin t\,\mathbf{j}) \cdot (\mathbf{i} + \arctan t\,\mathbf{j} - t^2\mathbf{k})$$
$$= \cos t + \sin t \arctan t.$$

$$(\mathbf{f} \cdot \mathbf{g})'(t) = -\sin t + \frac{\sin t}{1 + t^2} + \cos t \arctan t.$$

(d) $(\mathbf{f} \times \mathbf{g})'(t) = [\mathbf{f}(t) \times \mathbf{g}'(t)] + [\mathbf{f}'(t) \times \mathbf{g}(t)]$

$$= (\cos t\,\mathbf{i} + \sin t\,\mathbf{j}) \times \left(\frac{1}{1+t^2}\mathbf{j} - 2t\,\mathbf{k}\right)$$

$$+ (-\sin t\,\mathbf{i} + \cos t\,\mathbf{j}) \times (\mathbf{i} + \arctan t\,\mathbf{j} - t^2\,\mathbf{k})$$

$$= (\sin t)(-2t)\mathbf{i} - (\cos t)(-2t)\mathbf{j} + \frac{\cos t}{1+t^2}\mathbf{k}$$

$$+ (\cos t)(-t^2)\mathbf{i} - (-\sin t)(-t^2)\mathbf{j} + (-\sin t \arctan t - \cos t)\mathbf{k}$$

$$= [(-t)(2\sin t + t\cos t)]\mathbf{i} + [t(2\cos t - t\sin t)]\mathbf{j}$$

$$+ \left[\frac{\cos t}{1+t^2} - \sin t \arctan t - \cos t\right]\mathbf{k}.$$

Another method:

$(\mathbf{f} \times \mathbf{g})(t) = \mathbf{f}(t) \times \mathbf{g}(t)$

$$= (\cos t\,\mathbf{i} + \sin t\,\mathbf{j}) \times (\mathbf{i} + \arctan t\,\mathbf{j} - t^2\,\mathbf{k})$$

$$= (\sin t)(-t^2)\mathbf{i} - (\cos t)(-t^2)\mathbf{j} + (\cos t \arctan t - \sin t)\mathbf{k}.$$

$(\mathbf{f} \times \mathbf{g})'(t) = [-t^2\cos t - 2t\sin t]\mathbf{i} + [t^2(-\sin t) + 2t\cos t]\mathbf{j}$

$$+ \left[\frac{\cos t}{1+t^2} - \sin t \arctan t - \cos t\right]\mathbf{k}$$

$$= [(-t)(t\cos t + 2\sin t)]\mathbf{i} + [t(2\cos t - t\sin t)]\mathbf{j}$$

$$+ \left[\frac{\cos t}{1+t^2} - \sin t \arctan t - \cos t\right]\mathbf{k}.$$

(e) $(\mathbf{f} \circ u)'(t) = \mathbf{f}'(u(t))u'(t)$

$$= \left[-\sin\left(\log(t+1)\right)\mathbf{i} + \cos\left(\log(t+1)\right)\mathbf{j}\right]\frac{1}{t+1}.$$

Another method:

$(\mathbf{f} \circ u)(t) = \mathbf{f}(u(t))$

$$= \left[\cos\left(\log(t+1)\right)\right]\mathbf{i} + \left[\sin\left(\log(t+1)\right)\right]\mathbf{j}.$$

$(\mathbf{f} \circ u)'(t) = -\sin\left(\log(t+1)\right)\frac{1}{t+1}\mathbf{i} + \cos\left(\log(t+1)\right)\frac{1}{t+1}\mathbf{j}.$ □

Proofs for the differentiation formulas (1)–(6) can be derived by componentwise differentiation. As an example we verify Formula (3):

$$(u\mathbf{f})'(t) = u(t)\mathbf{f}'(t) + u'(t)\mathbf{f}(t).$$

Setting

$$\mathbf{f}(t) = f_1(t)\mathbf{i} + f_2(t)\mathbf{j} + f_3(t)\mathbf{k},$$

we have

$$(u\mathbf{f})(t) = u(t)\mathbf{f}(t) = u(t)f_1(t)\mathbf{i} + u(t)f_2(t)\mathbf{j} + u(t)f_3(t)\mathbf{k}$$

$$(u\mathbf{f})'(t) = [u(t)f_1'(t) + u'(t)f_1(t)]\mathbf{i} + [u(t)f_2'(t) + u'(t)f_2(t)]\mathbf{j}$$
$$+ [u(t)f_3'(t) + u'(t)f_3(t)]\mathbf{k}$$
$$= u(t)[f_1'(t)\mathbf{i} + f_2'(t)\mathbf{j} + f_3'(t)\mathbf{k}] + u'(t)[f_1(t)\mathbf{i} + f_2(t)\mathbf{j} + f_3(t)\mathbf{k}]$$
$$= u(t)\mathbf{f}'(t) + u'(t)\mathbf{f}(t). \quad \square$$

Componentwise verification of the other formulas we leave as exercises.

Let's take another look at Formula (3), this time from a different point of view. Instead of casting everything in terms of components, we will work with the vectors directly.

To evaluate $(u\mathbf{f})'(t)$ is to compute the limit of the difference quotient

$$\frac{1}{h}[u(t+h)\mathbf{f}(t+h) - u(t)\mathbf{f}(t)]$$

as h tends to 0. By subtracting and adding $u(t+h)\mathbf{f}(t)$ inside the brackets and then factoring, we can rewrite the difference quotient as

$$u(t+h)\frac{1}{h}[\mathbf{f}(t+h) - \mathbf{f}(t)] + \frac{1}{h}[u(t+h) - u(t)]\mathbf{f}(t).$$

It is clear that

$$u(t+h) \quad \text{tends to} \quad u(t),$$

$$\frac{1}{h}[\mathbf{f}(t+h) - \mathbf{f}(t)] \quad \text{tends to} \quad \mathbf{f}'(t),$$

$$\frac{1}{h}[u(t+h) - u(t)] \quad \text{tends to} \quad u'(t).$$

It can now be argued that the difference quotient tends to

$$u(t)\mathbf{f}'(t) + u'(t)\mathbf{f}(t). \quad \square$$

It is instructive to apply the same point of view to Formula (4):

$$(\mathbf{f} \cdot \mathbf{g})'(t) = [\mathbf{f}(t) \cdot \mathbf{g}'(t)] + [\mathbf{f}'(t) \cdot \mathbf{g}(t)].$$

On the left we have the limit of the difference quotient

$$\frac{1}{h}[\mathbf{f}(t+h) \cdot \mathbf{g}(t+h) - \mathbf{f}(t) \cdot \mathbf{g}(t)]$$

as h tends to 0. Subtracting and adding $\mathbf{f}(t+h) \cdot \mathbf{g}(t)$ inside the brackets and rearranging, we obtain

$$\mathbf{f}(t+h) \cdot \frac{1}{h}[\mathbf{g}(t+h) - \mathbf{g}(t)] + \frac{1}{h}[\mathbf{f}(t+h) - \mathbf{f}(t)] \cdot \mathbf{g}(t).$$

It probably seems obvious now that the limit is

$$[\mathbf{f}(t) \cdot \mathbf{g}'(t)] + [\mathbf{f}'(t) \cdot \mathbf{g}(t)]. \quad \square$$

In passing, we remark that all of these differentiation formulas can be derived in a component-free manner.

In Leibniz's notation the formulas become

---

(1) $\dfrac{d}{dt}[\mathbf{f}(t) + \mathbf{g}(t)] = \dfrac{d}{dt}[\mathbf{f}(t)] + \dfrac{d}{dt}[\mathbf{g}(t)].$

(2) $\dfrac{d}{dt}[\alpha\mathbf{f}(t)] = \alpha\dfrac{d}{dt}[\mathbf{f}(t)].$

(3) $\dfrac{d}{dt}[u(t)\mathbf{f}(t)] = u(t)\dfrac{d}{dt}[\mathbf{f}(t)] + \dfrac{d}{dt}[u(t)]\mathbf{f}(t).$

(4) $\dfrac{d}{dt}[\mathbf{f}(t) \cdot \mathbf{g}(t)] = \left(\mathbf{f}(t) \cdot \dfrac{d}{dt}[\mathbf{g}(t)]\right) + \left(\dfrac{d}{dt}[\mathbf{f}(t)] \cdot \mathbf{g}(t)\right).$

(5) $\dfrac{d}{dt}[\mathbf{f}(t) \times \mathbf{g}(t)] = \left(\mathbf{f}(t) \times \dfrac{d}{dt}[\mathbf{g}(t)]\right) + \left(\dfrac{d}{dt}[\mathbf{f}(t)] \times \mathbf{g}(t)\right).$

(6) $\dfrac{d}{dt}[\mathbf{f}(u(t))] = \dfrac{d\mathbf{f}}{du}[u(t)]\dfrac{du}{dt}(t).$

Exercises

Compute $\mathbf{f}'(t)$ and $\mathbf{f}''(t)$ for each of the following functions.

*1. $\mathbf{f}(t) = \mathbf{a} + t\mathbf{b}.$

2. $\mathbf{f}(t) = \mathbf{a} + t\mathbf{b} + t^2\mathbf{c}.$

*3. $\mathbf{f}(t) = e^{2t}\mathbf{i} - \sin t\,\mathbf{j}.$

4. $\mathbf{f}(t) = [(t^2\mathbf{i} - \mathbf{j}) \cdot (\mathbf{i} - t^2\mathbf{j})]\mathbf{i}.$

*5. $\mathbf{f}(t) = \mathbf{g}(t^2).$

6. $\mathbf{f}(t) = t\mathbf{g}(t^2).$

*7. $\mathbf{f}(t) = t\mathbf{g}(\sqrt{t}).$

8. $\mathbf{f}(t) = (\cos t\,\mathbf{i} - \mathbf{j}) \times (t\mathbf{j} + \sin t\,\mathbf{k}).$

Find

*9. $\dfrac{d}{dt}[e^{\cos t}\mathbf{i} + e^{\sin t}\mathbf{j}].$

10. $\dfrac{d^2}{dt^2}[e^t \cos t\,\mathbf{i} + e^t \sin t\,\mathbf{j}].$

*11. $\dfrac{d^3}{dt^3}[e^t\mathbf{i} - e^{-t}\mathbf{j}].$

12. $\dfrac{d}{dt}[(\mathbf{a} + t\mathbf{b}) \times (\mathbf{c} + t\mathbf{d})].$

*13. $\dfrac{d^2}{dt^2}[(\mathbf{a} + t\mathbf{b}) \times (\mathbf{c} + t\mathbf{d})].$

14. $\dfrac{d}{dt}[(\mathbf{a} + t\mathbf{b}) \cdot (\mathbf{a} + t\mathbf{b} + t^2\mathbf{c})].$

15. Show that

$$\frac{d}{dt}\left[u_1(t)\mathbf{r}_1(t) \times u_2(t)\mathbf{r}_2(t)\right] = u_1(t)u_2(t)\frac{d}{dt}\left[\mathbf{r}_1(t) \times \mathbf{r}_2(t)\right]$$

$$+ \frac{d}{dt}\left[u_1(t)u_2(t)\right]\left[\mathbf{r}_1(t) \times \mathbf{r}_2(t)\right].$$

Find $\mathbf{f}(t)$ given that:

16. $\mathbf{f}'(t) = \mathbf{b}$ for all real t, $\mathbf{f}(0) = \mathbf{a}$.

*17. $\mathbf{f}''(t) = \mathbf{c}$ for all real t, $\mathbf{f}'(0) = \mathbf{b}$, $\mathbf{f}(0) = \mathbf{a}$.

18. $\mathbf{f}''(t) = \mathbf{a} + t\mathbf{b}$ for all real t, $\mathbf{f}'(0) = \mathbf{c}$, $\mathbf{f}(0) = \mathbf{d}$.

19. Show that, if $\mathbf{f}$ is differentiable at t, then $\mathbf{f}$ is continuous at t.

20. Show that

$$\frac{d}{dt}\left[\mathbf{f}(t) \times \mathbf{f}'(t)\right] = \mathbf{f}(t) \times \mathbf{f}''(t).$$

21. Given that

$$\mathbf{f}(t) = \cos t\,\mathbf{i} + \sin t\,\mathbf{j},$$

show that

$$\mathbf{f}(t) \cdot \mathbf{f}'(t) = 0.$$

22. Given that $\mathbf{f}$ is differentiable, show that

$$\|\mathbf{f}(t)\| \text{ is constant} \qquad \text{iff} \qquad \mathbf{f}(t) \cdot \mathbf{f}'(t) = 0 \quad \text{identically.}$$

23. Given that

$$\mathbf{g}(t) = e^{kt}\,\mathbf{i} + e^{-kt}\,\mathbf{j},$$

show that $\mathbf{g}(t)$ and $\mathbf{g}''(t)$ have the same direction.

24. Give an informal derivation of the formula

$$(\mathbf{f} \times \mathbf{g})'(t) = \left[\mathbf{f}(t) \times \mathbf{g}'(t)\right] + \left[\mathbf{f}'(t) \times \mathbf{g}(t)\right]$$

without appealing to components.

25. Give a detailed proof of the chain rule for vector functions:

$$(\mathbf{f} \circ u)'(t) = \mathbf{f}(u(t))u'(t).$$

13.3 Curves and Tangents

A linear vector function

$$\mathbf{r}(t) = \mathbf{r}_0 + t\mathbf{d}$$

serves as a parametrization for a line l in the sense that, as t ranges over the set of real numbers, the tip of the position vector $\mathbf{r}(t)$ traces out the line l. The vector function

$$\mathbf{r}(t) = \cos t\,\mathbf{i} + \sin t\,\mathbf{j}, \qquad t \in [0, 2\pi]$$

serves as a parametrization for the unit circle in the xy-plane in the sense that, as t ranges over $[0, 2\pi]$, the tip of the vector $\mathbf{r}(t)$ traces out the unit circle in the xy-plane.

In general a vector function

$$\mathbf{r}(t) = x(t)\mathbf{i} + y(t)\mathbf{j} + z(t)\mathbf{k}$$

differentiable on some interval I serves as a *parametrization for a curve C* in the sense that, as t ranges over I, the tip of the vector $\mathbf{r}(t)$ traces out the curve C. (See Figure 13.3.1.)

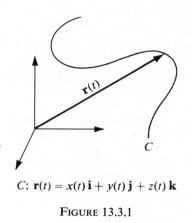

$$C:\ \mathbf{r}(t) = x(t)\,\mathbf{i} + y(t)\,\mathbf{j} + z(t)\,\mathbf{k}$$

FIGURE 13.3.1

It is time we gave a geometric interpretation to the derivative

$$\mathbf{r}'(t) = x'(t)\mathbf{i} + y'(t)\mathbf{j} + z'(t)\mathbf{k}.$$

Definition

Let

$$C:\ \mathbf{r}(t) = x(t)\mathbf{i} + y(t)\mathbf{j} + z(t)\mathbf{k}$$

be a differentiable curve. The vector $\mathbf{r}'(t)$, if not $\mathbf{0}$, is said to be *tangent* to C at the tip of $\mathbf{r}(t)$.

To see why, recall that we have

$$\mathbf{r}'(t) = \lim_{h \to 0} \frac{\mathbf{r}(t + h) - \mathbf{r}(t)}{h}.$$

With $\mathbf{r}'(t) \neq \mathbf{0}$, we can be sure that for $t + h$ close enough to t, the vector

$$\mathbf{r}(t + h) - \mathbf{r}(t)$$

will not be $\mathbf{0}$. Consequently, we can think of the vector $\mathbf{r}(t + h) - \mathbf{r}(t)$ as pictured in Figure 13.3.2.

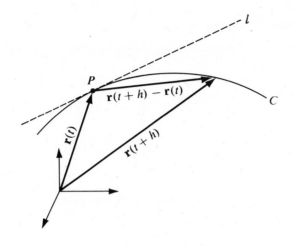

FIGURE 13.3.2

Let's agree that the line marked l in the figure corresponds to our intuitive notion of tangent. We know that l passes through the tip of $\mathbf{r}(t)$. To determine l, we need only find a direction vector for l. As h tends to zero, the vector

$$\mathbf{r}(t + h) - \mathbf{r}(t)$$

comes increasingly closer to serving as a direction vector for l. It may be tempting therefore to take the limiting case

$$\lim_{h \to 0} \left[\mathbf{r}(t + h) - \mathbf{r}(t) \right]$$

and call that a direction vector for l. The trouble here is that this limit vector is $\mathbf{0}$ and $\mathbf{0}$ has no direction. We can circumvent this difficulty by looking instead at the vector

$$\frac{1}{h} \left[\mathbf{r}(t + h) - \mathbf{r}(t) \right].$$

For each value of h, this vector is parallel to

$$\mathbf{r}(t + h) - \mathbf{r}(t)$$

and therefore its limit

$$\mathbf{r}'(t) = \lim_{h \to 0} \frac{1}{h} \left[\mathbf{r}(t + h) - \mathbf{r}(t) \right],$$

which by assumption is not $\mathbf{0}$, can be taken as a direction vector for l. This in a nutshell is why we call the vector $\mathbf{r}'(t)$ tangent to the curve C at the tip of $\mathbf{r}(t)$.

At the tip of $\mathbf{r}_0 = \mathbf{r}(t_0)$ the tangent vector is $\mathbf{r}'(t_0)$ and the *tangent line* is

$$\mathbf{R}(u) = \mathbf{r}_0 + u\,\mathbf{r}'(t_0).$$

Problem.   Show that in the case of the circle

$$C: \mathbf{r}(t) = \rho \cos t \, \mathbf{i} + \rho \sin t \, \mathbf{j},$$

the tangent vector is perpendicular to the radius vector.

SOLUTION.   As a tangent vector we can take

$$\mathbf{r}'(t) = -\rho \sin t \, \mathbf{i} + \rho \cos t \, \mathbf{j}.$$

Since

$$\mathbf{r}(t) \cdot \mathbf{r}'(t) = (\rho \cos t \, \mathbf{i} + \rho \sin t \, \mathbf{j}) \cdot (-\rho \sin t \, \mathbf{i} + \rho \cos t \, \mathbf{j})$$
$$= -\rho^2 \cos t \sin t + \rho^2 \cos t \sin t = 0,$$

we have

$$\mathbf{r}(t) \perp \mathbf{r}'(t). \quad \square$$

Problem.   The curve

$$\mathbf{r}(t) = t\mathbf{i} + t^2\mathbf{j} + t^3\mathbf{k}$$

is called the "twisted" cubic. Find a vector tangent to the twisted cubic at the point $P(2, 4, 8)$ and find the tangent line.

SOLUTION

$$\mathbf{r}'(t) = \mathbf{i} + 2t\mathbf{j} + 3t^2\mathbf{k}.$$

Since $P(2, 4, 8)$ is the tip of $\mathbf{r}(2)$, a suitable tangent vector is

$$\mathbf{r}'(2) = \mathbf{i} + 4\mathbf{j} + 12\mathbf{k}.$$

The tangent line is given by

$$\mathbf{R}(u) = (2\mathbf{i} + 4\mathbf{j} + 8\mathbf{k}) + u(\mathbf{i} + 4\mathbf{j} + 12\mathbf{k}). \quad \square$$

Problem.   Find the point P on the curve

$$\mathbf{r}(t) = (1 - 2t)\mathbf{i} + t^2\mathbf{j} + 2e^{2(t-1)}\mathbf{k}$$

at which the tangent vector $\mathbf{r}'(t)$ is parallel to $\mathbf{r}(t)$.

SOLUTION

$$\mathbf{r}'(t) = -2\mathbf{i} + 2t\mathbf{j} + 4e^{2(t-1)}\mathbf{k}.$$

For $\mathbf{r}'(t)$ to be parallel to $\mathbf{r}(t)$ there must exist a scalar α such that

$$\mathbf{r}(t) = \alpha \, \mathbf{r}'(t).$$

This implies the scalar equations

$$1 - 2t = -2\alpha, \qquad t^2 = 2\alpha t, \qquad 2e^{2(t-1)} = 4\alpha e^{2(t-1)}.$$

The last equation requires $\alpha = \frac{1}{2}$. This in turn tells us that $t = 1$. The point in question is $P(-1, 1, 2)$. $\quad \square$

Two curves

$$C_1: \mathbf{r}_1(t) = x_1(t)\mathbf{i} + y_1(t)\mathbf{j} + z_1(t)\mathbf{k} \quad \text{and} \quad C_2: \mathbf{r}_2(u) = x_2(u)\mathbf{i} + y_2(u)\mathbf{j} + z_2(u)\mathbf{k}$$

intersect iff there are numbers t and u such that

$$\mathbf{r}_1(t) = \mathbf{r}_2(u).$$

As the angle of intersection we take the acute angle $(0 \le \theta \le \tfrac{1}{2}\pi)$ between their tangent lines.

Problem. Show that the circles

$$C_1: \mathbf{r}_1(t) = \cos t\,\mathbf{i} + \sin t\,\mathbf{j} \quad \text{and} \quad C_2: \mathbf{r}_2(u) = \cos u\,\mathbf{j} + \sin u\,\mathbf{k} \qquad \text{(Figure 13.3.3)}$$

meet at right angles at the points $P(0, 1, 0)$ and $Q(0, -1, 0)$.

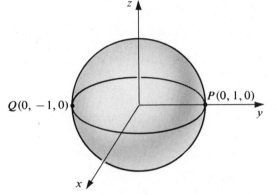

FIGURE 13.3.3

SOLUTION.   Since

$$\mathbf{r}_1(\tfrac{1}{2}\pi) = \mathbf{j} = \mathbf{r}_2(0),$$

the curves meet at the tip of $\mathbf{j}$, which is $P(0, 1, 0)$. Since

$$\mathbf{r}_1\left(\tfrac{3}{2}\pi\right) = -\mathbf{j} = \mathbf{r}_2(\pi),$$

the curves meet at the tip of $-\mathbf{j}$, which is $P(0, -1, 0)$. Obviously

$$\mathbf{r}_1'(t) = -\sin t\,\mathbf{i} + \cos t\,\mathbf{j} \quad \text{and} \quad \mathbf{r}_2'(u) = -\sin u\,\mathbf{j} + \cos u\,\mathbf{k}.$$

Since

$$\mathbf{r}_1'(\tfrac{1}{2}\pi) = -\mathbf{i} \quad \text{and} \quad \mathbf{r}_2'(0) = \mathbf{k},$$

we have

$$\mathbf{r}_1'(\tfrac{1}{2}\pi) \cdot \mathbf{r}_2'(0) = 0.$$

This tells us the curves are perpendicular at $P(0, 1, 0)$. Since

$$\mathbf{r}_1'(\tfrac{3}{2}\pi) = \mathbf{i} \quad \text{and} \quad \mathbf{r}_2'(\pi) = -\mathbf{k},$$

we have

$$\mathbf{r}_1'(\tfrac{3}{2}\pi) \cdot \mathbf{r}_2'(\pi) = 0.$$

This tells us the curves are perpendicular at $P(0, -1, 0)$. □

Exercises

1. Find the tangent vector and the tangent line at the indicated point.
   * (a) $\mathbf{r}(t) = \cos \pi t\, \mathbf{i} + \sin \pi t\, \mathbf{j} + t\mathbf{k}$ at $t = 2$.
     (b) $\mathbf{r}(t) = e^t\mathbf{i} + e^{-t}\mathbf{j} - \log t\, \mathbf{k}$ at $t = 1$.
   * (c) $\mathbf{r}(t) = \mathbf{a} + t\mathbf{b} + t^2\mathbf{c}$ at $t = -1$.
     (d) $\mathbf{r}(t) = (t + 1)\mathbf{i} + (t^2 + 1)\mathbf{j} + (t^3 + 1)\mathbf{k}$ at $P(1, 1, 1)$.
   * (e) $\mathbf{r}(t) = 2t^2\mathbf{i} + (1 - t)\mathbf{j} + (3 + 2t^2)\mathbf{k}$ at $P(2, 0, 5)$.

2. Show that

$$\mathbf{r}(t) = a \cos \omega t\, \mathbf{i} + b \sin \omega t\, \mathbf{j}$$

   traces out an ellipse. Find an equation for the ellipse in x and y.
* 3. Show that

$$\mathbf{r}(t) = at\mathbf{i} + bt^2\mathbf{j}$$

   traces out a parabola. Find an equation for the parabola in x and y.

4. Show that

$$\mathbf{r}(t) = \tfrac{1}{2}a(e^{\omega t} + e^{-\omega t})\mathbf{i} + \tfrac{1}{2}a(e^{\omega t} - e^{-\omega t})\mathbf{j}$$

   traces out the hyperbola

$$x^2 - y^2 = a^2.$$

* 5. Find (a) the points on the plane curve

$$\mathbf{r}(t) = t\mathbf{i} + (1 + t^2)\mathbf{j}$$

   at which $\mathbf{r}(t)$ and $\mathbf{r}'(t)$ are perpendicular; (b) the points at which they have the same direction; (c) the points at which they have opposite directions.
* 6. Find the curve given that

$$\mathbf{r}'(t) = \alpha\, \mathbf{r}(t) \quad \text{for all real } t \quad \text{and} \quad \mathbf{r}(0) = \mathbf{i} + 2\mathbf{j} + 3\mathbf{k}.$$

7. The curves

$$\mathbf{r}_1(t) = (e^t - 1)\mathbf{i} + 2 \sin t\, \mathbf{j} + \log (t + 1)\mathbf{k},$$
$$\mathbf{r}_2(u) = (u + 1)\mathbf{i} + (u^2 - 1)\mathbf{j} + (u^3 + 1)\mathbf{k}$$

   intersect at the origin. Find the angle of intersection.

8. Show that if $\mathbf{r}'(t)$ and $\mathbf{r}(t)$ are parallel, then the tangent line passes through the origin.

*9. Find the point at which the curves

$$\mathbf{r}_1(t) = e^t\mathbf{i} + 2\sin(t + \tfrac{1}{2}\pi)\mathbf{j} + (t^2 - 2)\mathbf{k},$$
$$\mathbf{r}_2(u) = u\mathbf{i} + 2\mathbf{j} + (u^2 - 3)\mathbf{k}$$

intersect and then find the angle of intersection.

10. The curve

$$\mathbf{r}(t) = a\cos\omega t\,\mathbf{i} + a\sin\omega t\,\mathbf{j} + b\omega t\mathbf{k} \qquad (a > 0, \quad b > 0, \quad \omega > 0)$$

is called a *circular helix*.

(a) Show that the circular helix climbs along the cylinder

$$x^2 + y^2 = a^2, \qquad z \text{ unrestricted.} \qquad \text{(Figure 13.4.2)}$$

(b) Show that it intersects lines parallel to the z-axis at a constant angle.

13.4   Velocity and Acceleration

A curve

$$C: \mathbf{r}(t) = x(t)\mathbf{i} + y(t)\mathbf{j} + z(t)\mathbf{k}, \qquad t \in I$$

can be thought of as the path traced out by a moving object. We can interpret I as a time interval and use $\mathbf{r}(t)$ to indicate the position of the object at time t. Let's assume that $\mathbf{r}$ is twice differentiable. In this context the derivatives, $\mathbf{r}'(t)$ and $\mathbf{r}''(t)$, have special names and special significance.

Definition

The vector $\mathbf{r}'(t)$ is called the *velocity* of the object at time t and $\mathbf{r}''(t)$ is called the *acceleration*.

In symbols, we have

$$\boxed{\mathbf{r}'(t) = \mathbf{v}(t) \quad \text{and} \quad \mathbf{r}''(t) = \mathbf{v}'(t) = \mathbf{a}(t).}$$

There should be nothing surprising about this. As before, we have velocity as the time rate of change of position and acceleration as the time rate of change of velocity. Since

$$\mathbf{v}(t) = \mathbf{r}'(t),$$

we know from the last section that the velocity vector, when not $\mathbf{0}$, is tangent to the path of the motion. The direction of the velocity vector gives us the direction of the motion. (Figure 13.4.1)

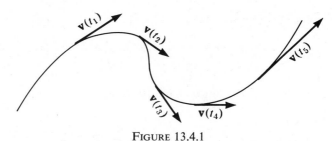

FIGURE 13.4.1

The magnitude of the velocity vector tells us how fast the object is moving. The details will be made clear in the next section. At this point consider it true by definition; namely, for an object moving with velocity $\mathbf{v}(t)$ we define

$$\|\mathbf{v}(t)\| = \text{speed at time } t.$$

Problem.   A particle moves along a circular helix (Figure 13.4.2)

$$\mathbf{r}(t) = a \cos \omega t \, \mathbf{i} + a \sin \omega t \, \mathbf{j} + bt\omega \, \mathbf{k}. \qquad (a > 0, \quad b > 0, \quad \omega > 0)$$

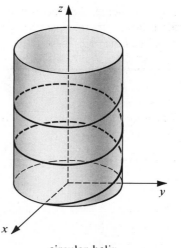

circular helix

FIGURE 13.4.2

For each time t, find

(a) the velocity.
(b) the speed.
(c) the acceleration.
(d) the magnitude of the acceleration vector.
(e) the angle between the velocity vector and the acceleration vector.

SOLUTION

(a) $\mathbf{v}(t) = \mathbf{r}'(t) = -a\omega \sin \omega t\, \mathbf{i} + a\omega \cos \omega t\, \mathbf{j} + b\omega\, \mathbf{k}$.

(b) speed at time $t = \|\mathbf{v}(t)\|$

$$= \sqrt{a^2\omega^2 \sin^2 \omega t + a^2\omega^2 \cos^2 \omega t + b^2\omega^2}$$
$$= \sqrt{a^2\omega^2 + b^2\omega^2} = \omega\sqrt{a^2 + b^2}.$$

The speed is thus constant.

(c) $\mathbf{a}(t) = \mathbf{v}'(t) = -a\omega^2 \cos \omega t\, \mathbf{i} - a\omega^2 \sin \omega t\, \mathbf{j}$
$$= -a\omega^2(\cos \omega t\, \mathbf{i} + \sin \omega t\, \mathbf{j}).$$

Since the speed is constant, the acceleration comes from the change in direction of the velocity vector.

(d) the magnitude of acceleration vector $= \|\mathbf{a}(t)\| = a\omega^2$.

(e) $\cos \theta = \dfrac{\mathbf{v}(t) \cdot \mathbf{a}(t)}{\|\mathbf{v}(t)\|\, \|\mathbf{a}(t)\|}$

$$= \frac{[-a\omega \sin \omega t\, \mathbf{i} + a\omega \cos \omega t\, \mathbf{j} + b\omega\, \mathbf{k}] \cdot [-a\omega^2 (\cos \omega t\, \mathbf{i} + \sin \omega t\, \mathbf{j})]}{a\omega^3\sqrt{a^2 + b^2}}$$

$$= \frac{a(\sin \omega t \cos \omega t - \cos \omega t \sin \omega t)}{\sqrt{a^2 + b^2}} = 0.$$

The angle is therefore $\frac{1}{2}\pi$. □

In classical mechanics the *momentum* $\mathbf{p}$ of an object is defined by the equation

$$\mathbf{p} = m\mathbf{v}$$

where m is the mass of the object and $\mathbf{v}$ is its velocity. To indicate the time dependence, we write

$$\mathbf{p}(t) = m\mathbf{v}(t).$$

The acceleration of an object is related to the *force* acting on it by Newton's second law of motion:

$$\mathbf{F}(t) = m\mathbf{a}(t). \qquad (\textit{force equals mass times acceleration})$$

Differentiating the momentum equation, we find that

$$\mathbf{p}'(t) = m\mathbf{v}'(t) = m\mathbf{a}(t).$$

Obviously then

$$\mathbf{F}(t) = \mathbf{p}'(t).$$

In other words, force is simply the time rate of change of momentum.

Problem. An object of mass m follows an elliptical path in such a manner that at each time t it has position

$$\mathbf{r}(t) = a \cos \omega t\, \mathbf{i} + b \sin \omega t\, \mathbf{j}.$$

Find the force acting on the object.

SOLUTION

$$v(t) = r'(t) = -a\omega \sin \omega t \, \mathbf{i} + b\omega \cos \omega t \, \mathbf{j}.$$

$$\mathbf{a}(t) = \mathbf{v}'(t) = \mathbf{r}''(t) = -a\omega^2 \cos \omega t \, \mathbf{i} - b\omega^2 \sin \omega t \, \mathbf{j} = -\omega^2 \mathbf{r}(t).$$

$$\mathbf{F}(t) = -m\omega^2 \mathbf{r}(t). \quad \square$$

This force is called a *central* force because it acts along the line of the radius vector. It is called an *attractive* force because its direction is opposite to that of the radius vector. In magnitude this force is proportional to the distance of the object from the origin. We illustrate all this in Figure 13.4.3.

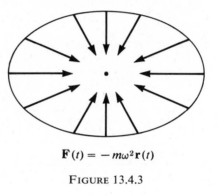

$$\mathbf{F}(t) = -m\omega^2 \mathbf{r}(t)$$

FIGURE 13.4.3

Problem. A force of magnitude α directed upward from the xy-plane is continually applied to an object of mass m. Given that the object starts at $P(0, y_0, z_0)$ with initial velocity $2\mathbf{j}$ find:

(a) the velocity of the object t seconds later.
(b) the speed of the object t seconds later.
(c) the momentum of the object t seconds later.
(d) the path followed by the object, both in vector form and in Cartesian coordinates.

SOLUTION. We begin with

$$\mathbf{F}(t) = \alpha \, \mathbf{k}.$$

Since in general

$$\mathbf{F}(t) = m \, \mathbf{a}(t),$$

we have here

$$\mathbf{a}(t) = \frac{\alpha}{m} \, \mathbf{k}.$$

Integration gives

$$\mathbf{v}(t) = C_1 \mathbf{i} + C_2 \mathbf{j} + \left(\frac{\alpha}{m} t + C_3 \right) \mathbf{k}.$$

Since
$$\mathbf{v}(0) = 2\mathbf{j},$$

you can see that
$$C_1 = 0, \qquad C_2 = 2, \qquad C_3 = 0$$

and consequently
$$\mathbf{v}(t) = 2\mathbf{j} + \frac{\alpha}{m}\, t\, \mathbf{k}.$$

Another integration gives
$$\mathbf{r}(t) = D_1\mathbf{i} + (2t + D_2)\mathbf{j} + \left(\frac{\alpha}{2m}\, t^2 + D_3\right)\mathbf{k}.$$

From the initial position
$$\mathbf{r}(0) = y_0\mathbf{j} + z_0\mathbf{k}$$

we find that
$$D_1 = 0, \qquad D_2 = y_0, \qquad D_3 = z_0$$

and thus that
$$\mathbf{r}(t) = (2t + y_0)\mathbf{j} + \left(\frac{\alpha}{2m}\, t^2 + z_0\right)\mathbf{k}, \qquad t \geq 0.$$

This path is a parabolic arc in the yz-plane. (See Figure 13.4.4.) To get an equation for this arc in Cartesian coordinates, we write out the components

$$x(t) = 0, \qquad y(t) = 2t + y_0, \qquad z(t) = \frac{\alpha}{2m}\, t^2 + z_0.$$

From the second equation we have
$$t = \tfrac{1}{2}[y(t) - y_0].$$

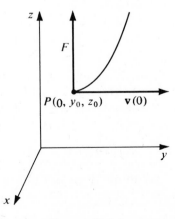

FIGURE 13.4.4

Substituting this into the third equation, we get

$$z(t) = \frac{\alpha}{8m}[y(t) - y_0]^2 + z_0.$$

Eliminating t altogether, we have

$$z = \frac{\alpha}{8m}(y - y_0)^2 + z_0.$$

Of course we must bear in mind the extra condition $x = 0$.

Answers to (a) through (d):

(a) velocity: $\mathbf{v}(t) = 2\mathbf{j} + \frac{\alpha}{m}t\,\mathbf{k}.$

(b) speed: $\|\mathbf{v}(t)\| = \frac{1}{m}\sqrt{4m^2 + \alpha^2 t^2}.$

(c) momentum: $m\mathbf{v}(t) = 2m\mathbf{j} + \alpha t\mathbf{k}.$

(d) path in vector form: $\mathbf{r}(t) = (2t + y_0)\mathbf{j} + \left(\frac{\alpha}{2m}t^2 + z_0\right)\mathbf{k}, \qquad t \geq 0.$

path in x, y, z: $z = \frac{\alpha}{8m}(y - y_0)^2 + z_0, \qquad x = 0.$ ☐

The scalar product

$$\mathbf{F}(t) \cdot \mathbf{v}(t)$$

is called the *power* (which is given to the object by the force). The *kinetic energy* of an object is defined as

$$\tfrac{1}{2}m\|\mathbf{v}(t)\|^2.$$

In Exercise 7 you are asked to show that power is the time rate of change of kinetic energy:

$$\frac{d}{dt}\left(\frac{1}{2}m\|\mathbf{v}(t)\|^2\right) = \mathbf{F}(t) \cdot \mathbf{v}(t).$$

We will return to these notions later. Now let's talk briefly about the dynamics of rotations. The central notions here are *torque*, usually denoted by τ (the Greek letter tau), and *angular momentum*, usually denoted by **L**. Torque, from the Latin word for "twist," is defined by the equation

(13.4.1) $\qquad\qquad \tau(t) = \mathbf{r}(t) \times \mathbf{F}(t).$

More precisely (13.4.1) gives the torque produced by the force $\mathbf{F}(t)$ about the origin Angular momentum is defined by setting

$$\mathbf{L}(t) = \mathbf{r}(t) \times m\mathbf{v}(t).$$

Just as force is the time rate of change of momentum, torque is the time rate of change of angular momentum:

$$\begin{aligned}
\mathbf{L}'(t) &= [\mathbf{r}(t) \times m\mathbf{v}'(t)] + [\mathbf{r}'(t) \times m\mathbf{v}(t)] \\
&= [\mathbf{r}(t) \times m\mathbf{a}(t)] + [\mathbf{v}(t) \times m\mathbf{v}(t)] \\
&= [\mathbf{r}(t) \times m\mathbf{a}(t)] + \mathbf{0} \\
&= \tau(t). \quad \square
\end{aligned}$$

Central forces, forces which act along the line of the radius vector, produce no torque. To verify this note that every central force has the form

$$\mathbf{F}(t) = u(t)\mathbf{r}(t).$$

Obviously then

$$\tau(t) = \mathbf{r}(t) \times \mathbf{F}(t) = \mathbf{r}(t) \times u(t)\mathbf{r}(t) = \mathbf{0}. \quad \square$$

Using these ideas, we can demonstrate Kepler's marvelous equal-time equal-area law for planetary motion.

Kepler's Second Law

As a planet moves around the sun, the radius vector sweeps out equal areas in equal times.

For convenience we picture the sun at the origin of our coordinate system. (See Figure 13.4.5.)

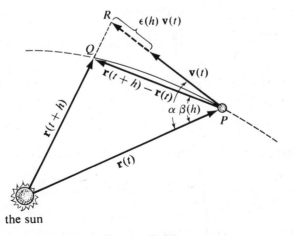

FIGURE 13.4.5

The force involved is the gravitational attraction of the sun. Since this is a central force, we know that the torque is zero and thus the angular momentum of the planet

$$\mathbf{L}(t) = \mathbf{r}(t) \times m\mathbf{v}(t)$$

is constant. In particular, its norm is constant and thus

$$\|\mathbf{r}(t)\| \, \|\mathbf{v}(t)\| \, |\sin \alpha|$$

is constant.

We will use $A(t)$ to indicate the area swept out by the radius vector from some fixed time up to time t. Our task is to show that $A'(t)$ is constant. This we will do by showing that

$$A'(t) = \tfrac{1}{2} \|\mathbf{r}(t)\| \, \|\mathbf{v}(t)\| \, |\sin \alpha|.$$

The area swept out during the time interval $[t, t + h]$ is simply

$$A(t + h) - A(t).$$

From obvious geometric considerations,

area of $\triangle OPQ \leq A(t + h) - A(t) \leq$ area of $\triangle OPQ$ + area of $\triangle PQR$†

and thus

(13.4.2) $\dfrac{\text{area of } \triangle OPQ}{h} \leq \dfrac{A(t + h) - A(t)}{h} \leq \dfrac{\text{area of } \triangle OPQ}{h} + \dfrac{\text{area of } \triangle PQR}{h}.$

Since

$$\frac{\text{area of } \triangle OPQ}{h} = \frac{1}{2} \|\mathbf{r}(t)\| \left\| \frac{\mathbf{r}(t + h) - \mathbf{r}(t)}{h} \right\| |\sin \beta(h)|,$$

we see that

$$\lim_{h \downarrow 0} \frac{\text{area of } \triangle OPQ}{h} = \frac{1}{2} \|\mathbf{r}(t)\| \, \|\mathbf{v}(t)\| \, |\sin \alpha|.$$

We turn our attention now to the other triangle.

$$\frac{\text{area of } \triangle PQR}{h} = \frac{1}{2} \left\| \frac{\mathbf{r}(t + h) - \mathbf{r}(t)}{h} \right\| \|\mathbf{v}(t) + t(h)\mathbf{v}(t)\| \, |\sin [\alpha - \beta(h)]|.$$

As h decreases to zero, $\beta(h)$ tends to α and thus $|\sin [\alpha - \beta(h)]|$ tends to zero. Since

$$\left\| \frac{\mathbf{r}(t + h) - \mathbf{r}(t)}{h} \right\| \quad \text{tends to} \quad \|\mathbf{v}(t)\|,$$

and $\|\mathbf{v}(t) + t(h)\mathbf{v}(t)\|$ certainly remains bounded, we see that

$$\lim_{h \downarrow 0} \frac{\text{area of } \triangle PQR}{h} = 0.$$

---

† Here we are assuming that the path curves toward the sun and not away from it. This is physically obvious: gravitational force is an attracting force, not a repelling force.

From (13.4.2)

$$\lim_{h \downarrow 0} \frac{A(t + h) - A(t)}{h} = \frac{1}{2} \|\mathbf{r}(t)\| \|\mathbf{v}(t)\| |\sin \alpha|,$$

which assures us that

$$A'(t) = \tfrac{1}{2}\|\mathbf{r}(t)\| \|\mathbf{v}(t)\| |\sin \alpha|$$

as asserted.   □

Exercises

1. An object moves so that

$$\mathbf{r}(t) = (a \cos t + b)\mathbf{i} + (a \sin t + c)\mathbf{j}.$$

   (a) Show that the speed remains constant.
   (b) What is the acceleration at time 0?
2. A moving point traces out the curve

$$\mathbf{r}(t) = at\mathbf{i} + b \sin at\, \mathbf{j}.$$

   Show that the magnitude of its acceleration remains proportional to its distance from the x-axis.
* 3. A point traces out the curve

$$\mathbf{r}(t) = 2\mathbf{i} + t^2\mathbf{j} + (t - 1)^2\mathbf{k}.$$

   At what time is the speed a minimum?
4. The path of a plane motion is the curve

$$\mathbf{r}(t) = 2 \cos 2t\, \mathbf{i} + 3 \cos t\, \mathbf{j}.$$

   * (a) Show that the object oscillates on an arc of the parabola

$$4y^2 - 9x - 18 = 0.$$

   (b) Draw the path.
   * (c) What are the acceleration vectors at the points of zero velocity?
   (d) Draw these vectors at the points in question.
5. An object of mass m moves along the path

$$\mathbf{r}(t) = \tfrac{1}{2}a(e^{\omega t} + e^{-\omega t})\mathbf{i} + \tfrac{1}{2}b(e^{\omega t} - e^{-\omega t})\mathbf{j}.$$

   (a) What is the velocity at $t = 0$?
   (b) Show that the acceleration vector is a constant positive multiple of the position vector.
   (c) What does (b) imply about the angular momentum and the torque? Verify your answers by direct calculation.
6. Show that a particle subject to no outside forces is either stationary or moves with constant speed along a straight line.
7. Show that power is the time rate of change of kinetic energy.
* 8. Find the force required to propel a particle of mass m so that

$$\mathbf{r}(t) = t\mathbf{j} + t^2\mathbf{k}.$$

*9. At each point $P(x(t), y(t), z(t))$ of its motion an object of mass m is subject to a force

$$\mathbf{F}(t) = m\pi^2[a \cos \pi t \, \mathbf{i} + b \sin \pi t \, \mathbf{j}]. \qquad (a > 0, b > 0)$$

Given that

$$\mathbf{v}(0) = -\pi b\mathbf{j} + \mathbf{k} \quad \text{and} \quad \mathbf{r}(0) = b\mathbf{j}$$

find the following at time $t = 1$.

(a) the velocity.   (b) the speed.
(c) the acceleration.   (d) the momentum.
(e) the power.   (f) the kinetic energy.
(g) the angular momentum.   (h) the torque.

13.5   Arc Length and Speed

Arc Length

In Figure 13.5.1 we have sketched a curve C and a polygonal approximation to it. The length of C is by definition the least upper bound of the lengths of such polygonal approximations.

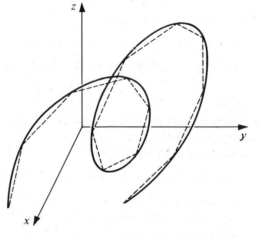

a space curve and a polygonal approximation

FIGURE 13.5.1

Earlier we considered plane curves and used the arc length formula

$$l(C) = \int_a^b \sqrt{[x'(t)]^2 + [y'(t)]^2} \; dt.$$

For space curves the formula becomes

$$l(C) = \int_a^b \sqrt{[x'(t)]^2 + [y'(t)]^2 + [z'(t)]^2} \, dt.$$

Both of these results can be cast in the form

$$l(C) = \int_a^b \|\mathbf{r}'(t)\| \, dt.$$

For plane curves,

$$\|\mathbf{r}'(t)\| = \|x'(t)\mathbf{i} + y'(t)\mathbf{j}\| = \sqrt{[x'(t)]^2 + [y'(t)]^2}$$

and for space curves,

$$\|\mathbf{r}'(t)\| = \|x'(t)\mathbf{i} + y'(t)\mathbf{j} + z'(t)\mathbf{k}\| = \sqrt{[x'(t)]^2 + [y'(t)]^2 + [z'(t)]^2}.$$

Theorem

The length of a continuously differentiable curve

$$C: \mathbf{r}(t), \qquad t \in [a, b]$$

is given by the formula

$$l(C) = \int_a^b \|\mathbf{r}'(t)\| \, dt.$$

PROOF.   First we show that

$$l(C) \le \int_a^b \|\mathbf{r}'(t)\| \, dt.$$

To do this, we begin with an arbitrary partition P of $[a, b]$

$$P = \{a = t_0, \dots, t_{i-1}, t_i, \dots, t_n = b\}.$$

Such a partition gives rise to a finite number of points of C

$$\mathbf{r}(a) = \mathbf{r}(t_0), \dots, \mathbf{r}(t_{i-1}), \mathbf{r}(t_i), \dots, \mathbf{r}(t_n) = \mathbf{r}(b)$$

and thus to a polygonal path of total length

$$l_P = \sum_{i=1}^n \|\mathbf{r}(t_i) - \mathbf{r}(t_{i-1})\|. \qquad\qquad \text{(Figure 13.5.2)}$$

For each i,

$$\mathbf{r}(t_i) - \mathbf{r}(t_{i-1}) = \int_{t_{i-1}}^{t_i} \mathbf{r}'(t) \, dt.$$

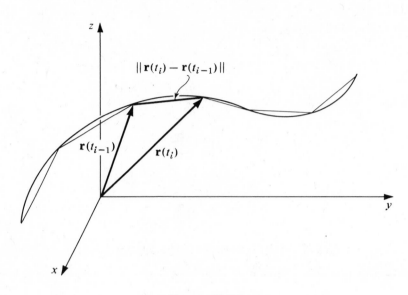

FIGURE 13.5.2

This gives

$$\|\mathbf{r}(t_i) - \mathbf{r}(t_{i-1})\| = \left\| \int_{t_{i-1}}^{t_i} \mathbf{r}'(t)\, dt \right\| \leq \int_{t_{i-1}}^{t_i} \|\mathbf{r}'(t)\|\, dt$$

and thus

$$l_P = \sum_{i=1}^{n} \|\mathbf{r}(t_i) - \mathbf{r}(t_{i-1})\| \leq \sum_{i=1}^{n} \int_{t_{i-1}}^{t_i} \|\mathbf{r}'(t)\|\, dt = \int_a^b \|\mathbf{r}'(t)\|\, dt.$$

From the arbitrariness of P, we know that the inequality

$$l_P \leq \int_a^b \|\mathbf{r}'(t)\|\, dt$$

must hold for all the l_P. This makes the integral on the right an upper bound for all the l_P. Since $l(C)$ is the *least* upper bound of all the l_P, we can conclude right now that

$$l(C) \leq \int_a^b \|\mathbf{r}'(t)\|\, dt.$$

To show equality, we need to know that arc length is additive; we need to know (see Figure 13.5.3) that the length of the curve from P to Q, plus the length of the curve from Q to R, is exactly equal to the length of the curve from P to R.† A proof of this is given in the supplement to this section.

---

† It is obvious that arc length should have this property. What we have to verify is that arc length, *as we have defined it,* has this property.

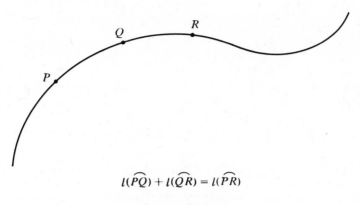

$$l(\widehat{PQ}) + l(\widehat{QR}) = l(\widehat{PR})$$

FIGURE 13.5.3

Let's go on. In Figure 13.5.4 we display the initial vector $\mathbf{r}(a)$, a general position vector $\mathbf{r}(t)$, and a nearby vector $\mathbf{r}(t + h)$. We now set

$$s(t) = \text{length of the curve from } \mathbf{r}(a) \text{ to } \mathbf{r}(t)$$

and

$$s(t + h) = \text{length of the curve from } \mathbf{r}(a) \text{ to } \mathbf{r}(t + h).$$

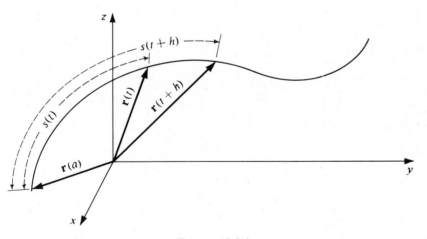

FIGURE 13.5.4

Assuming the additivity of arc length, we get

$$s(t + h) - s(t) = \text{length of the curve from } \mathbf{r}(t) \text{ to } \mathbf{r}(t + h).$$

From what we have shown already, we know that

$$\|\mathbf{r}(t + h) - \mathbf{r}(t)\| \le s(t + h) - s(t) \le \int_{t}^{t+h} \|\mathbf{r}'(u)\| \, du.$$

Dividing this inequality by h (which we are taking as positive), we get

$$\left\| \frac{r(t + h) - r(t)}{h} \right\| \leq \frac{s(t + h) - s(t)}{h} \leq \frac{1}{h} \int_t^{t+h} \|r'(u)\| \, du.$$

As $h \downarrow 0$, both extremes of the inequality tend to $\|r'(t)\|$ and consequently so does the middle term.† In short,

$$\lim_{h \downarrow 0} \frac{s(t + h) - s(t)}{h} = \|r'(t)\|.$$

By taking $h < 0$ and proceeding in a similar manner, one can show that

$$\lim_{h \uparrow 0} \frac{s(t + h) - s(t)}{h} = \|r'(t)\|.$$

It follows that

$$s'(t) = \lim_{h \to 0} \frac{s(t + h) - s(t)}{h} = \|r'(t)\|.$$

Recall now that

$$s(t) = \text{length of the curve from } r(a) \text{ to } r(t).$$

Obviously then $s(a) = 0$. This condition, together with the fact that

$$s'(t) = \|r'(t)\|,$$

implies that

$$s(t) = \text{length of the curve from } r(a) \text{ to } r(t) = \int_a^t \|r'(u)\| \, du.$$

The total length of C is therefore

$$s(b) = \int_a^b \|r'(t)\| \, dt. \quad \square$$

Problem.   Find the length of the curve

$$C: r(t) = 2 \cos t \, i + 2 \sin t \, j + t^2 \, k$$

from $t = 0$ to $t = 1$.

SOLUTION

$r'(t) = -2 \sin t \, i + 2 \cos t \, j + 2t \, k.$

$\|r'(t)\| = \sqrt{4 \sin^2 t + 4 \cos^2 t + 4t^2} = 2\sqrt{\sin^2 t + \cos^2 t + t^2} = 2\sqrt{1 + t^2}.$

$l(C) = \int_0^1 \|r'(t)\| \, dt = \int_0^1 2\sqrt{1 + t^2} \, dt = \left[t\sqrt{1 + t^2} + \log(t + \sqrt{1 + t^2}) \right]_0^1$

$= \sqrt{2} + \log(1 + \sqrt{2}). \quad \square$

---

† One way to show that the right-hand extreme of the inequality tends to $\|r'(t)\|$ is to use the first mean-value theorem for integrals. (Section 9.1)

Speed

In the last section we considered motions parametrized by time and, somewhat arbitrarily, called $\|\mathbf{r}'(t)\|$ the speed of the motion. We are now in a position to justify this. Imagine that an object moves so that at time t it has position

$$\mathbf{r}(t) = x(t)\mathbf{i} + y(t)\mathbf{j} + z(t)\mathbf{k}.$$

During a time interval $[0, t]$, the object moves along its path from $\mathbf{r}(0)$ to $\mathbf{r}(t)$ for a total distance which is equal to the length of the curve from $\mathbf{r}(0)$ to $\mathbf{r}(t)$. If, as before, we denote this distance by $s(t)$, then we have

$$s(t) = \int_0^t \|\mathbf{r}'(u)\|\ du$$

and

$$s'(t) = \|\mathbf{r}'(t)\|.$$

In other words, $\|\mathbf{r}'(t)\|$ is the *rate of change of distance with respect to time*. That is why we call it the *speed*.

In this setting, speed is commonly abbreviated by the letter v. In symbols, we have

$$\boxed{s'(t) = v(t)}$$

or, in the Leibniz notation,

$$\boxed{\dfrac{ds}{dt} = v.}$$

While integration of the velocity vector gives the position vector

$$\mathbf{r}(t) = \mathbf{r}(0) + \int_0^t \mathbf{v}(u)\ du,$$

integration of the speed gives the distance traveled:

$$s(t) = s(0) + \int_0^t v(u)\ du.$$

Exercises

Find the lengths of the following curves.

*1. $\mathbf{r}(t) = a \cos t\,\mathbf{i} + a \sin t\,\mathbf{j} + bt\mathbf{k}$ from $t = 0$ to $t = 2\pi$.

2. $\mathbf{r}(t) = t\mathbf{i} + \frac{2}{3}\sqrt{2}\,t^{3/2}\mathbf{j} + \frac{1}{2}t^2\mathbf{k}$ from $t = 0$ to $t = 2$.

*3. $\mathbf{r}(t) = t\mathbf{i} + \log(\sec t)\mathbf{j} + 3\mathbf{k}$ from $t = 0$ to $t = \frac{1}{4}\pi$.

4. $\mathbf{r}(t) = t^3\mathbf{i} + t^2\mathbf{j}$ from $t = 0$ to $t = 1$.

*5. $\mathbf{r}(t) = t\mathbf{i} + \mathbf{j} + (\frac{1}{6}t^3 + \frac{1}{2}t^{-1})\mathbf{k}$ from $t = 1$ to $t = 3$.

6. $\mathbf{r}(t) = e^t[\cos t\,\mathbf{i} + \sin t\,\mathbf{j}]$ from $t = 0$ to $t = \pi$.

*7. $\mathbf{r}(t) = 3t\cos t\,\mathbf{i} + 3t\sin t\,\mathbf{j} + 4t\,\mathbf{k}$ from $t = 0$ to $t = 4$.

8. $\mathbf{r}(t) = 2t\,\mathbf{i} + (t^2 - 2)\mathbf{j} + (1 - t^2)\mathbf{k}$ from $t = 0$ to $t = 2$.

(*Optional*) Estimate within 0.01 the lengths of the following curves from $t = 0$ to $t = \frac{1}{2}$.

*9. $\mathbf{r}(t) = t\mathbf{i} + \frac{1}{3}t^3\mathbf{j}.$ 10. $\mathbf{r}(t) = \frac{2}{5}t^{5/2}\mathbf{j} + t\mathbf{k}.$

(Evaluate the integrals by expanding the integrands in power series.)

Supplement

The Additivity of Arc Length

With P, Q, and R as in Figure 13.5.3 we wish to show that

$$l(\overparen{PQ}) + l(\overparen{QR}) = l(\overparen{PR}).$$

If L_1 is a polygonal path inscribed in $\overparen{PQ}$ and L_2 is a polygonal path inscribed in $\overparen{QR}$, then

$$L_1 \cup L_2 \quad \text{is inscribed in} \quad \overparen{PR}.$$

Since

$$l(L_1) + l(L_2) = l(L_1 \cup L_2) \quad \text{and} \quad l(L_1 \cup L_2) \leq l(\overparen{PR}),$$

we see that

$$l(L_1) + l(L_2) \leq l(\overparen{PR}) \quad \text{and thus} \quad l(L_1) \leq l(\overparen{PR}) - l(L_2).$$

Since L_1 is an arbitrary polygonal path inscribed in $\overparen{PQ}$, we conclude that

$$l(\overparen{PQ}) \leq l(\overparen{PR}) - l(L_2).$$

This last inequality we can rewrite as

$$l(L_2) \leq l(\overparen{PR}) - l(\overparen{PQ}).$$

From the arbitrariness of L_2 we conclude that

$$l(\overparen{QR}) \leq l(\overparen{PR}) - l(\overparen{PQ}),$$

which we rewrite as

$$l(\overparen{PQ}) + l(\overparen{QR}) \leq l(\overparen{PR}).$$

We now set out to prove that

$$l(\overparen{PR}) \leq l(\overparen{PQ}) + l(\overparen{QR}).$$

To do this, we need only take

$$L = \overline{T_0 T_1} \cup \cdots \cup \overline{T_{n-1} T_n}$$

as an arbitrary polygonal path inscribed in $\overset{\frown}{PR}$ and show that

$$l(L) \leq l(\overset{\frown}{PQ}) + l(\overset{\frown}{QR}).$$

If Q is one of the T_i, say $Q = T_k$, then

$$L_1 = \overline{T_0 T_1} \cup \cdots \cup \overline{T_{k-1} T_k} \quad \text{is inscribed in} \quad \overset{\frown}{PQ}$$

and

$$L_2 = \overline{T_k T_{k+1}} \cup \cdots \cup \overline{T_{n-1} T_n} \quad \text{is inscribed in} \quad \overset{\frown}{QR}.$$

Moreover, $l(L) = l(L_1) + l(L_2)$, so that

$$l(L) \leq l(\overset{\frown}{PQ}) + l(\overset{\frown}{QR}).$$

If Q is none of the T_i, then for some k

$$Q \text{ lies between } T_k \text{ and } T_{k+1}.$$

Since

$$d(T_k, T_{k+1}) \leq d(T_k, Q) + d(Q, T_{k+1}),$$

it follows that the polygonal path

$$L' = \overline{T_0 T_1} \cup \cdots \cup \overline{T_k Q} \cup \overline{Q T_{k+1}} \cup \cdots \cup \overline{T_{n-1} T_n}$$

satisfies

$$l(L) \leq l(L').$$

Proceeding as before, we see that

$$l(L') \leq l(\overset{\frown}{PQ}) + l(\overset{\frown}{QR}),$$

and thus again we have

$$l(L) \leq l(\overset{\frown}{PQ}) + l(\overset{\frown}{QR}). \quad \square$$

13.6 (Optional) Curvature

To define curvature we will use arc length as the parameter. Suppose that we have a curve

$$C: \mathbf{r}(s) = x(s)\mathbf{i} + y(s)\mathbf{j} + z(s)\mathbf{k}, \qquad s \in [0, l].$$

C is said to be *parametrized by arc length* iff for each $s \in [0, l]$, $\mathbf{r}(s)$ is that point of C which is at an arc distance s from the initial point $\mathbf{r}(0)$. The idea is illustrated in Figure 13.6.1.

There is a simple way to determine whether or not a given curve is being parametrized by arc length. In general

$$\int_0^s \|\mathbf{r}'(u)\| \, du = \text{the arc distance from } \mathbf{r}(0) \text{ to } \mathbf{r}(s).$$

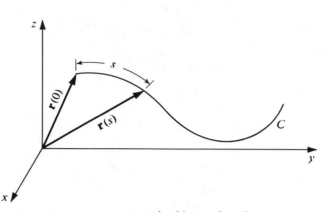

a curve parametrized by arc length

FIGURE 13.6.1

The parametrization is with respect to arc length iff

$$\int_0^s \|\mathbf{r}'(u)\| \, du = s$$

which, by differentiation, happens iff

$$\|\mathbf{r}'(s)\| = 1.$$

A curve is parametrized by arc length iff the tangent vector $\mathbf{r}'(s)$ has constant length 1.

Examples.   The curve most generally parametrized by arc length is the unit circle:

$$\mathbf{r}(s) = \cos s \, \mathbf{i} + \sin s \, \mathbf{j}, \qquad s \in [0, 2\pi].$$

A slightly more general arc length parametrization is

$$\mathbf{r}(s) = \rho[\cos (s/\rho)\mathbf{i} + \sin (s/\rho)\mathbf{j}], \qquad s \in [0, 2\pi\rho].$$

This gives the circle of radius ρ also centered at the origin.

To verify that this second parametrization is also with respect to arc length we need only verify that the tangent vector $\mathbf{r}'(s)$ has constant length 1:

$$\mathbf{r}'(s) = -\sin (s/\rho)\mathbf{i} + \cos (s/\rho)\mathbf{j}$$

and

$$\|\mathbf{r}'(s)\| = \sqrt{\sin^2 (s/\rho) + \cos^2 (s/\rho)} = 1. \quad \square$$

Take now any curve parametrized by arc length. Although the length of the tangent vector $\mathbf{r}'(s)$ remains constantly 1, the direction of this vector can change as s changes. The vector $\mathbf{r}''(s)$ signals this change of direction and is called the *curvature vector*. Its magnitude

(13.6.1)
$$\boxed{k(s) = \|\mathbf{r}''(s)\|}$$

is called the *curvature*. The curvature gives the change in direction per unit of arc length.

Example. In a circle of radius ρ the curvature is constantly $1/\rho$.

PROOF. For convenience we place the circle in the xy-plane and center it at the origin. In terms of arc length we have

$$C: \mathbf{r}(s) = \rho[\cos (s/\rho)\mathbf{i} + \sin (s/\rho)\mathbf{j}], \qquad s \in [0, 2\pi\rho].$$

This gives

$$\mathbf{r}'(s) = -\sin (s/\rho)\mathbf{i} + \cos (s/\rho)\mathbf{j}$$

and

$$\mathbf{r}''(s) = -(1/\rho)[\cos (s/\rho)\mathbf{i} + \sin (s/\rho)\mathbf{j}].$$

It follows then that

$$k(s) = \|\mathbf{r}''(s)\| = 1/\rho$$

as asserted. $\square$

That the curvature of a circle of radius ρ should be $1/\rho$ is intuitively obvious. It is geometrically and physically evident that in a circular path the change in direction takes place at a constant rate. Since a complete revolution entails a change in direction of 2π radians and this change is carried out on a path of length $2\pi\rho$, the change in direction per unit arc length must be $1/\rho$.

Returning now to

$$\|\mathbf{r}'(s)\| = 1,$$

we see that

$$\mathbf{r}'(s) \cdot \mathbf{r}'(s) = 1,$$
$$\mathbf{r}'(s) \cdot \mathbf{r}''(s) + \mathbf{r}''(s) \cdot \mathbf{r}'(s) = 0,$$
$$2[\mathbf{r}'(s) \cdot \mathbf{r}''(s)] = 0,$$

and therefore

$$\mathbf{r}'(s) \perp \mathbf{r}''(s).$$

This tells us that the *curvature vector is always perpendicular to the tangent vector.*

Curves other than the circle are seldom parametrized by arc length. It is therefore useful to know how to compute curvature in terms of an arbitrary parameter t, which for convenience we can interpret as time.

Theorem

Let

$$C: \mathbf{r}(t) = x(t)\mathbf{i} + y(t)\mathbf{j} + z(t)\mathbf{k}, \qquad t \in [a, b]$$

be a twice differentiable curve. If $\mathbf{r}'(t) \neq \mathbf{0}$, the curvature is given by the formula

(13.6.2)
$$k(t) = \left\| \frac{\mathbf{T}'(t)}{s'(t)} \right\|$$

where $\mathbf{T}$ is the unit tangent vector

$$\mathbf{T}(t) = \frac{\mathbf{r}'(t)}{\|\mathbf{r}'(t)\|}.$$

PROOF. We begin by assuming that $\mathbf{r}'(t) \neq \mathbf{0}$. By $s(t)$ we mean the length of the curve from $\mathbf{r}(a)$ to $\mathbf{r}(t)$. As t ranges from a to b, $s(t)$ ranges from 0 to l, the total length of the curve. For each number $s(t) \in [0, l]$, we define

$$\mathbf{g}(s(t)) = \mathbf{r}(t).\qquad\qquad\text{(see Figure 13.6.2)}$$

Since

$$\mathbf{g}'(s(t))s'(t) = \mathbf{r}'(t) \quad\text{and}\quad \mathbf{r}'(t) \neq \mathbf{0},$$

you can see that

$$s'(t) \neq 0.$$

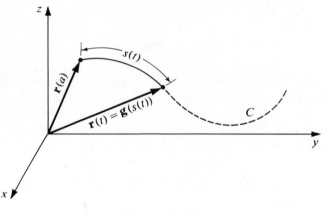

FIGURE 13.6.2

We defined the function $\mathbf{g}$ so that $\mathbf{g}(s)$ would give us an arc length parametrization for C. Consequently, we have as the unit tangent

$$\mathbf{T}(t) = \mathbf{g}'(s(t)).$$

Differentiating again, we get

$$\mathbf{T}'(t) = \mathbf{g}''(s(t))s'(t)$$

and thus

$$k(t) = \|g''(s(t))\| = \left\|\frac{\mathbf{T}'(t)}{s'(t)}\right\|. \quad\square$$

The curvature formula is often written

$$\boxed{k = \frac{1}{v}\left\|\frac{d\mathbf{T}}{dt}\right\|.}$$

Problem. Find the curvature of

$$\mathbf{r}(t) = t\mathbf{i} + t^2\mathbf{j} + \tfrac{2}{3}t^3\mathbf{k}\quad\text{at the point } P(-1, 1, -\tfrac{2}{3}).$$

SOLUTION

$$\mathbf{r}'(t) = \mathbf{i} + 2t\mathbf{j} + 2t^2\mathbf{k}$$

$$s'(t) = \|\mathbf{r}'(t)\| = \sqrt{1 + 4t^2 + 4t^4} = 1 + 2t^2$$

$$\mathbf{T}(t) = \frac{\mathbf{r}'(t)}{\|\mathbf{r}'(t)\|} = \frac{1}{1 + 2t^2}[\mathbf{i} + 2t\mathbf{j} + 2t^2\mathbf{k}]$$

$$\mathbf{T}'(t) = \frac{(1 + 2t^2)[2\mathbf{j} + 4t\mathbf{k}] - 4t[\mathbf{i} + 2t\mathbf{j} + 2t^2\mathbf{k}]}{(1 + 2t^2)^2}$$

$$= \frac{-4t\mathbf{i} + (2 - 4t^2)\mathbf{j} + 4t\mathbf{k}}{(1 + 2t^2)^2}.$$

The point $P(-1, 1, -\frac{2}{3})$ corresponds to the parameter value $t = -1$.

$$\mathbf{T}'(-1) = \tfrac{1}{9}[4\mathbf{i} - 2\mathbf{j} - 4\mathbf{k}], \qquad \|\mathbf{T}'(-1)\| = \tfrac{1}{9}\sqrt{16 + 4 + 16} = \tfrac{2}{3}, \qquad s'(-1) = 3.$$

It follows that

$$k = \left\| \frac{\mathbf{T}'(-1)}{s'(-1)} \right\| = \left(\frac{2}{3}\right)\left(\frac{1}{3}\right) = \frac{2}{9}. \quad \square$$

The quicker the turning, the larger the curvature. The slower the turning, the smaller the curvature. Along a straight line there is no turning and the curvature is 0.

PROOF. If

$$\mathbf{r}(t) = \mathbf{r}_0 + t\mathbf{d},$$

then

$$\mathbf{r}'(t) = \mathbf{d}$$

and the unit tangent

$$\mathbf{T}(t) = \frac{\mathbf{d}}{\|\mathbf{d}\|}$$

has derivative $\mathbf{0}$:

$$\mathbf{T}'(t) = \mathbf{0}.$$

It follows that

$$\|\mathbf{T}'(t)\| = 0 \quad \text{and hence that } k(t) = 0. \quad \square$$

For a plane curve

$$\mathbf{r}(t) = x(t)\mathbf{i} + y(t)\mathbf{j}$$

the curvature can be written

(13.6.3)

$$k(t) = \frac{|x'(t)y''(t) - y'(t)x''(t)|}{([x'(t)]^2 + [y'(t)]^2)^{3/2}}.$$

We can verify this directly from formula (13.6.2) but the calculations are laborious and we will omit them.

Problem. Find the curvature of

$$\mathbf{r}(t) = a \cos^3 t \, \mathbf{i} + a \sin^3 t \, \mathbf{j} \qquad \text{at } t = \tfrac{1}{4}\pi.$$

SOLUTION. Here

$$x'(t) = -3a \sin t \cos^2 t, \qquad x''(t) = 6a \sin^2 t \cos t - 3a \cos^3 t,$$
$$y'(t) = 3a \sin^2 t \cos t, \qquad y''(t) = -3a \sin^3 t + 6a \sin t \cos^2 t,$$

so that

$$x'(\tfrac{1}{4}\pi) = -\tfrac{3}{4}\sqrt{2}a, \qquad x''(\tfrac{1}{4}\pi) = \tfrac{3}{4}\sqrt{2}a,$$
$$y'(\tfrac{1}{4}\pi) = \tfrac{3}{4}\sqrt{2}a, \qquad y''(\tfrac{1}{4}\pi) = \tfrac{3}{4}\sqrt{2}a,$$

and, as you can check, formula (13.6.3) gives

$$k(\tfrac{1}{4}\pi) = \frac{2}{3|a|}. \quad \square$$

For the graph of a function

$$y = f(x)$$

the curvature formula is

(13.6.4)
$$\boxed{k(x) = \frac{|f''(x)|}{(1 + [f'(x)]^2)^{3/2}}.}$$

PROOF. We can parametrize the graph of f by setting

$$\mathbf{r}(t) = t\mathbf{i} + f(t)\mathbf{j}.$$

Here

$$x(t) = t, \qquad x'(t) = 1, \qquad x''(t) = 0,$$
$$y(t) = f(t), \qquad y'(t) = f'(t), \qquad y''(t) = f''(t),$$

so that (13.6.3) becomes

$$k(t) = \frac{|f''(t)|}{(1 + [f'(t)]^2)^{3/2}}.$$

Now replace t by x. $\quad \square$

Problem. Find the curvature of the graph at the point where

$$y = 3x - x^3$$

attains its maximum value.

SOLUTION. Here

$$f'(x) = 3 - 3x^2, \qquad f''(x) = -6x.$$

The tangent is horizontal both at $x = -1$ and at $x = 1$, but by the second derivative test only $x = 1$ gives rise to a maximum. The curvature at $x = 1$ is

$$k(1) = \frac{|f''(1)|}{(1 + [f'(1)]^2)^{3/2}} = 6. \quad \square$$

Problem. Find the curvature of the ellipse

$$x^2 + 4y^2 = 8 \qquad \text{at } x = 2, \ y = 1.$$

SOLUTION. We will use Leibniz's notation. In that notation

$$k = \frac{|d^2y/dx^2|}{[1 + (dy/dx)^2]^{3/2}}.$$

We begin by differentiating implicitly.

$$2x + 8y\,\frac{dy}{dx} = 0$$

so that

$$x + 4y\,\frac{dy}{dx} = 0.$$

A second differentiation gives

$$1 + 4y\,\frac{d^2y}{dx^2} + 4\left(\frac{dy}{dx}\right)^2 = 0.$$

From these last two equations you can see that when $x = 2$ and $y = 1$

$$\frac{dy}{dx} = -\frac{1}{2} \quad \text{and} \quad \frac{d^2y}{dx^2} = -\frac{1}{2}.$$

The curvature is $\frac{4}{25}\sqrt{5}$. $\square$

Exercises

Find the curvature.

*1. $y = x - x^2$.   2. $y = e^{-x}$.

*3. $y = x^3$.   4. $y = \log (\sec x)$.

* 5. Find the point of maximal curvature on the curve

$$y = \log x.$$

6. Find the curvature at the vertex of the parabola

$$x^2 = 4y.$$

* 7. Find the curvature of the hyperbola

$$x^2 - 4y^2 = 9 \qquad \text{at } x = 5, \ y = 2$$

by implicit differentiation.

8. Find the curvature of

$$x^2 + 4xy - 2y^2 = 10 \qquad \text{at } x = 2, \ y = 1$$

by implicit differentiation.

Find the curvature at the point indicated.

*9. $\mathbf{r}(t) = 2t\mathbf{i} + (t^2 - 1)\mathbf{j}$ at $t = 1$.

10. $\mathbf{r}(t) = t^2\mathbf{i} + (t^4 - t^5)\mathbf{j}$ at $t = 0$.

*11. $\mathbf{r}(t) = 2abt\mathbf{i} + b^2t^2\mathbf{j}$ at $t = 1$.

12. $\mathbf{r}(t) = (t - \sin t)\mathbf{i} + (1 - \cos t)\mathbf{j} + 4\sin \frac{1}{2}t\,\mathbf{k}$ at $t = \pi$.

*13. $\mathbf{r}(t) = e^t\mathbf{i} + e^{-t}\mathbf{j} + \sqrt{2}t\,\mathbf{k}$ at $t = 0$.

*14. Find a formula for the curvature of a polar curve

$$r = f(\theta).$$

15. Find the curvature of the logarithmic spiral

$$r = e^{a\theta}.$$

*16. Find the curvature of the spiral of Archimedes

$$r = a\theta.$$

17. Find the curvature of the cardioid

$$r = a(1 - \cos \theta).$$

Functions of Several Variables

14.1 What are They?

First we talk about functions of two variables. To create such a function we pick out a portion of the xy-plane and call it D. To each point (x, y) in D we assign a real number $f(x, y)$. The set of all these assignments is called a *function of two variables*. We call the function f and the set D its *domain*.

Example. Take D as the closed unit disk:

$$D = \{(x, y): x^2 + y^2 \leq 1\}.$$

To each point (x, y) in D assign the number

$$f(x, y) = \sqrt{1 - (x^2 + y^2)}. \quad \square$$

Example. Take D as the entire xy-plane and to each pair (x, y) assign the number

$$g(x, y) = xy. \quad \square$$

Example. Take D as the set of all points off the x-axis and to each such point (x, y) assign the number

$$h(x, y) = \arctan\left(\frac{x}{y}\right). \quad \square$$

Functions of three variables are just as easy to come by. To create such a function we pick out a portion of space and call it D. To each point (x, y, z) in D we assign a real number $f(x, y, z)$. The set of all these assignments is called a *function of three variables*. We call the function f and D its *domain*.

Example. Take D as the closed unit ball:

$$D = \{(x, y, z): x^2 + y^2 + z^2 \leq 1\}.$$

To each point (x, y, z) in D assign the number

$$f(x, y, z) = \sqrt{1 - (x^2 + y^2 + z^2)}. \quad \square$$

Example. Take D as all of three space and to each triple (x, y, z) assign the number

$$g(x, y, z) = xyz. \quad \square$$

Example. Take D as the set of all (x, y, z) such that $z \neq x + y$ and to each such triple assign the number

$$h(x, y, z) = \arctan\left(\frac{1}{x + y - z}\right). \quad \square$$

Functions of several variables arise naturally in very elementary settings. Here are some simple instances:

$$f(x, y) = xy, \qquad x > 0, \quad y > 0$$

gives the area of a rectangle of width x and length y;

$$f(x, y, z) = xyz, \qquad x > 0, \quad y > 0, \quad z > 0$$

gives the volume of a rectangular solid of dimensions x, y, z;

$$f(x, y) = \pi x^2 y, \qquad x > 0, \quad y > 0$$

gives the volume of a cylinder with base radius x and height y, while

$$f(x, y) = 2\pi x(x + y)$$

gives the total surface area;

$$f(x, y) = \sqrt{x^2 + y^2}$$

gives the norm of the vector $(x, y) = x\mathbf{i} + y\mathbf{j}$ and

$$f(x, y, z) = \sqrt{x^2 + y^2 + z^2}$$

gives the norm of the vector $(x, y, z) = x\mathbf{i} + y\mathbf{j} + z\mathbf{k}. \quad \square$

If the domain of a function of several variables is not explicitly given, it is to be understood that the domain is the maximal set for which the definition makes sense. Thus in the case of

$$f(x, y) = \frac{1}{x - y}$$

the domain is understood to be all the pairs (x, y) with $y \neq x$. That is, all points of the plane off the line $y = x$. In the case of

$$g(x, y, z) = \arcsin xyz$$

the domain is understood to be all the triples (x, y, z) such that

$$-1 \leq xyz \leq 1. \quad \square$$

The *range* of a function of several variables is the set of values that the function takes on. To say that a function of several variables is *bounded* is to say that its range is bounded. For example, the functions

$$f(x, y) = \frac{1}{x - y} \quad \text{and} \quad g(x, y, z) = \text{arc sin } xyz$$

have ranges

$$(-\infty, 0) \cup (0, \infty) \quad \text{and} \quad [-\tfrac{1}{2}\pi, \tfrac{1}{2}\pi]$$

respectively. The second function is bounded but the first one is not. $\square$

Problem.   Find the domain and range of

$$f(x, y) = \frac{1}{\sqrt{4x^2 - y^2}}.$$

SOLUTION.   A pair (x, y) is in the domain of f iff

$$4x^2 - y^2 > 0.$$

This occurs iff

$$y^2 < 4x^2$$

and thus iff

$$-2|x| < y < 2|x|.$$

The domain is shown in Figure 14.1.1. It consists of all points of the xy-plane which lie above $y = -2|x|$ but below $y = 2|x|$.

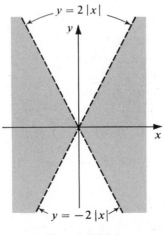

FIGURE 14.1.1

On this set $\sqrt{4x^2 - y^2}$ takes on all positive values and so does its reciprocal $f(x, y)$. The range of f is $(0, \infty)$. $\square$

Problem. Find the domain and range of

$$f(x, y, z) = \frac{\sqrt{a^2 - z^2}}{b^2 + \sqrt{r^2 - (x^2 + y^2)}}.$$

SOLUTION. Here we must have

$$x^2 + y^2 \le r^2 \quad \text{and} \quad z^2 \le a^2.$$

The domain is the solid circular cylinder of radius r and height $2a$ given in Figure 14.1.2.

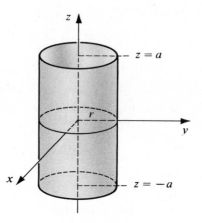

FIGURE 14.1.2

Since

$$\sqrt{a^2 - z^2}$$

takes on all values from 0 to a and

$$b^2 + \sqrt{r^2 - (x^2 + y^2)}$$

takes on all values from b^2 to $b^2 + r$, the quotient

$$f(x, y, z) = \frac{\sqrt{a^2 - z^2}}{b^2 + \sqrt{r^2 - (x^2 + y^2)}}$$

must take on all values from 0 to a/b^2. The range is the closed interval $[0, a/b^2]$.   □

Exercises

Find the domain and range of each of the following functions.

*1. $f(x, y) = \sqrt{xy}$.

2. $f(x, y) = \sqrt{1 - xy}$.

*3. $f(x, y) = \dfrac{1}{x + y}$.

4. $f(x, y) = \dfrac{1}{x^2 + y^2}$.

* 5. $f(x, y) = \dfrac{e^x - e^y}{e^x + e^y}$.

6. $f(x, y) = \dfrac{x^2}{x^2 + y^2}$.

* 7. $f(x, y) = \dfrac{1}{x} - \dfrac{1}{y}$.

8. $f(x, y) = \dfrac{y}{x} + \dfrac{x}{y}$.

* 9. $f(x, y, z) = \sqrt{x^2 + y^2 + z^2}$.

10. $f(x, y, z) = \cos x + \cos y + \cos z$.

11. $f(x, y, z) = \dfrac{1}{\sqrt{x^2 + y^2 + z^2}}$.

*12. $f(x, y) = \dfrac{1}{\sqrt{x^2 - y^2}}$.

*13. $f(x, y, z) = \dfrac{z^2}{x^2 - y^2}$.

14. $f(x, y, z) = \dfrac{z}{x - y}$.

14.2 A Brief Catalog of the Quadric Surfaces

We represented functions of one variable by curves in the xy-plane. We will picture functions of two variables by surfaces in three-dimensional space.

The simplest curved surfaces are those with equations of the second degree:

$$Ax^2 + By^2 + Cz^2 + Dxy + Exz + Fyz + Hx + Iy + Jz + K = 0.$$

Such surfaces are called *quadric surfaces*.

By suitable translations and rotations of the coordinate axes we can simplify such equations and thereby show that the nondegenerate† quadrics fall into nine distinct types:

1. the ellipsoid
2. the hyperboloid of one sheet
3. the hyperboloid of two sheets
4. the quadric cone
5. the elliptic paraboloid
6. the hyperbolic paraboloid
7. the parabolic cylinder
8. the elliptic cylinder
9. the hyperbolic cylinder.

As you go on with calculus you'll run across these surfaces time and time again. You will probably find it useful to familiarize yourself with each of these surfaces right now. In any case, we'll give you a picture of each one, together with its equation in standard form and some information about its special properties. These are some of the things to look for:

(a) the *intercepts* (the points at which the surface intersects the coordinate axes)
(b) the *traces* (the intersections with the coordinate planes)

---

† We are excluding such degenerate quadrics as
$$1 + x^2 + y^2 = -z^2 \quad \text{and} \quad x^2 + y^2 = -z^2.$$
The first one has no points and the second one consists of only one point, the origin.

(c) the *sections* (the intersections with planes in general)
(d) the *center* (some quadrics have a center; some do not)
(e) the *central line* (some quadrics have more than one center in which case they have
                infinitely many and they form a central line)
(f) *symmetry*
(g) *boundedness, unboundedness*

1. The Ellipsoid

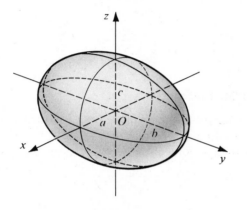

the ellipsoid: $\dfrac{x^2}{a^2} + \dfrac{y^2}{b^2} + \dfrac{z^2}{c^2} = 1$

FIGURE 14.2.1

All the sections parallel to the coordinate planes are ellipses. The numbers a, b, c are called the semiaxes of the ellipsoid. If two of the semiaxes are equal, then we have an *ellipsoid of revolution*. If the three semiaxes are equal, then we have a sphere. The surface is symmetric about the three coordinate planes.   □

2. The Hyperboloid of One Sheet

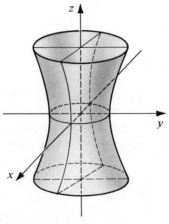

the hyperboloid of one sheet: $\dfrac{x^2}{a^2} + \dfrac{y^2}{b^2} - \dfrac{z^2}{c^2} = 1$

FIGURE 14.2.2

This surface is unbounded. It is symmetric about the three coordinate planes.

Sections parallel to the xy-plane are ellipses. Sections parallel to the other coordinate planes are hyperbolas. If $a = b$, we have a surface of revolution. ☐

3. The Hyperboloid of Two Sheets

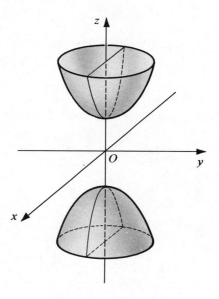

the hyperboloid of two sheets:

$$\frac{x^2}{a^2} + \frac{y^2}{b^2} - \frac{z^2}{c^2} = -1$$

FIGURE 14.2.3

The surface consists of two parts: one for which $z \geq c$, another for which $z \leq -c$. Sections parallel to the xy-plane are ellipses. Sections parallel to the other coordinate planes are hyperbolas. ☐

4. The Quadric Cone

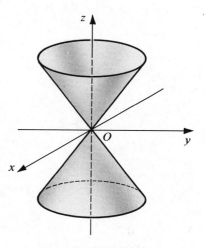

the quadric cone: $\dfrac{x^2}{a^2} + \dfrac{y^2}{b^2} = z^2$

FIGURE 14.2.4

Once again there is symmetry about the coordinate planes, the coordinate lines, and the origin. The surface is unbounded. A plane through the origin cuts the surface either only at that point or in a line or in a pair of lines. All other planes intersect the cone in an ellipse, a hyperbola, or a parabola depending on the slant of the plane. If $a = b$ we have the usual *circular cone*.   □

5.   The Elliptic Paraboloid

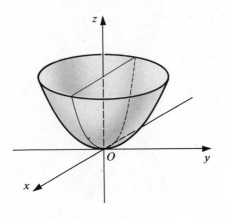

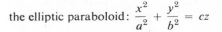

the elliptic paraboloid: $\dfrac{x^2}{a^2} + \dfrac{y^2}{b^2} = cz$

FIGURE 14.2.5

The surface does not extend below the xy-plane. The origin is called the *vertex*. Sections parallel to the xy-plane are ellipses; sections parallel to the other coordinate planes are parabolas; hence the name. The surface is symmetric about the xz-plane and about the yz-plane. It is also symmetric about the z-axis. If $a = b$, then the surface is a *paraboloid of revolution*.   □

6.   The Hyperbolic Paraboloid

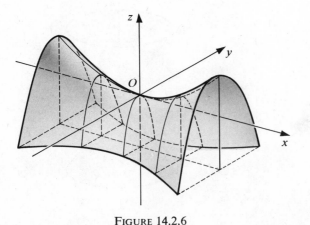

the hyperbolic paraboloid:

$$\frac{x^2}{a^2} - \frac{y^2}{b^2} = cz$$

FIGURE 14.2.6

As in the case of the elliptic paraboloid, the sections parallel to the xz-plane and the yz-plane are parabolas. Sections parallel to the xy-plane are hyperbolas. There is symmetry about the xz-plane and the yz-plane. The origin is a maximum point for the trace in the yz-plane but a minimum point for the trace in the xz-plane. Because of this, the origin is called a *minimax* or a *saddle point*. ☐

7.   The Parabolic Cylinder

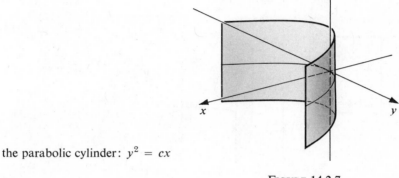

the parabolic cylinder: $y^2 = cx$

FIGURE 14.2.7

The surface is generated by a vertical line which runs along the parabola

$$y^2 = 4cx$$

of the xy-plane. ☐

8.   The Elliptic Cylinder

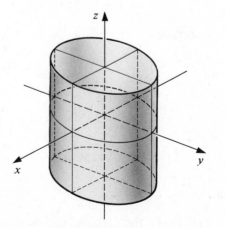

the elliptic cylinder: $\dfrac{x^2}{a^2} + \dfrac{y^2}{b^2} = 1$

FIGURE 14.2.8

The surface is generated by a vertical line which moves along the ellipse

$$\frac{x^2}{a^2} + \frac{y^2}{b^2} = 1$$

of the xy-plane. If $a = b$, we have the usual *circular cylinder*. □

9. The Hyperbolic Cylinder

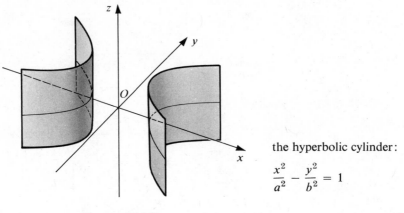

the hyperbolic cylinder:

$$\frac{x^2}{a^2} - \frac{y^2}{b^2} = 1$$

FIGURE 14.2.9

The surface is generated by a vertical line which moves along the hyperbola

$$\frac{x^2}{a^2} - \frac{y^2}{b^2} = 1$$

of the xy-plane. □

Exercises

Identify the following surfaces.

*1. $x^2 + 4y^2 - 16z^2 = 0.$ 2. $x^2 + 4y^2 + 16z^2 = 12.$

*3. $x - 4y^2 = 0.$ 4. $x^2 - 4y^2 - 2z = 0.$

*5. $5x^2 + 2y^2 - 6z^2 - 10 = 0.$ 6. $2x^2 + 4y^2 - 1 = 0.$

*7. $x^2 + y^2 + z^2 - 4 = 0.$ 8. $5x^2 + 2y^2 - 6z^2 + 10 = 0.$

*9. $x^2 + 2y^2 - 4z = 0.$ 10. $2x^2 - 3y^2 - 6 = 1.$

*11. $x - y^2 + 2z^2 = 0.$ 12. $x - y^2 - 6z^2 = 0.$

*13. Write the equation of an ellipsoid which is centered at the origin and has semiaxes 3, 4, 5. How many such ellipsoids are there?

14. Find the semiaxes of the ellipsoid

$$x^2 + 2y^2 + 3z^2 = 20.$$

*15. Describe the trace of

$$12x^2 + 9y^2 - 36z^2 = 36$$

in each of the coordinate planes.

16. Find the planes of symmetry of the cylinder

$$x - 4y^2 = 0.$$

*17. The parabola with equation

$$x^2 = 4z \quad \text{in the } xz\text{-plane}$$

is revolved about the z-axis. Find an equation for the resulting surface.

14.3   Level Curves and Level Surfaces

Here we take up the problem of representing functions of several variables by pictures. We begin with a function f of two variables defined on a subset D of the xy-plane. By the *graph of f* we mean the graph of the equation

$$z = f(x, y), \qquad (x, y) \in D.$$

Some Examples

In the case of

$$f(x, y) = x^2 + y^2$$

the domain is the entire plane. The graph of f is the paraboloid of revolution

$$z = x^2 + y^2. \quad \square \qquad\qquad \text{(Figure 14.3.1)}$$

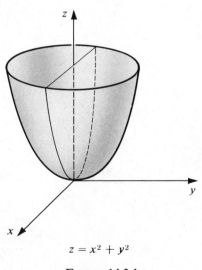

$$z = x^2 + y^2$$

FIGURE 14.3.1

In the case of

$$g(x, y) = ax + by + c$$

the domain is again the entire xy-plane. The graph of g is the plane

$$z = ax + by + c. \quad \square$$ (Figure 14.3.2)

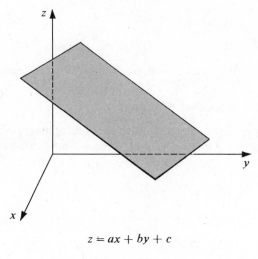

$$z \doteq ax + by + c$$

FIGURE 14.3.2

Another function with a simple graph is

$$f(x, y) = \sqrt{r^2 - (x^2 + y^2)}.$$

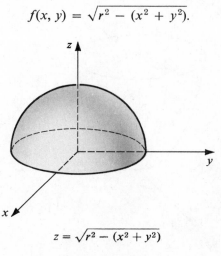

$$z = \sqrt{r^2 - (x^2 + y^2)}$$

FIGURE 14.3.3

The domain here is the disk of radius r centered at the origin:

$$\{(x, y) : x^2 + y^2 \leq r^2\}.$$

The graph has equation

$$z = \sqrt{r^2 - (x^2 + y^2)}$$

or equivalently

$$r^2 = x^2 + y^2 + z^2, \qquad z \geq 0.$$

It is the hemispherical surface depicted in Figure 14.3.3. □

Our next example

$$f(x, y) = xy$$

is more interesting. The function is simple enough, but its graph

$$z = xy$$

is quite difficult to draw. It is the "saddle-shaped" surface of Figure 14.3.4, a hyperbolic paraboloid. Rotate the x and y axes by $\frac{1}{4}\pi$ radians and you get

$$x^2 - y^2 = z,$$

surface 6 of the last section with a, b, c each 1. □

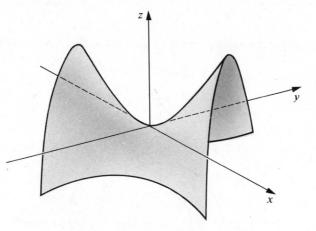

FIGURE 14.3.4

Level Curves

In practice the graph of a function of two variables is usually too difficult to draw. Moreover, if drawn, it is often difficult to interpret. Fortunately there is an easier way to convey the same information.

We take our cue from the art of the map maker. In mapping mountainous terrain it is a common practice to sketch curves joining points of constant elevation. A collection of such curves, properly labeled, gives a good idea of the altitude variations in a region. (See Figure 14.3.5.)

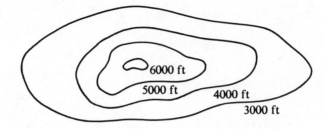

FIGURE 14.3.5

We can do the same thing to portray functions of two variables. Suppose that f is defined on some portion of the xy-plane. For a given value c that f takes on, we can sketch the curve

$$f(x, y) = c.$$

Such a curve is called a *level curve* for f. It lies in the domain of f and on it f is constantly c. A collection of such curves, properly drawn and labeled, can give a good idea of the behavior of f. Below we give some examples.

We begin with the function

$$f(x, y) = x^2 + y^2.$$

The level curves (see Figure 14.3.6) are circles centered at the origin:

$$x^2 + y^2 = c, \qquad c > 0.$$

The function takes on the value c on the circle of radius $\sqrt{c}$. At the origin the function is zero.   □

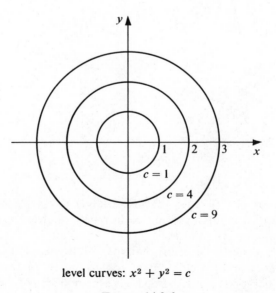

level curves: $x^2 + y^2 = c$

FIGURE 14.3.6

In the case of

$$g(x, y) = x + y + 1 \qquad \text{(Figure 14.3.7)}$$

the level curves are parallel lines

$$x + y + 1 = c. \quad \square$$

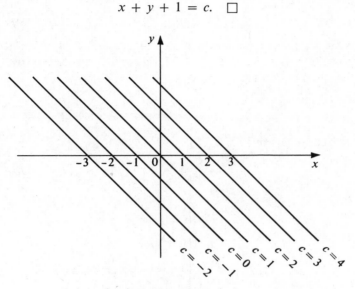

level curves: $x + y + 1 = c$

FIGURE 14.3.7

As our next example we take the function

$$h(x, y) = \begin{cases} \sqrt{x^2 + y^2}, & x \geq 0 \\ |y|, & x < 0 \end{cases}.$$

For $x \geq 0$, $h(x, y)$ is the distance from (x, y) to the origin. For $x < 0$, $h(x, y)$ is the distance from (x, y) to the x-axis. The level curves are pictured in Figure 14.3.8.

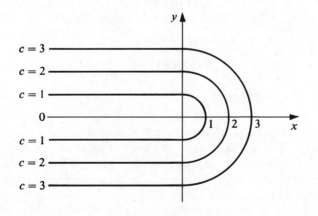

FIGURE 14.3.8

The zero level curve is the nonpositive x-axis. The other level curves are horseshoe-shaped: pairs of horizontal rays capped on the right by semicircles. □

We go back now to the function

$$f(x, y) = xy.$$

The graph of f is the hyperbolic paraboloid of Figure 14.3.4. This surface is readily understood in terms of its level curves. The 0-level curve consists of the two coordinate axes. The other level curves are simply hyperbolas:

$$xy = c.$$

Some are displayed in Figure 14.3.9. □

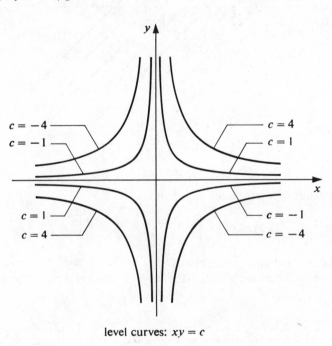

level curves: $xy = c$

FIGURE 14.3.9

Level Surfaces

Drawing graphs for functions of two variables is usually very difficult. Drawing graphs for functions of three variables is actually impossible. To draw such things, we would need four dimensions at our disposal. One way to try to visualize the behavior of a function of three variables is to examine its *level surfaces*. These are the sets with equations of the form

$$f(x, y, z) = c.$$

They are the counterparts of the level curves in the two-variable case. Unlike level curves, level surfaces are usually hard to draw. Nevertheless a knowledge of what they are is usually helpful. We shall restrict ourselves to a few very simple examples.

In the case of

$$f(x, y, z) = Ax + By + Cz$$

the level surfaces are parallel planes

$$Ax + By + Cz = c. \quad \square$$

For

$$g(x, y, z) = \sqrt{x^2 + y^2 + z^2}$$

the level surfaces are concentric spheres

$$x^2 + y^2 + z^2 = c^2. \quad \square$$

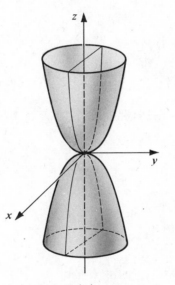

level surface: $|z| = c(x^2 + y^2)$

FIGURE 14.3.10

As our final example we take the function

$$f(x, y, z) = \frac{|z|}{x^2 + y^2},$$

which we extend to the origin by defining it there to be zero. At other points of the z-axis we leave f undefined. In the first place we note that f takes on only nonnegative values. Since it is zero only when $z = 0$, we see that the 0-level surface is the xy-plane. To get the other surfaces, we take $c > 0$ and set

$$f(x, y, z) = c.$$

This gives

$$\frac{|z|}{x^2 + y^2} = c$$

and thus

$$|z| = c(x^2 + y^2). \qquad \text{(Figure 14.3.10)}$$

Each of these surfaces is a double-paraboloid.† ☐

Exercises

Represent the following functions by level curves.

1. $f(x, y) = x - y.$ 2. $f(x, y) = x^2 - y.$
3. $f(x, y) = 2x - y.$ 4. $f(x, y) = x(y - 1).$

5. $f(x, y) = \dfrac{x}{x + y}.$ 6. $f(x, y) = \dfrac{y}{x^2}.$

7. $f(x, y) = x^3 - y.$ 8. $f(x, y) = e^{xy}.$
9. $f(x, y) = x^2 - y^2.$ 10. $f(x, y) = x^2.$

11. $f(x, y) = \dfrac{x^2}{x^2 + y^2}.$ 12. $f(x, y) = x^2 y^2.$

Identify the c-level surface and sketch it.

*13. $f(x, y, z) = x + 2y + 3z, \quad c = 0.$
14. $f(x, y, z) = x^2 + y^2, \quad c = 4.$

*15. $f(x, y, z) = \dfrac{z}{\sqrt{x^2 + y^2}}, \quad c = 1.$

16. $f(x, y, z) = \dfrac{x^2}{4} + \dfrac{y^2}{6} + \dfrac{z^2}{9}, \quad c = 1.$

*17. $f(x, y, z) = z^2 - 36x^2 - 9y^2, \quad c = 1.$
18. $f(x, y, z) = 4x^2 + 9y^2 - 72z, \quad c = 0.$

14.4 Partial Derivatives

Two Variables

The idea of partial differentiation is a very simple one. Let's start with a function f of two variables, say

$$f(x, y) = 3x^2 y - 5x \cos \pi y.$$

The *partial derivative of f with respect to x* is the function f_x obtained by differentiating f with respect to x, treating y as a constant; in this case

$$f_x(x, y) = 6xy - 5 \cos \pi y.$$

---

† Surface 5 of the last section together with its mirror image below the xy-plane.

The *partial derivative of f with respect to y* is the function f_y obtained by differentiating f with respect to y, treating x as a constant; in this case

$$f_y(x, y) = 3x^2 + 5\pi x \sin \pi y.$$

The limit definitions of these partial derivatives are as follows:

$$f_x(x, y) = \lim_{h \to 0} \frac{f(x + h, y) - f(x, y)}{h},$$

$$f_y(x, y) = \lim_{k \to 0} \frac{f(x, y + k) - f(x, y)}{k}.$$

Example. For

$$g(x, y) = x \arctan xy$$

we have

$$g_x(x, y) = x \frac{y}{1 + (xy)^2} + \arctan xy = \frac{xy}{1 + x^2 y^2} + \arctan xy$$

and

$$g_y(x, y) = x \frac{x}{1 + (xy)^2} = \frac{x^2}{1 + x^2 y^2}. \quad \square$$

In the one-variable case the derivative gives the rate of change of the function. In the two-variable case, f_x *gives the rate of change of f with respect to x* and f_y *gives the rate of change of f with respect to y*.

Example. In the case of

$$f(x, y) = x \tan y - y \tan x$$

the rate of change of f with respect to x is

$$f_x(x, y) = \tan y - y \sec^2 x$$

and the rate of change of f with respect to y is

$$f_y(x, y) = x \sec^2 y - \tan x.$$

At the point $(0, \frac{1}{4}\pi)$ we have

$$f_x(0, \tfrac{1}{4}\pi) = \tan \tfrac{1}{4}\pi - \tfrac{1}{4}\pi \sec^2 0 = 1 - \tfrac{1}{4}\pi$$

and

$$f_y(0, \tfrac{1}{4}\pi) = 0 \sec^2 \tfrac{1}{4}\pi - \tan 0 = 0. \quad \square$$

A Geometric Interpretation

In Figure 14.4.1 we have sketched a surface $z = f(x, y)$. Through this surface we have passed a plane $y = y_0$ perpendicular to the y-axis. This plane cuts the xy-plane in a line PQ and the surface above it in a curve C.

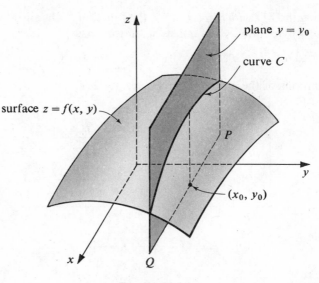

FIGURE 14.4.1

At each point (x, y_0) of PQ, $f(x, y_0)$ gives the height of C above (x, y_0). As x changes, $f(x, y_0)$ changes. The partial derivative $f_x(x_0, y_0)$ gives the rate of change when $x = x_0$.

The other partial derivative f_y can be given a similar interpretation. In Figure 14.4.2 you see the same surface $z = f(x, y)$, this time sliced by a plane $x = x_0$ perpendicular to the x-axis. On the line RS, x remains x_0 and only y changes. As y changes, $f(x_0, y)$ changes. The partial derivative gives the rate of this change when $y = y_0$.

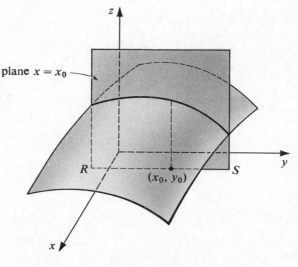

FIGURE 14.4.2

Three Variables

In the case of a function of three variables we can look for three partial derivatives: the partial with respect to x, the partial with respect to y, and also the partial with respect to z. These partials

$$f_x(x, y, z), \quad f_y(x, y, z), \quad f_z(x, y, z)$$

are obtained by differentiating with respect to the subscript variable keeping the other two variables constant. The function f_x *gives the rate of change with respect to* x, f_y *gives the rate of change with respect to* y, and f_z *gives the rate of change with respect to* z.

Example. For

$$f(x, y, z) = xy^2z^3$$

the rates of change are

$$f_x(x, y, z) = y^2z^3, \quad f_y(x, y, z) = 2xyz^3, \quad f_z(x, y, z) = 3xy^2z^2.$$

In particular

$$f_x(1, -2, -1) = -4, \quad f_y(1, -2, -1) = 4, \quad f_z(1, -2, -1) = 12. \quad \Box$$

Example. For

$$g(x, y, z) = x^2e^{y/z}$$

we have

$$g_x(x, y, z) = 2xe^{y/z}, \quad g_y(x, y, z) = \frac{x^2}{z} e^{y/z}, \quad g_z(x, y, z) = - \frac{x^2y}{z^2} e^{y/z}. \quad \Box$$

Example. If $T(x, y, z)$ gives the temperature at the point (x, y, z), then T_x gives the rate of change of T with respect to x, T_y the rate of change of T with respect to y, and T_z the rate of change of T with respect to z. $\quad \Box$

Example. For a function of the form

$$f(x, y, z) = F(x, y)G(y, z)$$

the partial derivatives are

$$f_x(x, y, z) = F_x(x, y)G(y, z),$$
$$f_y(x, y, z) = F(x, y)G_y(y, z) + F_y(x, y)G(y, z),$$
$$f_z(x, y, z) = F(x, y)G_z(y, z). \quad \Box$$

The geometric significance that we attached to the partials of a function of two variables has no analog in the case of three variables. Figures 14.4.1 and 14.4.2 do not generalize to three variables.

Other Notations

The idea of partial derivative is not tied to the letters x, y, z. In some cases it is more convenient to use other letters.

Problem.   Find the rate of change of the volume of a cylinder first with respect to its radius and then with respect to its height.

SOLUTION.   We could use the letter x for radius and the letter y for height, but it seems more natural to use r for radius and h for height. The volume then becomes

$$V(r, h) = \pi r^2 h.$$

The rate of change of V with respect to r is

$$V_r(r, h) = 2\pi r h.$$

The rate of change of V with respect to h is

$$V_h(r, h) = \pi r^2. \quad \square$$

Problem.   Let V be the volume of the frustum of a cone of lower base radius R, upper base radius r, and height h. See Figure 14.4.3. Find the rate of change of V with respect to each of these variables.

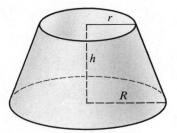

frustum of a cone

FIGURE 14.4.3

SOLUTION.   The volume is given by the formula

$$V = \tfrac{1}{3}\pi h(R^2 + Rr + r^2).$$

To emphasize the dependence of V on R, r, and h we write

$$V(R, r, h) = \tfrac{1}{3}\pi h(R^2 + Rr + r^2).$$

The rates of change we get by partial differentiation:

$$V_R(R, r, h) = \tfrac{1}{3}\pi h(2R + r),$$
$$V_r(R, r, h) = \tfrac{1}{3}\pi h(R + 2r),$$
$$V_h(R, r, h) = \tfrac{1}{3}\pi(R^2 + Rr + r^2). \quad \square$$

The subscript notation is not the only one used for partial differentiation. On the contrary, a variant of Leibniz's double-d notation is very common. In that notation the partials

$$f_x, \; f_y, \; f_z$$

are denoted by

$$\frac{\partial f}{\partial x}, \quad \frac{\partial f}{\partial y}, \quad \frac{\partial f}{\partial z}.$$

Thus for

$$f(x, y, z) = x^3 y^2 z + \sin xy$$

we have

$$\frac{\partial f}{\partial x}(x, y, z) = 3x^2 y^2 z + y \cos xy,$$

$$\frac{\partial f}{\partial y}(x, y, z) = 2x^3 yz + x \cos xy,$$

$$\frac{\partial f}{\partial z}(x, y, z) = x^3 y^2,$$

or more simply,

$$\frac{\partial f}{\partial x} = 3x^2 y^2 z + y \cos xy,$$

$$\frac{\partial f}{\partial y} = 2x^3 yz + x \cos xy,$$

$$\frac{\partial f}{\partial z} = x^3 y^2.$$

We can also write

$$\frac{\partial}{\partial x}(x^3 y^2 z + \sin xy) = 3x^2 y^2 z + y \cos xy,$$

$$\frac{\partial}{\partial y}(x^3 y^2 z + \sin xy) = 2x^3 yz + x \cos xy,$$

$$\frac{\partial}{\partial z}(x^3 y^2 z + \sin xy) = x^3 y^2.$$

Finally we remark that the "double-decker" notation is not restricted to the letters x, y, z. For instance,

$$\frac{\partial}{\partial r}(r^2 \cos \theta + e^{\theta r}) = 2r \cos \theta + \theta e^{\theta r}$$

and

$$\frac{\partial}{\partial \theta}(r^2 \cos \theta + e^{\theta r}) = -r^2 \sin \theta + re^{\theta r}.$$

For

$$\rho = \sin 2\theta \cos 3\phi$$

we have

$$\frac{\partial \rho}{\partial \theta} = 2 \cos 2\theta \cos 3\phi \quad \text{and} \quad \frac{\partial \rho}{\partial \phi} = -3 \sin 2\theta \sin 3\phi. \quad \square$$

Exercises

Find the partial derivatives.

*1. $f(x, y) = 3x^2 - xy + y$.

2. $g(x, y) = x^2 e^{-y}$.

*3. $z = Ax^2 + Bxy + Cy^2$.

4. $\rho = \sin^2 (\theta - \phi)$.

*5. $f(x, y) = e^{x-y} - x^2 y$.

6. $z = \sqrt{x^2 - 3y}$.

*7. $g(x, y) = \dfrac{Ax + By}{Cx + Dy}$.

8. $u = \dfrac{e^z}{xy^2}$.

*9. $u = xy + yz + zx$.

10. $f(x, y, z) = z \sin (x - y)$.

*11. $\rho = (\theta - \frac{1}{2}\pi) \sin (\phi - \frac{1}{2}\pi)$.

12. $g(u, v, w) = \log (u^2 + vw - w^2)$.

*13. $\rho = e^{\theta + \phi} \cos (\theta - \phi)$.

14. $f(x, y) = (x + y) \sin (x - y)$.

*15. For $f(x, y) = e^x \log y$, find $f_x(0, e)$ and $f_y(0, e)$.

16. For $f(x, y) = \dfrac{x}{x + y}$, find $f_x(1, 2)$ and $f_y(1, 2)$.

*17. For $g(x, y) = e^{-x} \sin (x + 2y)$, find $g_x(0, \frac{1}{4}\pi)$ and $g_y(0, \frac{1}{4}\pi)$.

18. Show that

$$u = Ax^4 + 2Bx^2 y^2 + Cy^4 \quad \text{satisfies the equation} \quad x \frac{\partial u}{\partial x} + y \frac{\partial u}{\partial y} = 4u.$$

19. Show that

$$u = \frac{x^2 y^2}{x + y} \quad \text{satisfies the equation} \quad x \frac{\partial u}{\partial x} + y \frac{\partial u}{\partial y} = 3u.$$

20. Show that

$$u = x^2 y + y^2 z + z^2 x \quad \text{satisfies the equation} \quad \frac{\partial u}{\partial x} + \frac{\partial u}{\partial y} + \frac{\partial u}{\partial z} = (x + y + z)^2.$$

*21. The area of a triangle is given by the formula

$$K = \tfrac{1}{2} bc \sin A.$$

Suppose that $b = 10$ inches, $c = 20$ inches, and $A = \frac{1}{3}\pi$ radians.

(a) Find the area of the triangle.

(b) Find the rate of change of the area with respect to the side b if c and A remain constant.

(c) Find the rate of change of the area with respect to the angle A if b and c remain constant.

(d) Using the rate found in (c), calculate (by differentials) the approximate change in area if the angle is increased by one degree.

(e) Find the rate of change of c with respect to b if the area and the angle remain constant.

22. The law of cosines for a triangle reads

$$a^2 = b^2 + c^2 - 2bc \cos A.$$

Suppose that $b = 10$ inches, $c = 15$ inches, and $A = \frac{1}{3}\pi$ radians.

(a) Find a.

(b) Find the rate of change of a with respect to b if c and A remain constant.

(c) Using the rate found in (b), calculate (by differentials) the approximate change in a if b is decreased by 1 inch.

(d) Find the rate of change of a with respect to A if b and c remain constant.

(e) Find the rate of change of c with respect to A if a and b remain constant.

14.5   Open Sets and Closed Sets

The domain of a function of several variables can be thought of as a set of points (or vectors) $\mathbf{x}$. In the two-variable case the points are points of the plane:

$$\mathbf{x} = (x, y).$$

In the three-variable case the points are in space:

$$\mathbf{x} = (x, y, z).$$

In this section we study some ways of describing such sets of points.

There are five simple notions that we wish to establish here:

(1) neighborhood of a point
(2) interior of a set
(3) boundary of a set
(4) open set
(5) closed set.

The fundamental notion here is the notion of neighborhood. The other four notions can be derived from it.

Definition of Neighborhood

A *neighborhood* of a point (or vector) $\mathbf{x}_0$ is a set of the form

$$\{\mathbf{x} : \|\mathbf{x} - \mathbf{x}_0\| < \delta\}.$$

In the plane a neighborhood of $\mathbf{x}_0 = (x_0, y_0)$ is the interior of a disk centered at (x_0, y_0). In three-space a neighborhood of $\mathbf{x}_0 = (x_0, y_0, z_0)$ is the interior of a ball centered at (x_0, y_0, z_0). (Figure 14.5.1)

Definition of Interior

A point $\mathbf{x}_0$ is said to be in the *interior* of the set S iff S contains some neighborhood of $\mathbf{x}_0$.

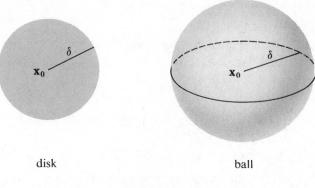

disk   ball

'FIGURE 14.5.1

Example. The points marked x_1 and x_2 in Figure 14.5.2 are interior points of the plane set marked Ω, but x_3 and x_4 are not. □

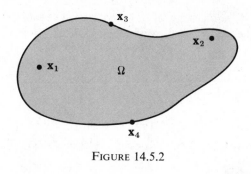

FIGURE 14.5.2

Definition of Boundary

A point x_0 is said to be on the *boundary* of the set S iff every neighborhood of x_0 contains points which are in S and points which are not in S.

Example. The points marked x_3 and x_4 in Figure 14.5.2 are boundary points. Each neighborhood of x_3 and of x_4 contains points in the set and points not in the set. □

It is now easy to phrase what we mean by open and closed.

Definition of Open Set

A set S is said to be *open* iff it contains a neighborhood of each of its points (iff each of its points is an interior point) (iff it contains no boundary points).

Definition of a Closed Set

A set S is said to be *closed* iff it contains its boundary.

Here are some examples of sets which are open, sets which are closed, and sets which are neither open nor closed:

Two-Dimensional Examples

The sets

$$S_1 = \{(x, y) : 1 < x < 2, \quad 1 < y < 2\},$$
$$S_2 = \{(x, y) : 3 \le x \le 4, \quad 1 \le y \le 2\},$$
$$S_3 = \{(x, y) : 5 \le x \le 6, \quad 1 < y < 2\}$$

are displayed in Figure 14.5.3. S_1 is the inside of the square with the dotted boundary. S_1 is open because it contains a neighborhood of each of its points. S_2 is the inside of the indicated square together with its boundary. S_2 is closed because it contains its entire boundary. S_3 is the inside of the square together with that part of the boundary which is indicated by a solid line. S_3 is not open because it contains part of its boundary and it is not closed because it does not contain all of its boundary.

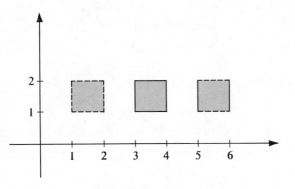

FIGURE 14.5.3

Three-Dimensional Examples

We now examine

$$S_1 = \{(x, y, z): z > x^2 + y^2\},$$
$$S_2 = \{(x, y, z): z \ge x^2 + y^2\},$$
$$S_2 = \left\{(x, y, z): 1 \ge \frac{x^2 + y^2}{z}\right\}.$$

Each of these sets has as its boundary the paraboloid

$$z = x^2 + y^2. \tag{Figure 14.3.1}$$

The first set consists of all points above this surface. This is an open set because if a point is above this surface then all points sufficiently close to it will also be above this surface. The second set is closed because it contains all of its boundary. The third set is neither open nor closed. It is not open because it contains some boundary points; for example, it contains the point $(\frac{1}{2}, \frac{1}{2}, \frac{1}{2})$. It is not closed because it fails to contain the boundary point $(0, 0, 0)$.

Exercises

Specify the interior and the boundary of the set. State whether the set is open, closed, or neither. Sketch the set.

* 1. $\{(x, y) : 2 \le x \le 4, \quad 1 \le y \le 3\}$.
  2. $\{(x, y) : 2 < x < 4, \quad 1 < y < 3\}$.
* 3. $\{(x, y) : 1 < x^2 + y^2 < 4\}$.
  4. $\{(x, y) : 1 \le x^2 \le 4\}$.
* 5. $\{(x, y) : 1 < x^2 \le 4\}$.
  6. $\{(x, y) : y < x^2\}$.
* 7. $\{(x, y) : y \le x^2\}$.
  8. $\{(x, y, z) : 1 \le x \le 2, \quad 1 \le y \le 2, \quad 1 \le z \le 2\}$.
* 9. $\{(x, y, z) : (x - 1)^2 + (y - 1)^2 + (z - 1)^2 < \frac{1}{4}\}$.
  10. $\{(x, y, z) : x^2 + y^2 \le 1, \quad 0 \le z \le 4\}$.

14.6 Limits and Continuity; Equality of Mixed Partials

The Notions

The limit process used in taking partial derivatives involved nothing new because in each instance all but one of the variables was held fixed. In this section we take up limits of the form

$$\lim_{(x,y) \to (x_0, y_0)} f(x, y) \quad \text{and} \quad \lim_{(x,y,z) \to (x_0, y_0, z_0)} f(x, y, z).$$

To avoid having to treat the two- and three-variable cases separately, we will write instead

$$\lim_{\mathbf{x} \to \mathbf{x}_0} f(\mathbf{x}).$$

This gives us both the two-variable case [set $\mathbf{x} = (x, y)$ and $\mathbf{x}_0 = (x_0, y_0)$] and the three-variable case [set $\mathbf{x} = (x, y, z)$ and $\mathbf{x}_0 = (x_0, y_0, z_0)$].

The essential idea of

$$\lim_{\mathbf{x} \to \mathbf{x}_0} f(\mathbf{x}) = l$$

is a simple one: as the point $\mathbf{x}$ gets close to the point $\mathbf{x}_0$ (as $\|\mathbf{x} - \mathbf{x}_0\|$ gets small), the number $f(\mathbf{x})$ gets close to the number l ($|f(\mathbf{x}) - l|$ gets small). For this idea to make sense we do not need that f be defined at $\mathbf{x}_0$ but we do need that f be defined at some points which are close to $\mathbf{x}_0$. To insure this, we shall assume that each neighborhood of $\mathbf{x}_0$ contains at least some points of the domain of f other than $\mathbf{x}_0$. We can now give the ϵ, δ definition of limit.

Definition of Limit

$$\lim_{\mathbf{x} \to \mathbf{x}_0} f(\mathbf{x}) = l \quad \text{iff} \quad \begin{cases} \text{for each } \epsilon > 0 \text{ there exists } \delta > 0 \text{ such that if} \\ \qquad 0 < \|\mathbf{x} - \mathbf{x}_0\| < \delta \quad \text{and} \quad \mathbf{x} \in \text{dom}\,(f) \\ \text{then} \\ \qquad\qquad |f(\mathbf{x}) - l| < \epsilon. \end{cases}$$

As in the one-variable case, a limit (if it exists) is unique. Moreover, if f and g are two functions with a common domain satisfying

$$\lim_{\mathbf{x} \to \mathbf{x}_0} f(\mathbf{x}) = l \quad \text{and} \quad \lim_{\mathbf{x} \to \mathbf{x}_0} g(\mathbf{x}) = m$$

then

$$\lim_{\mathbf{x} \to \mathbf{x}_0} [f(\mathbf{x}) + g(\mathbf{x})] = l + m,$$

$$\lim_{\mathbf{x} \to \mathbf{x}_0} [\alpha f(\mathbf{x})] = \alpha l,$$

$$\lim_{\mathbf{x} \to \mathbf{x}_0} [f(\mathbf{x})g(\mathbf{x})] = lm,$$

$$\lim_{\mathbf{x} \to \mathbf{x}_0} [f(\mathbf{x})/g(\mathbf{x})] = l/m \qquad \text{provided } m \neq 0.$$

These results are not hard to derive. You can do it simply by imitating the comparable arguments in the one-variable case.

The idea of continuity is the same as it is in other settings; namely, to say that f is *continuous* at $\mathbf{x}_0$, is to say that

$$\boxed{\lim_{\mathbf{x} \to \mathbf{x}_0} f(\mathbf{x}) = f(\mathbf{x}_0)} \quad \text{or equivalently that} \quad \boxed{\lim_{\mathbf{h} \to \mathbf{0}} f(\mathbf{x}_0 + \mathbf{h}) = f(\mathbf{x}_0).}$$

In two variables we can write

$$\lim_{(x,y) \to (x_0, y_0)} f(x, y) = f(x_0, y_0)$$

and in three variables

$$\lim_{(x,y,z) \to (x_0, y_0, z_0)} f(x, y, z) = f(x_0, y_0, z_0).$$

To say that f is *continuous on the set S* is to say that f is continuous at all points of S.

Some Examples of Continuous Functions

Probably the simplest functions which are everywhere continuous are the polynomials. These are simply finite linear combinations of finite products of x, y, and z: for example,

$$P(x, y) = x^2y + 3x^3y^4 - x + 2y,$$
$$Q(x, y, z) = 6x^3z - z^3y + 2xyz.$$

Rational functions, by which we mean quotients of polynomials, are continuous everywhere except where the denominator is zero. Thus

$$f(x, y) = \frac{2x - y}{x^2 + y^2}$$

is continuous at each point of the xy-plane except at the origin $(0, 0)$;

$$g(x, y) = \frac{x^4}{x - y}$$

is continuous except on the line $y = x$;

$$h(x, y) = \frac{1}{x^2 - y}$$

is continuous except on the parabola $y = x^2$;

$$F(x, y, z) = \frac{2x}{x^2 + y^2 + z^2}$$

is continuous at each point of space except the origin $(0, 0, 0)$;

$$G(x, y, z) = \frac{x^5 - y}{ax + by + cz}$$

is continuous except on the plane $ax + by + cz = 0$. $\square$

A simple way to construct more complicated continuous functions is to form composites:

$$\arctan\left(\frac{xz^2}{x + y}\right), \qquad \sqrt{x^2 + y^4 + z^6}, \qquad \sin xyz.$$

The first function is continuous except at those points where $x = -y$. The last two functions are continuous at each point of space. The continuity of these composites follows from a simple theorem which we state and prove below.

Theorem (Continuity of Composites)

If g is continuous at the point $\mathbf{x}_0$ and f is continuous at the number $g(\mathbf{x}_0)$, then the composition $f \circ g$ is continuous at the point $\mathbf{x}_0$.

PROOF. We begin with $\epsilon > 0$. We must show that there exists $\delta > 0$ such that

$$\text{if } \begin{Bmatrix} \|\mathbf{x} - \mathbf{x}_0\| < \delta \\ \text{and} \\ \mathbf{x} \in \text{dom} (f \circ g) \end{Bmatrix} \quad \text{then} \quad |f(g(\mathbf{x})) - f(g(\mathbf{x}_0))| < \epsilon.$$

From the continuity of f at $g(\mathbf{x}_0)$, we know that there exists $\delta_1 > 0$ such that

$$\text{if } \begin{Bmatrix} |y - g(\mathbf{x}_0)| < \delta_1 \\ \text{and} \\ y \in \text{dom} (f) \end{Bmatrix} \quad \text{then} \quad |f(y) - f(g(\mathbf{x}_0))| < \epsilon.$$

From the continuity of g at $\mathbf{x}_0$, we know that there exists $\delta > 0$ such that

$$\text{if } \begin{Bmatrix} \|\mathbf{x} - \mathbf{x}_0\| < \delta \\ \text{and} \\ \mathbf{x} \in \text{dom} (g) \end{Bmatrix} \quad \text{then} \quad |g(\mathbf{x}) - g(\mathbf{x}_0)| < \delta_1.$$

This last δ obviously works — for if

$$\|\mathbf{x} - \mathbf{x}_0\| < \delta \quad \text{and} \quad \mathbf{x} \in \text{dom} (f \circ g),$$

then

$$|g(\mathbf{x}) - g(\mathbf{x}_0)| < \delta_1 \quad \text{and} \quad g(\mathbf{x}) \in \text{dom} (f),$$

and consequently

$$|f(g(\mathbf{x})) - f(g(\mathbf{x}_0))| < \epsilon$$

as asserted. □

Continuity in Each Variable Separately

A continuous function of several variables is continuous in each of its variables separately. In the two-variable case this means that if

$$\lim_{(x,y) \to (x_0,y_0)} f(x, y) = f(x_0, y_0)$$

then

$$\lim_{x \to x_0} f(x, y_0) = f(x_0, y_0) \quad \text{and} \quad \lim_{y \to y_0} f(x_0, y) = f(x_0, y_0).$$

(This is not hard to prove.) The converse is false. *It is possible for a function to be continuous in each variable separately and yet fail to be continuous as a function of several variables.* You can see this kind of behavior in the next example.

Example

$$f(x, y) = \left. \begin{cases} \dfrac{2xy}{x^2 + y^2}, & (x, y) \neq (0, 0) \\ 0, & (x, y) = (0, 0) \end{cases} \right].$$

Since
$$f(x, 0) = 0 \text{ for all } x \quad \text{and} \quad f(0, y) = 0 \text{ for all } y,$$
you can see that
$$\lim_{x \to 0} f(x, 0) = f(0, 0) \quad \text{and} \quad \lim_{y \to 0} f(0, y) = f(0, 0).$$
Namely, at the point $(0, 0)$, f is continuous in x and continuous in y. However, as a function of two variables, f is not continuous at $(0, 0)$. An easy way to see this is to note that we can approach $(0, 0)$ as closely as we want by points of the form (t, t) with $t \neq 0$. At such points f takes on the value 1:
$$f(t, t) = \frac{2tt}{t^2 + t^2} = 1.$$
Hence f cannot tend to $f(0, 0) = 0$ as required. $\square$

Continuity and Partial Differentiability

The existence of partial derivatives also fails to guarantee continuity. To show this, we can use the same example:
$$f(x, y) = \begin{cases} \dfrac{2xy}{x^2 + y^2}, & (x, y) \neq (0, 0) \\ 0, & (x, y) = (0, 0) \end{cases}.$$
Since both $f(x, 0)$ and $f(0, y)$ are constantly zero, both partials exist (and are zero) at $(0, 0)$ and yet, as you saw, the function is discontinuous at $(0, 0)$. $\square$

It is not hard to understand why a function can have partial derivatives and yet fail to be continuous. The existence of
$$\frac{\partial f}{\partial x} \quad \text{at} \quad (x_0, y_0)$$
depends on the behavior of f only at points of the form
$$(x_0 + h, y_0).$$
Similarly, the existence of
$$\frac{\partial f}{\partial y} \quad \text{at} \quad (x_0, y_0)$$
depends on the behavior of f only at points of the form
$$(x_0, y_0 + k).$$
On the other hand, continuity at (x_0, y_0) depends on the behavior of f at points of the more general form
$$(x_0 + h, y_0 + k).$$
More briefly, we can put it this way: *The existence of a partial derivative depends on the behavior of the function along one linear path but continuity depends on the behavior of the function in all directions.*

Equality of Mixed Partials

There is nevertheless an interesting tie-in between partial differentiation and continuity. To describe it, we begin with a function f of x and y having two partial derivatives

$$\frac{\partial f}{\partial x} \quad \text{and} \quad \frac{\partial f}{\partial y}.$$

These are again functions of x and y and may themselves possess partial derivatives:

$$\frac{\partial}{\partial x}\left(\frac{\partial f}{\partial x}\right) = \frac{\partial^2 f}{\partial x^2}, \qquad \frac{\partial}{\partial y}\left(\frac{\partial f}{\partial x}\right) = \frac{\partial^2 f}{\partial y\,\partial x},$$

$$\frac{\partial}{\partial x}\left(\frac{\partial f}{\partial y}\right) = \frac{\partial^2 f}{\partial x\,\partial y}, \qquad \frac{\partial}{\partial y}\left(\frac{\partial f}{\partial y}\right) = \frac{\partial^2 f}{\partial y^2}.$$

These last functions are called the *second-order partials*. Note that there are two "mixed" partials

$$\frac{\partial^2 f}{\partial y\,\partial x} \quad \text{and} \quad \frac{\partial^2 f}{\partial x\,\partial y}.$$

The first of these is obtained by differentiating first with respect to x and then with respect to y. The second is obtained by differentiating first with respect to y and then with respect to x.

Example. For

$$f(x, y) = \sin x^2 y$$

we have

$$\frac{\partial f}{\partial x} = 2xy \cos x^2 y \quad \text{and} \quad \frac{\partial f}{\partial y} = x^2 \cos x^2 y.$$

The second-order partials are

$$\frac{\partial^2 f}{\partial x^2} = -4x^2 y^2 \sin x^2 y + 2y \cos x^2 y,$$

$$\frac{\partial^2 f}{\partial y\,\partial x} = -2x^3 y \sin x^2 y + 2x \cos x^2 y,$$

$$\frac{\partial^2 f}{\partial x\,\partial y} = -2x^3 y \sin x^2 y + 2x \cos x^2 y,$$

$$\frac{\partial^2 f}{\partial y^2} = -x^4 \sin x^2 y. \quad \square$$

Example. For

$$f(x, y) = \log (x^2 + y^3)$$

we have

$$\frac{\partial f}{\partial x} = \frac{2x}{x^2 + y^3} \quad \text{and} \quad \frac{\partial f}{\partial y} = \frac{3y^2}{x^2 + y^3}.$$

The second-order partials are

$$\frac{\partial^2 f}{\partial x^2} = \frac{(x^2 + y^3)2 - 2x(2x)}{(x^2 + y^3)^2} = \frac{2(y^3 - x^2)}{(x^2 + y^3)^2},$$

$$\frac{\partial^2 f}{\partial y \, \partial x} = \frac{-2x(3y^2)}{(x^2 + y^3)^2} = -\frac{6xy^2}{(x^2 + y^3)^2},$$

$$\frac{\partial^2 f}{\partial x \, \partial y} = \frac{-3y^2(2x)}{(x^2 + y^3)^2} = -\frac{6xy^2}{(x^2 + y^3)^2},$$

$$\frac{\partial^2 f}{\partial y^2} = \frac{(x^2 + y^3)6y - 3y^2(3y^2)}{(x^2 + y^3)^2} = \frac{3y(2x^2 - y^3)}{(x^2 + y^3)^2}. \quad \square$$

You may have noticed that in both examples we had

$$\frac{\partial^2 f}{\partial y \, \partial x} = \frac{\partial^2 f}{\partial x \, \partial y}.$$

Since in neither case is f symmetric in x and y, this equality of the mixed partials cannot be due to symmetry. Actually it is due to continuity. It can be proved that

---

(14.6.1)
$$\frac{\partial^2 f}{\partial y \, \partial x} = \frac{\partial^2 f}{\partial x \, \partial y}$$

on every open set U on which f and its partials

$$\frac{\partial f}{\partial x}, \quad \frac{\partial f}{\partial y}, \quad \frac{\partial^2 f}{\partial y \, \partial x}, \quad \frac{\partial^2 f}{\partial x \, \partial y}$$

are all continuous.†

In the case of a function of three variables you can look for three first partials

$$\frac{\partial f}{\partial x}, \quad \frac{\partial f}{\partial y}, \quad \frac{\partial f}{\partial z}$$

and nine second partials

$$\frac{\partial^2 f}{\partial x^2}, \quad \frac{\partial^2 f}{\partial x \, \partial y}, \quad \frac{\partial^2 f}{\partial x \, \partial z}$$

$$\frac{\partial^2 f}{\partial y \, \partial x}, \quad \frac{\partial^2 f}{\partial y^2}, \quad \frac{\partial^2 f}{\partial y \, \partial z}$$

$$\frac{\partial^2 f}{\partial z \, \partial x}, \quad \frac{\partial^2 f}{\partial z \, \partial y}, \quad \frac{\partial^2 f}{\partial z^2}.$$

---

† For a proof consult a text on advanced calculus.

Here again, there is equality of the mixed partials

$$\frac{\partial^2 f}{\partial y\, \partial x} = \frac{\partial^2 f}{\partial x\, \partial y}, \qquad \frac{\partial^2 f}{\partial z\, \partial x} = \frac{\partial^2 f}{\partial x\, \partial z}, \qquad \frac{\partial^2 f}{\partial y\, \partial z} = \frac{\partial^2 f}{\partial z\, \partial y}$$

provided that all the functions involved are continuous.

Example. For

$$f(x, y, z) = xe^y \sin \pi z$$

we have

$$\frac{\partial f}{\partial x} = e^y \sin \pi z, \qquad \frac{\partial f}{\partial y} = xe^y \sin \pi z, \qquad \frac{\partial f}{\partial z} = \pi xe^y \cos \pi z,$$

$$\frac{\partial^2 f}{\partial x^2} = 0, \qquad \frac{\partial^2 f}{\partial y^2} = xe^y \sin \pi z, \qquad \frac{\partial^2 f}{\partial z^2} = -\pi^2 xe^y \sin \pi z,$$

$$\frac{\partial^2 f}{\partial y\, \partial x} = e^y \sin \pi z = \frac{\partial^2 f}{\partial x\, \partial y},$$

$$\frac{\partial^2 f}{\partial z\, \partial x} = \pi e^y \cos \pi z = \frac{\partial^2 f}{\partial x\, \partial z},$$

$$\frac{\partial^2 f}{\partial y\, \partial z} = \pi xe^y \cos \pi z = \frac{\partial^2 f}{\partial z\, \partial y}. \quad \square$$

Exercises

Find the second partials.

*1. $f(x, y) = Ax^2 + 2Bxy + Cy^2$.

2. $f(x, y) = Ax^3 + Bx^2y + Cxy^2 + Dy^3$.

*3. $f(x, y) = Ax + By + Ce^{xy}$.

4. $f(x, y) = x^2 \cos y + y^2 \sin x$.

*5. $f(x, y) = \sqrt{x + y^2}$.

*6. $f(x, y, z) = (x + y^2 + z^3)^2$.

*7. $f(x, y) = \log\left(\dfrac{x}{x + y}\right)$.

8. $f(x, y) = \dfrac{Ax + By}{Cx + Dy}$.

*9. $f(x, y, z) = (x + y)(y + z)(z + x)$.

10. $f(x, y, z) = \arctan(xyz)$.

11. Show that

$$\text{if} \quad u = \frac{xy}{x + y} \quad \text{then} \quad x^2 \frac{\partial^2 u}{\partial x^2} + 2xy \frac{\partial^2 u}{\partial x\, \partial y} + y^2 \frac{\partial^2 u}{\partial y^2} = 0.$$

12. Show that

$$\text{if} \quad u = \log \sqrt{x^2 + y^2} \quad \text{then} \quad \frac{\partial^2 u}{\partial x^2} + \frac{\partial^2 u}{\partial y^2} = 0.$$

13. Show that

$$\text{if} \quad u = \frac{1}{\sqrt{x^2 + y^2 + z^2}} \quad \text{then} \quad \frac{\partial^2 u}{\partial x^2} + \frac{\partial^2 u}{\partial y^2} + \frac{\partial^2 u}{\partial z^2} = 0.$$

14.7 Differentiability and Gradient

The Notion of Differentiability

Our object here is to generalize the notion of differentiability from functions of one variable to functions of several variables. Partial derivatives in themselves do not fulfill this role for they take into account only behavior along a particular line.

In the one-variable case we formed the difference quotient

$$\frac{f(x + h) - f(x)}{h}$$

and called f differentiable at x provided that this quotient had a limit as h tended to zero. In the multivariable case we can still form the difference

$$f(\mathbf{x} + \mathbf{h}) - f(\mathbf{x})$$

but the "quotient"

$$\frac{f(\mathbf{x} + \mathbf{h}) - f(\mathbf{x})}{\mathbf{h}}$$

makes no sense because to divide by a vector makes no sense.

The notion of differentiability in the multivariable case is inspired not by difference quotients (we want to avoid $\mathbf{h}$'s in the denominator) but rather by the characterization of $f'(x)$ as the unique number satisfying

$$f(x + h) - f(x) = f'(x)h + o(h).$$

By $o(h)$, you will recall, we mean any function $g(h)$ such that

$$\lim_{h \to 0} \frac{g(h)}{h} = 0, \quad \text{or equivalently,} \quad \lim_{h \to 0} \frac{g(h)}{|h|} = 0.$$

To set things up for the multivariable case, let us agree to call $o(\mathbf{h})$ any function $g(\mathbf{h})$ which satisfies

$$\lim_{\mathbf{h} \to 0} \frac{g(\mathbf{h})}{\|\mathbf{h}\|} = 0.$$

We now consider a function f of several variables *defined at least in a neighborhood of* $\mathbf{x}$. In the three-variable case, $\mathbf{x} = (x, y, z)$ and in the two-variable case, $\mathbf{x} = (x, y)$.

Definition of Differentiability

The function f is said to be *differentiable* at $\mathbf{x}$ iff there exists a vector $\nabla f(\mathbf{x})$ such that

$$f(\mathbf{x} + \mathbf{h}) - f(\mathbf{x}) = \nabla f(\mathbf{x}) \cdot \mathbf{h} + o(\mathbf{h}).$$

The vector $\nabla f(\mathbf{x})$ is called the *gradient* of f at $\mathbf{x}$.

The similarities between the one-variable case,

$$f(x + h) - f(x) = f'(x)h + o(h),$$

and the multivariable case,

$$f(\mathbf{x} + \mathbf{h}) - f(\mathbf{x}) = \nabla f(\mathbf{x}) \cdot (\mathbf{h}) + o(\mathbf{h}),$$

are obvious. We point to the differences. There are essentially two of them:

(1) While the derivative $f'(x)$ is a number, the gradient $\nabla f(\mathbf{x})$ is a vector.
(2) While

$$f'(x)h$$

is the ordinary product of two real numbers,

$$\nabla f(\mathbf{x}) \cdot \mathbf{h}$$

is the dot product of two vectors.

Computing Gradients

First we compute some gradients by applying the definition directly. Then we bring in a theorem which makes such computations much easier. Finally we compute gradients by using the theorem. As for notation, in the two-variable case we will write

$$\nabla f(\mathbf{x}) = \nabla f(x, y)$$

and in the three-variable case

$$\nabla f(\mathbf{x}) = \nabla f(x, y, z).$$

Example. For

$$f(x, y, z) = ax + by + cz$$

we have

$$
\begin{aligned}
f(\mathbf{x} + \mathbf{h}) - f(\mathbf{x}) &= f(x + h_1, y + h_2, z + h_3) - f(x, y, z) \\
&= [a(x + h_1) + b(y + h_2) + c(z + h_3)] - [ax + by + cz] \\
&= ah_1 + bh_2 + ch_3 \\
&= [a\mathbf{i} + b\mathbf{j} + c\mathbf{k}] \cdot \mathbf{h}.
\end{aligned}
$$

The remainder $o(\mathbf{h})$ is here identically zero and

$$\nabla f(\mathbf{x}) = a\mathbf{i} + b\mathbf{j} + c\mathbf{k}. \quad \square$$

Example. For

$$f(x, y) = x^2 + y^2$$

we have

$$
\begin{aligned}
f(\mathbf{x} + \mathbf{h}) - f(\mathbf{x}) &= f(x + h_1, y + h_2) - f(x, y) \\
&= [(x + h_1)^2 + (y + h_2)^2] - [x^2 + y^2] \\
&= [2xh_1 + 2yh_2] + [h_1^2 + h_2^2] \\
&= [2x\mathbf{i} + 2y\mathbf{j}] \cdot \mathbf{h} + \|\mathbf{h}\|^2.
\end{aligned}
$$

As a remainder we have

$$o(\mathbf{h}) = \|\mathbf{h}\|^2$$

and as gradient

$$\nabla f(\mathbf{x}) = \nabla f(x, y) = 2x\mathbf{i} + 2y\mathbf{j}. \quad \square$$

Theorem 14.7.1

If f has continuous first partials in a neighborhood of $\mathbf{x}$, then f is differentiable at $\mathbf{x}$ and

$$\nabla f(\mathbf{x}) = \frac{\partial f}{\partial x}(\mathbf{x})\,\mathbf{i} + \frac{\partial f}{\partial y}(\mathbf{x})\,\mathbf{j} + \frac{\partial f}{\partial z}(\mathbf{x})\,\mathbf{k}.$$

In two variables,

$$\nabla f(\mathbf{x}) = \frac{\partial f}{\partial x}(\mathbf{x})\,\mathbf{i} + \frac{\partial f}{\partial y}(\mathbf{x})\,\mathbf{j}.$$

PROOF.   We prove the theorem in the two-variable case. The same ideas work for a proof in the three-variable case but the details are more burdensome.
   In the first place,

$$f(\mathbf{x} + \mathbf{h}) - f(\mathbf{x}) = f(x + h_1, y + h_2) - f(x, y).$$

We can rewrite this as

$$(1)\quad f(\mathbf{x} + \mathbf{h}) - f(\mathbf{x}) = [f(x + h_1, y + h_2) - f(x, y + h_2)] + [f(x, y + h_2) - f(x, y)].$$

By the mean-value theorem for functions of one variable, we know that there are numbers

$$0 < \theta_1, \theta_2 < 1$$

such that

$$f(x + h_1, y + h_2) - f(x, y + h_2) = \frac{\partial f}{\partial x}(x + \theta_1 h_1, y + h_2)h_1$$

and

$$f(x, y + h_2) - f(x, y) = \frac{\partial f}{\partial y}(x, y + \theta_2 h_2)h_2.$$

By the continuity of $\partial f/\partial x$,

$$\frac{\partial f}{\partial x}(x + \theta_1 h_1, y + h_2) = \frac{\partial f}{\partial x}(x, y) + \epsilon_1(\mathbf{h})$$

where

$$\epsilon_1(\mathbf{h}) \to 0 \quad \text{as} \quad \mathbf{h} \to 0.$$

By the continuity of $\partial f/\partial y$,

$$\frac{\partial f}{\partial y}(x, y + \theta_2 h_2) = \frac{\partial f}{\partial y}(x, y) + \epsilon_2(\mathbf{h})$$

where

$$\epsilon_2(\mathbf{h}) \to 0 \quad \text{as} \quad \mathbf{h} \to 0.$$

Substituting these expressions in Equation (1), we find that

$$f(\mathbf{x} + \mathbf{h}) - f(\mathbf{x}) = \left[\frac{\partial f}{\partial x}(x, y) + \epsilon_1(\mathbf{h})\right] h_1 + \left[\frac{\partial f}{\partial y}(x, y) + \epsilon_2(\mathbf{h})\right] h_2$$

and therefore

$$f(\mathbf{x} + \mathbf{h}) - f(\mathbf{x}) = \left[\frac{\partial f}{\partial x}(x, y)\mathbf{i} + \frac{\partial f}{\partial y}(x, y)\mathbf{j}\right] \cdot \mathbf{h} + \left[\epsilon_1(\mathbf{h})\mathbf{i} + \epsilon_2(\mathbf{h})\mathbf{j}\right] \cdot \mathbf{h}.$$

The proof will be complete once we show that

$$[\epsilon_1(\mathbf{h})\mathbf{i} + \epsilon_2(\mathbf{h})\mathbf{j}] \cdot \mathbf{h} = o(\mathbf{h}).$$

Using the Schwarz inequality,

$$|\mathbf{a} \cdot \mathbf{b}| \leq \|\mathbf{a}\| \|\mathbf{b}\|,$$

we see that

$$|[\epsilon_1(\mathbf{h})\mathbf{i} + \epsilon_2(\mathbf{h})\mathbf{j}] \cdot \mathbf{h}| \leq \|\epsilon_1(\mathbf{h})\mathbf{i} + \epsilon_2(\mathbf{h})\mathbf{j}\| \|\mathbf{h}\|$$

and thus

$$\frac{|[\epsilon_1(\mathbf{h})\mathbf{i} + \epsilon_2(\mathbf{h})\mathbf{j}] \cdot \mathbf{h}|}{\|\mathbf{h}\|} \leq \|\epsilon_1(\mathbf{h})\mathbf{i} + \epsilon_2(\mathbf{h})\mathbf{j}\| \leq \|\epsilon_1(\mathbf{h})\mathbf{i}\| + \|\epsilon_2(\mathbf{h})\mathbf{j}\|$$

$$= |\epsilon_1(\mathbf{h})| + |\epsilon_2(\mathbf{h})| \to 0,$$

since

$$\epsilon_1(\mathbf{h}) \to 0 \quad \text{and} \quad \epsilon_2(\mathbf{h}) \to 0. \quad \square$$

Example. For

$$f(x, y) = xe^y - ye^x$$

we have

$$\frac{\partial f}{\partial x}(x, y) = e^y - ye^x, \qquad \frac{\partial f}{\partial y}(x, y) = xe^y - e^x$$

and so

$$\nabla f(x, y) = (e^y - ye^x)\mathbf{i} + (xe^y - e^x)\mathbf{j}. \quad \square$$

When there is no reason to emphasize the point of evaluation we don't write

$$\nabla f(\mathbf{x}) \quad \text{or} \quad \nabla f(x, y) \quad \text{or} \quad \nabla f(x, y, z)$$

but simply ∇f. In this simpler notation, the function

$$f(x, y) = xe^y - ye^x$$

has partial derivatives

$$\frac{\partial f}{\partial x} = e^y - ye^x, \qquad \frac{\partial f}{\partial y} = xe^y - e^x$$

and gradient

$$\nabla f = (e^y - ye^x)\mathbf{i} + (xe^y - e^x)\mathbf{j}. \quad \square$$

Example.   For

$$f(x, y, z) = \sin xy^2z^3$$

we have

$$\frac{\partial f}{\partial x} = y^2z^3 \cos xy^2z^3,$$

$$\frac{\partial f}{\partial y} = 2xyz^3 \cos xy^2z^3,$$

$$\frac{\partial f}{\partial z} = 3xy^2z^2 \cos xy^2z^3,$$

and

$$\nabla f = (\cos xy^2z^3)[y^2z^3\mathbf{i} + 2xyz^3\mathbf{j} + 3xy^2z^2\mathbf{k}]. \quad \square$$

Example.   As a numerical example, we take

$$f(x, y, z) = x \sin \pi y + y \cos \pi z$$

and evaluate ∇f at $(0, 1, 2)$.

Here

$$\frac{\partial f}{\partial x} = \sin \pi y,$$

$$\frac{\partial f}{\partial y} = \pi x \cos \pi y + \cos \pi z,$$

$$\frac{\partial f}{\partial z} = -\pi y \sin \pi z.$$

At $(0, 1, 2)$

$$\frac{\partial f}{\partial x} = 0, \qquad \frac{\partial f}{\partial y} = 1, \qquad \frac{\partial f}{\partial z} = 0$$

and

$$\nabla f = \mathbf{j}. \quad \square$$

Uniqueness

We defined the gradient at $\mathbf{x}$ as *the* vector $\nabla f(\mathbf{x})$ which satisfies

$$f(\mathbf{x} + \mathbf{h}) - f(\mathbf{x}) = \nabla f(\mathbf{x}) \cdot \mathbf{h} + o(\mathbf{h}).$$

To make sure that this definition makes sense, we must make sure that no two vectors can satisfy that condition; namely, we must make sure that if

$$f(\mathbf{x} + \mathbf{h}) - f(\mathbf{x}) = \mathbf{y} \cdot \mathbf{h} + o(\mathbf{h})$$

and

$$f(\mathbf{x} + \mathbf{h}) - f(\mathbf{x}) = \mathbf{z} \cdot \mathbf{h} + o(\mathbf{h}),$$

then

$$\mathbf{y} = \mathbf{z}.$$

You are asked to prove this in Exercise 21.

Exercises

Find the gradient.

*1. $f(x, y) = xe^{xy}$.

2. $f(x, y) = x^2 + y^2$.

*3. $f(x, y) = 3x^2 - xy + y$.

4. $f(x, y) = Ax^2 + Bxy + Cy^2$.

*5. $f(x, y) = e^{x-y} - x^2y$.

6. $f(x, y) = x^2y^{-2}$.

*7. $f(x, y, z) = xy^2e^{-z}$.

8. $f(x, y, z) = z \sin (x - y)$.

*9. $f(x, y, z) = xy + yz + zx$.

10. $f(x, y, z) = \sqrt{x^2 + y^4 + z^6}$.

*11. $f(x, y) = (x + y)e^{x-y}$.

12. $f(x, y) = (x + y) \sin (x - y)$.

*13. $f(x, y) = e^x \log y$.

14. $f(x, y, z) = x^2y + y^2z + z^2x$.

*15. $f(x, y) = \dfrac{Ax + By}{Cx + Dy}$.

16. $f(x, y) = \dfrac{x - y}{x^2 + y^2}$.

Find the gradient vector at the indicated point.

*17. $f(x, y) = 2x^2 - 3xy + 4y^2$ at $(2, 3)$.

18. $f(x, y) = \dfrac{2x}{x - y}$ at $(3, 1)$.

*19. $f(x, y) = e^{-x} \sin (x + 2y)$ at $(0, \frac{1}{4}\pi)$.

20. $f(x, y, z) = (x - y) \cos \pi z$ at $(1, 0, \frac{1}{2})$.

21. (*Optional*) Show that if

$$f(\mathbf{x} + \mathbf{h}) - f(\mathbf{x}) = \mathbf{y} \cdot \mathbf{h} + o(\mathbf{h})$$

and

$$f(\mathbf{x} + \mathbf{h}) - f(\mathbf{x}) = \mathbf{z} \cdot \mathbf{h} + o(\mathbf{h}),$$

then

$$\mathbf{y} = \mathbf{z}.$$

14.8   Some Simple Properties of Gradients

Much of the calculus of functions of several variables is centered around gradients. Throughout the rest of this chapter we shall study them. In this section we begin.

Differentiability Implies Continuity

As in the one-variable case, differentiability implies continuity; namely, *if f is differentiable at* **x**, *then f is continuous at* **x**. To see this, we need only write

$$f(\mathbf{x} + \mathbf{h}) - f(\mathbf{x}) = \nabla f(\mathbf{x}) \cdot \mathbf{h} + o(\mathbf{h}).$$

As $\mathbf{h} \to \mathbf{0}$,

$$o(\mathbf{h}) \to 0$$

and

$$|\nabla f(\mathbf{x}) \cdot \mathbf{h}| \le \|\nabla f(x)\| \, \|\mathbf{h}\| \to 0.$$

It follows that

$$f(\mathbf{x} + \mathbf{h}) - f(\mathbf{x}) \to 0,$$

which means that

$$f(\mathbf{x} + \mathbf{h}) \to f(\mathbf{x}),$$

and thus f is continuous at **x**.   □

Some Formulas

In many respects gradients behave just like derivatives do in the one-variable case. In particular, if $\nabla f(\mathbf{x})$ and $\nabla g(\mathbf{x})$ exist, then so do

$$\nabla[f(\mathbf{x}) + g(\mathbf{x})], \qquad \nabla[\alpha f(\mathbf{x})], \qquad \nabla[f(\mathbf{x})g(\mathbf{x})].$$

Moreover

$$\begin{aligned}
\nabla[f(\mathbf{x}) + g(\mathbf{x})] &= \nabla f(\mathbf{x}) + \nabla g(\mathbf{x}), \\
\nabla[\alpha f(\mathbf{x})] &= \alpha \, \nabla f(\mathbf{x}), \\
\nabla[f(\mathbf{x})g(\mathbf{x})] &= f(\mathbf{x}) \, \nabla g(\mathbf{x}) + g(\mathbf{x}) \, \nabla f(\mathbf{x}).
\end{aligned}$$

The first two formulas are easy to derive. To derive the third formula, let's assume that $\nabla f(\mathbf{x})$ and $\nabla g(\mathbf{x})$ both exist. Our task is to show that

$$f(\mathbf{x} + \mathbf{h})g(\mathbf{x} + \mathbf{h}) - f(\mathbf{x})g(\mathbf{x}) = [f(\mathbf{x}) \, \nabla g(\mathbf{x}) + g(\mathbf{x}) \, \nabla f(\mathbf{x})] \cdot \mathbf{h} + o(\mathbf{h}).$$

A sketch of how this can be done is given below. See if you can justify each of the steps:

$$\begin{aligned}
f(\mathbf{x} + \mathbf{h})g(\mathbf{x} + \mathbf{h}) - f(\mathbf{x})g(\mathbf{x}) &= [f(\mathbf{x} + \mathbf{h})g(\mathbf{x} + \mathbf{h}) - f(\mathbf{x})g(\mathbf{x} + \mathbf{h})] \\
&\quad + [f(\mathbf{x})g(\mathbf{x} + \mathbf{h}) - f(\mathbf{x})g(\mathbf{x})] \\
&= [f(\mathbf{x} + \mathbf{h}) - f(\mathbf{x})]g(\mathbf{x} + \mathbf{h}) + f(\mathbf{x})[g(\mathbf{x} + \mathbf{h}) - g(\mathbf{x})] \\
&= [\nabla f(\mathbf{x}) \cdot \mathbf{h} + o(\mathbf{h})]g(\mathbf{x} + \mathbf{h}) + f(\mathbf{x})[\nabla g(\mathbf{x}) \cdot \mathbf{h} + o(\mathbf{h})] \\
&= g(\mathbf{x} + \mathbf{h}) \nabla f(\mathbf{x}) \cdot \mathbf{h} + f(\mathbf{x}) \nabla g(\mathbf{x}) \cdot \mathbf{h} + o(\mathbf{h}) \\
&= g(\mathbf{x}) \nabla f(\mathbf{x}) \cdot \mathbf{h} + f(\mathbf{x}) \nabla g(\mathbf{x}) \cdot \mathbf{h} \\
&\quad + [g(\mathbf{x} + \mathbf{h}) - g(\mathbf{x})] \nabla f(\mathbf{x}) \cdot \mathbf{h} + o(\mathbf{h}) \\
&= [g(\mathbf{x}) \nabla f(\mathbf{x}) + f(\mathbf{x}) \nabla g(\mathbf{x})] \cdot \mathbf{h} + o(\mathbf{h}). \quad □
\end{aligned}$$

Directional Derivatives

Here we take up an idea which generalizes the notion of partial derivative. Its connection to gradients will be made clear as we go on. The partial derivatives

$$\frac{\partial f}{\partial x}(x, y, z) = \lim_{h \to 0} \frac{f(x + h, y, z) - f(x, y, z)}{h},$$

$$\frac{\partial f}{\partial y}(x, y, z) = \lim_{h \to 0} \frac{f(x, y + h, z) - f(x, y, z)}{h},$$

$$\frac{\partial f}{\partial z}(x, y, z) = \lim_{h \to 0} \frac{f(x, y, z + h) - f(x, y, z)}{h}$$

can be expressed more elegantly in the form

$$\frac{\partial f}{\partial x}(\mathbf{x}) = \lim_{h \to 0} \frac{f(\mathbf{x} + h\mathbf{i}) - f(\mathbf{x})}{h},$$

$$\frac{\partial f}{\partial y}(\mathbf{x}) = \lim_{h \to 0} \frac{f(\mathbf{x} + h\mathbf{j}) - f(\mathbf{x})}{h},$$

$$\frac{\partial f}{\partial z}(\mathbf{x}) = \lim_{h \to 0} \frac{f(\mathbf{x} + h\mathbf{k}) - f(\mathbf{x})}{h}.$$

Each of the partials is therefore the limit of a quotient of the form

$$\frac{f(\mathbf{x} + h\mathbf{u}) - f(\mathbf{x})}{h},$$

where $\mathbf{u}$ is one of the unit coordinate vectors $\mathbf{i}$, $\mathbf{j}$, or $\mathbf{k}$. There is no reason to be so restrictive on $\mathbf{u}$. If f is defined in a neighborhood of $\mathbf{x}$, then, for small h, the difference quotient

$$\frac{f(\mathbf{x} + h\mathbf{u}) - f(\mathbf{x})}{h}$$

makes sense for any unit vector $\mathbf{u}$.

Definition of Directional Derivative

For any *unit vector* $\mathbf{u}$, the limit

$$f'_{\mathbf{u}}(\mathbf{x}) = \lim_{h \to 0} \frac{f(\mathbf{x} + h\mathbf{u}) - f(\mathbf{x})}{h},$$

if it exists, is called the *directional derivative* of f at $\mathbf{x}$ in the direction $\mathbf{u}$.

The partials are of course themselves directional derivatives:

$$\frac{\partial f}{\partial x}(\mathbf{x}) = f'_{\mathbf{i}}(\mathbf{x}), \qquad \frac{\partial f}{\partial y}(\mathbf{x}) = f'_{\mathbf{j}}(\mathbf{x}), \qquad \frac{\partial f}{\partial z}(\mathbf{x}) = f'_{\mathbf{k}}(\mathbf{x}).$$

As you know, the partials of f give the rates of change of f in the $\mathbf{i}, \mathbf{j}, \mathbf{k}$ directions. The directional derivative $f'_{\mathbf{u}}$ gives *the rate of change* of f *in the direction* $\mathbf{u}$.

There is a simple link between the gradient at $\mathbf{x}$ and the directional derivatives at $\mathbf{x}$.

Theorem 14.8.1

If f is differentiable at $\mathbf{x}$, then f has a directional derivative at $\mathbf{x}$ in every direction and

$$f'_{\mathbf{u}}(\mathbf{x}) = \nabla f(\mathbf{x}) \cdot \mathbf{u}.$$

PROOF.   We take $\mathbf{u}$ as a unit vector and assume that f is differentiable at $\mathbf{x}$. The differentiability at $\mathbf{x}$ tells us that $\nabla f(\mathbf{x})$ exists and

$$f(\mathbf{x} + h\mathbf{u}) - f(\mathbf{x}) = \nabla f(\mathbf{x}) \cdot h\mathbf{u} + o(h\mathbf{u}).$$

Division by h gives

$$\frac{f(\mathbf{x} + h\mathbf{u}) - f(\mathbf{x})}{h} = \nabla f(\mathbf{x}) \cdot \mathbf{u} + \frac{o(h\mathbf{u})}{h}.$$

It is not hard to see that the remainder term tends to zero with h and thus

$$\frac{f(\mathbf{x} + h\mathbf{u}) - f(\mathbf{x})}{h} \to \nabla f(\mathbf{x}) \cdot \mathbf{u},$$

which is what we were trying to show.   ☐

Earlier (Theorem 14.7.1) we proved that if f has *continuous first partials in a neighborhood of* $\mathbf{x}$, then f is differentiable at $\mathbf{x}$ and

$$\nabla f(\mathbf{x}) = \frac{\partial f}{\partial x}(\mathbf{x})\,\mathbf{i} + \frac{\partial f}{\partial y}(\mathbf{x})\,\mathbf{j} + \frac{\partial f}{\partial z}(\mathbf{x})\,\mathbf{k}.$$

Theorem 14.8.1 enables us to go the other way.

Theorem 14.8.2

If f is differentiable at $\mathbf{x}$ then all the first partials exist at $\mathbf{x}$ and

$$\nabla f(\mathbf{x}) = \frac{\partial f}{\partial x}(\mathbf{x})\,\mathbf{i} + \frac{\partial f}{\partial y}(\mathbf{x})\,\mathbf{j} + \frac{\partial f}{\partial z}(\mathbf{x})\,\mathbf{k}.$$

PROOF.   Assume that f is differentiable at $\mathbf{x}$. The existence of the first partials follows from Theorem 14.8.1 together with the observation that the first partials are directional derivatives:

$$\frac{\partial f}{\partial x}(\mathbf{x}) = f'_{\mathbf{i}}(\mathbf{x}), \qquad \frac{\partial f}{\partial y}(\mathbf{x}) = f'_{\mathbf{j}}(\mathbf{x}), \qquad \frac{\partial f}{\partial z}(\mathbf{x}) = f'_{\mathbf{k}}(\mathbf{x}).$$

Also from Theorem 14.8.1 we have

$$\frac{\partial f}{\partial x}(\mathbf{x}) = \nabla f(\mathbf{x}) \cdot \mathbf{i}, \qquad \frac{\partial f}{\partial y}(\mathbf{x}) = \nabla f(\mathbf{x}) \cdot \mathbf{j}, \qquad \frac{\partial f}{\partial z}(\mathbf{x}) = \nabla f(\mathbf{x}) \cdot \mathbf{k}.$$

Since

$$\nabla f(\mathbf{x}) = [\,\nabla f(\mathbf{x}) \cdot \mathbf{i}\,]\mathbf{i} + [\,\nabla f(\mathbf{x}) \cdot \mathbf{j}\,]\mathbf{j} + [\,\nabla f(\mathbf{x}) \cdot \mathbf{k}\,]\mathbf{k}$$

it follows that

$$\nabla f(\mathbf{x}) = \frac{\partial f}{\partial x}(\mathbf{x})\,\mathbf{i} + \frac{\partial f}{\partial y}(\mathbf{x})\,\mathbf{j} + \frac{\partial f}{\partial z}(\mathbf{x})\,\mathbf{k},$$

as asserted.   □

For an arbitrary unit vector **u**, we have

$$f'_{\mathbf{u}}(\mathbf{x}) = \nabla f(\mathbf{x}) \cdot \mathbf{u} = \frac{\nabla f(\mathbf{x}) \cdot \mathbf{u}}{\|\mathbf{u}\|} = \text{comp}_{\mathbf{u}}\,\nabla f(\mathbf{x}).$$

Namely, the *directional derivative is the component of the gradient vector in that direction.* We illustrate this in Figure 14.8.1.

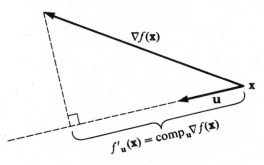

FIGURE 14.8.1

With

$$\nabla f(\mathbf{x}) = \frac{\partial f}{\partial x}(\mathbf{x})\,\mathbf{i} + \frac{\partial f}{\partial y}(\mathbf{x})\,\mathbf{j} + \frac{\partial f}{\partial z}(\mathbf{x})\,\mathbf{k},$$

the relation

$$f'_{\mathbf{u}}(\mathbf{x}) = \nabla f(\mathbf{x}) \cdot \mathbf{u}$$

gives

$$f'_{\mathbf{u}}(\mathbf{x}) = \frac{\partial f}{\partial x}(\mathbf{x})\,u_1 + \frac{\partial f}{\partial y}(\mathbf{x})\,u_2 + \frac{\partial f}{\partial z}(\mathbf{x})\,u_3.$$

Setting

$$\mathbf{x} = (x, y, z),$$

we have

$$\boxed{\,f'_{\mathbf{u}}(x, y, z) = \frac{\partial f}{\partial x}(x, y, z)\,u_1 + \frac{\partial f}{\partial y}(x, y, z)\,u_2 + \frac{\partial f}{\partial z}(x, y, z)\,u_3.\,}$$

In the two-variable case, the z-term drops out and we have

$$f_u'(x, y) = \frac{\partial f}{\partial x}(x, y)u_1 + \frac{\partial f}{\partial y}(x, y)u_2.$$

Below we carry out some computations.

Problem. Find the directional derivative of

$$f(x, y) = x^2 + y^2$$

at $(1, 2)$ in the direction of $2\mathbf{i} - 3\mathbf{j}$.

SOLUTION. In the first place, $2\mathbf{i} - 3\mathbf{j}$ is not a unit vector. Its norm is $\sqrt{13}$. The unit vector we want is

$$\mathbf{u} = \tfrac{1}{13}\sqrt{13}\,[2\mathbf{i} - 3\mathbf{j}].$$

Since we are in the two-variable case,

$$f_u'(x, y) = \frac{\partial f}{\partial x}(x, y)u_1 + \frac{\partial f}{\partial y}(x, y)u_2.$$

From

$$\frac{\partial f}{\partial x}(x, y) = 2x \quad \text{and} \quad \frac{\partial f}{\partial y}(x, y) = 2y,$$

we get

$$\frac{\partial f}{\partial x}(1, 2) = 2 \quad \text{and} \quad \frac{\partial f}{\partial y}(1, 2) = 4.$$

Since

$$u_1 = \tfrac{2}{13}\sqrt{13} \quad \text{and} \quad u_2 = -\tfrac{3}{13}\sqrt{13},$$

we have

$$f_u'(1, 2) = \tfrac{4}{13}\sqrt{13} - \tfrac{12}{13}\sqrt{13} = -\tfrac{8}{13}\sqrt{13}. \quad \square$$

Problem. Find the directional derivative of

$$f(x, y, z) = x \cos y \sin z$$

at $(1, \pi, \tfrac{1}{4}\pi)$ in the direction of $2\mathbf{i} - \mathbf{j} + 4\mathbf{k}$.

SOLUTION. The unit vector we want is

$$\mathbf{u} = \tfrac{1}{21}\sqrt{21}\,[2\mathbf{i} - \mathbf{j} + 4\mathbf{k}].$$

Here we are in the three-variable case:

$$f_u'(x, y, z) = \frac{\partial f}{\partial x}(x, y, z)u_1 + \frac{\partial f}{\partial y}(x, y, z)u_2 + \frac{\partial f}{\partial z}(x, y, z)u_3.$$

From

$$\frac{\partial f}{\partial x}(x, y, z) = \cos y \sin z, \quad \frac{\partial f}{\partial y}(x, y, z) = -x \sin y \sin z, \quad \frac{\partial f}{\partial z}(x, y, z) = x \cos y \cos z$$

we get

$$\frac{\partial f}{\partial x}(1, \pi, \tfrac{1}{4}\pi) = -\tfrac{1}{2}\sqrt{2}, \quad \frac{\partial f}{\partial y}(1, \pi, \tfrac{1}{4}\pi) = 0, \quad \frac{\partial f}{\partial z}(1, \pi, \tfrac{1}{4}\pi) = -\tfrac{1}{2}\sqrt{2}.$$

Since

$$u_1 = \tfrac{2}{21}\sqrt{21}, \quad u_2 = -\tfrac{1}{21}\sqrt{21}, \quad u_3 = \tfrac{4}{21}\sqrt{21},$$

it follows that

$$f_u'(1, \pi, \tfrac{1}{4}\pi) = -\tfrac{1}{21}\sqrt{42} + 0 - \tfrac{2}{21}\sqrt{42} = -\tfrac{3}{21}\sqrt{42}. \quad \square$$

The Gradient as the Most Rapid Increase Vector

The relation

$$f_u'(\mathbf{x}) = \nabla f(\mathbf{x}) \cdot \mathbf{u}$$

gives

$$f_u'(\mathbf{x}) = \| \nabla f(\mathbf{x}) \| \, \| \mathbf{u} \| \cos \theta$$

and thus

$$\boxed{f_u'(\mathbf{x}) = \| \nabla f(\mathbf{x}) \| \cos \theta}$$

where θ is the angle between $\nabla f(\mathbf{x})$ and the unit vector $\mathbf{u}$. It follows that to maximize $f_u'(\mathbf{x})$ we must choose $\mathbf{u}$ in the direction of $\nabla f(\mathbf{x})$. $[\theta = 0]$ To minimize $f_u'(\mathbf{x})$ we must choose $\mathbf{u}$ in the direction of $-\nabla f(\mathbf{x})$. $[\theta = \pi]$ In other words, the *gradient vector points in the direction of the most rapid increase of the function and the negative of the gradient points in the direction of the most rapid decrease.* At each point $\mathbf{x}$ the extreme rates of change are respectively

$$\| \nabla f(\mathbf{x}) \| \quad \text{and} \quad -\| \nabla f(\mathbf{x}) \|.$$

Example. In the case of

$$f(x, y) = \sqrt{1 - (x^2 + y^2)}$$

the graph is the upper hemisphere of radius one centered at the origin. The gradient at (x, y)

$$\nabla f = \frac{-x}{\sqrt{1 - (x^2 + y^2)}} \mathbf{i} + \frac{-y}{\sqrt{1 - (x^2 + y^2)}} \mathbf{j}$$

points from (x, y) through the origin. See Figure 14.8.2. The path of steepest ascent along the surface is the great circle route through the north pole. $\square$

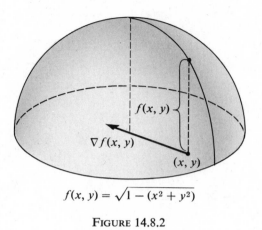

$$f(x, y) = \sqrt{1 - (x^2 + y^2)}$$

FIGURE 14.8.2

Problem. The temperature distribution on a metal plate is given by the function

$$T(x, y) = k(x^2 e^y + y^2 e^x),$$

where k is a positive constant.

(a) In what direction does the temperature increase most rapidly at the point $(1, 0)$? What is this rate of increase?
(b) In what direction does the temperature decrease most rapidly?

SOLUTION

$$\nabla T(x, y) = \frac{\partial T}{\partial x}(x, y)\,\mathbf{i} + \frac{\partial T}{\partial y}(x, y)\,\mathbf{j}$$

$$= k(2xe^y + y^2 e^x)\mathbf{i} + k(x^2 e^y + 2ye^x)\mathbf{j}.$$

(a) At $(1, 0)$ the temperature increases most rapidly in the direction of the gradient

$$\nabla T(1, 0) = 2k\mathbf{i} + k\mathbf{j}.$$

This rate of increase is

$$\|\nabla T(1, 0)\| = \|2k\mathbf{i} + k\mathbf{j}\| = \sqrt{5}\,k.$$

(b) The temperature decreases most rapidly in the direction of

$$-\nabla T(1, 0) = -2k\mathbf{i} - k\mathbf{j}. \quad \square$$

Problem. The temperature distribution of a metal ball centered at the origin is given by the function

$$T(x, y, z) = ke^{-(x^2 + y^2 + z^2)}, \quad k \text{ a positive constant.}$$

(a) In what direction does the temperature increase most rapidly at the point (x, y, z)? What is this rate of temperature increase?
(b) In what direction does the temperature decrease most rapidly?
(c) What is the rate of temperature change at (x, y, z) in the directions of $\mathbf{i}$, $\mathbf{j}$, and $\mathbf{k}$?

SOLUTION.   The gradient

$$\nabla T(x, y, z) = \frac{\partial T}{\partial x}(x, y, z)\mathbf{i} + \frac{\partial T}{\partial y}(x, y, z)\mathbf{j} + \frac{\partial T}{\partial z}(x, y, z)\mathbf{k}$$

$$= -2ke^{-(x^2+y^2+z^2)}[x\mathbf{i} + y\mathbf{j} + z\mathbf{k}]$$

$$= -2T(x, y, z)\mathbf{r}$$

points from (x, y, z) toward the origin.

(a) The temperature increases most rapidly in the direction toward the origin. The rate of increase is

$$\|\nabla T(x, y, z)\| = 2T(x, y, z)\|\mathbf{r}\| = 2T(x, y, z)\sqrt{x^2 + y^2 + z^2}.$$

(b) The temperature decreases most rapidly in the direction away from the origin.

(c) The rates of temperature change along the directions $\mathbf{i}$, $\mathbf{j}$, and $\mathbf{k}$ are given by the directional derivatives

$$T_\mathbf{i}'(x, y, z) = \nabla T(x, y, z)\cdot\mathbf{i} = -2xT(x, y, z),$$
$$T_\mathbf{j}'(x, y, z) = \nabla T(x, y, z)\cdot\mathbf{j} = -2yT(x, y, z),$$
$$T_\mathbf{k}'(x, y, z) = \nabla T(x, y, z)\cdot\mathbf{k} = -2zT(x, y, z). \quad \square$$

Problem.   The temperature distribution of a plate is given by the function

$$T(x, y) = A - 2x^2 - y^2.$$

(a) What is the hottest point?

(b) Find the path followed by a heat-seeking particle which starts out at the point $(-2, 1)$.

SOLUTION.   The hottest point is obviously the origin. The particle, starting out at $(-2, 1)$, will proceed at each point in the direction of the gradient. From

$$T(x, y) = A - 2x^2 - y^2$$

we have

$$\nabla T = -4x\mathbf{i} - 2y\mathbf{j}.$$

We now want the curve

$$C: \mathbf{r}(t) = x(t)\mathbf{i} + y(t)\mathbf{j}$$

which begins at $(-2, 1)$ and has at each point the tangent vector ∇T. We can satisfy the first condition by setting

$$x(0) = -2, \qquad y(0) = 1.$$

We can satisfy the second condition by setting

$$x'(t) = -4x(t), \qquad y'(t) = -2y(t). \qquad \qquad \text{(explain)}$$

These differential equations, together with initial conditions at $t = 0$, imply that

$$x(t) = -2e^{-4t}, \qquad y(t) = e^{-2t}.$$

We can eliminate the parameter t by noting that

$$x(t) = -2e^{-4t} = -2[e^{-2t}]^2 = -2[y(t)]^2.$$

In terms of just x and y, the path becomes

$$x = -2y^2.$$

This is a parabolic path. The part that concerns us begins at $(-2, 1)$ and ends at the origin. $\square$

Exercises

Find the directional derivative of each function at the given point in the indicated direction.

*1. $f(x, y) = x^2 + 3y^2$ at $(1, 1)$ in the direction of $\mathbf{i} - \mathbf{j}$.

2. $f(x, y) = x + \sin(x + y)$ at $(0, 0)$ in the direction of $2\mathbf{i} + \mathbf{j}$.

*3. $f(x, y) = xe^y - ye^x$ at $(1, 0)$ in the direction of $3\mathbf{i} + 4\mathbf{j}$.

4. $f(x, y) = \dfrac{2x}{x - y}$ at $(1, 0)$ in the direction of $\mathbf{i} - \sqrt{3}\,\mathbf{j}$.

*5. $f(x, y) = \dfrac{ax + by}{x + y}$ at $(1, 1)$ in the direction of $\mathbf{i} - \mathbf{j}$.

6. $f(x, y) = \dfrac{x + y}{cx + dy}$ at $(1, 1)$ in the direction of $c\mathbf{i} - d\mathbf{j}$.

*7. $f(x, y, z) = xy + yz + zx$ at $(1, -1, 1)$ in the direction of $\mathbf{i} + 2\mathbf{j} + \mathbf{k}$.

8. $f(x, y, z) = x^2y + y^2z + z^2x$ at $(1, 0, 1)$ in the direction of $3\mathbf{j} - \mathbf{k}$.

*9. $f(x, y, z) = (x + y^2 + z^3)^2$ at $(1, -1, 1)$ in the direction of $\mathbf{i} + \mathbf{j}$.

10. $f(x, y, z) = Ax^2 + Bxyz + Cy^2$ at $(1, 2, 1)$ in the direction of $A\mathbf{i} + B\mathbf{j} + C\mathbf{k}$.

*11. Find the directional derivative of

$$f(x, y) = \log \sqrt{x^2 + y^2}$$

in the direction toward the origin.

12. Find the directional derivative of

$$f(x, y) = (x - 1)y^2 e^{xy}$$

at $(0, 1)$ in the direction toward $(-1, 3)$.

*13. Find the directional derivative of

$$f(x, y) = Ax^2 + 2Bxy + Cy^2$$

at (a, b) in the direction toward (b, a)

(a) if $a > b$.   (b) if $a < b$.

14. Find the directional derivative of

$$f(x, y, z) = xy^2 + y^2z^3 + z^3x$$

at $(2, -1, 1)$ in the direction of the most rapid increase of f.

* 15. Find the directional derivative of

$$f(x, y, z) = xe^{y^2 - z^2}$$

at $(1, 2, -2)$ in the direction of the path

$$\mathbf{r}(t) = t\mathbf{i} + 2\cos(t - 1)\mathbf{j} - 2e^{t-1}\mathbf{k}.$$

16. Given the temperature distribution

$$T(x, y) = 48 - \tfrac{4}{3}x^2 - 3y^2$$

find the rate of temperature change
(a) at $(1, -1)$ in the direction of the most rapid cooling.
(b) at $(1, 2)$ in the direction $\mathbf{i}$.
(c) at $(2, 2)$ in the direction away from the origin.

* 17. The temperature distribution of a metal plate is given by the function

$$T(x, y) = A - x^2 - 3y^2.$$

Find the path followed by a heat-seeking particle which starts out at the point $(1, 1)$.

18. Same as 17 except that the particle starts out at $(1, -2)$.

* 19. The temperature distribution of a metal plate is given by the function

$$T(x, y) = 1 - a^2x^2 - b^2y^2.$$

Find the path followed by a heat-seeking particle which starts out at the point (a^2, b^2).

20. (*Optional*) A heat-fleeing particle starts out at the origin. Find the path followed if the temperature distribution is given by the function

$$T(x, y) = A + x + 2y - x^2 - 3y^2.$$

HINT: The resulting differential equations can be solved by the methods of Section 5.8.

14.9 Mean-Value Theorem and Chain Rule

You are already familiar with results bearing these names in the one-variable case. Here we take up the multivariable case.

Mean-Value Theorem

If f is differentiable at each point of the line segment $\mathbf{ab}$, then there exists on that line segment a point $\mathbf{c}$ between $\mathbf{a}$ and $\mathbf{b}$ such that

$$f(\mathbf{b}) - f(\mathbf{a}) = \nabla f(\mathbf{c}) \cdot (\mathbf{b} - \mathbf{a}).$$

PROOF. The idea of the proof is to apply the one-variable mean-value theorem to the function

$$g(t) = f(\mathbf{a} + t[\mathbf{b} - \mathbf{a}]), \qquad t \in [0, 1].$$

To get the differentiability of g, we take $t \in (0, 1)$ and form

$$
\begin{aligned}
g(t + h) - g(t) &= f(\mathbf{a} + (t + h)[\mathbf{b} - \mathbf{a}]) - f(\mathbf{a} + t[\mathbf{b} - \mathbf{a}]) \\
&= f(\mathbf{a} + t[\mathbf{b} - \mathbf{a}] + h[\mathbf{b} - \mathbf{a}]) - f(\mathbf{a} + t[\mathbf{b} - \mathbf{a}]) \\
&= \nabla f(\mathbf{a} + t[\mathbf{b} - \mathbf{a}]) \cdot h[\mathbf{b} - \mathbf{a}] + o(h[\mathbf{b} - \mathbf{a}]).
\end{aligned}
$$

Since

$$\nabla f(\mathbf{a} + t[\mathbf{b} - \mathbf{a}]) \cdot h(\mathbf{b} - \mathbf{a}) = \{\nabla f(\mathbf{a} + t[\mathbf{b} - \mathbf{a}]) \cdot (\mathbf{b} - \mathbf{a})\}h$$

and the $o(h(\mathbf{b} - \mathbf{a}))$ term is obviously $o(h)$, we can write

$$g(t + h) - g(t) = \{\nabla f(\mathbf{a} + t[\mathbf{b} - \mathbf{a}]) \cdot (\mathbf{b} - \mathbf{a})\}h + o(h).$$

Dividing both sides by h, we see that g is differentiable and

$$g'(t) = \nabla f(\mathbf{a} + t[\mathbf{b} - \mathbf{a}]) \cdot (\mathbf{b} - \mathbf{a}).$$

The function g is clearly continuous at 0 and at 1. Applying the one-variable mean-value theorem to g, we can conclude that there exists a number t_0 between 0 and 1 such that

(1)
$$g(1) - g(0) = g'(t_0)(1 - 0).$$

Since

$$g(1) = f(\mathbf{b}), \qquad g(0) = f(\mathbf{a}), \quad \text{and} \quad g'(t_0) = \nabla f(\mathbf{a} + t_0[\mathbf{b} - \mathbf{a}]) \cdot (\mathbf{b} - \mathbf{a}),$$

condition (1) gives us

$$f(\mathbf{b}) - f(\mathbf{a}) = \nabla f(\mathbf{a} + t_0[\mathbf{b} - \mathbf{a}]) \cdot (\mathbf{b} - \mathbf{a}).$$

Setting

$$\mathbf{c} = \mathbf{a} + t_0[\mathbf{b} - \mathbf{a}],$$

we have

$$f(\mathbf{b}) - f(\mathbf{a}) = \nabla f(\mathbf{c}) \cdot (\mathbf{b} - \mathbf{a}). \quad \square \qquad \text{(Figure 14.9.1)}$$

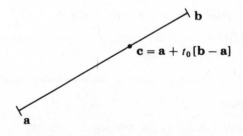

there exists a point $\mathbf{c}$ between $\mathbf{a}$ and $\mathbf{b}$ such that
$$f(\mathbf{b}) - f(\mathbf{a}) = \nabla f(\mathbf{c}) \cdot [\mathbf{b} - \mathbf{a}]$$

FIGURE 14.9.1

Before stating the chain rule we go over some terminology.

1. To say that a function f of several variables is *continuously differentiable* on a set U is to say that f is differentiable on U and its gradient ∇f is continuous on U.
2. To say that U is *open* is to say that U contains a neighborhood of each of its points.
3. To say that $\mathbf{r}$ is a *differentiable curve* in U is to say that $\mathbf{r}$ is a differentiable vector function defined on an interval of real numbers and its range lies entirely in U; namely,

$$\mathbf{r}'(t) \text{ exists} \quad \text{and} \quad \mathbf{r}(t) \in U.$$

If a curve $\mathbf{r}$ lies in the domain of a function f, then we can form the composition

$$(f \circ \mathbf{r})(t) = f(\mathbf{r}(t)).$$

The numbers $f(\mathbf{r}(t))$ are the values of f along the curve $\mathbf{r}$.

Chain Rule

If f is continuously differentiable on an open set U and $\mathbf{r}$ is a differentiable curve in U, then the composition $f \circ \mathbf{r}$ is differentiable and

$$\frac{d}{dt}[f(\mathbf{r}(t))] = \nabla f(\mathbf{r}(t)) \cdot \mathbf{r}'(t).$$

PROOF.   We will show that

$$\lim_{h \to 0} \frac{f(\mathbf{r}(t + h)) - f(\mathbf{r}(t))}{h} = \nabla f(\mathbf{r}(t)) \cdot \mathbf{r}'(t).$$

For $h \neq 0$ and sufficiently small, the line segment joining $\mathbf{r}(t)$ to $\mathbf{r}(t + h)$ lies entirely in U. This we get because U is open and $\mathbf{r}$ is continuous. (See Figure 14.9.2.) For such h, the mean-value theorem just proved tells us that there exists a point $\mathbf{c}(h)$ between $\mathbf{r}(t)$ and $\mathbf{r}(t + h)$ such that

$$f(\mathbf{r}(t + h)) - f(\mathbf{r}(t)) = \nabla f(\mathbf{c}(h)) \cdot [\mathbf{r}(t + h) - \mathbf{r}(t)].$$

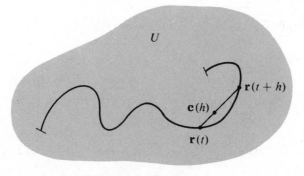

FIGURE 14.9.2

Dividing both sides by h,

$$\frac{f(\mathbf{r}(t + h)) - f(\mathbf{r}(t))}{h} = \nabla f(\mathbf{c}(h)) \cdot \left[\frac{\mathbf{r}(t + h) - \mathbf{r}(t)}{h} \right].$$

As h tends to zero, $\mathbf{c}(h)$ tends to $\mathbf{r}(t)$ and by the continuity of ∇f,

$$\nabla f(\mathbf{c}(h)) \rightarrow \nabla f(\mathbf{r}(t)).$$

Since

$$\frac{\mathbf{r}(t + h) - \mathbf{r}(t)}{h} \rightarrow \mathbf{r}'(t),$$

the result follows. $\square$

Problem.   Let

$$f(x, y) = \tfrac{1}{2}(x^2 + y^2) \quad \text{and} \quad \mathbf{r}(t) = a \cos t\, \mathbf{i} + b \sin t\, \mathbf{j}.$$

Find the rate of change of f with respect to t along the curve $\mathbf{r}$.

SOLUTION.   We'll use the chain rule

$$\frac{d}{dt} [f(\mathbf{r}(t))] = \nabla f(\mathbf{r}(t)) \cdot \mathbf{r}'(t).$$

Here

$$\nabla f = x\mathbf{i} + y\mathbf{j},$$
$$\nabla f(\mathbf{r}(t)) = a \cos t\, \mathbf{i} + b \sin t\, \mathbf{j},$$
$$\mathbf{r}'(t) = -a \sin t\, \mathbf{i} + b \cos t\, \mathbf{j}.$$
$$\frac{d}{dt} [f(\mathbf{r}(t))] = -a^2 \cos t \sin t + b^2 \sin t \cos t$$
$$= (b^2 - a^2) \sin t \cos t. \quad \square$$

Problem.   Find the rate of change of f with respect to t along the curve $\mathbf{r}$ if

$$f(x, y, z) = x^2 y + z \cos x \quad \text{and} \quad \mathbf{r}(t) = t\mathbf{i} + t^2\mathbf{j} + t^3\mathbf{k}.$$

SOLUTION

$$\nabla f = (2xy - z \sin x)\mathbf{i} + x^2\mathbf{j} + \cos x\, \mathbf{k},$$
$$\nabla f(\mathbf{r}(t)) = (2t^3 - t^3 \sin t)\mathbf{i} + t^2\mathbf{j} + \cos t\, \mathbf{k},$$
$$\mathbf{r}'(t) = \mathbf{i} + 2t\mathbf{j} + 3t^2\mathbf{k}.$$
$$\frac{d}{dt} [f(\mathbf{r}(t))] = 2t^3 - t^3 \sin t + 2t^3 + 3t^2 \cos t$$
$$= 4t^3 - t^3 \sin t + 3t^2 \cos t. \quad \square$$

Another View of the Chain Rule

In the one-variable case we can write the chain rule in two ways. We can write

(1) $$\frac{d}{dt}[f(g(t))] = f'(g(t))g'(t)$$

or, by setting $u = f(x)$, $x = g(t)$, we can write

(2) $$\frac{du}{dt} = \frac{du}{dx}\frac{dx}{dt}.$$

You can view the formula

$$\frac{d}{dt}[f(\mathbf{r}(t))] = \nabla f(\mathbf{r}(t)) \cdot \mathbf{r}'(t)$$

as the counterpart of (1). The counterpart of (2) can be obtained by setting

$$u = f(x, y, z); \qquad x = g(t), \quad y = h(t), \quad z = k(t).$$

The chain rule then reads

$$\boxed{\frac{du}{dt} = \frac{\partial u}{\partial x}\frac{dx}{dt} + \frac{\partial u}{\partial y}\frac{dy}{dt} + \frac{\partial u}{\partial z}\frac{dz}{dt}.}$$

In the two-variable case the z-term drops out and we have

$$\boxed{\frac{du}{dt} = \frac{\partial u}{\partial x}\frac{dx}{dt} + \frac{\partial u}{\partial y}\frac{dy}{dt}.}$$

Problem.   Find du/dt if

$$u = x^2 - y^2 \quad \text{and} \quad x = t^2 - 1, \quad y = 3\sin \pi t.$$

SOLUTION.   Here we are in the two-variable case

$$\frac{du}{dt} = \frac{\partial u}{\partial x}\frac{dx}{dt} + \frac{\partial u}{\partial y}\frac{dy}{dt}.$$

Since

$$\frac{\partial u}{\partial x} = 2x, \qquad \frac{\partial u}{\partial y} = -2y, \qquad \frac{dx}{dt} = 2t, \qquad \frac{dy}{dt} = 3\pi \cos \pi t,$$

we have

$$\frac{du}{dt} = (2x)(2t) + (-2y)(3\pi \cos \pi t)$$

$$= 2(t^2 - 1)(2t) + (-2)(3 \sin \pi t)(3\pi \cos \pi t)$$

$$= 4t^3 - 4t - 18 \sin \pi t \cos \pi t. \quad \square$$

Problem. The upper radius of the frustum of a cone is 10 inches, the lower radius is 12 inches, and the height is 18 inches. At what rate is the volume changing if the upper radius decreases at the rate of 2 inches per minute, the lower radius increases at the rate of 3 inches per minute, and the height decreases at the rate of 4 inches per minute?

SOLUTION. Let x be the upper radius, y the lower radius, and z the height. Then

$$V = \frac{1}{3} \pi z(x^2 + xy + y^2)$$

so that

$$\frac{\partial V}{\partial x} = \frac{1}{3} \pi z(2x + y), \qquad \frac{\partial V}{\partial y} = \frac{1}{3} \pi z(x + 2y), \qquad \frac{\partial V}{\partial z} = \frac{1}{3} \pi(x^2 + xy + y^2).$$

Since

$$\frac{dV}{dt} = \frac{\partial V}{\partial x}\frac{dx}{dt} + \frac{\partial V}{\partial y}\frac{dy}{dt} + \frac{\partial V}{\partial z}\frac{dz}{dt}$$

we have here

$$\frac{dV}{dt} = \frac{1}{3} \pi z(2x + y)\frac{dx}{dt} + \frac{1}{3} \pi z(x + 2y)\frac{dy}{dt} + \frac{1}{3} \pi(x^2 + xy + y^2)\frac{dz}{dt}.$$

Using

$$x = 10, \qquad y = 12, \qquad z = 18, \qquad \frac{dx}{dt} = -2, \qquad \frac{dy}{dt} = 3, \qquad \frac{dz}{dt} = -4,$$

you will find that

$$\frac{dV}{dt} = -\frac{772}{3} \pi.$$

The volume decreases at the rate of $\frac{772}{3}\pi$ cubic inches per minute. ☐

In Chapter 3 you saw that if $f'(x) = 0$ for all $x \in (a, b)$, then f is constant on (a, b). We have a similar result for functions of several variables.

Theorem

Let U be an open set with the property that any two points of U can be joined by a differentiable curve which lies entirely in U:

$$\text{if } \nabla f(\mathbf{x}) = \mathbf{0} \text{ for all } \mathbf{x} \text{ in } U, \text{ then } f \text{ is constant on } U.$$

PROOF. Let $\mathbf{a}$ and $\mathbf{b}$ be two points of U. We join these two points by a differentiable curve $\mathbf{r}(t)$ with $\mathbf{r}(a) = \mathbf{a}$ and $\mathbf{r}(b) = \mathbf{b}$. (See Figure 14.9.3.) Let f be the function in question. Since

$$\nabla f(\mathbf{r}(t)) = \mathbf{0} \qquad \text{for all } t \in [a, b],$$

we have

$$\int_a^b \left[\nabla f(\mathbf{r}(t)) \cdot \mathbf{r}'(t) \right] dt = 0.$$

On the other hand,

$$\int_a^b \left[\nabla f(\mathbf{r}(t)) \cdot \mathbf{r}'(t) \right] dt = \int_a^b \frac{d}{dt} \left[f(\mathbf{r}(t)) \right] dt$$

$$= f(\mathbf{r}(b)) - f(\mathbf{r}(a))$$

$$= f(\mathbf{b}) - f(\mathbf{a}).$$

This tells us that

$$f(\mathbf{b}) = f(\mathbf{a}).$$

Since $\mathbf{a}$ and $\mathbf{b}$ are arbitrary points of U, f must be constant on U.   □

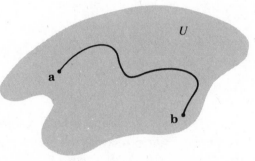

any two points of U can be joined by a
differentiable curve which lies entirely in U

FIGURE 14.9.3

Corollary

If $\nabla f(\mathbf{x}) = \nabla g(\mathbf{x})$ for all $\mathbf{x}$ in U, then f and g differ by a constant on U.

PROOF.   If

$$\nabla f(\mathbf{x}) = \nabla g(\mathbf{x}) \qquad \text{for all } \mathbf{x} \text{ in } U,$$

then

$$\nabla [f(\mathbf{x}) - g(\mathbf{x})] = \nabla f(\mathbf{x}) - \nabla g(\mathbf{x}) = \mathbf{0} \qquad \text{for all } \mathbf{x} \text{ in } U$$

and by the theorem $f - g$ must be constant on U.   □

CAUTION.   The last theorem and the corollary to it depend strongly on our being able to join any two points of U by a differentiable curve *which lies entirely in U*. If U is "disconnected," as in Figure 14.9.4, the results do not hold.

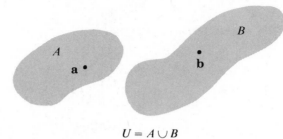

$$U = A \cup B$$

a and **b** cannot be joined by a
curve lying entirely in U

FIGURE 14.9.4

Exercises

Find $\dfrac{d}{dt} f(\mathbf{r}(t))$.

*1. $f(x, y) = x^2 y, \quad \mathbf{r}(t) = e^t \mathbf{i} + e^{-t}\mathbf{j}$.

2. $f(x, y) = x - y, \quad \mathbf{r}(t) = at\mathbf{i} + b \cos at\, \mathbf{j}$.

*3. $f(x, y, z) = xy - yz, \quad \mathbf{r}(t) = t\mathbf{i} + t^2\mathbf{j} + t^3\mathbf{k}$.

4. $f(x, y, z) = x^2 + y^2, \quad \mathbf{r}(t) = a \cos \omega t\, \mathbf{i} + b \sin \omega t\, \mathbf{j} + b\omega t\, \mathbf{k}$.

*5. $f(x, y, z) = x^2 + y^2 + z, \quad \mathbf{r}(t) = a \cos \omega t\, \mathbf{i} + b \sin \omega t\, \mathbf{j} + b\omega t\, \mathbf{k}$.

6. $f(x, y, z) = y^2 \sin (x + z), \quad \mathbf{r}(t) = 2t\mathbf{i} + \cos t\, \mathbf{j} + t^3\mathbf{k}$.

Find du/dt by the method of this section.

*7. $u = x^2 - 3xy + 2y^2; \quad x = \cos t, \quad y = \sin t$.

8. $u = x + 4\sqrt{xy} - 3y; \quad x = t^3, \quad y = t^{-1}. \qquad (t > 0)$

*9. $u = e^x \sin y + e^y \sin x; \quad x = \tfrac{1}{2}t, \quad y = 2t$.

10. $u = 2x^2 - xy + y^2; \quad x = \cos 2t, \quad y = \sin t$.

*11. $u = xy + yz + zx; \quad x = t^2, \quad y = t(1 - t), \quad z = (1 - t)^2$.

12. $u = x \sin \pi y - z \cos \pi x; \quad x = t^2, \quad y = 1 - t, \quad z = 1 - t^2$.

13. The height of a cone is 20 inches and the radius is 14 inches. Use the methods of this section to find the rate at which the volume is changing if the height decreases at the rate of 2 inches per second and the radius increases at the rate of 3 inches per second.

14. Find du/dt given that
$$u = f(x, y); \quad x = g(s), y = h(s); \quad s = k(t).$$

*15. Find $\partial u/\partial r$ and $\partial u/\partial s$ given that
$$u = f(x, y); \quad x = \phi(r, s), \quad y = \psi(r, s).$$

16. Let
$$u = f(x, y).$$

Show that the change of variables
$$x = r \cos \theta, \qquad y = r \sin \theta$$

gives
$$\left(\frac{\partial u}{\partial x}\right)^2 + \left(\frac{\partial u}{\partial y}\right)^2 = \left(\frac{\partial u}{\partial r}\right)^2 + \frac{1}{r^2}\left(\frac{\partial u}{\partial \theta}\right)^2.$$

17. Let
$$u = f(r), \qquad r = \sqrt{x^2 + y^2}.$$

Show that
$$\frac{\partial^2 u}{\partial x^2} + \frac{\partial^2 u}{\partial y^2} = \frac{d^2 u}{dr^2} + \frac{1}{r}\frac{du}{dr}.$$

*18. Find f, given that
$$\nabla f(x, y, z) = a_1 \mathbf{i} + a_2 \mathbf{j} + a_3 \mathbf{k} \quad \text{for all } (x, y, z).$$

19. What can you conclude about f and g given that
$$\nabla f(x, y, z) - \nabla g(x, y, z) = a_1 \mathbf{i} + a_2 \mathbf{j} + a_3 \mathbf{k} \quad \text{for all } (x, y, z)?$$

20. Let
$$u = f(x, y)$$

be a differentiable function. Assume that the equation
$$f(x, y) = 0$$

gives y implicitly as a differentiable function of x.
(a) Show that
$$\text{if } \frac{\partial f}{\partial y} \neq 0 \quad \text{then} \quad \frac{dy}{dx} = -\frac{\partial f/\partial x}{\partial f/\partial y}.$$

(b) Use this formula to find
$$\frac{dy}{dx} \text{ if } x^2 y^4 + \sin y = 0.$$

21. Show that if
$$\| \nabla f(\mathbf{x}) \| \leq M$$

for all $\mathbf{x}$ in some neighborhood of the origin, then
$$|f(\mathbf{x}) - f(\mathbf{y})| \leq M \|\mathbf{x} - \mathbf{y}\|$$

for all $\mathbf{x}$ and $\mathbf{y}$ in that neighborhood.

14.10 The Gradient as a Normal; Tangent Lines and Tangent Planes

Functions of Two Variables

You have seen that the gradient of a function points in the direction of the most rapid increase of the function. Here we show that *the gradient is at each point perpendicular to the level curve which passes through that point.*

We begin with a function f and a point (x_0, y_0) in its domain. We suppose that $f(x_0, y_0) = c$ and parametrize the level curve $f(x, y) = c$ by

$$\mathbf{r}(t) = x(t)\mathbf{i} + y(t)\mathbf{j}, \qquad t \in I.$$

We take t_0 so that

$$\mathbf{r}(t_0) = x_0\mathbf{i} + y_0\mathbf{j} = (x_0, y_0)$$

and assume that the tangent vector $\mathbf{r}'(t_0)$ exists and is not $\mathbf{0}$. Since on the level curve f is constantly c, we have

$$f(\mathbf{r}(t)) = c \qquad \text{for all } t \in I.$$

For such t

$$\frac{d}{dt}[f(\mathbf{r}(t))] = \nabla f(\mathbf{r}(t)) \cdot \mathbf{r}'(t) = 0.$$

In particular

$$\nabla f(\mathbf{r}(t_0)) \cdot \mathbf{r}'(t_0) = 0,$$

which means that

$$\nabla f(\mathbf{r}(t_0)) \perp \mathbf{r}'(t_0).$$

Since $\mathbf{r}'(t_0)$ is tangent to the level curve at (x_0, y_0), it follows that $\nabla f(\mathbf{r}(t_0)) = \nabla f(x_0, y_0)$ is perpendicular to the level curve at (x_0, y_0). (Figure 14.10.1)   □

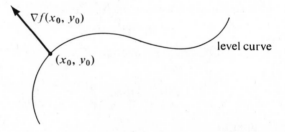

the gradient vector at (x_0, y_0) is perpendicular
to the level curve at (x_0, y_0)

FIGURE 14.10.1

Example.   For

$$f(x, y) = x^2 + y^2$$

the level curves are concentric circles

$$x^2 + y^2 = c.$$

At each point $(x, y) \neq (0, 0)$ the gradient

$$\nabla f(x, y) = 2x\mathbf{i} + 2y\mathbf{j}$$

points radially away from the origin and is thus perpendicular to the circle in question. At the origin the level curve is reduced to a point and the gradient is simply $\mathbf{0}$.   □

Normal Lines and Tangent Lines

We consider now a curve γ defined implicitly by an equation

$$f(x, y) = c.$$

We can think of this curve as the c-level curve of a function f. If f is differentiable, then the gradient $\nabla f(x_0, y_0)$ is perpendicular to the curve at (x_0, y_0). If $\nabla f(x_0, y_0) = \mathbf{0}$, there is not much more to say. Suppose therefore that $\nabla f(x_0, y_0) \neq \mathbf{0}$. In this case we can use $\nabla f(x_0, y_0)$ as a normal vector and from it derive an equation for the *normal line* at $\mathbf{x}_0 = (x_0, y_0)$. Let's proceed. Using $\nabla f(x_0, y_0)$ as a direction vector, we can parametrize the normal line by setting

$$\mathbf{R}(t) = \mathbf{x}_0 + t\nabla f(\mathbf{x}_0) = \left[x_0 + t\frac{\partial f}{\partial x}(x_0, y_0) \right]\mathbf{i} + \left[y_0 + t\frac{\partial f}{\partial y}(x_0, y_0) \right]\mathbf{j}.$$

In terms of components, we have

$$x(t) = x_0 + t\frac{\partial f}{\partial x}(x_0, y_0) \quad \text{and} \quad y(t) = y_0 + t\frac{\partial f}{\partial y}(x_0, y_0).$$

This gives

$$(x(t) - x_0)\frac{\partial f}{\partial y}(x_0, y_0) = (y(t) - y_0)\frac{\partial f}{\partial x}(x_0, y_0). \qquad \text{(explain)}$$

Eliminating the t altogether, we get

$$(x - x_0)\frac{\partial f}{\partial y}(x_0, y_0) = (y - y_0)\frac{\partial f}{\partial x}(x_0, y_0).$$

Usually the right side is transferred to the left so that the equation for the normal line becomes

$$\boxed{(x - x_0)\frac{\partial f}{\partial y}(x_0, y_0) - (y - y_0)\frac{\partial f}{\partial x}(x_0, y_0) = 0.}$$

The tangent line at (x_0, y_0) is perpendicular to the normal (Figure 14.10.2) and therefore has equation

$$\boxed{(x - x_0)\frac{\partial f}{\partial x}(x_0, y_0) + (y - y_0)\frac{\partial f}{\partial y}(x_0, y_0) = 0.}$$

Problem. Find equations for the lines normal and tangent to the ellipse

$$b^2x^2 + a^2y^2 = a^2b^2$$

at the point (x_0, y_0).

SOLUTION.   We can represent the ellipse by

$$f(x, y) = b^2x^2 + a^2y^2 - a^2b^2 = 0.$$

Note that

$$\frac{\partial f}{\partial x}(x, y) = 2b^2x, \qquad \frac{\partial f}{\partial y}(x, y) = 2a^2y,$$

so that

$$\frac{\partial f}{\partial x}(x_0, y_0) = 2b^2x_0, \qquad \frac{\partial f}{\partial y}(x_0, y_0) = 2a^2y_0.$$

For the normal, we have

$$(x - x_0)2a^2y_0 - (y - y_0)2b^2x_0 = 0,$$

which we can simplify to

$$a^2y_0(x - x_0) - b^2x_0(y - y_0) = 0.$$

For the tangent, we have

$$(x - x_0)2b^2x_0 + (y - y_0)2a^2y_0 = 0.$$

Dividing by 2 and noting that

$$b^2x_0^2 + a^2y_0^2 = a^2b^2, \qquad \text{(explain)}$$

we can simplify the equation for the tangent to read

$$b^2x_0x + a^2y_0y = a^2b^2. \quad \square$$

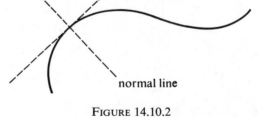

FIGURE 14.10.2

Functions of Three Variables

Here instead of level curves we have level surfaces, but the results are similar: *at each point the gradient is perpendicular to the level surface which passes through that point.* (See Figure 14.10.3.)

To verify this, we take a differentiable function f, of three variables this time, and fix our attention on a point $\mathbf{x}_0 = (x_0, y_0, z_0)$ of its domain. If $f(x_0, y_0, z_0) = c$, then the level surface through $\mathbf{x}_0 = (x_0, y_0, z_0)$ has equation

$$f(x, y, z) = c.$$

We suppose now that

$$\mathbf{r}(t) = x(t)\mathbf{i} + y(t)\mathbf{j} + z(t)\mathbf{k}, \qquad t \in I$$

is a differentiable curve which lies on the given surface and passes through the point $\mathbf{x}_0 = (x_0, y_0, z_0)$. Moreover, we suppose that

$$\mathbf{r}(t_0) = \mathbf{x}_0 = (x_0, y_0, z_0) \quad \text{and} \quad \mathbf{r}'(t_0) \neq \mathbf{0}.$$

Since $\mathbf{r}(t)$ lies on the surface,

$$f(\mathbf{r}(t)) = c \quad \text{for every } t.$$

Consequently,

$$\frac{d}{dt}\left[f(\mathbf{r}(t))\right] = \nabla f(\mathbf{r}(t)) \cdot \mathbf{r}'(t) = 0.$$

In particular,

$$\nabla f(\mathbf{r}(t_0)) \cdot \mathbf{r}'(t_0) = 0.$$

This tells us that the gradient

$$\nabla f(\mathbf{r}(t_0)) = \nabla f(\mathbf{x}_0) = \nabla f(x_0, y_0, z_0)$$

is perpendicular to the curve in question. This same argument applies to every differentiable curve which lies on the level surface and passes through the point $\mathbf{x}_0 = (x_0, y_0, z_0)$. Consequently, we view $\nabla f(\mathbf{x}_0)$ as perpendicular to the surface itself.

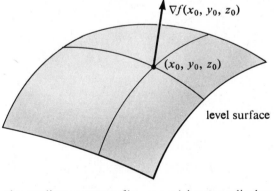

the gradient vector $\nabla f(x_0, y_0, z_0)$ is perpendicular
to the level surface at (x_0, y_0, z_0)

FIGURE 14.10.3

Tangent Planes

You have seen that the gradient vector ∇f is perpendicular to the surface

$$f(x, y, z) = c.$$

If the gradient vector $\mathbf{V}f$ is not zero, we can use it as a normal vector. We can then define the *tangent plane at* $\mathbf{x}_0 = (x_0, y_0, z_0)$ *as the plane through* $\mathbf{x}_0$ *with normal* $\mathbf{V}f(\mathbf{x}_0)$. (See Figure 14.10.4.) As a vector equation for this tangent plane, we have

$$\mathbf{V}f(\mathbf{x}_0) \cdot (\mathbf{x} - \mathbf{x}_0) = 0.$$

In Cartesian coordinates, this equation becomes

$$(x - x_0)\frac{\partial f}{\partial x}(x_0, y_0, z_0) + (y - y_0)\frac{\partial f}{\partial y}(x_0, y_0, z_0) + (z - z_0)\frac{\partial f}{\partial z}(x_0, y_0, z_0) = 0.$$

Geometrically, the tangent plane at a point is the plane that locally best approximates the surface. We'll come back to this idea later.

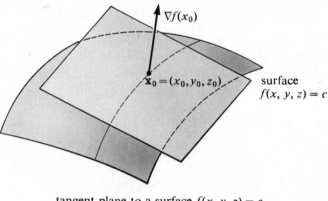

tangent plane to a surface $f(x, y, z) = c$

FIGURE 14.10.4

Problem. Find an equation for the plane tangent to the surface

$$xy + yz + zx = 11$$

at the point $(1, 2, 3)$.

SOLUTION. Here

$$f(x, y, z) = xy + yz + zx,$$

so that

$$\frac{\partial f}{\partial x}(x, y, z) = y + z, \qquad \frac{\partial f}{\partial y}(x, y, z) = x + z, \qquad \frac{\partial f}{\partial z}(x, y, z) = x + y$$

and

$$\frac{\partial f}{\partial x}(1, 2, 3) = 5, \qquad \frac{\partial f}{\partial y}(1, 2, 3) = 4, \qquad \frac{\partial f}{\partial z}(1, 2, 3) = 3.$$

For the tangent plane we have

$$5(x - 1) + 4(y - 2) + 3(z - 3) = 0,$$

which simplifies to

$$5x + 4y - 3z - 22 = 0. \quad \square$$

Problem.   The curve

$$\mathbf{r}(t) = \tfrac{1}{2}t^2\mathbf{i} + 4t^{-1}\mathbf{j} + (\tfrac{1}{2}t - t^2)\mathbf{k}$$

cuts the hyperbolic paraboloid

$$x^2 - 4y^2 - 4z = 0$$

at the point $(2, 2, -3)$. What is the angle of intersection?

SOLUTION.   First we need the value of t which corresponds to the point $(2, 2, -3)$. To find this, we set

$$\tfrac{1}{2}t^2 = 2, \qquad 4t^{-1} = 2, \qquad \tfrac{1}{2}t - t^2 = -3$$

and solve for t. An easy calculation shows that $t = 2$.

Since

$$\mathbf{r}'(t) = t\mathbf{i} - 4t^{-2}\mathbf{j} + (\tfrac{1}{2} - 2t)\mathbf{k},$$

we have

$$\mathbf{r}'(2) = 2\mathbf{i} - \mathbf{j} - \tfrac{7}{2}\mathbf{k}.$$

This is the tangent vector of the curve at the point of intersection.
   To represent the surface we will use the gradient. For

$$f(x, y, z) = x^2 - 4y^2 - 4z$$

we have

$$\nabla f = 2x\mathbf{i} - 8y\mathbf{j} - 4\mathbf{k}.$$

At the point $(2, 2, -3)$

$$\nabla f = 4\mathbf{i} - 16\mathbf{j} - 4\mathbf{k}.$$

Since

$$\frac{\mathbf{r}'(2) \cdot \nabla f}{\|\mathbf{r}'(2)\| \| \nabla f \|} = \frac{19}{414} \sqrt{138}, \qquad\qquad \text{(check this)}$$

the angle between the curve and the gradient is

$$\text{arc cos } (\tfrac{19}{414} \sqrt{138}).$$

Since the gradient is perpendicular to the surface, the angle between the curve and the surface itself is

$$\tfrac{1}{2}\pi - \text{arc cos } (\tfrac{19}{414}\sqrt{138}) \cong 1.57 - \text{arc cos } 0.539 \cong 0.57 \text{ radians.} \quad \square$$

   To find the tangent plane to a surface of the form

$$z = g(x, y)$$

we set

$$f(x, y, z) = g(x, y) - z = 0.$$

If g is differentiable, then so is f. Moreover,

$$\frac{\partial f}{\partial x}(x, y, z) = \frac{\partial g}{\partial x}(x, y), \qquad \frac{\partial f}{\partial y}(x, y, z) = \frac{\partial g}{\partial y}(x, y), \qquad \frac{\partial f}{\partial z}(x, y, z) = -1.$$

The tangent plane at (x_0, y_0, z_0) has equation

$$(x - x_0)\frac{\partial g}{\partial x}(x_0, y_0) + (y - y_0)\frac{\partial g}{\partial y}(x_0, y_0) + (z - z_0)(-1) = 0,$$

which we can rewrite as

$$\boxed{(x - x_0)\frac{\partial g}{\partial x}(x_0, y_0) + (y - y_0)\frac{\partial g}{\partial y}(x_0, y_0) = z - z_0.}$$

If $\nabla g(x_0, y_0) = 0$, then both partials of g are zero at (x_0, y_0) and the equation reduces to

$$\boxed{z = z_0.}$$

In this case the tangent plane is *horizontal*.

Problem. Find an equation for the plane tangent to the surface

$$z = \frac{x^2}{2p} + \frac{y^2}{2q}$$

at the point (x_0, y_0, z_0).

SOLUTION.   We set

$$g(x, y) = \frac{x^2}{2p} + \frac{y^2}{2q}$$

and note that

$$\frac{\partial g}{\partial x}(x, y) = \frac{x}{p} \quad \text{and} \quad \frac{\partial g}{\partial y}(x, y) = \frac{y}{q}.$$

As an equation for the tangent plane we have

$$\frac{x_0}{p}(x - x_0) + \frac{y_0}{q}(y - y_0) = z - z_0. \quad \square$$

Problem. At what points of the surface

$$z = 3xy - x^3 - y^3$$

is the tangent plane horizontal?

SOLUTION. We set

$$g(x, y) = 3xy - x^3 - y^3$$

and seek the pairs (x, y) for which

$$\frac{\partial g}{\partial x}(x, y) = 0 = \frac{\partial g}{\partial y}(x, y).$$

Since

$$\frac{\partial g}{\partial x}(x, y) = 3y - 3x^2 \quad \text{and} \quad \frac{\partial g}{\partial y}(x, y) = 3x - 3y^2,$$

we must solve the equations

$$3y - 3x^2 = 0 \quad \text{and} \quad 3x - 3y^2 = 0$$

simultaneously. This gives

$$x = 0, \quad y = 0 \quad \text{and} \quad x = 1, \quad y = 1.$$

Each pair gives rise to a point at which the tangent plane is horizontal. These points are $(0, 0, 0)$ and $(1, 1, 1)$. $\square$

Exercises

Find equations for the tangent and normal lines at the point indicated.

*1. $x^2 + xy + y^2 = 3$; $(-1, -1)$.
2. $(y - x)^2 = 2x$; $(2, 4)$.
*3. $(x^2 + y^2)^2 = 9(x^2 - y^2)$; $(\sqrt{2}, 1)$.
4. $x^3 + y^3 = 9$; $(1, 2)$.
*5. $xy^2 - 2x^2 + y + 5x = 6$; $(4, 2)$.
6. $x^5 + y^5 = 2x^3$; $(1, 1)$.
*7. $2x^3 - x^2y^2 - 3x + y + 7 = 0$; $(1, -2)$.

8. Find an equation for the tangent to the curve $x^3 + y^2 + 2x = 6$ at a point where $y = 3$.
*9. Find the slope of the tangent to the curve $x^2y = a^2(a - y)$ at the point $(0, a)$.

Find an equation for the tangent plane at the point indicated.

10. $z = (x^2 + y^2)^2$; $(1, 1, 4)$.
*11. $x^3 + y^3 = 3xyz$; $(1, 2, \frac{3}{2})$.
12. $xy^2 + 2z^2 = 12$; $(1, 2, 2)$.
*13. $\sqrt{x} + \sqrt{y} + \sqrt{z} = 4$; $(1, 4, 1)$.
14. $z = axy$; $(1, 1/a, 1)$.
*15. $z = \sin x + \sin y + \sin (x + y)$; $(0, 0, 0)$.
16. $z = x^2 + xy + y^2 - 6x + 2$; $(4, -2, -10)$.
*17. $b^2c^2x^2 - a^2c^2y^2 - a^2b^2z^2 = a^2b^2c^2$; (x_0, y_0, z_0).

Find the point(s) of the surface at which the tangent plane is horizontal.

18. $z = 4x + 2y - x^2 + xy - y^2$.

*19. $z - 2x^2 - 2xy + y^2 + 5x - 3y + 2 = 0$.

20. $xy + a^3/x + b^3/y - z = 0$.

*21. $z = xy$.

22. $x + y + z + xy - x^2 - y^2 = 0$.

*23. Given that the surfaces

$$F(x, y, z) = 0 \quad \text{and} \quad G(x, y, z) = 0$$

intersect at right angles in a curve γ, what condition must be satisfied by the partial derivatives of F and G on γ?

24. Show that in the case of a surface of the form

$$z = xf(x/y) \quad \text{with } f \text{ differentiable}$$

all the tangent planes share a point in common.

*25. Show that all the pyramids formed by the coordinate planes and a plane tangent to the surface $xyz = a^3$ have the same volume. What is this volume?

26. Show that the sum of the intercepts of a plane tangent to the surface

$$\sqrt{x} + \sqrt{y} + \sqrt{z} = \sqrt{a}$$

is constant.

27. Show that the sum of the squares of the intercepts of a plane tangent to the surface

$$x^{2/3} + y^{2/3} + z^{2/3} = a^{2/3}$$

is constant.

*28. The ellipsoid

$$x^2 + y^2 + 3z^2 = 25$$

and the curve

$$\mathbf{r}(t) = 2t\mathbf{i} + 3t^{-1}\mathbf{j} - 2t^2\mathbf{k}$$

intersect when $t = 1$. What is the angle of intersection?

29. Show that the ellipsoid

$$x^2 + 2y^2 + 3z^2 = 20$$

and the curve

$$\mathbf{r}(t) = \tfrac{3}{2}(t^2 + 1)\mathbf{i} + (t^4 + 1)\mathbf{j} + t^3\mathbf{k}$$

meet at right angles at the point $(3, 2, 1)$.

*30. The surfaces

$$x^2y^2 + 2x + z^3 = 16 \quad \text{and} \quad 3x^2 + y^2 - 2z = 9$$

intersect in a curve which passes through the point $(2, 1, 2)$. What are the equations of the respective tangent planes to the two surfaces at this point?

31. Show that the ellipsoid
$$x^2 + 3y^2 + 2z^2 = 9$$
and the sphere
$$x^2 + y^2 + z^2 - 8x - 8y - 6z + 24 = 0$$
are tangent to each other at the point $(2, 1, 1)$.

32. Show that the paraboloid
$$3x^2 + 2y^2 - 2z = 1$$
and the sphere
$$x^2 + y^2 + z^2 - 4y - 2z + 2 = 0$$
intersect at right angles at the point $(1, 1, 2)$.

33. Show that the surfaces
$$xy = az^2, \qquad x^2 + y^2 + z^2 = b, \qquad z^2 + 2x^2 = c(z^2 + 2y^2)$$
are mutually perpendicular.

14.11   Maximum and Minimum Values

In Chapter 3 we talked about the local extreme values of a function of one variable. Here we take up the subject for functions of several variables. The ideas are much the same.

Definition of Local Maximum and Local Minimum

A function f of several variables is said to have a *local maximum* at $\mathbf{x}_0$ iff

$$f(\mathbf{x}_0) \geq f(\mathbf{x}) \quad \text{for all } \mathbf{x} \text{ in some neighborhood of } \mathbf{x}_0.$$

It is said to have a *local minimum* at $\mathbf{x}_0$ iff

$$f(\mathbf{x}_0) \leq f(\mathbf{x}) \quad \text{for all } \mathbf{x} \text{ in some neighborhood of } \mathbf{x}_0.$$

As in the one-variable case, the local maxima and local minima together comprise the *local extreme values*.

In the one-variable case we know that if f is differentiable at x_0 and has a local extreme at x_0, then

$$f'(x_0) = 0.$$

We have a similar result for functions of several variables.

Theorem

If f is differentiable at $\mathbf{x}_0$ and has a local extreme at $\mathbf{x}_0$, then

$$\nabla f(\mathbf{x}_0) = \mathbf{0}.$$

PROOF.   Let's take f as a function of two variables and assume that f is differentiable at $\mathbf{x}_0 = (x_0, y_0)$. (The three-variable case is similar.) If f has a local extreme at $\mathbf{x}_0 = (x_0, y_0)$, then $g(x) = f(x, y_0)$ must have a local extreme at x_0 and, being differentiable there, must satisfy

$$g'(x_0) = \frac{\partial f}{\partial x}(x_0, y_0) = 0.$$

Similarly, $h(y) = f(x_0, y)$ must have a local extreme at y_0 and, being differentiable there, must satisfy

$$h'(y_0) = \frac{\partial f}{\partial y}(x_0, y_0) = 0.$$

The gradient is $\mathbf{0}$ since both partials are 0. (Theorem 14.8.2)   □

Although the ideas introduced so far are completely general, their application to functions of more than two variables is generally laborious. From now on we restrict ourselves to functions of two variables. Not only are the computations less formidable but also we can make use of our geometric intuition.

Two Variables

A continuously differentiable function of two variables can be viewed geometrically as a surface

$$z = f(x, y).$$

Where f has a local maximum, the surface has a high point. Where f has a local minimum, the surface has a low point. Where f has either a local maximum or a local minimum, the gradient is $\mathbf{0}$ and therefore the tangent plane is horizontal. (See Figures 14.11.1 and 14.11.2.)

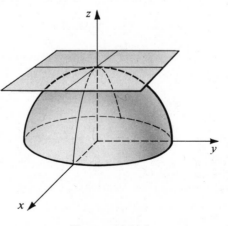

FIGURE 14.11.1

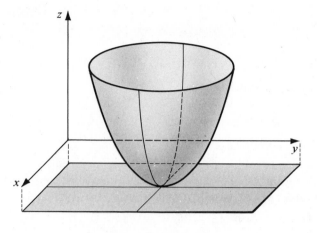

FIGURE 14.11.2

A zero gradient signals the possibility of an extreme value. It does not guarantee it. For example, in the case of the saddle-shaped surface of Figure 14.11.3, there is a horizontal tangent plane at the origin and therefore the gradient is zero there, but the origin produces neither a local maximum nor a local minimum.

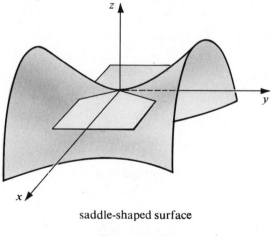

saddle-shaped surface

FIGURE 14.11.3

Points at which the gradient is zero are called *stationary points*. Those stationary points that give rise to extreme values are called *extreme points*. Those stationary points that do not give rise to extreme values are called *saddle points*.

Below we examine some differentiable functions for extreme values. In each case our first step is to seek out the stationary points.

Example.   For
$$f(x, y) = 2x^2 - xy + y^2 - 7y,$$
we have
$$\nabla f(x, y) = (4x - y)\mathbf{i} + (2y - x - 7)\mathbf{j}.$$
To find the stationary points, we set
$$\nabla f(x, y) = \mathbf{0}.$$
This gives
$$4x - y = 0 \quad \text{and} \quad 2y - x - 7 = 0.$$
The only simultaneous solution to these equations is given by $x = 1$, $y = 4$. The point $(1, 4)$ is therefore the only stationary point. Now we compare the value of f at $(1, 4)$ with the values of f at nearby points $(1 + h, 4 + k)$:
$$f(1, 4) = 2 - 4 + 16 - 28 = -14,$$
$$
\begin{aligned}
f(1 + h, 4 + k) &= 2(1 + h)^2 - (1 + h)(4 + k) + (4 + k)^2 - 7(4 + k) \\
&= 2 + 4h + 2h^2 - 4 - 4h - k - hk + 16 + 8k + k^2 - 28 - 7k \\
&= 2h^2 + k^2 - hk - 14,
\end{aligned}
$$
$$f(1 + h, 4 + k) - f(1, 4) = 2h^2 + k^2 - hk = h^2 + (h^2 - hk + k^2).$$
The difference
$$f(1 + h, 4 + k) - f(1, 4) = h^2 + (h^2 - hk + k^2)$$
is positive for small h and k (in fact for all h and k different from 0). It follows that f has a minimum value at $(1, 4)$. This minimum value is -14.   ☐

Example.   In the case of
$$f(x, y) = y^2 - xy + 2x + y + 1$$
we have
$$\nabla f(x, y) = (2 - y)\mathbf{i} + (2y - x + 1)\mathbf{j}.$$
Setting
$$\nabla f(x, y) = \mathbf{0}$$
gives
$$2 - y = 0 \quad \text{and} \quad 2y - x + 1 = 0.$$
The only simultaneous solution to these equations is $x = 5$, $y = 2$. The point $(5, 2)$ is the only stationary point. To determine its nature, we compare $f(5, 2)$ with $f(5 + h, 2 + k)$ for small h and k:
$$f(5, 2) = 4 - 10 + 10 + 2 + 1 = 7,$$
$$
\begin{aligned}
f(5 + h, 2 + k) &= (2 + k)^2 - (5 + h)(2 + k) + 2(5 + h) + (2 + k) + 1 \\
&= 4 + 4k + k^2 - 10 - 2h - 5k - hk + 10 + 2h + 2 + k + 1 \\
&= k^2 - hk + 7,
\end{aligned}
$$
$$f(5 + h, 2 + k) - f(5, 2) = k^2 - hk = k(k - h).$$
This last expression does not keep a constant sign for small h and k. It follows that $(5, 2)$ is a saddle point.   ☐

Example. For

$$f(x, y) = x^3 + y^3 - 3xy$$

we have

$$\nabla f(x, y) = (3x^2 - 3y)\mathbf{i} + (3y^2 - 3x)\mathbf{j}.$$

Setting

$$\nabla f(x, y) = \mathbf{0}$$

gives

$$x^2 - y = 0 \quad \text{and} \quad y^2 - x = 0.$$

The only simultaneous solutions to these equations are given by $x = 0$, $y = 0$ and $x = 1$, $y = 1$. The points $(0, 0)$ and $(1, 1)$ are the only stationary points. To investigate $(0, 0)$ we first note that $f(0, 0) = 0$. For small $h > 0$ and $k = 0$,

$$f(h, k) = h^3 > 0.$$

For small $h < 0$ and $k = 0$,

$$f(h, k) = h^3 < 0.$$

It follows that $f(0, 0) = 0$ is neither a local maximum nor a local minimum. The origin $(0, 0)$ is therefore a saddle point.

Now we investigate the function at $(1, 1)$ and at nearby points $(1 + h, 1 + k)$:

$$f(1, 1) = 1 + 1 - 3 = -1,$$

$$\begin{aligned}
f(1 + h, 1 + k) &= (1 + h)^3 + (1 + k)^3 - 3(1 + h)(1 + k) \\
&= 1 + 3h + 3h^2 + h^3 + 1 + 3k + 3k^2 + k^3 - 3 - 3h - 3k - 3hk \\
&= -1 + 3h^2 + 3k^2 - 3hk + h^3 + k^3,
\end{aligned}$$

$$f(1 + h, 1 + k) - f(1, 1) = 3h^2 + 3k^2 - 3hk + h^3 + k^3.$$

The idea now is to play around with the right-hand side until we can determine whether or not it keeps a constant sign for small h and k. Our problems are over once we realize that we can rewrite the equation to read

$$\begin{aligned}
f(1 + h, 1 + k) - f(1, 1) &= (h^2 + h^3) + (k^2 + k^3) + (h^2 + k^2 - hk) + (h^2 - 2hk + k^2) \\
&= h^2(1 + h) + k^2(1 + k) + (h^2 + k^2 - hk) + (h^2 - 2hk + k^2).
\end{aligned}$$

For small h and k the first three terms on the right are positive and the last is nonnegative. It follows that

$$f(1 + h, 1 + k) - f(1, 1) > 0.$$

We conclude that the point $(1, 1)$ gives rise to a local minimum. This minimum is -1. $\square$

The next example is more complicated. The function there has an infinite number of stationary points.

Example.   For

$$f(x, y) = \frac{x^2 + 2y^2}{(x + y)^2}, \qquad x \neq -y$$

we have

$$\nabla f(x, y) = 2y \frac{x - 2y}{(x + y)^3} \mathbf{i} + 2x \frac{2y - x}{(x + y)^3} \mathbf{j}.$$

Setting

$$\nabla f(x, y) = \mathbf{0}$$

gives

$$2y \frac{x - 2y}{(x + y)^3} = 0 \quad \text{and} \quad 2x \frac{2y - x}{(x + y)^3} = 0.$$

Solving these two equations simultaneously, we get

$$x = 2y.$$

The stationary points are all the points of the line $x = 2y$, except of course for $(0, 0)$ which was originally excluded. At each stationary point (x_0, y_0) the function takes on the value $\frac{2}{3}$:

$$f(x_0, y_0) = f(2y_0, y_0) = \frac{(2y_0)^2 + 2y_0^2}{(2y_0 + y_0)^2} = \frac{6y_0^2}{9y_0^2} = \frac{2}{3}.$$

In a moment we will show that for all (x, y) in the domain

$$f(x, y) = \frac{x^2 + 2y^2}{(x + y)^2} \geq \frac{2}{3}.$$

It will then follow that every stationary point is a minimum point and the number $\frac{2}{3}$ is a minimum value.

The inequality

$$\frac{x^2 + 2y^2}{(x + y)^2} \geq \frac{2}{3}$$

can be justified by the following sequence of equivalent inequalities, the last of which is obvious:

$$3(x^2 + 2y^2) \geq 2(x + y)^2,$$
$$3x^2 + 6y^2 \geq 2x^2 + 4xy + 2y^2,$$
$$x^2 - 4xy + 4y^2 \geq 0,$$
$$(x - 2y)^2 \geq 0. \quad \square$$

So far we have considered only functions which have gradients on their entire domain. A local extreme value can also occur at a point where the gradient does not exist. As an example we can take the function

$$f(x, y) = \sqrt{x^2 + y^2}.$$

Its graph is the upper half of a circular cone. (Figure 14.2.4 with $a = b = 1$.) Obviously f has a minimum at $(0, 0)$, but since the partials

$$\frac{\partial f}{\partial x} = \frac{x}{\sqrt{x^2 + y^2}}, \qquad \frac{\partial f}{\partial y} = \frac{y}{\sqrt{x^2 + y^2}}$$

are not defined at $(0, 0)$, the gradient is not defined at $(0, 0)$. (Theorem 14.8.2) The surface comes to a sharp point at $(0, 0)$ and there is no tangent plane. □

Exercises

Classify the stationary points and find the local extreme values.

*1. $f(x, y) = 2x - x^2 - y^2$.

 2. $f(x, y) = x^2 + 2y^2 - 4y$.

*3. $f(x, y) = 2x + 2y - x^2 + y^2 + 5$.

 4. $f(x, y) = x^2 - xy + y^2 + 1$.

*5. $f(x, y) = x^2 + xy + y^2 + 3x + 1$.

 6. $f(x, y) = x^3 - y^2 + 2y$.

*7. $f(x, y) = x^3 - 3x + y$.

 8. $f(x, y) = x^5 - 3x^2 - 2y$.

*9. $f(x, y) = x^2 + xy + y^2 - 3x - 3y$.

 10. $f(x, y) = x^2 - xy + y^2 + 2x + 2y - 6$.

*11. $f(x, y) = |x| + |y|$.

 12. $f(x, y) = 5xy - 7x^2 - y^2 + 3x - 6y$.

*13. $f(x, y) = x^2 + xy + 2x + 2y + 1$.

 14. $f(x, y) = x \sin y, \quad -\pi < y < \pi$.

*15. Find the point with the property that the sum of the squares of its distances from $P_1(x_1, y_1)$, $P_2(x_2, y_2)$, $P_3(x_3, y_3)$ is a minimum.

 16. A manufacturer finds that he can sell in two distinct markets. If he sells Q_1 units in the first market and Q_2 units in the second market, the revenue functions are

$$R_1 = A_1 Q_1 - B_1 Q_1^2, \qquad R_2 = A_2 Q_2 - B_2 Q_2^2$$

and the total cost function is

$$C = A_3 + B_3(Q_1 + Q_2).$$

Given that the A's and B's are known positive constants, find the quantities Q_1 and Q_2 that will maximize profit.

*17. Given that $0 < a < b$, find the two numbers x and y between a and b such that

$$\frac{xy}{(a + x)(x + y)(b + y)}$$

is a maximum.

14.12   Second-Partials Test

We start with a function g of one variable. Suppose now that $g'(x_0) = 0$. The second derivative test for extreme values can be phrased this way:

If $g''(x_0) > 0$, then g has a minimum at x_0.
If $g''(x_0) < 0$, then g has a maximum at x_0.
If $g''(x_0) = 0$, then anything can happen.

For functions of two variables we have a similar test. As you might expect, it is a little more complicated to state and definitely more difficult to prove.

Second-Partials Test

Suppose that f has continuous second-order partials in a neighborhood of (x_0, y_0) and suppose that $\nabla f(x_0, y_0) = \mathbf{0}$. Set

$$A = \frac{\partial^2 f}{\partial x^2}(x_0, y_0), \qquad B = \frac{\partial^2 f}{\partial y\, \partial x}(x_0, y_0), \qquad C = \frac{\partial^2 f}{\partial y^2}(x_0, y_0)$$

and form the discriminant
$$D = B^2 - AC.$$

1. If $D > 0$, (x_0, y_0) is a saddle point.
2. If $D < 0$, there is a maximum at (x_0, y_0) if $A < 0$ and a minimum if $A > 0$.
3. If $D = 0$, then anything can happen.

A detailed justification of this test would be extremely burdensome at this point.† We will not attempt to give one. Instead we move on directly to some examples.

Example.   For
$$f(x, y) = x^3 - y^3 + 3xy$$
we have
$$\frac{\partial f}{\partial x} = 3x^2 + 3y \quad \text{and} \quad \frac{\partial f}{\partial y} = -3y^2 + 3x.$$

Setting both partials equal to zero, we get
$$x^2 + y = 0 \quad \text{and} \quad -y^2 + x = 0.$$

The only simultaneous solutions to these equations are given by $x = 0$, $y = 0$ and $x = 1$, $y = -1$. The points $(0, 0)$ and $(1, -1)$ are the only stationary points. The second partials are
$$\frac{\partial^2 f}{\partial x^2} = 6x, \qquad \frac{\partial^2 f}{\partial y\, \partial x} = 3, \qquad \frac{\partial^2 f}{\partial y^2} = -6y.$$

---

† You can find one in most texts on advanced calculus. See for example Section 2-15 in Wilfred Kaplan's *Advanced Calculus* (Addison-Wesley, Reading, Mass., 1952).

At (0, 0) we have

$$A = 0, \quad B = 3, \quad C = 0$$

and thus

$$D = B^2 - AC = 9 > 0.$$

The test tells us that (0, 0) is a saddle point. At (1 − 1) we have

$$A = 6, \quad B = 3, \quad C = 6$$

and thus

$$D = B^2 - AC = 9 - 36 = -27 < 0.$$

From $D < 0$ and $A > 0$ we can conclude that $(1, -1)$ gives rise to a local minimum. This minimum is

$$f(1, -1) = 1 + 1 - 3 = -1. \quad \square$$

Example. For

$$f(x, y) = x^2 + xy + y^2 - 3x - 3y$$

we have

$$\frac{\partial f}{\partial x} = 2x + y - 3 \quad \text{and} \quad \frac{\partial f}{\partial y} = x + 2y - 3.$$

Setting both of these partials equal to zero, we have

$$2x + y - 3 = 0 \quad \text{and} \quad x + 2y - 3 = 0.$$

The only simultaneous solution to these equations is $x = 1$, $y = 1$. The point (1, 1) is the only stationary point. Since at this point

$$\frac{\partial^2 f}{\partial x^2} = 2, \quad \frac{\partial^2 f}{\partial y \, \partial x} = 1, \quad \frac{\partial^2 f}{\partial y^2} = 2,$$

we have

$$A = 2, \quad B = 1, \quad C = 2$$

and thus

$$D = B^2 - AC = 1 - 4 = -3 < 0.$$

With $D < 0$ and $A > 0$ we can conclude that (1, 1) gives a local minimum. This minimum is

$$f(1, 1) = 1 + 1 + 1 - 3 - 3 = -3. \quad \square$$

Example. We examine the function

$$f(x, y) = \sin x + \sin y + \sin (x + y)$$

on the square

$$0 < x < \pi, \quad 0 < y < \pi.$$

In the first place

$$\frac{\partial f}{\partial x} = \cos x + \cos (x + y), \quad \frac{\partial f}{\partial y} = \cos y + \cos (x + y).$$

Setting both of these partials equal to zero, we have

$$\cos x + \cos (x + y) = 0, \quad \cos y + \cos (x + y) = 0.$$

Together these equations imply that

$$\cos x = \cos y.$$

Since x and y must lie between 0 and π, we conclude that

$$x = y.$$

The condition

$$\cos x + \cos (x + y) = 0$$

now leads to

$$\cos x + \cos 2x = 0.$$

Into this last equation we substitute the identity

$$\cos 2x = 2 \cos^2 x - 1$$

and thereby obtain

$$2 \cos^2 x + \cos x - 1 = 0.$$

This is a quadratic in $\cos x$ with solutions

$$\cos x = \tfrac{1}{4}(-1 \pm \sqrt{1 + 8}) = -1 \text{ or } \tfrac{1}{2}.$$

We throw out the solution -1 because x has to lie between 0 and π. The other possibility, $\cos x = \tfrac{1}{2}$, leads to $x = \tfrac{1}{3}\pi$. Since $x = y$, we also have $y = \tfrac{1}{3}\pi$. This shows that $(\tfrac{1}{3}\pi, \tfrac{1}{3}\pi)$ is the only stationary point.

$$\frac{\partial^2 f}{\partial x^2} = -\sin x - \sin (x + y),$$

$$\frac{\partial^2 f}{\partial y \, \partial x} = -\sin (x + y),$$

$$\frac{\partial^2 f}{\partial y^2} = -\sin y - \sin (x + y).$$

At $(\tfrac{1}{3}\pi, \tfrac{1}{3}\pi)$ we have

$$A = -\sin \tfrac{1}{3}\pi - \sin \tfrac{2}{3}\pi = -\tfrac{1}{2}\sqrt{3} - \tfrac{1}{2}\sqrt{3} = -\sqrt{3},$$
$$B = -\sin \tfrac{2}{3}\pi = -\tfrac{1}{2}\sqrt{3},$$
$$C = -\sin \tfrac{1}{3}\pi - \sin \tfrac{2}{3}\pi = -\tfrac{1}{2}\sqrt{3} - \tfrac{1}{2}\sqrt{3} = -\sqrt{3},$$

and thus

$$D = B^2 - AC = \tfrac{3}{4} - 3 < 0.$$

From $D < 0$ and $A < 0$ we can conclude that $(\tfrac{1}{3}\pi, \tfrac{1}{3}\pi)$ gives a local maximum. This maximum is

$$f(\tfrac{1}{3}\pi, \tfrac{1}{3}\pi) = \sin \tfrac{1}{3}\pi + \sin \tfrac{1}{3}\pi + \sin \tfrac{2}{3}\pi = \tfrac{1}{2}\sqrt{3} + \tfrac{1}{2}\sqrt{3} + \tfrac{1}{2}\sqrt{3} = \tfrac{3}{2}\sqrt{3}. \quad \square$$

That anything can happen when $D = 0$ is illustrated by the behavior of

$$f(x, y) = x^4 + y^4,$$
$$g(x, y) = -(x^4 + y^4),$$
$$h(x, y) = 2x^2 - 3xy^2 + y^4.$$

As you can readily verify, for each of these functions $(0, 0)$ is a stationary point at which $D = 0$, and yet

(1) for f, $(0, 0)$ gives a minimum;
(2) for g, $(0, 0)$ gives a maximum;
(3) for h, $(0, 0)$ is a saddle point.

Statements (1) and (2) are obvious. To confirm (3), note that h can be written in the form

$$h(x, y) = (y^2 - x)(y^2 - 2x).$$

The parabolas

$$y^2 = x \quad \text{and} \quad y^2 = 2x$$

divide the plane in the manner depicted in Figure 14.12.1. The function is zero on the parabolas, negative between the parabolas and positive elsewhere. That the origin is a saddle point follows from the fact that, at the origin, the function is zero, but, in every neighborhood of the origin, the function takes on both positive and negative values.   ☐

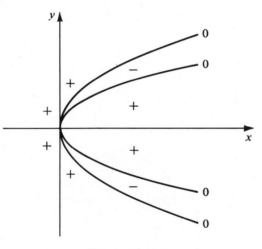

FIGURE 14.12.1

Exercises

Find the local extreme values.

*1. $f(x, y) = x^2 + xy + y^2 - 6x + 2$.
 2. $f(x, y) = x^2 + 2xy + 3y^2 + 2x + 10y + 1$.
*3. $f(x, y) = 4x + 2y - x^2 + xy - y^2$.
 4. $f(x, y) = x^2 - xy + y^2 + 2x + 2y - 1$.
*5. $f(x, y) = x^3 - 6xy + y^3$.
 6. $f(x, y) = 3x^2 + xy - y^2 + 5x - 5y + 4$.

* 7. $f(x, y) = x^3 + y^2 - 6xy + 6x + 3y - 2.$
  8. $f(x, y) = x^2 + xy + y^2 - 6y + 5.$
* 9. $g(x, y) = e^x \cos y.$
  10. $g(x, y) = x^2 - 2xy + 2y^2 - 3x + 5y.$
* 11. $g(x, y) = x \sin y.$
  12. $g(x, y) = y + x \sin y.$
* 13. $h(x, y) = (x + y)(xy + 1).$
  14. $h(x, y) = x/y - y/x.$
* 15. $F(x, y) = xy + 1/x + 8/y.$
  16. $G(x, y) = x^2 - 2xy - y^2 + 1.$
* 17. $f(x, y) = xy + x^{-1} + y^{-1}.$
  18. $g(x, y) = (x - y)(xy - 1).$

* 19. Find the maximum volume for a rectangular solid that has three faces on the coordinate planes and one vertex on the plane $x + y + z = 1$.

20. Describe the behavior of the function

$$f(x, y) = 2 \cos (x + y) + e^{xy}$$

at the origin.

* 21. Find the maximum volume for a rectangular solid inscribed in the ellipsoid

$$\frac{x^2}{a^2} + \frac{y^2}{b^2} + \frac{z^2}{c^2} = 1.$$

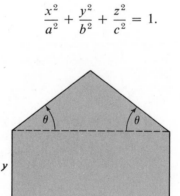

FIGURE 14.12.2

* 22. A pentagon is composed of a rectangle surmounted by an isosceles triangle. (Figure 14.12.2) Given that the perimeter of the pentagon has a fixed value P, find the dimensions for maximum area.

23. Find the distance between the lines

$$x = \tfrac{1}{2}y = \tfrac{1}{3}z \quad \text{and} \quad x = y - 2 = z.$$

24. A dairy produces two types of cheese at constant average costs of 50¢ and 60¢ per pound, respectively. If the selling price of the first type is x¢ per pound and of the second y¢ per pound, the number of pounds which can be sold each week is given by the formulas

$$N_1 = 250(y - x), \qquad N_2 = 32{,}000 + 250(x - 2y).$$

Show that for maximum profit the selling prices should be fixed at 89¢ and 94¢ per pound, respectively.

* 25. A manufacturer produces razors and blades at a constant average cost of 80¢ per razor and 60¢ per dozen blades. If the razors are sold at x¢ each and the blades at y¢ per dozen, the demand of the market each week is

$$\frac{4{,}000{,}000}{xy} \text{ razors} \quad \text{and} \quad \frac{8{,}000{,}000}{xy} \text{ dozen blades.}$$

Find the selling price for maximum profit.

* 26. Find the maximum value of

$$f(x, y) = \frac{(ax + by + c)^2}{x^2 + y^2 + 1}.$$

14.13   Maxima and Minima with Side Conditions

When we ask for the distance between a point $P(x_0, y_0)$ and a line $l\colon Ax + By + C = 0$, we are asking for the minimum value of

$$f(x, y) = \sqrt{(x - x_0)^2 + (y - y_0)^2}$$

with (x, y) subject to the side condition

$$Ax + By + C = 0.$$

When we ask for the distance between a point $P(x_0, y_0, z_0)$ and a plane

$$p\colon Ax + By + Cz + D = 0,$$

we are asking for the minimum value of

$$f(x, y, z) = \sqrt{(x - x_0)^2 + (y - y_0)^2 + (z - z_0)^2}$$

with (x, y, z) subject to the side condition

$$Ax + By + Cz + D = 0.$$

These particular problems we have already treated by special techniques. Our interest here is to present techniques for handling problems of this sort in general. In the two-variable case, the problems will take the form of maximizing (or minimizing) some function $f(x, y)$ subject to a side condition $g(x, y) = 0$. In the three-variable case, we will seek to maximize (or minimize) some function $f(x, y, z)$ subject to a side condition $g(x, y, z) = 0$. We begin with two simple examples.

Problem. Find the maximum value of xy subject to the side condition $x + y - 1 = 0$.

SOLUTION. The condition $x + y - 1 = 0$ gives us $y = 1 - x$. Our initial problem can therefore be solved simply by maximizing the product
$$h(x) = x(1 - x).$$
Obviously
$$h'(x) = 1 - 2x.$$
Setting
$$h'(x) = 0$$
gives
$$x = \tfrac{1}{2}.$$
Since
$$h''(x) = -2 < 0,$$
we know from the second-derivative test that
$$h(\tfrac{1}{2}) = \tfrac{1}{2}(1 - \tfrac{1}{2}) = \tfrac{1}{4}$$
is the desired maximum. □

Problem. Find the maximum volume of a rectangular solid given that the sum of the lengths of its edges is $12a$.

SOLUTION. We denote the dimensions of the solid by x, y, z. The volume then becomes
$$V = xyz.$$
The stipulation on the edges of the solid requires that
$$4(x + y + z) = 12a.$$
Solving this last equation for z, we get
$$z = 3a - (x + y).$$
If we now substitute this expression for z in the volume formula, we get
$$V = xy[3a - (x + y)].$$
Our problem is to maximize this function.

The partials are
$$\frac{\partial V}{\partial x} = xy(-1) + y[3a - (x + y)] = 3ay - 2xy - y^2,$$
$$\frac{\partial V}{\partial y} = xy(-1) + x[3a - (x + y)] = 3ax - x^2 - 2xy.$$

Setting both partials equal to zero, we get
$$(3a - 2x - y)y = 0,$$
$$(3a - x - 2y)x = 0.$$

The possibilities $x = 0$ and $y = 0$ do not interest us here. (Why?) We can therefore simplify the equations to read

$$3a - 2x - y = 0,$$
$$3a - x - 2y = 0.$$

Solving these equations simultaneously, we find that

$$x = y = a.$$

The corresponding value of V is a^3. The conditions of the problem make it clear that this is a maximum. $\square$

The last two problems were extremely easy. They were easy in part because the side conditions were such that we could solve for one of the variables in terms of the other(s). In general this is not possible and a more sophisticated approach is required.

The Method of Lagrange

We begin with what looks like a detour. To avoid having to make separate statements for the two- and three-variable cases, we will use vector notation.

Throughout the discussion, f will be a function of two or three variables continuously differentiable on some open set U. We take

$$C: \mathbf{r}(t), \quad t \in I$$

as a curve which lies entirely in U and has at each point a nonzero tangent vector $\mathbf{r}'(t)$. The basic result is this:

(14.13.1)

> If $\mathbf{x}_0$ maximizes (or minimizes) $f(\mathbf{x})$ on C, then $\nabla f(\mathbf{x}_0)$ is perpendicular to C at $\mathbf{x}_0$.

(Figure 14.13.1)

To show this, we suppose that

$$\mathbf{r}(t_0) = \mathbf{x}_0.$$

The composition

$$f(\mathbf{r}(t)), \quad t \in I$$

has a maximum (or minimum) at t_0. Consequently its derivative,

$$\frac{d}{dt}[f(\mathbf{r}(t))] = \nabla f(\mathbf{r}(t)) \cdot \mathbf{r}'(t),$$

must be zero at t_0:

$$0 = \nabla f(\mathbf{r}(t_0)) \cdot \mathbf{r}'(t_0) = \nabla f(\mathbf{x}_0) \cdot \mathbf{r}'(t_0).$$

This shows that

$$\nabla f(\mathbf{x}_0) \perp \mathbf{r}'(t_0).$$

Since $\mathbf{r}'(t_0)$ is tangent to C at $\mathbf{x}_0$, we have $\nabla f(\mathbf{x}_0)$ perpendicular to C at $\mathbf{x}_0$. $\square$

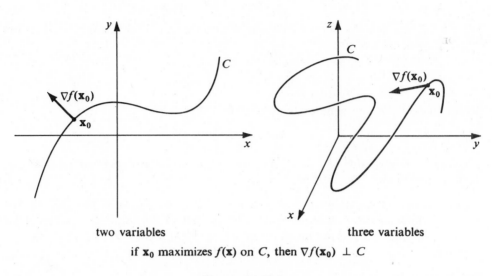

two variables   three variables

if $\mathbf{x}_0$ maximizes $f(\mathbf{x})$ on C, then $\nabla f(\mathbf{x}_0) \perp C$

FIGURE 14.13.1

We are now ready for the side condition problems. Suppose that g is a continuously differentiable function of two or three variables defined on a subset of the domain of f. Suppose furthermore that the gradient ∇g is never $\mathbf{0}$. Lagrange made the following observation:

> **(14.13.2)**   If $\mathbf{x}_0$ maximizes (or minimizes) $f(\mathbf{x})$ subject to the side condition $g(\mathbf{x}) = 0$, then $\nabla f(\mathbf{x}_0)$ and $\nabla g(\mathbf{x}_0)$ are collinear. Consequently there exists a scalar λ such that
> $$\nabla f(\mathbf{x}_0) = \lambda \, \nabla g(\mathbf{x}_0).$$

Such a scalar λ is now called a *Lagrange multiplier*.

A Geometric Argument for (14.13.2)

Suppose that $\mathbf{x}_0$ maximizes (or minimizes) $f(\mathbf{x})$ subject to the side condition $g(\mathbf{x}) = 0$. In the two-variable case we have

$$\mathbf{x}_0 = (x_0, y_0) \quad \text{and} \quad \text{the equation } g(x, y) = 0.$$

This equation defines a curve C, which, we assume, has everywhere a nonzero tangent. Since (x_0, y_0) maximizes (or minimizes) $f(x, y)$ on C, we know that $\nabla f(x_0, y_0)$ is perpendicular to C at (x_0, y_0). Since $\nabla g(x_0, y_0)$ is also perpendicular to C at (x_0, y_0) (Section 14.10), the two gradients must be collinear.

In the three-variable case we have

$$\mathbf{x}_0 = (x_0, y_0, z_0) \quad \text{and} \quad \text{the equation } g(x, y, z) = 0.$$

This equation defines a surface Γ that lies in the domain of f. Now let C be a curve which lies in Γ and passes through (x_0, y_0, z_0). As usual we assume that C has a nonzero tangent vector. We know that (x_0, y_0, z_0) maximizes (or minimizes) $f(x, y, z)$ on C. Consequently, $\nabla f(x_0, y_0, z_0)$ is perpendicular to C at (x_0, y_0, z_0). Since this is true for each such curve C, $\nabla f(x_0, y_0, z_0)$ must be perpendicular to Γ itself. But $\nabla g(x_0, y_0, z_0)$ is also perpendicular to Γ at (x_0, y_0, z_0). (Section 14.10) It follows that $\nabla f(x_0, y_0, z_0)$ and $\nabla g(x_0, y_0, z_0)$ are collinear. $\square$

It is time to apply Lagrange's idea to some problems.

Problem.   Maximize and minimize

$$f(x, y) = xy$$

subject to the side condition

$$x^2 + y^2 = 1.$$

SOLUTION.   Setting

$$g(x, y) = x^2 + y^2 - 1,$$

we can state the side condition as

$$g(x, y) = 0.$$

We are looking for those points (x, y) which satisfy a Lagrange condition

$$\nabla f(x, y) = \lambda \, \nabla g(x, y)$$

and, simultaneously, the side condition

$$g(x, y) = 0.$$

The gradients are

$$\nabla f(x, y) = y\mathbf{i} + x\mathbf{j},$$
$$\nabla g(x, y) = 2x\mathbf{i} + 2y\mathbf{j}.$$

Setting

$$\nabla f(x, y) = \lambda \, \nabla g(x, y),$$

we obtain

$$y = 2\lambda x, \qquad x = 2\lambda y.$$

Multiplying the first equation by y and the second equation by x, we find that

$$y^2 = 2\lambda xy, \qquad x^2 = 2\lambda xy$$

and thus

$$y^2 = x^2.$$

This equation together with the side condition

$$g(x, y) = x^2 + y^2 - 1 = 0$$

gives

$$2x^2 - 1 = 0,$$
$$x^2 = \tfrac{1}{2},$$
$$x = \pm\tfrac{1}{2}\sqrt{2}.$$

The points under consideration are therefore

$$(\tfrac{1}{2}\sqrt{2}, \tfrac{1}{2}\sqrt{2}), (\tfrac{1}{2}\sqrt{2}, -\tfrac{1}{2}\sqrt{2}), (-\tfrac{1}{2}\sqrt{2}, \tfrac{1}{2}\sqrt{2}), (-\tfrac{1}{2}\sqrt{2}, -\tfrac{1}{2}\sqrt{2}).$$

At the first and fourth points f takes on the value $\tfrac{1}{2}$. At the second and third points f takes on the value $-\tfrac{1}{2}$. The first is a maximum and the second a minimum. ☐

Problem.   Find the minimum value taken on by the function

$$f(x, y) = x^2 + (y - 2)^2$$

on the hyperbola

$$x^2 - y^2 = 1.$$

SOLUTION.   We set

$$g(x, y) = x^2 - y^2 - 1.$$

Our problem is to minimize $f(x, y)$ subject to the side condition

$$g(x, y) = x^2 - y^2 - 1 = 0.$$

The points of interest must satisfy a Lagrange condition

$$\nabla f(x, y) = \lambda \, \nabla g(x, y)$$

together with

$$g(x, y) = 0.$$

Here

$$\nabla f(x, y) = 2x\mathbf{i} + 2(y - 2)\mathbf{j}, \qquad \nabla g(x, y) = 2x\mathbf{i} - 2y\mathbf{j}.$$

The condition

$$\nabla f(x, y) = \lambda \, \nabla g(x, y)$$

gives

$$2x = 2\lambda x, \qquad 2(y - 2) = -2\lambda y,$$

which we can simplify to

$$x = \lambda x, \qquad y - 2 = -\lambda y.$$

The points of interest must therefore satisfy the following three equations:

$$x = \lambda x, \qquad y - 2 = -\lambda y, \qquad x^2 - y^2 - 1 = 0.$$

The last equation shows that x cannot be zero. Dividing $x = \lambda x$ by x, we get $\lambda = 1$. This means that $y - 2 = -y$ and therefore $y = 1$. With $y = 1$, the last equation gives $x = \pm\sqrt{2}$. The points to be checked are therefore $(-\sqrt{2}, 1)$ and $(\sqrt{2}, 1)$. At each of these points f takes on the value 3. This is the desired minimum. ☐

Problem.   Maximize

$$f(x, y, z) = xyz \qquad\qquad x \geq 0, y \geq 0, z \geq 0$$

subject to the side condition

$$x^3 + y^3 + z^3 = 1.$$

SOLUTION. We set
$$g(x, y, z) = x^3 + y^3 + z^3 - 1$$
so that the side condition becomes
$$g(x, y, z) = 0.$$
We seek those triples (x, y, z) which simultaneously satisfy a Lagrange condition
$$\nabla f(x, y, z) = \lambda \, \nabla g(x, y, z)$$
and the side condition
$$g(x, y, z) = 0$$
The gradients are
$$\nabla f(x, y, z) = yz\mathbf{i} + xz\mathbf{j} + xy\mathbf{k},$$
$$\nabla g(x, y, z) = 3x^2\mathbf{i} + 3y^2\mathbf{j} + 3z^2\mathbf{k}.$$
Setting
$$\nabla f(x, y, z) = \lambda \, \nabla g(x, y, z),$$
we find that
$$yz = \lambda 3x^2, \qquad xz = \lambda 3y^2, \qquad xy = \lambda 3z^2.$$
Multiplying the first equation by x, the second by y, and the third by z, we get
$$xyz = \lambda 3x^3, \qquad xyz = \lambda 3y^3, \qquad xyz = \lambda 3z^3$$
and consequently
$$\lambda x^3 = \lambda y^3 = \lambda z^3.$$

We can exclude $\lambda = 0$ because if $\lambda = 0$ then at least one of the other variables would have to be zero. That would force xyz to be 0 and that is obviously not a maximum. Having excluded $\lambda = 0$, we can divide by λ, and get
$$x^3 = y^3 = z^3.$$
This gives
$$x = y = z.$$
The side condition
$$x^3 + y^3 + z^3 = 1$$
now gives
$$z = (\tfrac{1}{3})^{1/3}, \qquad y = (\tfrac{1}{3})^{1/3}, \qquad z = (\tfrac{1}{3})^{1/3}.$$
The desired maximum is $\tfrac{1}{3}$. □

Problem. Show that of all the triangles inscribed in a fixed circle, the equilateral triangle has the largest perimeter.

SOLUTION. In Figure 14.13.2 we have marked the angles α, β, γ. The length of the side subtended by α is
$$2R \sin \tfrac{1}{2}\alpha.$$
Hence the perimeter is given by
$$f(\alpha, \beta, \gamma) = 2R(\sin \tfrac{1}{2}\alpha + \sin \tfrac{1}{2}\beta + \sin \tfrac{1}{2}\gamma).$$

As a side condition we have

$$g(\alpha, \beta, \gamma) = \alpha + \beta + \gamma - 2\pi = 0.$$

To maximize the perimeter we form the gradients

$$\nabla f(\alpha, \beta, \gamma) = R[\cos \tfrac{1}{2}\alpha \, \mathbf{i} + \cos \tfrac{1}{2}\beta \, \mathbf{j} + \cos \tfrac{1}{2}\gamma \, \mathbf{k}],$$
$$\nabla g(\alpha, \beta, \gamma) = \mathbf{i} + \mathbf{j} + \mathbf{k}.$$

The Lagrange condition

$$\nabla f(\alpha, \beta, \gamma) = \lambda \, \nabla g(\alpha, \beta, \gamma)$$

results in

$$\lambda = R \cos \tfrac{1}{2}\alpha, \qquad \lambda = R \cos \tfrac{1}{2}\beta, \qquad \lambda = R \cos \tfrac{1}{2}\gamma.$$

Under the special conditions of the problem this means that

$$\alpha = \beta = \gamma.$$

With the central angles equal, the sides are equal. The triangle is therefore equilateral. □

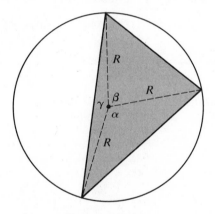

FIGURE 14.13.2

Exercises

*1. Minimize $x^2 + y^2$ subject to the condition $xy = 1$.

2. Maximize xy subject to the condition $\dfrac{x^2}{a^2} + \dfrac{y^2}{b^2} = 1$.

*3. Minimize xy subject to the condition $\dfrac{x^2}{a^2} + \dfrac{y^2}{b^2} = 1$.

4. Minimize xy^2 subject to the condition $x^2 + y^2 = 1$.

*5. Maximize xy^2 subject to the condition $\dfrac{x^2}{a^2} + \dfrac{y^2}{b^2} = 1$.

6. Maximize $x + y$ subject to the condition $x^4 + y^4 = 1$.

*7. Minimize xyz subject to the condition $x^2 + y^2 + z^2 = 1$.

8. Maximize xyz subject to the condition $\dfrac{x^2}{a^2} + \dfrac{y^2}{b^2} + \dfrac{z^2}{c^2} = 1$.

*9. Maximize xy^2z^3 subject to the condition $x + y + z = 1$.

10. Minimize $x + 2y + 4z$ subject to the condition $x^2 + y^2 + z^2 = 7$.

*11. Maximize $2x + 3y + 5z$ subject to the condition $x^2 + y^2 + z^2 = 19$.

12. Minimize $x^4 + y^4 + z^4$ subject to the side condition $x + y + z = 1$.

*13. Maximize the volume of a rectangular solid which has three faces on the coordinate planes and one vertex on the plane

$$\frac{x}{a} + \frac{y}{b} + \frac{z}{c} = 1. \qquad (a > 0, \quad b > 0, \quad c > 0)$$

14. Show that the square has the largest area of all the rectangles with a given perimeter.

*15. Find the distance between the point $(0, 1)$ and the parabola $x^2 = 4y$.

16. Find the distance between the point $(p, 4p)$ and the parabola $y^2 = 2px$.

*17. Find the largest volume of a rectangular solid which has the property that the sum of the areas of the six faces is $6a^2$.

18. Use the method of Lagrange multipliers to find the distance from the origin to the plane $Ax + By + Cz + D = 0$.

*19. Within a triangle, there is a point P such that the sum of the squares of its distances from the sides of the triangle is a minimum. Find this minimum.

20. Show that of all the triangles inscribed in a fixed circle the equilateral one has the largest
    (a) product of the lengths of the sides.
    (b) sum of the squares of the lengths of the sides.

21. A plane passes through the point (a, b, c). Find its intercepts with the coordinate axes if the volume of the solid bounded by the plane and the coordinate planes is a minimum.

22. Show that of all the triangles with a given perimeter the equilateral triangle has the largest area. HINT: Use

$$\text{area} = \sqrt{s(s - a)(s - b)(s - c)},$$

where s is the semiperimeter

$$s = \tfrac{1}{2}(a + b + c).$$

*23. A manufacturer can produce three distinct products in quantities Q_1, Q_2, Q_3 respectively and thereby derive a profit

$$P(Q_1, Q_2, Q_3) = 2Q_1 + 8Q_2 + 24Q_3.$$

Find the values of Q_1, Q_2, Q_3 that maximize profit if production is subject to the constraint

$$Q_1^2 + 2Q_2^2 + 4Q_3^2 = 4{,}500{,}000{,}000.$$

14.14   Differentials

We begin by reviewing the one-variable case. If f is differentiable at x, then

$$f(x + h) - f(x) = f'(x)h + o(h).$$

This tells us that for small h

$$f(x + h) - f(x) \quad \text{and} \quad f'(x)h$$

are approximately equal and consequently we can use the differential

$$df = f'(x)h$$

to approximate the increment

$$\Delta f = f(x + h) - f(x).$$

In symbols

$$\Delta f \cong df.$$

The differential of a function of several variables is defined in a similar manner and plays a similar approximating role. If f, now a function of several variables, is differentiable at $\mathbf{x}$, then

$$f(\mathbf{x} + \mathbf{h}) - f(\mathbf{x}) = \nabla f(\mathbf{x}) \cdot \mathbf{h} + o(\mathbf{h}).$$

This says that, for small $\mathbf{h}$,

$$f(\mathbf{x} + \mathbf{h}) - f(\mathbf{x}) \quad \text{and} \quad \nabla f(\mathbf{x}) \cdot \mathbf{h}$$

are approximately equal. In this setting, we call

$$\boxed{\Delta f = f(\mathbf{x} + \mathbf{h}) - f(\mathbf{x})}$$

the *increment* of f and

$$\boxed{df = \nabla f(\mathbf{x}) \cdot \mathbf{h}}$$

the *differential* (more formally, the *total differential*) at $\mathbf{x}$ with increment $\mathbf{h}$. The key point is that, as before,

$$\boxed{\Delta f \cong df.}$$

In the two-variable case we set

$$\mathbf{x} = (x, y) \quad \text{and} \quad \mathbf{h} = (\Delta x, \Delta y).$$

The increment

$$\Delta f = f(\mathbf{x} + \mathbf{h}) - f(\mathbf{x})$$

then becomes

$$\Delta f = f(x + \Delta x, y + \Delta y) - f(x, y)$$

and the differential

$$df = \nabla f(\mathbf{x}) \cdot \mathbf{h}$$

becomes

$$df = \frac{\partial f}{\partial x}(x, y)\,\Delta x + \frac{\partial f}{\partial y}(x, y)\,\Delta y.$$

By suppressing the point of evaluation, we can write

$$df = \frac{\partial f}{\partial x}\,\Delta x + \frac{\partial f}{\partial y}\,\Delta y.$$

The approximation

$$\Delta f \cong df$$

is illustrated in Figure 14.14.1. There we have represented f as a surface $z = f(x, y)$.

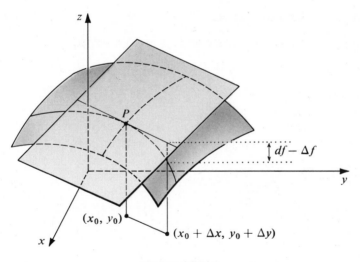

FIGURE 14.14.1

Through a point $P(x_0, y_0, f(x_0, y_0))$ we have drawn the tangent plane. The difference

$$df - \Delta f$$

is the vertical separation between this tangent plane and the surface as measured above the point $(x_0 + \Delta x, y_0 + \Delta y)$.

PROOF.   The plane tangent to the surface at P has equation

$$(x - x_0)\frac{\partial f}{\partial x}(x_0, y_0) + (y - y_0)\frac{\partial f}{\partial y}(x_0, y_0) = z - f(x_0, y_0).$$

The z-coordinate of this plane over the point $(x_0 + \Delta x, y_0 + \Delta y)$ is

$$f(x_0, y_0) + \frac{\partial f}{\partial x}(x_0, y_0)\,\Delta x + \frac{\partial f}{\partial y}(x_0, y_0)\,\Delta y. \qquad \text{(check this)}$$

The z-coordinate of the surface over this same point is

$$f(x_0 + \Delta x, y_0 + \Delta y).$$

The difference

$$\left[f(x_0, y_0) + \frac{\partial f}{\partial x}(x_0, y_0)\,\Delta x + \frac{\partial f}{\partial y}(x_0, y_0)\,\Delta y \right] - \left[f(x_0 + \Delta x, y_0 + \Delta y) \right]$$

can be written as

$$\left[\frac{\partial f}{\partial x}(x_0, y_0)\,\Delta x + \frac{\partial f}{\partial y}(x_0, y_0)\,\Delta y \right] - \left[f(x_0 + \Delta x, y_0 + \Delta y) - f(x_0, y_0) \right]$$

and that is just

$$df - \Delta f. \quad \square$$

For the three-variable case we set

$$\mathbf{x} = (x, y, z) \quad \text{and} \quad \mathbf{h} = (\Delta x, \Delta y, \Delta z).$$

The increment then becomes

$$\Delta f = f(x + \Delta x, y + \Delta y, z + \Delta z) - f(x, y, z)$$

and the approximating differential

$$df = \frac{\partial f}{\partial x}(x, y, z)\,\Delta x + \frac{\partial f}{\partial y}(x, y, z)\,\Delta y + \frac{\partial f}{\partial z}(x, y, z)\,\Delta z,$$

or more simply,

$$\boxed{df = \frac{\partial f}{\partial x}\,\Delta x + \frac{\partial f}{\partial y}\,\Delta y + \frac{\partial f}{\partial z}\,\Delta z.}$$

To illustrate the use of differentials we begin with a rectangle of sides x and y. The area is given by

$$A(x, y) = xy.$$

An increase in the dimensions of the rectangle to $x + \Delta x$ and $y + \Delta y$ produces a change in area

$$\begin{aligned}
\Delta A &= (x + \Delta x)(y + \Delta y) - xy \\
&= (xy + x\,\Delta y + y\,\Delta x + \Delta x\,\Delta y) - xy \\
&= x\,\Delta y + y\,\Delta x + \Delta x\,\Delta y.
\end{aligned}$$

The differential estimate for this change in area is

$$dA = \frac{\partial A}{\partial x}\,\Delta x + \frac{\partial A}{\partial y}\,\Delta y = y\,\Delta x + x\,\Delta y. \qquad \text{(Figure 14.14.2)}$$

The error of our estimate—the difference between the actual change and the estimated change—is the difference

$$\Delta A - dA = \Delta x \, \Delta y. \quad \square$$

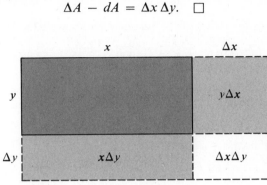

FIGURE 14.14.2

Problem. Given that

$$f(x, y) = yx^{2/5} + x\sqrt{y},$$

estimate by a differential the change in f from $(32, 16)$ to $(35, 18)$.

SOLUTION. Since

$$\frac{\partial f}{\partial x} = \frac{2}{5}\left(\frac{1}{x}\right)^{3/5} y + \sqrt{y} \quad \text{and} \quad \frac{\partial f}{\partial y} = x^{2/5} + \frac{x}{2\sqrt{y}},$$

we have

$$df = \left[\frac{2}{5}\left(\frac{1}{x}\right)^{3/5} y + \sqrt{y}\right] \Delta x + \left[x^{2/5} + \frac{x}{2\sqrt{y}}\right] \Delta y.$$

At $x = 32$, $y = 16$, $\Delta x = 3$, $\Delta y = 2$,

$$df = \left[\frac{2}{5}\left(\frac{1}{32}\right)^{3/5} 16 + \sqrt{16}\right] 3 + \left[32^{2/5} + \frac{32}{2\sqrt{16}}\right] 2 = 30.4.$$

The change increases the value of f by approximately 30.4. $\quad \square$

Problem. Use differentials to estimate

$$\sqrt{27} \sqrt[3]{1021}.$$

SOLUTION. We know $\sqrt{25}$ and $\sqrt[3]{1000}$. What we need is an estimate for the increase of

$$f(x, y) = \sqrt{x}\,\sqrt[3]{y} = x^{1/2}y^{1/3}$$

from $x = 25$, $y = 1000$ to $x = 27$, $y = 1021$. The differential is

$$df = \tfrac{1}{2}x^{-1/2}y^{1/3}\,\Delta x + \tfrac{1}{3}x^{1/2}y^{-2/3}\,\Delta y.$$

With $x = 25$, $y = 1000$, $\Delta x = 2$, $\Delta y = 21$, df becomes

$$(\tfrac{1}{2} \cdot 25^{-1/2} \cdot 1000^{1/3})2 + (\tfrac{1}{3} \cdot 25^{1/2} \cdot 1000^{-2/3})21 = 2.35.$$

The change increases the value of the function by about 2.35. It follows that

$$\sqrt{27} \sqrt[3]{1021} \cong \sqrt{25} \sqrt[3]{1000} + 2.35 = 52.35. \quad \square$$

Problem. Estimate the change in the volume of the frustum of a cone if the upper radius r is decreased from 3 to 2.7, the base radius R is increased from 8 to 8.1, and the height h is increased from 6 to 6.3.

SOLUTION

$$V(r, R, h) = \tfrac{1}{3}\pi h(R^2 + Rr + r^2),$$
$$dV = \tfrac{1}{3}\pi h(R + 2r)\,\Delta r + \tfrac{1}{3}\pi h(2R + r)\,\Delta R + \tfrac{1}{3}\pi(R^2 + Rr + r^2)\,\Delta h.$$

At $r = 3$, $R = 8$, $h = 6$, $\Delta r = -0.3$, $\Delta R = 0.1$, and $\Delta h = 0.3$,

$$dV = (28\pi)(-0.3) + (38\pi)(0.1) + \tfrac{1}{3}(97\pi)(0.3) = 5.1\pi \text{ cubic inches.}$$

The volume increases by about 5.1π cubic inches. $\quad \square$

Exercises

Find the differential df.

*1. $f(x, y) = x^3 y - x^2 y^2$.

2. $f(x, y) = xy + yz + xz$.

*3. $f(x, y) = x \cos y - y \cos x$.

4. $f(x, y, z) = x^2 y e^{2z}$.

*5. $f(x, y, z) = x - y \tan z$.

6. $f(x, y) = (x - y) \log (x + y)$.

*7. Compute Δu and du for $u = x^2 - 3xy + 2y^2$ at $x = 2$, $y = -3$, $\Delta x = -0.3$, $\Delta y = 0.2$.

*8. Compute du for $u = (x + y)\sqrt{x - y}$ at $x = 6$, $y = 2$, $\Delta x = \tfrac{1}{4}$, $\Delta y = -\tfrac{1}{2}$.

Use differentials to find the approximate value.

*9. $\sqrt{125}\sqrt[4]{17}$.

10. $(1 - \sqrt{10})(1 + \sqrt{24})$.

*11. $\sin \tfrac{6}{7}\pi \cos \tfrac{1}{5}\pi$.

12. $\sqrt{8} \tan \tfrac{5}{16}\pi$.

*13. Given that

$$z = \frac{x - y}{x + y}$$

use dz to find the approximate change in z if x is increased from 4 to $4\tfrac{1}{10}$ and y is increased from 2 to $2\tfrac{1}{10}$. What is the exact change?

*14. Estimate by a differential the change in the volume of a right circular cylinder if the height is increased from 12 inches to 12.2 inches and the radius is decreased from 8 inches to 7.7 inches.

15. Use a differential to estimate the change in

$$T = x^2 \cos \pi z - y^2 \sin \pi z$$

from $x = 2$, $y = 2$, $z = 2$ to $x = 2.1$, $y = 1.9$, $z = 2.2$.

16. Give bounding estimates for the volume and the lateral area of a right circular cone given that the radius is between 4.8 inches and 5.2 inches and the slant height is between 11.7 inches and 12.3 inches.

14.15   Reconstructing a Function from its Gradient

This section has essentially three parts. In the first part we show how to find $f(x, y)$, given its gradient

$$\nabla f(x, y) = \frac{\partial f}{\partial x}(x, y)\,\mathbf{i} + \frac{\partial f}{\partial y}(x, y)\,\mathbf{j}.$$

In the second part we show that, although all gradients $\nabla f(x, y)$ are of the form

$$P(x, y)\mathbf{i} + Q(x, y)\mathbf{j}$$

$\left(\text{set } P = \dfrac{\partial f}{\partial x} \text{ and } Q = \dfrac{\partial f}{\partial y} \right)$, not all combinations of the form

$$P(x, y)\mathbf{i} + Q(x, y)\mathbf{j}$$

are actually gradients. In the third part we tackle the problem of recognizing which combinations

$$P(x, y)\mathbf{i} + Q(x, y)\mathbf{j}$$

are gradients.

Problem.   Find f given that

$$\nabla f(x, y) = y^2\mathbf{i} + (2xy - 1)\mathbf{j}.$$

SOLUTION.   Since

$$\nabla f(x, y) = \frac{\partial f}{\partial x}(x, y)\mathbf{i} + \frac{\partial f}{\partial y}(x, y)\mathbf{j}$$

we must have

$$\frac{\partial f}{\partial x}(x, y) = y^2 \quad \text{and} \quad \frac{\partial f}{\partial y}(x, y) = 2xy - 1.$$

Integrating the first relation with respect to x, we have

$$f(x, y) = y^2 x + \phi(y)$$

where $\phi(y)$ is independent of x but may depend on y. Integrating the second relation with respect to y, we have

$$f(x, y) = xy^2 - y + \psi(x)$$

where $\psi(x)$ is independent of y but may depend on x. To reconcile

$$f(x, y) = xy^2 + \phi(y) \quad \text{with} \quad f(x, y) = xy^2 - y + \psi(x),$$

we set

$$xy^2 + \phi(y) = xy^2 - y + \psi(x).$$

This forces

$$\phi(y) = -y + \psi(x)$$

and thus

$$\phi(y) + y = \psi(x).$$

The left-hand side can depend at most on y and the right-hand side at most on x. This means that both sides are equal to some constant C. This gives

$$\phi(y) + y = C \quad \text{and} \quad \psi(x) = C.$$

The relation

$$f(x, y) = xy^2 + \phi(y)$$

now gives

$$f(x, y) = xy^2 - y + C.$$

The relation

$$f(x, y) = xy^2 - y + \psi(x)$$

also gives

$$f(x, y) = xy^2 - y + C. \quad \square$$

REMARK. Once we had derived

$$f(x, y) = xy^2 - y + C$$

from the relation

$$f(x, y) = xy^2 + \phi(y),$$

it was unnecessary to derive the same result from

$$f(x, y) = xy^2 - y + \psi(x).$$

We did it simply to check our computations. It's reassuring to get the same answer from both relations.

Problem. Find f given that

$$\nabla f(x, y) = (y^{1/2} - \tfrac{1}{2}yx^{-1/2} + 2x)\mathbf{i} + (\tfrac{1}{2}xy^{-1/2} - x^{1/2} + 1)\mathbf{j}.$$

SOLUTION. We begin by observing that

$$\frac{\partial f}{\partial x}(x, y) = y^{1/2} - \tfrac{1}{2}yx^{-1/2} + 2x \quad \text{and} \quad \frac{\partial f}{\partial y}(x, y) = \tfrac{1}{2}xy^{-1/2} - x^{1/2} + 1.$$

Integration of the first relation with respect to x, gives

$$f(x, y) = y^{1/2}x - x^{1/2}y + x^2 + \phi(y)$$

with $\phi(y)$ independent of x. Integration of the second relation with respect to y, gives

$$f(x, y) = xy^{1/2} - x^{1/2}y + y + \psi(x),$$

with $\psi(x)$ independent of y. These two characterizations of $f(x, y)$ can be reconciled only by having

$$x^2 + \phi(y) = y + \psi(x).$$

This in turn gives

$$\phi(y) - y = \psi(x) - x^2.$$

The left-hand side can depend at most on y, and the right-hand side at most on x. Both sides must therefore be constant. The relations

$$\phi(y) - y = C \quad \text{and} \quad \psi(x) - x^2 = C$$

lead to

$$\phi(y) = y + C \quad \text{and} \quad \psi(x) = x^2 + C.$$

Each of these relations gives

$$f(x, y) = xy^{1/2} - x^{1/2}y + x^2 + y + C. \quad \square$$

The function

$$f(x, y) = xy^{1/2} - x^{1/2}y + x^2 + y + C$$

that we first found is in its most general form in the sense that it contains an arbitrary constant C. If we want f to take on a particular value z_0 at some particular point (x_0, y_0) we need only adjust the constant C accordingly. Let's suppose for example that we require

$$f(1, 4) = 5.$$

Note that

$$f(1, 4) = 1\sqrt{4} - \sqrt{1} \cdot 4 + 1^2 + 4 + C = 3 + C.$$

To get

$$f(1, 4) = 5$$

we must have

$$3 + C = 5$$

and thus

$$C = 2. \quad \square$$

The next problem shows that not all linear combinations

$$P(x, y)\mathbf{i} + Q(x, y)\mathbf{j}$$

are actually gradients.

Problem. Show that there is no function f such that

$$\nabla f(x, y) = y\mathbf{i} - x\mathbf{j}.$$

SOLUTION. Suppose on the contrary that there is such a function. Then of course we must have

$$\frac{\partial f}{\partial x}(x, y) = y \quad \text{and} \quad \frac{\partial f}{\partial y}(x, y) = -x.$$

Integrating the first equation with respect to x, we get

$$f(x, y) = xy + \phi(y),$$

where $\phi(y)$ is independent of x. Integrating the second equation with respect to y, we get

$$f(x, y) = -xy + \psi(x),$$

where $\psi(x)$ is independent of y. To reconcile these two characterizations of f, we must have

$$xy + \phi(y) = -xy + \psi(x),$$

and consequently,

$$\psi(x) = 2xy + \phi(y).$$

Differentiation of this equation with respect to y gives

$$\frac{\partial}{\partial y}[\psi(x)] = 2x + \phi'(y),$$

$$0 = 2x + \phi'(y),$$

$$-2x = \phi'(y).$$

Differentiation of this last equation with respect to x gives

$$-2 = 0.$$

The assumption that

$$y\mathbf{i} - x\mathbf{j}$$

was a gradient has led to an absurdity.   □

We come now to the problem of recognizing which linear combinations

$$P(x, y)\mathbf{i} + Q(x, y)\mathbf{j}$$

are gradients.

Theorem 14.15.1

Let P and Q be functions of two variables each continuously differentiable in an open rectangle S, which we take to have sides parallel to the coordinate axes. The linear combination

$$P(x, y)\mathbf{i} + Q(x, y)\mathbf{j}$$

is a gradient in S iff

$$\frac{\partial P}{\partial y}(x, y) = \frac{\partial Q}{\partial x}(x, y) \quad \text{for all } (x, y) \in S.$$

PROOF.   Suppose that

$$P(x, y)\mathbf{i} + Q(x, y)\mathbf{j}$$

is a gradient in S, say

$$\nabla f(x, y) = P(x, y)\mathbf{i} + Q(x, y)\mathbf{j}.$$

Since

$$\nabla f(x, y) = \frac{\partial f}{\partial x}(x, y)\mathbf{i} + \frac{\partial f}{\partial y}(x, y)\mathbf{j},$$

we have

$$P = \frac{\partial f}{\partial x} \quad \text{and} \quad Q = \frac{\partial f}{\partial y}.$$

Since P and Q have continuous first partials, f has continuous second partials. Consequently, its mixed partials are equal (14.6.1), and

$$\frac{\partial P}{\partial y} = \frac{\partial^2 f}{\partial y\, \partial x} = \frac{\partial^2 f}{\partial x\, \partial y} = \frac{\partial Q}{\partial x}.$$

Conversely, suppose that

$$\frac{\partial P}{\partial y}(x, y) = \frac{\partial Q}{\partial x}(x, y) \quad \text{for all } (x, y) \in S.$$

We must show that

$$P(x, y)\mathbf{i} + Q(x, y)\mathbf{j}$$

is a gradient in S. To do this, we choose a point (x_0, y_0) and form the function

$$f(x, y) = C + \int_{x_0}^{x} P(u, y_0)\, du + \int_{y_0}^{y} Q(x, v)\, dv, \qquad (x, y) \in S.$$

[If you want to visualize f, you can refer to Figure 14.15.1. The function P is being integrated along the horizontal line segment joining (x_0, y_0) to (x, y_0) and Q is being integrated along the vertical line segment joining (x, y_0) to (x, y). Our assumptions on S guarantee that these line segments remain in S. C is simply an arbitrary constant.]

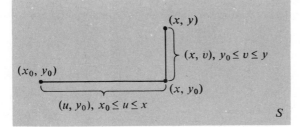

FIGURE 14.15.1

The first integral is independent of y. Hence

$$\frac{\partial f}{\partial y}(x, y) = \frac{\partial}{\partial y}\left(\int_{y_0}^{y} Q(x, v)\, dv \right) = Q(x, y).$$

The last equality holds because we are differentiating an integral with respect to its upper limit.† Differentiating f with respect to x is a little more complicated:

$$\frac{\partial f}{\partial x}(x, y) = \frac{\partial}{\partial x}\left(\int_{x_0}^{x} P(u, y_0)\, du \right) + \frac{\partial}{\partial x}\left(\int_{y_0}^{y} Q(x, v)\, dv \right).$$

---

† This is a fundamental notion. If you are fuzzy on it, see Theorem 4.3.2.

In the first term the variable x appears as an upper limit. In the second term the variable x appears under the sign of integration. Differentiation under the sign of integration is justified in more advanced texts.† Anticipating this result, we get

$$\frac{\partial f}{\partial x}(x, y) = P(x, y_0) + \int_{y_0}^{y} \frac{\partial Q}{\partial x}(x, v)\, dv$$

$$= P(x, y_0) + \int_{y_0}^{y} \frac{\partial P}{\partial y}(x, v)\, dv$$

$$= P(x, y_0) + P(x, y) - P(x, y_0)$$

$$= P(x, y).$$

With

$$P(x, y) = \frac{\partial f}{\partial x}(x, y) \quad \text{and} \quad Q(x, y) = \frac{\partial f}{\partial y}(x, y),$$

we have

$$P(x, y)\mathbf{i} + Q(x, y)\mathbf{j}$$

as the gradient of f.   ☐

The theorem we just proved for a rectangle can be extended to more general regions. The proof that we gave depended on our being able to join arbitrary points of S by a path consisting of a horizontal and a vertical line segment both of which remained in S. For more general regions a more complicated proof is required.

The test for gradients that is furnished by the theorem is easy to apply.

Examples

(1) To determine whether

$$xy\mathbf{i} + \tfrac{1}{2}(x + 1)^2 y^2 \mathbf{j}$$

is a gradient, we set

$$P(x, y) = xy \quad \text{and} \quad Q(x, y) = \tfrac{1}{2}(x + 1)^2 y^2.$$

Since

$$\frac{\partial P}{\partial y}(x, y) = x \quad \text{and} \quad \frac{\partial Q}{\partial x}(x, y) = (x + 1)y^2,$$

we have

$$\frac{\partial P}{\partial y}(x, y) \neq \frac{\partial Q}{\partial x}(x, y).$$

The expression is not a gradient.

(2) In the case of

$$2x \sin y \,\mathbf{i} + x^2 \cos y \,\mathbf{j}$$

---

† See, for example, Einar Hille, *Analysis*, Vol. 2 (Blaisdell, Waltham, Mass., 1966).

we have

$$P(x, y) = 2x \sin y, \qquad Q(x, y) = x^2 \cos y,$$

$$\frac{\partial P}{\partial y}(x, y) = 2x \cos y, \qquad \frac{\partial Q}{\partial x}(x, y) = 2x \cos y.$$

Since the two partials are equal, the expression is a gradient. □

Exercises

Find the function with the given gradient.

*1. $xy^2\mathbf{i} + x^2y\mathbf{j}$.

2. $x\mathbf{i} + y\mathbf{j}$.

*3. $y\mathbf{i} + x\mathbf{j}$.

4. $(x^2 + y)\mathbf{i} + (y^3 + x)\mathbf{j}$.

*5. $(y^3 + x)\mathbf{i} + (x^2 + y)\mathbf{j}$.

6. $(y^2e^x - y)\mathbf{i} + (2ye^x - x)\mathbf{j}$.

*7. $(\cos x - y \sin x)\mathbf{i} + \cos x\,\mathbf{j}$.

8. $(1 + e^y)\mathbf{i} + (xe^y + y^2)\mathbf{j}$.

*9. $ye^x(1 + x)\mathbf{i} + (xe^x - e^{-y})\mathbf{j}$.

10. $(x \tan y + \sec^2 x)\mathbf{i} + (\frac{1}{2}x^2 \sec^2 y + \pi y)\mathbf{j}$.

*11. $(x^2 \arcsin y)\mathbf{i} + \left(\dfrac{x^3}{3\sqrt{1 - y^2}} - \log y\right)\mathbf{j}$.

12. $\left(\dfrac{\arctan y}{\sqrt{1 - x^2}} + \dfrac{x}{y}\right)\mathbf{i} + \left(\dfrac{\arcsin x}{1 + y^2} - \dfrac{x^2}{2y^2} + 1\right)\mathbf{j}$.

*13. *(Optional)* $(\cosh x + \sinh y)\mathbf{i} + (x \cosh y)\mathbf{j}$.

14. *(Optional)* $(\cosh x - x \sinh y)\mathbf{i} + (2y \cosh x)\mathbf{j}$.

14.16   Work and Line Integrals, Part I

If an object moves along a straight line subject to a constant force $\mathbf{F}$, the work done is, by definition, the component of $\mathbf{F}$ in the direction of the displacement, multiplied by the length of the displacement:

$$W = (\text{comp}_{\mathbf{d}}\, \mathbf{F}) \cdot \|\mathbf{d}\|.$$

We can write this more briefly as a dot product.

$$\boxed{W = \mathbf{F} \cdot \mathbf{d}.}$$

This elementary notion of work may be useful, but it is not sufficient. Take for example the case of an object which moves through a magnetic field or a gravitational field. In such a situation the path of the motion is generally not a straight line but a curve, and the force does not remain constant but instead varies from point to point. What we want is a notion of work that applies also to this more general situation.

Suppose that an object moves through a force field and that at each time t it has position $\mathbf{r}(t)$ and is subject to a force $\mathbf{F}(\mathbf{r}(t))$. As the object moves from time a to time b, it traces out a curve

$$C: \mathbf{r}(t) = x(t)\mathbf{i} + y(t)\mathbf{j} + z(t)\mathbf{k}, \qquad t \in [a, b]$$

which we assume is *smooth*. (That is, we assume $\mathbf{r}'$ exists and is continuous.) What we want to do is *define* the total work done by $\mathbf{F}$ along the curve C.

To decide how to do this, we begin by focusing our attention on what happens during a short time interval $[t, t + h]$. As an estimate for the work done during this short time interval, we use the dot product

$$\mathbf{F}(\mathbf{r}(t)) \cdot [\mathbf{r}(t + h) - \mathbf{r}(t)].$$

In making this estimate (see Figure 14.16.1), we are using the force vector at time t and we are replacing the curved path from $\mathbf{r}(t)$ to $\mathbf{r}(t + h)$ by the corresponding line segment.

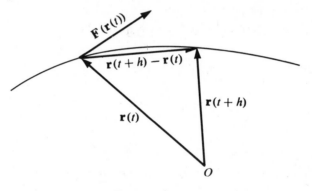

FIGURE 14.16.1

If we set

$$W(t) = \text{total work done up to time } t$$

and

$$W(t + h) = \text{total work done up to time } t + h,$$

then the work done from time t to time $t + h$ must be the difference

$$W(t + h) - W(t).$$

Bringing our estimate into play, we are led to the approximate equation

$$W(t + h) - W(t) \cong \mathbf{F}(\mathbf{r}(t)) \cdot [\mathbf{r}(t + h) - \mathbf{r}(t)],$$

which, upon division by h, becomes

$$\frac{W(t + h) - W(t)}{h} \cong \mathbf{F}(\mathbf{r}(t)) \cdot \frac{[\mathbf{r}(t + h) - \mathbf{r}(t)]}{h}.$$

The quotients here are average rates of change and the equation is only an approximate one, but, presumably, we want it to become increasingly accurate as h tends to zero.

The definition of work is made by requiring that, as h tends to zero, both sides have exactly the same limit; in other words, by requiring that

$$W'(t) = \mathbf{F}(\mathbf{r}(t)) \cdot \mathbf{r}'(t).$$

The rest is now predetermined. Since by definition

$$W(a) = 0 \quad \text{and} \quad W(b) = \text{total work done,}$$

we have

$$\text{total work done} = W(b) - W(a) = \int_a^b W'(t)\, dt = \int_a^b [\mathbf{F}(\mathbf{r}(t)) \cdot \mathbf{r}'(t)]\, dt.$$

In short, we have arrived at the following definition of work:

$$W = \int_a^b [\mathbf{F}(\mathbf{r}(t)) \cdot \mathbf{r}'(t)]\, dt.$$

The integral on the right is called a *line integral*. More precisely, it is called the *line integral of F over C* and is commonly denoted by

$$\int_C \mathbf{F}(\mathbf{r}) \cdot d\mathbf{r}.$$

In symbols, we have

$$W = \int_C \mathbf{F}(\mathbf{r}) \cdot d\mathbf{r} = \int_a^b [\mathbf{F}(\mathbf{r}(t)) \cdot \mathbf{r}'(t)]\, dt.$$

Problem. Find the total work done when an object traverses the parabolic arc

$$C: \mathbf{r}(t) = t\mathbf{i} + t^2\mathbf{j}, \qquad t \in [0, 1]$$

subject to a force

$$\mathbf{F}(x, y) = xy\mathbf{i} + y^2\mathbf{j}.$$

SOLUTION

$$\mathbf{F}(\mathbf{r}(t)) = \mathbf{F}(x(t), y(t)) = x(t)y(t)\mathbf{i} + [y(t)]^2\mathbf{j} = t^3\mathbf{i} + t^4\mathbf{j},$$

$$\mathbf{r}'(t) = \mathbf{i} + 2t\mathbf{j},$$

$$W = \int_C \mathbf{F}(\mathbf{r}) \cdot d\mathbf{r} = \int_0^1 [\mathbf{F}(\mathbf{r}(t)) \cdot \mathbf{r}'(t)]\, dt = \int_0^1 (t^3 + 2t^5)\, dt = \tfrac{7}{12}. \quad \square$$

Problem. Find the total work done when an object traverses the twisted cubic

$$C: \mathbf{r}(t) = t\mathbf{i} + t^2\mathbf{j} + t^3\mathbf{k}, \qquad t \in [-1, 1]$$

subject to a force

$$\mathbf{F}(x, y, z) = xy\mathbf{i} + yz\mathbf{j} + xz\mathbf{k}.$$

SOLUTION

$$\mathbf{F}(\mathbf{r}(t)) = \mathbf{F}(x(t), y(t), z(t)) = x(t)y(t)\mathbf{i} + y(t)z(t)\mathbf{j} + x(t)z(t)\mathbf{k}$$
$$= t^3\mathbf{i} + t^5\mathbf{j} + t^4\mathbf{k},$$

$$\mathbf{r}'(t) = \mathbf{i} + 2t\mathbf{j} + 3t^2\mathbf{k},$$

$$W = \int_C \mathbf{F}(\mathbf{r}) \cdot d\mathbf{r} = \int_{-1}^{1} [\mathbf{F}(\mathbf{r}(t)) \cdot \mathbf{r}'(t)]\, dt = \int_{-1}^{1} [t^3 + 2t^6 + 3t^6]\, dt$$

$$= \int_{-1}^{1} [t^3 + 5t^6]\, dt = \tfrac{10}{7}. \quad \square$$

If a curve C is not smooth but is made up of a finite number of smooth pieces $C_1, C_2, \ldots, C_n$, then we define the integral over C (and therefore the work done) as the sum of the integrals over the C_i:

$$\boxed{\int_C = \int_{C_1} + \int_{C_2} + \cdots + \int_{C_n}.}$$

A curve of this type is said to be *piecewise smooth*. Some pictorial examples are given in Figure 14.16.2.

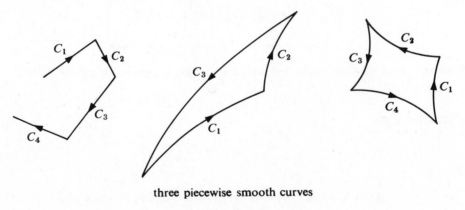

three piecewise smooth curves

FIGURE 14.16.2

Triangles are of course piecewise smooth curves. In the next problem we integrate over a triangle. We do this by integrating over each of the sides and then adding up the results. Keep in mind that the directed line segment which begins at $\mathbf{a} = (a_1, a_2)$ and ends at $\mathbf{b} = (b_1, b_2)$ can be parametrized by setting

$$\mathbf{r}(t) = (1 - t)\mathbf{a} + t\mathbf{b}, \qquad t \in [0, 1].$$

Problem. Evaluate the line integral

$$\int_C \mathbf{F}(\mathbf{r}) \cdot d\mathbf{r}$$

if

$$\mathbf{F}(x, y) = e^y\mathbf{i} - \sin \pi x \, \mathbf{j}$$

and C is the triangle with vertices $(1, 0)$, $(0, 1)$, $(-1, 0)$ traversed counterclockwise.

SOLUTION.   The path C (Figure 14.16.3) is made up of the three line segments

$$C_1: \mathbf{r}(t) = (1 - t)\mathbf{i} + t\mathbf{j}, \qquad t \in [0, 1],$$
$$C_2: \mathbf{r}(t) = (1 - t)\mathbf{j} + t(-\mathbf{i}) = -t\mathbf{i} + (1 - t)\mathbf{j}, \qquad t \in [0, 1],$$
$$C_3: \mathbf{r}(t) = (1 - t)(-\mathbf{i}) + t\mathbf{i} = (2t - 1)\mathbf{i}, \qquad t \in [0, 1].$$

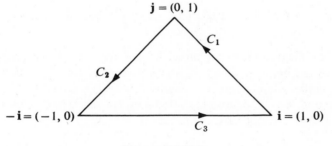

FIGURE 14.16.3

As you can verify,

$$\int_{C_1} \mathbf{F}(\mathbf{r}) \cdot d\mathbf{r} = \int_0^1 \left[e^{y(t)}x'(t) - \sin\left[\pi x(t)\right] y'(t) \right] dt$$

$$= \int_0^1 \left[-e^t - \sin\left[\pi(1 - t)\right] \right] dt$$

$$= \left[-e^t - \frac{1}{\pi} \cos\left[\pi(1 - t)\right] \right]_0^1$$

$$= 1 - e - \frac{2}{\pi};$$

$$\int_{C_2} \mathbf{F(r)} \cdot \mathbf{dr} = \int_0^1 \left[e^{y(t)} x'(t) - \sin\left[\pi x(t)\right] y'(t) \right] dt$$

$$= \int_0^1 \left[-e^{1-t} + \sin\left(-\pi t\right) \right] dt$$

$$= \left[e^{1-t} + \frac{1}{\pi} \cos\left(-\pi t\right) \right]_0^1$$

$$= 1 - e - \frac{2}{\pi};$$

and

$$\int_{C_3} \mathbf{F(r)} \cdot \mathbf{dr} = \int_0^1 \left[e^{y(t)} x'(t) - \sin\left[\pi x(t)\right] y'(t) \right] dt$$

$$= \int_0^1 2\, dt$$

$$= 2.$$

As the integral over the entire triangle, we have

$$\int_C \mathbf{F(r)} \cdot \mathbf{dr} = \left(1 - e - \frac{2}{\pi}\right) + \left(1 - e - \frac{2}{\pi}\right) + 2 = 4 - 2e - \frac{4}{\pi}. \quad \square$$

When we are integrating over a curve, we are integrating in a given direction. If we integrate in the opposite direction, our answer is altered by a factor of -1. To see how this comes about (and incidentally clarify this idea), let C be a piecewise smooth curve and let $-C$ denote the same path traversed in the opposite direction. See Figure 14.16.4. It's not hard to see that if

$$C\colon \mathbf{r}(t), \quad t \in [a, b]$$

then

$$-C\colon \mathbf{R}(u) = \mathbf{r}(a + b - u), \quad u \in [a, b].$$

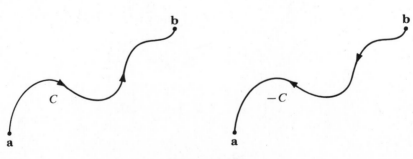

FIGURE 14.16.4

In a more formal treatment this is how $-C$ would be defined. Our assertion is that

$$\int_{-C} \mathbf{F}(\mathbf{R}) \cdot d\mathbf{R} = -\int_{C} \mathbf{F}(\mathbf{r}) \cdot d\mathbf{r}$$

or more briefly that

$$\int_{-C} = -\int_{C} .$$

PROOF.   We prove the result first in the case that C is smooth. Setting

$$t(u) = a + b - u,$$

we have

$$t(b) = a \quad \text{and} \quad t(a) = b.$$

Moreover,

$$\mathbf{R}(u) = \mathbf{r}(t(u)) \quad \text{and} \quad \mathbf{R}'(u) = \mathbf{r}'(t(u))t'(u).$$

The change of variables formula gives

$$\int_{a=t(b)}^{b=t(a)} f(t) \, dt = \int_{b}^{a} f(t(u))t'(u) \, du.$$

Applying this formula to

$$f(t) = \mathbf{F}(\mathbf{r}(t)) \cdot \mathbf{r}'(t),$$

we get

$$\int_{C} \mathbf{F}(\mathbf{r}) \cdot d\mathbf{r} = \int_{a=t(b)}^{b=t(a)} \left[\mathbf{F}(\mathbf{r}(t)) \cdot \mathbf{r}'(t)\right] dt$$

$$= \int_{b}^{a} \left[\mathbf{F}(\mathbf{r}(t(u))) \cdot \mathbf{r}'(t(u))\right]t'(u) \, du$$

$$= \int_{b}^{a} \left[\mathbf{F}(\mathbf{R}(u)) \cdot \mathbf{R}'(u)\right] du$$

$$= -\int_{a}^{b} \left[\mathbf{F}(\mathbf{R}(u)) \cdot \mathbf{R}'(u)\right] du$$

$$= -\int_{-C} \mathbf{F}(\mathbf{R}) \cdot d\mathbf{R}.$$

If C is only piecewise smooth, the result still holds because it holds for each of the smooth pieces. ☐

We will come back to these ideas in the next section.

Exercises

Compute the line integral of $\mathbf{F}$ on the indicated curve.

*1. $\mathbf{F}(x, y) = y\mathbf{i} + x\mathbf{j}$; $\quad \mathbf{r}(t) = t\mathbf{i} + t^2\mathbf{j}$, $\quad t \in [0, 1]$.

2. $\mathbf{F}(x, y) = (x^2 - y)\mathbf{i} + x\mathbf{j}$; $\quad \mathbf{r}(t) = (t - 2)\mathbf{i}$, $\quad t \in [0, 2]$.

*3. $\mathbf{F}(x, y) = y\mathbf{i} + x\mathbf{j}$; $\quad$ the unit circle traversed clockwise.

4. $\mathbf{F}(x, y) = y^2 x\mathbf{i} + 2\mathbf{j}$; $\quad \mathbf{r}(t) = e^t\mathbf{i} + e^{-t}\mathbf{j}$, $\quad t \in [0, 1]$.

*5. $\mathbf{F}(x, y) = (x - y)\mathbf{i} + xy\mathbf{j}$; $\quad$ the line segment from $(2, 3)$ to $(1, 2)$.

6. $\mathbf{F}(x, y) = (x - y)\mathbf{i} + xy\mathbf{j}$; $\quad$ the line segment from $(1, 2)$ to $(2, 3)$.

*7. $\mathbf{F}(x, y) = (\cos x - y \sin x)\mathbf{i} + \cos x\,\mathbf{j}$; $\quad$ the triangle with vertices $(0, 2)$, $(2, 3)$, $(1, 4)$ traversed counterclockwise.

8. $\mathbf{F}(x, y) = e^{x-y}\mathbf{i} + e^{x+y}\mathbf{j}$; $\quad$ the line segment from $(-1, 1)$ to $(1, 2)$.

*9. $\mathbf{F}(x, y) = \dfrac{1}{xy^2}\mathbf{i} + \dfrac{1}{x^2 y}\mathbf{j}$; $\quad \mathbf{r}(t) = \sqrt{t}\,\mathbf{i} + \sqrt{1 + t}\,\mathbf{j}$, $\quad t \in [1, 4]$.

10. $\mathbf{F}(x, y) = \dfrac{1}{xy^2}\mathbf{i} + \dfrac{1}{x^2 y}\mathbf{j}$; $\quad$ the curve of Exercise 9 traversed in the opposite sense.

*11. $\mathbf{F}(x, y) = (x + y)\mathbf{i} + (y^2 - x)\mathbf{j}$; $\quad$ the closed curve which begins at $(-1, 0)$, proceeds along the x-axis to $(1, 0)$, and returns to $(-1, 0)$ by the upper part of the unit circle.

12. $\mathbf{F}(x, y, z) = x\mathbf{i} + xy\mathbf{j} + xyz\mathbf{k}$; $\quad$ the line segment from $(0, 1, 4)$ to $(1, 0, -4)$.

*13. $\mathbf{F}(x, y, z) = yz\mathbf{i} + x^2\mathbf{j} + xz\mathbf{k}$; $\quad \mathbf{r}(t) = t\mathbf{i} + t^2\mathbf{j} + t^3\mathbf{k}$, $\quad t \in [0, 1]$.

14. $\mathbf{F}(x, y, z) = e^{\sqrt{y}}\mathbf{i} + yz\mathbf{j} + e^{xy}\mathbf{k}$; $\quad$ the curve of Exercise 13.

*15. An object orbits an ellipse in standard position subject to a force

$$\mathbf{F}(x, y) = -\tfrac{1}{2}[y\mathbf{i} - x\mathbf{j}].$$

Find a relation between the work done and the area of the ellipse.

16. Show that if C is a smooth curve which begins and ends at the same point, then

$$\int_C \nabla f(\mathbf{r}) \cdot d\mathbf{r} = 0.$$

14.17   Work and Line Integrals, Part II

There are many kinds of energy that a physicist recognizes. Among the most important of these is *kinetic energy*, which is the energy possessed by an object because of its motion. The kinetic energy of an object depends on its mass m and on its speed v:

$$\boxed{\text{K.E.} = \tfrac{1}{2}mv^2.}$$

There is an important relation between work and kinetic energy which we will now derive.

Suppose that an object of mass m moves along a curve

$$C: \mathbf{r}(t), \quad t \in [a, b]$$

subject to a force $\mathbf{F}(\mathbf{r})$. The total work done on the object along the curve is given by the formula

$$W = \int_C \mathbf{F}(\mathbf{r}) \cdot \mathbf{dr} = \int_a^b [\mathbf{F}(\mathbf{r}(t)) \cdot \mathbf{r}'(t)] \, dt.$$

From Newton's second law of motion, we know that at time t,

$$\mathbf{F}(\mathbf{r}(t)) = m\mathbf{a}(t) = m\mathbf{r}''(t).$$

It follows then that

$$\mathbf{F}(\mathbf{r}(t)) \cdot \mathbf{r}'(t) = m\mathbf{r}''(t) \cdot \mathbf{r}'(t)$$

$$= \frac{d}{dt}\left[\frac{1}{2} m(\mathbf{r}'(t) \cdot \mathbf{r}'(t))\right]$$

$$= \frac{d}{dt}\left[\frac{1}{2} m\|\mathbf{r}'(t)\|^2\right]$$

$$= \frac{d}{dt}\left[\frac{1}{2} m[v(t)]^2\right].$$

Substituting this last expression into the work integral, we find that

$$W = \int_a^b \frac{d}{dt}\left(\frac{1}{2} m[v(t)]^2\right) dt = \frac{1}{2} m[v(b)]^2 - \frac{1}{2} m[v(a)]^2.$$

This last equation says that the work done on an object along a curve C is exactly equal to the difference in the kinetic energy of the object as measured at the endpoints of C. In short, we have the following result:

$$\boxed{\text{work done} = \text{change in kinetic energy.}}$$

In general, if an object moves from one point to another, the work done depends on the path of the motion. There is however an important exception. If the force field is a *gradient field*,

$$\mathbf{F} = \nabla f,$$

then the work done depends only on the endpoints of the path and not on the path itself. Returning to pure mathematics for a moment, we have the following theorem.

The Fundamental Theorem for Line Integrals

Let

$$C: \mathbf{r}(t), \quad t \in [a, b]$$

be a piecewise smooth curve which begins at $\mathbf{a}\,[\mathbf{r}(a) = \mathbf{a}]$ and ends at $\mathbf{b}\,[\mathbf{r}(b) = \mathbf{b}]$. If ∇f is continuous (has continuous components) on an open set which contains C, then

$$\int_C \nabla f(\mathbf{r}) \cdot d\mathbf{r} = f(\mathbf{b}) - f(\mathbf{a}). \qquad \text{(Figure 14.17.1)}$$

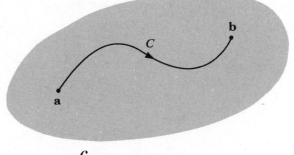

$$\int_C \nabla f(\mathbf{r}) \cdot d\mathbf{r} = f(\mathbf{b}) - f(\mathbf{a})$$

FIGURE 14.17.1

PROOF. If C is smooth,

$$\int_C \nabla f(\mathbf{r}) \cdot d\mathbf{r} = \int_a^b [\nabla f(\mathbf{r}(t)) \cdot \mathbf{r}'(t)]\, dt$$

$$= \int_a^b \frac{d}{dt} [f(\mathbf{r}(t))]\, dt$$

$$= f(\mathbf{r}(b)) - f(\mathbf{r}(a))$$

$$= f(\mathbf{b}) - f(\mathbf{a}).$$

If C is not actually smooth but only piecewise smooth, then we break up C into smooth pieces

$$C = C_1 \cup C_2 \cup \cdots \cup C_n.$$

With obvious notation,

$$\int_C \nabla f(\mathbf{r}) \cdot d\mathbf{r} = \int_{C_1} \nabla f(\mathbf{r}) \cdot d\mathbf{r} + \int_{C_2} \nabla f(\mathbf{r}) \cdot d\mathbf{r} + \cdots + \int_{C_n} \nabla f(\mathbf{r}) \cdot d\mathbf{r}$$

$$= [f(\mathbf{a}_1) - f(\mathbf{a}_o)] + [f(\mathbf{a}_2) - f(\mathbf{a}_1)] + \cdots + [f(\mathbf{a}_n) - f(\mathbf{a}_{n-1})]$$

$$= f(\mathbf{a}_n) - f(\mathbf{a}_o)$$

$$= f(\mathbf{b}) - f(\mathbf{a}). \quad \square$$

If the curve C is closed (if $\mathbf{a} = \mathbf{b}$), then of course

$$f(\mathbf{b}) = f(\mathbf{a})$$

and

$$\int_C \nabla f(\mathbf{r}) \cdot d\mathbf{r} = 0.$$

If a force field is a gradient field, then the work done along any closed path is 0. An object which starts at a point with a given kinetic energy returns to that same point with exactly the same kinetic energy.

We can use these ideas in computations.

Problem. Find the work done if an object traverses the helix

$$C: \mathbf{r}(t) = \cos t\, \mathbf{i} + \sin t\, \mathbf{j} + t\mathbf{k}, \qquad t \in [0, 2\pi]$$

subject to a force

$$\mathbf{F}(\mathbf{r}) = \frac{K}{\|\mathbf{r}\|^2}\, \mathbf{r}.$$

SOLUTION. The first thing we do is try to determine whether $\mathbf{F}$ is a gradient. In terms of x, y, z we have

$$\mathbf{F}(x, y, z) = \frac{K}{x^2 + y^2 + z^2}\, [x\mathbf{i} + y\mathbf{j} + z\mathbf{k}].$$

It is not hard to see that

$$\mathbf{F}(x, y, z) = \nabla f(x, y, z) \qquad \text{where} \quad f(x, y, z) = \tfrac{1}{2}K \log (x^2 + y^2 + z^2).$$

In terms of $\mathbf{r}$, we have

$$f(\mathbf{r}) = \tfrac{1}{2}K \log \|\mathbf{r}\|^2 = K \log \|\mathbf{r}\|.$$

To compute the work integral, we need only evaluate f at the endpoints of C. Since

$$\mathbf{r}(0) = \mathbf{i} \quad \text{and} \quad \mathbf{r}(2\pi) = \mathbf{i} + 2\pi\mathbf{k},$$

we have

$$W = \int_C \mathbf{F}(\mathbf{r}) \cdot d\mathbf{r} = f(\mathbf{i} + 2\pi\mathbf{k}) - f(\mathbf{i})$$

$$= K \log \|\mathbf{i} + 2\pi\mathbf{k}\| - K \log \|\mathbf{i}\|$$

$$= K \log \sqrt{1 + 4\pi^2} - K \log 1$$

$$= \tfrac{1}{2}K \log (1 + 4\pi^2). \quad \square$$

Problem. Evaluate the line integral

$$\int_C \mathbf{F}(\mathbf{r}) \cdot d\mathbf{r}$$

where C is the circular arc

$$\mathbf{r}(t) = \cos t\, \mathbf{i} + \sin t\, \mathbf{j}, \qquad t \in [0, \tfrac{1}{2}\pi]$$

and

$$\mathbf{F}(x, y) = y^2\mathbf{i} + (2xy - e^y)\mathbf{j}.$$

SOLUTION.   Here $\mathbf{F}(x, y)$ has the form

$$P(x, y)\mathbf{i} + Q(x, y)\mathbf{j}$$

with

$$P(x, y) = y^2 \quad \text{and} \quad Q(x, y) = 2xy - e^y.$$

Since

$$\frac{\partial P}{\partial y}(x, y) = 2y = \frac{\partial Q}{\partial x}(x, y),$$

we know from Section 14.15 that $\mathbf{F}(x, y)$ is a gradient. Since the integral then depends only on the endpoints of C and not on C itself, we may as well simplify our computations and integrate over the line segment C' that joins the same endpoints. (Figure 14.17.2) With

$$C': \mathbf{r}(t) = (1 - t)\mathbf{i} + t\mathbf{j}, \qquad t \in [0, 1]$$

we have

$$\int_C \mathbf{F}(\mathbf{r}) \cdot d\mathbf{r} = \int_{C'} \mathbf{F}(\mathbf{r}) \cdot d\mathbf{r}$$

$$= \int_0^1 \left[\mathbf{F}(\mathbf{r}(t)) \cdot \mathbf{r}'(t) \right] dt$$

$$= \int_0^1 \left[[y(t)]^2 x'(t) + [2x(t)y(t) - e^{y(t)}]y'(t) \right] dt$$

$$= \int_0^1 [2t - 3t^2 - e^t] \, dt$$

$$= \left[t^2 - t^3 - e^t \right]_0^1$$

$$= 1 - e.$$

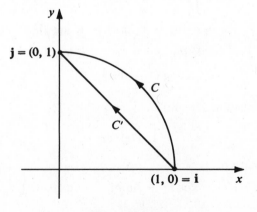

FIGURE 14.17.2

ALTERNATE SOLUTION.   Once we recognize

$$F(x, y) = y^2\mathbf{i} + (2xy - e^y)\mathbf{j}$$

as a gradient $\nabla f(x, y)$, we can apply the methods of Section 14.15 to find $f(x, y)$. With

$$\frac{\partial f}{\partial x}(x, y) = y^2 \quad \text{and} \quad \frac{\partial f}{\partial y}(x, y) = 2xy - e^y,$$

we have

$$f(x, y) = xy^2 + \phi(y) \quad \text{and} \quad f(x, y) = xy^2 - e^y + \psi(x).$$

Equating these two expressions for $f(x, y)$, we have

$$\phi(y) = -e^y + \psi(x)$$

and therefore

$$\phi(y) + e^y = \psi(x).$$

It is obvious then that $\psi(x)$ must be constant and therefore

$$f(x, y) = xy^2 - e^y + c.$$

Since the curve C begins at $(1, 0)$ and ends at $(0, 1)$, we have

$$\int_C \mathbf{F(r)} \cdot \mathbf{dr} = f(0, 1) - f(1, 0) = -e + 1 = 1 - e. \quad \square$$

Problem.   Evaluate the line integral

$$\int_C \mathbf{F(r)} \cdot \mathbf{dr}$$

where C is the unit circle

$$\mathbf{r}(t) = \cos t\,\mathbf{i} + \sin t\,\mathbf{j}, \qquad t \in [0, 2\pi]$$

and

$$\mathbf{F}(x, y, z) = (y^2 + y)\mathbf{i} + (2xy - e^y)\mathbf{j}.$$

SOLUTION.   Although

$$(y^2 + y)\mathbf{i} + (2xy - e^y)\mathbf{j}$$

is not a gradient, part of it

$$y^2\mathbf{i} + (2xy - e^y)\mathbf{j}$$

is a gradient. (See last problem.) Since we are integrating over a closed curve, the contribution of the gradient part is 0. All we need compute therefore is

$$\int_C \mathbf{G(r)} \cdot \mathbf{dr} \quad \text{where} \quad \mathbf{G}(x, y) = y\mathbf{i}.$$

This is easy to do:

$$\int_C \mathbf{G}(\mathbf{r}) \cdot d\mathbf{r} = \int_0^{2\pi} [\mathbf{G}(\mathbf{r}(t)) \cdot \mathbf{r}'(t)] \, dt = \int_0^{2\pi} y(t)x'(t) \, dt$$

$$= \int_0^{2\pi} -\sin^2 t \, dt = -\pi. \quad \Box$$

Exercises

Compute the line integral of $\mathbf{F}$ on the indicated curve.

*1. $\mathbf{F}(x, y) = x\mathbf{i} + y\mathbf{j}$; $\mathbf{r}(t) = a \cos t \, \mathbf{i} + b \sin t \, \mathbf{j}$, $t \in [0, 2\pi]$.

2. $\mathbf{F}(x, y) = (x + y)\mathbf{i} + y\mathbf{j}$; the curve of Exercise 1.

*3. $\mathbf{F}(x, y) = \cos \pi y \, \mathbf{i} - \pi x \sin \pi y \, \mathbf{j}$; $\mathbf{r}(t) = t^2\mathbf{i} - t^3\mathbf{j}$, $t \in [0, 1]$.

4. $\mathbf{F}(x, y) = (x^2 + \cos \pi y)\mathbf{i} + (x - \pi x \sin \pi y)\mathbf{j}$; the curve of Exercise 1.

*5. $\mathbf{F}(x, y) = xy^2\mathbf{i} + x^2 y\mathbf{j}$; $\mathbf{r}(t) = t \sin \pi t \, \mathbf{i} + \cos \pi t^2 \, \mathbf{j}$, $t \in [0, 1]$.

6. $\mathbf{F}(x, y) = (1 + e^y)\mathbf{i} + (xe^y - x)\mathbf{j}$; the square with vertices $(-1, -1)$, $(1, -1)$, $(1, 1)$, $(-1, 1)$ traversed counterclockwise.

*7. $\mathbf{F}(x, y) = (1 + e^y)\mathbf{i} + (xe^y - 4)\mathbf{j}$; $\mathbf{r}(t) = \cos t \, \mathbf{i} + \sin t \, \mathbf{j}$, $t \in [0, \pi]$.

8. (*Optional*) $\mathbf{F}(x, y) = 2xy \sinh x^2 y \, \mathbf{i} + x^2 \sinh x^2 y \, \mathbf{j}$; the curve of Exercise 1.

*9. (*Optional*) $\mathbf{F}(x, y) = [2x^2 y \sinh x^2 + y \cosh x^2]\mathbf{i} + [x(1 + \cosh x^2)]\mathbf{j}$; the square of Exercise 6.

10. Show that a central force field

$$\mathbf{F}(\mathbf{r}) = \frac{K}{\|\mathbf{r}\|^n} \, \mathbf{r}, \quad n \text{ a positive integer}$$

is a gradient field. Find a function f such that

$$\nabla f(\mathbf{r}) = \mathbf{F}(\mathbf{r})$$

*(a) if $n = 2$,

*(b) if $n \neq 2$.

Multiple Integrals

We began with ordinary integrals

$$\int_a^b f(x)\ dx.$$

Then we discussed line integrals

$$\int_C \mathbf{F(r)} \cdot \mathbf{dr}.$$

Now we take up double integrals

$$\iint_\Omega f(x,\ y)\ dxdy$$

and a little later triple integrals

$$\iiint_S f(x,\ y,\ z)\ dxdydz.$$

15.1 The Double Integral Over a Rectangle

We start with a function f continuous on a rectangle

$$R = \{(x, y): a \le x \le b, c \le y \le d\}.$$

See Figure 15.1.1. Our object is to define the double integral

$$\iint_R f(x,\ y)\ dxdy.$$

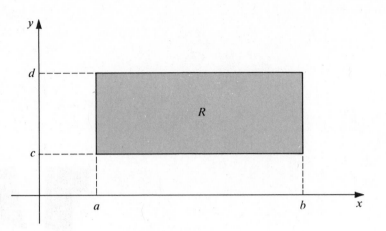

FIGURE 15.1.1

To define

$$\int_a^b f(x)\,dx$$

we first introduced some auxiliary notions: partition P, upper sum $U_f(P)$, and lower sum $L_f(P)$. We were then able to define

$$\int_a^b f(x)\,dx$$

as the unique number I which satisfies the inequality

$$L_f(P) \le I \le U_f(P) \quad \text{for all partitions } P \text{ of } [a, b]. \qquad \text{(see Section 4.2)}$$

We will follow exactly the same procedure to define the double integral

$$\iint_R f(x, y)\,dxdy.$$

First we explain what we mean by a partition of the rectangle R. To do this, we begin with a partition

$$P_1 = \{x_0, x_1, \ldots, x_m\} \quad \text{of} \quad [a, b]$$

and a partition

$$P_2 = \{y_0, y_1, \ldots, y_n\} \quad \text{of} \quad [c, d].$$

A set of the form

$$P = P_1 \times P_2 = \{(x_i, y_j): x_i \in P_1, y_j \in P_2\}$$

is called a *partition of R*. The idea is illustrated in Figure 15.1.2. P consists of all the grid points (x_i, y_j).

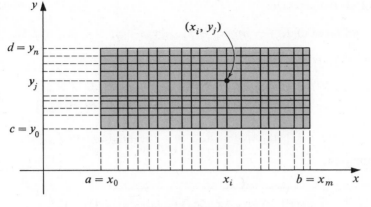

FIGURE 15.1.2

The partition P breaks up R into $m \times n$ nonoverlapping rectangles

$$R_{ij} = \{(x, y): x_{i-1} \leq x \leq x_i, y_{j-1} \leq y \leq y_j\}. \qquad \text{(Figure 15.1.3)}$$

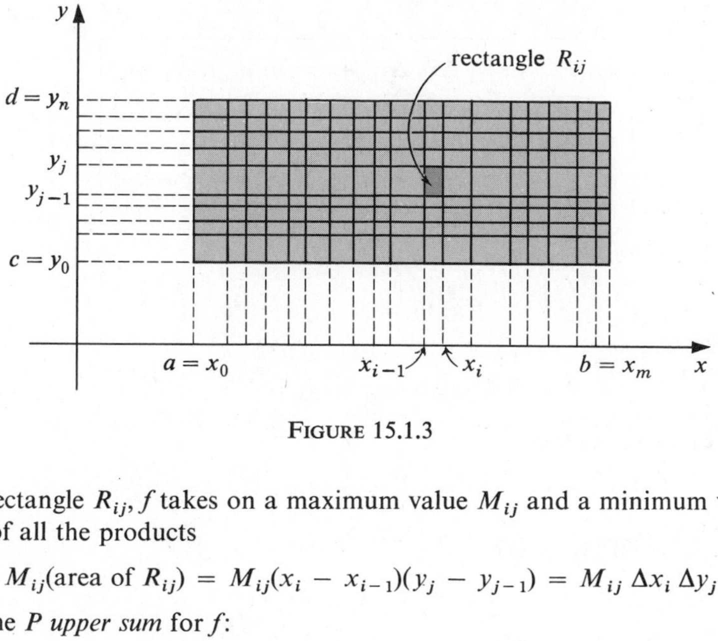

FIGURE 15.1.3

On each rectangle R_{ij}, f takes on a maximum value M_{ij} and a minimum value m_{ij}.† The sum of all the products

$$M_{ij}(\text{area of } R_{ij}) = M_{ij}(x_i - x_{i-1})(y_j - y_{j-1}) = M_{ij}\,\Delta x_i\,\Delta y_j$$

is called the P *upper sum* for f:

$$
\begin{aligned}
U_f(P) = \; & [M_{11}(\text{area of } R_{11}) + \cdots + M_{1n}(\text{area of } R_{1n})] \\
& + [M_{21}(\text{area of } R_{21}) + \cdots + M_{2n}(\text{area of } R_{2n})] \\
& \qquad\qquad\qquad \vdots \\
& + [M_{m1}(\text{area of } R_{m1}) + \cdots + M_{mn}(\text{area of } R_{mn})].
\end{aligned}
$$

---

† It can be shown that every continuous function takes on both a maximum and a minimum on every closed bounded set. Since R_{ij} is closed and bounded, we can be sure that f takes on both a maximum and a minimum there.

The sum of all the products

$$m_{ij}(\text{area of } R_{ij}) = m_{ij}(x_i - x_{i-1})(y_j - y_{j-1}) = m_{ij}\,\Delta x_i\,\Delta y_j$$

is called the *P lower sum* for *f*:

$$
\begin{aligned}
L_f(P) &= [m_{11}(\text{area of } R_{11}) + \cdots + m_{1n}(\text{area of } R_{1n})] \\
&\quad + [m_{21}(\text{area of } R_{21}) + \cdots + m_{2n}(\text{area of } R_{2n})] \\
&\qquad\qquad\qquad \vdots \\
&\quad + [m_{m1}(\text{area of } R_{m1}) + \cdots + m_{mn}(\text{area of } R_{mn})].
\end{aligned}
$$

In the Δ-notation

$$
\boxed{
\begin{aligned}
U_f(P) &= M_{11}\,\Delta x_1\,\Delta y_1 + \cdots + M_{1n}\,\Delta x_1\,\Delta y_n \\
&\quad + M_{21}\,\Delta x_2\,\Delta y_1 + \cdots + M_{2n}\,\Delta x_2\,\Delta y_n \\
&\qquad\qquad\qquad \vdots \\
&\quad + M_{m1}\,\Delta x_m\,\Delta y_1 + \cdots + M_{mn}\,\Delta x_m\,\Delta y_n
\end{aligned}
}
$$

and

$$
\boxed{
\begin{aligned}
L_f(P) &= m_{11}\,\Delta x_1\,\Delta y_1 + \cdots + m_{1n}\,\Delta x_1\,\Delta y_n \\
&\quad + m_{21}\,\Delta x_2\,\Delta y_1 + \cdots + m_{2n}\,\Delta x_2\,\Delta y_n \\
&\qquad\qquad\qquad \vdots \\
&\quad + m_{m1}\,\Delta x_m\,\Delta y_1 + \cdots + m_{mn}\,\Delta x_m\,\Delta y_n.
\end{aligned}
}
$$

Example.   Take the function

$$f(x, y) = x + y - 2$$

on the rectangle

$$R = \{(x, y): 1 \le x \le 4,\ 1 \le y \le 3\}.$$

As a partition of $[1, 4]$ take

$$P_1 = \{1, 2, 3, 4\}$$

and as a partition of $[1, 3]$ take

$$P_2 = \{1, \tfrac{3}{2}, 3\}.$$

The partition

$$P = P_1 \times P_2$$

breaks up the initial rectangle into the six rectangles marked in Figure 15.1.4. On each rectangle R_{ij}, f takes on its maximum value M_{ij} at the corner furthest away from the origin, (x_i, y_j); namely,

$$M_{ij} = f(x_i, y_j) = x_i + y_j - 2.$$

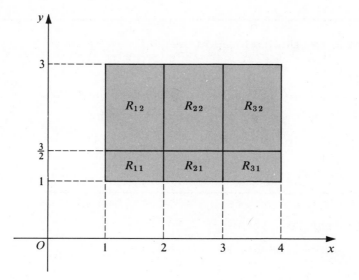

FIGURE 15.1.4

As the upper sum we have

$$U_f(P) = M_{11}(\text{area of } R_{11}) + M_{12}(\text{area of } R_{12}) + M_{21}(\text{area of } R_{21}) + M_{22}(\text{area of } R_{22})$$
$$+ M_{31}(\text{area of } R_{31}) + M_{32}(\text{area of } R_{32})$$
$$= \tfrac{3}{2}(\tfrac{1}{2}) + 3(\tfrac{3}{2}) + \tfrac{5}{2}(\tfrac{1}{2}) + 4(\tfrac{3}{2}) + \tfrac{7}{2}(\tfrac{1}{2}) + 5(\tfrac{3}{2}) = \tfrac{87}{4}.$$

On each rectangle R_{ij}, f takes on its minimum value m_{ij} at the corner closest to the origin, (x_{i-1}, y_{j-1}); namely,

$$m_{ij} = f(x_{i-1}, y_{j-1}) = x_{i-1} + y_{j-1} - 2.$$

The lower sum is given by

$$L_f(P) = m_{11}(\text{area of } R_{11}) + m_{12}(\text{area of } R_{12}) + m_{21}(\text{area of } R_{21}) + m_{22}(\text{area of } R_{22})$$
$$+ m_{31}(\text{area of } R_{31}) + m_{32}(\text{area of } R_{32})$$
$$= 0(\tfrac{1}{2}) + \tfrac{1}{2}(\tfrac{3}{2}) + 1(\tfrac{1}{2}) + \tfrac{3}{2}(\tfrac{3}{2}) + 2(\tfrac{1}{2}) + \tfrac{5}{2}(\tfrac{3}{2}) = \tfrac{33}{4}. \quad \square$$

We return now to the general situation. As in the one-variable case, it can be shown that, if f is continuous, then there exists one and only one number I which satisfies the inequality

$$L_f(P) \le I \le U_f(P) \qquad \text{for all partitions } P \text{ of } R.$$

Definition of the Double Integral over R

The unique number I which satisfies the inequality

$$L_f(P) \leq I \leq U_f(P) \quad \text{for all partitions } P \text{ of } R$$

is called the *double integral* of f over R and is denoted by

$$\iint_R f(x, y)\, dx dy.$$

The Double Integral as a Volume

If f is nonnegative on the rectangle R, the equation

$$z = f(x, y)$$

represents a surface that lies over R. The double integral

$$\iint_R f(x, y)\, dx dy$$

then gives the volume of the solid S which is bounded below by R and above by the surface $z = f(x, y)$. See Figure 15.1.5.

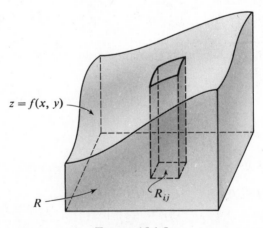

FIGURE 15.1.5

To see this, consider a partition P of R. P breaks up R into subrectangles R_{ij} and thus the entire solid S into parts S_{ij}. Since S_{ij} contains a rectangular solid with base R_{ij} and height m_{ij} (Figure 15.1.6), we must have

$$m_{ij}(\text{area of } R_{ij}) = \text{volume of } S_{ij}.$$

Since S_{ij} is contained in a rectangular solid with base R_{ij} and height M_{ij} (Figure 15.1.7), we must have

$$\text{volume of } S_{ij} \leq M_{ij}(\text{area of } R_{ij}).$$

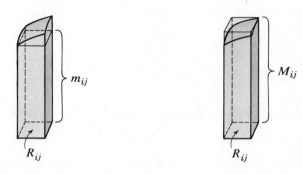

FIGURE 15.1.6   FIGURE 15.1.7

In short, for each pair of indices i and j, we must have

$$m_{ij}(\text{area of } R_{ij}) \leq \text{volume of } S_{ij} \leq M_{ij}(\text{area of } R_{ij}).$$

Adding up these inequalities, we are forced to conclude that

$$L_f(P) \leq \text{volume of } S \leq U_f(P).$$

Since P is arbitrary, the volume of S must be the double integral:

$$\boxed{\text{volume of } S = \iint_R f(x, y)\, dxdy.}$$

The integral

$$\iint_R 1\, dxdy = \iint_R dxdy$$

gives the volume of a solid of constant height 1 erected over R. This is simply the area of R:

$$\boxed{\text{area of } R = \iint_R dxdy.}$$

Some Computations

Double integrals are generally computed by techniques that we will take up later. It is, however, possible to evaluate simple double integrals directly from the definition.

Below we give some examples. In carrying out the calculations we will be using double sums,

$$\sum_{i=1}^{m} \sum_{j=1}^{n} a_{ij}.$$

Since you are already familiar with the sigma notation for ordinary sums,

$$\sum_{j=1}^{m} b_j = b_1 + b_2 + \cdots + b_m, \qquad \text{(Chapter 10)}$$

you will probably find the double sigma notation easy to understand:

$$\sum_{i=1}^{m} \sum_{j=1}^{n} a_{ij} = \sum_{i=1}^{m} \left(\sum_{j=1}^{n} a_{ij} \right) = \sum_{j=1}^{n} a_{1j} + \sum_{j=1}^{n} a_{2j} + \cdots + \sum_{j=1}^{n} a_{mj}$$

$$= (a_{11} + \cdots + a_{1n}) + (a_{21} + \cdots + a_{2n}) + \cdots + (a_{m1} + \cdots + a_{mn}).$$

Since constants α can be factored through single sums, they can also be factored through double sums:

$$\sum_{i=1}^{m} \sum_{j=1}^{n} \alpha a_{ij} = \sum_{i=1}^{m} \alpha \sum_{j=1}^{n} a_{ij} = \alpha \sum_{i=1}^{m} \sum_{j=1}^{n} a_{ij}.$$

Obviously also

$$\sum_{i=1}^{m} \sum_{j=1}^{n} (a_{ij} + b_{ij}) = \sum_{i=1}^{m} \sum_{j=1}^{n} a_{ij} + \sum_{i=1}^{m} \sum_{j=1}^{n} b_{ij}.$$

Problem.   Evaluate

$$\iint_R \alpha \, dx dy$$

where

$$R = \{(x, y): a \le x \le b, c \le y \le d\}.$$

SOLUTION.   Here

$$f(x, y) = \alpha \qquad \text{for all } (x, y) \in R.$$

We begin with

$$P_1 = \{x_0, x_1, \ldots, x_m\}$$

as an arbitrary partition of $[a, b]$ and

$$P_2 = \{y_0, y_1, \ldots, y_n\}$$

as an arbitrary partition of $[c, d]$. This gives

$$P = P_1 \times P_2 = \{(x_i, y_j): x_i \in P_1, y_j \in P_2\}$$

as an arbitrary partition of R. On each rectangle R_{ij}, f has constant value α. This gives $M_{ij} = \alpha$ and $m_{ij} = \alpha$ throughout. Thus

$$U_f(P) = \sum_{i=1}^{m} \sum_{j=1}^{n} \alpha(\text{area of } R_{ij}) = \alpha \sum_{i=1}^{m} \sum_{j=1}^{n} \text{area of } R_{ij}$$

$$= \alpha(\text{area of } R) = \alpha(b - a)(d - c)$$

and similarly

$$L_f(P) = \alpha(b - a)(d - c).$$

Since

$$L_f(P) \leq \alpha(b - a)(d - c) \leq U_f(P),$$

and P was chosen arbitrarily, the inequality must hold for all partitions P of R. This means that

$$\iint_R f(x, y) \, dxdy = \alpha(b - a)(d - c). \quad \square$$

REMARK. If $\alpha > 0$,

$$\iint_R \alpha \, dxdy = \alpha(b - a)(d - c)$$

gives the volume of the rectangular solid of constant height α erected over the rectangle R.

Problem. Evaluate

$$\iint_R (x + y) \, dxdy$$

where again R is the rectangle

$$\{(x, y): a \leq x \leq b, c \leq y \leq d\}.$$

SOLUTION. We take

$$P_1 = \{x_0, x_1, \ldots, x_m\}$$

as an arbitrary partition of $[a, b]$ and

$$P_2 = \{y_0, y_1, \ldots, y_n\}$$

as an arbitrary partition of $[c, d]$. This gives

$$P = P_1 \times P_2 = \{(x_i, y_j): x_i \in P_1, y_j \in P_2\}$$

as an arbitrary partition of R. On each rectangle

$$R_{ij} = \{(x, y): x_{i-1} \leq x \leq x_i, y_{j-1} \leq y \leq y_j\}$$

the function

$$f(x, y) = x + y$$

has a maximum

$$M_{ij} = x_i + y_j$$

and a minimum

$$m_{ij} = x_{i-1} + y_{j-1}.$$

Thus

$$U_f(P) = \sum_{i=1}^{m} \sum_{j=1}^{n} M_{ij}(\text{area of } R_{ij}) = \sum_{i=1}^{m} \sum_{j=1}^{n} (x_i + y_j)(x_i - x_{i-1})(y_j - y_{j-1})$$

and

$$L_f(P) = \sum_{i=1}^{m} \sum_{j=1}^{n} m_{ij}(\text{area of } R_{ij}) = \sum_{i=1}^{m} \sum_{j=1}^{n} (x_{i-1} + y_{j-1})(x_i - x_{i-1})(y_j - y_{j-1}).$$

For each pair of indices i and j,

$$x_{i-1} + y_{j-1} \le \tfrac{1}{2}(x_i + x_{i-1}) + \tfrac{1}{2}(y_j + y_{j-1}) \le x_i + y_j. \qquad \text{(explain)}$$

Consequently,

$$L_f(P) \le \sum_{i=1}^{m} \sum_{j=1}^{n} [\tfrac{1}{2}(x_i + x_{i-1}) + \tfrac{1}{2}(y_j + y_{j-1})](x_i - x_{i-1})(y_j - y_{j-1}) \le U_f(P).$$

The middle term can be written as

$$\sum_{i=1}^{m} \sum_{j=1}^{n} \tfrac{1}{2}(x_i + x_{i-1})(x_i - x_{i-1})(y_j - y_{j-1})$$

$$+ \sum_{i=1}^{m} \sum_{j=1}^{n} \tfrac{1}{2}(y_j + y_{j-1})(x_i - x_{i-1})(y_j - y_{j-1}).$$

The first double sum reduces to

$$\sum_{i=1}^{m} \tfrac{1}{2}(x_i^2 - x_{i-1}^2)\left(\sum_{j=1}^{n} (y_j - y_{j-1})\right) = \tfrac{1}{2} \sum_{i=1}^{m} (x_i^2 - x_{i-1}^2)(y_n - y_0)$$

$$= \tfrac{1}{2} \sum_{i=1}^{m} (x_i^2 - x_{i-1}^2)(d - c)$$

$$= \tfrac{1}{2}(d - c) \sum_{i=1}^{m} (x_i^2 - x_{i-1}^2)$$

$$= \tfrac{1}{2}(d - c)(x_m^2 - x_0^2)$$

$$= \tfrac{1}{2}(d - c)(b^2 - a^2).$$

In a similar manner, the second double sum reduces to

$$\tfrac{1}{2}(b - a)(d^2 - c^2).$$

The sum of these two numbers,

$$I = \tfrac{1}{2}(d - c)(b^2 - a^2) + \tfrac{1}{2}(b - a)(d^2 - c^2) = \tfrac{1}{2}(a + b + c + d)(b - a)(d - c),$$

satisfies the inequality

$$L_f(P) \le I \le U_f(P).$$

Since P is arbitrary, we have

$$I = \tfrac{1}{2}(a + b + c + d)(b - a)(d - c)$$

as the desired integral:

$$\iint_R (x + y)\, dx\, dy = \tfrac{1}{2}(a + b + c + d)(b - a)(d - c). \quad \square$$

REMARK.   If $x + y$ is positive on the rectangle

$$R = \{(x, y): a \le x \le b, c \le y \le d\},$$

then the double integral

$$\iint\limits_R (x + y)\, dxdy = \tfrac{1}{2}(a + b + c + d)(b - a)(d - c)$$

gives the volume of the prism bounded above by the plane $z = x + y$ and below by R.

Problem.   Evaluate

$$\iint\limits_R (y - 2x)\, dxdy$$

where

$$R = \{(x, y): 1 \le x \le 2, 3 \le y \le 5\}.$$

SOLUTION.   With

$$P_1 = \{x_0, x_1, \ldots, x_m\} \text{ as an arbitrary partition of } [1, 2]$$

and

$$P_2 = \{y_0, y_1, \ldots, y_n\} \text{ as an arbitrary partition of } [3, 5]$$

we have

$$P = P_1 \times P_2 = \{(x, y): x \in P_1, y \in P_2\}$$

as an arbitrary partition of R. On each rectangle

$$R_{ij} = \{(x, y): x_{i-1} \le x \le x_i, y_{j-1} \le y \le y_j\},$$

the function

$$f(x, y) = y - 2x$$

has a maximum

$$M_{ij} = y_j - 2x_{i-1} \qquad (y \text{ is maximized and } x \text{ is minimized})$$

and a minimum

$$m_{ij} = y_{j-1} - 2x_i. \qquad (y \text{ is minimized and } x \text{ is maximized})$$

Thus

$$U_f(P) = \sum_{i=1}^{m} \sum_{j=1}^{n} M_{ij}(\text{area of } R_{ij}) = \sum_{i=1}^{m} \sum_{j=1}^{n} (y_j - 2x_{i-1})(x_i - x_{i-1})(y_j - y_{j-1})$$

and

$$L_f(P) = \sum_{i=1}^{m} \sum_{j=1}^{n} m_{ij}(\text{area of } R_{ij}) = \sum_{i=1}^{m} \sum_{j=1}^{n} (y_{j-1} - 2x_i)(x_i - x_{i-1})(y_j - y_{j-1}).$$

For each pair of indices i and j,

$$y_{j-1} - 2x_i \le \tfrac{1}{2}(y_j + y_{j-1}) - (x_i + x_{i-1}) \le y_j - 2x_{i-1}.$$

This means that for arbitrary P we have

$$L_f(P) \le \sum_{i=1}^{m} \sum_{j=1}^{n} [\tfrac{1}{2}(y_j + y_{j-1}) - (x_i + x_{i-1})](x_i - x_{i-1})(y_j - y_{j-1}) \le U_f(P).$$

The middle term can be written as

$$\sum_{i=1}^{m} \sum_{j=1}^{n} \tfrac{1}{2}(y_j + y_{j-1})(x_i - x_{i-1})(y_j - y_{j-1})$$
$$+ \sum_{i=1}^{m} \sum_{j=1}^{n} - (x_i + x_{i-1})(x_i - x_{i-1})(y_j - y_{j-1}).$$

The first double sum reduces to

$$\tfrac{1}{2}(x_m - x_0)(y_n^2 - y_0^2).$$

The second double sum reduces to

$$-(x_m^2 - x_0^2)(y_n - y_0).$$

With

$$x_0 = 1, \qquad x_m = 2, \qquad y_0 = 3, \qquad y_n = 5$$

we have

$$\tfrac{1}{2}(x_m - x_0)(y_n^2 - y_0^2) - (x_m^2 - x_0^2)(y_n - y_0) = 8 - 3(2) = 2.$$

Since

$$L_f(P) \le 2 \le U_f(P) \quad \text{for arbitrary } P,$$

the integral is 2:

$$\iint\limits_{R} (y - 2x)\, dx dy = 2. \quad \square$$

REMARK. This last integral should not be interpreted as a volume. The expression $y - 2x$ does not keep a constant sign on

$$R = \{(x, y): 1 \le x \le 2, 3 \le y \le 5\}. \quad \square$$

Exercises

For Exercises 1–9 take

$$f(x, y) = x + 2y \quad \text{on} \quad R = \{(x, y): 0 \le x \le 2, 0 \le y \le 1\}$$

and P as the partition $P = P_1 \times P_2$.

*1. Find $L_f(P)$ if $P_1 = \{0, 1, 2\}$ and $P_2 = \{0, 1\}$.

*2. Find $U_f(P)$ with P_1 and P_2 as in Exercise 1.

3. Find $L_f(P)$ if $P_1 = \{0, 1, 2\}$ and $P_2 = \{0, \tfrac{1}{2}, 1\}$.

4. Find $U_f(P)$ with P_1 and P_2 as in Exercise 3.

*5. Find $L_f(P)$ if $P_1 = \{0, 1, \tfrac{3}{2}, 2\}$ and $P_2 = \{0, \tfrac{1}{2}, 1\}$.

*6. Find $U_f(P)$ with P_1 and P_2 as in Exercise 5.

7. Find $L_f(P)$ if $P_1 = \{x_0, x_1, \ldots, x_m\}$ is an arbitrary partition of $[0, 2]$ and $P_2 = \{y_0, y_1, \ldots, y_n\}$ is an arbitrary partition of $[0, 1]$.

8. Find $U_f(P)$ with P_1 and P_2 as in Exercise 7.

*9. (*Optional*) Using Exercises 7 and 8, evaluate the double integral

$$\iint_R (x + 2y)\,dxdy$$

and give a geometric interpretation to your answer.

For Exercises 10–14 take

$$f(x, y) = x - y \quad \text{on } R = \{(x, y): 0 \le x \le 1, 0 \le y \le 1\}$$

and P as the partition $P = P_1 \times P_2$.

*10. Find $L_f(P)$ if $P_1 = \{0, \frac{1}{2}, \frac{3}{4}, 1\}$ and $P_2 = \{0, \frac{1}{2}, 1\}$.

*11. Find $U_f(P)$ with P_1 and P_2 as in Exercise 10.

12. Find $L_f(P)$ if $P_1 = \{x_0, x_1, \ldots, x_m\}$ and $P_2 = \{y_0, y_1, \ldots, y_n\}$ are both arbitrary partitions of $[0, 1]$.

13. Find $U_f(P)$ with P_1 and P_2 as in Exercise 12.

*14. (*Optional*) Using Exercises 12 and 13, evaluate the double integral

$$\iint_R (x - y)\,dxdy.$$

*15. (*Optional*) Evaluate the double integral

$$\iint_R (x^2 + y)\,dxdy \quad \text{with } R = \{(x, y): 0 \le x \le 1, 0 \le y \le 2\}$$

by any method at your disposal. HINT: Think in terms of volume. You can use the method of this section and/or the method of parallel cross sections.

15.2 The Double Integral Over More General Regions

We start with a region Ω that we depict in Figure 15.2.1. We suppose that f is continuous on Ω. Our object is to define the double integral

$$\iint_\Omega f(x, y)\,dxdy.$$

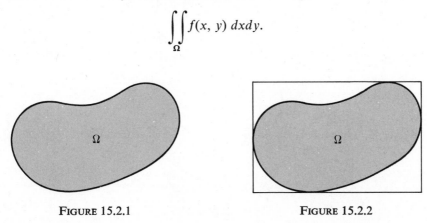

FIGURE 15.2.1   FIGURE 15.2.2

To do this, we surround Ω by a rectangle R in the manner of Figure 15.2.2. We now extend f to all of R by setting f equal to 0 outside of Ω. This extended function f is now continuous on all of R, except possibly at the boundary of Ω. If this boundary is not too wild (the details here are outside the scope of this course), the double integral

$$\iint_R f(x, y)\, dx dy$$

will still exist. We can then define the double integral over Ω by setting

$$\boxed{\iint_\Omega f(x, y)\, dx dy = \iint_R f(x, y)\, dx dy.}$$

If f is nonnegative over Ω, the extended f is nonnegative on all of R. The double integral gives the volume of the solid trapped between the surface $z = f(x, y)$ and the rectangle R. But since the surface has height 0 outside of Ω, the volume outside of Ω is 0. It follows therefore that

$$\iint_\Omega f(x, y)\, dx dy$$

gives the volume of the solid S bounded above by $z = f(x, y)$ and below by Ω.

$$\boxed{\text{volume of } S = \iint_\Omega f(x, y)\, dx dy.}$$

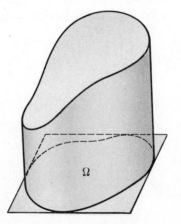

Ω

FIGURE 15.2.3

The double integral

$$\iint_{\Omega} 1 \, dxdy = \iint_{\Omega} dxdy$$

gives the volume of a solid of constant height 1 over Ω. This is of course just the area of Ω:

$$\boxed{\text{area of } \Omega = \iint_{\Omega} dxdy.}$$

Some Properties of the Double Integral

Like the ordinary integral, the double integral is linear:

$$\iint_{\Omega} [\alpha f(x, y) + \beta g(x, y)] \, dxdy = \alpha \iint_{\Omega} f(x, y) \, dxdy + \beta \iint_{\Omega} g(x, y) \, dxdy.$$

It preserves order:

$$\text{if} \quad f(x, y) \geq 0 \quad \text{for all } (x, y) \in \Omega, \qquad \text{then} \quad \iint_{\Omega} f(x, y) \, dxdy \geq 0;$$

$$\text{if} \quad f(x, y) \leq g(x, y) \quad \text{for all } (x, y) \in \Omega, \qquad \text{then}$$

$$\iint_{\Omega} f(x, y) \, dxdy \leq \iint_{\Omega} g(x, y) \, dxdy.$$

It is additive:

if Ω is broken up into two nonoverlapping sets Ω_1, Ω_2 over which f is still integrable, then

$$\iint_{\Omega_1} f(x, y) \, dxdy + \iint_{\Omega_2} f(x, y) \, dxdy = \iint_{\Omega} f(x, y) \, dxdy.$$

15.3   The Evaluation of Double Integrals by Repeated Integrals

The Reduction Formulas

We now discuss the problem of evaluating the double integral

$$\iint_{\Omega} f(x, y) \, dxdy.$$

For simplicity we restrict our attention to sets Ω of the form depicted in Figure 15.3.1.

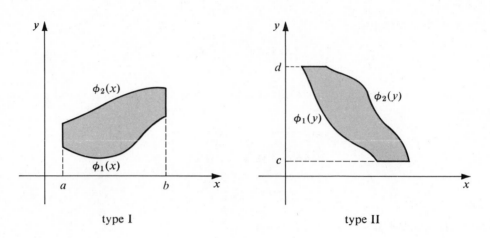

FIGURE 15.3.1

In both cases we assume that the bounding functions ϕ_1 and ϕ_2 are continuous. The fundamental idea is that double integrals over sets of this form can be reduced to "repeated" ordinary integrals: If Ω is of type I,

$$\Omega = \{(x, y): a \leq x \leq b, \quad \phi_1(x) \leq y \leq \phi_2(x)\},$$

then

$$\iint_{\Omega} f(x, y) \, dxdy = \int_a^b \left(\int_{\phi_1(x)}^{\phi_2(x)} f(x, y) \, dy \right) dx.$$

If Ω is of type II,

$$\Omega = \{(x, y): c \leq y \leq d, \quad \phi_1(y) \leq x \leq \phi_2(y)\},$$

then

$$\iint_{\Omega} f(x, y) \, dxdy = \int_c^d \left(\int_{\phi_1(y)}^{\phi_2(y)} f(x, y) \, dx \right) dy.$$

These formulas, although difficult to prove, are easy to understand geometrically.

The Reduction Formulas Viewed Geometrically

Let's take f as nonnegative and Ω of type I. The double integral over Ω gives the volume of the solid S which is trapped between Ω and the surface $z = f(x, y)$:

(1) $$\iint_{\Omega} f(x, y) \, dxdy = \text{volume of } S. \qquad \text{(Figure 15.3.2)}$$

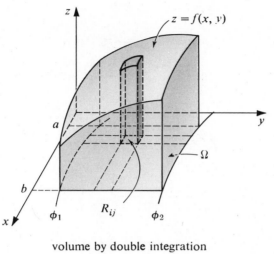

volume by double integration
$$V = \iint\limits_{\Omega} f(x, y)\, dxdy$$

FIGURE 15.3.2

We can also compute the volume of S by the method of parallel cross sections. Letting $A(x)$ be the area of that cross section of S which has first coordinate x, we can argue that

$$\int_a^b A(x)\, dx$$

also gives the volume of S. Recognizing that

$$A(x) = \int_{\phi_1(x)}^{\phi_2(x)} f(x, y)\, dy,$$

we are led to setting

(2) $$\int_a^b \left(\int_{\phi_1(x)}^{\phi_2(x)} f(x, y)\, dy \right) dx = \text{volume of } S. \qquad \text{(Figure 15.3.3)}$$

Combining (1) with (2), we have the first reduction formula

$$\iint\limits_{\Omega} f(x, y)\, dxdy = \int_a^b \left(\int_{\phi_1(x)}^{\phi_2(x)} f(x, y)\, dy \right) dx.$$

The other reduction formula can be obtained in a similar manner.

Our argument is of course a very loose one. How do we know for example that the "volume" obtained by double integration is the same as the "volume" obtained by the method of parallel cross sections? Intuitively it seems evident, but mathematically it is not clear. Certainly we have not proved it. We shall not pursue the matter further.

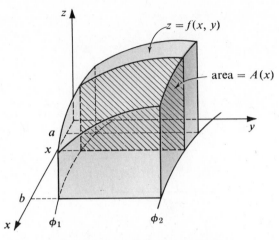

volume by parallel cross sections

$$V = \int_a^b A(x)\,dx = \int_a^b \left(\int_{\phi_1(x)}^{\phi_2(x)} f(x, y)\,dy \right) dx$$

FIGURE 15.3.3

Computations

Problem.   Evaluate

$$\iint_\Omega (x^2 - y)\,dxdy$$

with Ω as in Figure 15.3.4.

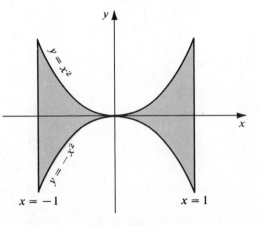

FIGURE 15.3.4

SOLUTION.   For each x from -1 to 1, y lies between $-x^2$ and x^2:

$$\Omega = \{(x, y): -1 \le x \le 1,\ -x^2 \le y \le x^2\}.$$

We integrate last between the constant limits:

$$\iint_\Omega (x^2 - y)\, dxdy = \int_{-1}^{1} \left(\int_{-x^2}^{x^2} (x^2 - y)\, dy \right) dx$$

$$= \int_{-1}^{1} \left[x^2 y - \tfrac{1}{2} y^2 \right]_{-x^2}^{x^2} dx$$

$$= \int_{-1}^{1} \left[(x^4 - \tfrac{1}{2} x^4) - (-x^4 - \tfrac{1}{2} x^4) \right] dx$$

$$= \int_{-1}^{1} 2x^4\, dx = \left[\tfrac{2}{5} x^5 \right]_{-1}^{1} = \tfrac{4}{5}. \quad \square$$

Problem.   Evaluate

$$\iint_\Omega (xy - y^3)\, dxdy$$

with Ω as in Figure 15.3.5.

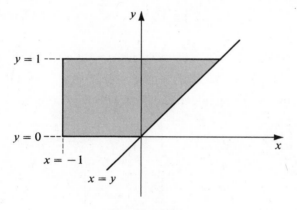

FIGURE 15.3.5

SOLUTION.   For each y from 0 to 1, x lies between -1 and y:

$$\Omega: \{(x, y): 0 \le y \le 1,\ -1 \le x \le y\}.$$

Again we integrate last between the constant limits:

$$\iint_\Omega (xy - y^3)\, dxdy = \int_{0}^{1} \left(\int_{-1}^{y} (xy - y^3)\, dx \right) dy$$

$$= \int_{0}^{1} \left[\tfrac{1}{2} x^2 y - x y^3 \right]_{-1}^{y} dy$$

$$= \int_{0}^{1} \left(-\tfrac{1}{2} y - \tfrac{1}{2} y^3 - y^4 \right) dy$$

$$= \left[-\tfrac{1}{4} y^2 - \tfrac{1}{8} y^4 - \tfrac{1}{5} y^5 \right]_{0}^{1} = -\tfrac{23}{40}. \quad \square$$

Problem. Evaluate

$$\iint_R (y - 2x) \, dx dy$$

where R is the rectangle

$$\{(x, y): 1 \leq x \leq 2, 3 \leq y \leq 5\}.$$

SOLUTION. Here both x and y range over constant limits and we can integrate in either order. We can integrate first with respect to y:

$$\iint_R (y - 2x) \, dx dy = \int_1^2 \left(\int_3^5 (y - 2x) \, dy \right) dx$$

$$= \int_1^2 \left[\tfrac{1}{2} y^2 - 2xy \right]_3^5 dx$$

$$= \int_1^2 \left[(\tfrac{25}{2} - 10x) - (\tfrac{9}{2} - 6x) \right] dx$$

$$= \int_1^2 (8 - 4x) \, dx = \left[8x - 2x^2 \right]_1^2 = 2.$$

We can get the same result by integrating first with respect to x:

$$\iint_R (y - 2x) \, dx dy = \int_3^5 \left(\int_1^2 (y - 2x) \, dx \right) dy$$

$$= \int_3^5 \left[yx - x^2 \right]_1^2 dy$$

$$= \int_3^5 \left[(2y - 4) - (y - 1) \right] dy$$

$$= \int_3^5 (y - 3) \, dy$$

$$= \left[\tfrac{1}{2} y^2 - 3y \right]_3^5 = 2. \quad \square$$

Problem. Evaluate

$$\iint_\Omega (x^{1/2} - y^2) \, dx dy$$

with Ω as in Figure 15.3.6.

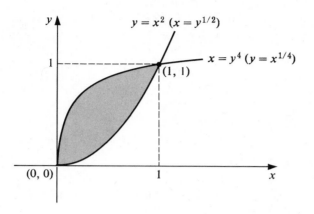

FIGURE 15.3.6

SOLUTION.   Here we also have a choice in the order of integration. We can take the point of view that for each x from 0 to 1, y lies between x^2 and $x^{1/4}$. This gives

$$\iint_{\Omega} (x^{1/2} - y^2)\, dxdy = \int_0^1 \left(\int_{x^2}^{x^{1/4}} (x^{1/2} - y^2)\, dy \right) dx$$

$$= \int_0^1 \left[x^{1/2}y - \tfrac{1}{3}y^3 \right]_{x^2}^{x^{1/4}} dx$$

$$= \int_0^1 (\tfrac{2}{3}x^{3/4} - x^{5/2} + \tfrac{1}{3}x^6)\, dx$$

$$= \left[\tfrac{8}{21}x^{7/4} - \tfrac{2}{7}x^{7/2} + \tfrac{1}{21}x^7 \right]_0^1$$

$$= \tfrac{8}{21} - \tfrac{2}{7} + \tfrac{1}{21} = \tfrac{1}{7}.$$

On the other hand, we can take the point of view that for each y from 0 to 1, x lies between y^4 and $y^{1/2}$. This gives

$$\iint_{\Omega} (x^{1/2} - y^2)\, dxdy = \int_0^1 \left(\int_{y^4}^{y^{1/2}} (x^{1/2} - y^2)\, dx \right) dy$$

$$= \int_0^1 \left[\tfrac{2}{3}x^{3/2} - y^2x \right]_{y^4}^{y^{1/2}} dy$$

$$= \int_0^1 (\tfrac{2}{3}y^{3/4} - y^{5/2} + \tfrac{1}{3}y^6)\, dy$$

$$= \left[\tfrac{8}{21}y^{7/4} - \tfrac{2}{7}y^{7/2} + \tfrac{1}{21}y^7 \right]_0^1$$

$$= \tfrac{8}{21} - \tfrac{2}{7} + \tfrac{1}{21} = \tfrac{1}{7}. \quad \square$$

Problem.   Calculate by double integration the area of the region Ω which lies between $\sqrt{x} + \sqrt{y} = \sqrt{a}$ and $x + y = a$.

SOLUTION.   The region Ω is pictured in Figure 15.3.7. Its area is given by the double integral

$$\iint_{\Omega} dx\,dy.$$

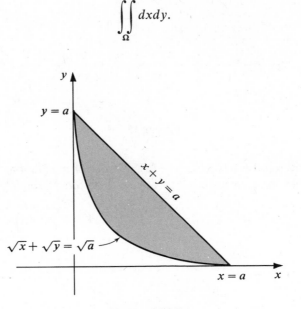

FIGURE 15.3.7

Here again we can integrate in either order. We can write the boundaries as functions of x,

$$y = (\sqrt{a} - \sqrt{x})^2 \quad \text{and} \quad y = a - x,$$

and note that for each x from 0 to a, y lies between $(\sqrt{a} - \sqrt{x})^2$ and $a - x$. This gives

$$\iint_{\Omega} dx\,dy = \int_0^a \left(\int_{(\sqrt{a}-\sqrt{x})^2}^{a-x} dy \right) dx$$

$$= \int_0^a [(a - x) - (\sqrt{a} - \sqrt{x})^2]\, dx$$

$$= \int_0^a (-2x + 2\sqrt{a}\,\sqrt{x})\, dx$$

$$= \left[-x^2 + \tfrac{4}{3}\sqrt{a}\,x\sqrt{x} \right]_0^a$$

$$= -a^2 + \tfrac{4}{3}a^2 = \tfrac{1}{3}a^2.$$

We can also take the point of view that for each y from 0 to a, x is between $(\sqrt{a} - \sqrt{y})^2$ and $a - y$. This gives

$$\iint_\Omega dx dy = \int_0^a \left(\int_{(\sqrt{a}-\sqrt{y})^2}^{a-y} dx \right) dy,$$

which is also $\frac{1}{3}a^2$.   □

Problem. Calculate the volume of that part of the cylinder $x^2 + y^2 = b^2$ which lies between the planes $y + z = a^2$ and $z = 0$.

SOLUTION.   See Figure 15.3.8. The solid in question is bounded below by the disk

$$\Omega = \{(x, y): x^2 + y^2 \le b^2\}$$

and above by the plane

$$z = a^2 - y.$$

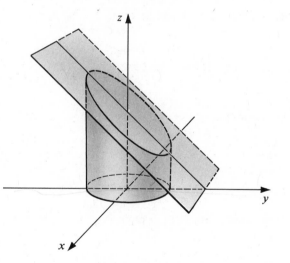

FIGURE 15.3.8

The volume is given by the double integral

$$\iint_\Omega (a^2 - y) \, dx dy.$$

Once again we have a choice in the order of integration. Taking the point of view that

$$\Omega = \{(x, y): -b \le x \le b, \ -\sqrt{b^2 - x^2} \le y \le \sqrt{b^2 - x^2}\},$$

we have

$$V = \iint_{\Omega} (a^2 - y)\, dx dy = \int_{-b}^{b} \left(\int_{-\sqrt{b^2 - x^2}}^{\sqrt{b^2 - x^2}} (a^2 - y)\, dy \right) dx$$

$$= \int_{-b}^{b} \left[a^2 y - \tfrac{1}{2} y^2 \right]_{-\sqrt{b^2 - x^2}}^{\sqrt{b^2 - x^2}} dx$$

$$= \int_{-b}^{b} 2a^2 \sqrt{b^2 - x^2}\, dx$$

$$= \pi a^2 b^2. \quad \square$$

Exercises

Evaluate.

*1. $\displaystyle\iint_{\Omega} x^2\, dx dy$ with $\Omega = \{(x, y): -1 \le x \le 1, -1 \le y \le 1\}$.

2. $\displaystyle\iint_{\Omega} e^{x+y}\, dx dy$ with Ω as in Exercise 1.

*3. $\displaystyle\iint_{\Omega} \cos \tfrac{1}{2}\pi y\, dx dy$ with Ω as in Exercise 1.

4. $\displaystyle\iint_{\Omega} x^3 y\, dx dy$ with $\Omega = \{(x, y): 0 \le x \le 1, 0 \le y \le x\}$.

*5. $\displaystyle\iint_{\Omega} (y + \sin \pi x^2)\, dx dy$ with Ω as in Exercise 4.

6. $\displaystyle\iint_{\Omega} \cos (x + y)\, dx dy$ with $\Omega = \{(x, y): 0 \le x \le \tfrac{1}{2}\pi, 0 \le y \le \tfrac{1}{2}\pi\}$.

*7. $\displaystyle\iint_{\Omega} \sin (x + y)\, dx dy$ with Ω as in Exercise 6.

8. $\displaystyle\iint_{\Omega} (x + y)\, dx dy$ with $\Omega = \{(x, y): x^2 + y^2 \le 1\}$.

*9. $\displaystyle\iint_{\Omega} (x^2 + 3y^2)\, dx dy$ with Ω as in Exercise 8.

10. $\displaystyle\iint_{\Omega} \sqrt{xy}\, dx dy$ with $\Omega = \{(x, y): 0 \le y \le 1, y^2 \le x \le y\}$.

*11. $\displaystyle\iint_\Omega ye^x \, dxdy$ with $\Omega = \{(x, y): 0 \leq y \leq 1, 0 \leq x \leq y^2\}$.

12. $\displaystyle\iint_\Omega (4 - y^2) \, dxdy$ with Ω the region between $y^2 = 2x$ and $y^2 = 8 - 2x$.

*13. $\displaystyle\iint_\Omega e^{x^2} \, dxdy$ where Ω is the triangle formed by $2y = x$, $x = 2$, and the x-axis.

Calculate by double integration the area of the bounded region determined by the following pairs of curves.

14. $x^2 = 4y$, $2y - x - 4 = 0$.
*15. $y = x$, $x = 4y - y^2$.
16. $y = x$, $4y^3 = x^2$.
*17. $x + y = 5$, $xy = 6$.

18. Find the volume of the solid bounded above by $z = x + y$ and below by the triangle with vertices $(0, 0)$, $(0, 1)$, $(1, 0)$.
*19. Find the volume of the solid bounded by $\frac{1}{2}x + \frac{1}{3}y + \frac{1}{4}z = 1$ and the coordinate planes.
20. Find the volume of the solid bounded above by the plane $z = 2x + 3y$ and below by the unit square $\{(x, y): 0 \leq x \leq 1, 0 \leq y \leq 1\}$.
*21. Find the volume of the solid bounded above by $z = x^3y$ and below by the triangle with vertices $(0, 0)$, $(2, 0)$, $(0, 1)$.
22. Find the volume of the solid bounded above by $z = 2x + 1$ and below by $\{(x, y): (x - 1)^2 + y^2 \leq 1\}$.
*23. Find the volume of the solid which lies between the paraboloid $z = x^2 + y^2$ and the unit disk $\{(x, y): x^2 + y^2 \leq 1\}$.
24. Find the volume of the solid bounded above by $z = 4 - y^2 - \frac{1}{4}x^2$ and below by the disk $\{(x, y): (y - 1)^2 + x^2 \leq 1\}$.
25. Show that if

$$R = \{(x, y): a \leq x \leq b, c \leq y \leq d\},$$

then

$$\iint_R f(x)g(y) \, dxdy = \left[\int_a^b f(x) \, dx\right] \cdot \left[\int_c^d g(y) \, dy\right].$$

15.4 Double Integrals in Polar Coordinates

We start with a region R given in polar coordinates by a pair of inequalities

$$r^* \leq r \leq r^{**}, \qquad \theta^* \leq \theta \leq \theta^{**}.\dagger \qquad\qquad \text{(Figure 15.4.1)}$$

---

† Throughout this chapter we take $r \geq 0$. Here $r^* \geq 0$ and $\theta^* \leq \theta^{**} \leq \theta^* + 2\pi$.

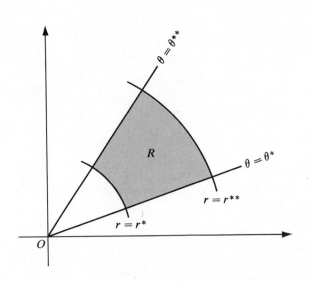

FIGURE 15.4.1

Such a set will be called a "polar rectangle." (It does not look like a rectangle but analytically it has the structure of a rectangle.) Over R we place a surface

$$z = f(r, \theta).$$ (Figure 15.4.2)

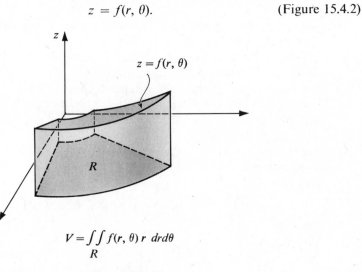

$$V = \iint_R f(r, \theta) \, r \, dr d\theta$$

FIGURE 15.4.2

The solid which lies between R and the surface $z = f(r, \theta)$ has volume

$$V = \iint_R f(r, \theta) r \, dr d\theta.$$ (note the extra r present)

To show how this comes about, we begin by partitioning the intervals $[r^*, r^{**}]$ and $[\theta^*, \theta^{**}]$:

$$r^* = r_0 < \cdots < r_{i-1} < r_i < \cdots < r_m = r^{**},$$
$$\theta^* = \theta_0 < \cdots < \theta_{j-1} < \theta_j < \cdots < \theta_n = \theta^{**}.$$

These partitions break up R into $m \times n$ "polar rectangles"

$$R_{ij}: r_{i-1} \leq r \leq r_i, \; \theta_{j-1} \leq \theta \leq \theta_j. \qquad \text{(Figure 15.4.3)}$$

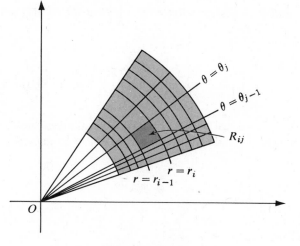

FIGURE 15.4.3

On each such "polar rectangle" f has a maximum value M_{ij} and a minimum value m_{ij}. (We are assuming as usual that f is continuous). The volume V that we seek is the only number that can be counted upon to satisfy the inequality

$$\sum_{i=1}^{m} \sum_{j=1}^{n} m_{ij}(\text{area of } R_{ij}) \leq V \leq \sum_{i=1}^{m} \sum_{j=1}^{n} M_{ij}(\text{area of } R_{ij}). \qquad \text{(explain)}$$

We now argue that the number

$$V = \iint\limits_{R} f(r, \theta) r \, dr d\theta$$

has this property. In the first place, it can be shown that the integral is additive over the R_{ij} and consequently

$$\iint\limits_{R} f(r, \theta) r \, dr d\theta = \sum_{i=1}^{m} \sum_{j=1}^{n} \iint\limits_{R_{ij}} f(r, \theta) r \, dr d\theta.$$

All we have to show therefore is that

$$m_{ij}(\text{area of } R_{ij}) \leq \iint\limits_{R_{ij}} f(r, \theta) r \, dr d\theta \leq M_{ij}(\text{area of } R_{ij}).$$

Observe now that

$$\iint\limits_{R_{ij}} r \, dr d\theta \underset{\substack{\uparrow \\ \text{from last section}}}{=} \int_{\theta_{j-1}}^{\theta_j} \left(\int_{r_{i-1}}^{r_i} r \, dr \right) d\theta = \tfrac{1}{2}(r_i^2 - r_{i-1}^2)(\theta_j - \theta_{j-1}) \underset{\substack{\uparrow \\ \text{from Section 9.5}}}{=} \text{area of } R_{ij}.$$

From this it follows that

$$\iint\limits_{R_{ij}} f(r, \theta) r \, dr d\theta \leq \iint\limits_{R_{ij}} M_{ij} r \, dr d\theta = M_{ij} \iint\limits_{R_{ij}} r \, dr d\theta = M_{ij}(\text{area of } R_{ij})$$

and

$$\iint\limits_{R_{ij}} f(r, \theta) r \, dr d\theta \geq \iint\limits_{R_{ij}} m_{ij} r \, dr d\theta = m_{ij} \iint\limits_{R_{ij}} r \, dr d\theta = m_{ij}(\text{area of } R_{ij}). \quad \square$$

The volume formula we just verified for "polar rectangles" holds for more general regions. In particular, it holds for regions Ω of the type depicted in Figure 15.4.4.

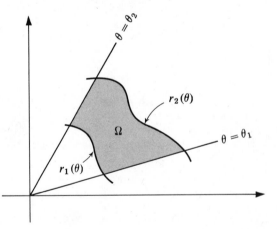

FIGURE 15.4.4

For such Ω we have

$$\textbf{(15.4.1)} \qquad V = \iint\limits_{\Omega} f(r, \theta) r \, dr d\theta = \int_{\theta_1}^{\theta_2} \left(\int_{r_1(\theta)}^{r_2(\theta)} f(r, \theta) r \, dr \right) d\theta.$$

If $f(r, \theta) \equiv 1$, the double integral takes the form

$$\iint\limits_{\Omega} r \, dr d\theta.$$

In cubic units, it gives the volume of a solid with base Ω and constant height 1. In square units, it gives the area of Ω.

Problem. Derive the formula for the volume of a sphere using the methods of this section.

SOLUTION. We take the sphere to have radius R and center at the origin. The upper spherical surface is given by

$$z = \sqrt{R^2 - r^2}.$$ (check this out)

As a base region we can take

$$\Omega: 0 \le r \le R, \, 0 \le \theta \le 2\pi.$$

Consequently,

$$V = 2 \iint_\Omega \sqrt{R^2 - r^2}\, r\, drd\theta$$

$$= \int_0^{2\pi} \left(\int_0^R \sqrt{R^2 - r^2}\, (2r)\, dr \right) d\theta$$

$$= \int_0^{2\pi} \left[-\tfrac{2}{3}(R^2 - r^2)^{3/2} \right]_0^R d\theta$$

$$= \int_0^{2\pi} \tfrac{2}{3} R^3\, d\theta$$

$$= \tfrac{4}{3}\pi R^3. \quad \square$$

Problem. Calculate by double integration the area of the region Ω that is enclosed by the cardioid

$$r = 1 - \cos\theta.$$

SOLUTION. We can express Ω by the inequalities

$$0 \le r \le 1 - \cos\theta, \, 0 \le \theta \le 2\pi.$$

Consequently,

$$\text{area of } \Omega = \iint_\Omega r\, drd\theta$$

$$= \int_0^{2\pi} \left(\int_0^{1-\cos\theta} r\, dr \right) d\theta$$

$$= \int_0^{2\pi} \left[\tfrac{1}{2} r^2 \right]_0^{1-\cos\theta} d\theta$$

$$= \int_0^{2\pi} \tfrac{1}{2}(1 - \cos\theta)^2\, d\theta$$

$$= \tfrac{1}{2} \int_0^{2\pi} (1 - 2\cos\theta + \cos^2\theta)\, d\theta.$$

Since

$$\tfrac{1}{2} \int_0^{2\pi} (1 - 2 \cos \theta) \, d\theta = \pi$$

and

$$\tfrac{1}{2} \int_0^{2\pi} \cos^2 \theta \, d\theta = \tfrac{1}{4} \int_0^{2\pi} (1 + \cos 2\theta) \, d\theta = \tfrac{1}{2}\pi,$$

we have

$$\text{area of } \Omega = \tfrac{3}{2}\pi. \quad \square$$

Problem. Calculate the volume of the solid bounded above by $z = 2 - r$ and below by the plane region

$$\Omega: 0 \le r \le 2 \cos \theta, \quad -\tfrac{1}{2}\pi \le \theta \le \tfrac{1}{2}\pi.$$

SOLUTION

$$V = \iint_\Omega (2 - r) r \, dr d\theta$$

$$= \int_{-\pi/2}^{\pi/2} \left(\int_0^{2 \cos \theta} (2r - r^2) \, dr \right) d\theta$$

$$= \int_{-\pi/2}^{\pi/2} \left[r^2 - \tfrac{1}{3} r^3 \right]_0^{2 \cos \theta} d\theta$$

$$= \int_{-\pi/2}^{\pi/2} (4 \cos^2 \theta - \tfrac{8}{3} \cos^3 \theta) \, d\theta$$

$$= 2 \int_0^{\pi/2} [2(1 + \cos 2\theta) - \tfrac{8}{3}(1 - \sin^2 \theta) \cos \theta] \, d\theta$$

$$= 2 \left[2\theta + \sin 2\theta - \tfrac{8}{3} \sin \theta + \tfrac{8}{9} \sin^3 \theta \right]_0^{\pi/2}$$

$$= 2(\pi - \tfrac{8}{3} + \tfrac{8}{9})$$

$$= \tfrac{2}{9}(9\pi - 16). \quad \square$$

Change of Variables Formula

It is possible to change a double integral

$$\iint_\Omega g(x, y) \, dxdy$$

from rectangular coordinates to polar coordinates. The change of variables rule is easy to state:

(15.4.2)
$$\boxed{\iint_\Omega g(x, y) \, dxdy = \iint_\Omega g(r \cos \theta, r \sin \theta) r \, dr d\theta.}$$

We shall verify this rule in the case that g is nonnegative.

VERIFICATION.   The surface given in rectangular coordinates by

$$z = g(x, y)$$

is given in polar coordinates by

$$z = g(r \cos \theta, r \sin \theta) = f(r, \theta).$$

The double integral

$$\iint_\Omega g(x, y) \, dxdy$$

gives the volume of the solid trapped between Ω and the given surface. By (15.4.1), the double integral

$$\iint_\Omega g(r \cos \theta, r \sin \theta) r \, drd\theta = \iint_\Omega f(r, \theta) r \, drd\theta$$

gives the volume of the same solid.   □

Problem.   Evaluate the double integral

$$\iint_\Omega e^{-(x^2 + y^2)} \, dxdy$$

where Ω is the unit disk

$$\{(x, y): x^2 + y^2 \le 1\}.$$

SOLUTION.   If you try to compute this integral by the usual methods, you will run into integrals that cannot be evaluated in closed form. By changing to polar coordinates, the calculations become simple. In polar coordinates we have

$$\Omega: 0 \le r \le 1, \, 0 \le \theta \le 2\pi.$$

The change of variables formula gives

$$\iint_\Omega e^{-(x^2 + y^2)} \, dxdy = \iint_\Omega e^{-r^2} r \, drd\theta$$

$$= \int_0^{2\pi} \left[\int_0^1 e^{-r^2} r \, dr \right] d\theta$$

$$= \int_0^{2\pi} \left[-\tfrac{1}{2} e^{-r^2} \right]_0^1 d\theta$$

$$= \int_0^{2\pi} \tfrac{1}{2} \left(1 - \frac{1}{e} \right) d\theta$$

$$= \pi \left(1 - \frac{1}{e} \right). \quad □$$

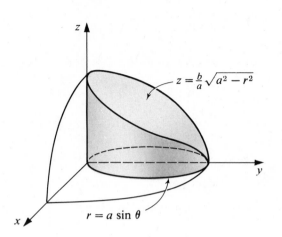

FIGURE 15.4.5

Problem. Find the volume of the solid bounded below by the xy-plane, above by the ellipsoid of revolution $b^2x^2 + b^2y^2 + a^2z^2 = a^2b^2$, and on the sides by the cylinder $x^2 + y^2 - ay = 0$.

SOLUTION. Setting

$$x = r \cos \theta, \qquad y = r \sin \theta$$

we can rewrite the upper boundary as

$$z = \frac{b}{a} \sqrt{a^2 - r^2}.$$

The boundary of the base region becomes

$$r = a \sin \theta. \qquad \text{(Figure 15.4.5)}$$

The base region itself can be written

$$\Omega : 0 \le r \le a \sin \theta,\ 0 \le \theta \le \pi.$$

The volume can now be computed as follows:

$$V = \iint_\Omega \frac{b}{a} \sqrt{a^2 - r^2}\, r\, dr d\theta$$

$$= \frac{b}{a} \int_0^\pi \left(\int_0^{a \sin \theta} r \sqrt{a^2 - r^2}\, dr \right) d\theta$$

$$= \tfrac{2}{9} a^2 b (3\pi - 4). \quad \square$$

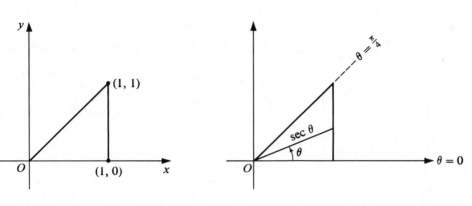

FIGURE 15.4.6   FIGURE 15.4.7

Problem.   Evaluate

$$I = \iint_{\Omega} \frac{1}{(1 + x^2 + y^2)^{3/2}} \, dxdy$$

where Ω is the triangle of Figure 15.4.6.

SOLUTION.   By the change of variables formula,

$$\iint_{\Omega} \frac{1}{(1 + x^2 + y^2)^{3/2}} \, dxdy = \iint_{\Omega} \frac{r}{(1 + r^2)^{3/2}} \, drd\theta.$$

In terms of polar coordinates, Ω is given by

$$0 \le \theta \le \tfrac{1}{4}\pi, \quad 0 \le r \le \sec \theta. \qquad \text{(Figure 15.4.7)}$$

Therefore

$$I = \int_0^{\pi/4} \left(\int_0^{\sec \theta} \frac{rdr}{(1 + r^2)^{3/2}} \right) d\theta$$

$$= - \int_0^{\pi/4} \left[(1 + r^2)^{-1/2} \right]_0^{\sec \theta} d\theta$$

$$= \int_0^{\pi/4} \left[1 - (1 + \sec^2 \theta)^{-1/2} \right] d\theta$$

$$= \int_0^{\pi/4} \left(1 - \frac{\cos \theta}{\sqrt{2 - \sin^2 \theta}} \right) d\theta$$

$$= \left[\theta - \arcsin \left(\tfrac{1}{2}\sqrt{2} \sin \theta \right) \right]_0^{\pi/4}$$

$$= \tfrac{1}{4}\pi - \tfrac{1}{6}\pi = \tfrac{1}{12}\pi. \quad \square$$

Exercises

*1. Find the area of the region which lies inside the circle $r = \frac{3}{2}$ but to the right of the line $4r \cos \theta = 3$.

2. Find the area of the region which lies inside the circle $r = 3 \cos \theta$ but outside the circle $r = \frac{3}{2}$.

*3. Find the area of the region which lies inside the circle $r = 3 \cos \theta$ but outside the circle $r = \cos \theta$.

4. Find the area of the region which lies inside the cardioid $r = 1 + \cos \theta$ but to the right of the line $4r \cos \theta = 3$.

*5. Find the area of the region which lies inside the cardioid $r = 1 + \cos \theta$ but outside the circle $r = 1$.

6. Find the area of the region which lies inside the circle $r = 1$ but outside the cardioid $r = 1 + \cos \theta$.

*7. Find the area of the region which lies inside the circle $r = 1$ but outside the parabola $r(1 + \cos \theta) = 1$.

8. Find the volume of the solid bounded above by the paraboloid $z = 1 - (x^2 + y^2)$ and below by the xy-plane.

*9. Find the volume of the solid bounded above by the plane $z = y + b$, below by the xy-plane, and on the sides by the circular cylinder $x^2 + y^2 = b^2$.

10. Find the volume of the solid bounded above by $z = 1 - (x^2 + y^2)$, below by the xy-plane, and on the sides by the cylinder $x^2 + y^2 - x = 0$.

*11. Find the volume of the solid bounded above by the spherical surface $x^2 + y^2 + z^2 = 4$, below by the xy-plane, and on the sides by the cylinder $x^2 + y^2 = 1$.

12. Find the volume of the solid bounded above by the cone $z^2 = x^2 + y^2$ and below by the region Ω which lies inside the curve $x^2 + y^2 = 2ax$.

15.5 Triple Integrals

Now that you are familiar with double integrals

$$\iint_\Omega f(x, y)\, dxdy,$$

you will find it easy to understand triple integrals

$$\iiint_S f(x, y, z)\, dxdydz.$$

Basically the only difference is this: instead of dealing with functions of two variables continuous on a plane region Ω, we will be dealing with functions of three variables continuous on some portion S of space.

Instead of beginning with a rectangle

$$R = \{(x, y): a_1 \leq x \leq a_2, \quad b_1 \leq y \leq b_2\}$$

we begin with a *box* (a rectangular solid)

$$\Pi = \{(x, y, z): a_1 \leq x \leq a_2, \quad b_1 \leq y \leq b_2, \quad c_1 \leq z \leq c_2\}. \qquad \text{(Figure 15.5.1)}$$

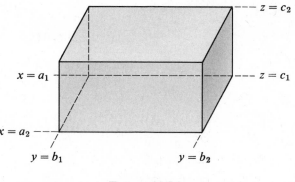

FIGURE 15.5.1

To partition this box, we first partition the edges. Taking

$$P_1 = \{x_0, \ldots, x_m\} \quad \text{as a partition of } [a_1, a_2],$$
$$P_2 = \{y_0, \ldots, y_n\} \quad \text{as a partition of } [b_1, b_2],$$
$$P_3 = \{z_0, \ldots, z_q\} \quad \text{as a partition of } [c_1, c_2],$$

we form the set

$$P = P_1 \times P_2 \times P_3 = \{(x_i, y_j, z_k): x_i \in P_1, \quad y_j \in P_2, \quad z_k \in P_3\}$$

and call this a *partition of* Π. P breaks up Π into $m \times n \times q$ nonoverlapping boxes

$$\Pi_{ijk} = \{(x, y, z): x_{i-1} \le x \le x_i, \quad y_{j-1} \le y \le y_j, \quad z_{k-1} \le z \le z_k\}.$$

A typical such box is pictured in Figure 15.5.2.

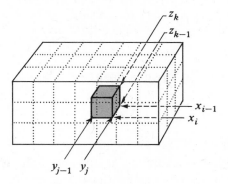

FIGURE 15.5.2

Taking

$$M_{ijk} \text{ as the maximum value of } f \text{ on } \Pi_{ijk}$$

and

$$m_{ijk} \text{ as the minimum value of } f \text{ on } \Pi_{ijk},$$

we form the *upper sum*

$$U_f(P) = \sum_{i=1}^{m} \sum_{j=1}^{n} \sum_{k=1}^{q} M_{ijk}(\text{volume of } \Pi_{ijk})$$

$$= \sum_{i=1}^{m} \sum_{j=1}^{n} \sum_{k=1}^{q} M_{ijk} \, \Delta x_i \, \Delta y_j \, \Delta z_k$$

and the *lower sum*

$$L_f(P) = \sum_{i=1}^{m} \sum_{j=1}^{n} \sum_{k=1}^{q} m_{ijk}(\text{volume of } \Pi_{ijk})$$

$$= \sum_{i=1}^{m} \sum_{j=1}^{n} \sum_{k=1}^{q} m_{ijk} \, \Delta x_i \, \Delta y_j \, \Delta z_k.$$

As in the case of functions of one and two variables, it turns out that, with f continuous on Π, there is one, and only one, number I which satisfies the inequality

$$L_f(P) \le I \le U_f(P) \quad \text{for all partitions } P \text{ of } \Pi.$$

Definition of the Triple Integral

The unique number I which satisfies the inequality

$$L_f(P) \le I \le U_f(P) \quad \text{for all partitions } P \text{ of } \Pi$$

is called the *triple integral* of f over Π and is denoted by

$$\iiint_{\Pi} f(x, y, z) \, dxdydz.$$

To obtain the integral over a more general solid S we first encase S in a rectangular box Π. We then extend f to all of Π by defining f to be 0 outside of S. The triple integral over S is then defined as the triple integral of f over Π:

$$\boxed{\iiint_{S} f(x, y, z) \, dxdydz = \iiint_{\Pi} f(x, y, z) \, dxdydz.}$$

Interpretation of the Triple Integral as a Mass

The density of a solid S is by definition its mass per unit volume. If the density ρ is constant throughout S, then we can get the total mass of S simply by multiplying the density by the volume:

$$\text{mass of } S = (\text{density of } S) \cdot (\text{volume of } S)$$

$$M = \rho V.$$

Suppose now that the density varies continuously from point to point; say

$$\rho = f(x, y, z).$$

What is the total mass of S then? The answer is simple:

$$\text{mass of } S = \iiint_S f(x, y, z) \, dxdydz.$$

To justify this assertion, we first encase S in a box Π assigning density 0 to points outside of S. The mass of S will then be the mass of Π. We now take P as a partition of Π, and use it to break up Π into little sub-boxes Π_{ijk} in the manner previously discussed. For each little box

$$\begin{pmatrix} \text{min density} \\ \text{on } \Pi_{ijk} \end{pmatrix} \cdot (\text{volume of } \Pi_{ijk}) \leq \text{mass of } \Pi_{ijk} \leq \begin{pmatrix} \text{max density} \\ \text{on } \Pi_{ijk} \end{pmatrix} \cdot (\text{volume of } \Pi_{ijk}).$$

In our previous notation,

$$m_{ijk}(\text{volume of } \Pi_{ijk}) \leq \text{mass of } \Pi_{ijk} \leq (\text{volume of } \Pi_{ijk}).$$

Adding up these inequalities, we see that the total mass of Π must satisfy the condition

$$L_f(P) \leq \text{mass of } \Pi \leq U_f(P).$$

Since P was chosen arbitrarily, this inequality must hold for all partitions P; consequently, we must have

$$\text{mass of } \Pi = \iiint_\Pi f(x, y, z) \, dxdydz.$$

Since the density outside of S was defined to be 0, we have

$$\iiint_S f(x, y, z) \, dxdydz = \iiint_\Pi f(x, y, z) \, dxdydz$$

and, since

$$\text{mass of } S = \text{mass of } \Pi,$$

we have

$$\text{mass of } S = \iiint_S f(x, y, z) \, dxdydz. \quad \square$$

The triple integral

$$\iiint_S 1 \, dxdydz = \iiint_S dxdydz$$

gives the mass of S when S has constant density 1. This is simply the volume of S:

$$\text{volume of } S = \iiint_S dxdydz.$$

15.6   Reduction to Repeated Integrals

As with double integrals, we usually evaluate triple integrals by first reducing them to repeated integrals. If S is given by inequalities of the form

$$a_1 \le x \le a_2, \qquad \phi_1(x) \le y \le \phi_2(x), \qquad \psi_1(x, y) \le z \le \psi_2(x, y)$$

(see Figure 15.6.1), then

$$\iiint_S f(x, y, z)\, dxdydz = \int_{a_1}^{a_2} \left[\int_{\phi_1(x)}^{\phi_2(x)} \left(\int_{\psi_1(x,y)}^{\psi_2(x,y)} f(x, y, z)\, dz \right) dy \right] dx,$$

which, by omitting the parentheses and brackets, can be written

$$\int_{a_1}^{a_2} \int_{\phi_1(x)}^{\phi_2(x)} \int_{\psi_1(x,y)}^{\psi_2(x,y)} f(x, y, z)\, dz\, dy\, dx.\dagger$$

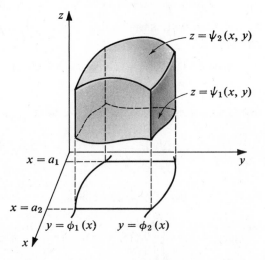

$$a_1 \le x \le a_2,\ \phi_1(x) \le y \le \phi_2(x),\ \psi_1(x, y) \le z \le \psi_2(x, y)$$

FIGURE 15.6.1

Here we integrate first with respect to z, then with respect to y, and then with respect to x. Other orders of integration are possible. If, for instance, S is of the form

$$b_1 \le y \le b_2, \qquad \phi_1(y) \le z \le \phi_2(y), \qquad \psi_1(y, z) \le x \le \psi_2(y, z),$$

---

† In this section we give no proofs. All the functions in question, f, ϕ_1, ϕ_2, ψ_1, ψ_2 are assumed to be continuous.

then the triple integral over S leads to the repeated integral

$$\int_{b_1}^{b_2}\left[\int_{\phi_1(y)}^{\phi_2(y)}\left(\int_{\psi_1(y,z)}^{\psi_2(y,z)}f(x,y,z)\,dx\right)dz\right]dy=\int_{b_1}^{b}\int_{\phi_1(y)}^{\phi_2(y)}\int_{\psi_1(y,z)}^{\psi_2(y,z)}f(x,y,z)\,dx\,dz\,dy.$$

In this case we integrate first with respect to x, then with respect to z, and then with respect to y. Still other orders of integration are of course possible.

We illustrate the technique in some problems.

Problem 1. Use triple integration to find the volume of the tetrahedron S of Figure 15.6.2.

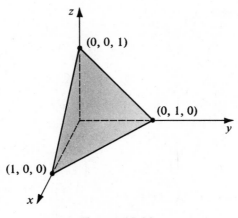

FIGURE 15.6.2

SOLUTION. We begin by referring to Figure 15.6.3. The inclined face of S has equation

$$x + y + z = 1.$$

It intersects the xy-plane in a line which has equation

$$x + y = 1.$$

As x ranges from 0 to 1 (see Figure 15.6.4), y ranges from 0 to $1 - x$, and z ranges from 0 to $1 - x - y$. In short, S is given by the inequalities

$$0 \le x \le 1, \qquad 0 \le y \le 1 - x, \qquad 0 \le z \le 1 - x - y.$$

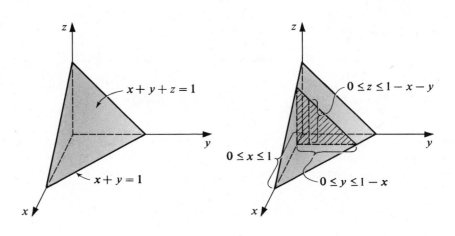

FIGURE 15.6.3   FIGURE 15.6.4

The volume of S is therefore

$$V = \iiint\limits_{S} dx\,dy\,dz$$

$$= \int_0^1 \int_0^{1-x} \int_0^{1-x-y} dz\,dy\,dx$$

$$= \int_0^1 \int_0^{1-x} (1 - x - y)\,dy\,dx$$

$$= \int_0^1 \left[(1 - x)y - \tfrac{1}{2}y^2\right]_0^{1-x} dx$$

$$= \int_0^1 \tfrac{1}{2}(1 - x)^2\,dx$$

$$= \left[-\tfrac{1}{6}(1 - x)^3\right]_0^1$$

$$= \tfrac{1}{6}. \quad \square$$

REMARK.   In the last problem we integrated first with respect to z, then with respect to y, and then with respect to x. This came from expressing the tetrahedron by the inequalities

$$0 \le x \le 1, \qquad 0 \le y \le 1 - x, \qquad 0 \le z \le 1 - x - y.$$

Other characterizations are also possible. If we use the inequalities

$$0 \le z \le 1, \qquad 0 \le x \le 1 - z, \qquad 0 \le y \le 1 - z - x,$$

then we obtain

$$V = \int_0^1 \int_0^{1-z} \int_0^{1-z-x} dy\, dx\, dz.$$

This formulation of the problem calls for us to integrate first with respect to y, then with respect to x, and then with respect to z. As you can check, this procedure leads to the same result. □

Problem 2. Find the total mass of the tetrahedron of Problem 1 if the density at each point (x, y, z) is xy.

SOLUTION. Using the first characterization of S, we have

$$\text{mass} = \iiint_S xy\, dxdydz = \int_0^1 \int_0^{1-x} \int_0^{1-x-y} xy\, dz\, dy\, dx.$$

Since

$$\int_0^{1-x-y} xy\, dz = xy(1 - x - y) = x(1 - x)y - xy^2,$$

we have

$$\int_0^{1-x} \int_0^{1-x-y} xy\, dz\, dy = \int_0^{1-x} [x(1 - x)y - xy^2]\, dy$$

$$= \left[\tfrac{1}{2}x(1 - x)y^2 - \tfrac{1}{3}xy^3 \right]_0^{1-x}$$

$$= \tfrac{1}{6}x(1 - x)^3$$

$$= \tfrac{1}{6}(x - 3x^2 + 3x^3 - x^4).$$

Consequently,

$$M = \int_0^1 \tfrac{1}{6}(x - 3x^2 + 3x^3 - x^4)\, dx$$

$$= \tfrac{1}{6}\left[\tfrac{1}{2}x^2 - x^3 + \tfrac{3}{4}x^4 - \tfrac{1}{5}x^5 \right]_0^1$$

$$= \tfrac{1}{6}(\tfrac{1}{2} - 1 + \tfrac{3}{4} - \tfrac{1}{5}) = \tfrac{1}{120}. \quad □$$

Problem 3. Find the mass of a cylindrical solid of base radius r and height h if the density is proportional to the distance from one of the bases.

SOLUTION. For convenience we place the cylinder as in Figure 15.6.5. The cylindrical solid, call it S, is then given by the three inequalities

$$-r \leq x \leq r, \quad -\sqrt{r^2 - x^2} \leq y \leq \sqrt{r^2 - x^2}, \quad 0 \leq z \leq h.$$

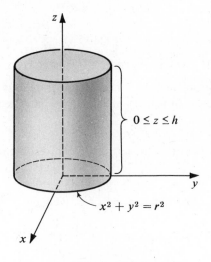

FIGURE 15.6.5

Taking k as the constant of proportionality, we have

$$M = \iiint_S kz \, dxdydz$$

$$= \int_{-r}^{r} \int_{-\sqrt{r^2-x^2}}^{\sqrt{r^2-x^2}} \int_{0}^{h} kz \, dz \, dy \, dx$$

$$= 4k \int_{0}^{r} \int_{0}^{\sqrt{r^2-x^2}} \int_{0}^{h} z \, dz \, dy \, dx \qquad \text{(explain)}$$

$$= 4k \int_{0}^{r} \int_{0}^{\sqrt{r^2-x^2}} \tfrac{1}{2}h^2 \, dy \, dx$$

$$= 2kh^2 \int_{0}^{r} \sqrt{r^2 - x^2} \, dx = 2kh^2 \cdot \tfrac{1}{4}\pi r^2 = \tfrac{1}{2}kh^2 r^2 \pi. \quad \square$$

Problem 4.   Integrate
$$f(x, y, z) = xy + yz + zx$$

over that portion of the first octant $x \geq 0$, $y \geq 0$, $z \geq 0$ which is cut off by the ellipsoid

$$\frac{x^2}{a^2} + \frac{y^2}{b^2} + \frac{z^2}{c^2} = 1.$$

SOLUTION.   See Figure 15.6.6. Call the solid S. The upper boundary of S has equation

$$z = \psi(x, y) = c\left(1 - \frac{x^2}{a^2} - \frac{y^2}{b^2}\right)^{1/2}.$$

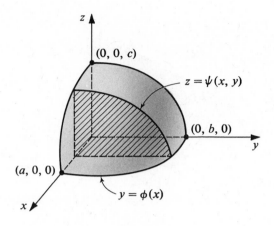

FIGURE 15.6.6

This surface intersects the xy-plane in the curve

$$y = \phi(x) = b\left(1 - \frac{x^2}{a^2}\right)^{1/2}.$$

As x ranges from 0 to a, y ranges from 0 to $\phi(x)$, and z ranges from 0 to $\psi(x, y)$. We are therefore led to the repeated integral

$$I = \int_0^a \int_0^{\phi(x)} \int_0^{\psi(x,y)} (xy + yz + zx) \, dz \, dy \, dx.$$

A straightforward but lengthy computation shows that

$$I = \tfrac{1}{15} abc(ab + bc + ca). \quad \square$$

Problem 5. Use triple integration to find the volume of the solid S bounded above by the parabolic cylinder $z = 4 - y^2$ and below by the elliptic paraboloid

$$z = x^2 + 3y^2.$$

SOLUTION. The two surfaces intersect at those points (x, y, z) where

$$4 - y^2 = x^2 + 3y^2$$

or, equivalently, at those points (x, y, z) where

$$x^2 + 4y^2 = 4.$$

Below these points on the xy-plane lies the ellipse

$$x^2 + 4y^2 = 4. \qquad \text{(Figure 15.6.7)}$$

As the base region we can take

$$-2 \le x \le 2, \qquad -\tfrac{1}{2}\sqrt{4 - x^2} \le y \le \tfrac{1}{2}\sqrt{4 - x^2}.$$

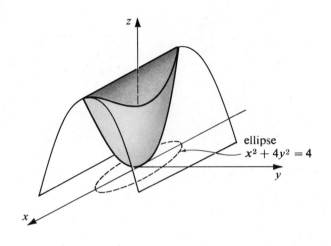

FIGURE 15.6.7

The solid S is then the set of all (x, y, z) such that

$$-2 \le x \le 2, \qquad -\tfrac{1}{2}\sqrt{4 - x^2} \le y \le \tfrac{1}{2}\sqrt{4 - x^2}, \qquad x^2 + 3y^2 \le z \le 4 - y^2.$$

Its volume is given by

$$V = \int_{-2}^{2} \int_{-(\sqrt{4-x^2})/2}^{(\sqrt{4-x^2})/2} \int_{x^2+3y^2}^{4-y^2} dz \, dy \, dx = 4 \int_{0}^{2} \int_{0}^{(\sqrt{4-x^2})/2} \int_{x^2+3y^2}^{4-y^2} dz \, dy \, dx = 4\pi. \quad \square$$

Exercises

Evaluate.

*1. $\displaystyle\int_{0}^{a} \int_{0}^{b} \int_{0}^{c} dx \, dy \, dz.$

2. $\displaystyle\int_{0}^{1} \int_{0}^{x} \int_{0}^{y} y \, dz \, dy \, dx.$

*3. $\displaystyle\int_{0}^{1} \int_{1}^{2y} \int_{0}^{x} (x + 2z) \, dz \, dx \, dy.$

4. $\displaystyle\int_{0}^{1} \int_{1-x}^{1+x} \int_{0}^{xy} 4z \, dz \, dy \, dx.$

*5. Use triple integration to find the volume of the tetrahedron of Figure 15.6.8.

6. Find the mass of the tetrahedron of Exercise 5 if $a = 1$, $b = 1$, $c = 1$, and the density at (x, y, z) is (a) $x^2 + y^2$. (b) $\tfrac{1}{2}(x + y + z)$.

*7. Find the mass of a block in the shape of a unit cube if its density at each point is proportional
   (a) to its distance from one of the faces.
   (b) to the square of its distance from one of the vertices.

8. Let S be the solid bounded above by the plane $z = y$, below by the xy-plane, and on the sides by the planes

$$x = 0, \qquad x = 1, \qquad y = 0, \qquad y = 1.$$

Find the mass of S given that the density at each point is proportional to the square of its distance from the origin.

* 9. Find the volume of the solid bounded above by the parabolic cylinder

$$y^2 + z = 4,$$

below by the plane $y + z = 2$, and on the sides by the planes $x = 0$ and $x = 2$.

10. Find the mass of the solid of Exercise 9 if the density at (x, y, z) is

  (a) x.   * (b) y^2.   (c) x^2y^2.

11. Find the mass of the solid discussed in Problem 5 (p. 807) if the density function is $f(x, y, z) = |x|$.

* 12. Use triple integration to find the volume of the solid which is bounded above by the plane $z - y = 0$ and below by the paraboloid $z = x^2 + y^2$.

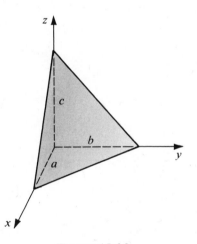

FIGURE 15.6.8

15.7 Averages and Centroids

In Chapter 9 we defined the average of a continuous function by setting

$$\text{average of } f \text{ on } [a, b] = \frac{1}{b - a} \int_a^b f(x)\, dx = \frac{1}{\text{length of } [a, b]} \int_a^b f(x)\, dx.$$

Averages for functions of several variables are defined in a similar manner. If f is continuous on a plane region Ω, its average on Ω is defined by setting

$$\text{average of } f \text{ on } \Omega = \frac{1}{\text{area of } \Omega} \iint_\Omega f(x, y)\, dxdy.$$

If f is continuous on a portion S of space, its average on S is defined by setting

$$\text{average of } f \text{ on } S = \frac{1}{\text{volume of } S} \iiint\limits_S f(x, y, z)\, dxdydz.$$

Problem. Calculate the average height (average z-coordinate) of that portion of the plane

$$\frac{x}{2} + \frac{y}{3} + \frac{z}{4} = 1$$

which lies above the first quadrant: $x \geq 0$, $y \geq 0$.

SOLUTION.   That portion of the plane which lies above the first quadrant is sketched in Figure 15.7.1. What we want to do here is average the function

$$z = 4(1 - \tfrac{1}{2}x - \tfrac{1}{3}y)$$

over the triangular region

$$\Omega: 0 \leq x \leq 2, \quad 0 \leq y \leq 3(1 - \tfrac{1}{2}x).$$

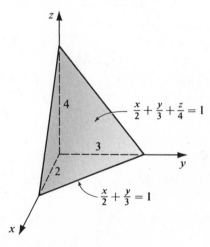

FIGURE 15.7.1

As you can verify,

$$\iint\limits_\Omega 4(1 - \tfrac{1}{2}x - \tfrac{1}{3}y)\, dxdy = 4 \quad \text{and} \quad \text{area of } \Omega = 3.$$

The average height is therefore

$$\frac{1}{\text{area of } \Omega} \iint_\Omega 4(1 - \tfrac{1}{2}x - \tfrac{1}{3}y) \, dx dy = \tfrac{4}{3}. \quad \square$$

Problem.   Find the average density of the cylindrical solid

$$S: -r \le x \le r, \quad -\sqrt{r^2 - x^2} \le y \le \sqrt{r^2 - x^2}, \quad 0 \le z \le h$$

if the density at each point (x, y, z) is $\rho = y^2\sqrt{r^2 - x^2}$.

SOLUTION.   The average we want is the quotient

$$\frac{\text{mass of } S}{\text{volume of } S} = \frac{1}{V} \iiint_S y^2 \sqrt{r^2 - x^2} \, dx dy dz.$$

In the first place,

$$\iiint_\Omega y^2 \sqrt{r^2 - x^2} \, dx dy dz = \int_{-r}^{r} \int_{-\sqrt{r^2-x^2}}^{\sqrt{r^2-x^2}} \int_0^h y^2 \sqrt{r^2 - x^2} \, dz \, dy \, dx$$

$$= 4 \int_0^r \int_0^{\sqrt{r^2-x^2}} \int_0^h y^2 \sqrt{r^2 - x^2} \, dz \, dy \, dx$$

$$= 4 \int_0^r \int_0^{\sqrt{r^2-x^2}} h y^2 \sqrt{r^2 - x^2} \, dy \, dx$$

$$= 4h \int_0^r \tfrac{1}{3}(r^2 - x^2)^2 \, dx$$

$$= \tfrac{4}{3}h \int_0^r (r^4 - 2r^2x^2 + x^4) \, dx$$

$$= \tfrac{4}{3}h \left[r^4x - \tfrac{2}{3}r^2x^3 + \tfrac{1}{5}x^5 \right]_0^r$$

$$= \tfrac{4}{3}h[r^5(1 - \tfrac{2}{3} + \tfrac{1}{5})]$$

$$= \tfrac{4}{3}hr^5(\tfrac{8}{15})$$

$$= \tfrac{32}{45}hr^5.$$

Since the volume of the cylindrical solid is $\pi r^2 h$, the average density is

$$\frac{1}{V} M = \frac{1}{\pi r^2 h} \cdot \frac{32}{45} hr^5 = \frac{32}{45\pi} r^3. \quad \square$$

Centroids

The *centroid* of a plane region Ω is the point $(\bar{x}, \bar{y})$ where $\bar{x}$ is the average x-coordinate of points in Ω and $\bar{y}$ is the average y-coordinate of points in Ω:

$$\bar{x} = \frac{1}{\text{area of }\Omega} \iint_\Omega x \, dxdy,$$

$$\bar{y} = \frac{1}{\text{area of }\Omega} \iint_\Omega y \, dxdy.$$

If you think of Ω as a thin distribution of matter of constant density, then the centroid of Ω becomes the center of gravity, the balance point.

For regions that have a geometric center, the center and the centroid coincide. For example, in the case of a rectangle

$$R: a \le x \le b, \, c \le y \le d$$

we have

$$(\bar{x}, \bar{y}) = (\tfrac{1}{2}[a + b], \tfrac{1}{2}[c + d]).$$

This follows from observing that

$$\iint_R x \, dxdy = \int_a^b \int_c^d x \, dy \, dx = \tfrac{1}{2}(b^2 - a^2)(d - c)$$

$$= \tfrac{1}{2}(a + b)(b - a)(d - c) = \tfrac{1}{2}(a + b)(\text{area of } R)$$

and

$$\iint_R y \, dxdy = \int_a^b \int_c^d y \, dy \, dx = \tfrac{1}{2}(d^2 - c^2)(b - a)$$

$$= \tfrac{1}{2}(c + d)(b - a)(d - c) = \tfrac{1}{2}(c + d)(\text{area of } R). \quad \square$$

A region may fail to have a center and yet be symmetric with respect to some line l. The centroid will then lie on l.

Problem. Find the centroid of the semicircular region bounded above by $y = \sqrt{r^2 - x^2}$ and below by the x-axis. (Figure 15.7.2)

SOLUTION.   Here

$$\Omega: -r \le x \le r, \, 0 \le y \le \sqrt{r^2 - x^2}.$$

Since Ω is symmetric with respect to the y-axis, we can expect $\bar{x} = 0$. You can verify this by showing that

$$\iint_\Omega x \, dxdy = 0.$$

Since

$$\iint_{\Omega} y \, dxdy = \int_{-r}^{r} \int_{0}^{\sqrt{r^2 - x^2}} y \, dy \, dx$$

$$= \int_{-r}^{r} \tfrac{1}{2}(r^2 - x^2) \, dx = \left[\tfrac{1}{2}r^2x - \tfrac{1}{6}x^3 \right]_{-r}^{r} = \tfrac{2}{3}r^3$$

and

$$\text{area of } \Omega = \tfrac{1}{2}\pi r^2,$$

we have

$$\bar{y} = \frac{1}{\text{area of } \Omega} \iint_{\Omega} y \, dxdy = \frac{4r}{3\pi}. \quad \Box$$

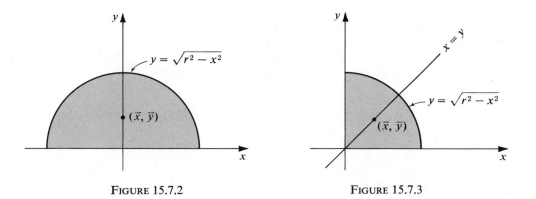

FIGURE 15.7.2   FIGURE 15.7.3

Problem. Find the centroid of the quarter disk depicted in Figure 15.7.3.

SOLUTION. Here we have symmetry with respect to the line $x = y$. We can therefore expect $\bar{x} = \bar{y}$. As in the last problem

$$\bar{y} = \frac{1}{\text{area of } \Omega} \iint_{\Omega} y \, dxdy = \frac{4r}{3\pi}.$$

The centroid is the point

$$(\bar{x}, \bar{y}) = \left(\frac{4r}{3\pi}, \frac{4r}{3\pi} \right). \quad \Box$$

Problem. Find the centroid of the region bounded above by the line $y = 2x$ and below by the parabola $y = x^2$.

SOLUTION. The line intersects the parabola at the origin and at the point (2, 4). The region Ω can be characterized by the inequalities

$$0 \leq x \leq 2, \quad x^2 \leq y \leq 2x.$$

Here

$$\text{area of } \Omega = \iint_{\Omega} dx\,dy = \int_0^2 \int_{x^2}^{2x} dy \, dx = \int_0^2 (2x - x^2) \, dx = \left[x^2 - \tfrac{1}{3}x^3 \right]_0^2 = \tfrac{4}{3},$$

$$\iint_{\Omega} x \, dx\,dy = \int_0^2 \int_{x^2}^{2x} x \, dy \, dx = \int_0^2 (2x^2 - x^3) \, dx = \left[\tfrac{2}{3}x^3 - \tfrac{1}{4}x^4 \right]_0^2 = \tfrac{4}{3},$$

$$\iint_{\Omega} y \, dx\,dy = \int_0^2 \int_{x^2}^{2x} y \, dy \, dx = \tfrac{1}{2} \int_0^2 (4x^2 - x^4) \, dx = \tfrac{1}{2}\left[\tfrac{4}{3}x^3 - \tfrac{1}{5}x^5 \right]_0^2 = \tfrac{32}{15},$$

$$\bar{x} = \frac{1}{\text{area of } \Omega} \iint_{\Omega} x \, dx\,dy = 1, \qquad \bar{y} = \frac{1}{\text{area of } \Omega} \iint_{\Omega} y \, dx\,dy = \tfrac{8}{5}.$$

The centroid is the point $(1, \tfrac{8}{5})$.   □

There is a simple relation between centroids and volumes of revolution, a relation which was first pointed out by Pappus of Alexandria (born about A.D. 340).

Theorem of Pappus

When a plane region Ω is revolved about an axis which lies in its plane but does not cross it, the volume of the resulting solid of revolution is the area of Ω multiplied by the circumference of the circle described by the centroid of Ω.

To confirm the theorem, we take Ω as in Figure 15.7.4 and see what happens when we revolve Ω about each of the coordinate axes.

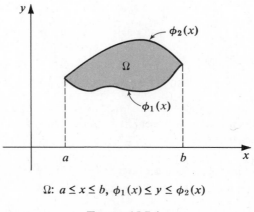

$\Omega: a \leq x \leq b, \, \phi_1(x) \leq y \leq \phi_2(x)$

FIGURE 15.7.4

The coordinates of the centroid are

$$\bar{x} = \frac{1}{\text{area of } \Omega} \iint_{\Omega} x \, dxdy, \qquad \bar{y} = \frac{1}{\text{area of } \Omega} \iint_{\Omega} y \, dxdy.$$

When Ω is revolved about the x-axis, the resulting solid has volume

$$V_x = \int_a^b \pi[\phi_2(x)]^2 \, dx - \int_a^b \pi[\phi_1(x)]^2 \, dx \qquad \text{(by 9.3.2)}$$

$$= \pi \int_a^b ([\phi_2(x)]^2 - [\phi_1(x)]^2) \, dx.$$

When Ω is revolved about the y-axis, the resulting solid has volume

$$V_y = \int_a^b 2\pi x \phi_2(x) \, dx - \int_a^b 2\pi x \phi_1(x) \, dx \qquad \text{(by 9.4.3)}$$

$$= 2\pi \int_a^b x[\phi_2(x) - \phi_1(x)] \, dx.$$

According to Pappus we must have

$$V_x = 2\pi\bar{y} \text{ (area of } \Omega\text{)} \quad \text{and} \quad V_y = 2\pi\bar{x} \text{ (area of } \Omega\text{)}.$$

Since

$$2\pi\bar{y} \cdot \text{area of } \Omega = 2\pi \iint_{\Omega} y \, dxdy$$

$$= 2\pi \int_a^b \int_{\phi_1(x)}^{\phi_2(x)} y \, dy \, dx$$

$$= 2\pi \int_a^b \left[\tfrac{1}{2} y^2 \right]_{\phi_1(x)}^{\phi_2(x)} dx$$

$$= 2\pi \int_a^b \tfrac{1}{2}([\phi_2(x)]^2 - [\phi_1(x)]^2) \, dx$$

$$= \pi \int_a^b ([\phi_2(x)]^2 - [\phi_1(x)]^2) \, dx$$

$$= V_x,$$

the first relation is confirmed. We now confirm the second relation:

$$2\pi\bar{x} \cdot \text{area of } \Omega = 2\pi \iint_{\Omega} x \, dxdy$$

$$= 2\pi \int_a^b \int_{\phi_1(x)}^{\phi_2(x)} x \, dy \, dx$$

$$= 2\pi \int_a^b x[\phi_2(x) - \phi_1(x)] \, dx$$

$$= V_y. \quad \square$$

The *centroid* of a solid S is the point $(\bar{x}, \bar{y}, \bar{z})$ where $\bar{x}$ is the average x-coordinate of points in S, $\bar{y}$ the average y-coordinate, and $\bar{z}$ the average z-coordinate. The defining formulas are:

$$\bar{x} = \frac{1}{\text{volume of } S} \iiint_S x \, dxdydz,$$

$$\bar{y} = \frac{1}{\text{volume of } S} \iiint_S y \, dxdydz,$$

$$\bar{z} = \frac{1}{\text{volume of } S} \iiint_S z \, dxdydz.$$

For a solid of constant density the centroid is the center of mass. Centers of mass are important in physics. The reason is roughly this. When a complicated object (or collection of objects) S of total mass M is subjected to an outside force $\mathbf{F}$, not every part of S reacts with the same acceleration. What happens then to Newton's famous $\mathbf{F} = M\mathbf{a}$? It still holds, not for every part of S, but for the center of mass. In short, Newton's $\mathbf{F} = M\mathbf{a}$ should be read

$$\begin{pmatrix} \text{total outside force} \\ \text{applied to } S \end{pmatrix} = \begin{pmatrix} \text{total mass} \\ \text{of } S \end{pmatrix} \cdot \begin{pmatrix} \text{acceleration of the} \\ \text{center of mass of } S \end{pmatrix}.$$

As in the two-dimensional case, if S has a geometric center, the center and centroid coincide. If S fails to have a center but has an axis or plane of symmetry, then the centroid will lie on that axis or on that plane.

Problem. Find the centroid of the tetrahedron with vertices $(0, 0, 0)$, $(1, 0, 0)$, $(0, 1, 0)$, $(0, 0, 1)$.

SOLUTION.   Call the tetrahedron S. S is characterized by the inequalities

$$0 \le x \le 1, \qquad 0 \le y \le 1 - x, \qquad 0 \le z \le 1 - x - y. \qquad \text{(Figure 15.6.4)}$$

As you have seen before,

$$V = \tfrac{1}{6}.$$

Consequently,

$$\bar{x} = \frac{1}{V} \iiint\limits_{S} x \, dx dy dz$$

$$= 6 \int_0^1 \int_0^{1-x} \int_0^{1-x-y} x \, dz \, dy \, dx$$

$$= 6 \int_0^1 \int_0^{1-x} [x(1 - x) - xy] \, dy \, dx$$

$$= 6 \int_0^1 \left[x(1 - x)y - \tfrac{1}{2}xy^2 \right]_0^{1-x} dx$$

$$= 6 \int_0^1 \tfrac{1}{2}x(1 - x)^2 \, dx$$

$$= 3 \int_0^1 (x - 2x^2 + x^3) \, dx$$

$$= 3 \left[\tfrac{1}{2}x^2 - \tfrac{2}{3}x^3 + \tfrac{1}{4}x^4 \right]_0^1$$

$$= 3(\tfrac{1}{2} - \tfrac{2}{3} + \tfrac{1}{4}) = 3(\tfrac{1}{12}) = \tfrac{1}{4}.$$

In a similar manner (or by symmetry), you can verify that $\bar{y}$ and $\bar{z}$ are also $\tfrac{1}{4}$. The centroid is the point

$$(\tfrac{1}{4}, \tfrac{1}{4}, \tfrac{1}{4}). \quad \square$$

Exercises

Find the centroid of the bounded region determined by the following curves.

* 1. $y = 6x - x^2$, $y = x$.   2. $y = 4x - x^2$, $y = 2x - 3$.
* 3. $x^2 = 4y$, $x - 2y + 4 = 0$.   4. $y = x^2$, $2x - y + 3 = 0$.
* 5. $y^3 = x^2$, $2y = x$.   6. $y^2 = 2x$, $y = x - x^2$.
* 7. $y = x^2 - 2x - 3$, $y = 6x - x^2 - 3$.
  8. $y = 6x - x^2$, $x + y = 6$.
* 9. $\sqrt{x} + \sqrt{y} = \sqrt{a}$, $x = 0$, $y = 0$.
 10. $x + 1 = 0$, $x + y^2 = 0$.

* 11. Use the theorem of Pappus to find the volume of the torus generated by revolving the circle

$$(x - b)^2 + y^2 = a^2, \qquad b > a$$

about the y-axis.

12. What is the centroid of the solid bounded above by the plane $z = 1 + x + y$, below by the plane $z = -2$, and on the sides by the planes $x = 1$, $x = 2$, $y = 1$, $y = 2$?

*13. Taking a, b, c as positive, find the centroid of the tetrahedron with vertices $(0, 0, 0)$, $(a, 0, 0)$, $(0, b, 0)$, and $(0, 0, c)$.

14. Find the centroid of the solid bounded above by the cylindrical surface $x^2 + z = 4$, below by the plane $x + z = 2$, and on the sides by the planes $y = 0$ and $y = 3$.

*15. The rectangle of Figure 15.7.5 is revolved about the line marked l. Find the volume of the resulting solid.

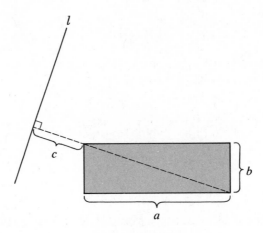

FIGURE 15.7.5

For a solid of varying density, $\rho = f(x, y, z)$, the center of mass is given by the equations

$$x_M = \frac{1}{M} \iiint_S xf(x, y, z)\, dxdydz, \qquad y_M = \frac{1}{M} \iiint_S yf(x, y, z)\, dxdydz$$

$$z_M = \frac{1}{M} \iiint_S zf(x, y, z)\, dxdydz.$$

*16. Find the center of mass of a unit cube

$$0 \le x \le 1, \qquad 0 \le y \le 1, \qquad 0 \le z \le 1$$

if the density is proportional to
(a) the square of the distance from the origin.
(b) the distance from the xy-plane.
(c) the square of the distance from the diagonal which joins $(0, 0, 0)$ to $(1, 1, 1)$.

17. Find the center of mass of the solid described in Exercise 8, Section 15.6.

18. Find the center of mass of the tetrahedron of Exercise 13 if the density at each point (x, y, z) is
*(a) yz.   (b) xyz.

19. Find the center of mass of the solid of Exercise 14 if the density at each point (x, y, z) is $1 + y$.

20. Show that in polar coordinates the formulas for the centroid become

$$\bar{x} = \frac{1}{\text{area of } \Omega} \iint_\Omega r^2 \cos \theta \, dr d\theta, \qquad \bar{y} = \frac{1}{\text{area of } \Omega} \iint_\Omega r^2 \sin \theta \, dr d\theta.$$

*21. Find the centroid of the region enclosed by the cardioid $r = a(1 + \cos \theta)$.

15.8 Cylindrical Coordinates

The cylindrical coordinates (r, θ, z) of a point P in space are illustrated in Figure 15.8.1. The first two coordinates r and θ are the usual polar coordinates. The third coordinate is the third rectangular coordinate z.

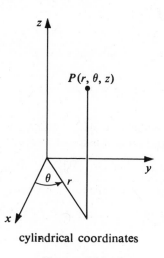

cylindrical coordinates

FIGURE 15.8.1

The fundamental solid in cylindrical coordinates is a set of the form

$$\Pi: a_1 \leq r \leq a_2, \quad b_1 \leq \theta \leq b_2, \quad d_1 \leq z \leq d_2.$$

Geometrically this is a *cylindrical wedge*. (See Figure 15.8.2.)

Suppose now that S is a solid given in cylindrical coordinates and suppose that $f(r, \theta, z)$ gives the density at each point (r, θ, z). We will sketch informally how the mass of S can be found. We begin by encasing S in a cylindrical wedge Π. We then extend the density function f to all of Π by defining f to be zero outside of S. The mass of S, whatever it is, must be the same as the mass of Π. The advantage of dealing with Π rather than with S is that we can break up Π in an obvious manner into little solids

$$\Pi_{ijk}: r_{i-1} \leq r \leq r_i, \quad \theta_{j-1} \leq \theta \leq \theta_j, \quad z_{k-1} \leq z \leq z_k,$$

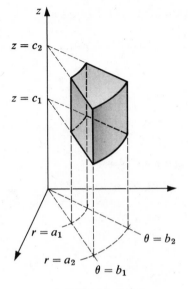

a cylindrical wedge

FIGURE 15.8.2

which are again cylindrical wedges. If M_{ijk} and m_{ijk} are respectively the maximum and minimum densities on Π_{ijk}, then obviously

$$m_{ijk}(\text{volume of } \Pi_{ijk}) \leq \text{mass of } \Pi_{ijk} \leq M_{ijk}(\text{volume of } \Pi_{ijk}).$$

Adding up all these inequalities, we have an estimate for the total mass:

$$\sum\sum\sum m_{ijk}(\text{volume of } \Pi_{ijk}) \leq \text{mass } \Pi \leq \sum\sum\sum M_{ijk}(\text{volume of } \Pi_{ijk}).$$

Since

$$\text{volume of } \Pi_{ijk} = \tfrac{1}{2}(r_i^2 - r_{i-1}^2)(\theta_j - \theta_{j-1})(z_k - z_{k-1})$$

$$= \int_{z_{k-1}}^{z_k} \int_{\theta_{j-1}}^{\theta_j} \int_{r_{i-1}}^{r_i} r \, dr \, d\theta \, dz$$

$$= \iiint_{\Pi_{ijk}} r \, dr d\theta dz,$$

we can rewrite this estimate as

$$\sum\sum\sum m_{ijk} \iiint_{\Pi_{ijk}} r \, dr d\theta dz \leq \text{mass of } \Pi \leq \sum\sum\sum M_{ijk} \iiint_{\Pi_{ijk}} r \, dr d\theta dz.$$

Much in the manner of Section 15.4, we can show that the triple integral

$$\iiint_{\Pi} f(r,\, \theta,\, z) \, r \, dr d\theta dz \qquad\qquad\text{(if it exists)}$$

does satisfy this estimate for every choice of the Π_{ijk}. As a consequence, we set

$$\text{mass of } \Pi = \iiint_\Pi f(r, \theta, z)r \, dr d\theta dz,$$

and, since Π has density 0 outside of S, we set

$$\boxed{\text{mass of } S = \iiint_S f(r, \theta, z)r \, dr d\theta dz.}$$

The triple integral

$$\iiint_S 1r \, dr d\theta dz = \iiint_S r \, dr d\theta dz$$

gives the mass of S when S has constant density 1. This is of course the volume of S:

$$\boxed{\text{volume of } S = \iiint_S r \, dr d\theta dz.}$$

We illustrate the use of these formulas in some sample problems.

Problem.   Find the mass of a cylindrical solid of radius R and height h if the density at each point is proportional to its distance from the axis of the cylinder.

SOLUTION.   We call the solid S and express it by the inequalities

$$0 \leq r \leq R, \qquad 0 \leq \theta \leq 2\pi, \qquad 0 \leq z \leq h.$$

The density function is of the form

$$f(r, \theta, z) = kr.$$

According to our formula,

$$\text{mass of } S = \iiint_S kr^2 \, dr d\theta dz$$

$$= \int_0^h \int_0^{2\pi} \int_0^R kr^2 \, dr \, d\theta \, dz$$

$$= \int_0^h \int_0^{2\pi} \tfrac{1}{3}kR^3 \, d\theta \, dz$$

$$= \int_0^h \tfrac{2}{3}\pi \, kR^3 \, dz$$

$$= \tfrac{2}{3}\pi \, kR^3 \, h. \quad \square$$

Problem. Use cylindrical coordinates to find the volume of the solid bounded above by the plane $z = y$ and below by the paraboloid $z = x^2 + y^2$.

SOLUTION. The two surfaces intersect at those points (x, y, z) where

$$y = x^2 + y^2.$$

The projection of this set onto the xy-plane is the circle

$$x^2 + y^2 - y = 0. \qquad\qquad \text{(Figure 15.8.3)}$$

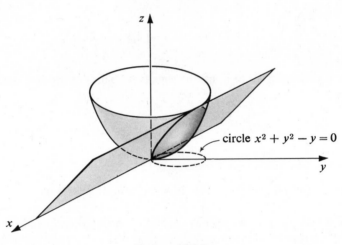

FIGURE 15.8.3

The region bounded by this circle constitutes the base region. In cylindrical coordinates the circle becomes

$$r = \sin \theta$$

and the base region becomes

$$0 \le \theta \le \pi, \qquad 0 \le r \le \sin \theta.$$

Since the upper boundary can be written as

$$z = r \sin \theta$$

and the lower boundary as

$$z = r^2,$$

the solid itself becomes

$$S: 0 \le \theta \le \pi, \quad 0 \le r \le \sin \theta, \quad r^2 \le z \le r \sin \theta.$$

The volume is then

$$V = \iiint_S r \, dr d\theta dz$$

$$= \int_0^\pi \int_0^{\sin\theta} \int_{r^2}^{r\sin\theta} r \, dz \, dr \, d\theta$$

$$= \int_0^\pi \int_0^{\sin\theta} (r^2 \sin\theta - r^3) \, dr \, d\theta$$

$$= \int_0^\pi \left[\tfrac{1}{3} r^3 \sin\theta - \tfrac{1}{4} r^4 \right]_0^{\sin\theta} d\theta$$

$$= \tfrac{1}{12} \int_0^\pi \sin^4\theta \, d\theta = \tfrac{1}{12}(\tfrac{3}{8}\pi) = \tfrac{1}{32}\pi. \quad \square$$

by Exercise 5, Section 7.3

Problem. Find the mass of the solid of the last problem if the density at each point is proportional to the square of its distance from the z-axis.

SOLUTION. The density function here is of the form

$$f(r, \theta, z) = kr^2.$$

According to the mass formula,

$$\text{mass of } S = \iiint_S kr^3 \, dr d\theta dz$$

$$= \int_0^\pi \int_0^{\sin\theta} \int_{r^2}^{r\sin\theta} kr^3 \, dz \, dr \, d\theta$$

$$= \int_0^\pi \int_0^{\sin\theta} k(r^4 \sin\theta - r^5) \, dr \, d\theta$$

$$= k \int_0^\pi \left[\tfrac{1}{5} r^5 \sin\theta - \tfrac{1}{6} r^6 \right]_0^{\sin\theta} d\theta$$

$$= k \int_0^\pi \tfrac{1}{30} \sin^6\theta \, d\theta$$

$$= \tfrac{1}{96} k\pi. \quad \square$$

Exercises

*1. Find the mass of a cylindrical solid of radius R and height h if the density at each point is proportional to
(a) the distance from one of the bases.
(b) the square root of the distance from the axis of the cylinder.
(c) the square of the distance from the curved boundary.

*2. Use cylindrical coordinates to show that a cone of base radius R and height h has volume
$$V = \tfrac{1}{3}\pi R^2 h.$$

3. Find the mass of the cone of Exercise 2 if the density is proportional to the distance
   (a) from the axis.   (b) from the base.

4. Use cylindrical coordinates to find the volume of the solid S which is bounded above by the paraboloid $z = 1 - (x^2 + y^2)$ and below by the xy-plane.

5. Find the mass of the solid of Exercise 4 if the density is proportional to
   (a) the distance from the xy-plane.
   (b) the square of the distance from the origin.

*6. Find the volume of the solid bounded by the paraboloid of revolution $x^2 + y^2 = az$, the xy-plane, and the cylinder $x^2 + y^2 = 2ax$.

*7. Find the volume of the solid bounded above by the cone $z^2 = x^2 + y^2$, below by the xy-plane, and on the sides by the cylinder $x^2 + y^2 = 2ax$.

*8. Find the volume of the solid bounded above by the plane $2z = 4 + x$, below by the xy-plane, and on the sides by the cylinder $x^2 + y^2 = 2x$.

*9. Find the volume of the solid bounded above by the cone $z = a - \sqrt{x^2 + y^2}$, below by the xy-plane, and on the sides by the cylinder $x^2 + y^2 = ax$.

10. Find the volume of the solid bounded above by $x^2 + y^2 + z^2 = 25$ and below by $z = \sqrt{x^2 + y^2} + 1$.

*11. Find the volume of the solid bounded by the plane $z = x$ and the paraboloid $z = x^2 + y^2$.

12. Find the volume of the solid bounded by the hyperboloid $z^2 = a^2 + x^2 + y^2$ and the upper nappe of the cone $z^2 = 2(x^2 + y^2)$.

15.9 Spherical Coordinates

The spherical coordinates (ρ, θ, ϕ) of a point P in space are illustrated in Figure 15.9.1. Rectangular coordinates (x, y, z) are related to spherical coordinates by the equations

$$x = \rho \sin \phi \cos \theta, \qquad y = \rho \sin \phi \sin \theta, \qquad z = \rho \cos \phi.$$

You can check out these formulas in terms of Figure 15.9.1. In spherical coordinates the fundamental solids are sets of the form

$$\Pi: a_1 \le r \le a_2, \quad b_1 \le \theta \le b_2, \quad c_1 \le \phi \le c_2.$$

Such solids are called *spherical wedges*. A typical one is pictured in Figure 15.9.2. The volume of a spherical wedge Π is given by the formula

$$V = \iiint_{\Pi} \rho^2 \sin \phi \, d\rho d\theta d\phi.$$

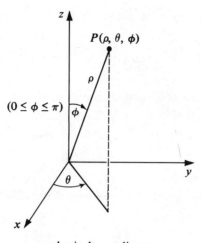

spherical coordinates

FIGURE 15.9.1

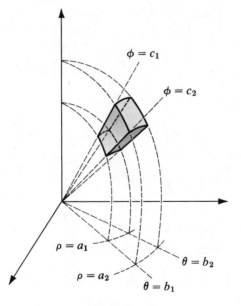

a spherical wedge

FIGURE 15.9.2

To see this, note first that Π is a solid of revolution. One way to obtain Π is to rotate the face

$$\Omega: a_1 \leq \rho \leq a_2, \quad \theta = b_1, \quad c_1 \leq \phi \leq c_2$$

about the z-axis for $b_2 - b_1$ radians. (See Figure 15.9.2.)

As you can verify (see Exercise 20 of Section 15.7), the centroid of Ω is at a distance

$$\frac{1}{\text{area of } \Omega} \iint_{\Omega} \rho^2 \sin \phi \, d\rho d\theta$$

from the z-axis. As the face Ω is rotated from $\theta = b_1$ to $\theta = b_2$, the centroid travels through a circular arc of length

$$l = (b_2 - b_1) \cdot \frac{1}{\text{area of } \Omega} \iint_{\Omega} \rho^2 \sin \phi \, d\rho d\phi.$$

From the theorem of Pappus we know that the volume of a solid of revolution is the length of the path traveled by the centroid multiplied by the area of Ω. In this case we have

$$\text{volume of } \Pi = l \; \text{area of } \Omega = (b_2 - b_1) \iint_{\Omega} \rho^2 \sin \phi \, d\rho d\phi.$$

Since

$$(b_2 - b_1) \iint_{\Omega} \rho^2 \sin \phi \, d\rho d\phi = (b_2 - b_1) \int_{a_1}^{a_2} \int_{c_1}^{c_2} \rho^2 \sin \phi \, d\phi \, d\rho$$

$$= \int_{b_1}^{b_2} \int_{a_1}^{a_2} \int_{c_1}^{c_2} \rho^2 \sin \phi \, d\phi \, d\rho \, d\theta$$

$$= \iiint_{\Pi} \rho^2 \sin \phi \, d\rho d\theta d\phi,$$

we have the desired formula

$$V = \iiint_{\Pi} \rho^2 \sin \phi \, d\rho d\theta d\phi. \quad \square$$

From the fact that a spherical wedge Π has volume

$$\iiint_{\Pi} \rho^2 \sin \phi \, d\rho d\theta d\phi$$

one can argue that the mass of Π is given by the integral

$$\iiint_{\Pi} f(\rho, \theta, \phi) \, \rho^2 \sin \phi \, d\rho d\theta d\phi$$

where, as usual, f is the density function. This formula for mass is then extended to a more general solid S in the usual manner. Encase S in a spherical wedge Π and define the density outside of S to be 0. Obviously then

$$\text{mass of } \Pi = \text{mass of } S$$

and, since

$$\iiint_{\Pi} f(\rho, \theta, \phi) \, \rho^2 \sin \phi \, d\rho d\theta d\phi = \iiint_{S} f(\rho, \theta, \phi) \, \rho^2 \sin \phi \, d\rho d\theta d\phi,$$

we have

$$
\text{mass of } S = \iiint\limits_{S} f(\rho,\, \theta,\, \phi)\, \rho^2 \sin \phi\, d\rho d\theta d\phi.
$$

If $f(\rho,\, \theta,\, \phi) = 1$, the integral on the right becomes simply

$$
\iiint\limits_{S} 1\, \rho^2 \sin \phi\, d\rho d\theta d\phi = \iiint\limits_{S} \rho^2 \sin \phi\, d\rho d\theta d\phi.
$$

This is the volume of S:

$$
\text{volume of } S = \iiint\limits_{S} \rho^2 \sin \phi\, d\rho d\theta d\phi.
$$

We illustrate the application of these formulas in some problems.

Problem. Find the mass of a solid ball of radius 1 if the density at each point d units from the center is

$$
\frac{1}{1 + d^2}.
$$

SOLUTION. In spherical coordinates the unit ball is the solid

$$
S: 0 \le \rho \le 1, \quad 0 \le \theta \le 2\pi, \quad 0 \le \phi \le \pi.
$$

The density function is

$$
f(\rho,\, \theta,\, \phi) = \frac{1}{1 + \rho^2}.
$$

We therefore have

$$
\begin{aligned}
\text{mass of } S &= \iiint\limits_{S} \frac{\rho^2}{1 + \rho^2} \sin \phi\, d\rho d\theta d\phi \\[2mm]
&= \int_0^\pi \int_0^{2\pi} \int_0^1 \frac{\rho^2}{1 + \rho^2} \sin \phi\, d\rho\, d\theta\, d\phi \\[2mm]
&= \int_0^\pi \int_0^{2\pi} \int_0^1 \left(1 - \frac{1}{1 + \rho^2}\right) \sin \phi\, d\rho\, d\theta\, d\phi \\[2mm]
&= \int_0^\pi \int_0^{2\pi} (1 - \tfrac{1}{4}\pi) \sin \phi\, d\theta\, d\phi \\[2mm]
&= (1 - \tfrac{1}{4}\pi) \int_0^\pi 2\pi \sin \phi\, d\phi \\[2mm]
&= 4\pi - \pi^2. \quad \square
\end{aligned}
$$

Problem. Find the volume of the solid enclosed by the surface

$$(x^2 + y^2 + z^2)^2 = 2z(x^2 + y^2).$$

SOLUTION. We begin by expressing the surface in spherical coordinates. Since z remains nonnegative, ϕ must remain less than or equal to $\frac{1}{2}\pi$. (Figure 15.9.1) With $x = \rho \sin \phi \cos \theta$, $y = \rho \sin \phi \sin \theta$, $z = \rho \cos \phi$, we find that the bounding surface becomes

$$\rho = 2 \sin^2 \phi \cos \phi. \qquad\qquad \text{(check this out)}$$

We can express the solid inside this surface by

$$S: 0 \le \theta \le 2\pi, \quad 0 \le \phi \le \tfrac{1}{2}\pi, \quad 0 \le \rho \le 2 \sin^2 \phi \cos \phi.$$

Consequently,

$$V = \iiint_S \rho^2 \sin \phi \, d\rho d\theta d\phi$$

$$= \int_0^{2\pi} \int_0^{\pi/2} \int_0^{2 \sin^2 \phi \cos \phi} \rho^2 \sin \phi \, d\rho \, d\phi \, d\theta$$

$$= \int_0^{2\pi} \int_0^{\pi/2} \tfrac{8}{3} \sin^7 \phi \cos^3 \phi \, d\phi \, d\theta$$

$$= \tfrac{8}{3} \int_0^{2\pi} \int_0^{\pi/2} (\sin^7 \phi \cos \phi - \sin^9 \phi \cos \phi) \, d\phi \, d\theta$$

$$= \tfrac{8}{3} \int_0^{2\pi} \tfrac{1}{40} \, d\theta$$

$$= \tfrac{2}{15} \pi. \quad \square$$

Exercises

* 1. Find the spherical coordinates (ρ, θ, ϕ) of the point with rectangular coordinates $(1, 1, 1)$.
2. Find the rectangular coordinates of the point with spherical coordinates $(2, \frac{1}{6}\pi, \frac{1}{4}\pi)$.
* 3. Find the spherical coordinates of the point with cylindrical coordinates $(2, \frac{2}{3}\pi, 6)$.

The following equations are given in spherical coordinates. Interpret each one geometrically.

* 4. $\rho = \rho_0$.   * 5. $\theta = \theta_0$.   * 6. $\phi = \phi_0$.

7. $\tan \theta = 1$.   * 8. $\rho \cos \phi = 1$.   9. $\rho = \cos \phi$.

10. Derive the formula for the volume of a sphere of radius a using spherical coordinates.

*11. A wedge is cut from a ball of radius a by two planes which meet at a diameter. Find the volume of the wedge if the angle between the planes is α radians.

12. Find the mass of a ball of radius a if the density at each point is proportional to its distance from the boundary.

*13. Use spherical coordinates to derive the formula for the volume of a circular cone of base radius r and height h.

*14. Find the mass of a circular cone of base radius r and height h if the density at each point is proportional to its distance from the vertex.

*15. Find the volume of the solid common to the sphere $\rho = a$ and the cone $\phi = \alpha$. Take $\alpha \in (0, \frac{1}{2}\pi)$.

*16. Find an equation for the sphere

$$x^2 + y^2 + (z - b)^2 = b^2$$

in spherical coordinates.

17. Find the mass of the ball

$$\rho \leq 2b \cos \phi$$

if the density at each point (ρ, θ, ϕ) is

(a) ρ.   *(b) $\rho \sin \phi$.   (c) $\rho \cos^2 \theta \sin \phi$.

*18. Find the volume of the solid common to the two spheres

$$\rho = 2 \quad \text{and} \quad \rho = 2\sqrt{2} \cos \phi.$$

19. Find the volume of the solid enclosed by the surface

$$\rho = 1 - \cos \phi.$$

Appendix A

A.1 Sets

By a *set* we mean a collection of objects. We are not particular about the nature of the objects. However, we do require that a set be *well defined*; namely, given any object x, the statement "x is in the set" must be either true or false. The question "Is x in the set?" must be answerable in terms of an unqualified yes or an unqualified no. This excludes collections that involve highly subjective judgments, such as "all men of good will" and "all pleasant days."

The objects which are in the set are known as the *elements* or *members* of the set. To indicate "x is a member of the set A," we write

$$x \in A.$$

This is also read "x is an element of A" or "x belongs to A" or "x is in A." The notation $x \notin A$ is used to indicate that x is not an element of A. For example,

$$-1 \notin \text{ set of positive numbers.}$$

A set B is called a *subset* of a set A if and only if

$$x \in B \quad \text{implies} \quad x \in A.$$

Thus, for B to fail to be a subset of A, there must be an object in B which is not in A. To indicate that B is a subset of A, we write $B \subseteq A$ or $A \supseteq B$. (See Figure A.1.1.) These expressions are also read "B is contained in A" or "A contains B."

Example. The set of equilateral triangles $\subseteq$ the set of triangles.

Example. The set of integers $\supseteq$ the set of even integers.

Example. The set of college freshmen $\subseteq$ the set of college students.

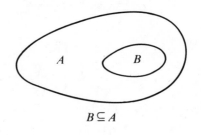

$$B \subseteq A$$

FIGURE A.1.1

Equality between sets is reserved for identity. That is, $A = B$ if and only if A and B have exactly the same membership. More formally, $A = B$ if and only if $A \subseteq B$ and $B \subseteq A$.

We distinguish between the element a of A and the subset $\{a\}$ of A which has a as the only element. Thus $a \in \{a\}$ is valid, but $a = \{a\}$ is not. The set consisting of the elements a and b is denoted by $\{a, b\}$. That consisting of a, b, and c is denoted by $\{a, b, c\}$, and so on.

The set of elements common to two sets A and B is called the *intersection* of A and B, and is denoted by $A \cap B$. (See Figure A.1.2.) Thus,

$$x \in A \cap B \qquad \text{if and only if} \qquad x \in A \quad \text{and} \quad x \in B.$$

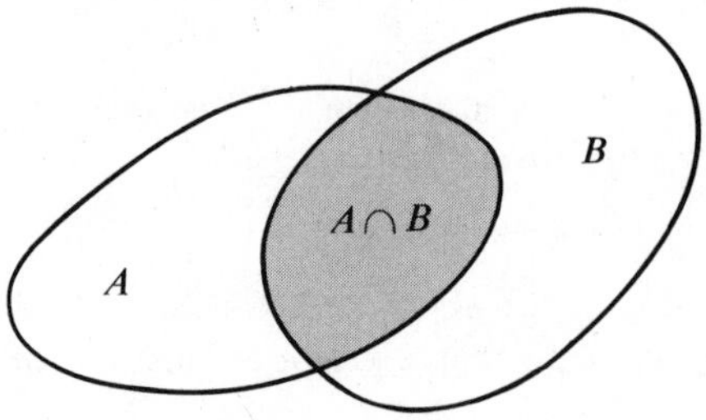

FIGURE A.1.2

Example. If A is the set of nonnegative numbers and B is the set of nonpositive numbers, then $A \cap B = \{0\}$.

Example. If A is the set of all multiples of 3 and B is the set of all multiples of 4, then $A \cap B$ is the set of all multiples of 12.

If the sets A and B have no elements in common, we say that A and B are disjoint and write $A \cap B = \emptyset$. We regard $\emptyset$ as a set having no elements and refer to it as an *empty, vacuous,* or *void* set.

Example.   The set of all real numbers whose square is negative is an empty set.

The *union* of sets A and B, written $A \cup B$, is the set of elements which are either in A or in B. (See Figure A.1.3.) This does not exclude objects which are elements of both A and B. In symbols,

$$x \in A \cup B \quad \text{if and only if} \quad x \in A \quad \text{or} \quad x \in B.$$

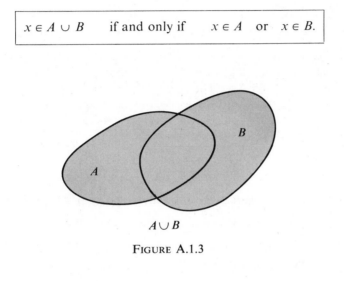

$A \cup B$

FIGURE A.1.3

Example.   If A is the set of all even numbers between 1 and 9 and B is the set of all even numbers greater than 5, then $A \cup B$ is the set of all positive even numbers.

Example.   If A is the set of all college students and B is the set of all college freshmen, then $A \cup B$ is the set of all college students.

Set theory is now a highly developed discipline, but it is not our intention to develop it here. The notions just introduced are enough for our purposes.

A.2   Induction

Suppose that one were asked to verify that a certain set S contains the set of positive integers. One could in sequence start to verify that

$$1 \in S, \quad 2 \in S, \quad 3 \in S, \quad 4 \in S, \quad 5 \in S, \quad \text{etc.}$$

Even if for each positive integer n it required only one second to verify that $n \in S$, such a procedure could not in any finite amount of time yield a verification that S contains the entire set of positive integers. To avoid such difficulties, mathematicians have invented a special procedure. That this procedure works is an *assumption* we make. We therefore state the result as an axiom.

Axiom of Induction

Let S be a set of integers. If

$$(A) \; 1 \in S \quad \text{and} \quad (B) \; k \in S \text{ implies } k + 1 \in S,$$

then all the positive integers are in S.

To show the *plausibility* of this axiom, we suppose that S satisfies conditions (A) and (B). Let's choose a positive integer m and "argue" that $m \in S$.

From (A) we know that $1 \in S$. Thus from (B) we know that $1 + 1 \in S$, and thus that $(1 + 1) + 1 \in S$, and so on. Since m can be obtained from 1 by adding 1, $(m - 1)$ successive times, it "seems obvious" that $m \in S$.

As an example of the above procedure, we note that

$$5 = ([(1 + 1) + 1] + 1) + 1$$

and thus $5 \in S$.

As an application of the axiom of induction, we prove that, if $x \geq -1$, then

$$(1 + x)^n \geq (1 + nx) \quad \text{for all positive integers } n.$$

Let S be the set of integers for which the inequality holds. We have to show that S contains the set of positive integers. Since

$$(1 + x)^1 \geq (1 + 1x),$$

we see that $1 \in S$. If $k \in S$, then

$$(1 + x)^k \geq (1 + kx).$$

Thus

$$(1 + x)^{k+1} = (1 + x)^k(1 + x) \geq (1 + kx)(1 + x),$$

the inequality holding since $1 + x \geq 0$. Since

$$(1 + kx)(1 + x) = 1 + (k + 1)x + kx^2 \geq 1 + (k + 1)x,$$

it follows that

$$(1 + x)^{k+1} \geq 1 + (k + 1)x,$$

and thus $k + 1 \in S$. Having shown that $1 \in S$ and that

$$k \in S \quad \text{implies} \quad k + 1 \in S,$$

we conclude that S contains the set of positive integers. $\square$

Exercises

1. Show that the following statements hold for each positive integer n:
   (a) $n(n + 1)$ is divisible by 2. HINT: $(k + 1)(k + 2) = k(k + 1) + 2(k + 1)$.
   (b) $n(n + 1)(n + 2)$ is divisible by 6.

(c) $1 + 2 + 3 + \cdots + n = \frac{1}{2}n(n + 1)$.

(d) $1 + 3 + 5 + \cdots + (2n - 1) = n^2$.

(e) $1^2 + 2^2 + 3^2 + \cdots + n^2 = \frac{1}{6}n(n + 1)(2n + 1)$.

(f) $1^3 + 2^3 + 3^3 + \cdots + n^3 = (1 + 2 + 3 + \cdots + n)^2$. HINT: use part (c).

(g) $3^{2n+1} + 2^{n+2}$ is divisible by 7.

2. For what integers n is $9^n - 8n - 1$ divisible by 64? Prove that your answer is correct.

In the rest of the Appendix we prove some important results that we did not prove in the main body of the text. Many details are omitted. These are left as exercises.

A.3　The Intermediate-Value Theorem (Bernard Bolzano, 1781–1848)

Lemma

Let f be continuous on $[a, b]$. If

$$f(a) < 0 < f(b) \quad \text{or} \quad f(b) < 0 < f(a),$$

then there is a number c between a and b such that

$$f(c) = 0.$$

PROOF.　Suppose that $f(a) < 0 < f(b)$. (The other case can be treated in a similar manner.) Since $f(a) < 0$, we know from the continuity of f that there exists a number ξ such that f is negative on $[a, \xi)$. Let

$$c = \text{lub } \{\xi : f \text{ is negative on } [a, \xi)\}.$$

Clearly, $c \leq b$. We cannot have $f(c) > 0$, for then f would be positive on some interval extending to the left of c, and we know that, to the left of c, f is negative. Incidentally this argument excludes the possibility $c = b$ and means that $c < b$. We cannot have $f(c) < 0$, for then there would be an interval $[a, t)$, with $t > c$, on which f is negative, and this would contradict the definition of c. It follows that $f(c) = 0$. □

The Intermediate-Value Theorem

If f is continuous on $[a, b]$ and C is a number between $f(a)$ and $f(b)$, then there is at least one number c between a and b such that

$$f(c) = C.$$

PROOF. Suppose for example that

$$f(a) < C < f(b).$$

(The other possibility can be handled in a similar manner.) The function

$$g(x) = f(x) - C$$

is continuous on $[a, b]$. Since

$$g(a) = f(a) - C < 0 \quad \text{and} \quad g(b) = f(b) - C > 0,$$

we know from the lemma that there is a number c between a and b such that

$$g(c) = 0.$$

Obviously then

$$f(c) = C. \quad \square$$

A.4 The Maximum-Minimum Theorem (Karl Weierstrass, 1815–1897)

Lemma

If f is continuous on $[a, b]$, then f is bounded on $[a, b]$.

PROOF. Consider

$$\{x: x \in [a, b] \text{ and } f \text{ is bounded on } [a, x]\}.$$

It is easy to see that this set is nonempty and bounded above by b. Thus we can set

$$c = \text{lub } \{x: f \text{ is bounded on } [a, x]\}.$$

Now we argue that $c = b$. To do so, we suppose that

$$c < b.$$

From the continuity of f at c, it is easy to see that f is bounded on $[c - \epsilon, c + \epsilon]$ for some $\epsilon > 0$. Being bounded on $[a, c - \epsilon]$ and on $[c - \epsilon, c + \epsilon]$, it is obviously bounded on $[a, c + \epsilon]$. This contradicts our choice of c. We can therefore conclude that $c = b$. This tells us that

$$f \text{ is bounded on } [a, x] \text{ for all } x < b.$$

We are now almost through. From the continuity of f, we know that f is bounded on some interval of the form $[b - \epsilon, b]$. Since $b - \epsilon < b$, we know from what we just proved that f is bounded on $[a, b - \epsilon]$. Being bounded on $[a, b - \epsilon]$ and on $[b - \epsilon, b]$, it is bounded on $[a, b]$. $\square$

The Maximum-Minimum Value Theorem

If f is continuous on $[a, b]$, then, somewhere on $[a, b]$, f takes on a maximum value M and a minimum value m.

PROOF. By the lemma, f is bounded on $[a, b]$. Set

$$M = \text{lub } \{f(x): x \in [a, b]\}.$$

We must show that there exists c in $[a, b]$ such that

$$f(c) = M.$$

To do this, we set

$$g(x) = \frac{1}{M - f(x)}.$$

If f does not take on the value M, then g is continuous on $[a, b]$ and thus, by the lemma, bounded on $[a, b]$. A look at the definition of g makes it clear that g cannot be bounded on $[a, b]$. The assumption that f does not take on the value M has led to a contradiction. (That f takes on a minimum value m can be proved in a similar manner.) □

A.5 Inverses

Continuity Theorem

Let f be a one-to-one function defined on an interval (a, b). If f is continuous, then its inverse f^{-1} is also continuous.

PROOF. If f is continuous, then, being one-to-one, f either increases throughout (a, b) or it decreases throughout (a, b). The proof of this assertion we leave to you.

Let's suppose now that f increases throughout (a, b). Let's take c in the domain of f^{-1} and show that f^{-1} is continuous at c.

We first observe that $f^{-1}(c)$ lies in (a, b) and choose $\epsilon > 0$ sufficiently small so that $f^{-1}(c) - \epsilon$ and $f^{-1}(c) + \epsilon$ also lie in (a, b). We seek $\delta > 0$ such that

$$\text{if} \quad c - \delta < x < c + \delta \quad \text{then} \quad f^{-1}(c) - \epsilon < f^{-1}(x) < f^{-1}(c) + \epsilon.$$

This condition can be met by choosing δ to satisfy

$$f(f^{-1}(c) - \epsilon) < c - \delta \quad \text{and} \quad c + \delta < f(f^{-1}(c) + \epsilon)$$

for then, if

$$c - \delta < x < c + \delta,$$

then

$$f(f^{-1}(c) - \epsilon) < x < f(f^{-1}(c) + \epsilon)$$

and, since f^{-1} also increases,

$$f^{-1}(c) - \epsilon < f^{-1}(x) < f^{-1}(c) + \epsilon.$$

The case where f decreases throughout (a, b) can be handled in a similar manner. $\quad\square$

Differentiability Theorem

Let f be a one-to-one function defined on an interval (a, b). If f is differentiable and its derivative does not take on the value 0, then f^{-1} is differentiable and

$$(f^{-1})'(x) = \frac{1}{f'(f^{-1}(x))}.$$

PROOF. Let x be in the domain of f^{-1}. We take $\epsilon > 0$ and show that there exists $\delta > 0$ such that

$$\text{if} \quad 0 < |t - x| < \delta \quad \text{then} \quad \left| \frac{f^{-1}(t) - f^{-1}(x)}{t - x} - \frac{1}{f'(f^{-1}(x))} \right| < \epsilon.$$

Since f is differentiable at $f^{-1}(x)$ and $f'(f^{-1}(x)) \neq 0$, there exists $\delta_1 > 0$ such that if

$$0 < |y - f^{-1}(x)| < \delta_1$$

then

$$\left| \frac{1}{\dfrac{f(y) - f(f^{-1}(x))}{y - f^{-1}(x)}} - \frac{1}{f'(f^{-1}(x))} \right| < \epsilon$$

and therefore

$$\left| \frac{y - f^{-1}(x)}{f(y) - f(f^{-1}(x))} - \frac{1}{f'(f^{-1}(x))} \right| < \epsilon.$$

By the previous theorem, f^{-1} is continuous at x and therefore there exists $\delta > 0$ such that

$$\text{if} \quad 0 < |t - x| < \delta \quad \text{then} \quad 0 < |f^{-1}(t) - f^{-1}(x)| < \delta_1.$$

It follows from the special property of δ_1 that

$$\left| \frac{f^{-1}(t) - f^{-1}(x)}{t - x} - \frac{1}{f'(f^{-1}(x))} \right| < \epsilon. \quad\square$$

A.6   The Integrability of Continuous Functions (G. F. B. Riemann, 1826–1866)

The aim here is to prove that, if f is continuous on $[a, b]$, then there is one and only one number I which satisfies the inequality

$$L_f(P) \leq I \leq U_f(P) \quad \text{for all partitions } P \text{ of } [a, b].$$

Definition

A function f is said to be *uniformly continuous* on $[a, b]$ iff for each $\epsilon > 0$ there exists $\delta > 0$ such that, if

$$x, y \in [a, b] \quad \text{and} \quad |x - y| < \delta,$$

then

$$|f(x) - f(y)| < \epsilon.$$

For convenience, let's agree to say that *the interval $[a, b]$ has the property P_ϵ* iff there exist sequences $\{x_n\}$, $\{y_n\}$ satisfying

$$x_n, y_n \in [a, b], \quad |x_n - y_n| < 1/n, \quad |f(x_n) - f(y_n)| \geq \epsilon.$$

Lemma

If f is not uniformly continuous on $[a, b]$, then $[a, b]$ has the property P_ϵ for some $\epsilon > 0$.

The proof is easy. □

Lemma

Let f be continuous on $[a, b]$. If $[a, b]$ has the property P_ϵ, then at least one of the subintervals $[a, \frac{1}{2}(a + b)]$, $[\frac{1}{2}(a + b), b]$ has the property P_ϵ.

PROOF. Let's suppose that the lemma is false. For convenience, we let

$$c = \tfrac{1}{2}(a + b),$$

so that the halves become

$$[a, c] \quad \text{and} \quad [c, b].$$

Since $[a, c]$ fails to have the property P_ϵ, there exists an integer p such that, if

$$x, y \in [a, c] \quad \text{and} \quad |x - y| < 1/p,$$

then

$$|f(x) - f(y)| < \epsilon.$$

Since $[c, b]$ fails to have the property P_ϵ, there exists an integer q such that, if

$$x, y \in [c, b] \quad \text{and} \quad |x - y| < 1/q,$$

then

$$|f(x) - f(y)| < \epsilon.$$

Since f is continuous at c, there exists an integer r such that if $|x - c| < 1/r$, then $|f(x) - f(c)| < \frac{1}{2}\epsilon$. Set

$$s = \max\{p, q, r\}$$

and suppose that

$$x, y \in [a, b], \qquad |x - y| < 1/s.$$

If x, y are both in $[a, c]$ or both in $[c, b]$, then

$$|f(x) - f(y)| < \epsilon.$$

The only other possibility is that

$$x \in [a, c] \quad \text{and} \quad y \in [c, b].$$

In this case we have

$$|x - c| < 1/r, \qquad |y - c| < 1/r,$$

and thus

$$|f(x) - f(c)| < \tfrac{1}{2}\epsilon, \qquad |f(y) - f(c)| < \tfrac{1}{2}\epsilon.$$

By the triangle inequality, we again have

$$|f(x) - f(y)| < \epsilon.$$

In summary, we have obtained the existence of an integer s with the property that

$$x, y \in [a, b], \quad |x - y| < 1/s \quad \text{implies} \quad |f(x) - f(y)| < \epsilon.$$

Hence $[a, b]$ does not have the property P_c. This is a contradiction and proves the lemma. □

Theorem

If f is continuous on $[a, b]$, then f is uniformly continuous on $[a, b]$.

PROOF. We suppose that f is not uniformly continuous on $[a, b]$ and base our argument on a mathematical version of the classical maxim "Divide and Conquer."

By the first lemma of this section, we know that $[a, b]$ has the property P_c for some $\epsilon > 0$. We bisect $[a, b]$ and note by the second lemma that one of the halves, say $[a_1, b_1]$, has the property P_c. We then bisect $[a_1, b_1]$ and note that one of the halves, say $[a_2, b_2]$, has the property P_c. Continuing in this manner, we obtain a sequence of intervals $[a_n, b_n]$, each with the property P_ϵ. For each n, we can choose $x_n, y_n \in [a_n, b_n]$ such that

$$|x_n - y_n| < 1/n \quad \text{and} \quad |f(x_n) - f(y_n)| \geq \epsilon.$$

Since

$$a \leq a_n \leq a_{n+1} < b_{n+1} \leq b_n \leq b,$$

we see that the sequences

$$\{a_n\} \quad \text{and} \quad \{b_n\}$$

are both bounded and monotonic. Thus they are convergent. Since $b_n - a_n \to 0$, we see that $\{a_n\}$ and $\{b_n\}$ both converge to the same limit, say l. From the inequality

$$a_n \leq x_n \leq y_n \leq b_n,$$

we conclude that

$$x_n \to l \quad \text{and} \quad y_n \to l.$$

This tells us that

$$|f(x_n) - f(y_n)| \to |f(l) - f(l)| = 0,$$

which contradicts the statement that for all n

$$|f(x_n) - f(y_n)| \geq \epsilon. \quad \square$$

Lemma

If P and Q are partitions of $[a, b]$, then

$$L_f(P) \leq U_f(Q).$$

PROOF. $P \cup Q$ is a partition of $[a, b]$ which contains both P and Q. It's obvious then that

$$L_f(P) \leq L_f(P \cup Q) \leq U_f(P \cup Q) \leq U_f(Q). \quad \square$$

From the last lemma it follows that the set of all lower sums is bounded above and has a least upper bound L. The number L satisfies the inequality

$$L_f(P) \leq L \leq U_f(P) \quad \text{for all partitions } P$$

and is clearly the least of such numbers. Similarly, we find that the set of all upper sums is bounded below and has a greatest lower bound U. The number U satisfies the inequality

$$L_f(P) \leq U \leq U_f(P) \quad \text{for all partitions } P$$

and is clearly the greatest of such numbers.

We are now ready to prove the basic theorem.

The Integrability Theorem

If f is continuous on $[a, b]$, then there exists one and only one number I which satisfies

$$L_f(P) \leq I \leq U_f(P) \quad \text{for all partitions } P \text{ of } [a, b].$$

PROOF. We know that

$$L_f(P) \leq L \leq U \leq U_f(P) \quad \text{for all } P,$$

so that existence is no problem. We will have uniqueness if we can prove that

$$L = U.$$

To do this, we let $\epsilon > 0$ and note that f, being continuous on $[a, b]$, is uniformly continuous on $[a, b]$. Thus there exists $\delta > 0$ such that, if

$$x, y \in [a, b] \quad \text{and} \quad |x - y| < \delta,$$

then

$$|f(x) - f(y)| < \frac{\epsilon}{b - a}.$$

We now choose a partition

$$P = \{x_0, x_1, \ldots, x_n\}$$

for which

$$\max \Delta x_i < \delta.$$

For this partition P, we have

$$U_f(P) - L_f(P) = \sum_{i=1}^{n} M_i \, \Delta x_i - \sum_{i=1}^{n} m_i \, \Delta x_i$$

$$= \sum_{i=1}^{n} (M_i - m_i) \, \Delta x_i$$

$$< \sum_{i=1}^{n} \frac{\epsilon}{b - a} \Delta x_i$$

$$= \frac{\epsilon}{b - a} \sum_{i=1}^{n} \Delta x_i$$

$$= \frac{\epsilon}{b - a} (b - a)$$

$$= \epsilon.$$

Since

$$U_f(P) - L_f(P) < \epsilon \text{ and } 0 \leq U - L \leq U_f(P) - L_f(P),$$

you can see that

$$0 \leq U - L < \epsilon.$$

Since ϵ was chosen arbitrarily, we must have

$$U - L = 0 \quad \text{and} \quad L = U. \quad \square$$

A.7   A Variant of Bliss's Theorem (G. A. Bliss, 1876–1951)

For the notation we refer to Section 9.4.

Theorem

If f and g are continuous and nonnegative on $[a, b]$, then the definite integral

$$\int_a^b f(x)g(x)\, dx$$

is the only number I which satisfies the inequality

$$L_{f,g}^{\star}(P) \leq I \leq U_{f,g}^{\star}(P) \quad \text{for all partitions } P \text{ of } [a, b].$$

PROOF.   Take f and g as continuous and nonnegative on $[a, b]$ and take P as an arbitrary partition. It's easy to show first of all that

$$L_{f,g}^{\star}(P) \leq L_{fg}(P) \quad \text{and} \quad U_{fg}(P) \leq U_{f,g}^{\star}(P).$$

Since in general

$$L_{fg}(P) \leq \int_a^b f(x)g(x)\, dx \leq U_{fg}(P),$$

it follows that

$$L_{f,g}^{\star}(P) \leq \int_a^b f(x)g(x)\, dx \leq U_{f,g}^{\star}(P).$$

We have shown that the integral lies between $L_{f,g}^{\star}(P)$ and $U_{f,g}^{\star}(P)$ for all partitions P of $[a, b]$. We show now that the integral is the *only* such number. We assume that there are two such numbers I_1 and I_2 and obtain a contradiction by showing the existence of a partition P for which

$$U_{f,g}^{\star}(P) - L_{f,g}^{\star}(P) < |I_1 - I_2|.$$

Since f and g are continuous on $[a, b]$ they are there uniformly continuous. It follows that there exists $\delta > 0$ such that if

$$x, y \in [a, b] \quad \text{and} \quad |x - y| < \delta$$

then

$$|f(x) - f(y)| < \frac{|I_1 - I_2|}{2(b - a)[1 + M(g)]} \quad \text{and} \quad |g(x) - g(y)| < \frac{|I_1 - I_2|}{2(b - a)[1 + M(f)]},$$

where $M(f)$ and $M(g)$ are the maximum values of f and g on all of $[a, b]$. We now choose a partition

$$P = \{x_0, x_1, \ldots, x_n\}$$

for which

$$\max \Delta x_i < \delta.$$

For this partition P we have

$$U^{\star}_{f,g}(P) - L^{\star}_{f,g}(P) = \sum_{i=1}^{n} M_i(f)M_i(g)\,\Delta x_i - \sum_{i=1}^{n} m_i(f)m_i(g)\,\Delta x_i$$

$$= \sum_{i=1}^{n} [M_i(f)M_i(g) - m_i(f)m_i(g)]\,\Delta x_i$$

$$= \sum_{i=1}^{n} [M_i(f)M_i(g) - M_i(f)m_i(g) + M_i(f)m_i(g) - m_i(f)m_i(g)]\,\Delta x_i$$

$$= \sum_{i=1}^{n} \{M_i(f)[M_i(g) - m_i(g)] + m_i(g)[M_i(f) - m_i(g)]\}\,\Delta x_i$$

$$\leq \sum_{i=1}^{n} \left\{ \frac{M(f)|I_1 - I_2|}{2(b-a)[1+M(f)]} + \frac{M(g)|I_1 - I_2|}{2(b-a)[1+M(g)]} \right\}\,\Delta x_i$$

$$< \sum_{i=1}^{n} \left(\frac{|I_1 - I_2|}{b-a} \right)\,\Delta x_i$$

$$= \frac{|I_1 - I_2|}{b-a} \sum_{i=1}^{n} \Delta x_i$$

$$= \frac{|I_1 - I_2|}{b-a}\,(b-a)$$

$$= |I_1 - I_2|. \quad \square$$

Appendix B. Tables

TABLE 1. Natural logs

x	$\log x$	x	$\log x$	x	$\log x$
		4.0	1.386	8.0	2.079
0.1	-2.303	4.1	1.411	8.1	2.092
0.2	-1.609	4.2	1.435	8.2	2.104
0.3	-1.204	4.3	1.459	8.3	2.116
0.4	-0.916	4.4	1.482	8.4	2.128
0.5	-0.693	4.5	1.504	8.5	2.140
0.6	-0.511	4.6	1.526	8.6	2.152
0.7	-0.357	4.7	1.548	8.7	2.163
0.8	-0.223	4.8	1.569	8.8	2.175
0.9	-0.105	4.9	1.589	8.9	2.186
1.0	0.000	5.0	1.609	9.0	2.197
1.1	0.095	5.1	1.629	9.1	2.208
1.2	0.182	5.2	1.649	9.2	2.219
1.3	0.262	5.3	1.668	9.3	2.230
1.4	0.336	5.4	1.686	9.4	2.241
1.5	0.405	5.5	1.705	9.5	2.251
1.6	0.470	5.6	1.723	9.6	2.262
1.7	0.531	5.7	1.740	9.7	2.272
1.8	0.588	5.8	1.758	9.8	2.282
1.9	0.642	5.9	1.775	9.9	2.293
2.0	0.693	6.0	1.792	10	2.303
2.1	0.742	6.1	1.808	20	2.996
2.2	0.788	6.2	1.825	30	3.401
2.3	0.833	6.3	1.841	40	3.689
2.4	0.875	6.4	1.856	50	3.912
2.5	0.916	6.5	1.872	60	4.094
2.6	0.956	6.6	1.887	70	4.248
2.7	0.993	6.7	1.902	80	4.382
2.8	1.030	6.8	1.917	90	4.500
2.9	1.065	6.9	1.932	100	4.605
3.0	1.099	7.0	1.946		
3.1	1.131	7.1	1.960		
3.2	1.163	7.2	1.974		
3.3	1.194	7.3	1.988		
3.4	1.224	7.4	2.001		
3.5	1.253	7.5	2.015		
3.6	1.281	7.6	2.028		
3.7	1.308	7.7	2.041		
3.8	1.335	7.8	2.054		
3.9	1.361	7.9	2.067		

TABLE 2. Exponentials (0.01 to 0.99)

x	e^x	e^{-x}	x	e^x	e^{-x}	x	e^x	e^{-x}
0.01	1.010	0.990	0.34	1.405	0.712	0.67	1.954	0.512
0.02	1.020	0.980	0.35	1.419	0.705	0.68	1.974	0.507
0.03	1.030	0.970	0.36	1.433	0.698	0.69	1.994	0.502
0.04	1.041	0.961	0.37	1.448	0.691	0.70	2.014	0.497
0.05	1.051	0.951	0.38	1.462	0.684	0.71	2.034	0.492
0.06	1.062	0.942	0.39	1.477	0.677	0.72	2.054	0.487
0.07	1.073	0.932	0.40	1.492	0.670	0.73	2.075	0.482
0.08	1.083	0.923	0.41	1.507	0.664	0.74	2.096	0.477
0.09	1.094	0.914	0.42	1.522	0.657	0.75	2.117	0.472
0.10	1.105	0.905	0.43	1.537	0.651	0.76	2.138	0.468
0.11	1.116	0.896	0.44	1.553	0.644	0.77	2.160	0.463
0.12	1.127	0.887	0.45	1.568	0.638	0.78	2.181	0.458
0.13	1.139	0.878	0.46	1.584	0.631	0.79	2.203	0.454
0.14	1.150	0.869	0.47	1.600	0.625	0.80	2.226	0.449
0.15	1.162	0.861	0.48	1.616	0.619	0.81	2.248	0.445
0.16	1.174	0.852	0.49	1.632	0.613	0.82	2.270	0.440
0.17	1.185	0.844	0.50	1.649	0.607	0.83	2.293	0.436
0.18	1.197	0.835	0.51	1.665	0.600	0.84	2.316	0.432
0.19	1.209	0.827	0.52	1.682	0.595	0.85	2.340	0.427
0.20	1.221	0.819	0.53	1.699	0.589	0.86	2.363	0.423
0.21	1.234	0.811	0.54	1.716	0.583	0.87	2.387	0.419
0.22	1.246	0.803	0.55	1.733	0.577	0.88	2.411	0.415
0.23	1.259	0.795	0.56	1.751	0.571	0.89	2.435	0.411
0.24	1.271	0.787	0.57	1.768	0.566	0.90	2.460	0.407
0.25	1.284	0.779	0.58	1.786	0.560	0.91	2.484	0.403
0.26	1.297	0.771	0.59	1.804	0.554	0.92	2.509	0.399
0.27	1.310	0.763	0.60	1.822	0.549	0.93	2.535	0.395
0.28	1.323	0.756	0.61	1.840	0.543	0.94	2.560	0.391
0.29	1.336	0.748	0.62	1.859	0.538	0.95	2.586	0.387
0.30	1.350	0.741	0.63	1.878	0.533	0.96	2.612	0.383
0.31	1.363	0.733	0.64	1.896	0.527	0.97	2.638	0.379
0.32	1.377	0.726	0.65	1.916	0.522	0.98	2.664	0.375
0.33	1.391	0.719	0.66	1.935	0.517	0.99	2.691	0.372

TABLE 3.   Exponentials (1.0 to 4.9)

x	e^x	e^{-x}	x	e^x	e^{-x}
1.0	2.718	0.368	3.0	20.086	0.050
1.1	3.004	0.333	3.1	22.198	0.045
1.2	3.320	0.301	3.2	24.533	0.041
1.3	3.669	0.273	3.3	27.113	0.037
1.4	4.055	0.247	3.4	29.964	0.033
1.5	4.482	0.223	3.5	33.115	0.030
1.6	4.953	0.202	3.6	36.598	0.027
1.7	5.474	0.183	3.7	40.447	0.025
1.8	6.050	0.165	3.8	44.701	0.024
1.9	6.686	0.150	3.9	49.402	0.020
2.0	7.389	0.135	4.0	54.598	0.018
2.1	8.166	0.122	4.1	60.340	0.017
2.2	9.025	0.111	4.2	66.686	0.015
2.3	9.974	0.100	4.3	73.700	0.014
2.4	11.023	0.091	4.4	81.451	0.012
2.5	12.182	0.082	4.5	90.017	0.011
2.6	13.464	0.074	4.6	99.484	0.010
2.7	14.880	0.067	4.7	109.947	0.009
2.8	16.445	0.061	4.8	121.510	0.008
2.9	18.174	0.055	4.9	134.290	0.007

TABLE 4.   Sines, cosines, tangents (radian measure)

x	$\sin x$	$\cos x$	$\tan x$	x	$\sin x$	$\cos x$	$\tan x$
0.00	0.000	1.000	0.000	0.46	0.444	0.896	0.495
0.01	0.010	1.000	0.010	0.47	0.453	0.892	0.508
0.02	0.020	1.000	0.020	0.48	0.462	0.887	0.521
0.03	0.030	1.000	0.030	0.49	0.471	0.882	0.533
0.04	0.040	0.999	0.040	0.50	0.479	0.878	0.546
0.05	0.050	0.999	0.050	0.51	0.488	0.873	0.559
0.06	0.060	0.998	0.060	0.52	0.497	0.868	0.573
0.07	0.070	0.998	0.070	0.53	0.506	0.863	0.586
0.08	0.080	0.997	0.080	0.54	0.514	0.858	0.599
0.09	0.090	0.996	0.090	0.55	0.523	0.853	0.613
0.10	0.100	0.995	0.100	0.56	0.531	0.847	0.627
0.11	0.110	0.994	0.110	0.57	0.540	0.842	0.641
0.12	0.120	0.993	0.121	0.58	0.548	0.836	0.655
0.13	0.130	0.992	0.131	0.59	0.556	0.831	0.670
0.14	0.140	0.990	0.141	0.60	0.565	0.825	0.684
0.15	0.149	0.989	0.151	0.61	0.573	0.820	0.699
0.16	0.159	0.987	0.161	0.62	0.581	0.814	0.714
0.17	0.169	0.986	0.172	0.63	0.589	0.808	0.729
0.18	0.179	0.984	0.182	0.64	0.597	0.802	0.745
0.19	0.189	0.982	0.192	0.65	0.605	0.796	0.760
0.20	0.199	0.980	0.203	0.66	0.613	0.790	0.776
0.21	0.208	0.978	0.213	0.67	0.621	0.784	0.792
0.22	0.218	0.976	0.224	0.68	0.629	0.778	0.809
0.23	0.228	0.974	0.234	0.69	0.637	0.771	0.825
0.24	0.238	0.971	0.245	0.70	0.644	0.765	0.842
0.25	0.247	0.969	0.255	0.71	0.652	0.758	0.860
0.26	0.257	0.966	0.266	0.72	0.659	0.752	0.877
0.27	0.267	0.964	0.277	0.73	0.667	0.745	0.895
0.28	0.276	0.961	0.288	0.74	0.674	0.738	0.913
0.29	0.286	0.958	0.298	0.75	0.682	0.732	0.932
0.30	0.296	0.955	0.309	0.76	0.689	0.725	0.950
0.31	0.305	0.952	0.320	0.77	0.696	0.718	0.970
0.32	0.315	0.949	0.331	0.78	0.703	0.711	0.989
0.33	0.324	0.946	0.343	0.79	0.710	0.704	1.009
0.34	0.333	0.943	0.354	0.80	0.717	0.697	1.030
0.35	0.343	0.939	0.365	0.81	0.724	0.689	1.050
0.36	0.352	0.936	0.376	0.82	0.731	0.682	1.072
0.37	0.362	0.932	0.388	0.83	0.738	0.675	1.093
0.38	0.371	0.929	0.399	0.84	0.745	0.667	1.116
0.39	0.380	0.925	0.411	0.85	0.751	0.660	1.138
0.40	0.389	0.921	0.423	0.86	0.758	0.652	1.162
0.41	0.399	0.917	0.435	0.87	0.764	0.645	1.185
0.42	0.408	0.913	0.447	0.88	0.771	0.637	1.210
0.43	0.417	0.909	0.459	0.89	0.777	0.629	1.235
0.44	0.426	0.905	0.471	0.90	0.783	0.622	1.260
0.45	0.435	0.900	0.483	0.91	0.790	0.614	1.286

TABLE 4   (continued)

x	$\sin x$	$\cos x$	$\tan x$	x	$\sin x$	$\cos x$	$\tan x$
0.92	0.796	0.606	1.313	1.26	0.952	0.306	3.113
0.93	0.802	0.598	1.341	1.27	0.955	0.296	3.224
0.94	0.808	0.590	1.369	1.28	0.958	0.287	3.341
0.95	0.813	0.582	1.398	1.29	0.961	0.277	3.467
0.96	0.819	0.574	1.428	1.30	0.964	0.267	3.602
0.97	0.825	0.565	1.459	1.31	0.966	0.258	3.747
0.98	0.830	0.557	1.491	1.32	0.969	0.248	3.903
0.99	0.836	0.549	1.524	1.33	0.971	0.238	4.072
1.00	0.841	0.540	1.557	1.34	0.973	0.229	4.256
1.01	0.847	0.532	1.592	1.35	0.976	0.219	4.455
1.02	0.852	0.523	1.628	1.36	0.978	0.209	4.673
1.03	0.857	0.515	1.665	1.37	0.980	0.199	4.913
1.04	0.862	0.506	1.704	1.38	0.982	0.190	5.177
1.05	0.867	0.498	1.743	1.39	0.984	0.180	5.471
1.06	0.872	0.489	1.784	1.40	0.985	0.170	5.798
1.07	0.877	0.480	1.827	1.41	0.987	0.160	6.165
1.08	0.882	0.471	1.871	1.42	0.989	0.150	6.581
1.09	0.887	0.462	1.917	1.43	0.990	0.140	7.055
1.10	0.891	0.454	1.965	1.44	0.991	0.130	7.602
1.11	0.896	0.445	2.014	1.45	0.993	0.121	8.238
1.12	0.900	0.436	2.066	1.46	0.994	0.111	8.989
1.13	0.904	0.427	2.120	1.47	0.995	0.101	9.887
1.14	0.909	0.418	2.176	1.48	0.996	0.091	10.983
1.15	0.913	0.408	2.234	1.49	0.997	0.081	12.350
1.16	0.917	0.399	2.296	1.50	0.997	0.071	14.101
1.17	0.921	0.390	2.360	1.51	0.998	0.061	16.428
1.18	0.925	0.381	2.427	1.52	0.999	0.051	19.670
1.19	0.928	0.372	2.498	1.53	0.999	0.041	24.498
1.20	0.932	0.362	2.572	1.54	1.000	0.031	32.461
1.21	0.936	0.353	2.650	1.55	1.000	0.021	48.078
1.22	0.939	0.344	2.733	1.56	1.000	0.011	92.620
1.23	0.942	0.334	2.820	1.57	1.000	0.001	1255.770
1.24	0.946	0.325	2.912	1.58	1.000	-0.009	-108.649
1.25	0.949	0.315	3.010				

TABLE 5.   Sines, cosines, tangents (degree measure)

x	$\sin x$	$\cos x$	$\tan x$	x	$\sin x$	$\cos x$	$\tan x$
0°	0.00	1.00	0.00	45°	0.71	0.71	1.00
1	0.02	1.00	0.02	46	0.72	0.69	1.04
2	0.03	1.00	0.03	47	0.73	0.68	1.07
3	0.05	1.00	0.05	48	0.74	0.67	1.11
4	0.07	1.00	0.07	49	0.75	0.66	1.15
5	0.09	1.00	0.09	50	0.77	0.64	1.19
6	0.10	0.99	0.11	51	0.78	0.63	1.23
7	0.12	0.99	0.12	52	0.79	0.62	1.28
8	0.14	0.99	0.14	53	0.80	0.60	1.33
9	0.16	0.99	0.16	54	0.81	0.59	1.38
10	0.17	0.98	0.18	55	0.82	0.57	1.43
11	0.19	0.98	0.19	56	0.83	0.56	1.48
12	0.21	0.98	0.21	57	0.84	0.54	1.54
13	0.22	0.97	0.23	58	0.85	0.53	1.60
14	0.24	0.97	0.25	59	0.86	0.52	1.66
15	0.26	0.97	0.27	60	0.87	0.50	1.73
16	0.28	0.96	0.29	61	0.87	0.48	1.80
17	0.29	0.96	0.31	62	0.88	0.47	1.88
18	0.31	0.95	0.32	63	0.89	0.45	1.96
19	0.33	0.95	0.34	64	0.90	0.44	2.05
20	0.34	0.94	0.36	65	0.91	0.42	2.14
21	0.36	0.93	0.38	66	0.91	0.41	2.25
22	0.37	0.93	0.40	67	0.92	0.39	2.36
23	0.39	0.92	0.42	68	0.93	0.37	2.48
24	0.41	0.91	0.45	69	0.93	0.36	2.61
25	0.42	0.91	0.47	70	0.94	0.34	2.75
26	0.44	0.90	0.49	71	0.95	0.33	2.90
27	0.45	0.89	0.51	72	0.95	0.31	3.08
28	0.47	0.88	0.53	73	0.96	0.29	3.27
29	0.48	0.87	0.55	74	0.96	0.28	3.49
30	0.50	0.87	0.58	75	0.97	0.26	3.73
31	0.52	0.86	0.60	76	0.97	0.24	4.01
32	0.53	0.85	0.62	77	0.97	0.22	4.33
33	0.54	0.84	0.65	78	0.98	0.21	4.70
34	0.56	0.83	0.67	79	0.98	0.19	5.14
35	0.57	0.82	0.70	80	0.98	0.17	5.67
36	0.59	0.81	0.73	81	0.99	0.16	6.31
37	0.60	0.80	0.75	82	0.99	0.14	7.12
38	0.62	0.79	0.78	83	0.99	0.12	8.14
39	0.63	0.78	0.81	84	0.99	0.10	9.51
40	0.64	0.77	0.84	85	1.00	0.09	11.43
41	0.66	0.75	0.87	86	1.00	0.07	14.30
42	0.67	0.74	0.90	87	1.00	0.05	19.08
43	0.68	0.73	0.93	88	1.00	0.03	28.64
44	0.69	0.72	0.97	89	1.00	0.02	57.29
45	0.71	0.71	1.00	90	1.00	0.00	—

TABLE 6.   Integrals

1. $\int \alpha\,dx = \alpha x + C.$

2. $\int x^n\,dx = \dfrac{x^{n+1}}{n+1} + C, \quad n \neq -1.$

3. $\int \dfrac{dx}{x} = \log|x| + C.$

4. $\int e^x\,dx = e^x + C.$

5. $\int p^x\,dx = \dfrac{p^x}{\log p} + C.$

6. $\int \log x\,dx = x\log x - x + C.$

7. $\int \cos x\,dx = \sin x + C.$

8. $\int \sin x\,dx = -\cos x + C.$

9. $\int \sec^2 x\,dx = \tan x + C.$

10. $\int \operatorname{cosec}^2 x\,dx = -\cot x + C.$

11. $\int \sec x \tan x\,dx = \sec x + C.$

12. $\int \operatorname{cosec} x \cot x\,dx = -\operatorname{cosec} x + C.$

13. $\int \tan x\,dx = \log|\sec x| + C.$

14. $\int \cot x\,dx = \log|\sin x| + C.$

15. $\int \sec x\,dx = \log|\sec x + \tan x| + C.$

16. $\int \operatorname{cosec} x\,dx = \log|\operatorname{cosec} x - \cot x| + C.$

17. $\int \arcsin x\,dx = x\arcsin x + \sqrt{1 - x^2} + C.$

18. $\int \arctan x\,dx = x\arctan x - \tfrac{1}{2}\log(1 + x^2) + C.$

19. $\int \dfrac{dx}{x^2 + a^2} = \dfrac{1}{a}\arctan\left(\dfrac{x}{a}\right) + C.$ (take $a > 0$ throughout)

20. $\int \dfrac{dx}{x^2 - a^2} = \dfrac{1}{2a}\log\left|\dfrac{x-a}{x+a}\right| + C.$

21. $\int \dfrac{dx}{\sqrt{x^2 \pm a^2}} = \log(x + \sqrt{x^2 \pm a^2}) + C.$

22. $\int \dfrac{dx}{\sqrt{a^2 - x^2}} = \arcsin\left(\dfrac{x}{a}\right) + C.$

23. $\int \sqrt{x^2 \pm a^2}\,dx = \dfrac{x}{2}\sqrt{x^2 \pm a^2} \pm \dfrac{a^2}{2}\log(x + \sqrt{x^2 \pm a^2}) + C.$

24. $\int \sqrt{a^2 - x^2}\,dx = \dfrac{x}{2}\sqrt{a^2 - x^2} + \dfrac{a^2}{2}\arcsin\left(\dfrac{x}{a}\right) + C.$

TABLE 6.   (continued)

Hyperbolic Functions

25. $\displaystyle\int \cosh x \, dx = \sinh x + C.$

26. $\displaystyle\int \sinh x \, dx = \cosh x + C.$

27. $\displaystyle\int \operatorname{sech}^2 x \, dx = \tanh x + C.$

28. $\displaystyle\int \operatorname{cosech}^2 x \, dx = -\coth x + C.$

29. $\displaystyle\int \operatorname{sech} x \tanh x \, dx = -\operatorname{sech} x + C.$

30. $\displaystyle\int \operatorname{cosech} x \coth x \, dx = -\operatorname{cosech} x + C.$

31. $\displaystyle\int \frac{dx}{\sqrt{x^2 + a^2}} = \sinh^{-1}\left(\frac{x}{a}\right) + C.$ (take $a > 0$ throughout)

32. $\displaystyle\int \frac{dx}{\sqrt{x^2 - a^2}} = \cosh^{-1}\left(\frac{x}{a}\right) + C.$

33. $\displaystyle\int \sqrt{x^2 + a^2} \, dx = \frac{x}{2}\sqrt{x^2 + a^2} + \frac{a^2}{2}\sinh^{-1}\left(\frac{x}{a}\right) + C.$

34. $\displaystyle\int \sqrt{x^2 - a^2} \, dx = \frac{x}{2}\sqrt{x^2 - a^2} - \frac{a^2}{2}\cosh^{-1}\left(\frac{x}{a}\right) + C.$

Answers to Starred Exercises

Section 1.3

1. $x = 2$

3. $x = a$ and $x = b$

4. $(1, 3)$

6. $(-\infty, 1) \cup (3, \infty)$

7. $(-\infty, \frac{1}{3})$

9. $(-1, 1)$

10. $(-\infty, \infty)$

12. $[0, \infty)$

13. $(-\infty, 0)$

15. $(1, \infty)$

16. $[-2, 2]$

18. $(0, \infty)$

19. $(-\infty, 0] \cup (5, \infty)$

Section 1.4

1. dom $(f) = (-\infty, \infty)$; ran $(f) = [0, \infty)$

3. dom $(F) = (-\infty, \infty)$; ran $(F) = (-\infty, \infty)$

5. dom $(f) = (-\infty, 0) \cup (0, \infty)$; ran $(f) = (0, \infty)$

7. dom $(f) = (-\infty, 1]$; ran $(f) = [0, \infty)$

9. dom $(f) = (-\infty, 1]$; ran $(f) = [-1, \infty)$

11. dom $(h) = (-\infty, 1)$; ran $(h) = (0, \infty)$

13. one-to-one

15. not one-to-one

17. not one-to-one

19. not one-to-one

21. one-to-one

23. one-to-one

25. not one-to-one

27. one-to-one

29. odd

31. neither

33. even

35. even

36. even

37. odd

38. symmetric about the y-axis

39. symmetric about the origin

Section 1.5

1. $(f \circ g)(x) = 2x^2 + 5$

3. $(f \circ g)(x) = \sqrt{x^2 + 5}$

5. $(f \circ g)(x) = x, \quad x \neq 0$

7. $(f \circ g)(x) = \dfrac{1}{x} - 1$

9. $(f \circ g)(x) = \dfrac{1}{x^2 + 1}$

11. $(f \circ g \circ h)(x) = 4(x^2 - 1)$

13. $(f \circ g \circ h)(x) = (x^4 - 1)^2$

15. $(f \circ g \circ h)(x) = 2x^2 + 1$

17. (a) $f(x) = \dfrac{1}{x}$ (c) $f(x) = \sqrt{a^2 + x}$

18. (b) $g(x) = a^2 x^2$

Section 1.6

1. $f^{-1}(x) = \frac{1}{5}(x - 3)$

3. $f^{-1}(x) = \dfrac{1}{m}(x - b)$

5. $f^{-1}(x) = (x - 1)^{1/5}$

7. $f^{-1}(x) = \left(\dfrac{x - 1}{3}\right)^{1/3}$

9. $f^{-1}(x) = (x - 2)^{1/3} - 1$

11. $f^{-1}(x) = x^{5/3}$

13. $f^{-1}(x) = (1 - x^{1/5})^{1/3}$

15. $f^{-1}(x) = \dfrac{1}{x}$

17. $f^{-1}(x) = \left(\dfrac{1}{x} - 1\right)^{1/3}$

Section 2.1

1. 3

3. -3

5. does not exist

7. 4

9. 2

11. does not exist

13. 2

15. 3

17. 0

18. does not exist

Section 2.2

1. $\frac{1}{2}$

3. 0

5. $\frac{4}{3}$

7. does not exist

9. does not exist

11. only δ_1

13. $\frac{1}{5}\epsilon$

15. 5ϵ

17. Take $\delta = \frac{1}{3}\epsilon$. If $0 < |x - 2| < \frac{1}{3}\epsilon$, then $|3x - 6| < \epsilon$ and therefore $|(3x - 1) - 5| < \epsilon$.

19. Take $\delta = \frac{1}{5}\epsilon$. If $0 < |x - 0| < \frac{1}{5}\epsilon$, then $|5x| < \epsilon$ and so $|(2 - 5x) - 2| < \epsilon$.

21. Take $\delta = \epsilon$. If $0 < |x - 2| < \epsilon$, then $||x - 2| - 0| < \epsilon$.

23. Take $\delta = $ minimum of 9 and ϵ. If $0 < |x - 9| < \delta$, then $x > 0$ and $|x - 9| < \epsilon$. It follows that $|\sqrt{x} - 3| \, |\sqrt{x} + 3| < \epsilon$ and, since $|\sqrt{x} + 3| > 1$, that $|\sqrt{x} - 3| < \epsilon$.

25. Take $\delta = $ minimum of 4 and ϵ. If $0 < |x - 5| < \delta$, then $x > 1$ and

$|(x - 1) - 4| < \epsilon$. It follows that $|\sqrt{x - 1} - 2| \, |\sqrt{x - 1} + 2| < \epsilon$ and, since $|\sqrt{x - 1} + 2| > 1$, that $|\sqrt{x - 1} - 2| < \epsilon$.

Section 2.3

1. (b) $\frac{1}{4}$ (c) 4   (e) no limit   (h) 1
2. (b) 2   (c) no limit
3. (a) 3   (d) 0   (e) no limit
4. (a) 27   (c) 0

Section 2.4

1. (a) 0   (c) 0   (e) 0   (h) no limit   (j) -1 (l) 1
3. (a) no limit   (b) 1
5. (b) $\min\{f(x), g(x)\} = \frac{1}{2}\left[[f(x) + g(x)] - |f(x) - g(x)|\right]$.

Section 2.5

1. (b) (i) -1 (iii) no limit
2. (b) (i) 1   (iii) 1   (v) 2   (vii) 4   (ix) no limit
3. (a) 1   (c) -1 (e) no limit   (g) 0   (i) 1   (k) -1 (m) does not exist
4. (b) (i) 0   (ii) 1   (iii) 1   (iv) 1   (v) no limit   (vi) 1
8. (a) 0   (c) 1

Section 2.6

1. (a) continuous at 2   (b) continuous at 2
    (c) continuous at 2   (d) not defined at 2
    (e) continuous at 2   (f) discontinuous at 2
    (g) continuous at 2   (h) continuous at 2
    (i) discontinuous at 2   (j) discontinuous at 2
2. (a) no discontinuities   (b) no discontinuities
    (c) no discontinuities   (d) no discontinuities
    (e) not defined at 3   (f) discontinuous at 2, not defined at 3
5. (a) nowhere continuous   (b) continuous only at 0

Section 3.1

1. $f'(x) = 0$ 3. $f'(x) = 3$ 5. $f'(x) = -1/x^2$
7. $f'(x) = x$ 9. $f'(x) = 6x^2$ 11. $f'(x) = 2(x - 1)$
13. $f'(x) = -\frac{1}{2}x^{-3/2}$ 15. $f'(x) = -\frac{1}{2}(x - 1)^{-3/2}$
17. $f'(2) = -\frac{1}{9}$ 19. $f'(2) = \frac{1}{4}$

Section 3.2

1. $F'(x) = -1$

3. $F'(x) = 55x^4 - 18x^2$

5. $F'(x) = 2ax + b$

7. $F'(x) = \dfrac{2}{x^3}$

9. $F'(x) = \dfrac{bc - ad}{(cx - d)^2}$

11. $G'(x) = \dfrac{3x^2 - 2x^3}{(1 - x)^2}$

13. $G'(x) = 3x^2 - 6x - 1$

15. $G'(x) = 2x - 3$

17. $G'(x) = -\dfrac{2(3x^2 - x + 1)}{x^2(x - 2)^2}$

19. $f'(x) = (9x^8 - 8x^9)\left(1 - \dfrac{1}{x^2}\right) + \left(x + \dfrac{1}{x}\right)(72x^7 - 72x^8)$

$\qquad = -80x^9 + 81x^8 - 64x^7 + 63x^6$

21. $f'(0) = -\frac{1}{4}, \ f'(1) = -1$

23. $f'(0) = 0, \ f'(1) = -1$

25. $f'(0) = \dfrac{ad - bc}{d^2}, \ f'(1) = \dfrac{ad - bc}{(c + d)^2}$

27. $f'(0) = 1$

29. $f'(0) = -4$

Section 3.3

1. $\dfrac{dy}{dx} = 12x^3 - 2x$

3. $\dfrac{dy}{dx} = 1 + \dfrac{1}{x^2}$

5. $\dfrac{dy}{dx} = \dfrac{1 - x^2}{(1 + x^2)^2}$

7. $\dfrac{dy}{dx} = \dfrac{2x - x^2}{(1 - x)^2}$

9. $\dfrac{dy}{dx} = \dfrac{-6x^2}{(x^3 - 1)^2}$

11. $\dfrac{d}{dx}(2x - 5) = 2$

13. $21x^2$

15. $\dfrac{-4t}{(t^2 - 1)^2}$

17. $\dfrac{2t^3(t^3 - 2)}{(2t^3 - 1)^2}$

19. $\dfrac{2}{(1 - 2u)^2}$

21. $-\left[\dfrac{1}{(u - 1)^2} + \dfrac{1}{(u + 1)^2}\right] = -2\left[\dfrac{u^2 + 1}{(u^2 - 1)^2}\right]$

23. $2x\left[\dfrac{1}{(1 - x^2)^2} + \dfrac{1}{x^4}\right]$

25. $\dfrac{-2}{(x - 1)^2}$

27. 47

29. $\frac{1}{4}$

Section 3.4

1. 8   3. 4   5. $-\frac{5}{36}$

8. $\dfrac{dA}{dr} = 8\pi r,\ 8\pi r_0, r_0 = \dfrac{1}{8\pi}$ 10. $-\dfrac{h}{b}$

12. -4

Section 3.5

1. $f'(x) = 2(1 - 2x)^{-2}$

3. $f'(x) = 20(x^5 - x^{10})^{19}(5x^4 - 10x^9)$

5. $f'(x) = 4\left(x - \dfrac{1}{x}\right)^3 \left(1 + \dfrac{1}{x^2}\right)$

7. $f'(x) = 4(x - x^3 - x^5)^3(1 - 3x^2 - 5x^4)$

9. $f'(t) = 200t(t^2 - 1)^{99}$

11. $f'(t) = -4(t^{-1} + t^{-2})^3(t^{-2} + 2t^{-3})$

13. $f'(x) = 324x^3\ \dfrac{1 - x^2}{(x^2 + 1)^5}$

15. $f'(x) = 2(x^4 + x^2 + x)(4x^3 + 2x + 1)$

17. $f'(x) = -\left(\dfrac{x^3}{3} + \dfrac{x^2}{2} + \dfrac{x}{1}\right)^{-2}(x^2 + x + 1)$

19. $f'(x) = 3\left(\dfrac{1}{x + 2} - \dfrac{1}{x - 2}\right)^2 \left[\dfrac{-1}{(x + 2)^2} + \dfrac{1}{(x - 2)^2}\right] = \dfrac{384x}{(x + 2)^4(x - 2)^4}$

21. $\dfrac{dy}{dx} = -\dfrac{2x + 1}{(2x^2 + 2x + 1)^2}$ 23. $\dfrac{dy}{dx} = \dfrac{80x(5x^2 + 1)^3}{[1 - 4(5x^2 + 1)^4]^2}$

27. 1   29. 1   31. 1

33. 2   35. 2   37. 0

39. $2xf'(x^2 + 1)$ 41. $2f(x)f'(x)$

Section 3.6

1. $42x - 120x^3$ 3. $2 - \dfrac{6}{x^4}$

5. 48   7. $\dfrac{4}{(1 + x)^3}$

9. $\dfrac{2c(bc - ad)}{(cx + d)^3}$ 11. $\dfrac{2(1 - 2x)}{(1 + x)^4}$

13. 0   15. 2

17. $2\left[\dfrac{(x^2 - 1)(-3x^2 + 36x - 73) + 6x(x - 6)(x^2 - 12x + 1)}{(x^2 - 1)^4}\right]$

19. $n!$

Section 3.7

1. $\dfrac{dy}{dx} = \frac{3}{2}x^2(x^3 + 1)^{-1/2}$

3. $\dfrac{dy}{dx} = \dfrac{2x^2 + 1}{\sqrt{x^2 + 1}}$

5. $\dfrac{dy}{dx} = x(2x^2 + 1)^{-3/4}$

7. $\dfrac{dy}{dx} = \dfrac{x(2x^2 - 5)}{\sqrt{3 - x^2}\sqrt{2 - x^2}}$

9. $\frac{1}{2}(x^{-1/2} - x^{-3/2})$

11. $(x^2 + 1)^{-3/2}$

13. $\frac{1}{3}(x^{-2/3} - x^{-4/3})$

15. $\dfrac{(ad - bc)x}{(cx^2 + d)^2}\sqrt{\dfrac{cx^2 + d}{ax^2 + b}}$

17. $\dfrac{f'(\sqrt{x} + 1)}{2\sqrt{x}}$

19. $\dfrac{xf'(x^2 + 1)}{\sqrt{f(x^2 + 1)}}$

21. $(f^{-1})'(x) = \dfrac{1}{x}$

23. $(f^{-1})'(x) = \dfrac{1}{\sqrt{1 - x^2}}$

24. (a) elastic (c) unitary
(e) inelastic for $0 < P < 20$; unitary at $P = 20$; elastic for $20 < P < 40$.

Section 3.8

1. (a) $y = 11x + 16$ (c) $y = -4x$
2. (a) $(\frac{1}{4}, \frac{1}{2})$ (c) $(2, 6)$
3. (a) $(-2, -10)$ (c) $(0, 1)$
4. (a) $(1, \frac{2}{3})$ (c) $(\frac{1}{3}, \frac{2}{27}\sqrt{3})$
7. $C = \dfrac{1}{B}(1 - 2\sqrt{AB}), \left(\sqrt{\dfrac{B}{A}}, \dfrac{1}{B}(1 - \sqrt{AB})\right)$ (assuming $x > 0$)
8. $A = -1$, $B = 0$, $C = 4$

Section 3.9

2. $\dfrac{dy}{dx} = \frac{3}{2}x^{-1/4} - x^{-5/4}$

4. $\dfrac{dy}{dx} = \dfrac{1}{4\sqrt{x}} + \dfrac{1}{x\sqrt{x}}$

6. $\dfrac{ds}{dt} = -\dfrac{a}{2t\sqrt{t}} + \dfrac{b}{2\sqrt{t}} + \dfrac{3c\sqrt{t}}{2}$

8. $f'(t) = -18t(2 - 3t^2)^2$

10. $\dfrac{dy}{dx} = \dfrac{2b}{x^2}\left(a - \dfrac{b}{x}\right)$

12. $\dfrac{dy}{dx} = \dfrac{2a + 3bx}{2\sqrt{a + bx}}$

14. $\dfrac{dy}{dx} = -\dfrac{2a}{(a + x)^2}$

16. $\dfrac{dy}{dx} = -\dfrac{a^2}{x^2\sqrt{a^2 + x^2}}$

18. $\dfrac{dy}{dx} = \dfrac{a^2}{(a^2 - x^2)^{3/2}}$

20. $\dfrac{dr}{d\theta} = \dfrac{60 - 100\theta^2}{\sqrt{3 - 40}}$

22. $f'(x) = x(2 + 3x)^{-2/3} + (2 + 3x)^{1/3}$

24. $\dfrac{ds}{dt} = \dfrac{4}{(2 + 3t)^{2/3}(2 - 3t)^{4/3}}$

26. $\dfrac{ds}{dt} = \dfrac{t^3 - 1}{t^2\sqrt{2t^3 - 1}}$

28. $\dfrac{dy}{dx} = \dfrac{(x + 2)(3x^2 + 2x + 4)}{\sqrt{x^2 + 2}}$

30. $\dfrac{dy}{dx} = -\dfrac{(a^{3/5} - x^{3/5})^{2/3}}{x^{2/5}}$

31. 540

33. $\frac{1}{12}$

35. $\frac{5}{6}$

37. $\frac{3}{64}$

39. $-\frac{41}{90}$

41. 0

43. 20

Section 3.10

1. $c = \frac{3}{2}$

3. $c = \sqrt{\frac{13}{3}}$

Section 3.11

1. increasing on $(-\infty, -1]$ and $[1, \infty)$, decreasing on $[-1, 1]$
3. increasing on $(-\infty, -1]$ and $[1, \infty)$, decreasing on $[-1, 0)$ and $(0, 1]$
5. increasing on $(-\infty, -1 - \frac{1}{3}\sqrt{3}]$ and $[-1 + \frac{1}{3}\sqrt{3}, \infty)$, decreasing on $[-1 - \frac{1}{3}\sqrt{3}, -1 + \frac{1}{3}\sqrt{3}]$
7. increasing on $(-\infty, 2)$, decreasing on $(2, \infty)$
9. increasing on $(-\infty, -1)$ and $(-1, 0]$, decreasing on $[0, 1)$ and $(1, \infty)$
11. increasing on $[-\sqrt{5}, 0]$ and $[\sqrt{5}, \infty)$, decreasing on $(-\infty, -\sqrt{5}]$ and $[0, \sqrt{5}]$
13. increasing on $(-\infty, -1)$ and $(-1, \infty)$
15. decreasing on $[0, \infty)$
17. increasing on $[0, \infty)$, decreasing on $(-\infty, 0]$
19. $f(x) = \frac{1}{3}x^3 - x + 1$
21. $f(x) = x^2 - 5x + 10$
23. $f(x) = -2x^{-2} + 2$
25. $f(x) = \frac{1}{5}(x^3 - 1)^5$

Section 3.12

1. no critical pts; no local extreme values
3. critical pts ± 1; local max $f(-1) = -2$, local min $f(1) = 2$
5. critical pts $-1 \pm \frac{1}{3}\sqrt{3}$; local max $f(-1 - \frac{1}{3}\sqrt{3}) = \frac{2}{9}\sqrt{3}$, local min $f(-1 + \frac{1}{3}\sqrt{3}) = -\frac{2}{9}\sqrt{3}$
7. no critical pts; no local extreme values
9. no critical pts; no local extreme values
11. critical pts $-4, 0, 4$; local max $f(0) = 16$, local min $f(-4) = f(4) = 0$
13. critical pt 2; no local extreme values
15. critical pts $-1, \frac{1}{2}$; local max $f(\frac{1}{2}) = \frac{27}{16}$
17. critical pt 0; local min $f(0) = 0$
19. critical pts $-2, -\frac{1}{2}, 1$; local max $f(-\frac{1}{2}) = \frac{9}{4}$, local min $f(-2) = f(1) = 0$
21. critical pts $-0, \frac{3}{2}$; local max $f(-\frac{3}{2}) = -6\frac{3}{4}$
23. critical pt $-\frac{3}{2}$; local max $f(-\frac{3}{2}) = -4$

Section 3.13

1. critical pt -2; $f(-2) = 0$ endpt min and absolute min
3. critical pts 0, 2, 3; $f(0) = 1$ endpt max and absolute max, $f(2) = -3$ local min and absolute min, $f(3) = -2$ endpt max
5. critical pt $2^{-1/3}$; $f(2^{-1/3}) = 3 \cdot 2^{-2/3}$ local min
7. critical pts $\frac{1}{10}$, $2^{-1/3}$, 2; $f(\frac{1}{10}) = 10\frac{1}{100}$ endpt max and absolute max; $f(2^{-1/3}) = 2^{-2/3} + 2^{1/3} = 3 \cdot 2^{-2/3}$ local min and absolute min, $f(2) = 4\frac{1}{2}$ endpt max
9. critical pts 0, $\frac{3}{2}$, 2; $f(0) = 2$ endpt max and absolute max, $f(\frac{3}{2}) = -\frac{1}{4}$ local min and absolute min, $f(2) = 0$ endpt max
11. critical pt 0; $f(0) = \frac{1}{3}$ endpt max
13. critical pts 0, $\frac{1}{4}$, 1; $f(0) = 0$ endpt min and absolute min, $f(\frac{1}{4}) = \frac{1}{16}$ local max, $f(1) = 0$ local min and absolute min
15. critical pts 2, 3; $f(2) = 2$ local max and absolute max, $f(3) = 0$ endpt min
17. critical pt 1; no extreme values
19. critical pts 1, 4; $f(1) = -2$ local min, $f(4) = 1$ local max

Section 3.14

1. 20, 20   3. 20 by 10   5. $2r^2$ 7. $\dfrac{\beta - b}{2(a + \alpha)}$ tons

9. $l = \frac{1}{3}\sqrt{3}\, d, w = \frac{1}{3}\sqrt{6}\, d$
11. the equilateral triangle
13. half of the area of the given triangle
15. length of side $\frac{1}{6}(a + b - \sqrt{a^2 - ab + b^2})$
16. $\frac{3}{4}\sqrt{3}r^2$

21. (a) use it all for the circle   (b) use $\dfrac{9l}{\sqrt{3\pi} + 9}$ for the triangle

22. (a) base radius $\frac{2}{3}r$ and height $\frac{1}{3}h$
23. (a) base radius $\frac{1}{3}\sqrt{6}R$ (c) base radius $\frac{2}{3}\sqrt{2}R$
25. 18

Section 3.15

1. concave down on $(-\infty, 0)$, concave up on $(0, \infty)$
3. concave down on $(-\infty, 0]$, concave up on $[0, \infty)$; pt of inflection $(0, 2)$
5. concave up on $(-\infty, -1/\sqrt{3}]$, concave down on $[-1/\sqrt{3}, 1/\sqrt{3}]$, concave up on $[1/\sqrt{3}, \infty)$; pts of inflection $(-1/\sqrt{3}, -\frac{5}{36})$, $(1/\sqrt{3}, -\frac{5}{36})$
7. concave down on $(-\infty, -1)$ and on $[0, 1)$, concave up on $(-1, 0]$ and on $(1, \infty)$; pt of inflection $(0, 0)$
9. concave up on $(-\infty, -1/\sqrt{3}]$, concave down on $[-1/\sqrt{3}, 1/\sqrt{3}]$, concave up on $[1/\sqrt{3}, \infty)$; pts of inflection $(-1/\sqrt{3}, \frac{4}{9})$, $(1/\sqrt{3}, \frac{4}{9})$

11. concave up on $(0, \infty)$

13. $d = \frac{1}{3}(a + b + c)$ 15. $a = -\frac{1}{2}, \quad b = \frac{1}{2}$

Section 3.16

1. critical pt 2; $f(2) = 0$ local min and absolute min; decreasing on $(-\infty, 2]$, increasing on $[2, \infty)$; concave up everywhere; no pts of inflection

3. critical pts $\frac{1}{3}$, 1; $f(\frac{1}{3}) = 1\frac{4}{27}$ local max, $f(1) = 1$ local min; increasing on $(-\infty, \frac{1}{3}]$, decreasing on $[\frac{1}{3}, 1]$, increasing on $[1, \infty)$; concave down on $(-\infty, \frac{2}{3}]$, concave up on $[\frac{2}{3}, \infty)$; pt of inflection $(\frac{2}{3}, 1\frac{2}{27})$

5. no critical pts; no extreme values; decreasing on $(-\infty, \frac{1}{3})$ and on $(\frac{1}{3}, \infty)$; concave down on $(-\infty, \frac{1}{3})$, concave up on $(\frac{1}{3}, \infty)$; no pts of inflection

7. critical pts $-\frac{2}{3}$, 0; $f(-\frac{2}{3}) = -\frac{4}{9}$ local max, $f(0) = 0$ local min; increasing on $(-\infty, -\frac{2}{3}]$, decreasing on $[-\frac{2}{3}, -\frac{1}{3})$, decreasing on $(-\frac{1}{3}, 0]$, increasing on $[0, \infty)$; concave down on $(-\infty, -\frac{1}{3})$, concave up on $(-\frac{1}{3}, \infty)$; no pts of inflection

9. critical pt $\frac{1}{3}$; $f(\frac{1}{3}) = \frac{1}{12}$ local max and absolute max; decreasing on $(-\infty, -\frac{1}{3})$, increasing on $(-\frac{1}{3}, \frac{1}{3}]$, decreasing on $[\frac{1}{3}, \infty)$; concave down on $(-\infty, -\frac{1}{3})$ and on $(-\frac{1}{3}, \frac{2}{3}]$, concave up on $[\frac{2}{3}, \infty)$; pt of inflection $(\frac{2}{3}, \frac{2}{27})$

11. critical pt 0; no extreme values; increasing everywhere; concave down on $(-\infty, 0]$, concave up on $[0, \infty)$; pt of inflection $(0, 0)$

13. critical pt 2; $f(2) = 1$ local min and absolute min; decreasing on $(-\infty, 2]$, increasing on $[2, \infty)$; concave up everywhere; no pts of inflection

15. critical pts -1, $-\frac{1}{2}$, 0; $f(-1) = f(0) = 0$ local and absolute min, $f(-\frac{1}{2}) = \frac{1}{16}$ local max; decreasing on $(-\infty, -1]$, increasing on $[-1, -\frac{1}{2}]$, decreasing on $[-\frac{1}{2}, 0]$, increasing on $[0, \infty)$; concave up on $(-\infty, -\frac{1}{2} - \frac{1}{6}\sqrt{3}]$, concave down on $[-\frac{1}{2} - \frac{1}{6}\sqrt{3}, -\frac{1}{2} + \frac{1}{6}\sqrt{3}]$, concave up on $[-\frac{1}{2} + \frac{1}{6}\sqrt{3}, \infty)$; pts of inflection $(-\frac{1}{2} - \frac{1}{6}\sqrt{3}, \frac{1}{36})$, $(-\frac{1}{2} + \frac{1}{6}\sqrt{3}, \frac{1}{36})$

17. critical pts $\frac{2}{3}$, 1; $f(\frac{2}{3}) = \frac{2}{9}\sqrt{3}$ local max and absolute max, $f(1) = 0$ endpt min; increasing on $(-\infty, \frac{2}{3}]$, decreasing on $[\frac{2}{3}, 1]$; concave down on $(-\infty, 1]$; no pts of inflection

19. critical pts $\pm\sqrt{2}$; $f(-\sqrt{2}) = -\frac{1}{2}\sqrt{2}$ local min and absolute min, $f(\sqrt{2}) = \frac{1}{2}\sqrt{2}$ local max and absolute max; decreasing on $(-\infty, -\sqrt{2}]$, increasing on $[-\sqrt{2}, \sqrt{2}]$, decreasing on $[\sqrt{2}, \infty)$; concave down on $(-\infty, -\sqrt{6}]$, concave up on $[-\sqrt{6}, 0]$, concave down on $[0, \sqrt{6}]$, concave up on $[\sqrt{6}, \infty)$; pts of inflection $(-\sqrt{6}, -\frac{1}{4}\sqrt{6})$, $(0, 0)$, $(\sqrt{6}, \frac{1}{4}\sqrt{6})$

21. critical pt -1; $f(-1) = 2$ local min and absolute min; decreasing on $(-\infty, -1]$, increasing on $[-1, \infty)$; concave up everywhere; no pts of inflection

Section 3.17

1. (a) -2 units per second   (b) 4 units per second

3. increasing $\frac{8}{3}$ units per second

5. $\dfrac{24}{\sqrt{13}}$ (about 6.66) feet per second

8. (a) $-\dfrac{5}{2\sqrt{2}}$ (b) $-\dfrac{7}{8\sqrt{2}}$ (c) $\dfrac{5}{2\sqrt{2}}$ (d) $\dfrac{7}{8\sqrt{2}}$

   (e) max velocity at $t = 0$ (f) no max speed

9. $\dfrac{2r}{h}$ inches per minute

11. (a) $\dfrac{1}{\pi}$ inches per minute   (b) $\dfrac{1}{2\sqrt[3]{2\pi}}$ inches per minute

13. (a) 2 seconds   (b) 16 feet   (c) 48 feet per second

15. (a) $\frac{1625}{16}$ feet   (b) $\frac{6475}{64}$ feet   (c) 100 feet

Section 3.18

3. $10\frac{1}{30}$ taking $x = 1000$ 5. 0.212 taking $x = 25$

7. 1.975 taking $x = 32$ 9. 8.15 taking $x = 32$

13. $2\pi rht$ 14. error ≤ 0.01 feet

15. (a) $x > 2500$ 16. 0.00153 seconds

Section 3.19

1. $\dfrac{dy}{dx} = -\dfrac{x}{y}$ 3. $\dfrac{dy}{dx} = -\dfrac{b^2 x}{a^2 y}$ 5. $\dfrac{dy}{dx} = -\left(\dfrac{y}{x}\right)^{1/3}$

7. $\dfrac{dy}{dx} = -\dfrac{x^3 + 3x^2 y}{x^3 + y^3}$ 9. $\dfrac{dy}{dx} = -\dfrac{2y + 1}{2x + 1}$

11. $m = -\frac{2}{3}$ 13. $m = -\frac{1}{2}$ 15. at right angles

18. $\dfrac{d^2 y}{dx^2} = \dfrac{16}{(x + y)^3}$ 21. $-\dfrac{4a^2}{y^3}$ 23. $\dfrac{d^2 y}{dx^2} = -\dfrac{b^4}{a^2 y^3}$

24. $\dfrac{dy}{dx} = \dfrac{5}{8}, \; \dfrac{d^2 y}{dx^2} = -\dfrac{9}{128}$ 25. $\dfrac{dy}{dx} = 0, \; \dfrac{d^2 y}{dx^2} = -\dfrac{1}{3}$

Section 3.20

1. (a) $x_0 = 2$ (b) $x_0 = \pm\sqrt{2}$ (c) $x_0 = \frac{1}{4}$

3. 54 inches by 72 inches

4. $A = 1, \; B = -3, \; C = 5$ 6. $A = 9, \; B = 1$

8. max of $\frac{3}{2}$ at $x = -2$; min of $\frac{5}{6}$ at $x = 2$

10. max of $\frac{1}{8}$ at $x = \frac{4}{3}$ 12. 20 inches

14. $(3, 6)$ 16. 120 feet per minute

17. (a) decreasing 2.8 miles per hour

(b) increasing $\dfrac{86}{\sqrt{97}}$ (about 8.73) miles per hour

18. increasing 15 miles per hour or $15\sqrt{3}$ miles per hour
20. base $= 2a\sqrt{3}$, altitude $= 3b$
22. decreasing 4 inches per minute   23. 30 units
26. $\frac{1}{2}(\beta - b)$
28. increasing 7 pounds per square inch per second

29. $\dfrac{2}{3\sqrt{3r^2}}$ 31. $(\frac{1}{3}, (\frac{1}{3})^{3/2})$ 32. (a) 1   (b) $\frac{4}{3}$

33. $\frac{1}{10}$ foot per minute   35. $\frac{3}{2}ka^2$ square units per second
36. decreasing 4 units per second   38. \$5400

39. $x = \dfrac{a^{1/3}l}{a^{1/3} + b^{1/3}}$

Partial Derivatives

1. $\dfrac{\partial f}{\partial x} = 6x - y$, $\dfrac{\partial f}{\partial y} = 1 - x$ 2. $\dfrac{\partial g}{\partial x} = 2x\sqrt{y}$, $\dfrac{\partial g}{\partial y} = \dfrac{x^2}{2\sqrt{y}}$

3. $\dfrac{\partial h}{\partial x} = 2Ax + By + D$, $\dfrac{\partial h}{\partial y} = Bx + 2Cy + E$

4. $\dfrac{\partial f}{\partial x} = \dfrac{x}{\sqrt{x^2 - 3y}}$, $\dfrac{\partial f}{\partial y} = -\dfrac{3}{2\sqrt{x^2 - 3y}}$

5. $\dfrac{\partial g}{\partial x} = \dfrac{(AD - BC)y}{(Cx + Dy)^2}$, $\dfrac{\partial g}{\partial y} = \dfrac{(BC - AD)x}{(Cx + Dy)^2}$

6. $\dfrac{\partial f}{\partial x} = y + z$, $\dfrac{\partial f}{\partial y} = x + z$, $\dfrac{\partial f}{\partial z} = x + y$

7. $\dfrac{\partial g}{\partial x} = 2x$, $\dfrac{\partial g}{\partial y} = 2y$, $\dfrac{\partial g}{\partial z} = 2z$

8. $\dfrac{\partial f}{\partial x} = \dfrac{(1 - y)(1 - z^2)}{2\sqrt{x(1 - y)(1 - z^2)}}$, $\dfrac{\partial f}{\partial y} = -\dfrac{x(1 - z^2)}{2\sqrt{x(1 - y)(1 - z^2)}}$,

$\dfrac{\partial f}{\partial z} = -\dfrac{x(1 - y)z}{\sqrt{x(1 - y)(1 - z^2)}}$

Section 4.2

1. $L_f(P) = \frac{5}{8}$, $U_f(P) = 1\frac{3}{8}$ 3. $L_f(P) = \frac{9}{64}$, $U_f(P) = \frac{37}{64}$
5. $L_f(P) = 1\frac{1}{16}$, $U_f(P) = 1\frac{9}{16}$
8. (a) $L_f(P) = -3x_1(x_1 - x_0) - 3x_2(x_2 - x_1) - \cdots - 3x_n(x_n - x_{n-1})$
$ U_f(P) = -3x_0(x_1 - x_0) - 3x_1(x_2 - x_1) - \cdots - 3x_{n-1}(x_n - x_{n-1})$
(b) $-\frac{3}{2}(b^2 - a^2)$

Section 4.3

1. (a) 5   (b) -2 (c) -1 (d) 0   (e) -4 (f) 1
2. (a) Since $P_1 \subseteq P_2$, it follows from the first theorem of this section that

$$U_f(P_2) \le U_f(P_1).$$

3. (b) $F'(x) = x\sqrt{x + 1}$ (d) $F(2) = \displaystyle\int_0^2 t\sqrt{t + 1}\, dt$

9. (a) $\frac{1}{10}$ (c) $\frac{4}{37}$

Section 4.4

1. -2 3. 1   5. $4\frac{2}{3}$ 7. $\frac{32}{3}$ 9. $8\frac{5}{8}$
11. $-\frac{4}{15}$ 13. $\frac{1}{18}(2^{18} - 1)$ 15. $\frac{1}{6}a^2$ 17. $1\frac{3}{4}$ 19. $-\frac{1}{12}$

Section 4.5

1. part of the motion is in the negative direction
3. $\frac{2}{3}$ 5. 2   7. $\frac{1}{3}$ 9. $21\frac{1}{3}$ 11. $\frac{1}{3}a + \frac{1}{2}b + c$
13. (a) $-285\frac{1}{3}$ (b) $283\frac{2}{3}$ 15. $(x_0 + \frac{1}{4}, y_0 + \frac{2}{3})$

Section 4.7

1. $-\dfrac{1}{3x^3} + C$ 3. $\frac{1}{2}ax^2 + bx + C$

5. $2\sqrt{1 + x} + C$ 7. $\frac{2}{3}x^{3/2} - 2x^{1/2} + C$

9. $\frac{1}{3}t^3 - \frac{1}{2}(a + b)t^2 + abt + C$ 11. $-\dfrac{1}{g(x)} + C$

13. $\dfrac{d}{dx}\left(\displaystyle\int f(x)\, dx\right) = f(x),\quad \displaystyle\int \dfrac{d}{dx}\,[f(x)]\, dx = f(x) + C$

14. (a) $x(t) = x_0 + v_0 t + At^2 + Bt^3$

Section 4.8

1. $\dfrac{1}{3(2 - 3x)} + C$ 3. $\frac{1}{3}(2x + 1)^{3/2} + C$

5. $\dfrac{4(ax + b)^{7/4}}{7a} + C$ 7. $-\dfrac{1}{8(4x^2 + 9)} + C$

9. $\frac{1}{75}(5x^3 + 9)^5 + C$ 11. $\frac{4}{15}(1 + x^3)^{5/4} + C$

13. $-\dfrac{1}{4(1 + s^2)^2} + C$ 15. $\sqrt{x^2 + 1} + C$

17. $-\dfrac{b^3}{2a^4}\sqrt{1 - a^4x^4} + C$

19. $\frac{15}{8}$

21. $\frac{31}{2}$

23. 0

25. $\frac{1}{3}a^3$

Section 4.9

1. yes; $\displaystyle\int_a^b [f(x) - g(x)]\,dx = \int_a^b f(x)\,dx - \int_a^b g(x)\,dx > 0.$

2. no; take for example the functions

$$f(x) = x \quad \text{and} \quad g(x) = 0 \quad \text{on} \quad [-\tfrac{1}{2}, 1]$$

3. yes; otherwise we would have

$$f(x) \le g(x) \qquad \text{for all} \quad x \in [a, b]$$

and it would follow that

$$\int_a^b f(x)\,dx \le \int_a^b g(x)\,dx.$$

4. no; take for example

$$f(x) = 0 \quad \text{and} \quad g(x) = -1 \quad \text{on} \quad [0, 1].$$

5. yes; $\displaystyle\int_a^b |f(x)|\,dx > \int_a^b f(x)\,dx$ and we are assuming that

$$\int_a^b f(x)\,dx > \int_a^b g(x)\,dx.$$

6. no; take

$$f(x) = 0, g(x) = -1 \quad \text{on} \quad [0, 1].$$

16. $\dfrac{2}{x}$

17. $\dfrac{2x}{\sqrt{2x^2 + 7}}$

21. $\dfrac{1}{x}$

Section 4.10

1. $\frac{2}{3}[(x - a)^{3/2} - (x - b)^{3/2}] + C$

3. $\frac{1}{2}(x^{2/3} - 1)^3 + C$

5. $x + \frac{8}{3}x^{3/2} + 2x^2 + C$

7. $\dfrac{2(a + b\sqrt{x + 1})^3}{3b} + C$

9. $-\dfrac{1}{2[g(x)]^2} + C$

11. $\frac{13}{3}$

13. $\frac{2}{9}$

15. (a) $\displaystyle\int_0^{x_0} [g(x) - y_0]\,dx$

(b) $\displaystyle\int_0^{x_0} [y_0 - f(x)]\,dx$

(c) $\displaystyle\int_0^{x_0} f(x)\,dx + \int_{x_0}^a g(x)\,dx$

(d) $\displaystyle\int_0^{x_0} \left[\frac{c-b}{a}x + b - g(x)\right] dx + \int_{x_0}^a \left[\frac{c-b}{a}x + b - f(x)\right] dx$

(e) $\displaystyle\int_{x_0}^a [f(x) - g(x)]\,dx$

16. necessarily holds: $L_g(P) \leq \displaystyle\int_a^b g(x)\,dx < \int_a^b f(x)\,dx \leq U_f(P)$

17. need not hold: take for example

$$f(x) = -x^2 + 1 \quad \text{and} \quad g(x) = \tfrac{1}{10} \quad \text{on} \quad [-1, 1]$$

and use $P = \{-1, 1\}$.

20. necessarily holds: $U_f(P) \geq \displaystyle\int_a^b f(x)\,dx > \int_a^b g(x)\,dx$

Section 5.2

1. $\log 2 + \log 10 \cong 2.99$

3. $2 \log 4 - \log 10 \cong 0.48$

4. $4 \log 3 \cong 4.40$

6. $2 \log 5 - \log 10 \cong 0.92$

7. $\log 8 + \log 9 - \log 10 \cong 1.98$

9. $\tfrac{1}{2} \log 2 \cong 0.35$

11. for $x \in (0, 1)$

$$\log x = \int_1^x \frac{dt}{t} = -\int_x^1 \frac{dt}{t} = -\text{ area of } \Omega$$

where Ω is the region below the graph of

$$g(t) = \frac{1}{t}, \quad t \in [x, 1].$$

13. (a) 1.65   (c) 1.71

14. (b) 2.26

15. $x = e^2$

18. $x = 1$

Section 5.3

1. domain $(0, \infty)$, $f'(x) = \dfrac{1}{x}$

3. domain $(-1, \infty)$, $f'(x) = \dfrac{3x^2}{x^3 + 1}$

5. domain $(-\infty, \infty)$, $f'(x) = \dfrac{x}{1 + x^2}$

7. domain all $x \neq \pm 1$, $f'(x) = \dfrac{4x^3}{x^4 - 1}$

9. domain $(0, 1) \cup (1, \infty)$, $f'(x) = -\dfrac{1}{x(\log x)^2}$

11. domain $(0, \infty)$, $f'(x) = 1 + \log x$

13. domain $(-1, \infty)$, $f'(x) = \dfrac{1 - \log(x + 1)}{(x + 1)^2}$

15. $\log |x + 1| + C$

17. $-\frac{1}{2} \log |3 - x^2| + C$

19. $\dfrac{1}{2(3 - x^2)} + C$

21. $\log \left| \dfrac{x - a}{x - b} \right| + C$

23. $\dfrac{1}{2} \log \left| \dfrac{x^2 - a^2}{x^2 - b^2} \right| + C$

25. $\frac{2}{3} \log |1 + x\sqrt{x}| + C$

27. 1   29. 1

31. $\frac{1}{2} \log \frac{8}{5}$ 33. $\frac{1}{2}(\log 2)^2$

35. $g'(x) = (x^2 + 1)^2(x - 1)^5 x^3 \left(\dfrac{4x}{x^2 + 1} + \dfrac{5}{x - 1} + \dfrac{3}{x} \right)$

37. $g'(x) = \dfrac{x^4(x - 1)}{x + 2} \left(\dfrac{4}{x} + \dfrac{1}{x - 1} - \dfrac{1}{x + 2} \right)$

39. $g'(x) = \dfrac{1}{2} \sqrt{\dfrac{(x - 1)(x - 2)}{(x - 3)(x - 4)}} \left(\dfrac{1}{x - 1} + \dfrac{1}{x - 2} - \dfrac{1}{x - 3} - \dfrac{1}{x - 4} \right)$

41. (a) $\frac{15}{8} - \log 4$ 42. (a) $\log 5$ feet

44. (a) domain $(0, \infty)$; increasing everywhere; no extreme values; concave down everywhere; no pts of inflection

(c) domain $(0, \infty)$; decreasing on $(0, 1/e]$, increasing on $[1/e, \infty)$; $f(1/e) = -1/e$ local and absolute min; concave up throughout; no points of inflection

(e) domain $(0, \infty)$; increasing on $(0, 1]$, decreasing on $[1, \infty)$; $f(1) = -\log 2$ local max and absolute max; concave down on $(0, \sqrt{2 + \sqrt{5}}]$, concave up on $[\sqrt{2 + \sqrt{5}}, \infty)$; pt of inflection at $x = \sqrt{2 + \sqrt{5}}$

Section 5.4

1. $\dfrac{dy}{dx} = -2e^{-2x}$

3. $\dfrac{dy}{dx} = 2xe^{x^2 - 1}$

5. $\dfrac{dy}{dx} = e^x \left(\dfrac{1}{x} + \log x \right)$

7. $\dfrac{dy}{dx} = -(x^{-1} + x^{-2})e^{-x}$

9. $\dfrac{dy}{dx} = \frac{1}{2}(e^x - e^{-x})$

11. $\dfrac{dy}{dx} = \frac{1}{2}e^{\sqrt{x}} \left(\dfrac{1}{x} + \dfrac{\log \sqrt{x}}{\sqrt{x}} \right)$

13. $\dfrac{dy}{dx} = -\dfrac{xe^{\sqrt{1 - x^2}}}{\sqrt{1 - x^2}}$

15. $\dfrac{dy}{dx} = \dfrac{2e^x}{(e^x + 1)^2}$

17. $\frac{1}{2}e^{2x} + C$

19. $\dfrac{1}{k} e^{kx} + C$

21. $\frac{1}{2}e^{x^2} + C$

23. $-e^{1/x} + C$

25. $-\frac{1}{2}e^{-2x} + 2e^{-x} + x + C$

27. $\log (e^x + 1) + C$

29. $3(e^x + 1)^{2/3} + C$

31. $e - 1$

33. $\frac{1}{6}(1 - \pi^{-6})$

35. $\frac{1}{2}(e + e^{-1})$

37. $3 - 4e^{-1}$

39. $\frac{1}{2}e + \frac{1}{2}$

41. $e^{-0.4} = \dfrac{1}{e^{0.4}} \cong \dfrac{1}{1.49} \cong 0.67$

43. $e^{2.8} = (e^2)(e^{0.8}) \cong (7.39)(2.23) \cong 16.48$

45. 7.61   47. 23.10   51. $\dfrac{2}{a}(e^{a^2} + e^{-a^2} - 2)$

53. (i) domain $(-\infty, \infty)$ (ii) increasing on $[0, \infty)$, decreasing on $(-\infty, 0]$
    (iii) $f(0) = 1$ local min and absolute min   (iv) concave up everywhere

55. (i) domain $(-\infty, \infty)$ (ii) increasing on $[0, \infty)$, decreasing on $(-\infty, 0]$
    (iii) $f(0) = 1$ local min and absolute min   (iv) concave up everywhere

57. (i) domain $(-\infty, \infty)$ (ii) increasing on $[-1, \infty)$, decreasing on $(-\infty, -1]$
    (iii) $f(-1) = -1/e$ local min and absolute min   (iv) concave up on $[-2, \infty)$,
    concave down on $(-\infty, -2]$, point of inflection at $x = -2$

59. (i) domain $(-\infty, 0) \cup (0, \infty)$ (ii) increasing on $(-\infty, 0)$, decreasing on
    $(0, \infty)$ (iii) no extreme values   (iv) concave up on $(-\infty, 0)$ and on $(0, \infty)$

Section 5.5

1. (a) 6   (d) -2 (f) -1

2. (a) $\dfrac{3^x}{\log 3} + C$ (c) $-\dfrac{2^{-x}}{\log 2} + C$

   (d) $\dfrac{10^{x^2}}{2 \log 10} + C$ (f) $\dfrac{2^x - 2^{-x}}{\log 2} + C$

   (g) $\log_5 x + C$

3. (a) $\dfrac{1}{e \log 3}$ (c) $\dfrac{1}{e}$

4. (a) $(x + 1)^x \left[\dfrac{x}{x + 1} + \log (x + 1) \right]$

   (c) $[(x^2 + 2)^{\log x}] \left[\dfrac{2x \log x}{x^2 + 2} + \dfrac{\log (x^2 + 2)}{x} \right]$

   (d) $-\left(\dfrac{1}{x} \right)^x (1 + \log x)$ (f) $[(\log x)^{\log x}] \left[\dfrac{1 + \log (\log x)}{x} \right]$

8. (a) 0   (c) $e^{\pm \sqrt{\log 2 \cdot \log 3}}$ (d) 2   (e) $e^{(\log 2)^2/(\log 2 - 1)}$

9. (a) $t_1 < \log a < t_2$ (b) $x_1 < e^b < x_2$

10. (a) domain $(-\infty, \infty)$; increasing on $(-\infty, 0]$, decreasing on $[0, \infty)$; $f(0) = 10$
    local and absolute max
    (b) domain all $x \neq \pm 1$; decreasing on $(-\infty, -1)$ and on $(-1, 0]$, increasing on
    $[0, 1)$ and on $(1, \infty)$; $f(0) = 10$ local min
    (c) domain $[-1, 1]$; increasing on $[-1, 0]$, decreasing on $[0, 1]$; $f(0) = 10$
    local max and absolute max, $f(-1) = f(1) = 1$ endpt min and absolute min

11. (a) $\dfrac{2^{2\pi+1}-1}{2\pi+1}$ (c) $\dfrac{3}{\log 4}$

    (e) 2   (g) $\dfrac{2(p-1)}{\log p}$

13. (a) $\log_{10}4 = \dfrac{\log 4}{\log 10} \cong \dfrac{1.39}{2.30} \cong 0.60$

    (c) $\log_{10}12 = \dfrac{\log 12}{\log 10} = \dfrac{\log 3 + \log 4}{\log 10} \cong \dfrac{1.10 + 1.39}{2.30} \cong 1.08$

Section 5.6

1. (a) $f(x) = 2e^{3x}$ (c) $f(x) = e^{3x-5}$
2. (a) about \$325 ($e^{0.5} \cong 1.65$)   (c) about \$505 ($e^{0.7} \cong 2.01$)
3. (a) about \$820 ($e^{0.2} \cong 1.22$)   (c) about \$671 ($e^{0.4} \cong 1.49$)
4. (a) about 17 years and 3 months   (c) about 8 years and $7\frac{1}{2}$ months
6. (a) about $5\frac{1}{2}\%$ 7. $10{,}000 \log 2$
9. (a) $15e^{-(1/2)\log 1.5}$ (about 12.2) pounds
   (b) $15e^{-(3/2)\log 1.5}$ (about 8.2) pounds
11. 640 pounds
12. (a) (i) $\dfrac{4\log 2}{\log 4 - \log 3}$ (about $9\frac{1}{2}$) years
    (b) (i) $\frac{16}{3}$ pounds   (iii) $3(\frac{3}{4})^{10}$ pounds
14. about $16\frac{4}{5}$ years

Section 5.7

1. $-xe^{-x} - e^{-x} + C$
3. $\frac{1}{2}(x^2 - 1)\log(x+1) - \frac{1}{4}x^2 + \frac{1}{2}x + C$
5. $\frac{1}{3}x^3 \log x - \frac{1}{9}x^3 + C$ 7. $\frac{2}{3}x^{3/2}\log x - \frac{4}{9}x^{3/2} + C$
9. $-\frac{1}{2}x^2 e^{-x^2} - \frac{1}{2}e^{-x^2} + C$ 11. $2\sqrt{x+1}\log(x+1) - 4\sqrt{x+1} + C$
13. $\frac{2}{3}x(x+1)^{3/2} - \frac{4}{15}(x+1)^{5/2} + C$ 15. $x(\log x)^2 - 2x\log x + 2x + C$
17. $-\dfrac{1}{2(\log x)^2} + C$ 19. $\dfrac{1 + ne^{n+1}}{(n+1)^2}$

Section 5.8

1. $y(x) = -\frac{1}{2} + Ce^{2x}$ 3. $y(x) = \frac{2}{5} + Ce^{-(5/2)x}$
5. $y(x) = x + Ce^{2x}$ 7. $y(x) = nx + Cx^3$
9. $y(x) = \frac{2}{11}(x+1)^{7/2} + C(x+1)^{-2}$
11. $y(x) = e^x + Cx^{-1}$ 13. $y(x) = x + 2e^{-x} - 1$
15. $y(x) = e^{-x}\left[\log(1 + e^x) + e - \log 2\right]$

17. $y(x) = x^2(e^x - e)$

19. $y(x) = \left(\dfrac{5}{e} - 1\right)\dfrac{1}{x^2} - e^{-x}\left(1 + \dfrac{2}{x} + \dfrac{2}{x^2}\right)$

21. (a) $200(\frac{4}{5})^{t/5}$ gallons   (b) $200(\frac{4}{5})^{t^2/25}$ gallons

23. $20 + 5e^{-0.03t}$ pounds

Section 5.9

1. $2\sqrt{ab}$ 3. $\left(\pm\dfrac{1}{\sqrt{2}}, \dfrac{1}{\sqrt{e}}\right)$ 5. $x = \dfrac{1}{\sqrt{e}}$ 7. 1

9. (a) $y = \frac{1}{2}(e^{ax} - e^{-ax})$ (c) $y = e^{ax}$

11. $a^2 \log 2$ 13. $\dfrac{5280}{30 \log 2}$ feet (about 255 feet)

Section 6.1

1. $3 \cos 3x$ 3. $x \sec^2 x + \tan x$ 4. $\cos x - \sin x$

6. $\dfrac{1}{2\sqrt{x}} \cos \sqrt{x}$ 7. $\pi[\cos^2 \pi x - \sin^2 \pi x]$ 9. $\pi x^2 \cos \pi x + 2x \sin \pi x$

10. $\sin x \cos x$ 12. $6 \tan^2 2x \sec^2 2x$ 13. $e^x \cos x + e^x \sin x$

15. $-\dfrac{x \sin x^2}{\sqrt{\cos x^2}}$ 16. $(\cos \pi x)^x[-\pi x \tan \pi x + \log (\cos \pi x)]$

18. $ke^{-kx}[\cos kx - \sin kx]$ 19. $\pi 10^x \sec^2 \pi x + 10^x \log 10 \tan \pi x$

21. $\dfrac{1}{2}\left(\dfrac{\sec^2 x}{\tan x}\right) = \operatorname{cosec} 2x$ 22. $e^x(\frac{1}{2}\pi \cos \frac{1}{2}\pi x + \sin \frac{1}{2}\pi x)$

24. $\pi \tan^3 \frac{1}{4}\pi x \sec^2 \frac{1}{4}\pi x$ 25. $(\cos x)e^{\sin x}$

27. $2 \tan x \sec^2 x \left[\sec^2 x + \tan^2 x\right]$

28. $n \sin^{n-1} x \left[\sin nx \cos x + \sin x \cos nx\right]$

30. $-\tan 2x$

31. $2 \tan \theta \sec^2 \theta$ 33. $-6 \operatorname{cosec}^2 3\theta \cot 3\theta$

34. $-k^2 \sin kx$ 36. $-x \cos x - 2 \sin x$

37. $e^{2t}(3 \cos t - 4 \sin t)$ 39. $-e^{-t}(3 \sin 2t + 4 \cos 2t)$

40. $\dfrac{\sin (x - y)}{\sin (x - y) - 1}$ 42. $-\dfrac{1}{1 + (x + y) \sin y}$

44. $\tan \theta = 2\sqrt{2}, \quad \theta \cong 1.23$ radians

46. 0.868   48. 0.248

49. 0.884   51. 0.647

52. 3   54. $\frac{3}{2}$ 55. 0   57. does not exist

59. increasing on $[0, 2\pi]$; $f(0) = 0$ endpt min and absolute min, $f(2\pi) = 2\pi$ endpt max and absolute max; concave down on $[0, \pi]$, concave up on $[\pi, 2\pi]$; point of inflection (π, π)

61. decreasing on $(-\frac{1}{2}\pi, 0]$, increasing on $[0, \frac{1}{2}\pi)$; $f(0) = 0$ local min and absolute min; concave up on $(-\frac{1}{2}\pi, \frac{1}{2}\pi)$

63. increasing on $[0, \frac{1}{2}\pi]$ and on $[\frac{3}{2}\pi, 2\pi]$, decreasing on $[\frac{1}{2}\pi, \frac{3}{2}\pi]$; $f(0) = 1$ endpt min, $f(\frac{1}{2}\pi) = e^{\pi/2}$ local max, $f(\frac{3}{2}\pi) = -e^{3\pi/2}$ local min and absolute min, $f(2\pi) = e^{2\pi}$ endpt max and absolute max; concave up on $[0, \frac{1}{4}\pi]$ and on $[\frac{5}{4}\pi, 2\pi]$, concave down on $[\frac{1}{4}\pi, \frac{5}{4}\pi]$; pts of inflection $(\frac{1}{4}\pi, \sqrt{2}\, e^{\pi/4})$, $(\frac{5}{4}\pi, -\sqrt{2}\, e^{5\pi/4})$

Section 6.2

1. $\frac{1}{3} \sin (3x - 1) + C$

3. $\sin^3 x + C$

4. $\frac{1}{2} \sec 2x + C$

6. $-\frac{2}{3}(1 + \cos x)^{3/2} + C$

7. $-e^{-\sin x} + C$

9. $-2 \cos x^{1/2} + C$

10. $-2\sqrt{1 - \sin x} + C$

12. $-\log (1 + \cos x) + C$

13. $-\dfrac{1}{3\pi} \cos^3 \pi x + C$

15. $\dfrac{1}{\pi} \log |\sin \pi x| + C$

16. $\frac{1}{2}\log |1 + 2 \sin x| + C$

18. $-\frac{1}{2} \log (2 - \sin 2x) + C$

19. $\frac{1}{2} \tan^2 x + C$

21. $\frac{2}{3}(1 + \tan x)^{3/2} + C$

22. $\frac{1}{2}(x + \sin x) + C$

23. $\frac{1}{2}(x + \frac{1}{2} \sin 2x) + C$

25. $\dfrac{1}{2}\left(x + \dfrac{1}{2\pi} \sin 2\pi x\right) + C$

26. $x \sin x + \cos x + C$

28. $\frac{1}{2}e^x (\sin x - \cos x) + C$

Section 6.3

1. $\frac{1}{2}\pi$

3. 0

4. $\frac{1}{2}\pi$

6. $\frac{1}{4}\pi$

7. $\frac{1}{4}\pi$

9. $\frac{1}{3}\pi$

10. $-\frac{1}{6}\pi$

12. 0

13. $\dfrac{1}{x^2 + 2x + 2}$

15. $\dfrac{2x}{\sqrt{1 - x^4}}$

17. $\dfrac{2x}{\sqrt{1 - 4x^2}} + \arcsin 2x$

19. $\dfrac{2 \arcsin x}{\sqrt{1 - x^2}}$

21. $\dfrac{x - (1 + x^2) \arctan x}{x^2(1 + x^2)}$

23. $\dfrac{1}{(1 + 4x^2)\sqrt{\arctan 2x}}$

25. $\dfrac{1}{x[1 + (\log x)^2]}$

27. $-\dfrac{r}{|r|\sqrt{1 - r^2}}$

29. $\dfrac{1}{1 + r^2}$

31. $\sqrt{\dfrac{c - x}{c + x}}$ $(c > 0)$

33. $\dfrac{x^2}{(c^2 - x^2)^{3/2}}$ $(c > 0)$

38. $\frac{1}{4}\pi$

40. $\frac{1}{4}\pi$

41. $\frac{1}{6}\pi$

42. $\frac{1}{20}\pi$

43. $\frac{1}{10}\pi$

44. (b) $\dfrac{d}{dx}$ (arc cos x) $= -\dfrac{1}{\sqrt{1-x^2}}$

(c) $\dfrac{d}{dx}$ (arc cot x) $= -\dfrac{1}{1+x^2}$

(d) $\dfrac{d}{dx}$ (arc cosec x) $= -\dfrac{1}{x\sqrt{x^2-1}}$

Section 6.4

1. $\frac{1}{2}\log 2$ 3. $\frac{1}{6}\pi$ 5. $\frac{1}{6}(\pi\sqrt{3}-3)$ 7. $\sqrt{a^2+b^2}$
9. (a) $\frac{1}{4}\pi$ (c) $1-\frac{1}{32}\pi^2$ (e) $-\frac{1}{2}$
11. decreasing 0.04 radians per minute
12. increasing 0.06 radians per minute
13. 5π miles per minute   14. 12 feet
15. $r=2\sqrt{6}$ inches, $h=4\sqrt{3}$ inches   17. $\theta=$ arc tan m
18. $\theta=\frac{1}{4}\pi+\frac{1}{2}\alpha$ radians

Section 6.5

1. $x(t)=\sin(8t+\frac{1}{2}\pi)$; $a=1$, $f=4/\pi$

3. $\pm\dfrac{2\pi a}{p}$ 5. $x(t)=\dfrac{15}{\pi}\sin\dfrac{\pi}{3}t$

7. $x=\pm\frac{1}{2}\sqrt{3}\,C_2$ 9. $x(t)=x_0\sin(\sqrt{k/m}\,t+\frac{1}{2}\pi)$

Section 6.6

1. $2x\cosh x^2$ 3. $\dfrac{a\sinh ax}{2\sqrt{\cosh ax}}$

5. $\dfrac{1}{1-\cosh x}$ 7. $ab(\cosh bx-\sinh ax)$

9. $\dfrac{a\cosh ax}{\sinh ax}$ 16. minimum value of 3

18. no extreme values   21. $y=2\cosh 3x+\frac{1}{3}\sinh 3x$

Section 6.7

1. $2\tanh x\,\text{sech}^2 x$ 3. $\dfrac{\text{sech}^2 x}{\tanh x}$ 5. $\dfrac{2e^{2x}\cosh(\text{arc tan }e^{2x})}{1+e^{4x}}$

7. $\dfrac{-x\,\text{cosech}^2(\sqrt{x^2+1})}{\sqrt{x^2+1}}$ 9. $\dfrac{-\text{sech }x(\tanh x+2\sinh x)}{(1+\cosh x)^2}$

15. (a) $\frac{3}{5}$ (b) $\frac{5}{3}$ (c) $\frac{4}{3}$ (d) $\frac{5}{4}$ (e) $\frac{3}{4}$

Section 7.1

1. $-e^{2-x} + C$ 3. $\frac{3}{2} \sin \frac{2}{3}x + C$ 4. $\dfrac{\pi}{4c}$ 6. $e - \sqrt{e}$

7. $\frac{1}{2}(x^2 - 1) \log(x + 1) - \frac{1}{4}x^2 + \frac{1}{2}x + C$ 9. $-\dfrac{1}{5^x \log 5} + C$

10. $-\log(2\sqrt{3} - 3)$ 12. 2 13. $-\tan(1 - x) + C$

15. $\dfrac{x5^x}{\log 5} - \dfrac{5^x}{(\log 5)^2} + C$ 16. $\frac{1}{4} \log 2$ 18. $\frac{1}{2} \log 3$

19. $-\frac{1}{5} e^{-x} (\sin 2x + 2 \cos 2x) + C$ 21. $\frac{1}{2} \arcsin x^2 + C$

22. $\frac{1}{4}(2x^2 \arcsin 2x^2 + \sqrt{1 - 4x^4}) + C$

24. $-\frac{1}{2} \cot x + C$ 25. $\frac{2}{3}\sqrt{3 \tan \theta + 1} + C$

27. $\dfrac{\log |ae^x - b|}{a} + C$ 28. $\dfrac{a^x e^x}{1 + \log a} + C$

30. $2 - \sqrt{3}$ 31. $\frac{1}{4} \sinh 2x - \frac{1}{2}x + C$

33. $\frac{1}{2}(\cos x \cosh x + \sin x \sinh x) + C$

34. $\log(\cosh x) - \frac{1}{2} \tanh^2 x + C$ 36. $\dfrac{1}{4a} e^{2ax} + \dfrac{x}{2} + C$

Section 7.2

1. $\log \left| \dfrac{x - 2}{x + 5} \right| + C$ 3. $\log \left| \dfrac{x^2 - 1}{x} \right| + C$

4. $5 \log |x - 1| - 3 \log |x| + \dfrac{3}{x} + C$

6. $\dfrac{x^4}{4} + \dfrac{4x^3}{3} + 6x^2 + 32x - \dfrac{32}{x - 2} + 80 \log |x - 2| + C$

7. $5 \log |x - 2| - 4 \log |x - 1| + C$

9. $\dfrac{-1}{2(x - 1)^2} + C$

10. $\dfrac{3}{4} \log |x - 1| - \dfrac{1}{2(x - 1)} + \dfrac{1}{4} \log |x + 1| + C$

12. $\dfrac{1}{3} \left(\dfrac{1}{x + 1} - \dfrac{1}{x - 2} \right) + C$ 13. $\dfrac{1}{32} \log \left| \dfrac{x - 2}{x + 2} \right| - \dfrac{1}{16} \arctan \dfrac{x}{2} + C$

15. $\dfrac{1}{16} \log \left| \dfrac{x^2 + 2x + 2}{x^2 - 2x + 2} \right| + \dfrac{1}{8} [\arctan(x + 1) + \arctan(x - 1)] + C$

16. $\frac{1}{16}\pi$

20. (a) $C(t) = 2A_0 \left(\dfrac{t}{t+1} \right)$ (b) $C(t) = 4A_0 \dfrac{3^t - 2^t}{2(3^t) - 2^t}$

(c) $C(t) = A_0(m+n) \left(\dfrac{m^t - n^t}{m^{t+1} - n^{t+1}} \right)$

Section 7.3

1. $\frac{1}{3} \cos^3 x - \cos x + C$ 3. $-\frac{1}{5} \cos^5 x + \frac{1}{7} \cos^7 x + C$
5. $\frac{3}{8} \pi$ 7. $\frac{3}{16} \pi$
9. $\frac{1}{5} \cos^5 x - \frac{1}{3} \cos^3 x + C$ 11. $\frac{1}{2} \sin x + \frac{1}{10} \sin 5x + C$
13. $-\frac{1}{3} \cos^3 x + \frac{2}{5} \cos^5 x - \frac{1}{7} \cos^7 x + C$
15. $\frac{1}{16} x - \frac{1}{64} \sin 4x + \frac{1}{48} \sin^3 2x + C$
17. 1   19. 0   21. 0

Section 7.4

1. $\frac{1}{2} \tan^2 x + \log |\cos x| + C$ 3. $\tan(x+1) + C$
5. $\frac{1}{3} \tan^3 x + C$ 7. $\frac{1}{7} \tan^7 x + \frac{1}{5} \tan^5 x + C$
9. $\frac{1}{12} \tan^4 3x - \frac{1}{6} \tan^2 3x + \frac{1}{3} \log |\sec 3x| + C$
11. $-\frac{1}{4} \cot^2 2x - \frac{1}{2} \log |\sin 2x| + C$
13. $-\frac{1}{2} \operatorname{cosec} x \cot x + \frac{1}{2} \log |\operatorname{cosec} x - \cot x| + C$

Section 7.5

1. $\operatorname{arc\,sin} \left(\dfrac{x}{a} \right) + C$ 3. $\dfrac{x}{5\sqrt{5 - x^2}} + C$
5. $\sqrt{3} - \frac{1}{2} \log(2 + \sqrt{3})$
7. $\frac{1}{2} x \sqrt{x^2 - 4} + 2 \log |x + \sqrt{x^2 - 4}| + C$
9. $\frac{1}{2} x \sqrt{x^2 + 4} - 2 \log |x + \sqrt{x^2 + 4}| + C$
11. $\dfrac{x}{\sqrt{a^2 - x^2}} - \operatorname{arc\,sin} \left(\dfrac{x}{a} \right) + C$ 13. $\frac{1}{3}(16\sqrt{2} - 8)$
15. $\operatorname{arc\,sin} \left(\frac{1}{3} e^x \right) + C$ 17. $-\dfrac{(1 - x^2)^{3/2}}{3x^3} + C$
19. $\log |x + \sqrt{x^2 + a^2}| + C$ 21. $\dfrac{x}{2} \sqrt{a^2 - x^2} + \dfrac{a^2}{2} \operatorname{arc\,sin} \left(\dfrac{x}{a} \right) + C$
23. $\dfrac{1}{a} \log \left| \dfrac{x}{a + \sqrt{a^2 - x^2}} \right| + C$ 25. $-\dfrac{1}{a^2 x} \sqrt{a^2 - x^2} + C$
27. $\dfrac{1}{a} \log \left| \dfrac{x}{a + \sqrt{a^2 + x^2}} \right| + C$

Section 7.6

1. $\tan \frac{1}{2} x + C$ 3. $-\cot \frac{1}{2} x + C$
5. $\frac{1}{2}\sqrt{2} \operatorname{arc\,tan} \left(\frac{1}{2}\sqrt{2} \right)$ 7. $\frac{2}{3} \operatorname{arc\,tan} \left[\frac{1}{3} \tan \frac{1}{2} x \right] + C$

9. $-\frac{1}{3}x + \frac{5}{6}$ arc tan $[2 \tan \frac{1}{2}x] + C$

11. $-2[\tan \frac{1}{2}x + 1]^{-1} - \log(1 + \sin x) + C$

Section 7.7

1. $-2(\sqrt{x} + \log|1 - \sqrt{x}|)$ 3. $2\sqrt{x} - 2$ arc tan $\sqrt{x} + C$

5. $2 \log(\sqrt{1 + e^x} - 1) - x + 2\sqrt{1 + e^x} + C$

7. $\frac{17}{400}$ 9. $x + 2\sqrt{x} + 2 \log|\sqrt{x} - 1| + C$

11. $x + 4\sqrt{x - 1} + 4 \log|\sqrt{x - 1} - 1| + C$

13. $\log\left(\dfrac{\sqrt{e^x + 1} - 1}{\sqrt{e^x + 1} + 1}\right) + C$ 15. $\frac{2}{3}(x - 8)\sqrt{x + 4} + C$

17. $\dfrac{13\sqrt{5} - 25}{150}$ 19. $\dfrac{4b + 2ax}{a^2\sqrt{ax + b}} + C$

Section 7.8

1. (a) 506   (b) 650   (c) 572   (d) 578   (e) 576

3. (a) $\frac{1}{4}\pi \cong 0.78$ (b) $\frac{1}{4}\pi \cong 0.78$

5. (a) 1.23   (b) 1.24

Section 8.1

1. (a) $8/\sqrt{13}$ (c) $1/\sqrt{5}$ 3. (a) $\frac{17}{2}$

4. decreasing $\frac{8}{5}b^2\pi$ units per minute

Section 8.3

1. (a) $y^2 = 8x$ (b) $y^2 = -8x$

   (d) $(y - 2)^2 = 8(x - 1)$ (f) $(x - 1)^2 = 4y$

   (h) $y^2 + 4y + 6x - 17 = 0$

2. (a) vertex $(0, 0)$, focus $(\frac{1}{2}, 0)$, axis $y = 0$, directrix $x = -\frac{1}{2}$

   (c) vertex $(1, 0)$, focus $(\frac{3}{2}, 0)$, axis $y = 0$, directrix $x = \frac{1}{2}$

   (e) vertex $(-2, \frac{3}{2})$, focus $(-2, \frac{7}{2})$, axis $x = -2$, directrix $y = -\frac{1}{2}$

   (g) vertex $(-\frac{1}{2}, \frac{3}{4})$, focus $(-\frac{1}{2}, 1)$, axis $x = -\frac{1}{2}$, directrix $y = \frac{1}{2}$

3. (a) $(x - y)^2 = 6x + 10y - 9$ (c) $(x + y)^2 = -12x + 20y + 28$

7. $2y = x^2 - 4x + 7$, $18y = x^2 - 4x + 103$

9. $y = -\dfrac{16}{v_0^2}(\sec^2 \theta) x^2 + (\tan \theta) x$ 11. $\frac{1}{16}v_0^2 \cos \theta \sin \theta$ 13. $\frac{1}{4}\pi$

15. vertex $\left(-\dfrac{B}{2A}, \dfrac{4AC - B^2}{4A}\right)$, focus $\left(-\dfrac{B}{2A}, \dfrac{4AC - B^2 + 1}{4A}\right)$,

   directrix $y = \dfrac{4AC - B^2 - 1}{4A}$

Section 8.4

1. foci $(\pm\sqrt{5}, 0)$, major axis has length 6, minor axis has length 4
3. foci $(0, \pm\sqrt{2})$, major axis has length $2\sqrt{6}$, minor axis has length 4
5. foci $(\pm 1, 0)$, major axis has length 4, minor axis has length $2\sqrt{3}$
7. foci $(1, \pm 4\sqrt{3})$, major axis has length 16, minor axis has length 8
9. $\dfrac{x^2}{9} + \dfrac{y^2}{8} = 1$ 11. $\dfrac{(x-6)^2}{25} + \dfrac{(y-1)^2}{16} = 1$
13. $\dfrac{(x-1)^2}{21} + \dfrac{(y-3)^2}{25} = 1$
16. the ellipse approaches a circle of radius a
18. $\dfrac{x^2}{25} + \dfrac{4y^2}{75} = 1$

Section 8.5

1. $\dfrac{x^2}{9} - \dfrac{y^2}{16} = 1$ 3. $\dfrac{y^2}{25} - \dfrac{x^2}{144} = 1$
5. $\dfrac{x^2}{9} - \dfrac{(y-1)^2}{16} = 1$ 7. $xy = \frac{1}{2}$

8. transverse axis 2, vertices $(\pm 1, 0)$, foci $(\pm\sqrt{2}, 0)$, asymptotes $y = \pm x$
10. transverse axis 6, vertices $(\pm 3, 0)$, foci $(\pm 5, 0)$, asymptotes $y = \pm\frac{4}{3}x$
12. transverse axis 8, vertices $(0, \pm 4)$, foci $(0, \pm 5)$, asymptotes $y = \pm\frac{4}{3}x$
14. transverse axis 6, vertices $(4, 3)$ and $(-2, 3)$, foci $(6, 3)$ and $(-4, 3)$, asymptotes
    $y = \frac{4}{3}(x-1) + 3$ and $y = -\frac{4}{3}(x-1) + 3$
16. transverse axis 2, vertices $(0, 0)$ and $(-2, 0)$, foci $(1, 0)$ and $(-3, 0)$, asymptotes
    $y = \sqrt{3}(x+1)$ and $y = -\sqrt{3}(x+1)$
18. transverse axis $2\sqrt{2}$, vertices $(1, 1)$ and $(-1, -1)$, foci $(\sqrt{2}, \sqrt{2})$ and $(-\sqrt{2}, -\sqrt{2})$,
    asymptotes $x = 0$ and $y = 0$

Section 8.6

9. $(-2, 0)$ 11. $(-\frac{3}{2}, \frac{3}{2}\sqrt{3})$ 13. $(1, 0)$ 15. $(0, 3)$
17. $[1, \frac{1}{2}\pi + 2n\pi], [-1, \frac{3}{2}\pi + 2n\pi]$ 19. $[3, \pi + 2n\pi], [-3, 2n\pi]$
21. $[\sqrt{8}, \frac{7}{4}\pi + 2n\pi], [-\sqrt{8}, \frac{3}{4}\pi + 2n\pi]$
23. $[8, \frac{1}{6}\pi + 2n\pi], [-8, \frac{7}{6}\pi + 2n\pi]$
25. symmetry about the x-axis
27. symmetry about both axes and the origin
29. symmetry about the origin
31. $r^2 \sin 2\theta = 1$ 33. $r = 2$
35. $r = a(1 - \cos\theta)$ 37. the horizontal line $y = 4$

39. the line $y = \sqrt{3}x$

41. the parabola $y^2 = 4(x + 1)$

43. circle of radius $\frac{1}{2}$ centered at $(\frac{1}{2}, 0)$

45. Figure 8.6.15

47. Figure 8.6.16

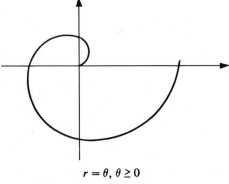

$r = \theta, \theta \geq 0$

spiral of Archimedes

FIGURE 8.6.15

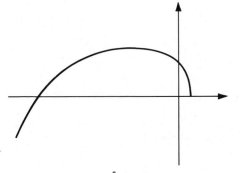

$r = e^\theta, \theta \geq 0$

exponential spiral

FIGURE 8.6.16

49. yes; $[1, \pi] = [-1, 0]$ and the pair $r = -1, \theta = 0$ satisfies the equation

51. yes; the pair $r = \frac{1}{2}, \theta = \frac{1}{2}\pi$ satisfies the equation

53. $[2, \pi] = [-2, 0]$; the coordinates of $[-2, 0]$ satisfy $r^2 = 4 \cos \theta$ and the coordinates of $[2, \pi]$ satisfy $r = 3 + \cos \theta$

54. (0, 0), (1, 1)

56. (0, 1)

57. Figure 8.6.17

59. Figure 8.6.18

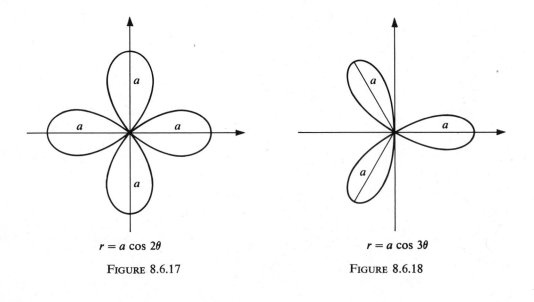

$r = a \cos 2\theta$

FIGURE 8.6.17

$r = a \cos 3\theta$

FIGURE 8.6.18

Section 8.7

1. $(y - 1)^2 = 4x$ 3. $a^3 y = (x - b)^3$
5. $y = \log \left[\frac{1}{4}(x - 1)^2 + 1\right]$ 7. 3
9. 0   11. -3 13. -1
15. 0   17. $\frac{1}{2}(1 - 2\sqrt{3})$ 19. $\frac{1}{3}$
22. no horizontal tangent; vertical tangents at $(2, 2), (-2, 0)$
24. horizontal tangents at $(0, 1), (0, -1)$; vertical tangents at $(1, 1/\sqrt{2}), (-1, 1/\sqrt{2}),$
     $(1, -1/\sqrt{2}), (-1, -1/\sqrt{2})$
26. (a) the paths intersect at $(2, 10)$ and $(5, 3)$
     (b) the particles collide at $(5, 3)$

Section 8.8

1. $\alpha = \frac{1}{4}\pi, \; \frac{1}{2}(X^2 - Y^2) = 1$ 3. $\alpha = \frac{1}{6}\pi, \; 4X^2 - Y^2 = 1$
5. $\alpha = \frac{1}{4}\pi, \; 2Y^2 + \sqrt{2}X = 0$ 7. $\alpha = -\frac{1}{6}\pi, \; Y^2 + X = 0$
9. $\alpha = \frac{1}{8}\pi, \; \cos\alpha = \frac{1}{2}\sqrt{2 + \sqrt{2}}, \; \sin\alpha = \frac{1}{2}\sqrt{2 - \sqrt{2}}$

Section 9.1

1. $\frac{1}{2}\alpha a + \beta$ 3. $\frac{1}{4}a^3$ 5. $\dfrac{\log a}{a - 1}$
7. $\frac{1}{4}\pi a$ 9. 0
14. (a) the terminal velocity is twice the average velocity
     (b) the average velocity during the first $\frac{1}{2}x$ seconds is one-third of the average
       velocity during the next $\frac{1}{2}x$ seconds
16. $\frac{1}{2}\pi b$ and $\frac{1}{2}\pi a$

Section 9.2

1. $10\frac{2}{3}$ 3. $10\frac{2}{3}$ 5. 10   7. $\frac{3}{16}$
9. $\frac{37}{12}$ 11. πab 13. $2\frac{1}{2}$ 15. $5\frac{3}{4}$

Section 9.3

1. $\frac{1}{5}\pi$ 3. $\frac{1}{2}\pi(e^2 - 1)$ 5. $\frac{12}{7}(2^{1/3})\pi$ 7. $\frac{1}{2}\pi$
9. $\frac{3}{10}\pi$ 11. $\frac{1}{3}\pi^4 - \frac{1}{2}\pi^2$ 13. $\frac{1}{4}\pi a^3$
14. $\frac{1}{3}\pi h(R^2 + Rr + r^2)$
16. (a) $\frac{4}{3}ab^2$ (b) $\frac{16}{3}ab^2$ (c) πab
17. (a) $31\frac{1}{4}\%$ (b) $14\frac{22}{27}\%$
19. (a) $4h$ (b) $\frac{3}{2}\sqrt{3}$ (c) $\frac{3}{2}$
20. (a) 64π (b) $\frac{1024}{35}\pi$ (c) $\frac{704}{5}\pi$ (d) $\frac{512}{7}\pi$
21. $\frac{16}{3}r^3$

Section 9.4

1. $\frac{1}{2}\pi$ 3. 2π 5. $3(4^{1/3})\pi$ 7. 2π

9. $\frac{4}{5}\pi$ 11. $\frac{3}{10}\pi$ 13. $\frac{2}{3}\pi^4 - 2\pi^2$ 15. $\frac{1}{4}\sqrt{3}\pi a^3$

Section 9.5

1. $\frac{1}{4}\pi a^2$ 3. $\frac{1}{2}a^2$ 5. $\frac{1}{2}\pi a^2$

7. $\frac{1}{4} - \frac{1}{16}\pi$ 9. $\frac{3}{16}\pi + \frac{3}{8}$ 11. $\frac{5}{2}a^2$

13. $\frac{1}{12}(3e^{2\pi} - 3 - 2\pi^3)$ 15. $\frac{1}{4}(e^{2\pi} + 1 - 2e^\pi)$

Section 9.6

1. 2   3. 2   5. $\sqrt{2}$

7. e 8. 0   11. 2

12. $\sqrt{m^2 + 1}(d - c)$ 14. $2\sqrt{3}$ 16. $2\log(\sqrt{2} + 1)$

18. $2 - \sqrt{2} - \frac{1}{2}\log 3 + \log(\sqrt{2} + 1)$ 20. $\frac{2}{3}[2\sqrt{2} - 1]$

22. $\frac{1}{3}\pi + \frac{1}{2}\sqrt{3}$ 24. $\frac{1}{2}[\sqrt{2} + \log(1 + \sqrt{2})]$

26. $\sqrt{2}(e^\pi - 1)$ 28. $\sqrt{2}(e^{4\pi} - 1)$

30. $\frac{1}{2}\sqrt{5}(e^{4\pi} - 1)$ 34. $\log(1 + \sqrt{2})$

35. $f(x) = \cosh x = \frac{1}{2}(e^x + e^{-x})$

Section 9.7

1. $2\pi r$ 3. $2\pi[\sqrt{2} + \log(1 + \sqrt{2})]$

5. $\frac{1}{9}\pi(17^{3/2} - 1)$ 7. $\frac{61}{432}\pi$

9. $3\pi\sqrt{13}$ 11. $2\pi r^2$ 13. $\frac{6}{5}\pi a^2$

15. $2\pi \displaystyle\int_{\theta_1}^{\theta_2} \rho(\theta) \sin\theta \sqrt{[\rho(\theta)]^2 + [\rho'(\theta)]^2}\, d\theta$

17. $\pi s(R + r)$ where s is the slant height

19. (a) $2\pi b^2 + \dfrac{2\pi ab}{e}$ arc sin e (b) $2\pi a^2 + \dfrac{\pi b^2}{e}\log\left|\dfrac{1 + e}{1 - e}\right|$,

where e is the eccentricity $\dfrac{\sqrt{a^2 - b^2}}{a}$

Section 9.8

1. (a) $\frac{1}{4}$ foot-pounds   (b) $\frac{9}{16}$ foot-pounds   3. $\frac{3}{2}$ feet

5. (a) $6480\pi + 8640$ foot-pounds   (b) $15,120\pi + 8640$ foot-pounds

7. (a) $\frac{1}{2}\sigma l^2$ foot-pounds   (b) $\frac{3}{2}\sigma l^2$ foot-pounds

Section 9.9

1. 2160 pounds
3. $\frac{8000}{3}\sqrt{2}$ pounds
5. 2560 pounds
7. (a) 41,250 pounds   (b) 11 feet

Section 9.10

1. (a) \$3700   (b) \$3425
4. (a) \$1008.40   (b) \$897

Section 9.11

1. 2
3. $\frac{9}{2}$
5. $\frac{5}{12}$
7. 3.6
9. $\frac{9}{2}\pi$
11. $\sqrt{2} - \frac{1}{8}\pi$
12. $a^2 \log 2$
13. 36
14. 2.7
15. $\frac{3}{2}$
17. 5
19. $\dfrac{2}{\pi - 2}$
21. $\frac{1}{4}\pi a^3$
23. $\frac{1}{15}\pi a^3$
25. $\frac{32}{35}\pi ab^2$
27. $\frac{64}{5}\pi$
29. $\frac{4}{5}\pi a^2 b$
31. $\frac{19}{27}$
33. $\frac{335}{27}a$
35. $6a$
37. $\dfrac{a^2 + ab + b^2}{a + b}$
38. $\frac{52}{3}\pi p^2$
40. $\frac{47}{16}\pi a^2$
43. $\frac{2}{3}r^3 \tan\theta$

Section 10.1

1. decreasing; bounded below by 0 and above by 2
3. increasing; bounded below by 1 but not bounded above
5. nonincreasing; bounded below by 0 and above by $\frac{1}{2}$
7. increasing; bounded below by $\frac{1}{2}$ but not bounded above
9. increasing; bounded below by $\frac{4}{5}\sqrt{5}$ and above by 2
11. decreasing; bounded below by 0 and above $\frac{2}{5}$
13. not monotonic; not bounded below and not bounded above
15. increasing; bounded below by 0 and above by log 2
17. decreasing; bounded below by 1 and above by 4
19. increasing; bounded below by $\frac{1}{2}$ and above by 1
21. decreasing; bounded below by 0 and above by $\frac{1}{2}$
23. decreasing; bounded below by 0 and above by $\frac{1}{3}\log 3$
25. not monotonic; bounded below by $-\frac{1}{3}$ and above by $\frac{4}{9}$
27. not monotonic; bounded below by -1 and above by 1

Section 10.2

1. converges to 0
3. diverges
5. converges to 1
7. diverges
9. converges to 4
11. converges to 0
13. diverges
15. converges to log 2
17. converges to 1

19. converges to 1

21. converges to 1

23. converges to 0

25. converges to 0

27. converges to 0

28. converges to $\frac{1}{2}$

30. converges to e^2

Section 10.3

1. converges to 1

2. converges to 1

3. converges to 0

5. converges to 0

7. converges to 0

8. converges to 1

9. converges to 0

11. converges to 1

13. converges to 0

15. converges to 1

17. converges to 0

19. converges to π

21. converges to 2

22. converges to $1/e$

24. diverges

27. $\lim\limits_{n \to \infty} 2n \sin (\pi/n) = 2\pi$; the number $2n \sin (\pi/n)$ is the perimeter of a regular polygon of n sides inscribed in the unit circle; as n tends to ∞, the perimeter of the polygon tends to the circumference of the circle

28. $\frac{1}{2}$ 30. $\frac{1}{8}$ 33. converges to $1/e$ 35. converges to 1

37. converges to 0 if $t < 1$, converges to e^x if $t = 1$, diverges if $t > 1$

Section 10.5

1. $\frac{1}{4}$

3. $\frac{1}{2}$

5. $\frac{400}{33}$

7. $\frac{5}{6}$

9. $\frac{15}{2}$

11. $\frac{2150}{99}$

13. $\frac{1}{4}$

15. $\sum\limits_{k=1}^{\infty} \dfrac{7}{10^k} = \dfrac{7}{9}$

17. $\sum\limits_{k=1}^{\infty} \dfrac{24}{100^k} = \dfrac{8}{33}$

19. $\sum\limits_{k=1}^{\infty} \dfrac{112}{1000^k} = \dfrac{112}{999}$

21. $\dfrac{62}{100} + \dfrac{1}{100} \sum\limits_{k=1}^{\infty} \dfrac{45}{100^k} = \dfrac{687}{1100}$

27. (a) $\sum\limits_{k=0}^{\infty} (-1)^k x^{k+1}$ (c) $\sum\limits_{k=0}^{\infty} x^{2k}$

29. $100 \sum\limits_{k=1}^{\infty} \left(\dfrac{1}{1.05}\right)^k = 2{,}000$ dollars

Section 10.6

1. converges, comparison with $\sum (1/k^2)$ 3. converges, root test

5. converges, ratio test   7. converges, comparison with $\sum (1/k^2)$

9. diverges, comparison with $\sum (1/\sqrt{k})$ 11. diverges, ratio test

13. diverges, comparison with $\sum (1/3k)$ 15. converges, root test

17. diverges, ratio test   19. converges, root test

21. converges, ratio test   23. converges, integral test

25. converges, comparison with $\sum (2/k^{3/2})$

27. converges, ratio test

29. diverges, kth term does not tend to zero

31. diverges, comparison with $\sum(1/k)$

Section 10.7

1. diverges

3. diverges

5. converges but not absolutely

7. diverges

9. converges but not absolutely

11. diverges

13. converges absolutely (terms > 0)

15. converges but not absolutely

17. converges absolutely

19. diverges

Section 10.8

1. $1 - x + \dfrac{x^2}{2!} - \dfrac{x^3}{3!} + \dfrac{x^4}{4!} - \dfrac{x^5}{5!}$

3. $x - \frac{1}{3}x^3 + \frac{1}{5}x^5$

5. $1 - x + x^2 - x^3 + x^4 - x^5$

7. $x^3 + \frac{1}{3}x^5$

9. $1 + \frac{1}{2}x^2 + \frac{5}{24}x^4$

11. $x - \frac{1}{2}x^2 + \frac{1}{3}x^3 - \frac{1}{4}x^4$

13. $-\frac{1}{2}x^2 - \frac{1}{12}x^4$

15. $x + x^2 + \frac{5}{6}x^3 + \frac{1}{2}x^4$

17. Table 2

19. Table 4

21. Table 4

23. Table 3

30. $\sin x = \sin a + (x - a)\cos a - \dfrac{(x - a)^2}{2!}\sin a - \dfrac{(x - a)^3}{3!}\cos a + \cdots$

32. $\log x = \log a + \dfrac{1}{a}(x - a) - \dfrac{1}{2a^2}(x - a)^2 + \dfrac{1}{3a^3}(x - a)^3 - \cdots$

33. $x + \dfrac{x^3}{3!} + \dfrac{x^5}{5!}$

34. $1 + \dfrac{x^2}{2!} + \dfrac{x^4}{4!}$

Section 10.9

1. Table 1

3. Table 1

5. Table 4

7. $\log x = \log a + \dfrac{x}{a} - \dfrac{x^2}{2a^2} + \dfrac{x^3}{3a^3} - \cdots + (-1)^{n-1}\dfrac{x^n}{na^n} + \cdots$

9. $\frac{1}{4}$

11. $\frac{120}{119}$

Section 10.10

1. $(-1, 1)$

3. $(-\infty, \infty)$

5. converges only at 0

7. $[-2, 2)$

9. $[-1, 1]$

11. $[-\frac{1}{2}, \frac{1}{2})$

13. $(-1, 1)$

15. $(-10, 10)$

17. $[-1/|a|, 1/|a|]$

19. $[-1/\pi, 1/\pi]$

21. $(-2, 2)$

Section 10.11

1. $\tan x = x + \frac{1}{3}x^3 + \frac{2}{15}x^5 + \frac{17}{315}x^7 + \cdots$
   and therefore
   $\sec^2 x = 1 + x^2 + \frac{2}{3}x^4 + \frac{17}{45}x^6 + \cdots$

3. $\dfrac{1}{1-x} = 1 + x + x^2 + x^3 + \cdots$
   and therefore
   $\log(1-x) = -x - \frac{1}{2}x^2 - \frac{1}{3}x^3 - \frac{1}{4}x^4 - \cdots$

4. $0.80 \le I \le 0.81$ 6. $0.60 \le I \le 0.61$

7. $0.29 \le I \le 0.30$ 9. $0.60 \le I \le 0.61$

11. $0.35 \le I \le 0.36$

Section 10.12

6. (a) $1 + \frac{1}{2}x - \frac{1}{8}x^2 + \frac{1}{16}x^3 - \frac{5}{128}x^4$
   (c) $1 + \frac{1}{2}x^2 - \frac{1}{8}x^4$

7. (a) 9.90   (c). 4.99   (e) 0.49

Section 10.13

1. 1   3. 0   5. 1
7. 0   9. diverges   11. 0
13. 0   15. 0   16. $\frac{4}{3}$
18. $\frac{2}{3}$ 20. 1   22. diverges
24. converges absolutely   26. converges but not absolutely
28. converges absolutely   30. converges but not absolutely
32. $[-1, 1)$ 34. $(-1, 1)$

36. $1 + (\log a)\, x + \dfrac{(\log a)^2}{2!}\, x^2 + \cdots + \dfrac{(\log a)^n}{n!}\, x^n + \cdots$

38. $x^2 - \dfrac{1}{2}x^4 + \dfrac{1}{3}x^6 - \cdots + \dfrac{(-1)^{n-1}}{n}\, x^{2n} + \cdots$

41. $0.49 \le I \le 0.50$ 43. $3.91 \le \sqrt[3]{60} \le 3.92$

45. $\tanh^{-1} x = x + \dfrac{1}{3}x^3 + \dfrac{1}{5}x^5 + \cdots + \dfrac{x^{2n-1}}{2n-1} + \cdots$

Section 11.1

1. 0   3. 1   4. 0   6. 1   7. 0
9. a^2 10. 0   12. 1   13. $\frac{1}{2}\pi$ 15. 0
17. $\frac{1}{2}$

Section 11.2

1. 0   3. 1   4. $\frac{1}{n} a^{1-n}$ 6. 1   7. 0

9. 2   10. $\log \dfrac{a}{a+1}$ 12. $\dfrac{1+\pi}{1-\pi}$ 13. $\dfrac{1}{\sqrt{a}}$ 15. -2

16. $\frac{1}{2}$ 18. 0   19. $-\frac{1}{2}$ 21. e^2 22. $\frac{1}{2}$

24. $2/n$ 25. $-\frac{1}{6}$

Section 11.3

1. ∞ 3. $-\infty$ 4. -1 6. 0   7. $\frac{1}{5}$ 9. 1

10. 0   12. 0   13. 1   15. 1   16. 1   18. $\frac{1}{2}$

19. e 21. 1   22. $\frac{1}{3}$ 24. ∞ 25. $-\infty$

Section 11.4

1. 1   3. $1/p$ 4. $\frac{1}{2}\pi$ 6. 2

7. 6   9. $\frac{1}{2}\pi$ 10. 2   12. $\frac{1}{2}\pi$

13. diverges   15. 1   16. $-\frac{1}{4}$ 18. $\frac{1}{2}\log 3$

19. -1 21. $\frac{3}{2}$ 22. $\log 2$ 24. $\frac{1}{4}$

25. (b) 1   (c) $\frac{1}{2}\pi$ (d) 2π (e) 2π

27. (b) $\frac{4}{3}$ (c) 2π (d) $\frac{8}{7}\pi$

29. converges by comparison with $\displaystyle\int_0^{\infty} \frac{dx}{x^{3/2}}$

31. diverges since for x large the integrand is greater than $\dfrac{1}{x}$ and $\displaystyle\int_1^{\infty} \dfrac{1}{x}\, dx$ diverges

32. converges by comparison with $\displaystyle\int_{\pi}^{\infty} \frac{1}{x^2}\, dx$

34. diverges by comparison with $\displaystyle\int_e^{\infty} \frac{dx}{(x+1)\log(x+1)}$

35. \$2000

Section 12.3

1. $3\mathbf{i} - 4\mathbf{j} + 6\mathbf{k}$ 3. $-3\mathbf{i} - \mathbf{j} + 8\mathbf{k}$

5. 5   7. 3   9. $\sqrt{6}$

11. (a) $\mathbf{a}, \mathbf{c}, \mathbf{d}$ (b) $\mathbf{a}, \mathbf{c}$
    (c) $\mathbf{a}$ and $\mathbf{c}$ both have direction opposite to $\mathbf{d}$

12. (i) $\mathbf{a} + \mathbf{b}$ (ii) $-(\mathbf{a} + \mathbf{b})$ (iii) $\mathbf{a} - \mathbf{b}$ (iv) $\mathbf{b} - \mathbf{a}$

14. (a) $6\mathbf{i} + 3\mathbf{j} + 12\mathbf{k}$ (b) $A = \frac{26}{7}$, $B = -\frac{11}{7}$, $C = \frac{3}{7}$

16. $\alpha = \pm 3$ 19. $\frac{1}{3}\sqrt{6}\,(\mathbf{i} + 2\mathbf{j} - \mathbf{k})$

Section 12.4

1. $\mathbf{a} \cdot \mathbf{b} = 5$, $\mathbf{a} \cdot \mathbf{c} = 8$, $\mathbf{b} \cdot \mathbf{c} = 18$

3. $\text{comp}_\mathbf{b}\,\mathbf{a} = \frac{5}{14}\sqrt{14}$, $\text{comp}_\mathbf{c}\,\mathbf{a} = \frac{8}{5}$

5. $\sqrt{2}\mathbf{i} + \sqrt{2}\mathbf{j}$ 7. (a) $\mathbf{a} \perp \mathbf{b}$ (b) $\mathbf{a} \perp \mathbf{b}$

8. $\mathbf{a} \cdot \mathbf{b} = \mathbf{a} \cdot \mathbf{c}$ iff $\mathbf{a} \cdot (\mathbf{b} - \mathbf{c}) = 0$ iff $\mathbf{a} \perp \mathbf{b} - \mathbf{c}$

13. (a) $W = \mathbf{F} \cdot \mathbf{r}$ (c) (i) $W_2 = -W_1$ (ii) $W_2 = \sqrt{3}W_1$

14. $x = 3$, $x = \frac{5}{2}$

16. $\theta = \text{arc cos}\,(\frac{1}{3}\sqrt{3}) \cong \text{arc cos}\,0.346 \cong 1.22$ radians

Section 12.5

1. P and Q 2. l_1, l_3, l_4 are parallel

3. all nonzero scalar multiples of $\mathbf{d}$

4. (a) $\mathbf{r}(t) = (3\mathbf{i} + \mathbf{j}) + t\mathbf{k}$
   (d) $\mathbf{r}(t) = x_0\mathbf{i} + y_0\mathbf{j} + z_0\mathbf{k} + t[(x_1 - x_0)\mathbf{i} + (y_1 - y_0)\mathbf{j} + (z_1 - z_0)\mathbf{k}]$

5. (a) $x(t) = 1 + t$, $y(t) = -t$, $z(t) = 3 + t$
   (c) $x(t) = 2$, $y(t) = -2 + t$, $z(t) = 3$

7. $\dfrac{x}{x_0} = \dfrac{y}{y_0} = \dfrac{z}{z_0}$ or $\dfrac{x - x_0}{x_0} = \dfrac{y - y_0}{y_0} = \dfrac{z - z_0}{z_0}$ if $x_0 y_0 z_0 \neq 0$

9. (b) $P(6, \sqrt{3}, 0)$, $\frac{1}{3}\pi$
   (d) $P(1, 3, 1)$, arc cos $(\frac{1}{6}\sqrt{3}) \cong$ arc cos $0.289 \cong 1.28$ radians

10. the lines meet at right angles at $P(x_0, y_0, z_0)$

11. the lines are parallel

13. (a) $t_0 = -\dfrac{\mathbf{r}_0 \cdot \mathbf{d}}{\|\mathbf{d}\|^2}$

    (b) $\mathbf{R}(t) = \mathbf{r}(t_0) \pm t\dfrac{\mathbf{d}}{\|\mathbf{d}\|}$ where t_0 is as in part (a)

14. (a) $\frac{1}{2}\sqrt{2}$ (b) $\sqrt{2}$

Section 12.6

1. Q 2. R, S 3. $x - 4y + 3z - 2 = 0$

5. $3x - 2y + 5z - 9 = 0$ 9. at $x = -\frac{2}{3}$, at $y = 2$, at $z = -\frac{1}{2}$

10. $\pm\frac{1}{62}\sqrt{62}(2\mathbf{i} - 3\mathbf{j} + 7\mathbf{k})$ 11. $\frac{1}{2}\pi$

13. $\cos\theta = \frac{2}{21}\sqrt{42} \cong 0.617,\ \theta \cong 0.91$ radians

15. coplanar   17. not coplanar   19. $\dfrac{|D_2 - D_1|}{\sqrt{A^2 + B^2 + C^2}}$

Section 12.7

1. $-2\mathbf{k}$ 3. $\mathbf{i} + \mathbf{j} + \mathbf{k}$ 5. $-3\mathbf{i} - \mathbf{j} - 2\mathbf{k}$

7. -1 9. $\mathbf{0}$ 11. $\mathbf{i} + \mathbf{j} - 2\mathbf{k}$

13. -3 15. $5\mathbf{i} - 4\mathbf{j} - \mathbf{k}$ 22. (a) $\frac{3}{2}$ (b) $2\sqrt{3}$

23. (a) 1   (b) 2   25. (a) $x + z = 2$ (b) $x + y - z = 1$

Section 13.1

1. $\mathbf{f}'(t) = 2\mathbf{i} - \mathbf{j} + 3\mathbf{k}$

3. $\mathbf{f}'(t) = -\dfrac{1}{2\sqrt{1 - t}}\mathbf{i} + \dfrac{1}{2\sqrt{1 + t}}\mathbf{j} - \dfrac{2}{(1 + t)^2}\mathbf{k}$

5. $\mathbf{f}'(t) = \cos t\,\mathbf{i} - \sin t\,\mathbf{j} + \sec^2 t\,\mathbf{k}$ 7. $\mathbf{f}'(t) = \dfrac{1}{\sqrt{1 - t^2}}\mathbf{i} + \dfrac{2}{1 + 2t}\mathbf{j} + 2t\mathbf{k}$

9. $2\mathbf{i} + \frac{1}{2}\pi^2\mathbf{k}$

11. $\dfrac{1}{e}\left\{(2e - 5)\mathbf{i} + \sqrt{2}(e - 2)\mathbf{j} + (e - 1)\mathbf{k}\right\}$

15. $\mathbf{f}(t) = (\frac{1}{2}t^2 + 1)\mathbf{i} + (\sqrt{1 + t^2} + 1)\mathbf{j} + (te^t - e^t + 4)\mathbf{k}$

16. $\mathbf{f}(t) = e^{2t}\mathbf{i} - e^{2t}\mathbf{k}$

Section 13.2

1. $\mathbf{f}'(t) = \mathbf{b},\quad \mathbf{f}''(t) = \mathbf{0}$

3. $\mathbf{f}'(t) = 2e^{2t}\mathbf{i} - \cos t\,\mathbf{j},\quad \mathbf{f}''(t) = 4e^{2t}\mathbf{i} + \sin t\,\mathbf{j}$

5. $\mathbf{f}'(t) = 2t\mathbf{g}'(t^2),\quad \mathbf{f}''(t) = 4t^2\mathbf{g}''(t^2) + 2\mathbf{g}'(t^2)$

7. $\mathbf{f}'(t) = \frac{1}{2}\sqrt{t}\,\mathbf{g}'(\sqrt{t}) + \mathbf{g}(\sqrt{t}),\quad \mathbf{f}''(t) = \frac{1}{4}\mathbf{g}''(\sqrt{t}) + \dfrac{3}{4\sqrt{t}}\mathbf{g}'(\sqrt{t})$

9. $(-\sin t)e^{\cos t}\mathbf{i} + (\cos t)e^{\sin t}\mathbf{j}$ 11. $e^t\mathbf{i} + e^{-t}\mathbf{j}$

13. $2(\mathbf{b} \times \mathbf{d})$ 17. $\mathbf{f}(t) = \mathbf{a} + t\mathbf{b} + \frac{1}{2}t^2\mathbf{c}$

Section 13.3

1. (a) $\pi\mathbf{j} + \mathbf{k},\quad \mathbf{R}(u) = (\mathbf{i} + 2\mathbf{k}) + u(\pi\mathbf{j} + \mathbf{k})$
   (c) $\mathbf{b} - 2\mathbf{c},\quad \mathbf{R}(u) = (\mathbf{a} - \mathbf{b} + \mathbf{c}) + u(\mathbf{b} - 2\mathbf{c})$
   (e) $4\mathbf{i} - \mathbf{j} + 4\mathbf{k},\quad \mathbf{R}(u) = (2\mathbf{i} + 5\mathbf{k}) + u(4\mathbf{i} - \mathbf{j} + 4\mathbf{k})$

3. The scalar components

$$x(t) = at, \qquad y(t) = bt^2$$

satisfy

$$a^2y(t) = a^2bt^2 = b(at)^2 = b[x(t)]^2$$

and generate the parabola

$$a^2y = bx^2.$$

5. (a) $P(0, 1)$ (b) $P(1, 2)$ (c) $P(-1, 2)$

6. $\mathbf{r}(t) = e^{at}(\mathbf{i} + 2\mathbf{j} + 3\mathbf{k})$

9. $P(1, 2, -2)$, arc cos $(\frac{1}{5}\sqrt{5}) \cong$ arc cos $0.447 \cong 1.11$ radians

Section 13.4

3. $t = \frac{1}{2}$

4. (a) The scalar components

$$x(t) = 2 \cos 2t, \; y(t) = 3 \cos t$$

satisfy

$$
\begin{aligned}
4[y(t)]^2 - 9x(t) - 18 &= 4(9 \cos^2 t) - 9(2 \cos 2t) - 18 \\
&= 36 \cos^2 t - 18(\cos^2 t - \sin^2 t) - 18 \\
&= 18 (\sin^2 t + \cos^2 t) - 18 = 0.
\end{aligned}
$$

(c) $-8\mathbf{i} - 3\mathbf{j}$ at $P(2, 3)$, $-8\mathbf{i} + 3\mathbf{j}$ at $Q(2, -3)$

8. $\mathbf{F}(t) = 2m\mathbf{k}$

9. (a) $\pi b\mathbf{j} + \mathbf{k}$ (b) $\sqrt{\pi^2b^2 + 1}$ (c) $-\pi^2a\mathbf{i}$

(d) $m(\pi b\mathbf{j} + \mathbf{k})$ (e) 0   (f) $\frac{1}{2}m(\pi^2b^2 + 1)$

(g) $m[b(1 - \pi)\mathbf{i} - 2a\mathbf{j} + 2\pi ab\mathbf{k}]$

(h) $-m\pi a^2[\mathbf{j} - b\mathbf{k}]$

Section 13.5

1. $2\pi\sqrt{a^2 + b^2}$ 3. log $(1 + \sqrt{2})$ 5. $\frac{14}{3}$

7. $26 + \frac{25}{6}$ log 5   9. 0.50

Section 13.6

1. $\dfrac{\sqrt{2}}{2(1 - 2x + 2x^2)^{3/2}}$ 3. $\dfrac{6|x|}{(1 + 9x^4)^{3/2}}$

5. $(\frac{1}{2}\sqrt{2}, \frac{1}{2}$ log $\frac{1}{2})$ 7. $\frac{36}{7921}\sqrt{89}$

9. $\frac{1}{8}\sqrt{2}$ 11. $\dfrac{|a|}{2(a^2 + b^2)^{3/2}}$ 13. $\frac{1}{4}\sqrt{2}$

14. $k(\theta) = \dfrac{[f(\theta)]^2 + 2[f'(\theta)]^2 - f(\theta)f''(\theta)}{([f(\theta)]^2 + [f'(\theta)]^2)^{3/2}}$

16. $\dfrac{\theta^2 + 2}{a(\theta^2 + 1)^{3/2}}$

Section 14.1

1. dom (f) = the first and third quadrants, including the axes
ran $(f) = [0, \infty)$

3. dom (f) = the set of all points (x, y) except those that lie on the line $y = -x$
ran $(f) = (-\infty, 0) \cup (0, \infty)$

5. dom (f) = the entire plane
ran $(f) = (-1, 1)$

7. dom (f) = the set of all points (x, y) except for those on the coordinate axes
ran $(f) = (-\infty, \infty)$

9. dom (f) = all of space
ran $(f) = [0, \infty)$

12. dom (f) = the set of all points (x, y) such that $x^2 > y^2$; in other words, all the points of the plane which lie between the graph of $y = |x|$ and the graph of $y = -|x|$
ran $(f) = (0, \infty)$

13. dom (f) = the set of all points (x, y, z) such that $x^2 \neq y^2$; that is, all points of space except those which lie on the planes $x + y = 0$ and $x - y = 0$
ran $(f) = (-\infty, \infty)$

Section 14.2

1. a quadric cone

3. a parabolic cylinder

5. a hyperboloid of one sheet

7. sphere of radius 2 centered at the origin

9. an elliptic paraboloid

11. a hyperbolic paraboloid

13. there are six such ellipsoids, among them

$$\frac{x^2}{9} + \frac{y^2}{16} + \frac{z^2}{25} = 1;$$

the others can be obtained by permuting the denominators.

15. in the xy-plane, the ellipse $12x^2 + 9y^2 = 36$
in the xz-plane, the hyperbola $12x^2 - 36z^2 = 36$
in the yz-plane, the hyperbola $9y^2 - 36z^2 = 36$

17. $4z = x^2 + y^2$, a paraboloid of revolution

Section 14.3

13. $x + 2y + 3z = 0$, plane through the origin

15. $z = \sqrt{x^2 + y^2}$, the upper half of the cone $z^2 = x^2 + y^2$ (Figure 14.2.4)

17. the hyperboloid of two sheets

$$\frac{x^2}{(\frac{1}{6})^2} + \frac{y^2}{(\frac{1}{3})^2} - \frac{z^2}{(1)^2} = -1 \qquad \text{(Figure 14.2.3)}$$

Section 14.4

1. $\dfrac{\partial f}{dx} = 6x - y, \quad \dfrac{\partial f}{\partial y} = 1 - x$

3. $\dfrac{\partial z}{\partial x} = 2Ax + By, \quad \dfrac{\partial z}{\partial y} = Bx + 2Cy$

5. $\dfrac{\partial f}{\partial x} = e^{x-y} - 2xy, \quad \dfrac{\partial f}{\partial y} = -e^{x-y} - x^2$

7. $\dfrac{\partial g}{\partial x} = \dfrac{(AD - BC)y}{(Cx + Dy)^2}, \quad \dfrac{\partial g}{\partial y} = \dfrac{(BC - AD)x}{(Cx + Dy)^2}$

9. $\dfrac{\partial u}{\partial x} = y + z, \quad \dfrac{\partial u}{\partial y} = x + z, \quad \dfrac{\partial u}{\partial z} = x + y$

11. $\dfrac{\partial \rho}{\partial \theta} = \sin(\phi - \frac{1}{2}\pi), \quad \dfrac{\partial \rho}{\partial \phi} = (\theta - \frac{1}{2}\pi)\cos(\phi - \frac{1}{2}\pi)$

13. $\dfrac{\partial \rho}{\partial \theta} = e^{\theta + \phi}[\cos(\theta - \phi) - \sin(\theta - \phi)]$

$\dfrac{\partial \rho}{\partial \phi} = e^{\theta + \phi}[\cos(\theta - \phi) + \sin(\theta - \phi)]$

15. $f_x(0, e) = 1, \quad f_y(0, e) = e^{-1}$

17. $g_x(0, \frac{1}{4}\pi) = 0, \quad g_y(0, \frac{1}{4}\pi) = 0$

21. (a) $50\sqrt{3}$ square inches   (b) $5\sqrt{3}$ (c) 50
    (d) $\frac{5}{18}\pi$ square inches   (e) -2

Section 14.5

1. interior $= \{(x, y): 2 < x < 4, 1 < y < 3\}$ (the inside of the rectangle)
   boundary $=$ the union of the four line segments that bound the rectangle
   set is closed

3. interior $=$ the entire set (region between two concentric circles)
   boundary $= \{(x, y): x^2 + y^2 = 1 \text{ or } x^2 + y^2 = 4\}$ (the two circles)
   set is open

5. interior $= \{(x, y): 1 < x^2 < 4\}$
   $\qquad\quad = \{(x, y): -2 < x < 1\} \cup \{(x, y): 1 < x < 2\}$
   $\qquad\qquad$ (two vertical strips without the boundary lines)
   boundary $= \{(x, y): x = -2, x = -1, x = 1, \text{ or } x = 2\}$
   $\qquad\qquad$ (four vertical lines)
   set is neither open nor closed

7. interior $= \{(x, y): y < x^2\}$ (region below the parabola)
   boundary $= \{(x, y): y = x^2\}$ (the parabola)
   set is closed

9. interior $=$ the entire set [the inside of the ball of radius $\frac{1}{2}$ centered at $(1, 1, 1)$]
   boundary $= \{(x, y, z): (x - 1)^2 + (y - 1)^2 + (z - 1)^2 = \frac{1}{4}\}$
              (the spherical surface)
   set is open

Section 14.6

1. $\dfrac{\partial^2 f}{\partial x^2} = 2A, \quad \dfrac{\partial^2 f}{\partial y\, \partial x} = \dfrac{\partial^2 f}{\partial x\, \partial y} = 2B, \quad \dfrac{\partial^2 f}{\partial y^2} = 2C$

3. $\dfrac{\partial^2 f}{\partial x^2} = Cy^2 e^{xy}, \quad \dfrac{\partial^2 f}{\partial y\, \partial x} = \dfrac{\partial^2 f}{\partial x\, \partial y} = Ce^{xy}(xy + 1), \quad \dfrac{\partial^2 f}{\partial y^2} = Cx^2 e^{xy}$

5. $\dfrac{\partial^2 f}{\partial x^2} = -\dfrac{1}{4(x + y^2)^{3/2}}, \quad \dfrac{\partial^2 f}{\partial y^2} = \dfrac{x}{(x + y^2)^{3/2}}$

$\dfrac{\partial^2 f}{\partial y\, \partial x} = \dfrac{\partial^2 f}{\partial x\, \partial y} = -\dfrac{y}{2(x + y^2)^{3/2}}$

7. $\dfrac{\partial^2 f}{\partial x^2} = \dfrac{1}{(x + y)^2} - \dfrac{1}{x^2}, \quad \dfrac{\partial^2 f}{\partial y^2} = \dfrac{1}{(x + y)^2}$

$\dfrac{\partial^2 f}{\partial y\, \partial x} = \dfrac{\partial^2 f}{\partial x\, \partial y} = \dfrac{1}{(x + y)^2}$

9. $\dfrac{\partial^2 f}{\partial x^2} = 2(y + z), \quad \dfrac{\partial^2 f}{\partial y^2} = 2(x + z), \quad \dfrac{\partial^2 f}{\partial z^2} = 2(x + y);$ all the second mixed

partials are $2(x + y + z)$

Section 14.7

1. $\nabla f = e^{xy}[(xy + 1)\mathbf{i} + x^2 \mathbf{j}]$
3. $\nabla f = (6x - y)\mathbf{i} + (1 - x)\mathbf{j}$
5. $\nabla f = (e^{x-y} - 2xy)\mathbf{i} - (e^{x-y} + x^2)\mathbf{j}$
7. $\nabla f = e^{-z}[y^2 \mathbf{i} + 2xy \mathbf{j} - xy^2 \mathbf{k}]$
9. $\nabla f = (y + z)\mathbf{i} + (x + z)\mathbf{j} + (x + y)\mathbf{k}$
11. $\nabla f = e^{x-y}[(1 + x + y)\mathbf{i} + (1 - x - y)\mathbf{j}]$

13. $\nabla f = e^x \left[\log y\, \mathbf{i} + \dfrac{1}{y}\, \mathbf{j} \right]$

15. $\nabla f = \dfrac{AD - BC}{(Cx + Dy)^2}\, [y\mathbf{i} - x\mathbf{j}]$

17. $\nabla f = -\mathbf{i} + 18\mathbf{j}$

19. $\nabla f = -\mathbf{i}$

Section 14.8

1. $-2\sqrt{2}$

3. $\frac{1}{5}(7 - 4e)$

5. $\frac{1}{4}\sqrt{2}\,(a - b)$

7. $\frac{2}{3}\sqrt{6}$

9. $-3\sqrt{2}$

11. $-\dfrac{1}{\sqrt{x^2 + y^2}}$

13. (a) $\sqrt{2}[a(B - A) + b(C - B)]$ (b) $\sqrt{2}[a(A - B) + b(B - C)]$

15. $-\frac{7}{5}\sqrt{5}$ 17. $y = x^3$ from $(1, 1)$ to $(0, 0)$

19. $(b^2)^{a^2}x^{b^2} = (a^2)^{b^2}y^{a^2}$ from (a^2, b^2) to $(0, 0)$

Section 14.9

1. e^t 3. $3t^2 - 5t^4$

5. $2\omega(b^2 - a^2)\sin\omega t\cos\omega t + b\omega$ 7. $\sin 2t - 3\cos 2t$

9. $e^{t/2}(\frac{1}{2}\sin 2t + 2\cos 2t) + e^{2t}(2\sin\frac{1}{2}t + \frac{1}{2}\cos\frac{1}{2}t)$

11. $1 - 4t + 6t^2 - 4t^3$

15. $\dfrac{\partial u}{\partial r} = \dfrac{\partial u}{\partial x}\dfrac{\partial x}{\partial r} + \dfrac{\partial u}{\partial y}\dfrac{\partial y}{\partial r}, \quad \dfrac{\partial u}{\partial s} = \dfrac{\partial u}{\partial x}\dfrac{\partial x}{\partial s} + \dfrac{\partial u}{\partial y}\dfrac{\partial y}{\partial s}$

18. $f(x, y, z) = a_1 x + a_2 y + a_3 z + C$

Section 14.10

1. tangent $x + y + 2 = 0$; normal $x - y = 0$

3. tangent $\sqrt{2}x - 5y + 3 = 0$; normal $5x + \sqrt{2}y - 6\sqrt{2} = 0$

5. tangent $7x - 17y + 6 = 0$; normal $17x + 7y - 82 = 0$

7. tangent $x - y - 3 = 0$; normal $x + y + 1 = 0$

9. 0   11. $4x - 5y + 4z = 0$

13. $2x + y + 2z - 8 = 0$ 15. $z = 2(x + y)$

17. $b^2c^2x_0 x - a^2c^2y_0 y - a^2b^2z_0 z = a^2b^2c^2$

19. $(\frac{1}{3}, \frac{11}{6}, -\frac{1}{12})$ 21. $(0, 0, 0)$

23. the tangent planes meet at right angles and therefore the normals ∇F and ∇G must meet at right angles:

$$\frac{\partial F}{\partial x}\frac{\partial G}{\partial x} + \frac{\partial F}{\partial y}\frac{\partial G}{\partial y} + \frac{\partial F}{\partial z}\frac{\partial G}{\partial z} = 0$$

25. $\frac{9}{2}a^3$ $(V = \frac{1}{3}Bh)$

28. $\frac{1}{2}\pi - \text{arc cos}(\frac{19}{203}\sqrt{29}) \cong 1.57 - \text{arc cos } 0.504 \cong 0.53$ radians

30. $3x + 4y + 6z = 22, \quad 6x + y - z = 11$

Section 14.11

1. $(1, 0)$ gives max of 1
3. $(1, -1)$ is a saddle point
5. $(-2, 1)$ gives min of -2
7. no stationary points and no extreme values
9. $(1, 1)$ gives min of -3
11. no stationary points but 0 is min
13. $(-2, 2)$ is a saddle point
15. $(\frac{1}{3}[x_1 + x_2 + x_3], \frac{1}{3}[y_1 + y_2 + y_3])$
17. $x = (a^2b)^{1/3}, \quad y = (ab^2)^{1/3}$

Section 14.12

1. $(4, -2)$ gives min of -10
3. $(\frac{10}{3}, \frac{8}{3})$ gives max of $\frac{28}{3}$
5. $(0, 0)$ is a saddle point; $(2, 2)$ gives a local min of -8
7. $(1, \frac{3}{2})$ is a saddle point; $(5, \frac{27}{2})$ gives a local min of $-\frac{117}{4}$
9. no extreme values
11. no extreme values
13. no extreme values
15. $(\frac{1}{2}, 4)$ gives a local min of 6
17. $(1, 1)$ gives a local min of 3
19. $\frac{1}{27}$
21. $\frac{8}{9}\sqrt{3}\, abc$
22. $\theta = \frac{1}{6}\pi, \quad x = \dfrac{P}{2 + 2 \sec \theta - \tan \theta}, \quad y = \frac{1}{2}[P - x(1 + \sec \theta)]$
25. \$1 per dozen blades, \$2 per razor
26. $a^2 + b^2 + c^2$

Section 14.13

1. 2
3. $-\frac{1}{2}ab$
5. $\frac{2}{9}\sqrt{3}\, ab^2$
7. $-\frac{1}{9}\sqrt{3}$
9. no maximum exists
11. $19\sqrt{2}$
13. $\frac{1}{27}abc$
15. 1
17. a^3
19. $4A^2(a^2 + b^2 + c^2)$ where A is the area of the triangle and $a, b, c,$ are the sides.
23. $Q_1 = 10,000; Q_2 = 20,000; Q_3 = 30,000$

Section 14.14

1. $df = (3x^2y - 2xy^2)\,\Delta x + (x^3 - 2x^2y)\,\Delta y$
3. $df = (\cos y + y \sin x)\,\Delta x - (x \sin y + \cos x)\,\Delta y$
5. $df = \Delta x - (\tan z)\,\Delta y - (y \sec^2 z)\,\Delta z$
7. $\Delta u = -7.15, \quad du = -7.5$
8. $du = 1$
9. $22\frac{249}{352}$ taking $u = x^{1/2}y^{1/4}, \quad x = 121, \quad y = 16, \quad \Delta x = 4, \quad \Delta y = 1$
11. $\frac{1}{14}\sqrt{2}\pi$ taking $u = \sin x \cos y, \quad x = \pi, \quad y = \frac{1}{4}\pi, \quad \Delta x = -\frac{1}{7}\pi,$
    $\Delta y = -\frac{1}{20}\pi$
13. $dz = -\frac{1}{90}, \Delta z = -\frac{1}{93}$
14. decreases by about 44.8π cubic inches

Section 14.15

1. $f(x, y) = \frac{1}{2}x^2y^2 + C$

3. $f(x, y) = xy + C$

5. no such function exists

7. $f(x, y) = \sin x + y \cos x + C$

9. $f(x, y) = xe^xy + e^{-y} + C$

11. $f(x, y) = \frac{1}{3}x^3 \arcsin y + y - y \log y + C$

13. $f(x, y) = \sinh x + x \sinh y + C$

Section 14.16

1. 1   3. 0   5. $-\frac{17}{6}$ 7. 0

9. $\log \frac{8}{5}$ 11. $-\pi$ 13. $\frac{23}{21}$ 15. $|W| = A$

Section 14.17

1. 0   3. -1 5. 0   7. $-\pi$ 9. 4

10. (a) $f(\mathbf{r}) = K \log \|\mathbf{r}\| + C$ (b) $f(\mathbf{r}) = -\left(\dfrac{K}{n-2}\right)\dfrac{1}{\|\mathbf{r}\|^{n-2}} + C$

Section 15.1

1. $L_f(P) = 1$ 2. $U_f(P) = 7$ 5. $L_f(P) = 2\frac{1}{4}$ 6. $U_f(P) = 5\frac{3}{4}$

9. $I = 4$; the volume of the prism bounded above by the plane $z = x + 2y$ and below by R

10. $L_f(P) = -\frac{7}{16}$ 11. $U_f(P) = \frac{7}{16}$ 14. $I = 0$ 15. $I = 2\frac{2}{3}$

Section 15.3

1. $\displaystyle\int_{-1}^{1}\left(\int_{-1}^{1} x^2\, dy\right) dx = \int_{-1}^{1}\left(\int_{-1}^{1} x^2\, dx\right) dy = \frac{4}{3}$

3. $\displaystyle\int_{-1}^{1}\left(\int_{-1}^{1} \cos \tfrac{1}{2}\pi y\, dy\right) dx = \int_{-1}^{1}\left(\int_{-1}^{1} \cos \tfrac{1}{2}\pi y\, dx\right) dy = 8/\pi$

5. $\displaystyle\int_{0}^{1}\left(\int_{0}^{x} (y + \sin \pi x^2)\, dy\right) dx = \frac{1}{6} + 1/\pi$

7. $\displaystyle\int_{0}^{\pi/2}\left(\int_{0}^{\pi/2} \sin(x + y)\, dy\right) dx = \int_{0}^{\pi/2}\left(\int_{0}^{\pi/2} \sin(x + y)\, dx\right) dy = 2$

9. $\displaystyle\int_{-1}^{1}\left(\int_{-\sqrt{1-x^2}}^{\sqrt{1-x^2}} (x^2 + 3y^2)\, dy\right) dx = 4\int_{0}^{1}\left(\int_{0}^{\sqrt{1-x^2}} (x^2 + 3y^2)\, dy\right) dx = \pi$

$\displaystyle\int_{-1}^{1}\left(\int_{-\sqrt{1-y^2}}^{\sqrt{1-y^2}} (x^2 + 3y^2)\, dx\right) dy = 4\int_{0}^{1}\left(\int_{0}^{\sqrt{1-y^2}} (x^2 + 3y^2)\, dx\right) dy = \pi$

11. $\int_0^1 \left(\int_0^{y^2} ye^x \, dx \right) dy = \frac{1}{2}e - 1$

13. $\int_0^2 \left(\int_0^{x/2} e^{x^2} \, dy \right) dx = \frac{1}{4}(e^4 - 1)$

15. $\int_0^3 \left(\int_y^{4y-y^2} dx \right) dy = \frac{9}{2}$

17. $\int_2^3 \left(\int_{6/x}^{5-x} dy \right) dx = \int_2^3 \left(\int_{6/y}^{5-y} dx \right) dy = \frac{5}{2} + 6 \log \frac{2}{3}$

19. $\int_2^2 \left(\int_0^{-(3x/2)+3} (4 - 2x - \frac{4}{3}y) \, dy \right) dx$

$$= \int_0^3 \left(\int_0^{-(2y/3)+2} (4 - 2x - \frac{4}{3}y) \, dx \right) dy = 4$$

21. $\int_0^2 \left(\int_0^{-(x/2)+1} x^3y \, dy \right) dx = \int_0^1 \left(\int_0^{-2y+2} x^3y \, dx \right) dy = \frac{2}{15}$

23. $4 \int_0^1 \left(\int_0^{\sqrt{1-x^2}} (x^2 + y^2) \, dy \right) dx = 4 \int_0^1 \left(\int_0^{\sqrt{1-y^2}} (x^2 + y^2) \, dx \right) dy = \frac{1}{2}\pi$

Section 15.4

1. $\int_{-\pi/3}^{\pi/3} \left(\int_{(3/4)\sec\theta}^{3/2} r \, dr \right) d\theta = \frac{3}{4}\pi - \frac{9}{16}\sqrt{3}$

3. $\int_{-\pi/2}^{\pi/2} \left(\int_{\cos\theta}^{3\cos\theta} r \, dr \right) d\theta = 2\pi$

5. $\int_{-\pi/2}^{\pi/2} \left(\int_1^{1+\cos\theta} r \, dr \right) d\theta = \frac{1}{4}\pi + 2$

7. $\int_{-\pi/2}^{\pi/2} \left(\int_{(1+\cos\theta)^{-1}}^1 r \, dr \right) d\theta = \frac{1}{2}\pi - \frac{2}{3}$

9. $\int_0^{2\pi} \left(\int_0^b (r^2 \sin\theta + br) \, dr \right) d\theta = \int_0^b \left(\int_0^{2\pi} (r^2 \sin\theta + br) \, d\theta \right) dr = b^3\pi$

11. $\int_0^{2\pi} \left(\int_0^1 r\sqrt{4 - r^2} \, dr \right) d\theta = \int_0^1 \left(\int_0^{2\pi} r\sqrt{4 - r^2} \, d\theta \right) dr = \frac{2}{3}(8 - 3\sqrt{3})\pi$

Section 15.6

1. abc 3. $\frac{2}{3}$

5. $V = \int_0^a \int_0^{\phi(x)} \int_0^{\psi(x,y)} dz \, dy \, dx = \frac{1}{6}abc$

with $\phi(x) = b\left(1 - \dfrac{x}{a}\right)$, $\psi(x, y) = c\left(1 - \dfrac{x}{a} - \dfrac{y}{b}\right)$

7. (a) $M = \displaystyle\int_0^1 \int_0^1 \int_0^1 kx\, dz\, dy\, dx = \tfrac{1}{2}k$ (where k is the constant of proportionality)

(b) $M = \displaystyle\int_0^1 \int_0^1 \int_0^1 k(x^2 + y^2 + z^2)\, dz\, dy\, dx = k$

9. $V = \displaystyle\int_0^2 \int_{-1}^2 \int_{2-y}^{4-y^2} dz\, dy\, dx = 9$

10. (b) $M = \displaystyle\int_0^2 \int_{-1}^2 \int_{2-y}^{4-y^2} y^2\, dz\, dy\, dx = \tfrac{63}{10}$

12. $V = \displaystyle\int_{-1/2}^{1/2} \int_{\phi_1(x)}^{\phi_2(x)} \int_{x^2+y^2}^{y} dz\, dy\, dx = \tfrac{1}{32}\pi$

with $\phi_1(x) = \tfrac{1}{2} - \tfrac{1}{2}\sqrt{1 - 4x^2}$, $\phi_2(x) = \tfrac{1}{2} + \tfrac{1}{2}\sqrt{1 - 4x^2}$

Section 15.7

1. $\bar{x} = \tfrac{6}{125} \displaystyle\int_0^5 \int_x^{6x-x^2} x\, dy\, dx = \tfrac{5}{2}$

   $\bar{y} = \tfrac{6}{125} \displaystyle\int_0^5 \int_x^{6x-x^2} y\, dy\, dx = 5$

3. $\bar{x} = \tfrac{1}{9} \displaystyle\int_{-2}^4 \int_{x^2/4}^{(x/2)+2} x\, dy\, dx = 1$

   $\bar{y} = \tfrac{1}{9} \displaystyle\int_{-2}^4 \int_{x^2/4}^{(x/2)+2} y\, dy\, dx = \tfrac{8}{5}$

5. $\bar{x} = \tfrac{5}{16} \displaystyle\int_0^8 \int_{x/2}^{x^{2/3}} x\, dy\, dx = \tfrac{10}{3}$

   $\bar{y} = \tfrac{5}{16} \displaystyle\int_0^8 \int_{x/2}^{x^{2/3}} y\, dy\, dx = \tfrac{40}{21}$

7. $\bar{x} = \tfrac{3}{64} \displaystyle\int_0^4 \int_{x^2-2x-3}^{6x-x^2-3} x\, dy\, dx = 2$

   $\bar{y} = \tfrac{3}{64} \displaystyle\int_0^4 \int_{x^2-2x-3}^{6x-x^2-3} y\, dy\, dx = 1$

9. $\bar{x} = \dfrac{6}{a^2} \displaystyle\int_0^a \int_0^{(\sqrt{a}-\sqrt{x})^2} x\, dy\, dx = \tfrac{1}{5}a$

   $\bar{y} = \dfrac{6}{a^2} \displaystyle\int_0^a \int_0^{(\sqrt{a}-\sqrt{x})^2} y\, dy\, dx = \tfrac{1}{5}a$

11. $V = (\pi a^2)(2\pi b) = 2\pi^2 a^2 b$

13. $(\frac{1}{4}a, \frac{1}{4}b, \frac{1}{4}c)$ with volume and limits of integration as in Exercise 5, Section 15.6

15. $V = \pi ab(2c + \sqrt{a^2 + b^2})$ (theorem of Pappus)

16. (a) $(\frac{7}{12}, \frac{7}{12}, \frac{7}{12})$ (b) $(\frac{1}{2}, \frac{1}{2}, \frac{2}{3})$ (c) $(\frac{1}{2}, \frac{1}{2}, \frac{1}{2})$

18. (a) $(\frac{1}{2}a, \frac{2}{3}b, \frac{2}{3}c)$ with limits of integration as in Exercise 5, Section 15.6

21. $\bar{x} = \dfrac{2}{3\pi a^2} \displaystyle\int_0^{2\pi} \int_0^{a(1+\cos\theta)} r^2 \cos\theta \, dr \, d\theta = \frac{5}{6}a$

$\bar{y} = 0$ since the region is symmetric about the y-axis

Section 15.8

1. (a) $M = \displaystyle\int_0^{2\pi} \int_0^{R} \int_0^{h} kzr \, dz \, dr \, d\theta = \frac{1}{2}k\pi R^2 h^2$

(b) $\frac{4}{5}k\pi R^{5/2}h$ (c) $\frac{1}{6}k\pi R^4 h$

2. $V = \displaystyle\int_0^{2\pi} \int_0^{R} \int_0^{(h/r)(R-r)} r \, dz \, dr \, d\theta = \frac{1}{3}\pi R^2 h$

6. $V = \displaystyle\int_{-\pi/2}^{\pi/2} \int_0^{2a\cos\theta} \int_0^{r^2/a} r \, dz \, dr \, d\theta = \frac{3}{2}\pi a^3$

7. $V = \displaystyle\int_{-\pi/2}^{\pi/2} \int_0^{2a\cos\theta} \int_0^{r} r \, dz \, dr \, d\theta = \frac{32}{9}a^3$

8. $V = \displaystyle\int_{-\pi/2}^{\pi/2} \int_0^{2\cos\theta} \int_0^{2+(r/2)\cos\theta} r \, dz \, dr \, d\theta = \frac{5}{2}\pi$

9. $V = \displaystyle\int_{-\pi/2}^{\pi/2} \int_0^{a\cos\theta} \int_0^{a-r} r \, dz \, dr \, d\theta = \frac{1}{36}a^3(9\pi - 16)$

11. $V = \displaystyle\int_{-\pi/2}^{\pi/2} \int_0^{\cos\theta} \int_{r^2}^{r\cos\theta} r \, dz \, dr \, d\theta = \frac{1}{32}\pi$

Section 15.9

1. $(\sqrt{3}, \frac{1}{4}\pi, \arccos \frac{1}{3}\sqrt{3})$ 3. $(2\sqrt{10}, \frac{2}{3}\pi, \arccos \frac{3}{10}\sqrt{10})$

4. sphere of radius ρ_0 centered at the origin

5. vertical half-plane hinged at the z-axis and making an angle θ_0 with the positive x-axis

6. for $0 < \phi_0 < \frac{1}{2}\pi$ and $\frac{1}{2}\pi < \phi_0 < \pi$, the cone generated by revolving about the z-axis a ray which starts out at the origin and makes an angle of ϕ_0 radians with the positive z-axis; for $\phi_0 = \frac{1}{2}\pi$, the xy-plane; for $\phi_0 = 0$ and $\phi_0 = \pi$, the surface degenerates: in the first case, to the nonnegative z-axis, and in the second case, to the nonpositive z-axis

8. horizontal plane one unit above the xy-plane

11. $V = \displaystyle\int_0^\alpha \int_0^\pi \int_0^a \rho^2 \sin\phi \, d\rho \, d\phi \, d\theta = \frac{2}{3}\alpha a^3$

13. $V = \displaystyle\int_0^{2\pi} \int_0^{\arctan (r/h)} \int_0^{h\sec\phi} \rho^2 \sin\phi \, d\rho \, d\phi \, d\theta = \frac{1}{3}\pi r^2 h$

14. $M = \displaystyle\int_0^{2\pi} \int_0^{\arctan (r/h)} \int_0^{h\sec\phi} k\rho^3 \sin\phi \, d\rho \, d\phi \, d\theta = \frac{1}{6}k\pi h[h^3 - (r^2 + h^2)^{3/2}]$

15. $V = \displaystyle\int_0^{2\pi} \int_0^\alpha \int_0^a \rho^2 \sin\phi \, d\rho \, d\phi \, d\theta = \frac{2}{3}\pi(1 - \cos\alpha)a^3$

16. $\rho = 2b \cos\phi$

17. (b) $M = \displaystyle\int_0^{2\pi} \int_0^{\pi/2} \int_0^{2b\cos\phi} \rho^3 \sin^2\phi \, d\rho \, d\phi \, d\theta = \frac{1}{4}\pi^2 b^4$

18. $V = \displaystyle\int_0^{2\pi} \int_0^{\pi/4} \int_0^2 \rho^2 \sin\phi \, d\rho \, d\phi \, d\theta + \int_0^{2\pi} \int_{\pi/4}^{\pi/2} \int_0^{2\sqrt{2}\cos\phi} \rho^2 \sin\phi \, d\rho \, d\phi \, d\theta$

$= \frac{1}{3}(16 - 6\sqrt{2})\pi$

Index